Artificial Intelligence, Data and Robotics

Edward Curry • Philip Piatkiewicz
Fredrik Heintz • Heike Vornhagen
Ahmed Nabil Belbachir • Emanuela Girardi
Marc Schoenauer • Juha Röning
Editors

Artificial Intelligence, Data and Robotics

Foundations, Transformations and Future Directions

 Springer

Editors
Edward Curry
Insight Research Ireland Centre for Data
Analytics, Data Science Institute
University of Galway
Galway, Ireland

Fredrik Heintz
Linköping University
Linköping, Sweden

Ahmed Nabil Belbachir
Norwegian Research Centre (NORCE)
Bergen, Norway

Marc Schoenauer
INRIA (National Institute for Research in
Digital Science and Technology)
Rocquencourt, France

Philip Piatkiewicz
The AI, Data and Robotics
Association (ADRA)
Brussels, Belgium

Heike Vornhagen
Insight Research Ireland Centre for Data
Analytics, Data Science Institute
University of Galway
Galway, Ireland

Emanuela Girardi
The AI, Data and Robotics
Association (ADRA)
Brussels, Belgium

Juha Röning
University of Oulu
Oulu, Finland

ISBN 978-3-032-10560-8   ISBN 978-3-032-10561-5   (eBook)
https://doi.org/10.1007/978-3-032-10561-5

This work was supported by the University of Galway.

This Springer imprint is published by the registered company Springer Nature Switzerland AG
The registered company address is: Gewerbestrasse 11, 6330 Cham, Switzerland

If disposing of this product, please recycle the paper.

Preface

The most transformative technologies in human history have been those that have fundamentally reshaped societies, economies, and human capabilities. Agriculture enabled permanent human settlements and the rise of civilizations, writing systems facilitated record-keeping and cultural transmission, and electricity revolutionized industry and communication. Today, the convergence of artificial intelligence, data, and robotics (ADR) technology marks a significant step change, with the potential to transform society. It offers opportunities to automate cognition and physical interaction, as well as to foster new forms of creativity and decision-making. However, it also presents challenges in translating core ADR strengths into advances in industrial competitiveness, societal well-being, and environmental sustainability.

The aim of this book is to educate readers about techniques and technologies related to AI, data, and robotics. It explores cutting-edge theories, tools, methodologies, and best practices within the context of both industrial and public sector scenarios. The book provides a foundation for understanding the scientific principles underlying areas such as digital twins, generative AI, and hybrid intelligence, as well as exploring the transformational potential of ADR technologies across domains, including manufacturing, mobility, safety-critical systems, and healthcare. Furthermore, it discusses future research challenges and roadmaps for ADR.

The contributions within this book are rooted in the European Partnership on AI, data, and robotics, which has mobilized the ADR ecosystem across Europe to provide strong leadership in science, innovation, and deployment. The Partnership's work addresses fundamental issues around deployment and citizen trust in ADR, fostering a vibrant innovation ecosystem built on Europe's strengths—its academic excellence, skilled workforce, global companies, and supportive regulations and standards, along with best practices. This positions Europe to lead the world in researching, developing, and deploying trustworthy, value-driven ADR based on European rights, principles, and values.

The book is intended for two main audiences. First, researchers and students in ADR and related fields such as computer science, information technology, and information systems. Second, industrial practitioners seeking practical guidelines based on rigorous studies, offering insights across various technological and innovative areas. Additionally, policymakers at local, national, and international levels will find this book relevant.

The structure of the book is divided into three parts: Foundations, Transformations, and Future Directions. Chapter "AI, Data, and Robotics: A Structured Overview of Contributions and Priorities" introduces the scope of the book, highlighting how each chapter contributes to the objectives of the European Partnership on AI, Data, and Robotics.

Part "Foundations"—covers the core scientific and technical principles of AI, data, and robotics. Chapter "Shaping Europe's Future: The Convergence of AI, Data, and Robotics" investigates the convergence of these technologies and their impact on Europe's landscape. Chapter "SRIDA: Charting the Future of a Progressive, Inclusive, and Sustainable European ADR Ecosystem" discusses the strategic research, innovation, and deployment plans aimed at establishing a progressive, inclusive, and sustainable ecosystem. Chapter "Digital Twins in Europe: Driving Sustainable Innovation and Sovereignty" offers an in-depth look at digital twin technologies, their applications, and their role in Europe's digital and green transitions. Chapter "The Cross-Fertilization Between the Human-in-the-Loop Approach and the Explainable AI Techniques Toward Trustworthiness" explores the synergy between the human-in-the-loop approach and explainable AI techniques. Chapter "Generative Artificial Intelligence to Tackle Visual Data Accessibility Challenges" focuses on generative AI and its role in addressing challenges related to visual data accessibility. Chapter "Hybrid Intelligence: The Fusion of Science-Based and Machine Learning Models" details machine learning models underpinning hybrid intelligence. The final chapter in this section examines how frugal machine learning can reduce the energy footprint of AI.

Part "Transformations"—illustrates how ADR technologies can disrupt various domains and applications. Chapter "The Role of Digital Transformation in Manufacturing Under the Light of AI Integration" discusses AI's role in transforming manufacturing. Chapter "AI-Based Management Approaches for Production Processes: Identifying Requirements for Frameworks and Architectures" analyzes AI-driven management approaches for production processes, emphasizing frameworks and architectures. Chapter "Data, AI, Robotics Transformative Power in Industry 5.0" examines Industry 5.0, including sustainable and circular applications and pilot cases. Chapter "Toward the Irish Mobility Data Space: Challenges, Opportunities, and Requirements" considers data-sharing requirements and opportunities for mobility data spaces. Chapter "Toward a Holistic Framework for Human-AI Collaboration in Safety-Critical Systems" presents a framework for human-AI collaboration in safety-critical systems. Chapter "CyclOps: Leveraging Semantic Technologies for AI and Data Life Cycle Management and Governance" highlights the use of semantic technologies in AI and data life cycle management and governance. Chapter "Intelligent Underwater Perception: Current Trends and Future Directions" explores current and future trends in intelligent underwater perception. Chapter "Brain-to-Speech: Prosody Feature Engineering and Transformer-Based Reconstruction" investigates feature engineering and reconstruction in brain-to-speech systems. Chapter "A Companion Robot Platform for Exploring Technical and Ethical Aspects in Elderly Care" describes a companion robot platform designed for elderly care. The section concludes with chapter "Robotics in Healthcare: IoT and Deep Learning Perspectives" on robotics, deep learning, and the healthcare Internet of Things.

Part "Future Directions"—discusses opportunities and challenges for ADR. It begins with chapter "Roadmaps and Agendas for Research and Innovation in Artificial Intelligence, Data, and Robotics", which sets out roadmaps and research agendas. Chapter "The Future of Digital Twins in Europe" considers the future of digital twins in Europe. Chapter "Advancing Industrial Collaboration: The Next Generation of Human-Robot Interaction" explores the next phase of human-robot interaction for advancing industrial cooperation. The book concludes with chapter "Human-AI Interaction and Visualization Perspectives on ADR", offering perspectives on human-AI interaction and visualization.

Galway, Ireland   Edward Curry
September 2025

Acknowledgments

The editors would like to extend their sincere thanks to Ralf Gerstner, Ramya Prakash, and the entire team at Springer for their professionalism and invaluable support throughout the preparation of this book.

This book was made possible through funding from the European Union's Horizon 2021 Research and Innovation Programme under grant agreement no. 101070336.

We are deeply grateful to all the contributing authors for sharing their expertise and research through the chapters in this volume. Special thanks are due to the reviewers, whose time, effort, and thoughtful feedback significantly enhanced the quality and coherence of the book.

We would particularly like to acknowledge the dedication and commitment of Majjed Al-Qatf, Fergus O'Donoghue, Nakul Mehta, Gaurav Negi, Muhammad Asif Razzaq, Al Waskow, and Yang Yang for their exceptional contributions as reviewers.

Edward Curry
Philip Piatkiewicz
Fredrik Heintz
Heike Vornhagen
Ahmed Nabil Belbachir
Emanuela Girardi
Marc Schoenauer
Juha Röning

Contents

Editors and Contributors

About the Editors

Edward Curry is the Established Professor of Data Science and Director of the Insight Research Ireland Centre for Data Analytics and Data Science Institute at the University of Galway. Edward has made substantial contributions to semantic technologies, incremental data management, event processing middleware, software engineering, distributed systems, and information systems. He combines strong theoretical results with high-impact practical applications. The excellence and impact of his research have been acknowledged by numerous awards, including best paper awards and the University of Galway President's Award for Societal Impact in 2017. His team's technology enables intelligent systems for smart environments in collaboration with several industrial partners. He is the organizer and program co-chair of major international conferences, including CIKM 2020, ECML 2018, IEEE Big Data Congress, ESWC 2025, and the European Big Data Value Forum. Edward is co-founder and elected Vice President of the AI, Data and Robotics Association (ADRA) and the Big Data Value Association (BDVA). He has built consensus on joint European research and innovation agendas and influenced European data and AI innovation policy to deliver on these agendas. In 2023, the Science Foundation Ireland Best International Engagement Award recognized his contributions to the European research and innovation ecosystem. He is a member of the Data Architecture & Technical Committee of the Irish Government Data Governance Board.

Philip Piatkiewicz is a seasoned leader in European affairs and project management, currently serving as the Secretary General of ADRA—the AI, Data and Robotics Association. ADRA is a leading strategic technology network of stakeholders from academia, industry, and the public sector, dedicated to advancing and adopting AI, data, and robotics technologies across Europe. ADRA acts as the private side for the AI, Data, and Robotics public-private partnership in Horizon

Europe. Philip's expertise encompasses strategic planning, stakeholder engagement, policy analysis, and project execution. He has successfully managed and coordinated complex collaborative projects that support the research, innovation, and industrial agendas of regional, national, and EU level, with a particular focus on key enabling technologies and digital industries.

Fredrik Heintz   is a Professor of Computer Science at Linköping University, where he leads the Division of Artificial Intelligence and Integrated Computer Systems (AIICS) and the Reasoning and Learning (ReaL) lab. His research focus is artificial intelligence, especially Trustworthy AI and the intersection between machine reasoning and machine learning. He is also Director of the Wallenberg AI and Transformative Technologies Education Development Program (WASP-ED); Co-director of the Wallenberg AI, Autonomous Systems and Software Program (WASP); Coordinator of the TrustLLM project; Vice President for AI Research ADRA, the AI, Data, and Robotics partnership; Member of the Swedish AI Commission; and Fellow of the Royal Swedish Academy of Engineering Sciences (IVA).

Heike Vornhagen   is a researcher and educator whose work bridges the fields of design, data visualization, and technology with a strong emphasis on human-centered approaches. With an academic background in digital media and a PhD in computer science from the University of Galway, her research has consistently focused on how people make sense of complex data, particularly through the design of interactive dashboards. Currently based at the Insight SFI Centre for Data Analytics, Dr. Vornhagen contributes to several interdisciplinary research initiatives that aim to make data more accessible and actionable, especially in the context of public services. Her broader research interests include data literacy, design thinking, and ethical AI—fields in which she frequently collaborates with both academic and public-sector partners.

Ahmed Nabil Belbachir   has nearly 30 years of experience in computer vision, AI, and robotics as an engineer, scientist, and strategist, driving innovative solutions and sustainable technologies. He is a Research Director at NORCE, Director at euRobotics aisbl, and Strategy Advisor for ADRA aisbl, contributing to Europe's AI, Data, and Robotics agenda 2030 and the Horizon Europe 2025–2027 program. Holding a PhD in Computer Science from TU Vienna, he previously led Austria's technical role in ESA-Herschel infrared observatory and co-invented a 360° panoramic 3D camera at AIT. Author of *Smart Cameras* (Springer, 2009; Chinese edition 2014), he has over 140 publications and 3 patents and has featured on Euronews and TV2 Norway. He coordinates major EU projects, COGNIMAN (€11M) and iBot4CRMs (€9.5M), and advises companies on advanced AI-driven robotics. Dr. Belbachir is also Vice General Chair of ADRA Forum 2024, General Chair of ADRA Forum 2025, and Co-chair of euRobotics Forum 2026.

Emanuela Girardi   is President of ADRA, the AI, Data and Robotics Association, and founder of *Pop AI*, a nonprofit raising public awareness of AI and its societal impact. She co-authored Italy's national AI strategy in 2019 as part of the expert group appointed by the Ministry of Economic Development. She also serves on the Advisory Board of CLAIRE, the Confederation of AI Research Laboratories in Europe, and on ADRA's Board of Directors, supporting the Horizon Europe 2021–2027 agenda for AI, data and robotics. Emanuela holds a business administration degree from Bocconi University and a CEMS master's from Université catholique de Louvain. She has studied and worked across Europe and the USA, beginning her career in mobile telecom, broadband, and video streaming. Later, she became an entrepreneur, managing an Italian automotive company and driving its digital transformation.

Marc Schoenauer   is Principal Senior Researcher Emeritus at INRIA, where he founded the TAO team on Machine Learning and Optimization in 2003. A graduate of École Normale Supérieure, he earned his PhD in Applied Mathematics from Paris 6 University and began his career at CNRS before joining INRIA. He later served as Head of Research at INRIA Saclay (2010–2016) and Deputy Research Director for AI (2020–2024). Since the 1990s, he has worked at the intersection of Evolutionary Computation and Machine Learning, publishing over 200 papers and advising 42 PhD students. He has held leading roles including Chair of ACM-SIGEVO, Founding President of SPECIES and Evolution Artificielle, and President of AFIA. He has also served as Editor-in-Chief of the *Evolutionary Computation Journal* and on multiple editorial boards. In 2018, he seconded Cédric Villani in writing his report on the French Strategy for AI delivered to President Macron.

Juha Röning   is Professor of Embedded System at the University of Oulu. He serves also as Visiting Professor of Tianjin University of Technology, P. R. China. He is principal investigator of the Biomimetics and Intelligent Systems Group (BISG). In 1985, he received ASLA/Fulbright scholarship. From 1985 to 1986, he was a visiting research scientist in the Center for Robotic Research at the University of Cincinnati. From 1986 to 1989, he held a Young Researcher Position in the Finnish Academy. In 2000, he was nominated as Fellow of SPIE. He has three patents and has published more than 400 papers in the areas of computer vision, robotics, intelligent signal analysis, and software security. He is currently serving as a Board of Director for euRobotics aisbl (Vice President Research) and ADRA (Robotics Research Vice President).

Contributors

Alberto Abella FIWARE Foundation, Berlin, Germany

Seyed Enayatallah Alavi Department of Computer Engineering, Faculty of Engineering, Shahid Chamran University of Ahvaz, Ahvaz, Iran

Carlos Agostinho UNINOVA–Centre of Technology and Systems (CTS), FCT Campus, Caparica, Portugal

Gorka Aguirre IDEKO, Elgoibar, Spain

Kosmas Alexopoulos Laboratory for Manufacturing Systems and Automation, Department of Mechanical Engineering and Aeronautics, University of Patras, Rio Patras, Greece

Department of Digital Industry Technologies, National and Kapodistrian University of Athens, Psachna, Greece

Jesus Alonso Innovalia Association, Bilbao, Spain

Rubén Alonso ICT and Robotics, R2M Solution s.r.l., Pavia, Italy

Majjed Al-Qatf Insight Research Ireland Centre for Data Analytics, Data Science Institute, University of Galway, Galway, Ireland

Mohammed Salah Al-Radhi Department of Telecommunications and Artificial Intelligence, Budapest University of Technology and Economics, Budapest, Hungary

John Angelopoulos Laboratory for Manufacturing Systems and Automation, Department of Mechanical Engineering and Aeronautics, University of Patras, Rio Patras, Greece

Zoi Arkouli Laboratory for Manufacturing Systems and Automation, University of Patras, Patras, Greece

Juanan Arreta IDEKO, Elgoibar, Spain

Magnus Bång Department of Computer and Information Science, Linköping University, Linköping, Sweden

Iddo Bante University of Twenty, Enschede, The Netherlands

Alex Barceló Barcelona Supercomputing Center, Barcelona, Spain

Marielena Marquez Barreiro Gradiant Technology Center, Pontevedra, Vigo, Spain

Leire Bastida eServices, TECNALIA, Basque Research and Technology Alliance (BRTA), Derio, Spain

Ahmed Nabil Belbachir Norwegian Research Centre (NORCE), Bergen, Norway

Ricardo J. Bessa INESC TEC, Porto, Portugal

Dimitris Bibikas Zenith Gas & Light SA, Thessaloniki, Greece

Dimitrios Bimpikas Zenith Gas Supply Company, Thessaloniki, Greece

Lorenzo Boi Department of Mathematics and Computer Science, University of Cagliari, Cagliari, Italy

Daniel Boos SBB Swiss Federal Railways, Bern, Switzerland

Clark Borst Delft University of Technology, Delft, The Netherlands

Nefeli Bountouni Suite5 Data Intelligence Solutions Ltd, Limassol, Cyprus

Dimitris Bouras UBITECH, Chalandri, Greece

Monica Caballero NTT DATA, Barcelona, Spain

Diego Calvanese Free University of Bozen-Bolzano, Ontopic s.r.l., Bolzano, Italy

Piero Campalani Eurac Research—Center for Climate Change and Transformation, Bolzano, Italy

Davide Dalle Carbonare Engineering Ingegneria Informatica, Rome, Italy

Michele Cardinali Institute of Law, Politics and Development, Sant'Anna School of Advanced Studies, Pisa, Italy

Department of Human Studies, University of Macerata, Macerata, Italy

Ludes Campus, Lugano, Switzerland

Gabriel Guimarães Carvalho imec, Kapeldreef, Leuven, Belgium

Alberto Castagna EnliteAI, Wien, Austria

Fernando Castaño Centre for Automation and Robotics (CAR), Spanish National Research Council (CSIC), Arganda del Rey, Spain

Ricardo Chavarriaga Zurich University of Applied Sciences, Zurich, Switzerland

Theodora Chrysoula Athena Research Center, Marousi, Greece

Benjamin Cogrel Ontopic s.r.l., Bolzano, Italy

Diarmuid Ó. Conchubhair Future Mobility Campus Ireland, Shannon, Clare, Ireland

Javier Conejero Barcelona Supercomputing Center, Barcelona, Spain

Daniele Crippa Consorzio Intellimech, Bergamo, Italy

Edward Curry Insight Research Ireland Centre for Data Analytics, Data Science Institute, University of Galway, Galway, Ireland

Theodore Dalamagas Athena Research Center, Marousi, Greece

Viktor Daropoulos domx IoT Technologies, Thessaloniki, Greece

Mireya de Diego Industry 4.0 Research Area, Fundacion CARTIF, Boecillo, Spain

Wassim Derguech Future Mobility Campus Ireland, Shannon, Clare, Ireland

Tom De Schepper imec, Leuven, Belgium
imec, Kapeldreef, Leuven, Belgium

Konstantina-Christina Diamanti Information Technologies Institute, Centre for Research and Technology Hellas, Thessaloniki, Greece

Duarte Dias INESC TEC, Porto, Portugal

Marios Dikaiakos Department of Computer Science, University of Cyprus, Nicosia, Cyprus

Nikos Dimitropoulos Laboratory for Manufacturing Systems and Automation, University of Patras, Patras, Greece

Adrian Egli SBB Swiss Federal Railways, Bern, Switzerland

Andrina Eisenegger University of Applied Sciences and Arts Northwestern Switzerland, Olten, Switzerland

Joost Ellerbroek Delft University of Technology, Delft, The Netherlands

Damiano Falcioni Research Group, BOC Products & Services AG, Vienna, Austria

Anna Fedorova Zurich University of Applied Sciences, Zurich, Switzerland

Cristina Felix NAV Portugal, Lisbon, Portugal

Paulo Figueiras UNINOVA—Centre of Technology and Systems (CTS), FCT Campus, Caparica, Portugal

Maria Font NTT DATA, Barcelona, Spain

Anton Fuxjäger EnliteAI, Wien, Austria

Gema Antequera García CTAG Automotive Technology Center of Galicia, Pontevedra, Vigo, Spain
CTAG Automotive Technology Center of Galicia, O Porriño, Spain

Amaya Garmendia AI, Data and Robotics Association (ADRA), Brussels, Belgium

Joaquim Geraldes NAV Portugal, Lisbon, Portugal

Emanuela Girardi AI, Data and Robotics Association (ADRA), Brussels, Belgium

Vasileios Gkolemis Athena Research Center, Marousi, Greece

Christos Gkrizis Laboratory for Manufacturing Systems and Automation, University of Patras, Patras, Greece

Lore Goetschalckx imec, Leuven, Belgium

Sergio Gusmeroli Politecnico di Milano, Milan, Italy

Rodolfo E. Haber Centre for Automation and Robotics (CAR), Spanish National Research Council (CSIC), Arganda del Rey, Spain

Samira Hamouche University of Applied Sciences and Arts Northwestern Switzerland, Olten, Switzerland

Carl Hans OHS Engineering GmbH, Bremen, Germany

Rafiqul Haque Insight Research Ireland Centre for Data Analytics, Data Science Institute, University of Galway, Galway, Ireland

Mohamed Hassouna Fraunhofer IEE, Kassel, Germany

Fredrik Heintz Linköping University, Linköping, Sweden

Department of Computer and Information Science, Artificial Intelligence and Integrated Computer Systems, Faculty of Science & Engineering, Linköping University, Linköping, Sweden

Robert Hellbach BIBA–Bremer Institut für Produktion und Logistik GmbH, University of Bremen, Bremen, Germany

BIBA—Bremer Institut für Produktion und Logistik GmbH, Bremen, Germany

Marina Da Bormida in Cugurra S&D Consulting Europe, Milan, Italy

Giorgos Ioannou Department of Computer Science, University of Cyprus, Nicosia, Cyprus

Morteza Jaderyan Department of Computer Engineering, Faculty of Engineering, Shahid Chamran University of Ahvaz, Ahvaz, Iran

Ricardo Jardim-Gonçalves UNINOVA–Centre of Technology and Systems (CTS), FCT Campus, Caparica, Portugal

Admela Jukan Technische Universität Braunschweig, Braunschweig, Germany

Alexandros Kalafatelis Four Dot Infinity, Athens, Greece

Stratos Keranidis domx IoT Technologies, Thessaloniki, Greece

Mila Koeva University of Twenty, Enschede, The Netherlands

Ioannis Kompatsiaris Information Technologies Institute, Centre for Research and Technology Hellas, Thessaloniki, Greece

Sjoerd Kop TenneT, Arnhem, The Netherlands

Sotiris Koussouris Suite5 Data Intelligence Solutions Ltd, Limassol, Cyprus

Katalin Kovacs Innomine, Budapest, Hungary

Kostiantyn Kucher Department of Science and Technology, Linköping University, Norrköping, Sweden

Ignacio Lacalle Universitat Politècnica de València, València, Spain

Charalambos Lambri Department of Computer Science, University of Cyprus, Nicosia, Cyprus

Eleni Lavasa Athena Research Center, Marousi, Greece

Bruno Lemetayer Réseau de Transport d'Électricité, Paris, France

Giulia Leto Delft University of Technology, Delft, The Netherlands

Milad Leyli-Abadi IRT SystemX, Paris, France

Roman Liessner Deutsche Bahn, Berlin, Germany

Katerina Linden Department of Computer and Information Science, Artificial Intelligence and Integrated Computer Systems, Faculty of Science & Engineering, Linköping University, Linköping, Sweden

Ross Little ATOS, Madrid, Spain

Jonas Lundberg Department of Science and Technology, Linköping University, Norrköping, Sweden

Sotiris Makris Laboratory for Manufacturing Systems and Automation, University of Patras, Patras, Greece

Antoine Marot Réseau de Transport d'Électricité, Paris, France

Simone Martin Marotta Expert.AI, Naples, Italy

Silvia M. Massa Department of Mathematics and Computer Science, University of Cagliari, Cagliari, Italy

Marlene Mayr Research Group, BOC Products & Services AG, Vienna, Austria

Maroua Meddeb IRT SystemX, Paris, France

Francisco Melendez FIWARE Foundation, Berlin, Germany

Manuel Meyer Flatland Association, Bern, Switzerland

Wael M. Mohammed FAST-Lab, Faculty of Engineering and Natural Sciences, Tampere University, Tampere, Finland

Gabriella Monteleone Politecnico di Milano, Milan, Italy

Elena Mossali Consorzio Intellimech, Bergamo, Italy

Santiago Muños-Landin AIMEN Technology Centre, Pontevedra, Vigo, Spain

Chukwuemeka Muonagor Technische Universität Braunschweig, Braunschweig, Germany

Sergi Nadal Universitat Politècnica de Catalunya, Barcelona Tech, Barcelona, Spain

Géza Németh Department of Telecommunications and Artificial Intelligence, Budapest University of Technology and Economics, Budapest, Hungary

Martina Imarisio Neviani Consorzio Intellimech, Bergamo, Italy

Nikolaos Nikolakis Laboratory for Manufacturing Systems, University of Patras, Patras, Greece

Alexandros Nizamis Centre for Research and Technology, Hellas—Information Technologies Institute (CERTH/ITI), Thessaloniki, Greece

Philip O'Brien Walton Institute, South East Technological University, Waterford, Ireland

Ibon Ocaña Ceit-IK4, San Sebastian, Spain

Alexandros Oikonomidis Centre for Research and Technology, Hellas—Information Technologies Institute (CERTH/ITI), Thessaloniki, Greece

Peio Oiz AnySolution SL, Palma de Mallorca, Spain

Dolores Ordóñez AnySolution SL, Palma de Mallorca, Spain

George Pallis Department of Computer Science, University of Cyprus, Nicosia, Cyprus

Julia Palma CeADAR—Ireland's Centre for Artificial Intelligence, University College Dublin, Dublin, Ireland

Symeon Papadopoulos Information Technologies Institute, Centre for Research and Technology Hellas, Thessaloniki, Greece

Konstantinos Perakis UBITECH, Chalandri, Greece

Florencia Pérez Avoris Corporación Empresarial, Palma de Mallorca, Spain

Philip Piatkiewicz AI, Data and Robotics Association (ADRA), Brussels, Belgium

Alberto Pirni Institute of Law, Politics and Development, Sant'Anna School of Advanced Studies, Pisa, Italy

Pietro Pittaro Prima Industrie, Collegno, Torino, Italy

Carmen Polcaro Innovalia Association, Bilbao, Spain

Anna Queralt Universitat Politècnica de Catalunya, Barcelona Tech, Barcelona, Spain

Victor Alonso Ramos CTAG Automotive Technology Center of Galicia, O Porriño, Spain

Muhammad Asif Razzak Insight Research Ireland Centre for Data Analytics, Data Science Institute, University of Galway, Galway, Ireland

Diego Reforgiato Recupero Department of Mathematics and Computer Science, University of Cagliari, Cagliari, Italy

ICT and Robotics, R2M Solution s.r.l., Pavia, Italy

Phil Reiter imec, Kapeldreef, Leuven, Belgium

Sangheeta Reji Digital Public Services, Fraunhofer Institute for Open Communication Systems, Berlin, Germany

Anibal Reñones CARTIF Technology Center, Valladolid, Spain
Industry 4.0 Research Area, Fundacion CARTIF, Boecillo, Spain

Celia Minguet Requeni CTAG Automotive Technology Center of Galicia, O Porriño, Spain

Daniele Riboni Department of Mathematics and Computer Science, University of Cagliari, Cagliari, Italy

Elena Ricci Institute of Law, Politics and Development, Sant'Anna School of Advanced Studies, Pisa, Italy
Department of Human Studies, European University of Rome, Rome, Italy

Nicoletta Risi imec, Kapeldreef, Leuven, Belgium

Iria Galiñanes Romero CTAG Automotive Technology Center of Galicia, O Porriño, Spain

Juha Röning University of Oulu, Oulu, Finland

Cinzia Rubattino Engineering Ingegneria Informatica, Rome, Italy

Hélio Sales NAV Portugal, Lisbon, Portugal

A'adel Sayyahi Department of Computer Engineering, Faculty of Engineering, Shahid Chamran University of Ahvaz, Ahvaz, Iran

Viola Schiaffonati Politecnico di Milano, Milan, Italy

Manuel Schneider Flatland Association, Bern, Switzerland

Marc Schoenauer INRIA (National Institute for Research in Digital Science and Technology), Rocquencourt, France

Jose Ramón Sierra LiveM, Zaragoza, Spain

Ricardo Simón-Carbajo CeADAR—Ireland's Centre for Artificial Intelligence, University College Dublin, Dublin, Ireland

Charalambos Sinnis Athena Research Center, Marousi, Greece

Irene Sturm Deutsche Bahn, Berlin, Germany

Anna Sumereder Research Group, BOC Products & Services AG, Vienna, Austria

Andre Tabone Department of R&D, MCS DataLabs, Berlin, Germany
MCS Data Labs, Berlin, Germany

Saeed Talebzadeh DataCalculus, Tallinn, Estonia

Gabriele Tassi Expert.AI, Naples, Italy

Andon Tchechmedjiev EuroMov Digital Health in Motion, University of Montpellier, IMT Mines Alès, Montpellier, France

Julia Usher University of Applied Sciences and Arts Northwestern Switzerland, Olten, Switzerland

Herke Van Hoof University of Amsterdam, Amsterdam, The Netherlands

Olivier Vasseur Faculty of Applied Engineering, IDLab—imec—University of Antwerp, Antwerp, Belgium

Jan Viebahn TenneT, Arnhem, The Netherlands

John Violos Information Technologies Institute, Centre for Research and Technology Hellas, Thessaloniki, Greece

Heike Vornhagen Insight Research Ireland Centre for Data Analytics, Data Science Institute, University of Galway, Galway, Ireland

Toni Wäfler University of Applied Sciences and Arts Northwestern Switzerland, Olten, Switzerland

Kaili Wang imec, Leuven, Belgium

Siri Willems imec, Leuven, Belgium

Robert Woitsch Research Group, BOC Products & Services AG, Vienna, Austria

Binbin Xu EuroMov Digital Health in Motion, University of Montpellier, IMT Mines Alès, Montpellier, France

Mouadh Yagoubi IRT SystemX, Paris, France

Giacomo Zanotti Politecnico di Milano, Milan, Italy

Fatemeh Ahmadi Zeleti Insight Research Ireland Centre for Data Analytics, Data Science Institute, University of Galway, Galway, Ireland

AI, Data, and Robotics: A Structured Overview of Contributions and Priorities

Heike Vornhagen, Philip Piatkiewicz, Fredrik Heintz, Marc Schoenauer, Ahmed Nabil Belbachir, Emanuela Girardi, Juha Röning, and Edward Curry

Abstract This chapter provides a comprehensive overview of the rapidly evolving fields of artificial intelligence, data, and robotics (ADR), introducing the foundational concepts, transformative applications, and forward-looking frameworks that structure the book. Positioned within Europe's broader digital innovation landscape, it highlights how ADR technologies are reshaping key sectors (including manufacturing, healthcare, mobility, and environmental sustainability) while underscoring the importance of ethical, human-centric, and sustainable approaches to their development. Organized around three core themes, Foundations, Transformations, and Future Directions, the chapter situates the contributions of each subsequent section, tracing current research priorities, emerging technological trajectories, and real-world deployments. By emphasizing multidisciplinary collaboration and strategic foresight, it outlines how integrating technical excellence with societal values can foster trustworthy, equitable, and resilient digital ecosystems. Serving as a roadmap

H. Vornhagen (✉) · E. Curry
Insight Research Ireland Centre for Data Analytics, Data Science Institute, University of Galway, Galway, Ireland
e-mail: hcike.vornhagen@insight-centre.org; edward.curry@insight-centre.org

P. Piatkiewicz · E. Girardi
AI, Data and Robotics Association (ADRA), Brussels, Belgium
e-mail: secretary-general@adr-association.eu; emanuela.girardi@popai.me

F. Heintz
Linköping University, Linköping, Sweden
e-mail: fredrik.heintz@liu.se

M. Schoenauer
INRIA (National Institute for Research in Digital Science and Technology), Rocquencourt, France
e-mail: marc.schoenauer@inria.fr

A. N. Belbachir
Norwegian Research Centre (NORCE), Bergen, Norway
e-mail: nabe@norceresearch.no

J. Röning
University of Oulu, Oulu, Finland
e-mail: juha.roning@oulu.fi

E. Curry et al. (eds.), *Artificial Intelligence, Data and Robotics*,
https://doi.org/10.1007/978-3-032-10561-5_1

for the chapters that follow, this overview aims to inspire coordinated efforts among researchers, industry leaders, policymakers, and citizens to harness the full potential of AI, data, and robotics for the benefit of society.

Keywords Artificial intelligence (AI) · Data · Robotics · Digital transformation · Human-centric design · Ethics · Sustainability · European innovation · Emerging technologies · Trustworthy AI

1   Introduction

This book explores the rapidly evolving landscape of artificial intelligence (AI), data, and robotics (ADR), highlighting their foundational principles, transformative applications, and future directions. As Europe positions itself at the forefront of digital innovation, these technologies are driving profound changes across industries—from manufacturing and healthcare to mobility and environmental sustainability. The chapters presented herein provide a comprehensive overview of the key research agendas, innovative frameworks, and real-world deployments that define the current and emerging state of ADR. Emphasizing ethical considerations, human-centric design, and sustainability, this introduction sets the stage for understanding how multidisciplinary collaboration and strategic foresight are critical to unlocking the full potential of AI, data, and robotics in creating a competitive, inclusive, and resilient digital society.

2   AI, Data, and Robotics: Foundations

This part introduces the strategic, technical, and conceptual foundations that underpin Europe's AI, data, and robotics (ADR) ecosystem. It begins with a chapter on the **AI, Data, and Robotics Association (ADRA)**, highlighting its formation, mission, and evolving role as the private-side partner in the European Partnership on AI, data, and robotics under Horizon Europe. Positioned at the intersection of research, industry, and policy, ADRA plays a central role in ensuring technological progress aligns with European values—ethics, inclusivity, and sustainability. Key focus areas include the **Strategic Research, Innovation, and Deployment Agenda (SRIDA)** and the emerging convergence of **generative AI and robotics**, both of which are shaping the roadmap for Europe's digital future.

Building on this strategic context, a dedicated chapter on **SRIDA** outlines its function as a dynamic and inclusive framework for aligning ADR research with societal and industrial goals. It emphasizes the collaborative processes and iterative structures that ensure SRIDA remains responsive and representative of Europe's diverse stakeholder base.

Following on from this comes an update from the ADRA topic group on **digital twins (DTs)**. This chapter explores in detail the multiple domains that are utilizing DTs to simulate scenarios such as crisis responses and to provide effective real-time monitoring.

Subsequent chapters explore foundational approaches to responsible and effective AI development. One focuses on the integration of **human-in-the-loop (HITL)** methods with **explainable AI (XAI)** to improve transparency, reduce bias, and increase trustworthiness in AI decision-making, especially in critical domains like healthcare and robotics. Another examines the use of **generative AI** to create synthetic visual data for machine learning applications, addressing challenges of data scarcity, privacy, and cost—particularly in sectors such as autonomous driving and medical imaging.

The part continues with a chapter on **hybrid intelligence**, which combines science-based (white-box) models with machine learning (black-box) approaches to enhance predictive performance, interpretability, and robustness in complex systems across industries. Finally, the chapter on **frugal machine learning (FML)** addresses the growing need for energy-efficient, resource-aware AI. It surveys techniques for minimizing computational and data demands, which are crucial for AI systems operating in edge and IoT environments.

Collectively, these contributions establish a robust foundation for the ADR field—strategically guided by ADRA's vision and practically informed by emerging technologies and methodologies. They offer a roadmap for building trustworthy, efficient, and human-centric AI systems that can drive both innovation and public value across Europe.

3 AI, Data, and Robotics: Transformations

The second part explores the transformative impact of AI, data, and robotics technologies across diverse industrial and societal domains, organized to reflect a natural progression from foundational industrial applications to advanced human-centric systems.

We begin by examining **digital transformation in manufacturing**, where AI integration is revolutionizing production processes. The initial chapter opens with an analysis of **digital transformation in manufacturing** underpinned by AI integration. Here, Industry 4.0 and 5.0 paradigms emphasize not only optimization and automation but also resilience, sustainability, and adaptability—especially important for small and medium-sized enterprises navigating complex technological and societal challenges such as climate change. The chapter stresses the importance of tailored strategies and organizational alignment with digital platforms to advance manufacturing maturity.

Building on this foundation, the discussion moves to **AI-based management approaches for production processes**, advocating hybrid symbolic and subsymbolic AI models to enhance transparency, explainability, and decision support within modular and dynamic production environments. Real-world European projects exemplify how such hybrid approaches can improve maintenance, compliance, and workforce allocation.

Industry 5.0's vision of human-centric, sustainable, and circular manufacturing is explored through the lens of **data, AI, and robotics technologies**, illustrating their decisive role in embedding sustainability principles within manufacturing ecosystems. Multiple European initiatives demonstrate how data spaces, AI testing facilities, and digital innovation hubs contribute to this transformative agenda.

Following this industrial focus, the discussion shifts to **mobility and data infrastructures**, exemplified by the challenges and opportunities in establishing interoperable data-sharing frameworks such as the Irish Mobility Data Space. This illustrates the critical importance of robust data governance and collaboration platforms to enable smart, sustainable mobility solutions.

Next, the part delves into **human-AI collaboration and data governance**, addressing the complex interplay between transparency, trust, and safety in critical systems.

Addressing safety-critical contexts, an interdisciplinary chapter proposes a **holistic framework for human-AI collaboration** that balances transparency, trust, and robust decision-making. Drawing from cognitive science, engineering, and AI, this framework supports effective integration of human expertise and automation across critical infrastructures like energy and mobility.

Data life cycle management is tackled through the **CyclOps project**, which leverages semantic technologies and knowledge graphs to automate governance and maintenance across distributed, heterogeneous data sources. This approach enhances reproducibility, traceability, and explainability in AI applications, enabling scalable, trustworthy data management aligned with FAIR principles.

Building on these foundations, the focus moves toward **specialized, frontier applications** of AI and robotics. This includes a chapter on **intelligent underwater perception** technologies that face unique challenges such as limited lighting and connectivity. Advances in sensing, processing, and simulation promise next-generation autonomous underwater systems for inspection and rescue operations.

Bridging neuroscience and AI, the **brain-to-speech synthesis** chapter presents a transformer-based model integrating prosody-aware feature engineering to decode speech from intracranial brain signals. This novel approach advances neuroprosthetics with implications for restoring communication in individuals with speech impairments.

The part concludes with a focus on **robotics in social and healthcare contexts**, highlighting the integration of advanced AI models in companion robots for elderly care and the synergy of robotics, IoT, and deep learning in healthcare delivery. A chapter on **companion robots for elderly care** showcases the integration of large language models with robotics to enhance interaction quality, personalization, and empathy. This project critically examines technological and ethical aspects, aiming to create natural reassuring human-robot relationships that improve well-being.

The final chapter examines **robotics in healthcare IoT and deep learning**, illustrating how this convergence revolutionizes surgical assistance, rehabilitation, and patient care through real-time connectivity, intelligent analytics, and autonomous decision-making—addressing resource constraints and enabling better patient outcomes.

Together, these chapters provide a comprehensive and nuanced view of how AI, data, and robotics are reshaping industries, infrastructure, and human-centered applications. They underline the importance of interdisciplinary approaches, ethical considerations, and technological innovation in driving forward a sustainable, inclusive, and impactful future for AI and robotics across Europe and beyond.

4 AI, Data, and Robotics: Future Directions

The final part explores the critical pathways for advancing AI, data, and robotics (ADR) technologies through strategic planning, applied industrial innovation, and human-centered approaches.

The opening chapter lays out comprehensive **roadmaps and research agendas** that define Europe's strategic priorities for ADR development. Emphasizing the necessity of ethical, explainable, and adaptive AI, it advocates for cross-disciplinary collaboration to address complex challenges such as bias, fairness, and accountability. This foundational vision aligns public and private sector investments to strengthen Europe's competitiveness while ensuring technologies remain trustworthy, inclusive, and aligned with societal values.

Building on this strategic vision, the following chapter explores how digital twin technologies can shape Europe's industrial and societal transformation by 2030 and beyond. It outlines a forward-looking roadmap developed within the ADRA community, highlighting the role of AI, data governance, frugal computing, and human-centric design in enabling trusted, sustainable, and globally competitive digital twin ecosystems.

The third chapter focuses on **industrial human-robot collaboration** as a key driver for digital competitiveness and sustainability. Highlighting pioneering EU-funded projects—ARISE, FORTIS, and JARVIS—it showcases innovative solutions for scalable, safe, and adaptive human-robot interaction in manufacturing and logistics. These initiatives demonstrate how integrating AI, data, and robotics can enhance task efficiency and multimodal collaboration, particularly supporting small and medium enterprises in their digital transformation and green transition efforts.

The part concludes with a deep dive into **human-AI interaction and visualization**, emphasizing the importance of safety, trust, and explainability in mission-critical and complex operational environments. By drawing on interdisciplinary insights from human factors, cognitive engineering, and data science, it advocates for embedding interactive visualization and human-centered design principles within ADR systems. This integration is vital for fostering robust decision-making, ensuring operator trust and enabling effective human-AI collaboration across diverse domains

Together, these chapters present a coherent and forward-looking narrative: setting strategic priorities, translating them into transformative industrial applications, and addressing the nuanced human factors essential for the responsible and effective deployment of AI, data, and robotics technologies.

5 Thematic Overview of Chapter Content

To better understand the thematic distribution and focus areas of the book, the following content analysis table (Table 1) maps each chapter against key topics and themes derived from their abstracts. This overview allows readers to quickly grasp how the chapters complement each other, identify thematic clusters, and ensure balanced coverage across the diverse landscape of artificial intelligence, data, and robotics applications. The analysis highlights recurring themes such as industrial transformation, healthcare, human-robot interaction, and AI methodologies, providing a structured lens through which the entire collection can be viewed and navigated.

Table 1 Overview of themes, technologies, application domains, and addressed challenges for each chapter

	Chapter title	Key themes/ topics covered	Technologies/ methods highlighted	Application domains	Strategic focus/ challenges addressed
2	Shaping Europe's Future	Overview of AI, data, robotics convergence	Strategic frameworks, technology integration	Cross-industry	Roadmapping, interdisciplinary collaboration
3	Strategic Research and Innovation Agenda (SRIDA)	Research priorities, policy alignment	AI strategy, innovation agendas	European tech ecosystem	Ethics, competitiveness, public-private cooperation
4	Digital Twins in Europe	Digital twins	Real-time monitoring, predictive analysis, simulation	Robotics, healthcare, manufacturing, transport, construction	Digital twin innovation, integrating AI, robotics, and data governance
5	The Cross-Fertilization Between the Human-in-the-Loop Approach and the Explainable AI Techniques	Human-in-the-loop (hitl), explainable ai (xai), ethical ai, human oversight and interaction, cognitive load and safety	Human-in-the-loop systems, explainable ai techniques, wearable sensors	Robotics, healthcare	Ethics, cognitive load, safety, human-centered automation, trust, interaction effectiveness

Table 1 (continued)

	Chapter title	Key themes/ topics covered	Technologies/ methods highlighted	Application domains	Strategic focus/ challenges addressed
6	Generative AI to Tackle Visual Data Challenges	Visual data processing, generative AI applications	Generative AI models, computer vision	Imaging, media, autonomous systems, healthcare	Data efficiency, complex visual interpretation
7	Hybrid Intelligence: Integrating Symbolic and Sub-symbolic AI	Combining symbolic and neural AI	Hybrid AI approaches, explainability	Knowledge management, decision-making	Transparency, hybrid reasoning, user trust
8	Frugal Machine Learning for Resource-Constrained Environments	Efficient ML under limited resources	Lightweight models, edge computing	IoT, mobile, embedded systems	Energy efficiency, accessibility, scalability
9	Digital Transformation in Manufacturing under the Light of AI Integration	Industry 4.0/5.0, digital maturity, sustainability	Industrial internet platforms, TOE framework	Manufacturing SMEs	Low-carbon transition, resilience, organizational change
10	AI-Based Management Approaches for Production Processes	Process transformation, hybrid AI for transparency	Hybrid symbolic and sub-symbolic AI, process knowledge	Manufacturing, maintenance	Process optimization, decision support
11	Data, AI, Robotics Transformative Power in Industry 5.0	Sustainability, circular economy	AI, data spaces, digital twins	Manufacturing	Circular manufacturing, sustainability, resilience
12	Toward the Irish Mobility Data Space	Mobility data frameworks, interoperability, governance	Data spaces, secure data sharing	Mobility and transportation (Ireland)	Standardization, governance, smart mobility
13	Holistic Framework for Human-AI Collaboration in Safety-Critical Systems	Human-AI interaction, trust, transparency in critical systems	Cognitive engineering, decision theory	Energy, mobility, critical infrastructure	Safety, explainability, human-AI collaboration
14	CyclOps: Leveraging Semantic Technologies	Data life cycle governance, semantic technologies	Knowledge graphs, FAIR principles, automation	Large-scale AI/ data applications	Automated governance, traceability, reproducibility
15	Intelligent Underwater Perception	Autonomous sensing, AI in underwater systems	Robotics, AI, underwater simulation	Underwater inspection, rescue	Connectivity, power constraints, autonomy

(continued)

Table 1 (continued)

	Chapter title	Key themes/ topics covered	Technologies/ methods highlighted	Application domains	Strategic focus/ challenges addressed
16	Brain-to-Speech: Prosody Feature Engineering and Transformer-Based Reconstruction	Neural decoding, speech synthesis	Transformer models, prosody feature extraction	Neuro-prosthetics, assistive tech	Speech naturalness, expressiveness, real-time processing
17	A Companion Robot Platform for Exploring Technical and Ethical Aspects in Elderly Care	LLMs in robotics, ethics, human-robot interaction	Large language models, empathetic interaction systems	Elderly care, healthcare robotics	Ethical AI, personalized interaction
18	Robotics in Healthcare IoT	Robotics, IoT integration, deep learning	IoT, deep learning, real-time analytics	Healthcare	Precision, chronic disease management, efficiency
19	Roadmaps and Agendas for Research and Innovation	Research priorities, ethical AI	AI strategy, adaptive and explainable AI	Cross-sector	Ethical AI, trust, strategic vision
20	The Future of Digital twins in Europe	Digital twins	Real-time monitoring, predictive analysis, simulation	Robotics, healthcare, manufacturing, transport, construction	Digital twin innovation, integrating AI, robotics, and data governance
21	The Next Generation of Human-Robot Interaction	Human-centric robotics, collaborative systems	Middleware, multimodal interfaces	Manufacturing, healthcare, logistics	Human-robot collaboration, SME support, sustainability
22	Human-AI Interaction and Visualisation Perspectives	Explainability, human-centered AI, visualization	Visual analytics, human-computer interaction	Mission-critical systems	Safety, trust, robust AI decision support

A second analysis (Table 2) provides an overview of the key themes covered by each chapter, thereby contextualizing the book's contributions within a broader European research and innovation agenda. This perspective aims to support readers in understanding the strategic relevance of each chapter and their collective impact on shaping future research and applications.

Table 2 Key priorities addressed by each chapter

Chapter	Ethical AI	Human-centric robotics	Sustainable manufacturing	Data governance	Industrial transformation	Healthcare	AI methodologies	Digital transformation	Mobility and infrastructure
AI, data, and robotics: Foundations									
2	+	+			+		+	+	
3	+	+	+	+	+	+	+	+	+
4	+		+	+		+			+
5	+	+		+		+	+		
6	+					+	+	+	+
7	+	+					+	+	
8	+		+				+		
AI, data, and robotics: Transformations									
9			+		+				
10					+		+	+	
11		+	+		+			+	
12				+				+	+
13	+				+		+	+	+
14	+			+			+	+	
15					+		+		+
16						+			
17	+	+				+	+	+	
18		+				+		+	
AI, data, and robotics: Future directions									
19	+	+		+	+		+	+	
20	+		+	+		+			+
21	+	+	+	+	+	+	+	+	
22	+	+		+	+		+		
Total	14	10	7	9	10	9	14	13	7

6 Conclusion

The convergence of AI, data, and robotics presents unprecedented opportunities and challenges that call for coordinated research, innovation, and deployment efforts. This introductory chapter has outlined the foundational technologies, ongoing transformations across sectors, and forward-looking frameworks that will shape the trajectory of ADR development in Europe and beyond. By integrating technical excellence with ethical principles and human-centered perspectives, the field is moving toward solutions that are not only powerful and efficient but also trustworthy and equitable. As the subsequent chapters delve deeper into specific domains and applications, this overview provides a cohesive roadmap, inspiring collaboration among researchers, industry leaders, policymakers, and society at large to build a future where AI, data, and robotics can serve the greater good.

Foundations

Shaping Europe's Future:
The Convergence of AI, Data, and Robotics

Heike Vornhagen, Philip Piatkiewicz, Fredrik Heintz, Marc Schoenauer,
Amaya Garmendia, Fatemeh Ahmadi Zeleti, Ahmed Nabil Belbachir,
Emanuela Girardi, Juha Röning, and Edward Curry

Abstract This chapter explores the convergence of generative AI and robotics within the broader context of Europe's AI, data, and robotics (ADR) strategy. It examines how these powerful general-purpose technologies are reshaping innovation across strategic sectors while also presenting complex challenges in safety, trustworthiness, and sustainability. The chapter highlights Europe's distinct approach anchored in values-driven innovation, technological sovereignty, and strategic autonomy and outlines the role of the ADRA partnership in steering this transformation. Through initiatives such as the SRIDA, high-impact pilots, and challenge-based innovation, ADRA is fostering a resilient European ecosystem that supports responsible, human-centric development of GenAI and Robotics. The chapter concludes by emphasizing the need for coordinated investment, policy

H. Vornhagen (✉) · F. A. Zeleti · E. Curry
Insight Research Ireland Centre for Data Analytics, Data Science Institute,
University of Galway, Galway, Ireland
e-mail: heike.vornhagen@insight-centre.org; fatemeh.ahmadizeleti@insight-centre.org;
edward.curry@insight-centre.org

P. Piatkiewicz · A. Garmendia · E. Girardi
AI, Data and Robotics Association (ADRA), Brussels, Belgium
e-mail: secretary-general@adr-association.eu; amaya.garmendia@adr-association.eu;
emanuela.girardi@popai.me

F. Heintz
Linköping University, Linköping, Sweden
e-mail: fredrik.heintz@liu.se

M. Schoenauer
INRIA (National Institute for Research in Digital Science and Technology),
Rocquencourt, France
e-mail: marc.schoenauer@inria.fr

A. N. Belbachir
Norwegian Research Centre (NORCE), Bergen, Norway
e-mail: nabe@norceresearch.no

J. Röning
University of Oulu, Oulu, Finland
e-mail: juha.roning@oulu.fi

E. Curry et al. (eds.), *Artificial Intelligence, Data and Robotics*,
https://doi.org/10.1007/978-3-032-10561-5_2

13

alignment, and bold action to ensure Europe remains at the forefront of this rapidly evolving technological landscape.

Keywords Artificial intelligence (AI) · Robotics · European technological sovereignty · Innovation ecosystem · Generative AI (GenAI)

1   Introduction

This chapter outlines the formation and strategic direction of the AI, Data, and Robotics Association (ADRA), a key player in Europe's digital innovation landscape. Established to unify the efforts of industry, academia, and policymakers, ADRA serves as the private side of the European Partnership on AI, Data, and Robotics under Horizon Europe. As technological capabilities accelerate, ADRA plays a central role in ensuring these advancements align with European values and priorities by driving progress that is not only competitive but also ethical, inclusive, and sustainable.

The chapter delves into the development of ADRA's vision and mission, offering insight into how they guide the organization's activities and influence the broader research and innovation ecosystem. It also explores the structure and diversity of the ADRA community, its objectives, and the core activities that support its mission. Special attention is given to the Strategic Research, Innovation, and Deployment Agenda (SRIDA), which forms the backbone of ADRA's long-term roadmap, and to the emerging convergence of generative AI (that stands on the shoulders of Giant Data) and robotics, a frontier with transformative potential and significant societal implications. Both of these are summarized later in this chapter. This technological intersection represents an opportunity and a challenge for ADRA as it works to shape a trustworthy and forward-looking AI, data, and robotics agenda.

Finally, the chapter concludes with a look toward the future of ADRA, reflecting on potential directions for growth, collaboration, and impact. Through strategic foresight and stakeholder engagement, ADRA intends to help Europe remain a leader in developing and deploying cutting-edge technologies that serve industry and the public good.

2   Introduction to ADRA

The AI, Data, and Robotics Association (ADRA) was founded on May 21, 2021, to bring together European communities in artificial intelligence (AI), data, and robotics. It was created through a joint effort by five founding organizations: Big Data Value Association (BDVA), European Association for Artificial Intelligence (EurAI), European Lab for Learning and Intelligent Systems (ELLIS), Confederation of Laboratories for Artificial Intelligence Research in Europe (CLAIRE, now CAIRNE), and euRobotics. ADRA acts as the private partner in a public-private partnership with the European Commission.

2.1   Vision and Mission

ADRA envisions Europe leading a revolution in AI, data, and robotics, developing the next generation of systems that are seamless, safe, and trustworthy, thereby becoming essential parts of everyday life and European society. This vision goes beyond technology, aiming for Europe to shape a future where AI-driven tools are reliable and deeply integrated. Achieving this requires significant technological advancements and new value chains within Europe.

ADRA's mission is to unite the European ADR communities to:

- Increase Europe's strategic autonomy in safe, trustworthy critical technologies
- Boost competitiveness through digital and green transformations
- Achieve excellence in global research
- Deliver solutions for climate, health, energy, and security challenges

As of April 2025, ADRA has 180 members from 27 countries, including industry leaders, researchers, academics, and government representatives.

2.2   Community and Ecosystem

ADRA acts as a key link between the European Commission and the ADR community, influencing EU initiatives, policies, and funding calls. The association prioritizes stakeholder engagement and ecosystem building to ensure long-term impact and scalability of projects within Europe's AI ecosystem. ADRA's ecosystem connects Europe's leading researchers and industry players in ADR technologies, working together to turn research into market-ready solutions. ADRA strives to build a sovereign, competitive European AI, data, and robotics landscape aligned with EU values and regulations.

2.3   Objectives and Key Activities

ADRA aims to speed up the development and integration of AI, data, and robotics to tackle societal and industrial challenges. The community contributes to key topics like the SRIDA, the roadmap on generative AI for robotics, and other policy documents, helping shape EU calls and initiatives. Key activities include:

- **Establishing a single point of contact**: ADRA is the EC's primary partner for advancing AI, data, and robotics R&I in Europe.
- **Integrated ecosystem**: Connecting AI, data, and robotics industry, academia, and government expertise across Europe.
- **Research excellence**: Corralling Europe's top minds in ADR technologies.
- **Bridge to industry**: Transforming research into market-ready solutions.
- **Geographic reach**: Established and active presence across Member States.
- **Rapid mobilization**: Providing ready expertise for strategic initiatives.

ADRA's community is active across numerous sectors, including agrifood, digital twin automation, energy, education, future of work, generative AI for robotics and manufacturing, healthcare, innovation, inspection and maintenance, legal, Strategic Research Innovation and Deployment Agenda (SRIDA), and policy and standardization, highlighting use cases that shape European strategies.

2.4 International Collaborations

ADRA works with international partners who share European values, such as Canada's Information and Communications Technology Council (ICTC), to strengthen bilateral economic cooperation, grow AI ecosystems, and compare regulatory approaches. The association believes ethical AI progress requires collaboration across government, academia, industry, and civil society.

2.5 Horizon Europe Projects

ADRA is involved in several Horizon Europe projects, e.g., AIOLIA, COORDINATEF, ROBO-KNOT, and LLM-BRIDGE, that focus on AI ethics guidelines, boosting AI innovation, accelerating robotics knowledge transfer, and strengthening Europe's role in generative AI to ensure European digital sovereignty.

2.6 Flagship Events

ADRA organizes major events like the AI, Data and Robotics Forum (ADRF) and the European Convergence Summit (ECS). These bring together the ADR community, stakeholders, and the European Commission to share use cases and address socioeconomic issues.

The 2025 ADRF in Norway brought together industry leaders, researchers, and policymakers to discuss infrastructure, cybersecurity, and regulatory frameworks supporting Europe's technological independence.

The 2025 ECS addressed resilience in crises, featuring diverse perspectives from industry, research, innovation, civil society, legal, ethical, regulatory, and policy experts. ECS events target ADR decision-makers to shape research and innovation directions.

3 Strategic Context for AI, Data, and Robotics

In the face of ongoing global crises, Europe's sustainable development and welfare are under increasing pressure. Climate change, recent pandemics, and geopolitical conflicts have further strained industries already grappling with rapid technological transformation. As vulnerabilities in critical value chains and essential resources become more apparent, resilience and technological sovereignty have become renewed priorities for Europe. Strengthening Europe's position in AI, data, and robotics (ADR) is vital for supporting societal welfare and accelerating the digital and green transitions across industries and the economy.

While ADR technologies offer significant potential, they also present risks that must be managed through a combination of regulation and innovation. A key objective of ADRA is ensuring the safe and ethical use of ADR technologies.

Innovation is progressing at an unprecedented pace, and ADR technologies are already transforming daily life. It is crucial that European companies and the public sector not only reap the benefits of these advancements but also maintain strategic autonomy over the development and deployment of ADR technologies. Achieving this goal requires a well-connected, resilient innovation ecosystem that empowers European stakeholders.

To enhance global competitiveness, particularly in industrial sectors, the EU must pivot from a predominantly regulatory focus to making substantial investments in innovation. This shift will enable the development of trustworthy, human-centered ADR tools and applications made in Europe, and their deployment worldwide will contribute to the dissemination of European values. Investments must span the entire technology stack—from computing infrastructure and foundational AI models to sector-specific applications. Much of the technology stack is sourced externally, which must change through targeted investments at all levels to create viable sovereign European alternatives.

A strategic approach involves investing in ADR technology tailored to industry sectors where Europe has existing strengths. Rather than attempting to predict future breakthroughs and risk falling behind, Europe must proactively support high-potential initiatives and commit to their success. Maintaining strategic autonomy requires Europe to possess the necessary competencies and capabilities by learning and adapting existing technologies and developing unique innovations.

A systematic approach to research funding and technology transfer will be essential in both advancing new technologies and deepening Europe's expertise in existing ones. European companies should actively engage in international research collaborations to acquire and integrate leading-edge knowledge. Additionally, bridging the gap between education, research, and industry through joint infrastructures and real-world training programs will strengthen domain-specific competencies.

To transition from research to widespread innovation and deployment, Europe must elevate the Technology Readiness Level (TRL) of ADR innovations, making them mature and easily integrable into existing products and services. Combining AI technologies with traditional European industrial strengths such as manufacturing, agrifood, energy, and healthcare could offer a significant competitive advantage.

ADR technologies developed under the ADR Partnership will also serve as key enablers for addressing major societal challenges such as climate resilience, energy security, and food sustainability. European leadership in ADR is a matter of industrial competitiveness and is essential for societal welfare and sustainability. This vision aligns with the principle of "AI for Good" and "AI for All," ensuring that technological progress benefits everyone.

For example, AI-enhanced and data-driven robotics can contribute to the green economy (e.g., recycling, remanufacturing, agrifood, and inspection), support healthy aging (e.g., healthcare and assisted living), improve workplace safety (e.g., human-robot collaboration, handling hazardous tasks), and reinforce ethical standards in robotics. Furthermore, strengthening Europe's industrial policy through ADR investments will fortify strategic sectors.

Europe's ADR vision for 2030 is ambitious yet attainable through coordinated research, investment, and policymaking. By integrating AI, data, and robotics while adhering to ethical AI standards, Europe can establish itself as a global leader in digital transformation.

Lastly, ensuring that the public sector has access to cutting-edge ADR tools is imperative for maintaining high-quality public services that reflect the values and standards of the European Union. Investing in ADR technology today is essential for securing Europe's technological sovereignty, fostering sustainable economic growth and positioning Europe at the forefront of the global AI revolution.

4 The Role of ADRA's Strategic Research, Innovation, and Deployment Agenda (SRIDA)

The ADRA Strategic Research, Innovation, and Deployment Agenda (SRIDA) results from dedicated collaboration within ADRA's community. It presents a bold, long-term vision for developing and deploying trustworthy AI, data, and robotics (ADR) technologies in Europe. This agenda addresses the most pressing challenges of our time, ensuring that Europe achieves strategic autonomy while driving global leadership in ADR.

While ADRA's vision and mission describe who the community is and why it acts, SRIDA defines what needs to be done to advance Europe's leadership in ADR by outlining the concrete steps required to develop sovereign technologies, foster innovation, and deliver societal impact.

SRIDA articulates a vision for 2030 in which Europe leads a responsible AI-powered green digital transformation that supports a sustainable, prosperous, secure multicultural society rooted in European values. A shared, secure data infrastructure will be established, balancing privacy with interoperability. Autonomous robotic systems will optimize efficiency and safety in critical sectors such as agriculture, transport, and healthcare. AI-based systems will enhance trustworthy decision-making, while European ADR enterprises will expand their global market share, attracting top talent and fostering an ecosystem of AI-first enterprises.

Achieving this vision demands a strong, multi-stakeholder ecosystem. SRIDA charts a course for the rapid deployment of next-generation ADR technologies while ensuring alignment with societal values. It reinforces the role of the ADR Partnership in bridging research, innovation, and implementation, positioning Europe as a technological leader while reducing fragmentation and securing a competitive edge in the global market.

The agenda outlines a set of missions to guide this effort. The ADR Partnership is committed to establishing a cohesive, high-performing ADR ecosystem that enhances Europe's industrial leadership in these transformative technologies. This entails uniting Europe's research landscape, attracting global talent, and developing workforce capabilities in ADR. Crucially, ADR technologies must yield high socioeconomic impact while maintaining public trust and regulatory compliance. Through SRIDA, these missions are translated into targeted actions designed to deliver both economic and social impact while maintaining public trust and regulatory compliance. Although Europe has a strong foundation in AI, robotics, and computer vision, it faces increasing competition from North America and Asia. Strategic investments such as the EU Chips Act are vital for sustaining leadership. SRIDA also addresses the need for more integrated research frameworks to reduce duplication, increase efficiency, and reinforce Europe's technological sovereignty.

A key focus is tackling the growing ADR skills gap. SRIDA calls for a coordinated education strategy that can support talent retention and foster innovation ecosystems capable of attracting and developing expertise. Without decisive action, Europe risks falling behind in the global technological race.

Public skepticism toward ADR must also be addressed. By encouraging transparency, ethical AI development, and proactive public engagement, SRIDA supports the cultivation of trust and paves the way for large-scale ADR adoption.

To achieve ADR leadership, SRIDA sets out ambitious goals: enhancing industry competitiveness, accelerating digital transformation, and achieving strategic autonomy in ADR technologies. This requires strengthening Europe's global research impact in AI, data, and robotics, advancing ADR-driven solutions for environmental, food, energy, health, and security challenges and reducing reliance on external suppliers by securing critical technological resources and expanding domestic innovation.

SRIDA calls for investment in AI-driven manufacturing, sustainable ADR initiatives, and greater control over essential digital infrastructure. It also emphasizes the importance of large-scale AI and robotics testbeds, AI-driven sustainability solutions, trustworthy ADR development with ethical safeguards, and industry-wide collaboration to accelerate ADR adoption.

In mapping major trends and challenges, SRIDA identifies emerging technologies such as generative AI, large-scale models, neuro-symbolic AI, AI systems that learn from human feedback, AutoML for SME uptake, and AI-enabled compliance tools.

Despite these advancements, Europe must address key gaps such as the lack of standardized ADR infrastructure; insufficient European participation in large-scale AI projects; challenges in ADR regulation, governance, and compliance; limited access to high-quality data and federated computing infrastructure; and fragmented robotics innovation across industries. To close these gaps, Europe must prioritize investments in AI regulation, AI-enhanced robotics, and cross-industry ADR deployment.

SRIDA also emphasizes the strategic role of ADR technologies in responding to global challenges such as resource security, workforce shortages, and technological dependency. ADR can mitigate these risks by enhancing food production and agricultural efficiency, securing energy resources through AI-powered grid management, strengthening technological sovereignty via semiconductor advancements, and improving healthcare and industrial efficiency with autonomous robotics.

Future ADR missions should focus on AI-powered climate adaptation and sustainability solutions, AI for advanced healthcare and medical research, robotics-driven industrial automation and safety enhancements, AI-enhanced security for cyber and physical threats, and sustainable ADR policies for long-term European prosperity.

Looking ahead to 2025–2027, SRIDA sets out a strategic plan to strengthen collaboration between academia, industry, and policymakers. The development of large-scale AI models tailored to European needs must be prioritized, ensuring that Europe remains competitive in the race for AI dominance. These models should be built on secure, interoperable infrastructures, allowing for seamless integration across industries while adhering to the highest ethical standards. Open ADR research initiatives must be expanded to ensure accessibility and transparency, facilitating cross-sector collaboration and accelerating innovation.

The creation of a sustainable AI infrastructure is imperative. Europe must develop AI technologies that minimize energy consumption and environmental impact while maximizing efficiency and reliability. Investment in general-purpose, trustworthy AI models will enable scalable applications that benefit both industry and society. Furthermore, a standardized and interoperable data ecosystem must support AI training and deployment, ensuring that European enterprises can access high-quality, ethically sourced data.

Advancements in next-generation autonomous robotics must be accelerated. These systems should be designed to enhance industrial automation, healthcare, and security while ensuring safety and compliance with European values. AI-driven scientific research and discovery should also be prioritized, harnessing ADR capabilities to revolutionize fields such as drug discovery, climate modelling, and materials science.

Regulatory and compliance tools are essential for ensuring ethical ADR adoption. Europe must invest in developing robust legal frameworks that protect citizens while enabling innovation. This includes establishing high-performance AI research infrastructure and regulatory sandboxes, which will provide safe environments for testing novel AI and robotics solutions before full-scale deployment. Bridging the gap between theoretical research and industrial application will be key, as will improving data governance and AI ethics compliance frameworks. Targeted funding and training programs must support SMEs in integrating ADR technologies, ensuring that smaller enterprises can compete in the evolving digital economy.

A final priority identified by SRIDA is addressing the shortage of AI-related skills, which continues to limit adoption across Europe. To address this, Europe must raise ADR awareness and engagement through accessible education initiatives, integrate ADR literacy into primary and secondary education, develop a dedicated European ADR curriculum at the bachelor's and master's levels, scale up education capacity in ADR technology while maintaining high standards, expand ADR education across multiple disciplines and professions, and invest in re-skilling and upskilling professionals already in the workforce who were trained before the advent of ADR.

Strengthening Europe's AI talent pipeline, as laid out in SRIDA, ensures that companies can harness ADR technology for value creation and global competitiveness.

4.1 Big Ticket Items: Strategic Focus for Maximum Impact

As part of the SRIDA roadmap, six **Big Ticket Items** were identified—high-priority areas where coordinated investment can unlock the greatest value for Europe. These were selected based on stakeholder consultations, alignment with European policy priorities, and potential for global impact. They serve as focal points for public-private collaboration and strategic investment.

#1: Groundbreaking technological foundations in ADR
#2: Effective and trustworthy general-purpose ADR
#3: An interoperable and integrated framework for data and model ecosystems
#4: Next-generation smart embodied robotic systems
#5: Developing ADR technology for the sciences
#6: Research, innovation, and tools for compliance

For more details on the SRIDA, download the latest version at https://adr-association.eu/news/adra-releases-strategic-research-innovation-and-deployment-agenda.

5 From Strategy to Action: The Convergence of Generative AI and Robotics

While the SRIDA sets out a comprehensive framework for a sovereign European AI, data, and robotics (ADR) landscape, one area merits particular attention for its disruptive potential: the convergence of generative artificial intelligence (GenAI) and robotics. These two general-purpose technologies are increasingly merging into a new paradigm that tightly couples perception, reasoning, and physical action.

This fusion enables unprecedented capabilities by integrating physical action and perception with cognitive adaptability, leading to transformative applications across AI, data, and robotics (ADR).

GenAI enhances robotics, and robotics, in turn, expands the potential of GenAI. Robots provide unique capabilities in action and perception, enabling GenAI to interact with the physical world in a dynamic way. Researchers view robotics as a means to push the boundaries of AI, fostering advancements toward general intelligence through embodied AI, a model that integrates perception, reasoning, and action. In return, GenAI enhances robots' ability to comprehend their environments, communicate naturally with humans, and execute tasks they were not explicitly trained for, creating an adaptive and intelligent system.

These advancements impact core technological components, including multimodal perception and cognition, control systems, simulation of the physical world, human-robot and robot-robot interactions, mapping, navigation, and the development of foundational models for 3D objects. Focusing on robotics, it is essential to research and develop new mechatronics components and versatile robotics platforms to ensure progress in ADR.

A significant divide exists between industrial end-users seeking mature AI solutions for practical implementation and the broader research and development community. Many industries struggle to integrate these technologies effectively due to a lack of accessible tools and expertise. Integrators serve as the crucial link, translating research advancements into real-world applications. Dedicated programs should be established to support these integrators, fostering pilot initiatives and upskilling programs that bridge the knowledge gap. By doing so, SMEs and other industrial players can leverage cutting-edge AI and robotics solutions to drive innovation and competitiveness using ADR technologies.

The convergence of GenAI and robotics has broad applications across numerous strategic ecosystems, including manufacturing, logistics, laboratory automation, inspection and maintenance, agriculture and agrifood, assistive robotics and healthcare, security, defense, public services, construction, hospitality, retail, biotechnologies, medical, aerospace, and more. However, this technological convergence also presents critical challenges that must be addressed, such as safety, explainability, reliability, cybersecurity, embodiment and real-time interaction, sustainability beyond energy efficiency, and alignment with European values.

5.1 A European Strategy for AI, Data, and Robotics

While the future of GenAI and robotics remains uncertain, their potential to reshape economic, social, and environmental landscapes is undeniable. Europe must act now to harness the opportunities presented by these transformative technologies. However, Europe should not compete with the USA and China on the same terms, where massive investments, vast data resources, and immense computing power dominate. Instead, Europe should carve its path by aligning the convergence of GenAI and robotics with its unique objectives and values. This approach should focus on Europe's strengths and weaknesses, considering our commitment to economic growth, social preservation of culture and values, and environmental sustainability—especially pursuing the UN Sustainable Development Goals. By doing so, Europe can seize the opportunity to lead in emerging markets and shape a future that reflects its ideals and priorities.

Seven key areas of action to guide this strategy are as follows:

1. Adopt a Blue Ocean Strategy for innovation.
2. Educate, train, and retain talent.
3. Build a resilient and high-performance AI infrastructure.
4. Integrate AI and robotics across the technology stack.
5. Advance mechatronics for adaptive AI.
6. Explore novel AI approaches.
7. Promote open source and digital commons.

These are described in more detail below.

1. **Adopt a Blue Ocean Strategy for Innovation**

   The Blue Ocean Strategy is a business framework that emphasizes creating new, uncontested markets rather than competing in existing, saturated ones. For Europe, this means focusing on value-driven, sustainable innovation that aligns with social, economic, and environmental objectives. European companies can establish a competitive edge in previously inaccessible markets by refining this approach and ensuring its transfer into real business applications.

   A critical component of this strategy is ensuring safety and trustworthiness in GenAI and robotics. While end-to-end neural networks offer extraordinary capabilities, they lack robust safety guarantees. Europe can lead in pioneering new safety-centric designs, including safety-by-design frameworks, rigorous testing across diverse generated scenarios, and learning methods that inherently incorporate formal (e.g., physical or biological) constraints. Additionally, explainability, guarantees, and certifications should be prioritized to establish trust in AI-driven robotics.

   Ethical considerations and social impact must also be embedded in this strategy. Europe has the opportunity to lead in ethical and inclusive AI development, ensuring transparency, privacy protection, and democratic inclusiveness in human-robot interactions. By incorporating social and human sciences expertise, Europe can address real societal needs and mitigate the unintended consequences of uncontrolled AI deployment.

Sustainability and resilience are also central to Europe's competitive positioning. Enforcing the "do not significantly harm" principle, as aligned with the EU Green Deal, and committing to climate-conscious AI development will be essential. Investments should be directed toward frugal AI models that minimize resource consumption while maximizing efficiency. Moreover, modular and upgradable robotics systems will ensure longevity and circularity, reducing environmental impact.

2. **Educate, Train, and Retain Talent**

To support Europe's leadership in the rapidly evolving fields of GenAI and robotics, it is essential to build on existing strategic frameworks such as SRIDA by expanding targeted education and training programs. A GenAI and robotics literacy program should be developed for the general public, fostering awareness, curiosity, and engagement with these technologies. Such initiatives would demystify AI and robotics, ensuring that society is well informed about their potential benefits, their limitations, and the challenges they address.

Alongside the foundational work laid out in SRIDA, there is a need to develop specialized training tailored specifically to the unique demands of GenAI and robotics. This can be achieved by creating high-quality, interactive courses and other training initiatives that incorporate the core elements of Europe's Blue Ocean Strategy. Making these courses widely accessible across Europe will ensure that professionals have the necessary skills to drive innovation in this domain. Additionally, upskilling robotics integrators should be an urgent priority, as they are the critical link between cutting-edge research and real-world industrial applications.

Beyond education, these training programs must also elevate the visibility of existing European tools and solutions while addressing their constraints and limitations. By strengthening communication between stakeholders and highlighting ongoing technological advancements, Europe can encourage the adoption of homegrown solutions rather than defaulting to non-European alternatives. European corporations should reduce their dependence on US-based GenAI and robotics tools where viable European alternatives exist, reinforcing strategic autonomy. Structures such as digital innovation hubs (e-DIHs), industry clusters, and technology transfer platforms should be leveraged to enhance collaboration, streamline knowledge sharing, and accelerate the adoption of European technologies.

To counteract the brain drain caused by the dominance of global tech actors, Europe must invest in moonshot projects that attract and retain top talent. This requires a commitment to ambitious, long-term research initiatives, access to state-of-the-art equipment, and fostering stronger partnerships between private sector experts and public researchers. By creating an environment where cutting-edge innovation is possible and actively supported, Europe can retain its brightest minds and establish itself as a global leader in GenAI and robotics.

3. **Build a Resilient and High-Performance AI Infrastructure**

Developing a robust and future-proof infrastructure for computation, data, and connectivity is essential to strengthen Europe's technological independence. This means ensuring easy access to high-performance computing (HPC) centers, AI factories, and European common data spaces and leveraging existing initiatives like EuroHPC supercomputers and next-generation cloud infrastructure projects. As AI models evolve to incorporate 3D and multimodal data, Europe must scale up computational power and data storage to support increasingly complex foundation models.

Ensuring fair and affordable access to HPC, GPUs, TPUs, and NPUs will be key to fostering innovation. While major tech firms dominate advanced AI hardware, short-term measures should support European SMEs and start-ups in securing partnerships and resources. In the long term, Europe should invest in its own AI hardware ecosystem to reduce dependency and enhance technological sovereignty.

Foundation models require vast amounts of data, yet accessing high-quality datasets remains a challenge. Europe must facilitate secure, privacy-compliant data sharing through regulatory frameworks and EU-controlled cloud storage, enabling researchers and businesses to leverage diverse datasets for AI training. Simulators and synthetic data generation should also be explored to enhance data availability while safeguarding privacy.

Europe should establish sandbox environments and testbeds for GenAI-integrated robotics, ensuring safe real-world applications through controlled trials. A full-stack simulator is crucial for training and validating AI models, offering physical simulation, photorealistic rendering, and scalability to generate high-quality synthetic training data. These platforms will accelerate AI and robotics development while ensuring safety, reliability, and real-world applicability.

4. **Integrate AI and Robotics Across the Technology Stack**

GenAI and robotics rely on a complex, interconnected value chain spanning hardware and software. Each component plays a crucial role, but they must be developed in tandem to manage the growing computational demands of AI models, which are expanding at unprecedented rates. Europe must foster a holistic approach supporting the entire ecosystem, ensuring strategic investments in software innovation and hardware advancements.

Hardware and software must be co-optimized for AI to be scalable and sustainable. AI-robotics accelerators tailored for perception, planning, and reasoning must be developed, while algorithms must be adapted for efficient execution on specialized processors. Investing in novel chip technologies, low-energy hardware, and real-time secure applications will ensure that Europe remains competitive in AI-driven robotics.

To sustain innovation, Europe must:

- Develop energy-efficient AI hardware for training and inference
- Create robust digital environments for AI development, including simulation and validation tools

- Improve sensor and actuator systems for safer and more adaptive robotics
- Enhance data accessibility by fostering large-scale, high-quality datasets combining movement, language, and vision

  Ensuring safety, trustworthiness, and efficiency requires a coordinated effort. Europe should expand Digital Innovation Hubs to support AI-robotics integration, coordinate research laboratories and industries, and promote large-scale data-sharing initiatives like euROBIN. Establishing a legal accountability and risk-sharing framework will also incentivize innovation, ensuring that integrators can adopt cutting-edge technologies without undue burden. Dedicated calls for projects uniting data, AI, and robotics companies will accelerate breakthroughs and drive European leadership in this domain.

5. **Advance Mechatronics for Adaptive AI**

   For robots to effectively carry out tasks, they must be capable of interacting with their surroundings. A GenAI-powered robot will be able to make decisions regarding navigation, manipulation, and operations in unpredictable situations and unstructured environments. Therefore, it is crucial that robots not only develop cognitive capabilities but also possess the necessary physical attributes to engage with diverse tasks. Achieving this will require ongoing investment and innovation in the field of mechatronics.

6. **Explore Novel AI Approaches**

   Europe needs to harness the full potential of GenAI. Dedicated efforts should also focus on investigating innovative AI methodologies. Particular emphasis should be placed on approaches that enhance explainability, optimize data usage, and improve energy efficiency. These emerging AI advancements have the potential to strengthen Europe's technological sovereignty and position it as a leader in the field.

7. **Promote Open Source and Digital Commons**

   To strengthen European technological sovereignty, Europe must invest in open-source AI and robotics platforms, ensuring accessibility, interoperability, and long-term sustainability. This includes developing digital commons consisting of shared datasets, software, and models to accelerate innovation.

   Key initiatives should focus on:

- Open-source robotic platforms and libraries for GenAI embodiment, leveraging European successes like Linux, Mistral AI, and Hugging Face
- European-led licensing models and governance structures to maintain sovereignty and competitiveness
- Collaborative projects for large-scale AI model training, ensuring alignment with European values of trustworthiness and efficiency
- In parallel, Europe must establish test beds and experimentation centers to support start-ups, industry, and researchers. These facilities should be equipped with EU-manufactured robotics and skilled engineers, fostering collaboration and enabling rapid deployment of new AI-driven robotic solutions. Existing networks like the Edge AI TEF and European Digital Innovation Hubs can be expanded to support these initiatives, ensuring Europe's leadership in AI and robotics.

For more details on the GenAI and robotics roadmap, download the full report at https://adr-association.eu/news/adra-releases-policy-paper-and-technology-roadmap-genai-and-robotics-4eu.

6 Future Directions

Building on the activities and priorities already outlined, ADRA is now focused on translating its strategic vision into action.

Central to this is the continued implementation of the SRIDA, which offers not only a guiding vision but a practical roadmap for shaping a resilient, sovereign, and innovation-driven ecosystem. While the SRIDA's core principles have been addressed earlier, its role now is to steer the delivery of concrete outcomes, from the development of foundational and frontier AI models to their rapid deployment in domains such as healthcare, mobility, and manufacturing.

A key enabler of this ambition is the launch of high-impact pilots that showcase the convergence of AI software with robotic hardware and control systems. These pilots are intended to demonstrate Europe's ability to lead across the full AI technology stack while also providing real-world validation of new capabilities. By accelerating the integration of AI into intelligent machines, these pilots will help bridge the gap between research and deployment, creating momentum across industry and public services alike.

To support the emergence of technology leaders, ADRA is also adopting a challenge-driven, stage-gate approach to innovation. This structured process enables the most promising ideas to advance through successive development stages, encouraging agility, early application discovery, and quicker paths to deployment. It also reinforces the alignment between research outcomes and industry needs, strengthening Europe's innovation pipeline and fostering a new generation of ADR champions.

Equally critical is Europe's effort to lead in the development of international standards, particularly for generative AI in robotics. With a growing number of use cases and a rapidly expanding market, there is a clear opportunity for Europe to shape the global rules of the game. Establishing harmonized, high-quality standards will not only enhance trust and safety but also support European manufacturers in building value chains that meet and define global expectations.

Finally, achieving these ambitions will require greater coordination among Member States. Policy, funding, and priority fragmentation remain persistent barriers. ADRA is working to create stronger alignment across European governance layers from national to regional to EU-wide with a particular focus on increasing private investment and boosting public procurement. This harmonization is vital to unlocking the full potential of the European research and innovation ecosystem, enabling shared progress toward strategic autonomy.

7 Conclusion

ADRA stands at the forefront of a transformative era, where the convergence of AI, data, and robotics reshapes how societies function, economies grow, and values are embedded in technology. As this chapter has shown, ADRA is not only responding to this moment but is helping define it. Through bold strategic initiatives like the SRIDA, high-impact pilots, and a challenge-driven innovation approach, the foundation is being laid for a sovereign, inclusive, and globally competitive European ecosystem.

At the heart of this effort is a vision of Europe as a world leader in responsible, human-centric technological advancement. ADRA is committed to enabling a green and digital future where ADR technologies drive prosperity, resilience, and shared progress. The rapid evolution of generative AI and large-scale models brings both unprecedented opportunities and challenges. Europe must seize this moment by investing ambitiously, advancing trustworthy and transparent models such as neuro-symbolic AI, and building technological capabilities that reflect its deepest democratic values.

To achieve strategic autonomy, Europe must balance regulation with significant investments in ADR innovation. Proactive investment and calculated risks will secure Europe's technological sovereignty and global leadership, ensuring a future where ADR technologies drive prosperity, sustainability, and security.

Realizing this future will demand courage, creativity, and unity. It means aligning national policies, mobilizing resources, and taking bold steps to secure leadership in the global tech landscape. With momentum already building, ADRA is poised to help Europe shape the next wave of digital transformation: one that empowers people, protects the planet, and reaffirms Europe's place as a guiding force in the world.

SRIDA: Charting the Future of a Progressive, Inclusive, and Sustainable European ADR Ecosystem

Fredrik Heintz and Katerina Linden

Abstract The Strategic Research, Innovation, and Deployment Agenda (SRIDA) is a defining framework for the AI, Data, and Robotics Association (Adra) and a forward-looking roadmap for the advancement of AI, data, and robotics (ADR) in Europe. It aims to align ADR research and development efforts, address societal challenges, enhance economic competitiveness, and inform the European Commission's Horizon Europe work program. This chapter outlines the SRIDA's purpose and the stakeholders involved and proposes a collaborative methodology based on our practical experience preparing the SRIDA for the past 3 years. It describes the operational framework, including tools, contribution processes, development phases, and iterative updates that ensure the SRIDA evolves in alignment with diverse priorities and voices. The SRIDA's dynamically evolving nature reflects its commitment to transparency, inclusivity, and balancing technical and non-technical goals. Building on previous SRIDA editions, the chapter examines constraints such as balancing agendas, fostering equitable representation within the ADR community, and aligning immediate objectives with long-term strategies. By addressing open challenges and providing actionable recommendations for future SRIDA iterations, this chapter demonstrates SRIDA's potential as a driver of Europe's ADR leadership, paving the way for sustainable innovation.

Keywords SRIDA · ADR · European research and development · Economic competitiveness · Transparency and inclusivity · Sustainable innovation · Horizon Europe

F. Heintz · K. Linden (✉)
Department of Computer and Information Science, Artificial Intelligence and Integrated Computer Systems, Faculty of Science & Engineering, Linköping University, Linköping, Sweden
e-mail: fredrik.heintz@liu.se; katerina.linden@liu.se

© The Author(s) 2026
E. Curry et al. (eds.), *Artificial Intelligence, Data and Robotics*,
https://doi.org/10.1007/978-3-032-10561-5_3

1   Introduction

1.1   A Vision for Europe's AI, Data, and Robotics Future

This book is filled with scientific insights from various AI, data, and robotics (ADR) researchers who shared their knowledge with the AI, Data, and Robotics Association (Adra). But it would be incomplete without describing one of Adra's most significant instruments, the Strategic Research, Innovation, and Deployment Agenda (SRIDA), which is a collaborative blueprint for guiding Europe's ambitions in ADR. It has been developed through a broad collaboration initiated and navigated by Adra to spur the motivation and aspirations of the European ADR community. The SRIDA outlines priorities, identifies challenges, and establishes a consensual framework for research and innovation development within the fields of AI, data, and robotics in the European Union.

ADR research and development are central to Europe's ongoing technological transformation, rapidly influencing industries, public services, and everyday life. As computing power continues to grow day by day, data becomes increasingly available, and machine learning algorithms advance at an unprecedented pace, the potential for ADR to redefine the future of innovation appears almost limitless. It brings extraordinary opportunities for researchers, businesses, and governments, promising to enhance productivity across various sectors. However, these technologies also introduce complex challenges related to ethics, transparency, and the responsible deployment of ADR.

Alongside this technological acceleration, globalization is undergoing a shift, with disruptions in supply chains and growing concerns over economic dependencies. The European Union is responding to these challenges by reinforcing its sovereignty, resilience, and strategic autonomy in key technological areas. But establishing true leadership in ADR requires a comprehensive strategy that not only supports resilience and security but also facilitates the digital and green transitions across industries and the broader economy. Achieving balance is essential for leveraging Europe's technological capabilities while ensuring industrial competitiveness, societal well-being, and environmental sustainability.

Europe already possesses the critical components necessary for success: a highly skilled workforce, competitive industries, established regulatory frameworks, and globally recognized academic institutions all provide a strong foundation for future advancements in ADR. But the main challenge lies in integrating these strengths into a coordinated and forward-looking strategy that allows the European Union to be a key player in the global ADR landscape. The overarching objective of the SRIDA, to which this chapter is devoted, is to provide a roadmap for the research, development, and deployment of trustworthy ADR technologies that align with European values, ethical principles, and societal needs.

This final book chapter provides a comprehensive exploration of the objectives and foundational principles of the SRIDA, the stakeholders involved in its development, and the contribution mechanisms in alignment with the goals of the AI, Data, and Robotics Partnership [1].

Specifically, Sect. 1 introduces the SRIDA as a strategic framework for European technological advancement, emphasizing its alignment with European values, regulations, and industrial aims, including fostering innovation, ethical development, and global competitiveness. It also touches upon its integration with broader policies and its adaptive nature through ongoing engagement. Section 2 examines the diverse stakeholders shaping the SRIDA, from internal contributors like researchers and policymakers to external entities, highlighting the importance of balanced representation, including SMEs and emerging voices. Section 3 outlines the collaborative methodology behind the SRIDA's development, detailing the tools and structured processes ensuring a transparent and inclusive evolution. Section 4 addresses key challenges in the SRIDA's creation and implementation, such as aligning diverse interests, ensuring equitable representation, balancing strategic goals with immediate needs, and enhancing transparency. Finally, Sect. 5 reflects on the SRIDA's impact on Europe's ADR strategy, summarizing lessons learned and outlining future directions for this crucial instrument in fostering responsible, innovative, and globally competitive ADR ecosystems.

By analyzing these aspects and offering recommendations for future iterations of SRIDA, this chapter highlights its role as a critical instrument for reinforcing Europe's ADR leadership and fostering sustainable, ethically grounded innovation.

1.2 Why the SRIDA Matters: Strategic Goals and Impact

According to research, a technology roadmap should provide three main benefits:

> First, it creates a forecasting mechanism for the developed technology; second, it produces a unified collection of requirements, needs, and values along with the developed technology; and third, it helps construct a framework to plan and assist the development of the technology or the product. [2]

The SRIDA was created with similar goals in mind. It aimed to mobilize diverse stakeholders and provide a framework that could help them find consensus on shared challenges and even develop a shared strategy. Such a practical strategic document could enable Europe to remain at the forefront of ethical and innovative ADR development. As a result of years of work, Adra succeeded in fostering collaboration across academia, industry, and policymakers and delivering the SRIDA—a consensual framework for the responsible development and deployment of ADR technologies that not only empowers stakeholders to navigate the complexities of an evolving technological environment but also offers key recommendations to shape future European research and innovation programs.

Essentially, the SRIDA, as a roadmap for the future of ADR in Europe, is designed to foster collaboration, strengthen economic resilience, and ensure technological sovereignty. Its primary goals include:

- Strengthening the European ADR ecosystem, by facilitating synergies between various stakeholders
- Enhancing global competitiveness, by supporting European leadership in ADR solutions
- Promoting ethical and trustworthy ADR, by ensuring that technological advancements are aligned with human-centered values, privacy considerations, and robust governance frameworks
- Accelerating societal and economic benefits, by leveraging ADR to address pressing global challenges

The SRIDA also highlights the importance of ADR technologies in supporting the twin digital and green transitions—essential transformations for securing a competitive and sustainable economy. While creating this roadmap, we knew that the responsible adoption of ADR technologies requires a proactive, dynamic approach, one that not only follows but anticipates rapid changes and adapts to new environments, whether they are economic, political, or technological.

The long-term impact of the SRIDA will be seen in the future. For now, it serves as a reference for stakeholders across the ADR ecosystem, offering a structured, consensus-driven approach for advancing European ADR. By harnessing the transformative potential of ADR while ensuring alignment with societal values, economic imperatives, and regulatory frameworks, the SRIDA will help maintain Europe's competitive edge and reinforce its role as a global leader in responsible and sustainable technological development.

1.3   Aligning with European Priorities: From Policy to Practice

The SRIDA was designed to function within a broader European policy framework, aligning with the continent's strategic and regulatory priorities. Adra ensures the engagement of national representatives from across Europe, which allows for the alignment of diverse perspectives with overarching policy objectives. Moreover, collaboration with European Commission representatives further strengthens this alignment, as these officials provide regulatory insight, strategic direction, and a connection to broader EU initiatives. On top of that, the SRIDA is strongly aligned with EU's key funding programs for research and innovation, such as Horizon Europe, by setting research priorities to drive European innovation in ADR. It follows up on European AI and Data Regulations and highlights the necessity of ensuring that technological developments comply with evolving legal and ethical frameworks. As a result of actively bridging policy, research, and industry, the SRIDA promotes a holistic approach to technological advancement. Thus, it ensures that ADR solutions are not only innovative and scalable but also responsible and aligned with European values.

As the region continues to strengthen its role in the global ADR landscape, the SRIDA will help translate policy objectives into real-world impact, shaping a future where technological progress serves both economic and societal goals.

1.4 Conceptual Methodology: Transparency, Collaboration, and Evolution

This section describes how the SRIDA is created, step by step, to ensure that this process is transparent, explainable, and easy to participate in, for all kinds of ADR stakeholders. One of the defining characteristics of the SRIDA is its openness, adaptability, and, most importantly, continuous refinement. Due to the rapid pace of technological change, a static document would quickly become outdated. Thus, the SRIDA functions as a living strategy, one that evolves dynamically through ongoing stakeholder engagement, expert consultation, and iterative updates. This approach ensures maximum relevance as well as the ability to address emerging challenges and opportunities in a timely manner.

A crucial element of the SRIDA's ability to dynamically adapt to new factors lies in its inclusive and highly organized development process. Various contributors shape its direction, fostering broad and representative perspectives. While integrating these diverse viewpoints, the SRIDA ensures equal opportunities for all voices and provides numerous channels for feedback and contributions at different stages.

To maintain both its relevance and collaborative spirit, the SRIDA undergoes regular reviews, allowing for updates that reflect the latest advancements in ADR. The key mechanisms supporting this process include:

- Inclusive development, by engaging a wide range of stakeholders ensuring a balanced representation and informed decision-making
- Continuous feedback, by implementing regular review cycles
- Regular updates, through periodic revisions that incorporate technological breakthroughs and evolving regulatory landscapes

It is important to point out that the SRIDA's dynamic nature is not only a response to technological and economic changes but also a means of fostering transparency and collaboration. By maintaining an open, structured, and repeatable framework, it provides a clear methodology for its continuous evolution. This methodology includes refining procedures for establishing priority areas known as "Big Tickets," as well as defining effective stakeholder engagement strategies and setting forth a structured timeline for implementation.

By embracing transparency, inclusivity, and adaptability, the SRIDA establishes its usefulness for Europe's ADR ecosystem. As the technological landscape continues to shift, the SRIDA's living strategy approach will be instrumental in ensuring that technological advancements are not only cutting-edge but also aligned with European values, global economic and social challenges, and ethical values.

2 Stakeholder Engagement: Who Is Creating the SRIDA?

A consensual framework like the SRIDA could not be developed by a small group of experts. On the contrary, its strength lies in its collaborative foundation, as it brings together a diverse network of stakeholders who set the priorities and outcomes of ADR in Europe. These stakeholders represent the breadth of the ADR ecosystem, from research institutions and industries to policymakers and broader societal actors. Their active engagement ensures that the SRIDA is as dynamic and inclusive as possible and, as a result, aligned with national and European goals, in line with the expert-based roadmapping approach [3]. This section defines the stakeholders and their involvement in the collaborative SRIDA creation process.

2.1 Internal Contributors: The Core Drivers of the SRIDA

The SRIDA's creation is a collaborative effort that involves internal and external contributors. While external bodies provide valuable feedback during designated events and sessions, it is the internal contributors who play a central role in shaping the document. These individuals and organizations work closely within Adra, its affiliated projects, and partner networks, engaging in continuous dialogue that refines the SRIDA's structure, content, and strategic direction. Their regular contributions include brainstorming, writing, editing, and iterative feedback, as well as alignment with the goals of the AI, Data, and Robotics Partnership [1]. The key internal contributors include different groups of stakeholders.

Adra Members provide critical strategic input, shaping research priorities and influencing funding directions. This diverse group includes academics, industry experts, and business representatives who collaborate with Adra to guide the SRIDA's development. Their involvement ensures that the agenda remains reflective of real-world needs, scientific findings, and industry trends. As Adra continues to expand, so does its membership, bringing in fresh perspectives and expertise.

The Adra Board of Directors plays a crucial oversight role, reviewing and approving the progress of the SRIDA at key stages. By ensuring that the document remains aligned with Adra's mission and strategic objectives, the Board provides governance and direction, helping maintain coherence and focus throughout the drafting process.

Adra-e is the coordination and support action team that facilitates Adra's operations. Its involvement ensures that the SRIDA is well integrated into the broader European policy and funding landscape.[1]

The Editing Committee and Coordination Team are instrumental in structuring and refining the SRIDA, ensuring clarity, accuracy, and strategic alignment. The Editing Committee consists of three dedicated editors, each representing one of the

---

[1]Adra-e project is finished in June 2025.

key constituencies—AI, data, and robotics—ensuring balanced expertise across the ADR domains.[2] Their weekly online meetings foster continuous collaboration and iterative refinement. This team plays a key role in shaping the SRIDA by formulating key themes, structuring the document, and engaging in ongoing scientific discussions to maintain a forward-looking and relevant agenda. The committee is led by a lead editor, appointed by the Board of Directors, who holds the responsibility of reporting to the Board.

The Coordination Team plays an equally crucial role in harmonizing stakeholder contributions, actively participating in Editing Committee meetings, and maintaining close collaboration with the Adra-e support team.[3] Their efforts streamline processes and contribute to a coherent final document, ensuring that diverse inputs are effectively integrated into the SRIDA while upholding its strategic and scientific integrity.

Topic Groups and Reference Groups—a diverse network of experts consults with the Editing Committee throughout the SRIDA's development, offering insights, feedback, and domain-specific expertise. These contributors help refine research objectives, align the document with evolving technological trends, and ensure that the SRIDA addresses practical challenges faced by the European ADR ecosystem.

Through their collective expertise, these internal contributors create the backbone of the SRIDA, ensuring that it is both ambitious and feasible within the broader European landscape. But while internal contributors' engagement and consultation are present on every level of work, external bodies provide their independent feedback during specific events and sessions. The external contributors' engagement guarantees that the SRIDA maintains balance and remains objective and unbiased.

2.2 External Stakeholders: Aligning National and International Priorities

While internal contributors provide the structure and operational backbone of the SRIDA, external stakeholders bring diversity and a broader strategic perspective. These stakeholders play a crucial role in aligning the SRIDA with European policy frameworks, national strategies, as well as international research efforts, fostering a truly integrated and forward-looking approach.

The European Commission, as the primary policymaker and funder, plays a central role in ensuring that the SRIDA aligns with the European Union's overarching initiatives. EC representatives actively contribute to the SRIDA's development by providing regulatory insight, giving feedback on the SRIDA iterations, and

---

[2] The current Editing Committee includes Fredrik Heintz (lead editor, AI), Nabil Belbachir (robotics), and Edward Curry (data).

[3] It currently consists of Ana Garcia (BDVA Secretary General), Philip Piatkiewicz (Adra Secretary General), Jozef Geurts (interim Adra Secretary General), and Katerina Linden.

embedding the agenda within their broader objectives. Their involvement ensures that the SRIDA is implementable within the EU's legal and financial frameworks.

Member States—national governments influence how SRIDA recommendations translate into concrete policies at the national level, facilitating alignment between EU-wide ADR strategies and country-specific research and industrial priorities. While the European Commission oversees strategic coordination, it is the individual Member States who are responsible for local implementation, ensuring that the aspirations outlined in the SRIDA lead to actionable national initiatives. It is important to note that although each EU country appoints a European Commissioner, these commissioners do not represent their respective nations but work collectively in the interest of all Member States, reinforcing the pan-European nature of the SRIDA.

Founding organizations such as the Big Data Value Association (BDVA), Confederation of Laboratories for AI Research in Europe (CAIRNE), European Laboratory for Learning and Intelligent Systems (ELLIS), the European Association for Artificial Intelligence (EurAI), and European Robotics (euRobotics) played a central role in establishing Adra in May 2021. These organizations continue to contribute their expertise and cutting-edge research perspectives and help shape the SRIDA and ensure its alignment with Europe's technological and scientific priorities.

Finally, various EU projects and initiatives, such as Networks of Excellence (NoEs[4]) and the AI-on-Demand Platform (AIoD[5]), provide strategic guidance on emerging technological and scientific trends. NoEs unite over 1000 researchers and 100 industry organizations, fostering a cohesive research community. The European Commission has invited NoEs to identify gaps and challenges within the European AI research landscape, ensuring that the SRIDA incorporates the latest developments and remains adaptive to new breakthroughs. AIoD serves as a key infrastructure for NoEs, enhancing its capabilities in tools, competencies, and services. Representatives of these expert groups, involved in AI, data, and robotics research and on-demand AI services, contributed their expertise to the SRIDA development process during various events, workshops, and conferences. These consultations have been instrumental in integrating research, education, and innovation into the roadmap, strengthening the link between Adra's SRIDA and broader European AI strategies (Fig. 1).

By incorporating these diverse voices, the SRIDA ensures that Europe's ADR agenda remain both globally competitive and locally relevant. However, input from scientific, governmental, and industrial communities alone is not enough to claim that the SRIDA truly represents the entire ADR landscape. To fully capture the breadth of Europe's AI, data, and robotics ambitions, the SRIDA must also engage with underrepresented stakeholders, including small and medium-sized enterprises (SMEs), startups, and emerging innovators.

---

[4] https://digital-strategy.ec.europa.eu/en/policies/networks-excellence-ai-robotics-researchers
[5] https://www.ai4europe.eu/

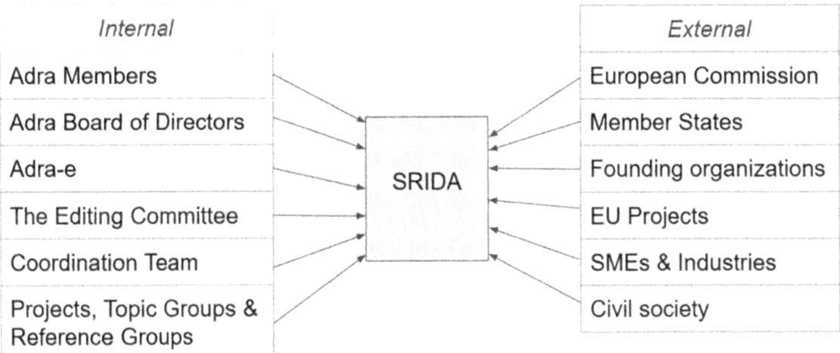

Fig. 1 Stakeholders' contribution to the SRIDA

2.3 Addressing Underrepresented Voices: SMEs and Emerging Stakeholders

We believe that one of the SRIDA's most significant challenges and, at the same time, one of its greatest strengths lies in its ability to balance a wide array of interests. National governments prioritize sovereignty, regulatory frameworks, and public policy objectives, while industry stakeholders focus on economic viability, market competitiveness, and technological leadership. Research institutions, in their own turn, emphasize scientific advancements, pushing the boundaries of innovation, whereas policymakers must carefully navigate the ethical, legal, and societal implications of ADR technologies. All these divergent priorities, if not properly reconciled, risk creating a roadmap that is fragmented and far from perfection.

To balance these perspectives, the SRIDA employed a highly democratic and inclusive approach, fostering open dialogue through a variety of engagement mechanisms, such as workshops, public consultations, collaborative writing sessions, and many others. This approach ensures that this document is not dominated by any single group but reflects a shared vision that accounts for the diversity of regional, economic, and sectoral priorities across Europe.

Here the role of smaller actors, such as startups, small and medium-sized enterprises (SMEs), and civil society organizations, becomes indispensable. These stakeholders contribute to the ecosystem in several critical ways. First of all, startups and SMEs are often at the forefront of disruptive innovation, pioneering novel applications of ADR technologies that challenge existing paradigms and push the industry forward. Aside from this, civil society organizations and smaller industry players are able to provide valuable insights into the ethical, local, and community-driven dimensions of ADR, ensuring that technological progress aligns with societal needs. Finally, by incorporating voices from diverse sectors such as healthcare, education, and social services, the SRIDA is not confined to traditional industry silos, instead paying attention to solutions that serve a broader range of users and applications.

To ensure genuine inclusivity, Adra actively engages these underrepresented groups in the SRIDA creation through dedicated participation channels, targeted outreach, and tailored consultation mechanisms.

3 Operational Methodology: How Does Creating the SRIDA Work?

Continuous work on the new editions of the SRIDA has helped us formulate a well-structured operational methodology, through which we can more easily ensure inclusivity, transparency, and accountability in every process. A clear methodology is essential, as it defines the mandate of the editors, establishes oversight mechanisms to ensure alignment with broader objectives, and enables the involvement of key stakeholders.

Research has shown that effective technology roadmapping plays a crucial role in strategic planning, functioning not merely as a static document but as an evolving process. This process can be broadly divided into three key phases: preliminary activities, the development of the roadmap, and follow-up actions [4]. Planning—or preliminary activities—is the most resource-intensive phase [5], as it involves numerous parallel, interdependent processes, including constant feedback loops and iterative refinement. According to the research findings in R&D management [6], these processes include:

1. Identifying stakeholders
2. Asking the right questions
3. Collecting and analyzing feedback
4. Categorizing and prioritizing input
5. Defining the basic requirements for the next-generation ADR strategy
6. Identifying dependencies and leverage points to manage multiple development initiatives simultaneously
7. Creating the strategic roadmap
8. Continuously updating the plans to keep the roadmap relevant and effective

The SRIDA follows a methodology rooted in these established practices, ensuring meticulous planning, collaborative engagement, and iterative refinement. Such principle, based on a cyclical process of drafting, gathering feedback, revising, and soliciting further input, also fosters inclusivity by default, as it draws upon a wide range of viewpoints to guarantee comprehensive coverage and accuracy, promotes a sense of community, aligns with academic standards, and establishes a sustainable framework for ongoing refinement. Finally, this iterative process also establishes a sustainable collaborative framework for ongoing refinement and enables the formulation of precise and overarching ideas, aims, and goals.

The whole process of writing SRIDA can be divided into the following three broad stages:

1. Defining the vision, mission, and goals. The initial phase centers on establishing the overarching vision and specific strategic research objectives for ADR development in Europe. Extensive consultations with stakeholders, ranging from policymakers to industry leaders and researchers, ensure that SRIDA's goals align with the European ADR community's broader ambitions. This collaborative effort results in a well-defined list of missions and goals aimed at achieving a shared European ADR vision.
2. Identifying global challenges and emerging trends. The second stage involves mapping global challenges and identifying major technological and societal trends that will shape research, innovation, and deployment efforts. The goal is to strike a careful balance between cutting-edge technological progress and the need to address pressing societal concerns. Through expert consultations and collaborative analysis, key ADR trends are highlighted, and gaps in research and policy are identified, setting the stage for targeted innovation strategies.
3. "Big Tickets"—in the final stage, editors and contributors work together on formulating the Research, Innovation, and Deployment Roadmap. This phase translates strategic objectives into actionable steps, detailing the initiatives required to achieve the defined goals. The roadmap is developed iteratively, incorporating continuous feedback from stakeholders to ensure its effectiveness and feasibility. A central aspect of this phase is the identification of six "Big Tickets"—high-priority focus areas in AI, data, and robotics for the new edition of the SRIDA. These strategic priorities are determined through extensive collaboration with stakeholders, ensuring that they address both technological advancements and societal needs.

As the contributing stakeholders were identified and described in the previous section of this chapter, we now focus on the process of collaboration. Specifically, how the feedback is collected, evaluated, and prioritized. Then we clarify how the process works in defining the basic requirements for the next-generation product and identifying dependencies and leverage points.

3.1　Tools and Processes for Collaboration

Technology roadmaps may appear straightforward in format, yet their development involves intricate challenges, particularly due to the broad scope they encompass, which integrates complex conceptual and human interactions [7]. Similarly, the SRIDA development process is built upon a robust framework of collaborative tools and methodologies that foster community engagement and facilitate input collection. These mechanisms play a crucial role in structuring discussions, refining strategic objectives, and achieving consensus among diverse stakeholders and are shortly described below.

Active engagement with the ADR community, both online and offline, is fundamental to the SRIDA's development. Every interaction, whether formal or informal, presents an opportunity for knowledge exchange, discussion, and consensus building. Adra is actively involved in the community engagement, organizing, and participating in workshops, Topic Groups, consultations, and various events. Most of these activities are listed on the Adra Web site: https://adr-association.eu/.

Workshops, webinars, and seminars serve as interactive platforms where stakeholders deliberate on emerging trends, challenges, and opportunities. By fostering collaborative ideation and facilitating in-depth discussions, workshops enhance engagement with Member States, industry representatives, and research institutions. They also provide the foundation for formulating the "Big Tickets"—key focus areas deemed essential for shaping Europe's ADR landscape.

Topic Groups serve as specialized working bodies, operating under the Adra umbrella.[6] They engage in continuous dialogue, exploring technological advancements and identifying their potential advantages for Europe. These groups investigate relevant use cases, develop sustainable solutions, and explore emerging technologies. Through regular online meetings and workshops, they strive to build long-term societal, market, and political impact.

Consultations, both structured and informal, provide targeted avenues for stakeholders to contribute insights and feedback. These sessions are designed to gather diverse perspectives, ensuring that input from industry leaders, policymakers, researchers, and the broader community is effectively integrated into the roadmap.

Events such as conferences, forums, symposiums, and panel discussions further enrich the dialogue. These gatherings foster networking opportunities, enable the exchange of best practices, and enhance visibility for key strategic initiatives. The culmination of these engagement activities leads to the iterative development of SRIDA, ensuring that it remains relevant, forward-looking, and aligned with community priorities.

3.2 Input and Feedback Mechanisms

To uphold a democratic and transparent process, the SRIDA integrates robust input and feedback mechanisms. The iterative nature of its development allows for continuous refinement, incorporating insights from various stakeholders at multiple stages.

It is known that the process of roadmapping is "somewhat paradoxical in that the appropriate expertise must be employed to develop a roadmap, but the appropriate expertise becomes fully known only after a complete roadmap has been constructed. An iterative roadmap development process is, therefore, essential" [3]. This approach aligns with best practices in technology roadmapping, where stakeholder involvement is crucial in shaping a comprehensive and dynamic strategy [7].

---

[6] https://adr-association.eu/topic-groups

By embracing multiple feedback loops and iterative revisions, the SRIDA remains adaptable and reflective of evolving technological and societal landscapes. Stakeholders are encouraged to contribute at different stages, ensuring that the roadmap evolves in response to new developments and changing priorities.

3.3 Role of Editors: Ensuring Transparency and Neutrality

The Editing Committee plays a key role in shaping the SRIDA, ensuring that diverse inputs are effectively transformed into a coherent and balanced document. Editors do not simply compile the contributions; they facilitate discussions, encourage broad participation, and navigate differing viewpoints to produce a unified consensual framework.

A key responsibility of the editors is maintaining neutrality. No single stakeholder's perspective should dominate and set the agenda; instead, the SRIDA must represent the most balanced and inclusive vision possible. Transparency of all processes is paramount, and the editors must ensure that decision-making processes are as open as possible, all voices are considered, and the final document remains an objective representation of collective priorities.

In addition to compiling content, the editors oversee the development of the "Big Tickets," ensuring that they are substantiated with clear goals, socioeconomic impact analyses, and practical applications. Throughout the process, together with the coordination team, they facilitate consultations, integrate feedback, and refine the document to uphold coherence, clarity, and strategic alignment.

Ultimately, the SRIDA's strength lies in its collaborative foundation. By integrating structured engagement mechanisms, iterative feedback loops, and a transparent editorial process, it continues to evolve and support Europe's ambitions in ADR development.

3.4 Development Phases and Timeline

The development of the SRIDA follows a structured, iterative approach informed by 3 years of practical experience in preparing previous editions. This process aligns with established roadmapping methodologies, which emphasize clear phases, milestones, and scheduled updates to ensure that strategic goals remain relevant in an evolving technological and societal landscape [7].

The SRIDA development process consists of three primary phases: drafting, refining, and finalizing. These phases unfold within a predefined cycle, ensuring a systematic approach to continuous improvement. Based on practical experience, we established a 1-year framework for developing the SRIDA from scratch, as described below. However, future iterations of the SRIDA will be published as needed, with the release timeline determined by the pace of ADR innovation and the needs or requests of the European Commission. This means that new editions may be released sooner or later than 1 year, with the cycle structure adjusted accordingly. Despite variations in cycle length, the three principal phases of development will remain unchanged. Below, we outline the gold standard—the optimal timeline structure that has proven effective for an annual SRIDA release—followed by a reference timeline for a biennial (2-year) release.

1. Drafting (Month 1–Month 6)

   The development of the SRIDA begins at the start of the year. The Editing Committee is formed, with responsibilities distributed among its members. The most recent SRIDA version serves as the foundational draft for the upcoming iteration. Weekly meetings take place throughout winter and early spring, during which editors analyze political, economic, and technological developments, identifying overarching strategic goals for the new edition. To ensure inclusivity and objectivity, specific questions are formulated for industry experts and researchers, helping shape discussions on the key themes for the year. During this initial phase, a preliminary document—a condensed version of key ideas, or "Big Tickets"—is distributed among stakeholders to solicit early feedback. In 2023, for example, a 10-page draft[7] was distributed as an internal document to Adra members to solicit their initial feedback before it was made public to the larger ADR community. This consultation provided a platform for an exchange of ideas on objectives and priorities, enabling further effective feedback gathering from various stakeholders across the ADR ecosystem. This iterative approach ensures that the foundational ideas are continuously refined in response to expert insights and emerging trends.

2. Refining (Month 3–Month 9)

   In parallel with the drafting process, the refinement phase runs throughout winter and spring. Contributions are actively collected from a broad spectrum of stakeholders through discussions at conferences, seminars, workshops, and dedicated Adra and Adra-e events. These interactions provide insights into recent technological advancements and shifting industry priorities while incorporating considerations of global and local economic and social challenges. By April and May, editors analyze the accumulated feedback to determine which revisions to the overarching Vision, Mission, and Goals of the SRIDA are necessary. Special attention is given to refining the "Big Tickets". Once these priorities are established, in May, they are assigned to the Adra "Big Tickets" Groups, which work collaboratively throughout the summer to develop detailed short-, medium-, and long-term roadmaps for their respective areas of expertise.

   By early September, the contributions from these expert groups are integrated into the evolving SRIDA draft. The updated document is then distributed to all stakeholder levels for further consultation and feedback. This process ensures transparency and broad democratic involvement, aligning the roadmap with the diverse perspectives within the ADR community.

---

[7] More information: https://adra-e.eu/events/adra-srida-deep-dive-workshop

3. Finalizing (Month 10–Month 12)

Throughout October and November, the Editing Committee carefully reviews and incorporates the final round of stakeholder feedback. This phase focuses on ensuring balance, coherence, and clarity, refining the SRIDA to meet both strategic objectives and practical implementation needs. Once the final revisions are made, the completed edition of SRIDA is officially presented to the public in December. This marks the culmination of a year-long process of collaboration, iteration, and refinement. Such a structured approach to development and regular updates ensures that SRIDA remains a dynamic and adaptive roadmap, providing strategic guidance that reflects both technological progress and societal needs (Fig. 2).

Research has demonstrated that the value of roadmapping is maximized only when the information it contains is continually updated to reflect new developments [7]. In practice, this means that revising the SRIDA regularly is necessary to maintain its relevance and alignment with the latest technological and policy advancements. Roadmapping serves as a framework to plan and coordinate technology developments [3], and the SRIDA follows this principle by ensuring a structured and iterative review cycle. By adhering to this predefined schedule of regular updates, the SRIDA functions as an evolving document rather than a static one. This commitment to ongoing refinement strengthens its role as a foundational strategy

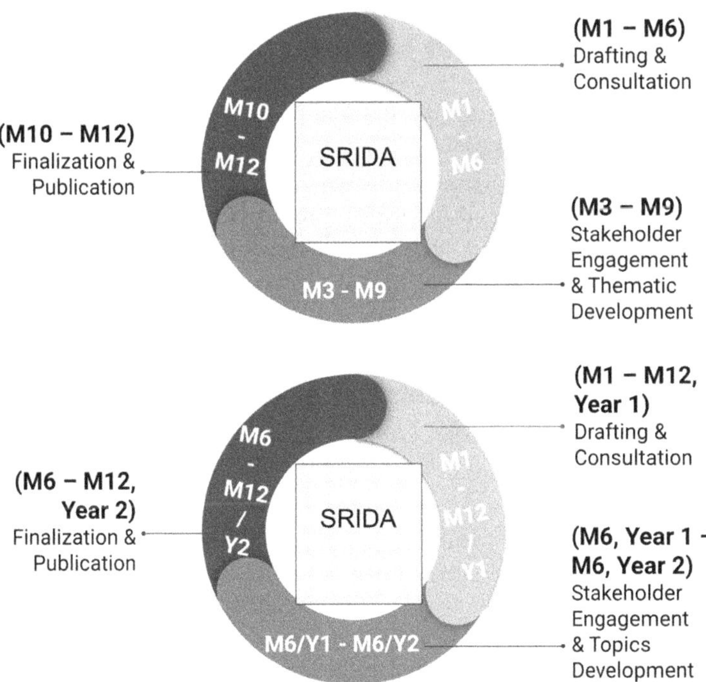

Fig. 2 SRIDA annual update timeline; SRIDA biennial update timeline

document for Europe's AI, data, and robotics ecosystem, ensuring that its guidance remains relevant, forward-thinking, and aligned with broader societal and industrial transformations.

3.5 Technical Aspects: Managing Feedback Loops and Document Iterations

Effectively managing feedback and document iterations is a special task for the coordination team. They are utilizing online interaction tools, collaborative online editing platforms allowing for version control - all to ensure that all inputs are systematically captured and integrated. Clear mechanisms for feedback submission and review help maintain order and prevent information overload. Such meticulous management aligns with best practices in roadmapping, where structured processes are essential for integrating diverse inputs and maintaining coherence [6].

The SRIDA's operational methodology, as described above, is an example of efficient structured collaboration and iterative refinement. By following established roadmapping practices and fostering an inclusive approach, the SRIDA serves as a dynamic blueprint for the European ADR future.

4 Challenges and Lessons Learned

The development of the SRIDA remains an ambitious and complex undertaking that requires extensive coordination among a broad spectrum of stakeholders. While the process of its creation fosters innovation and strategic alignment, it also presents several significant challenges that must be carefully managed to guarantee that the SRIDA remains an effective and representative roadmap for Europe's ADR ecosystem.

4.1 Navigating Potential Conflicts of Interest: Aligning Agendas with Broader Goals

With stakeholders ranging from government agencies and industry leaders to academic institutions and civil society organizations, differences in their priorities are inevitable. Industry representatives often focus on commercial viability and the rapid deployment of technology, policymakers emphasize ethical considerations and regulatory frameworks, and academic institutions tend to prioritize long-term scientific advancement and fundamental research. Striking a balance between these perspectives requires careful negotiation and structured dialogue to ensure consensus and alignment.

To mitigate potential conflicts, the SRIDA incorporates structured discussion strategies, transparent decision-making processes, and iterative stakeholder engagement, as was demonstrated earlier. It accommodates diverse perspectives while maintaining strategic coherence by fostering inclusive consultations and providing multiple avenues for feedback. As highlighted in technology roadmapping literature, successful frameworks must integrate multiple perspectives while maintaining coherence and strategic focus [7]. Following these ideas, the SRIDA remains aligned with broader European objectives while addressing stakeholder-specific concerns—through continuous dialogue and compromise-driven decision-making.

4.2 Ensuring Equitable Representation Across Stakeholders

A well-known challenge in strategic roadmapping is ensuring that all relevant actors, regardless of their size, sector, or geographical location, have meaningful opportunities to contribute. This issue remains critical for the SRIDA as well. Large corporations and well-established research institutions tend to dominate discussions, overshadowing the voices of smaller enterprises, startups, and regional stakeholders. Therefore, achieving a more balanced representation is a critical consideration for future SRIDA iterations.

Attracting contributors from outside academia has proven challenging, particularly given the time constraints and competing priorities of industry representatives. To counteract this imbalance, the SRIDA employs inclusive consultation mechanisms, targeted outreach efforts, and dedicated engagement channels for underrepresented stakeholders. As research has shown, broad representation leads to more resilient and widely accepted roadmaps [8]. Raising awareness about the agenda's impact on smaller entities is essential to encourage their participation. By ensuring an equitable distribution of influence, we aim for a representative and democratically structured SRIDA.

4.3 Fostering Closer Industry Collaboration

Strengthening ties with industry stakeholders is important for ensuring that the SRIDA remains relevant and responsive to market needs. In practice, industry engagement always provides valuable insights into emerging trends, technological challenges, and evolving expectations, helping shape a roadmap that is both ambitious and practical. To facilitate industry involvement, the SRIDA prioritizes direct engagement with key stakeholders through structured feedback mechanisms and working groups. Their real-world insights and practical considerations help the SRIDA's objectives remain aligned with market demands.

Additionally, maintaining a strong connection with policymakers helps bridge the gap between technological innovation and regulatory requirements. With this goal in mind, encouraging active participation from Member States and national representatives throughout the whole development process further enhances the SRIDA's practical applicability.

Another important aspect of the SRIDA's development is its commitment to interdisciplinary collaboration. The involvement of different contributors is particularly valuable in ensuring that various expertise is effectively integrated into the roadmap. Moreover, social and ethical considerations are fundamental to the responsible deployment of ADR technologies, and the future iterations of the SRIDA should actively incorporate expertise from the humanities, law, and social sciences, fostering deeper connections between technical advancements and societal needs. By embedding these perspectives into the development process, the SRIDA can ensure that ethical, legal, and societal implications are fully addressed.

4.4 Balancing Long-Term Vision with Immediate Objectives

As was highlighted before, the SRIDA is designed to address both immediate research needs and the broader trajectory of ADR development in Europe. However, maintaining this balance is a challenge, as short-term funding cycles, shifting political priorities, and evolving technological landscapes can impact long-term strategic planning.

An iterative approach allows us addressing these uncertainties for periodic reassessment and recalibration, as was described earlier. Although time intensive, this refinement process aligns with best practices in technology foresight, where "roadmaps provide flexibility while maintaining a coherent long-term vision" [3]. Regular updates aim to ensure that the SRIDA remains relevant, forward-looking, and responsive to new developments.

4.5 Strengthening Transparency in Decision-Making

Transparency is a cornerstone of the SRIDA's methodology. Given the diverse and sometimes competing interests involved, maintaining open communication and clear decision-making processes remains essential in building trust among stakeholders. To enhance transparency in the future, the SRIDA's coordination team and the Editing Committee will incorporate structured documentation, more public consultations and open discussions, and comprehensive reporting on stakeholder engagements.

A well-documented and publicly accessible process of the SRIDA creation fosters accountability and ensures that all contributors have the opportunity to participate in the decision-making process. Regular online meetings, events, and workshops, organized by Adra and other ADR stakeholders, serve as essential forums for collecting external feedback, validating concepts, and refining proposed strategies. These interactions not only strengthen the credibility of the SRIDA but also ensure that it remains inclusive, adaptable, and community driven.

5   Conclusion

The development of the SRIDA has marked a significant milestone in shaping Europe's ADR strategy. By encouraging a collaborative approach, it successfully integrated diverse perspectives, providing a balanced, inclusive, and forward-looking roadmap. The SRIDA serves as an effective strategic document by balancing long-term and short-term objectives and ensuring transparency. At the same time, being more than just a technological roadmap, the SRIDA presents a dynamic and constantly evolving process that aligns technological advancements with societal needs, ethical considerations, and long-term sustainability. As the ADR landscape continues to develop, the SRIDA will adapt, responding to emerging challenges while upholding its foundational principles of democratic engagement and responsible innovation.

5.1   Reflecting on the SRIDA's Impact on European ADR Strategy

Garcia and Bray [4] highlighted that technology roadmaps serve as essential tools for planning and coordinating scientific and technological developments across multiple levels, from individual organizations to entire industries and even national and international initiatives. In this regard, the SRIDA is establishing a structured mechanism for aligning research, industry, and policy objectives. More than that, by fostering collaboration among policymakers, researchers, industry leaders, and civil society, Adra essentially creates an ADR discussion platform where emerging trends, societal needs, and ethical considerations can be integrated into research and development efforts. By addressing the challenges and incorporating lessons learned, the SRIDA will continue to evolve as a robust, representative, and strategically focused roadmap for Europe's ADR ecosystem. Nevertheless, its success ultimately depends on successful collaboration, openness, and a commitment to inclusivity, which would be impossible without the active involvement of the whole ADR community—ensuring that all voices are heard and that the SRIDA truly reflects the collective vision of the European AI, data, and robotics.

5.2 Future Directions for SRIDA

Looking ahead, the SRIDA will continue evolving to remain responsive to emerging technological shifts, policy changes, and stakeholder needs. One of its existing key challenges is ensuring equitable representation across different stakeholder communities. It is essential that not only scientists from the AI, data, and robotics fields remain actively involved in its creation but also industry representatives, including large corporations, small and medium-sized enterprises (SMEs), and startups. A truly representative SRIDA will require Adra's ongoing efforts to engage underrepresented groups, ensuring that Europe's ADR ecosystem benefits from a diversity of insights and expertise.

Balancing long-term strategic research goals with immediate technological applications remains another critical priority. On one hand, the SRIDA must maintain its ambitious objective of fostering a European society powered by safe, reliable, and beneficial ADR technologies. On the other, it must address short-term imperatives such as accelerating collaboration between ADR stakeholders, facilitating technology deployment, and addressing pressing societal challenges.

Future iterations of the SRIDA will, we hope, further refine mechanisms for stakeholder engagement, particularly focusing on SMEs and emerging innovators who may lack the resources to participate in high-level policy discussions. Additionally, adapting to evolving regulatory frameworks, strengthening interdisciplinary collaborations, and aligning with global AI and data governance structures will be critical. By reinforcing its commitment to adaptability, transparency, and inclusivity, the SRIDA aims to continue guiding the responsible development of ADR in Europe while consistently following the established processes outlined in the conceptual and operational methodologies presented in this chapter.

References

1. Curry, E., Heintz, F., Irgens, M., Smeulders, A. W., & Stramigioli, S. (2022). Partnership on AI, data, and robotics. *Communications of the ACM, 65*(4), 54–55.
2. Khalifa, Z., Burgan, M., Bregaj, T., & Almallak, M. (2014). Technology roadmap for next generation PC: Hybrid PC. In T. Daim, M. Pizarro, & R. Talla (Eds.), *Planning and roadmapping technological innovations. Innovation, technology, and knowledge management.* Springer. https://doi.org/10.1007/978-3-319-02973-3_8
3. Kostoff, R. N., & Schaller, R. R. (2001). Science and technology roadmaps. *IEEE Transactions on Engineering Management, 48*(2), 132–143. https://doi.org/10.1109/17.922473
4. Garcia, M., & Bray, O. (1997). *Fundamentals of technology road mapping.* Strategic Business Development Department, Sandia National Laboratories.
5. Abudawod, B., Bregaj, N., Khalifa, Z., & Pizarro, M. (2014). Critical elements of R&D management: Case study of three firms from different sectors of industry. In T. U. Daim et al. (Eds.), *Planning and roadmapping technological innovations: Cases and tools* (pp. 67–82). Springer. https://doi.org/10.1007/978-3-319-02973-3_4
6. Daim, T. U., Pizarro, M., & Talla, R. (Eds.). (2014). *Planning and roadmapping technological innovations: Cases and tools.* Springer. https://doi.org/10.1007/978-3-319-02973-3

7. Phaal, R., Farrukh, C. J. P., & Probert, D. R. (2004). Technology roadmapping—A planning framework for evolution and revolution. *Technological Forecasting and Social Change, 71*(1–2), 5–26. https://doi.org/10.1016/S0040-1625(03)00072-6
8. Chakraborty, S., Nijssen, E. J., & Valkenburg, R. (2022). A systematic review of industry-level applications of technology roadmapping: Evaluation and design propositions for roadmapping practitioners. *Technological Forecasting and Social Change, 179*, 121141. https://doi.org/10.1016/j.techfore.2021.121141

Digital Twins in Europe: Driving Sustainable Innovation and Sovereignty

A Comprehensive Overview of Digital Twin Technologies, Strategic Applications, and Their Role in Europe's Digital and Green Transitions

Victor Alonso Ramos, Gema Antequera García, Iria Galiñanes Romero,
Celia Minguet Requeni, Sergio Gusmeroli, Iddo Bante, Jose Ramón Sierra,
Nikolaos Nikolakis, Mila Koeva, Katalin Kovacs, Gorka Aguirre,
and Ibon Ocaña

Abstract Digital twins are emerging as a key technological enabler for Europe's industrial future. As high-fidelity virtual replicas of physical systems, digital twins harness advances in artificial intelligence, connectivity, and the growing availability

V. A. Ramos (✉) · G. A. García · I. G. Romero · C. M. Requeni
CTAG Automotive Technology Center of Galicia, O Porriño, Spain
e-mail: victor.alonso@ctag.com; gema.antequera@ctag.com; iria.galinanes@ctag.com;
celia.minguet3892@ctag.com

S. Gusmeroli
Politecnico di Milano, Milan, Italy
e-mail: sergio.gusmeroli@polimi.it

I. Bante · M. Koeva
Univerity of Twente, Enschede, The Netherlands
e-mail: i.bante@utwente.nl; m.n.koeva@utwente.nl

J. R. Sierra
LiveM, Zaragoza, Spain
e-mail: jrsierra@livem.eu

N. Nikolakis
Laboratory for Manufacturing Systems, University of Patras, Patras, Greece
e-mail: nikolakis@lms.mech.upatras.gr

K. Kovacs
Innomine, Budapest, Hungary
e-mail: katalin.kovacs@innomine.com

G. Aguirre
IDEKO, Elgoibar, Spain
e-mail: gaguirre@ideko.es

I. Ocaña
Ceit-IK4, San Sebastian, Spain
e-mail: iocana@ceit.es

E. Curry et al. (eds.), *Artificial Intelligence, Data and Robotics*,
https://doi.org/10.1007/978-3-032-10561-5_4

of data to enable real-time monitoring, simulation, and optimization. Their impact spans diverse domains, including manufacturing, healthcare, energy, logistics, and smart cities, playing a pivotal role in driving the green and digital transition.This chapter explores the foundational principles and strategic relevance for digital twins within the European context, with particular emphasis on the contributions of the AI, Data, and Robotics Association (ADRA). It outlines state-of-the-art technical approaches and highlights prominent use cases in this critical field.

Keywords Artificial intelligence · Manufacturing · Advanced robotics · Internet of Things · Data analytics

1   Introduction

In today's fast-paced digital environment, digital twin (DT) technology has become a pivotal paradigm for the analysis, optimization, and transformation of complex systems and processes. By generating high-fidelity virtual representations of physical assets, digital twins facilitate real-time monitoring, predictive analytics, and scenario-based simulations, effectively bridging the physical and digital realms. This capability empowers stakeholders to anticipate system behaviors, improve operational efficiency, and support evidence-based decision-making across diverse domains [1–6].

The concept of the DT was first articulated by Dr. Michael Grieves in the context of aerospace [6]. Since then, DTs have expanded their influence on a wide array of sectors, including healthcare [7, 8], construction [9, 10], logistics [11, 12], and manufacturing [2, 13, 14]. This transformative potential is driven by the convergence of artificial intelligence (AI), Internet of Things (IoT), advanced robotics, and data analytics, which provide the foundation for their operation.

For Europe, digital twins are not just technological tools but strategic enablers of competitiveness and innovation. As highlighted in the *Draghi report* on EU competitiveness [15], since the early 2000s, the EU has experienced sluggish economic growth and a growing productivity gap compared to the United States of Ameria. As a consequence, Europe faces an urgent need to consolidate its leadership in this domain.

Through initiatives such as Horizon Europe and leading associations including the AI, Data, and Robotics Association (ADRA), the European Factories of the Future Research Association (EFFRA), and the European Institute of Innovation and Technology (EIT), Europe is well positioned to align technological advancement with societal values, such as sustainability, trustworthiness, and ethical AI [16].

By making strategic investments in DTs and embedding ethical, regulatory, and environmental considerations from the outset, Europe can strengthen its position at the forefront of digital transformation. DT technology plays a pivotal role in supporting Europe's technological sovereignty and in contributing to the Sustainable Development Goals [17].

This chapter provides an analysis of the state of the art in DT technology, with emphasis on its implications for European industries and society. In particular:

- **Section 2** analyses the state of the art, identifies opportunities and challenges, and presents global and European perspectives on the development and adoption of DTs.
- **Section 3** explores their transformative potential in sectors such as manufacturing, logistics, healthcare, and construction.

2   A Technology Perspective into Digital Twins

This chapter aims to describe how digital twins are embedded in society at all levels (Sect. 2.2), the benefits for European industry (Sect. 2.3), the impact of digital twins worldwide and how they are developing in other continents (Sect. 2.4), and a SWOT analysis (Sect. 2.5), drawn from our discussions in the ADRA digital twin group. We include in this chapter key definitions (Sect. 2.1) to establish a common understanding.

2.1   *Definitions*

Digital twin (DT): It is a dynamic, virtual representation of a physical system, process, or object, which is continuously updated with real-time data collected from sensors, IoT devices, and other sources to reflect the current state of its physical counterpart. Going beyond simulation models, DTs capture the structural and functional aspects of the physical entity but also model its behavior, enabling high-fidelity simulation, predictive analysis, and operational optimization. Through this integration, DTs bridge the gap between the physical and digital worlds, supporting informed decision-making and self-adaptive control [2, 6, 18–21]. Key components include:

- **Physical entity**: Real-world object, system, or environment being represented.
- **Digital model**: The virtual counterpart designed to mimic the physical entity.
- **Data connectivity**: Real-time integration of data streams from the physical entity to the digital model.
- **Feedback loop**: The capability to influence the physical system through insights generated by the digital twin.
- **Closed loop**: A system where data is continuously collected, analyzed, and used to automatically control or influence the physical environment without human intervention. In a closed-loop system, the feedback generated by the DT or AI model is directly used to autonomously adjust operations in real time, creating a self-regulating process (Fig. 1).

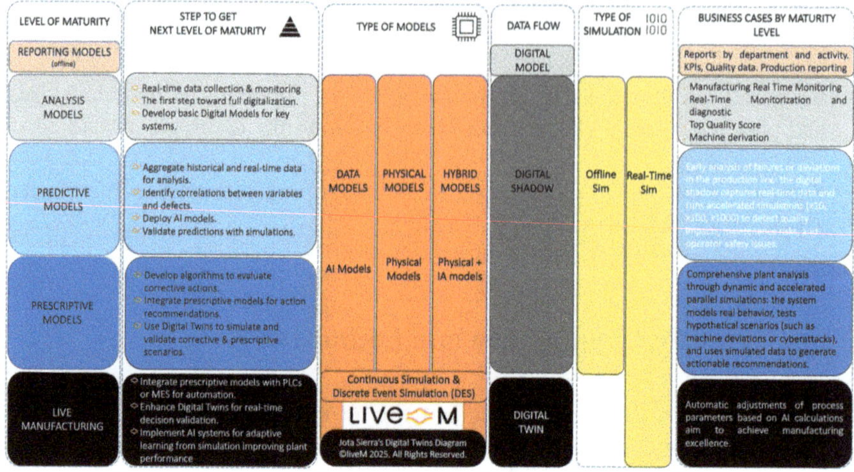

Fig. 1 Digital twins diagram. (Source: Jose Ramón Sierra)

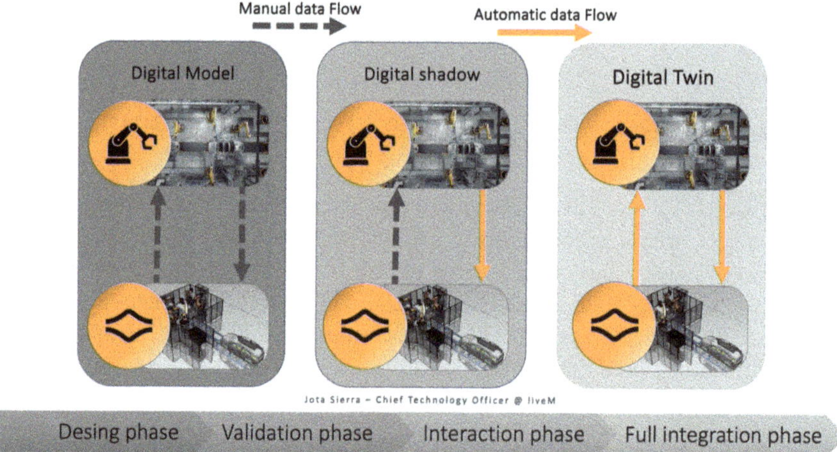

Fig. 2 Digital model, digital shadow, and digital twin and the steps of maturity. (Source: Jose Ramón Sierra)

The distinction between a digital model, digital shadow, and digital twin depends on the bidirectionality of information flow between the real world and the digital world, as illustrated in Fig. 2.

- **Generative artificial intelligence (GenAI)**: GenAI refers to a class of machine learning models capable of creating new content, such as text, images, or simulations, based on patterns in existing data. When integrated with DTs, GenAI enhances predictive capabilities and decision-making by generating plausible scenarios and optimized solutions for specific industrial challenges [22, 23].
- **Internet of Things (IoT)**: The IoT is a network of interconnected physical devices embedded with sensors and software, enabling them to collect, process,

and exchange data. IoT infrastructure provides the real-time data streams that are foundational to DT systems, supporting continuous monitoring and synchronization between digital and physical entities [24, 25].

- **Artificial intelligence (AI)**: AI comprises computational technologies that enable machines to perform tasks typically requiring human intelligence, such as learning, reasoning, and problem-solving. Within DTs, AI powers predictive analytics, automated optimization, and advanced decision support systems, increasing their added value and autonomy [26, 27].
- **Cyber-physical systems (CPS)**: CPS integrate computational (digital) and physical processes through embedded computers and networks that monitor and control physical environments. CPS form the architectural backbone for DTs by enabling seamless interaction and feedback between physical assets and their digital representations [28–30].

Key Associated Terms

- **Predictive maintenance**: The use of DTs to anticipate equipment failures and schedule maintenance proactively, thereby minimizing unplanned downtime and operational costs [31].
- **Simulation**: The use of a DT to test various scenarios, optimize designs, and predict outcomes without impacting the physical asset, enabling virtual experimentation and validation [32].
- **Edge computing**: A distributed computing paradigm that processes data close to its source (e.g., IoT sensors), rather than relying solely on centralized cloud servers. Edge computing reduces latency and bandwidth usage and is essential for real-time DT applications requiring immediate responsiveness.
- **Interoperability**: The ability of DTs to seamlessly integrate and communicate with other systems, platforms, or devices, ensuring cross-industry compatibility and data exchange [33, 34].
- **Data governance**: The set of policies and procedures that ensure the security, quality, and ethical use of data within DT systems, addressing compliance, privacy, and data integrity concerns [35–37].

2.2 How Digital Twins Are Being Adopted in Society at All Levels

DTs have moved beyond niche applications and are now widely adopted across numerous sectors, transforming how organizations interact with technology, manage resources, and address complex challenges [38, 39].

In government and urban planning, DTs support the development of smarter, sustainable, and more efficient cities. By constructing virtual models of urban environments, municipalities can simulate traffic flows, monitor energy consumption, and assess the environmental impact of new policies before implementation. Pioneering examples include Herrenberg's [40] and Helsinki's [41] city-wide DT initiatives, which are used to improve infrastructure, waste management, and public transport, thus advancing sustainable urban development.

Healthcare benefits from DTs by creating personalized digital replicas of patients using real medical data. These virtual models enable providers to simulate treatment options, predict outcomes, and optimize resource allocation, which improves decision-making, efficiency, and patient outcomes [42].

In agriculture, DTs empower farmers to monitor crop growth, soil conditions, water usage, and plant health in real time, promoting efficient and sustainable resource management. By simulating different scenarios, DTs help farmers prepare for unpredictable weather or market fluctuations [43].

DTs are also instrumental in environmental monitoring and disaster management. Virtual models of ecosystems and cities allow researchers and governments to simulate the impacts of climate change, natural disasters, or industrial accidents. For instance, DT-based flood management systems enable real-time risk prediction and enhance early warning capabilities [19].

In manufacturing and, more general, in industry, DTs optimize production lines, reduce waste, and improve product quality by allowing changes to be tested virtually before physical implementation. DTs also enhance system interoperability, enabling diverse factory technologies to communicate and collaborate [44].

Sectors such as energy and logistics are also adopting DTs to increase efficiency and safety, such as simulating energy distribution to minimize outages or optimizing logistics routes and intralogistics flows by simulation.

DTs are expected to become more prevalent in everyday life. Future applications may include homes using DTs to monitor and maintain appliances or the retail sector leveraging DTs to personalize shopping experiences and optimize inventory.

In summary, the broad and growing adoption of DTs across diverse sectors demonstrates their strategic significance for industry and society. As technological capabilities continue to advance, DTs are expected to become even more integral to data-driven decision-making and sustainable development. Ongoing research, investment, and cross-sector collaboration will be critical to fully realizing their transformative potential.

2.3 Benefits for Competitiveness and Development in Europe

DT technologies are rapidly becoming foundational tools in Europe's pursuit of a more competitive, resilient, and technologically sovereign economy [15, 16]. A principal advantage of DTs lies in the industrial sector. By enabling the creation of dynamic, real-time digital counterparts to physical assets and systems, DTs empower manufacturers to monitor and optimize operations with unprecedented precision. They facilitate not only predictive maintenance but also deep analysis and correction of inefficiencies. The result is improved efficiency, reduced downtime, and enhanced product quality—factors that directly reinforce the global competitiveness of European industry [19].

Beyond manufacturing, DTs are increasingly valuable for addressing complex societal challenges. In urban environments, DTs enable better planning and infrastructure management, allowing cities to test sustainable development strategies and anticipate the impact of policy decisions before implementation. For example, the city of Aachen, Germany, developed a comprehensive digital model of its transport infrastructure, which enhanced interdepartmental coordination and citizen engagement, supporting broader EU sustainability and smart urban development objectives [2, 45]. Another example is the Netherlands, where multiple cities are united behind the idea of Netherlands 3D. In healthcare, DTs support more accurate forecasting of service needs, leading to improved resource allocation and patient outcomes [42]. These applications contribute not only to efficiency but also to quality of life for European citizens.

A critical enabler of DT uptake in Europe is the network of European Digital Innovation Hubs (EDIHs) [46]. EDIHs serve as regional catalysts, connecting companies, especially SMEs, with the expertise, test environments, and funding needed to adopt advanced digital technologies such as DTs. Through EDIHs, businesses access training, skills development, and practical support, lowering barriers to entry for complex systems. By fostering ecosystems of collaboration between academia, industry, and public administration, EDIHs ensure DT solutions are technically robust and aligned with local and regional needs [47–49]. Complementary to this, organizations such as the Digital Twin Geohub (DTG), the Digital Twin City Center (DTCC), and Open & Agile Smart Cities (OASC) provide cross-sectoral knowledge exchange, international cooperation, and scalable urban DT frameworks.

These developments align closely with the conclusions of the Draghi report on EU competitiveness, which emphasizes the need for ambitious investment in innovation, infrastructure, and talent retention. The report highlights the urgency of strengthening the Single Market, removing barriers to cross-border cooperation, and building strategic technological capabilities. DTs, situated at the intersection of AI, data, robotics, and system integration, embody many of the report's key recommendations and represent a concrete step toward high-impact innovation [15].

Moreover, DTs reinforce Europe's digital sovereignty. As dependence on foreign platforms and technologies grows, the ability to develop and govern critical digital infrastructure becomes increasingly important. DTs promote the creation of secure, sovereign solutions that reduce reliance on external providers and foster European innovation.

In this context, DTs should be considered as a strategic enabler of Europe's broader economic and policy ambitions. Continued investment and integration of DTs into both public and private sector initiatives will be essential for building a competitive, sustainable, and sovereign Europe.

2.4 Landscape

The adoption of DT technologies is accelerating globally, driven by their potential to enhance efficiency, reduce costs, and foster innovation across diverse sectors [50]. The global DT market is projected to grow from €16.55 billion in 2024 to €242.11 billion by 2032, representing a compound annual growth rate (CAGR) of 39.8% [51].

2.4.1 United States of America

In the United States of Ameria, DT technology has gained significant traction, particularly within the manufacturing and aerospace sectors. Major companies such as Boeing, Lockheed Martin, and General Electric are utilizing DTs to optimize production lines, enhance predictive maintenance, and improve supply chain efficiency. These innovations are helping companies reduce downtime, improve operational efficiency, and cut operational costs.

Several SMEs (small and medium-sized enterprises) and industry associations are also leading the charge in DT innovation, driving forward frameworks, tools, and use cases across a wide array of industries:

- **Digital Twin Consortium (DTC)**: A global partnership of industry, government, and academia, the DTC is focused on developing best practices and accelerating the adoption of DT technologies. Notable members such as Microsoft, Dell, and GE Digital are collaborating to set standards and drive the future of DT applications. Their work spans industrial IoT, smart cities, and healthcare solutions, focusing on creating interoperable and scalable models.
- **National Institute of Standards and Technology (NIST)**: is collaborating with SMEs and research institutions to improve DT interoperability and cybersecurity, particularly in critical infrastructure like energy grids and transportation networks. Their focus is on ensuring that DT technology can be implemented securely and efficiently across various sectors.
- **Manufacturing United States of Ameria**: Through public-private partnerships, Manufacturing United States of Ameria promotes the integration of DT technologies to boost productivity and drive innovation in advanced manufacturing. These efforts are intended to modernize manufacturing processes, improve supply chain efficiency, and ultimately ensure global competitiveness for US manufacturers.

2.4.2 Europe

Europe, while facing great competition from both the United States of Ameria and China, has made significant strides in advancing DT technology. European countries are increasingly leveraging DTs for a wide range of applications, from the development of smart buildings and energy-efficient systems to digital replicas of entire cities.

Notably, the European Union has made DT technology a central component of its digital transformation agenda. Numerous research and innovation projects are aimed at boosting the adoption of DTs across sectors such as manufacturing, mobility, and energy. Europe's strong emphasis on sustainability is driving the development of DTs that focus on environmental impact, energy optimization, and resource efficiency.

Several initiatives underscore Europe's strategic approach to DT development, with an emphasis on collaboration, sustainability, and digital innovation:

- **Horizon Europe Programme**: This European funding initiative supports a wide range of projects in healthcare, transportation, urban sustainability, and so on. DT technology is being integrated into these sectors to improve operational efficiencies and enhance the quality of services. There are some important associations under the Horizon Europe Programme:

  - **ADRA (AI, Data, and Robotics Association)**: the main private representative in Europe's public-private partnership on artificial intelligence, data, and robotics, working alongside the European Commission. Created in 2021, the association brings together companies, research institutions, and academic organizations to help shape the future of these key technologies in Europe.
  - **Made in Europe**: a key European partnership under Horizon Europe focused on strengthening the EU's manufacturing sector. Its goal is to support the development of sustainable, digital, and resilient production technologies that keep European industry competitive on a global scale.

This partnership combines public investment from the European Commission with the technological and strategic expertise of the private sector, represented by the European Factories of the Future Research Association (EFFRA). EFFRA plays a key role in defining the partnership's research priorities, ensuring that industrial needs are effectively embedded within the innovation agenda. Together, these actors are advancing the modernization of manufacturing across Europe, with a particular emphasis on green technologies, digitalization, and strategic autonomy.

In parallel, Europe is spearheading a series of targeted initiatives across critical sectors such as manufacturing, energy, mobility, and urban development. These efforts, led by the European Union in collaboration with innovation hubs and major industry stakeholders, seek to accelerate digital transformation through the deployment of advanced virtual modelling frameworks. The following projects exemplify this strategic direction:

1. **DTs in Industry, SMEs, and Smart Manufacturing**

    Europe has launched multiple initiatives focused on the implementation of DTs in manufacturing and automation, with a strong presence in Germany, France, and Italy. Some key initiatives include:

    - **Plattform Industrie 4.0 (Germany)**: Germany's Industrie 4.0 initiative is a key driver of DT adoption in manufacturing. The platform focuses on integrating DTs into smart factories for real-time optimization of production and supply chain processes. This integration is helping boost productivity and establish a more agile manufacturing environment.

- **Industry 4.0 National Technology Platform (Hungary)**: Unites industrial players, research institutions, and government bodies to develop and implement Industry 4.0 technologies; aims on the formulation of strategic guidelines, support for research and development projects, and integration of DT technologies into industrial processes.
- **Change2Twin**: This project is dedicated to helping small and medium-sized enterprises (SMEs) in Europe digitize their operations by adopting DT solutions. Working in collaboration with local Digital Innovation Hubs (DIHs), Change2Twin provides tailored mentoring plans and DT implementation strategies, allowing SMEs to digitally transform their processes and improve their competitive edge.
- **ASHVIN Project**: Developing DTs for the **construction and infrastructure sectors** to enhance efficiency, safety, and sustainability.
- **DT Manufacturing EU**: An initiative focused on **virtualizing industrial processes** to optimize factory performance across Europe.
- **Booster Manufacturing Lab (Spain)**: An industrial TEF for human-machine collaboration, developing robotic applications, logistics, and collaborative environments for people with exoskeletons that allow the development of industrial experiments in which the digital twin developed supports all these types of processes. It has equipment that allows the development of applications with GenAI to support the processes developed. It is a space for co-creation in which companies, startups, and the technology center can scale the applications until they are taken to an industrial plant.
- **Digital Fuel Twin**: This cloud-based data platform developed by Bosch enables those maintaining and operating fleets of vehicles to track fuel properties across the entire supply chain. Through this solution, stakeholders can track on live what fuel is being used, carbon emissions, and fuel properties. This will help companies in reaching their sustainability targets.
- **Smart Digital Reality**: This is a platform developed by Hexagon AB that allows users to create DTs that blend physical with digital using data, creating immersive DTs for clients.
- **EcoStruxure Machine Expert Twin**: The digital automation company Schneider Electric provides customers with this tool that enables them to build twins that capture and simulate the entire life cycle of real machines. According to Schneider Electric, this approach cuts commissioning time by 60% and time to market by 50% and increase cost savings by 20%.
- **RAMI 4.0 (Reference Architecture Model for Industry 4.0)**: The RAMI 4.0 is an important framework that integrates DT technology into the broader context of Industry 4.0 in Europe. Developed in Germany, this model provides a comprehensive reference architecture that helps organizations structure and implement DT solutions within the framework of smart factories and industrial IoT.

2. **DTs in the Energy and Environment Sector**

   The European energy sector key initiatives include:

   - **ENTSO-E Digital Twin Grid**: A project developing a **DT of Europe's electricity grid**, enabling real-time simulation of energy generation, transmission, and demand.
   - **Digital Twins for the Green Transition**: Applications in **wind and solar farm management, energy storage optimization, and carbon footprint reduction**.
   - **Digital Twin of the Ocean (DT4O)**: This European project aims to map and simulate the oceans to support environmental protection and marine research. It highlights the growing use of DTs in addressing large-scale, global challenges, including climate change and marine conservation.
   - **Destination Earth (DestinE): A DT of the Earth**: One of the EU's most ambitious projects is **Destination Earth (DestinE)**, which aims to develop a high-precision DT of the planet to simulate, predict, and monitor climate changes, natural disasters, and human-environment interactions.

3. **Smart Cities and Urban Planning**

   European cities are integrating DTs for mobility, infrastructure, and sustainability planning. Some of the most notable projects include:

   - **Madrid Digital Twin**: The development of a virtual model of the city to simulate mobility scenarios, infrastructure planning, and emergency response.
   - **Twins4Europe Project**: A platform aimed at implementing DTs in urban communities to enhance quality of life and sustainable planning.

2.4.3   Asia

Asia has emerged as a leader in large-scale DT implementations, with several countries making substantial investments in this technology.

- **China** has heavily invested in DT capabilities, particularly in smart city initiatives and industrial applications. **The Chinese government has included DT technology in its national strategic plans**, viewing it as a key tool for urban management and sustainability. Several Chinese cities, such as Beijing and Shanghai, have adopted DT models to simulate and optimize urban systems, from transportation networks to energy grids.

  Additionally, China has successfully deployed DTs for large-scale infrastructure projects, enabling better planning, real-time monitoring, and more efficient decision-making processes. The government's vision for a "smart China" leverages these digital replicas to improve environmental sustainability, reduce resource consumption, and enhance public services.
- One specific example of DTs' encouraged actions in China is **Smart City Alliance**, a collaboration between local governments, universities, and tech com-

panies aimed at promoting the use of DTs for urban planning and efficient resource management in Chinese cities.

- **Japan**: Under the Society 5.0 initiative, Japan is integrating DTs into its approach to smart manufacturing, logistics, and healthcare innovation. This forward-thinking initiative seeks to balance technological advancement with societal needs, using DTs to enhance the quality of life, improve productivity, and address societal challenges.

2.4.4　Middle East

The Middle East, particularly the United Arab Emirates (UAE), has also embraced DT technology for urban management and sustainability. One notable example is the **Dubai Digital Twin initiative**, which aims to optimize urban management, resource allocation, and energy consumption through real-time data simulations. By replicating the city's infrastructure digitally, authorities are able to monitor and manage public services, traffic flow, and energy distribution more efficiently, contributing to a more sustainable and resilient urban environment.

2.4.5　Digital Twins in Universities and Research Institutions

Leading universities and research institutions are playing a pivotal role in advancing DT research and application. These institutions are focusing on how to integrate DTs with other cutting-edge technologies, such as AI, machine learning, and IoT, to create more dynamic, responsive systems:

- **The Massachusetts Institute of Technology (MIT, United States of Ameria)**: conducts groundbreaking research on smart manufacturing, leveraging DTs for predictive maintenance, system optimization, and real-time monitoring. Their work is helping drive the next generation of intelligent manufacturing solutions, ensuring greater efficiency and reduced downtime in industrial processes.
- **Technical University of Munich (Germany)**: explores how DTs can facilitate intelligent supply chain management and real-time decision-making in the context of Industry 4.0. Their research aims to create more adaptive, resilient systems that respond quickly to changes in demand or disruptions.

2.5　European SWOT in Digital Twin

A SWOT analysis evaluates Europe's position in the DT ecosystem, identifying strengths, weaknesses, opportunities, and threats (Table 1). **This framework highlights Europe's unique advantages while addressing the challenges and external pressures shaping its future.**

Table 1 SWOT analysis

Strengths	Weaknesses
1. **Robust Research and Innovation Ecosystem:** • Europe hosts leading research institutions, universities, and innovation clusters driving advancements in DTs, AI, and robotics • Strong public-private partnerships foster collaborative development, exemplified by initiatives like Horizon Europe and ADRA 2. **Regulatory Framework and Ethical Standards:** • Europe is a global leader in establishing regulations for data security, privacy (e.g., GDPR), and AI ethics, ensuring trusted adoption of DTs • Initiatives like Gaia-X promote data sovereignty and interoperability across industries 3. **Industrial Strength:** Europe's manufacturing and engineering sectors, particularly in Germany, France, Spain and Italy, are integrating DT technology to optimize processes and improve competitiveness 4. **Focus on Sustainability:** European strategies emphasize using DTs for climate resilience, sustainable urban planning, and green energy transitions	1. **Fragmentation Across Member States:** Diverse regulatory environments and technological disparities between EU countries hinder the unified adoption of DT technologies 2. **Skills Gap:** A shortage of skilled professionals in AI, robotics, and data analytics limits the pace of DT adoption and innovation 3. **Limited Private Investment:** Compared to the USA. and China, Europe faces a funding gap, particularly in scaling startups and SMEs specializing in DT technology 4. **Slow Pace of Standardization:** Despite progress, Europe lags in establishing universal technical standards for DT implementation, creating barriers to interoperability
Opportunities	Threats
1. **Global Leadership in Ethical AI and DTs:** Europe can leverage its ethical frameworks and focus on safety to become a global standard-bearer for trustworthy DT applications 2. **DTs as AI models for training:** By simulating real-world conditions with high precision, DTs enable the generation of high-quality synthetic data, which can be used to train AI systems. This allows AI to learn and adapt in safe, controlled settings, accelerating development while reducing the risks associated with deploying AI in real-life environments 3. **Integration of AI using LLMs (Large Language Models):** These tools can simplify how users interact with complex systems, allowing them to ask questions or explore data in plain language. LLMs also help process unstructured information and support faster, smarter decision-making. This makes DTs more accessible and useful across different sectors, not just for technical experts 4. **Expansion of Use Cases:** Emerging sectors such as personalized healthcare, autonomous systems, and smart cities provide vast opportunities for DT deployment 5. **Green Transition and Climate Adaptation:** DTs are pivotal in advancing Europe's Green Deal goals by optimizing energy systems, infrastructure, and resource management 6. **Strategic Alliances and International Collaboration:** Strengthening partnerships with global leaders and developing alliances in underserved regions can expand market reach and influence 7. **Integration of Generative AI and Robotics:** Combining DTs with GenAI and advanced robotics opens pathways for innovation in manufacturing, logistics, and autonomous systems 8. **Optimizing natural resources through DT and AI:** DTs combined with AI offer a powerful way to manage natural resources more efficiently. By creating accurate, real-time models of ecosystems, energy grids, or water systems, it becomes easier to predict demand, reduce waste, and plan sustainable use	1. **Competition from Global Players:** The USA. and China are heavily investing in DT ecosystems, supported by significant funding and rapid technological advancements 2. **Cybersecurity and Data Risks:** As DTs rely on vast amounts of real-time data, vulnerabilities to cyberattacks and breaches could undermine trust and adoption 3. **Economic Uncertainty:** Fluctuating economic conditions and limited resources in some EU member states may delay investments in DT infrastructure 4. **Technological Dependency:** Heavy reliance on non-European technologies (e.g., cloud services, semiconductors) threatens Europe's digital sovereignty and resilience 5. **Resistance to Change:** Industries with entrenched practices may resist adopting disruptive technologies like DTs, slowing the transformation process

Strategic Directions for Europe Based on the SWOT Analysis

- **Unify DT policies across EU member states** to reduce fragmentation and improve interoperability.
- **Increase investment in DT startups and SMEs** to bridge the funding gap and accelerate innovation.
- **Expand workforce training programs** in AI, data analytics, and DT applications to address the skills shortage.
- **Leverage Europe's leadership in ethical AI and sustainability** to set global standards and drive adoption.
- **Enhance cybersecurity frameworks** to protect DT infrastructures from data breaches and cyberthreats.
- **Strengthen international partnerships** to expand market influence and promote European DT standards.

By addressing these key challenges and opportunities, Europe can secure its leadership in DT technology, ensuring long-term competitiveness, sustainability, and digital sovereignty in an increasingly interconnected world.

3 Use Cases

DT technology has become an essential tool in various industries, providing innovative solutions to optimize processes and improve efficiency. DTs are helping businesses and governments tackle challenges and seize new opportunities. In this section, we will take a look at how DTs are being applied in different and more specific fields, showcasing their potential and impact.

3.1 Manufacturing

In manufacturing and engineering, DT technology plays a crucial role in improving production processes, predicting maintenance needs, and enhancing system performance. By creating virtual replicas of physical assets, systems, and processes, manufacturers can monitor real-time data, perform predictive analysis, and optimize efficiency. This way, DTs help European businesses be more competitive.

One of the main advantages of DTs in this sector is predictive maintenance. By continuously monitoring equipment and machinery, DTs help identify early signs of wear or potential breakdowns, allowing for proactive maintenance and reducing downtime. This not only helps avoid costly disruptions but also extends the lifespan of assets. Apart from that, DTs can help not only predict failures but correct current mistakes and avoid future problems. Furthermore, by running simulations and exploring different operational scenarios, DTs provide prescriptive insights that support decision-making and optimize operations in real time. As a result, they will be capable of commanding the factories of the future in a semi-autonomous or even fully autonomous way.

The following applications show how DTs are progressively making a huge impact in manufacturing:

- **Use of Robots and Devices in Factories (Humanoids, AMRs, AFTs)**: DT technology optimizes human-robot collaboration in industrial settings. By modelling robot performance and workflows, it enhances efficiency, reduces downtime, and ensures safe and effective integration of robots into manufacturing processes.
- **Battery Assembly for Electric Vehicles (EVs)**: DT technology can significantly optimize the battery assembly process for electric vehicles (EVs) by simulating and improving the assembly workflow, predicting failures, and integrating AI-driven predictive maintenance. This leads to cost reduction, fewer assembly failures, and a more efficient use of materials, making the entire process more sustainable and flexible in adapting to different battery designs.
- **Battery Disassembly for Electric Vehicles (EVs)**: Similarly, DT technology enhances the disassembly process for EV batteries. By simulating disassembly steps, it ensures safe handling of components, improves material recovery, and minimizes hazardous waste. The use of predictive analysis to detect potential failure points increases safety while promoting compliance with environmental sustainability goals.
- **Battery Cell Manufacturing**: In battery cell manufacturing, DT technology is employed to optimize the production process by modeling and simulating energy use, minimizing defects, and improving quality control. Real-time monitoring through IoT sensors ensures consistency, and process simulations lead to more durable and efficient battery cells.
- **Circular Manufacturing Data Spaces and AI-Enabled DTs (Circular TwAIn)**: DTs also enable the creation of Digital Product Passports (DPPs), which track the environmental footprint of products from creation through their entire life cycle. AI-driven data analysis helps optimize material reuse and supports the development of circular economy models by simulating various waste reduction strategies.
- **Steel Parts Manufacturing**: In steel parts manufacturing, digital twin technology is used for real-time simulation, quality control, and predictive maintenance. This leads to more accurate production, lower energy consumption, and reduced CO_2 emissions, contributing to sustainability in steel manufacturing.
- **Mixed Packaging in the Food Industry**: In the food industry, DTs are used to model packaging processes, reduce waste, and ensure food safety. AI-based detection corrects faults in the packaging line, improving efficiency and compliance with food safety regulations.
- **Beverage Packaging Optimization**: As part of industrial process digitization, a DT of the can packaging line at the Pécs Brewery in Hungary was developed using *Plant Simulation* software. This virtual model enabled the simulation and analysis of real production behavior, allowing the identification of operational inefficiencies and synchronization of machinery to reduce frequent stop-and-go

scenarios. Moreover, the DT significantly enhanced the understanding of production dynamics, providing a solid foundation for informed decision-making.

Benefits Achieved

- Increased packaging line efficiency from 71% to 76%
- Reduced utility consumption and maintenance costs
- Improved operational efficiency and final product quality

3.2 Logistics

In logistics, DTs optimize the entire supply chain, from inventory management to shipment tracking. By creating real-time models of supply chain networks, businesses can monitor goods in transit, predict potential disruptions, and improve efficiency.

- **Global Logistic Ship Transport**: DTs are used to model and optimize shipping routes based on variables such as weather, port congestion, and fuel costs. Real-time monitoring through IoT sensors ensures better tracking of cargo conditions, while predictive maintenance helps prevent downtime, and energy-efficient cargo load optimization reduces environmental impact.
- **Complete Supply Chain in the Automotive Industry**: DTs optimize end-to-end supply chain processes in the automotive industry. From raw materials to finished products, they allow for real-time tracking and predictive analysis of disruptions. Integration with Digital Product Passports (DPPs) ensures better traceability, and AI-driven models help forecast demand, improving efficiency and sustainability in the global supply chain.

3.3 Agriculture and Agrifood

- **Smart Farming and Precision Agriculture**: In agriculture, DTs enable farmers to monitor crops and soil conditions in real time through IoT sensors, improving yield while minimizing waste. AI-based predictive analytics help in proactive pest and disease management, while automated irrigation systems optimize water use, promoting sustainability.
- **Livestock Management and Animal Welfare**: DT technology also aids in livestock management by tracking health and productivity. With real-time monitoring of animal movements and health, predictive models can prevent disease outbreaks, improving animal welfare and farm productivity.
- **Agrifood Supply Chain and Food Safety**: DTs enable end-to-end traceability in the food supply chain. From farm to consumer, IoT sensors monitor temperature and humidity to ensure optimal storage and transport conditions, minimiz-

ing spoilage and waste. AI-driven demand forecasting optimizes logistics, improving food safety and sustainability.

3.4 Urban Planning

- **Urban Traffic Simulation**: In Sofia, Bulgaria, a LiDAR-based digital twin framework was developed to simulate traffic scenarios and improve congestion management [52], showcasing the potential of DTs for urban mobility planning.
- **Urban Heat Island (UHI) Mitigation**: DT-based Planning Support Systems (DT-PSS) using Physiological Equivalent Temperature (PET) simulations and real-time data have been developed to assess urban planning strategies and mitigate heat in Wuppertal, Germany [53]; Enschede, The Netherlands [54, 55]; and Padua, Italy [56], demonstrating the role of digital twins in climate-adaptive design.
- **Water Management**: In Enschede, The Netherlands, a DT-driven system for dynamic groundwater monitoring was implemented [57], integrating 3D geospatial data, IoT protocols, and predictive models for adaptive pump control in response to floods and droughts.

3.5 Social Robotics: Assistive Robots and Healthcare

In healthcare, DT technology is revolutionizing personalized medicine by creating digital replicas of patients. These virtual models simulate a patient's response to different treatments, enabling more accurate and effective medical interventions. This personalized approach helps healthcare providers deliver care tailored to each patient's specific needs, improving outcomes.

In the healthcare sector, DT technology also supports the use of assistive robots and wearable devices. By analyzing real-time motion and posture data, DTs can improve ergonomics in healthcare settings, reduce injuries, and optimize workflows for healthcare professionals.

3.6 Security, Defense, and Public Services

By simulating systems in real time, predicting failures, and optimizing resources, DTs offer organizations the ability to enhance resilience, mitigate risks, and improve decision-making in critical situations.

3.6.1 Public Safety and Emergency Management

DTs are being used to simulate and model emergency scenarios such as natural disasters, industrial accidents, or health crises. This allows emergency teams to assess different situations and make better-informed decisions on how to respond. By integrating IoT sensors and data platforms, digital models provide an accurate representation of risks and conditions in real time, aiding in the prevention and mitigation of incidents.

- **Improved emergency response**: Proactive planning and crisis simulations for better resource coordination and personnel deployment in emergencies.
- **Reduced response time**: Scenario modeling enables emergency services to act more quickly, minimizing damage and loss of life.
- **Real-time monitoring**: Integration of sensors and data platforms for continuous tracking of events and emergency conditions.

3.6.2 Critical Infrastructure Security

In defense and security sectors, DTs are used to monitor critical infrastructures such as nuclear plants, power facilities, and telecommunications systems. By creating digital replicas of these systems, they can be continuously monitored in real time to predict failures or imminent breakdowns, enabling proactive management and rapid intervention when needed [9].

Benefits

- **Predictive maintenance**: Identifying potential failures before they occur allows for component replacement or repair before critical breakdowns.
- **Enhanced security**: Continuous vulnerability assessments of infrastructures enable targeted security improvements.
- **Efficient resource management**: Better allocation of operational resources and personnel for maintaining secure infrastructures.

3.6.3 Defense and Scenario Simulation

In defense, DTs are applied to simulate battlefield scenarios and military training exercises. By creating precise digital models of military equipment, vehicles, and terrains, defense forces can analyze system behavior under realistic combat conditions. These simulations also provide valuable insight into testing new strategies and tactics without any risk in the real world.

Benefits

- **Advanced training**: Detailed simulations provide training opportunities without the associated risks of live exercises, enhancing readiness.

- **Optimized equipment and tactics**: Analyzing different military strategies in a controlled digital environment improves defense tactics' effectiveness.
- **System testing**: Virtual models of new defense technologies and equipment enable performance assessments before real-world deployment.

3.6.4 Public Health Management

DTs are revolutionizing public health management, especially in pandemic preparedness and healthcare planning. Digital models simulate disease spread, evaluate healthcare system capacities, and forecast resource needs in real time. Additionally, these models assist in managing medical supply chains, staff deployment, and hospital readiness.

Benefits

- **Disease outbreak prediction**: Identifying disease spread patterns enables quicker and more effective responses.
- **Optimized medical resource allocation**: Planning and distributing medical supplies and staff based on projected needs.
- **Better healthcare facility management**: Evaluating hospital capacity to handle surge demands, such as during a pandemic.

3.6.5 Cybersecurity and Data Protection

In the field of cybersecurity, DTs are used to simulate computer networks and systems, enabling organizations to perform penetration testing and identify vulnerabilities before they are exploited by malicious actors. These digital models help assess the resilience of technological infrastructures and security protocols.

Benefits

- **Cyberattack simulations**: DTs can simulate potential cyberattacks on critical systems, helping organizations identify and mitigate threats proactively.
- **Advanced data protection**: Continuous monitoring of network security ensures the integrity of systems and prevents data breaches.
- **Contingency planning**: Improved response strategies during cyber crises through detailed scenario simulations.

3.7 Construction

The construction industry, known for its labor-intensive nature, is also benefiting from the integration of DT technology. By creating digital models of construction sites and buildings, stakeholders can improve project visualization, monitor progress, and ensure projects stay on track in terms of time and budget.

Beyond project management, DTs are enhancing **safety on construction sites**. By simulating different scenarios and analyzing real-time data from machinery and workers, DTs can detect potential hazards and provide early warnings, ensuring better adherence to safety protocols and reducing accidents.

The integration of DTs with Building Information Modeling (BIM) has also proven useful in **designing and managing buildings**. With this technology, construction professionals can make informed decisions at every stage of a building's life cycle, from planning and design to construction and operation.

Employing a BIM-integrated DT during the construction phase allows companies to track construction progress as it happens on site. This reduces the time needed to detect a design issue and correct it, streamlining site operations and avoiding costly and complex changes.

DT technology is also making a significant impact on **energy consumption**. By integrating IoT sensors and smart meters, it enables real-time monitoring of energy use in buildings and construction sites, helping optimize consumption and reduce waste.

Additionally, regarding **sustainability**, DTs simulate energy-saving strategies, such as renewable energy integration, HVAC optimization, and demand response programs, allowing for the testing of solutions before implementation. This helps in choosing the most effective strategies, like evaluating the impact of solar, wind, and battery storage.

Predictive maintenance is another key benefit. By detecting faults in power grids, HVAC systems, and machinery, DTs can alert maintenance teams to issues before they lead to failures, reducing downtime and enhancing system reliability.

3.8 Transport

DT technology is revolutionizing the transportation industry. By speeding up decision-making, simplifying data analysis, and automating issue resolution, DTs are improving efficiency while minimizing resource usage across various transportation systems.

One significant advantage is their ability to provide real-time, reliable data that helps companies and cities stay ahead of changing Environmental, Social, and Governance (ESG) regulations. DTs track energy consumption, emissions, and sustainability metrics, making it easier to adapt to new requirements.

These technologies also facilitate autonomous problem-solving, from managing navigation to optimizing cargo and public transport systems. With real-time insights, transportation managers can anticipate and address potential issues, enhancing both reliability and efficiency. DTs also improve traffic flow and scheduling by offering better situational awareness, helping reduce congestion and delays.

In the realm of public transport, they enhance safety and the overall passenger experience. By continuously monitoring vehicle conditions, DTs enable proactive maintenance, minimizing breakdowns and ensuring smoother operations. Additionally, they help cut down on carbon emissions by optimizing routes and supporting energy-efficient operations, contributing to more sustainable transportation systems.

Some examples of "Smart Cities" in Europe using DTs are Madrid (developing a virtual model of the city to simulate mobility scenarios, infrastructure planning, and emergency response), Logroño (using AI and DTs to optimize traffic, water management, and energy efficiency in public buildings), and Aachen (developing a comprehensive digital model of its transport infrastructure).

3.9   Other DT Markets

The following examples of uses of DTs illustrate the expanding scope of DTs, highlighting their potential to transform beyond "traditional sectors" by providing innovative solutions and enhancing operational efficiency.

3.9.1   DTs in Energy and Utilities

Across the energy and utilities sector, DTs are quietly reshaping the way systems are managed and optimized. In smart grids, they provide operators with real-time visibility by collecting data from IoT sensors. This helps balance loads, detect faults early, and keep things running smoothly, even under pressure.

One of the most impactful applications lies in renewable energy. With more solar, wind, and hydro sources being added to the grid, DTs make it possible to simulate how these variable sources behave, helping planners avoid disruptions. They are also being used to fine-tune battery storage systems and run scenario testing for grid stability.

Water and waste services are seeing benefits too. Utility managers can use DTs to predict usage patterns, detect leaks before they become major problems, and optimize waste collection routes. Some cities are even exploring ways to model circular economy systems, like waste-to-energy conversion, to improve sustainability. An example of a DT-based solution for waste management is shown for the city of Johannesburg [58].

Perhaps one of the most valuable contributions of DTs is in **preparing for the unexpected.** Whether it's a heatwave, flood, or cyberattack, these systems can simulate the impact on infrastructure and help teams prepare for, and respond to, crises far more effectively.

3.9.2 Retail: Enhancing Customer Experience

Retailers are using DTs to create virtual storefronts, allowing customers to explore products in a simulated environment. Apex Imaging Services, for instance, has developed DTs for retail clients to stand out in the market.

3.9.3 Sports: Performance Optimization

In sports, DTs are being used to simulate and analyze athletes' performances. Coaches and trainers can test different training regimens and strategies in a virtual setting to enhance performance and prevent injuries.

3.9.4 Education

Universities are simulating campus layouts and virtual laboratories, offering more interactive learning experiences. One great example of application of DTs in education are projects like DT Lab (DTLab). This project seeks to develop new innovative methods for education in biomedical science and aquaculture.

4 Conclusions

This conclusion also sets the stage for the next chapter, which looks beyond the current state to examine the future evolution of DTs in Europe: emerging paradigms, speculative applications, and policy frameworks that will shape the next decade.

The advancement and implementation of DT technology presents an opportunity for Europe to assert leadership in the ongoing digital transformation. As discussed throughout this chapter, the implications of DTs extend well beyond technical advancement. They offer concrete pathways to address key European objectives, including climate resilience, healthcare efficiency, industrial competitiveness, and digital sovereignty.

DTs have the potential to serve as the essential interface bridging the digital and physical realms. By enabling systems that are more intelligent, adaptive, and predictive, especially through integration with advanced artificial intelligence, DTs may unlock new prospects for innovation and operational excellence across a wide array of sectors.

Realizing the full potential of DTs in Europe will require sustained investment in foundational research, talent development, and the ethical deployment of scalable, secure systems. Priorities should include building interoperable infrastructures, establishing robust standards, and fostering the integration of AI and robotics into DT ecosystems. Moreover, embedding data security, transparency, and sustainability as core principles will be crucial for building trust and ensuring long-term societal benefit.

Despite this progress, challenges remain. Fragmentation across Member States, talent shortages, and slower private investment continue to hinder scale-up. However, these can be mitigated through unified policies, strategic investment in skills, and the establishment of interoperable standards and data governance frameworks. The SWOT analysis provided in this chapter offers a clear roadmap for action.

The integration of DTs into critical sectors such as energy, mobility, and public services is poised to accelerate progress toward the European Green Deal, the Digital Decade, and the Sustainable Development Goals. By aligning technological advancement with European values and priorities, the region can distinguish itself as a global leader in innovation but also in ethical and responsible digital transformation.

For Europe, the strategic relevance of DTs extends beyond efficiency gains. They represent a critical enabler for achieving digital sovereignty, aligning innovation with sustainability goals, and strengthening competitiveness in a rapidly evolving global landscape. Initiatives such as Horizon Europe, ADRA, and the European Digital Innovation Hubs (EDIHs) are positioning the continent at the forefront of this transformation, reinforcing collaboration between industry, academia, and policy.

Acknowledgments The content of this chapter originates from the last years of workshops, discussions, and webinars inside the ADRA's Digital Twin Topic Group, where experts, projects, and industrial pilots have actively participated.

References

1. Qi, Q., Tao, F., Hu, T., Anwer, N., Liu, A., Wei, Y., Wang, L., & Nee, A. Y. C. (2019). Enabling technologies and tools for digital twin. *Journal of Manufacturing Systems, 58*. https://doi.org/10.1016/j.jmsy.2019.10.001
2. Nikolakis, N., Alexopoulos, K., Xanthakis, E., & Chryssolouris, G. (2019). The digital twin implementation for linking the virtual representation of human-based production tasks to their physical counterpart in the factory-floor. *International Journal of Computer Integrated Manufacturing, 32*(1), 1–12.
3. Schleich, B., Anwer, N., Mathieu, L., & Wartzack, S. (2017). Shaping the digital twin for design and production engineering. *CIRP Annals, 66*(1), 141–144.
4. Anwer, N., Stark, R., Tao, F., & Erkoyuncu, J. A. (2025). Developing and leveraging digital twins in engineering design. *CIRP Annals, 74*, 843.
5. Zhang, J., Li, C., Deng, C., Luo, T., Deng, R., Luo, D., Tao, G., & Cao, H. (2025). Toward digital twins for intelligence manufacturing: Self-adaptive control in assisted equipment through multi-sensor fusion smart tool real-time machine condition monitoring. *Journal of Manufacturing Systems, 82*, 301–318.
6. Grieves, M. (2014). Digital twin: Manufacturing excellence through virtual factory replication. *White Paper, 1*, 1–7.

7. Zheng, R., Ng, S. T., Shao, Y., Li, Z., & Xing, J. (2025). Leveraging digital twin for healthcare emergency management system: Recent advances, critical challenges, and future directions. *Reliability Engineering & System Safety, 261*, 111079.
8. Li, T., Shen, Y., Li, Y., Zhang, Y., & Wu, S. (2024). The status quo and future prospects of digital twins for healthcare. *EngMedicine, 1*(3), 100042.
9. Tuhaise, V. V., Tah, J. H. M., & Abanda, F. H. (2023). Technologies for digital twin applications in construction. *Automation in Construction, 152*, 104931.
10. Saif, W., RazaviAlavi, S., & Kassem, M. (2024). Construction digital twin: A taxonomy and analysis of the application-technology-data triad. *Automation in Construction, 167*, 105715.
11. Zhu, Y., Cheng, J., Liu, Z., Cheng, Q., Zou, X., Xu, H., Wang, Y., & Tao, F. (2023). Production logistics digital twins: Research profiling, application, challenges and opportunities. *Robotics and Computer-Integrated Manufacturing, 84*, 102592.
12. Liu, Y., & Pan, S. (2024). Unveiling the potential of digital twins in logistics and supply chain management: Services, capabilities, and research opportunities. *Digital Engineering, 3*, 100025.
13. Garcia, F. A., Devriendt, H., Metin, H., Özer, M., & Naets, F. (2025). Physics-informed digital twin design for supporting the selection of process settings in continuous manufacturing, with a focus in fiberboard production. *Computers in Industry, 168*, 104267.
14. Jin, L., Zhai, X., Wang, K., Zhang, K., Wu, D., Nazir, A., Jiang, J., & Liao, W. H. (2024). Big data, machine learning, and digital twin assisted additive manufacturing: A review. *Materials & Design, 244*, 113086.
15. https://commission.europa.eu/topics/eu-competitiveness/draghi-report_en
16. https://research-and-innovation.ec.europa.eu/funding/funding-opportunities/funding-programmes-and-open-calls/horizon-europe/strategic-plan_en
17. https://www.undp.org/sustainable-development-goals
18. Kritzinger, W., Karner, M., Traar, G., Henjes, J., & Sihn, W. (2018). Digital Twin in manufacturing: A categorical literature review and classification. *Ifac-PapersOnline, 51*(11), 1016–1022.
19. Tao, F., Zhang, H., Liu, A., & Nee, A. Y. (2018). Digital twin in industry: State-of-the-art. *IEEE Transactions on Industrial Informatics, 15*(4), 2405–2415.
20. Grieves, M., & Vickers, J. (2017). Digital twin: Mitigating unpredictable, undesirable emergent behavior in complex systems. In *Transdisciplinary perspectives on complex systems* (pp. 85–113). Springer.
21. Rasheed, A., San, O., & Kvamsdal, T. (2020). Digital Twin: Values, challenges and enablers from a modeling perspective. *IEEE Access, 8*, 21980–22012.
22. Goodfellow, I. J., Pouget-Abadie, J., Mirza, M., Xu, B., Warde-Farley, D., Ozair, S., Courville, A., & Bengio, Y. (2014). Generative adversarial nets. *Advances in Neural Information Processing Systems, 27*.
23. Mukhuty, S., Dixon, R., & Upadhyay, A. (2025). Industry 5.0 era of digital supply chain: A generative artificial intelligence (GenAI) action model for workforce engagement. In *Technological innovations and industry 5.0* (pp. 37–53). Elsevier.
24. Afrin, S., Rafa, S. J., Kabir, M., Farah, T., Alam, M. S. B., Lameesa, A., Ahmed, S. F., & Gandomi, A. H. (2025). Industrial Internet of Things: Implementations, challenges, and potential solutions across various industries. *Computers in Industry, 170*, 104317.
25. Hamdan, M., Hassan, E., Abdelaziz, A., Elhigazi, A., Mohammed, B., Khan, S., Vasilakos, A. V., & Marsono, M. N. (2021). A comprehensive survey of load balancing techniques in software-defined network. *Journal of Network and Computer Applications, 174*, 102856.
26. Kreuzer, T., Papapetrou, P., & Zdravkovic, J. (2024). Artificial intelligence in digital twins: A systematic literature review. *Data & Knowledge Engineering, 151*, 102304.
27. Samuel, P., Saini, A., Poongodi, T., & Nancy, P. (2023). Artificial intelligence–driven digital twins in Industry 4.0. In *Digital Twin for smart manufacturing* (pp. 59–88). Academic Press.
28. Napoleone, A., Macchi, M., & Pozzetti, A. (2020). A review on the characteristics of cyber-physical systems for the future smart factories. *Journal of Manufacturing Systems, 54*, 305–335.

29. Monostori, L., Kádár, B., Bauernhansl, T., Kondoh, S., Kumara, S., Reinhart, G., Sauer, O., Schuh, G., Sihn, W., & Ueda, K. (2016). Cyber-physical systems in manufacturing. *CIRP Annals, 65*(2), 621–641.
30. Nikolakis, N., Maratos, V., & Makris, S. (2019). A cyber physical system (CPS) approach for safe human-robot collaboration in a shared workplace. *Robotics and Computer-Integrated Manufacturing, 56*, 233–243.
31. Cerquitelli, T., Nikolakis, N., O'Mahony, N., Macii, E., Ippolito, M., & Makris, S. (2021). *Predictive maintenance in smart factories*. Springer Singapore.
32. Zhang, X., Zheng, J., Li, P., Yang, Y., Ye, Y., Causone, F., & Shi, X. (2025). A review of building digital twins: Framework and enabling technologies. *Journal of Building Engineering, 111*, 113117.
33. Raza, S. M., Minerva, R., Crespi, N., Alvi, M., Herath, M., & Dutta, H. (2025). A comprehensive survey of Network Digital Twin architecture, capabilities, challenges, and requirements for Edge-Cloud Continuum. *Computer Communications, 236*, 108144.
34. Liu, Y., Feng, J., Lu, J., & Zhou, S. (2024). A review of digital twin capabilities, technologies, and applications based on the maturity model. *Advanced Engineering Informatics, 62*, 102592.
35. Khatri, V., & Brown, C. V. (2010). Designing data governance. *Communications of the ACM, 53*(1), 148–152.
36. Juddoo, S., George, C., Duquenoy, P., & Windridge, D. (2018). Data governance in the health industry: Investigating data quality dimensions within a big data context. *Applied System Innovation, 1*(4), 43.
37. Serrano, J. Y., & Zorrilla, M. (2021). A data governance framework for Industry 4.0. *IEEE Latin America Transactions, 19*(12), 2130–2138.
38. Mohsen, M., & Gokhan Celik, B. (2023). Digital Twin: Benefits, use cases, challenges, and opportunities. *Decision Analytics Journal, 6*, 100165. https://doi.org/10.1016/j.dajour.2023.100165
39. https://www.sphereinc.com/blogs/digital-twins-use-cases
40. Dembski, F., Wössner, U., Letzgus, M., Ruddat, M., & Yamu, C. (2020). Urban digital twins for smart cities and citizens: The case study of Herrenberg, Germany. *Sustainability, 12*(6), 2307.
41. Hämäläinen, M. (2021). Urban development with dynamic digital twins in Helsinki city. *IET Smart Cities, 3*(4), 201–210.
42. Bruynseels, K., Santoni de Sio, F., & Van den Hoven, J. (2018). Digital twins in health care: Ethical implications of an emerging engineering paradigm. *Frontiers in Genetics, 9*, 31.
43. Purcell, W., & Neubauer, T. (2023). Digital twins in agriculture: A state-of-the-art review. *Smart Agricultural Technology, 3*, 100094.
44. Cimino, C., Negri, E., & Fumagalli, L. (2019). Review of digital twin applications in manufacturing. *Computers in Industry, 113*, 103130.
45. https://urban-mobility-observatory.transport.ec.europa.eu/resources/case-studies/digital-twins-lessons-learned-city-aachen_en
46. https://digital-strategy.ec.europa.eu/en/policies/edihs
47. Kalpaka, A., Rissola, G., De Nigris, S., & Nepelski, D. (2023). *Digital Maturity Assessment (DMA) framework & questionnaires for SMEs/PSOs. A guidance document for EDIHs*. European Commission.
48. De Nigris, S., Kalpaka, A., & Nepelski, D. (2023). *Characteristics and regional coverage of the European digital innovation hubs network*. Publications Office of the European Union.
49. Romo, A. M. (2020) Twinning SA II of Central Bohemia.
50. https://www.marketresearchfuture.com/reports/digital-twin-market-4504
51. https://www.statista.com/statistics/1296187/global-digital-twin-market-by-industry/
52. Wibisana, M. I., Koeva, M., Nourian, P., Petrova-Antonova, D., & Karamitov, K. (2024). A LiDAR-based digital twinning workflow for traffic monitoring and simulation. *ISPRS Annals of the Photogrammetry, Remote Sensing and Spatial Information Sciences, X-4-2024*(4), 411–418. https://doi.org/10.5194/isprs-annals-X-4-2024-411-2024

53. Afzalinezhad, A. (2024). *Digital Twin-based planning support system for urban heat island mitigation* (Master's thesis, University of Twente). University of Twente Student Theses. https://essay.utwente.nl/104873/

54. Cárdenas-León, I., Morales-Ortega, L. R., Koeva, M., Atun, F., & Pfeffer, K. (2024). *Digital Twin-based framework for heat stress calculation*. SSRN. https://doi.org/10.2139/ssrn.4861693

55. Tripathy, A., Koeva, M. N., Cárdenas-León, I., & Nourian, P. (2024). Planning Support System Development for analysing environmental performance of form-based design decisions: Case Study of Enschede. *Copernicus, 48*, 137–144. https://doi.org/10.5194/isprs-archives-XLVIII-4-W11-2024-137-2024

56. Sukma, A. I., Koeva, M. N., Reckien, D., Bočkarjova, M., Da Silva Mano, A., Canilli, G., Vicentini, G., & Kerle, N. (2024). 3D city Digital Twin simulation to mitigate heat risk of urban heat islands. *Copernicus, 48*, 129–136. https://doi.org/10.5194/isprs-archives-XLVIII-4-W11-2024-129-2024

57. Morales Ortega, L. R. (2023). *A digital twin for ground water table monitoring* (Master's thesis, University of Twente). University of Twente Student Theses. https://essay.utwente.nl/97168/

58. Cárdenas, I., Koeva, M., Davey, C., & Nourian, P. (2024). Solid waste in the virtual world: A digital twinning approach for waste collection planning. In T. H. Kolbe, A. Donaubauer, & C. Beil (Eds.), *Recent advances in 3D geoinformation science—Proceedings of the 18th 3D GeoInfo Conference* (Lecture Notes in Geoinformation and Cartography) (pp. 61–74). Springer. https://doi.org/10.1007/978-3-031-43699-4_4

The Cross-Fertilization Between the Human-in-the-Loop Approach and the Explainable AI Techniques Toward Trustworthiness

Marina Da Bormida in Cugurra, Sotiris Koussouris, Daniele Crippa, Carl Hans, Robert Hellbach, Andre Tabone, and Dimitris Bibikas

Abstract One of the primary concerns related to the AI systems regards the potential for them to perpetuate biases and discriminatory practices, since they learn from historical data that may contain inherent prejudices. The human-in-the-loop (HITL) approach has emerged as a possible solution to tackle the risks of biased decision-making and related challenges associated with AI adoption. Such an approach relies on the assumption that AI decisions and actions should be supervised and, if necessary, modified or validated by humans: the machine can learn human knowledge and experience during the loop for improving the transparency, accountability, and performance of AI systems. On the other hand, the HITL approach is helpful for adapting and reconceptualizing the users' role, valorizing their interaction with the

M. D. B. in Cugurra (✉)
S&D Consulting Europe, Milan, Italy
e-mail: marina.cugurra@sdconsulting-eu.com

S. Koussouris
Suite5 Data Intelligence Solutions Ltd, Limassol, Cyprus
e-mail: sotiris@suite5.eu

D. Crippa
Consorzio Intellimech, Bergamo, Italy
e-mail: daniele.crippa@intellimech.it

C. Hans
OHS Engineering GmbH, Bremen, Germany
e-mail: carl.hans@ohs-engineering.de

R. Hellbach
BIBA—Bremer Institut für Produktion und Logistik GmbH, University of Bremen, Bremen, Germany
e-mail: hel@biba.uni-bremen.de

A. Tabone
Department of R&D, MCS DataLabs, Berlin, Germany
e-mail: andre.tabone@mcs-datalabs.com

D. Bibikas
Zenith Gas & Light SA, Thessaloniki, Greece
e-mail: d.bimpikas@zenith.gr

© The Author(s) 2026
E. Curry et al. (eds.), *Artificial Intelligence, Data and Robotics*,
https://doi.org/10.1007/978-3-032-10561-5_5

learning system, as well as their provision of feedback, guidance, or input when needed. Although the HITL method brings several benefits, it might also convey some challenges. These include, for instance, the potential slowing down of decision-making processes and the risk of human error. The explainable AI (XAI) techniques facilitate a more informed and efficient HITL process, including human oversight. XAI is key to adopt an effective human-centered perspective and make the process more accessible and efficient. XAI gives rise to a bidirectional communication channel between human operators and the intelligent system. The users are prioritized as the primary driver of this interactive bi-directional process: they are empowered to collaborate with the system for enhancing the effectiveness of the interaction and the outcomes of the system itself, thereby contributing to adaptability. This chapter explores the HITL approach and the XAI techniques from a double perspective—technical and ethics driven—and outlines how they can be effectively combined with automation to make all AI models and their associated results transparent and ethically sound. On the other hand, the synergic use of the HITL approach and XAI can contribute to prevent or minimize the so-called hallucination effect. Such an effect occurs when AI systems generate inaccurate or logically inconsistent responses, outcomes, or data, relying on patterns learned during training, which can lead to incorrect or far-fetched conclusions. These kinds of conclusions are not acceptable, when it comes to accountability, e.g., ethical or contractual situations, for instance, in health or manufacturing applications. The chapter also outlines the initial expected outcomes of the validation of this cross-intersection of the HITL and XAI in different domain-specific demo cases, in alignment with the AI, Data, and Robotics Partnership. Once completed, such demo cases will demonstrate how this joint action of the HITL and XAI is expected to be paramount in different scenarios in view of prioritizing the role of humans toward fairness, accountability, and trustworthiness. All the demo cases will showcase high real-world relevance. For instance, the robotics demo case will showcase real-world relevance by using wearable sensors to provide real-time feedback, enhancing safety and efficiency through effective cognitive load management.

Keywords Human-in-the-loop · Explainable AI · Data · Robotics · Ethics

1 Introduction

The artificial intelligence (AI) systems, processing vast amount of data to generate automated recommendations and prediction and make decisions, are increasingly integral to our daily lives. It is expected that in the near future automated decision-making processes will be even more widespread [1].

Their advantages are unquestionable and range from the speeding up of several work processes and operations to the cost reduction and the possible identification of hidden relationships and patterns that humans might overlook. Nevertheless, sometimes, the operation of such systems might not be flawless: they might raise some concerns, such as of adopting unintended biases or inherent prejudices from the data they are trained on [2].

The involvement of humans in the AI-related process, the so-called human-in-the-loop, could contribute to tackle these challenges, since the AI decisions and actions are supervised and, if necessary, modified or validated by humans [3, 4].

The human participation is directed to optimize the output of an automated system, oversee its decisions, ensure fair and transparent decision-making processes, as well as facilitate successful human-machine interactions.

The importance of the human oversight in relation to the high-risk systems is emphasized also by the AI Act [5] (Art. 14), in order to prevent or minimize the risks to health, safety, or fundamental rights that could emerge.

To ensure that the humans involved can have meaningful opportunities to supervise or influence the automation processes, they should be properly empowered to accurately assess the quality of automated outputs and intervene as necessary. In other words, humans should be equipped with the tools necessary for the effective supervision and intervention, to be able to effectively contribute to the improvement of the AI models and their related outcomes [6].

This chapter explores the various facets of the human-in-the-loop approach and its benefits, which, however, are also accompanied by some possible issues and concerns. The role played by the adoption of explainable AI (XAI) techniques is paramount to tackle them and to make the AI-based decision-making processes and outcomes more transparent and understandable: XAI techniques are essential for facilitating a more informed and efficient HITL process, as well as for enabling a bidirectional communication channel between the human operators and the intelligent system. The chapter also investigates how the combination of the HITL approach and XAI can contribute to minimize the risk that the AI system generate inaccurate or logically inconsistent responses, outcomes, or data, relying on patterns learned during training, which can lead to incorrect or far-fetched conclusions. It will delve into the potentialities of the joint use of HITL and XAI to ensure that the AI systems are used in an ethically responsible manner, relying on a practically feasible process. Such potentialities are explored both through high-level insights and in four application domains currently running in the AI-DAPT Project: the health sector, the manufacturing sector, the robotics sector, and the energy sector.

2 The Promise of the Human-in-the-Loop Approach to Ensure Unbiased and Fair Decision-Making Outcomes: Opportunities and Challenges

2.1 Opportunities Offered by the Human-in-the Loop Approach

One of the primary concerns related to the AI system regards the potential for them to perpetuate biases and discriminatory practices, since they learn from historical data, which may contain inherent prejudices. The human-in-the-loop (HITL) approach, incorporating into the AI algorithms the human expertise and values, in conjunction with the human judgment and oversight, has emerged as a potential solution that addresses the risks of biased or opaque decision-making and related concerns.

The HITL approach [6], acknowledging the importance of human intervention in AI-based decision-making processes, consists of a method where humans are involved and interpret the output of AI models. The users are engaged to improve and personalize automatic AI-based solutions: the AI decisions and actions are supervised and, if necessary, modified or validated by humans, avoiding a fully automated process.

The HITL approach connects humans to the model loop in a worthwhile way, so that the machine can learn human knowledge and experience during the loop. The transparency, accountability, and performance of AI systems are enhanced, ultimately contributing to build trust with end users and foster the adoption of the AI system.

This paradigm is helpful for adapting and reconceptualizing the users' role in the AI system [6]: the role of the user in his/her interaction with the learning system is valorized, including feedback, guidance, or input when necessary.

The HITL method brings a number of benefits [7], such as:

- The valorization of **human intuition in the contextual understanding**: human beings still retain an irreplaceable advantage in understanding complex and variable contexts, thanks to their ability to assess new situations and interpret cultural or emotional nuances.
- The **human ability to manage exception**: in unforeseen situations or in cases falling outside standard AI learning patterns, the human intervention provides greater flexibility, together with a more adaptive and personalized response to specific needs.
- The **alignment with the ethical, social, and legal values of the society**: the human presence guarantees that the decisions made by the AI system—which doesn't possess yet a sense of ethics or an understanding of the moral implications of its actions—are consistent with such values. This feature is expected to foster trust and a better acceptance of the AI system itself. Thanks to the involvement of the human perspectives in the AI development process, biases might often be more easily identified and mitigated, thereby ensuring that AI algorithms are more representative and equitable. The HITL approach is therefore useful to **prevent and/or mitigate bias**.
- **Customization and fine-tuning of the AI system** by end users (e.g., through re-training and debugging of models/classifiers), to suite their individual needs and expectations in different contexts: it is possible to address a wider diversity of user profiles and expectations and, at the same time, to increase the flexibility of the system, by adapting most of its functionality to the users' needs. By enabling the users to tailor the system to their own preferences and daily routines, as well as to modify the system's functionalities to better suit their own context and evolving feelings toward the system, the **perceived utility of the**

system itself is enhanced. It is relevant to prevent future disengagement and to enhance the self-efficacy and user satisfaction [6]. Hence, the HITL enhances the **adaptability**, enabling humans to assist the intelligent systems in adjusting to evolving/changing conditions, goals, or preferences, with advantages in terms of robustness and personalization of the system itself.

- The **continuous monitoring, evaluation, and improvement of AI actions**, thanks to the integration of human intelligence, which promotes a feedback loop for enhancing the accuracy, system performance, and reliability of automated systems: by allowing humans to provide feedback, such paradigm increases accuracy and enables error detection and correction in cases where the data are noisy, sparse, or ambiguous. At the same time, the machine-learning (ML) model is assisted in focusing on relevant data, thereby reducing the amount of data that needs processing or labeling. Therefore, the HITL approach also improves the **AI accuracy**: the incorporation of the human expertise and judgment into the AI decision-making process supports in identifying and correcting errors, biases, and limitations in AI models. The human operators are also able to offer valuable **domain-specific knowledge and contextual understanding** that AI algorithms may lack. Such aspect enables them to catch and rectify mistakes that might otherwise go unnoticed: the incorporation of human expertise during the training phase can help guide the learning process, resulting in models that are better aligned with the nuances and complexities of the situation. Similarly, the involvement of human experts in the evaluation phase helps identify and correct the model errors, leading to more accurate and reliable predictions. In conclusion, the human feedback loop makes possible the **refinement of the AI models**, ensuring that they capture the relevant patterns and relationships within the data. As regards data, data processing methods based on human-in-the-loop might be useful for data preprocessing, data annotation, and iterative labeling.

2.2 The Human-in-the-Loop Approach Within the Ethical and Regulatory Landscape

The **Ethics Guidelines for Trustworthy AI** [3] connect the HITL approach with the human oversight, directed to prevent that an AI system undermines human autonomy or causes other adverse effects. The HITL is one of the governance mechanisms aimed at achieving such a human oversight. The Guidelines distinguish between human-in-the-loop (HITL), human-on-the-loop (HOTL), or human-in-command (HIC) approach.

The HITL regards "the capability for human intervention in every decision cycle of the system, which in many cases is neither possible nor desirable." The HOTL consists in "the capability for human intervention during the design cycle of the system and monitoring the system's operation." HIC refers to the capability to oversee the overall activity of the AI system and the ability to decide when and how to use the system in any particular situation, including the decision "not to use an AI system in a particular situation, to establish levels of human discretion during the use of the system, or to ensure the ability to override a decision made by a system." This distinction and this classification are not always adopted in the discourse on human oversight and related governance mechanisms, where the locution HITL is often used to refer to all these approaches, without distinction regarding the varying degrees of the oversight itself. In this chapter, we adopt this wider meaning.

The **Artificial Intelligence Act** [5] highlights the importance of adopting artificial intelligence (AI) systems with a human-centric approach to ensure their safe deployment. One of the pillars of the human-centric approach is keeping a "human-in-the-loop." Under the AI Act, AI designers of high-risk systems are required to allow human control or interference with an AI system to achieve effective human oversight (**Art. 14, "Human Oversight"** of the AI Act).

Recital 73 clarifies that it is necessary that high-risk AI systems are:

> Designed and developed in such a way that natural persons can oversee their functioning, ensure that they are used as intended and that their impacts are addressed over the system's lifecycle. To that end, appropriate human oversight measures should be identified by the provider of the system before its placing on the market or putting into service. It is also essential, as appropriate, to ensure that high-risk AI systems include mechanisms to guide and inform a natural person to whom human oversight has been assigned to make informed decisions if, when and how to intervene in order to avoid negative consequences or risks, or stop the system if it does not perform as intended.

2.3 The Combination of the Human-in-the-Loop Approach with Automation Through the Underlying Use of Explainable AI Techniques

There might also be some challenges in integrating the human element into AI-based systems and, more in general, in the HITL approach.[1] They range from the **potential slowing down of decision-making processes**, due to the time needed for the human intervention, which might reduce the speed and efficiency of the automation process, to the **risk of human error** (due to misjudgment, lack of knowledge, or misinterpretation of data). In the case of human-in-the-loop topic modeling task, some existing HITL techniques might allow **malicious individuals to efficiently train models that serve their purpose**, with possible consequent damages to the society. Another possible lowlight regards the **need for continuous training** for operators interacting with increasingly complex AI systems, to keep them up-to-date and competent in such interaction. Furthermore, the scalability of the AI system might be limited by the need for human intervention, unless the supervisory personnel are proportionately increased, which implies a cost.

---

[1] https://www.holisticai.com

Explainable AI techniques make AI-based decision-making processes and outcomes more transparent and understandable, thereby facilitating more informed and efficient HITL, including human oversight.

The HITL approach is paramount in the **AI-DAPT[2] Project**, whose methodology rotates around the effective combination of such approach with automation through the underlying use of explainable AI techniques to explain all AI models, for (1) supporting the automation of certain pipeline steps or (2) providing the expected prediction/inference results for the problem at hand and their associated results or (3) contributing to filtering out (training) data that may hurt the model performance because of poor quality or biases.

In the AI-DAPT Project, the HITL approach is effectively combined with automation through the underlying, implicit application of explainable AI techniques to explain all AI models (that either support the automation of certain pipeline steps or provide the expected prediction/inference results for the problem at hand) and their associated results. Such techniques also contribute to filtering out (training) data that may hurt the model performance because of poor quality or biases.

Explainable artificial intelligence (XAI) is essential to interact with and understand decisions made by the AI systems, making their processes transparent and understandable to humans: it is paramount to adopt a human-centered perspective and, as highlighted hereabove, to provide human operators with the tools necessary for effective supervision and intervention. In addition, XAI is essential to consider the human viewpoint within the human-machine interaction, allowing the users to contribute and improve the AI model, giving rise to a bidirectional communication channel between the human operators and the intelligent systems [6]. This bidirectional process prioritizes the users as the primary driver to achieve the desired system behavior and empowers them to collaborate with intelligent systems in view of enhancing the effectiveness of the interaction and the outcomes of the system itself, ensuring adaptability. The adaptability is essential to ensure human comfort, which, in turn, increases trust and acceptance of AI systems. There are several aspects in which XAI adds value to human-machine collaboration. First, XAI, by providing insights into the "how" and "why" behind decisions made by the AI system, helps the human operators understand the underlying patterns and decision-making processes: it makes it easier to identify and correct any errors or biases in the system, as well as facilitates a more effective collaboration between humans and machines. In fact, the operators, thanks to the information provided by XAI, can make informed decisions, taking full advantage of AI systems' data analysis capabilities, and/or provide more accurate feedback for improving and refining the AI models (continuous feedback loop). Second, a better understanding of such patterns and processes improves the trust in the decision of the AI system.

However, also the implementation of the XAI might present some challenges. For instance, it might be difficult, in some complex cases, to create adequate

---

[2]"AI-Ops Framework for Automated, Intelligent and Reliable Data/AI Pipelines Lifecycle with Humans-in-the-Loop and Coupling of Hybrid Science-Guided and AI Models" Project (GA 101135826).

explanations, especially for non-expert users. In addition, there is a challenge related to the **perceived trade-off between AI explainability and performance/accuracy**, with the need to seek for a balance: it is often assumed that the pursuit of XAI may come at the cost of sacrificing some degree of accuracy.

Moreover, the so-called illusion of explanation [8] occurs when the users find emotional satisfaction in explanations and believe they grasp the processes, ending up with only a superficial understanding of the explanations provided: the explanations fail to faithfully reflect the realities of decision-making, leading to a false sense of security, misinterpretation, misusage, or overreliance on the explanations themselves. The illusion of explanation might be originated by different factors, such as the overestimation of human understanding. It is a phenomenon known as the illusion of explanatory depth (IOED), which gives rise to misinterpretations of explanations; it might be facilitated by the non-transparency and lack of interpretability of the ML[3] systems. However, the users, deployers, and developers need to remain vigilant and enable to discern the actual reasons behind the decisions in question. To mitigate the risk of illusion of explanation, it is key to remind that the explainability is intricately linked to the human-in-the-loop approach: their combination is critical to equip the users with comprehensible insights into the outcomes generated by complex models, crafting a comprehensible and reliable explanation, beneficial for both developers and users. Nevertheless, a cautious approach is recommended, rooted in a deep understanding of human-in-the-loop responsibilities and integrating the robustness of probabilistic reasoning with the clarity of logical rules. In this way, it is possible to prevent overreliance and misinterpretation of AI-generated explanations, balancing the need for interpretability and the ability to handle complex, uncertain scenarios.

Another challenge posed by some AI systems, which the HITL approach and XAI can contribute to prevent or minimize, regards the so-called hallucination effect [7]. It is a phenomenon where the AI systems, like text/images generators or neural networks, can produce inaccurate or logically inconsistent responses or data, relying on patterns learned during training that can lead to incorrect or far-fetched conclusions: the machine produces an outcome that is not based on real or logical data but rather on a kind of "digital fantasy"/imaginary elements created by its own learning network. In these situations, the human intervention is important to guide, correct, and improve decisions made by machines, while XAI helps in navigating and tackling the complexities and challenges posed by the hallucination effect.

The adoption of the **hybrid science-guided AI model** approach might be very helpful to prevent the "hallucination effect." More in general, it is useful to improve the accuracy and explainability of the AI pipelines and to solve complex problems, advance data analysis, prediction, and decision-making.

---

[3] Machine learning.

3 Explainable AI Techniques: State of the Art and Role for Facilitating a More Informed and Efficient HITL, Relying on an Interactive Bidirectional Process

The exponential proliferation of sophisticated artificial intelligence (AI) systems, particularly deep neural networks (DNNs), has precipitated a critical need for explainability. Empirical evidence consistently demonstrates that increased architectural depth in DNNs correlates with enhanced predictive performance; however, this augmentation in capability is often accompanied by a commensurate reduction in model transparency, yielding "black box" systems. This opacity poses a significant impediment to the establishment of trust and the validation of AI-driven decisions, thereby hindering widespread adoption. XAI research endeavors to mitigate this challenge by developing methodologies that elucidate the internal mechanisms of these complex systems. As such, it is promoting human comprehension and fostering confidence in AI-derived outcomes. This need is amplified in high-stakes domains like healthcare and finance, where opaque AI decisions can have profound consequences.

The methodological landscape of XAI can be systematically categorized along four orthogonal axes, aligned with the AI life cycle and the specific objectives of explainability [9], as explained in the following lines.

The Data-Centric Explainability axis focuses on enhancing the interpretability of input data, facilitating informed feature engineering, data normalization, and model selection. Understanding the statistical properties and inherent structures within the dataset enables practitioners to select appropriate algorithmic paradigms, such as decision trees for linearly separable data or deep learning architectures for high-dimensional, non-linear representations. Techniques like data visualization, statistical analysis, and feature importance assessment are crucial.

Intrinsic Model Explainability is the axis that addresses the inherent interpretability of AI models. This axis is encompassing the development and application of transparent, "white box" models or the augmentation of complex models with mechanisms that enhance their inherent transparency. Models like decision trees, linear regression, and rule-based systems offer inherent interpretability. For complex models, techniques like attention mechanisms and disentangled representations contribute to transparency.

The third category is Post-hoc Explanatory Techniques, which encompasses methods that extract salient features and generate explanations for pre-trained, opaque models. Techniques such as Shapley Additive Explanations (SHAP) and Local Interpretable Model-Agnostic Explanations (LIME) provide insights into the causal factors influencing model predictions. These techniques are particularly valuable for understanding the behavior of complex models that lack inherent interpretability.

Finally, Evaluation and Validation of Explanations focuses on the rigorous assessment of explanation quality and efficacy. Evaluation metrics include cognitive psychological assessments, user satisfaction surveys, measures of trust and transparency, and computational validation of explanation fidelity and robustness. Methodologies include human-subject studies, algorithmic evaluation of explanation consistency, and the assessment of explanation completeness [9].

Recognizing the fundamental link between XAI and human cognitive processing, research has increasingly focused on the development of concept-based XAI (C-XAI) methodologies. C-XAI aims to bridge the semantic gap between AI representations and human understanding by expressing explanations in terms of high-level, human-interpretable concepts, rather than raw input features [10]. This approach leverages cognitive psychology principles, acknowledging that humans naturally reason using abstract concepts. C-XAI methods can be broadly classified into Post-hoc Concept-Based Models and Explainable-by-Design Concept-Based Models. The former category includes methods that generate conceptual explanations for pre-trained models using supervised or unsupervised learning techniques. For example, identifying "bird species" as a high-level concept in an image classification task. The latter category includes approaches that integrate concept prediction directly into the model architecture, utilizing supervised, unsupervised, or hybrid learning paradigms. This ensures that the model's internal representations are aligned with human-understandable concepts.

Beyond concept-based methodologies, the integration of generative AI [7, 11, 12], particularly large language models (LLMs), is gaining momentum within XAI research. LLMs facilitate the transformation of abstract AI explanations into coherent, natural language narratives, tailored to individual user preferences. By leveraging prompt engineering with key contributing features extracted via methods like SHAP, LLMs can generate contextually relevant and readily comprehensible explanations [12]. Furthermore, iterative feedback mechanisms enable continuous refinement of explanations, ensuring alignment with evolving user needs and AI capabilities. The ability of LLMs to generate counterfactual explanations and provide "what-if" scenarios further enhance their utility in XAI.

As it is turning out and is visible when seeing the evolution of LLMs, in dynamic, real-world environments characterized by heterogeneous user expertise, static, one-shot explanations may prove inadequate. Interactive XAI systems, which emphasize responsive feedback and user-centric interfaces, facilitate the establishment of dynamic trust. Iterative exploration of AI decision-making processes reinforces trust when model reliability is consistently demonstrated while appropriately inducing skepticism in instances of inconsistency. In this way, interactive XAI allows users to probe the model, ask follow-up questions, and explore alternative scenarios, fostering a deeper understanding of the AI's behavior.

In individualized learning contexts, AI-driven personalization plays a pivotal role. AI-driven adaptive learning leverages techniques such as recommender systems, educational adaptive hypermedia systems, and dynamic content adjustment to personalize learning pathways based on user interactions and learning preferences. Ensuring transparency in AI-driven personalization is paramount. Research in educational AI has explored the concept of "scrutability," which enables learners to understand and critique AI-generated recommendations. XAI can provide explanations for personalized learning recommendations, enabling learners to understand why specific content was selected and to provide feedback. Furthermore, in a direct analogy to digital twins in manufacturing, learner modeling is fundamental to adaptive learning research. Techniques such as knowledge space theory enable AI-driven customization by capturing user knowledge and learning needs. Integrating these models into XAI facilitates the generation of personalized explanations, enhancing learning efficacy and trust. By understanding a learner's knowledge gaps and learning style, XAI can provide explanations that are tailored to their individual needs.

Summarizing, it becomes evident that as AI continues to permeate diverse domains, explainability is posing even more boldly as a critical factor in fostering trust, usability, and effective decision-making. XAI research is evolving to accommodate the growing complexity of AI systems and the expanding diversity of end users. By ensuring transparency, comprehensibility, and adaptability, XAI can facilitate a future where AI-driven insights are not only accurate but also genuinely useful to human stakeholders. Empirical studies have demonstrated that users who receive explanations for AI decisions exhibit higher levels of trust, satisfaction, and task performance. For instance, research in medical AI has shown that clinicians who understand the rationale behind AI-driven diagnoses are more likely to accept and act upon the recommendations. Furthermore, studies in human-computer interaction have highlighted the importance of user-centered design principles in the development of effective XAI systems. As the future of AI hinges on its ability to seamlessly integrate with human decision-making processes, prioritizing the development of transparent, comprehensible, and adaptable XAI systems could be the key to unlock a great share of the potential of AI, ensuring that it remains an ethical tool that empowers, rather than replaces, human judgment.

4   HITL and Explainable AI Domain-Specific Demo Cases

4.1   Robotics and Cognitive Ergonomics Demo Case: Human-Centered Automation

4.1.1   Brief Description of the Demo Case

In the context of Industry 5.0, the Robotics demonstrator focuses on integrating human-centered automation into digital representations and simulations to enhance the flexibility and responsiveness of manufacturing processes. The main goal is to provide information to workers about their workload conditions to monitor their safety and increase productivity. By leveraging wearable devices and AI models, the demonstrator enables real-time monitoring of workers' physical and mental states, such as stress and fatigue levels. This approach not only enhances worker well-being but also optimizes productivity by adapting automated systems based on real-time data. Alerting them of their critical stress conditions, for example, they could take a break, move on less-demanding tasks, or simply be aware of safety issues.

This initiative aims to bridge the gap between advanced digital technologies and human factors, ensuring that workers remain central to the value generation process. For this reason, the inclusion of worker subjectivity is a key aspect, as it allows for the consideration of individual skills, medical conditions, and emotional states. This personalized data is crucial for creating adaptive automation systems that can respond dynamically to the needs and conditions of each worker. Furthermore, the demonstrator's ability to fine-tune AI models for different and more specific use cases highlights its potential for broad industrial applications.

To achieve these goal, two scenarios were designed, each leveraging cognitive ergonomics technology: the first one focuses on operator efficiency and human-robot collaboration, while the second one emphasizes operator safety in dynamic environments.

Scenario 1: Human-Robot Collaboration for Enhanced Quality Control

The first scenario demonstrates how a human operator can perform visual quality control with the support of a robot and a 3D vision system. Worker data will be acquired through four different devices (Fig. 1):

- *Environmental sensor*: air quality and other environmental measurements (e.g., ozone level, humidity)
- *Noisemeter*: working environment noise
- *Thermocamera*: worker head temperature, converted in time-series signal
- *Smartwatch (Garmin Venu 2)*: other physiological signals (e.g., heart rate)

Fig. 1 Robotics demonstrator—first scenario sensors: AI-generated image created with OpenAI, ChatGPT (GPT-4o, May 2024 version), 2024

Moreover, specific surveys will allow to evaluate mental state of the worker and context relevant information (e.g., expertise) a posteriori. This system will adapt to the operator's skills and experience, potentially upskilling them and improving overall efficiency. The integration of human factors into the digital representation allows for dynamic adaptation of the automation technologies, ensuring that the system can respond to the operator's needs and conditions.

Scenario 2: Worker Safety Enhancement Through Cognitive Load Monitoring

The second scenario focuses on the integration of wearable devices and AI models to monitor workers' physical and mental states, possibly in real time. The goal is to predict mental stress and fatigue, allowing for adaptive automation that optimizes the behavior of collaborative robotics cells based on the workers' conditions. To reach this goal, a set of three wearable devices will be exploited to gather physiological and relevant signals from the operator (Fig. 2):

- *Smartwatch (Empatica EmbracePlus)*: wrist physiological signals (e.g., electro-dermal activity, temperature)
- *Smartshirt (Hexoskin)*: chest physiological signals (e.g., electrocardiogram, respiration)
- *Headband (Muse S)*: mental signals (e.g., electroencephalogram)

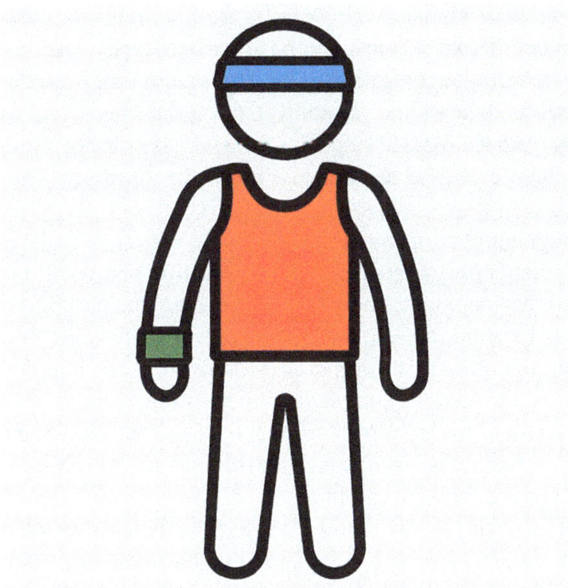

Fig. 2 Robotics Demonstrator—second scenario sensors: AI-generated image created with OpenAI, ChatGPT (GPT-4o, May 2024 version), 2024

These signals will be processed through dedicated pipelines and utilized to develop predictive AI models, such as anomaly detection or time-series classification models, to increase worker safety, productivity, and well-being by providing feedback to operators about their fatigue and, possibly, distraction levels. The system's ability to adapt to changes in the ergonomics conditions or the physical assets ensures high mobility and safety even in potentially hazardous environments.

Both scenarios include an initial setup phase for each involved operator, applicable during both AI model development and deployment. This phase involves defining operator-specific baselines, which will be incorporated into the AI models to enhance their adaptability to individual operators.

4.1.2 Human-in-the-Loop and Explainable AI

Both demonstrator scenarios involve the collection of a substantial number of signals. This is due to the decision to sensorize the operators as much as possible (within reasonable and legally compliant limits), aiming to identify a minimal subset of signals that is functional for developing AI models capable of predicting or detecting the desired target, such as cognitive overload or stress. While there are several references in the literature to experimental applications in controlled laboratory environments for these tasks [13–17], there are few real-world examples [12, 18, 19], especially in industrial context, where the physical components of tasks significantly affect the physiological signals collected.

In particular, many guidelines [20, 21] exist in the literature for extracting numerous features from these signals, which can be used as inputs for the development of a black-box model. These features can have various types, such as those derived from frequency filtering, aggregation of the signal into time windows using appropriate techniques (e.g., heart rate variability), or considering signal shape characteristics (e.g., R-R distance for ECG signals). Most of these transformations are based on medical domain expertise or derived empirically from other medical research studies.

Given the large number of signals and the wide variety of features suggested by the literature for laboratory cases, the feature selection phase will be crucial during the demonstrator development. This phase will ensure that the data used for training AI models is as informative as possible, excluding misleading contributions that could affect the algorithm's performance. Moreover, as the number of considered features increases, the model's complexity must also increase to maintain accuracy, which raises the risk of overfitting on training data and increases the need for model tuning. Therefore, finding ways to reduce the dimensionality of the problem, removing less useful features, will be essential.

This kind of analysis usually requires an iterative process, in which the data scientist is part of the development loop, applying statistical and data analysis techniques, such as identifying correlations between available features or studying their intrinsic properties (e.g., cyclicity and trends), before selecting the most useful features and starting the development of the actual predictive AI models.

Traditional feature selection and dimensionality reduction techniques, however, have some limitations, particularly regarding assumptions about the type of correlation between features and the target to be predicted (e.g., PCA).

Explainable AI (XAI) techniques could prove very useful in this scenario. By using them, it is possible to analyze the influence of features on the model's performance, integrating the model development process into the feature selection loop, thus greatly simplifying the developer's task. A possible approach would involve using all available features to train a simplified prediction model (e.g., a random forest with a small number of trees) as a first attempt. By analyzing this model, it would be possible to identify the most influential features on the model's predictions and select only those for training the final model.

Furthermore, XAI techniques can provide additional insights beyond feature importance, highlighting how features influence the predictions, such as different behaviors depending on the feature value, complex nonlinear correlations, or features that are highly correlated with the target but not significantly used by the model. All this information can be useful during feature selection and can significantly accelerate the development process. Finally, analyzing potential biases and quantifying model fairness can help identify situations where models are influenced by specific data conditions, thus supporting the generalization of algorithm performance.

For the Robotics demonstrator, both supervised classification approaches and unsupervised anomaly detection methods will be considered. These methods will enable labelling specific time windows with labels indicative of the operator's mental state. Training data will come from different subjects and will include metadata describing individual characteristics (e.g., weight, age, gender). This data will make it possible to use XAI techniques to study the fairness properties of the model and evaluate whether it performs equally well across all subjects or if there are specific situations where individual characteristics influence performance and generate biases in prediction.

Additionally, the results of the predictions will be analyzed to study any different impacts of features depending on the subject, thus investigating potential correlations between individual characteristics and variables—information that could be very useful in defining subject-specific baselines. Moreover, XAI techniques will allow for the quantification of feature influence (e.g., using SHAP or LIME) to exclude less significant features, such as those derived from literature in laboratory contexts, that are less relevant in a dynamic industrial environment. Global XAI techniques can evaluate the overall performance of the model, while local XAI techniques can focus on analyzing cases where the model outputs false positives and evaluate whether certain relevant features should be removed to improve model generalization.

Finally, XAI techniques will also be used to validate domain knowledge extracted from literature, examining the effects of known correlations in the medical domain between physiological signals and expected outcomes.

Once the feature selection and the final model training phases are complete, a feedback collection system will be prepared for the final testing phase, where operators will use the developed AI system. This will allow operators to provide real-time feedback on the correctness or errors in the model's predictions, enabling a new fine-tuning cycle that will use the previously discussed XAI techniques to further refine the model, ultimately enhancing its performance in real-world contexts through data collected on the field.

4.2 The Manufacturing Demo Case: Predictive Maintenance of Production Assets

4.2.1 Brief Description of the Demo Case

The Manufacturing demonstrator at OHS and BIBA is defined by the relationship of an OEM[4] manufacturer and its outsourced maintenance activities concerning manufacturing equipment—with specific domain-relevant constraints—to a maintenance service provider. The use case is designed to enhance reliability of manufacturing equipment toward the OEM through *AI-driven monitoring, to ensure data quality*, and *predictive analytics, to support a forecast-based decision process*. The system integrates sensor data, historical maintenance records, and machine learning models to anticipate failures—not machine failure, such as in classical predictive maintenance scenarios, but maintenance process inefficiency or errors—minimizing unplanned downtime and optimizing maintenance processes [22]. A well-structured maintenance strategy, including preventive and predictive approaches, ensures operational efficiency, cost reduction, and improved productivity. The predictive approach leverages real-time data from sensors to detect anomalies and forecast potential errors or bottlenecks in maintenance processes, whereas preventive maintenance is the standard in the given use case setup and follows scheduled inspections to maintain equipment reliability and working safety. By incorporating AI into maintenance processes, the demonstrator seeks to improve overall reliability in manufacturing process invoked by maintenance, optimize operational costs, and ensure enhanced diagnostics and planning.

4.2.2 Human-in-the-Loop and Explainable AI

Human-in-the-Loop (HITL) Implementation

The HITL approach in this demonstrator is essential for refining machine learning models and improving the accuracy of anomaly detection. Human operators act as a critical interface between AI-driven analytics and real-world operational processes, ensuring that AI-generated insights are not only technically sound but also

---

[4] Original Equipment Manufacturer.

aligned with the contextual realities of manufacturing maintenance. Unlike fully automated AI systems, which may lack the ability to distinguish between irrelevant noise and critical anomalies [23], HITL empowers operators to validate, interpret, and refine AI outputs dynamically. Through iterative engagement, human expertise enhances AI models by providing contextual feedback that improves the accuracy of maintenance strategies. The process begins with AI detecting potential anomalies and predicting failures based on sensor data and historical trends. Human operators then assess these predictions, verifying their relevance, adjusting thresholds, and refining algorithms to better accommodate domain-specific conditions. This interplay fosters a symbiotic relationship where AI augments human decision-making, while human validation ensures the reliability and credibility of AI-driven recommendations [24]. One of the key benefits of HITL in predictive maintenance is its ability to reduce both false positives and false negatives. False positives, where AI incorrectly flags a functional component as faulty, can lead to unnecessary maintenance interventions, increasing costs and operational disruptions. Conversely, false negatives, where AI fails to detect an impending failure, can result in unexpected equipment breakdowns and costly downtime. By incorporating human expertise, predictive models are continuously refined to strike the right balance between sensitivity and specificity, ultimately enhancing the overall efficiency of maintenance workflows [25]. However, the integration of HITL into AI-driven maintenance solutions presents challenges. One of the primary barriers is the need for continuous operator training [26]. As AI models evolve, maintenance personnel must stay informed about changes in data interpretation methodologies and algorithmic decision-making. Additionally, HITL implementation must be seamlessly integrated into existing workflows to avoid bottlenecks. If human validation introduces delays or inefficiencies, it may undermine the advantages of AI automation. Therefore, organizations must develop robust training programs and user-friendly AI interfaces that facilitate intuitive operator interactions with machine learning models. The effectiveness of HITL is further amplified through the integration of explainable AI (XAI) techniques. XAI ensures that AI decision-making processes are transparent and interpretable, enabling operators to understand why certain anomalies are flagged and how predictive analysis recommendations are generated. This level of explainability fosters trust and acceptance of AI-driven maintenance solutions among human operators, bridging the gap between complex AI computations and real-world decision-making.

Explainable AI (XAI) Techniques and Hybrid AI Approaches in Maintenance

To support transparency and enrich the HITL approach, the demonstrator incorporates explainable AI (XAI) techniques, for example, in intelligent spare part management or optimized logistics for maintenance activities. Feature Importance Analysis will be used to determine which predictive factors contribute the most to AI-generated decisions, enabling maintenance service provider to better understand why specific anomalies are flagged or are encountered [27]. The SHAP (Shapley

Additive Explanations) technique further enhances interpretability by quantifying the influence of each input feature on predictive outcomes [28]. Additionally, Counterfactual Explanations allow operators to explore alternative scenarios by showing how small changes in input data would alter the AI-generated predictions [29]. These XAI methodologies ensure that AI recommendations remain interpretable, fostering trust and effective decision-making within the maintenance workflow [30]. Hybrid AI approaches, which combine physical models with AI-driven insights, are particularly relevant in the maintenance sector. These approaches address limitations of purely data-driven techniques by integrating domain-specific models, e.g., quality control theories, reliability engineering principles, as well as logistics optimization theories, to improve prediction accuracy and reliability. Forecasting methods benefits significantly from hybrid AI by leveraging system behavior models alongside AI-driven anomaly detection [31], ensuring a more holistic understanding of machine conditions. Digital twins, virtual models continuously updated with real-time sensor data, further enhance maintenance planning by providing a simulated representation of maintenance process performance and conditional information about maintained manufacturing equipment. Robust Condition Monitoring [32] of processes and maintained manufacturing equipment is another crucial application, where hybrid models ensure that predictive analysis remains effective even in scenarios with limited training data or unexpected operating conditions.

4.3 The Health Demo Case: Personalized Medicine Based on Noninvasive Glucose Monitoring

4.3.1 Brief Description of the Demo Case

The Health Demonstrator represents a significant shift in personalized diabetes management, focusing on noninvasive glucose monitoring using wearable sensors and hybrid AI models. Traditional glucose monitoring relies on fingerprick blood samples or continuous glucose monitoring (CGM) systems, both of which can be cumbersome for patients. In contrast, this demonstrator employs photoplethysmography (PPG) sensors integrated into wearable devices to estimate blood glucose levels indirectly [33]. By analyzing fluctuations in blood volume and vascular responses, these sensors capture physiological signals that, when processed through advanced AI pipelines, provide real-time glucose estimates. The challenge, however, lies in the inherent variability of PPG signals, which are susceptible to motion artifacts, skin tone differences, and environmental noise. This variability necessitates a robust AI framework that not only corrects for inconsistencies but also incorporates human oversight to refine predictions and ensure clinical applicability.

Background and Rationale

Historically, diabetes management has relied on invasive techniques that often deter a patient's adherence to prescribed care. Recent advancements in optical sensor technology have motivated research into noninvasive methods that harness physiological signals. PPG, which measures blood volume fluctuations using light, has shown promising correlations with glucose levels when combined with sophisticated analytical models. A growing body of literature supports the shift toward noninvasive methodologies, with several studies highlighting that integrating signal processing with data-driven approaches can overcome limitations inherent in conventional sensors [33, 34]. Such hybrid techniques not only improve the reliability of sensor data but also enable dynamic correction of anomalies. This is an essential feature given that PPG signals are susceptible to motion artifacts and environmental noise. For this reason, the demonstrator is building their dataset around both wrist-based PPG signals and fingertip-based PPG since the latter are less prone to motion artifacts.

4.3.2 Human-in-the-Loop and Explainable AI

Human-in-the-Loop (HITL) Approaches

In the context of noninvasive glucose monitoring, human-in-the-loop (HITL) methodologies serve as a critical component in balancing automation with clinical oversight. Despite advances in machine learning for diabetes care, fully autonomous glucose prediction systems remain unreliable without human supervision [36]. The demonstrator implements HITL through a multi-level interaction framework, where clinicians and patients provide continuous feedback to refine the AI's decision-making process. Clinicians review model predictions, flagging inconsistencies arising from sensor noise or anomalous physiological states, while patients contribute through active self-reporting, validating glucose trends against fingerprick reference values [37, 38]. By integrating these expert corrections into the AI's training loop, the system becomes progressively adaptive, refining its predictions to align with real-world conditions [38].

Existing closed-loop insulin delivery systems, such as artificial pancreas technologies, have demonstrated the necessity of human intervention, particularly for handling unexpected variations due to meal intake, stress, or exercise [27]. These systems rely on users to announce meals, calibrate sensors, and manually adjust dosing thresholds, reinforcing the importance of human involvement even in highly automated frameworks. The AI-DAPT Health Demonstrator proposes a similar approach by embedding clinician-driven corrections into the AI pipeline, allowing for continuous adaptation. Unlike conventional glucose monitoring, where sensor data is passively interpreted, this system can actively learn and benefit from human interactions, improving long-term reliability and patient safety. The inclusion of reinforcement learning techniques based on HITL ensures that as the system is exposed to diverse physiological responses and fine-tunes its adaptive glucose prediction model accordingly.

Explainable AI (XAI) Techniques in Glucose Monitoring

The effectiveness of AI-driven glucose monitoring depends not only on prediction accuracy but also on its interpretability for clinicians and patients. The black-box nature of deep learning models remains a fundamental challenge in healthcare AI, where trust and explainability are essential for clinical adoption. Explainable AI (XAI) methodologies help bridge this gap by providing transparency into how and why the model arrives at its glucose predictions. Techniques such as Local Interpretable Model-Agnostic Explanations (LIME) and SHapley Additive exPlanations (SHAP) allow users to visualize the influence of specific physiological features on glucose estimates [35, 39]. These methods ensure that clinicians can identify potential biases, assess feature significance, and verify whether the model's predictions align with known physiological mechanisms. Complementing this, the work by Doshi-Velez offers robust guidelines for integrating explainability into high-stakes decision-making processes, ensuring that model outputs are not only accurate but also transparent to clinical operators [40]. Furthermore, hybrid XAI models have been explored for biomarker discovery in diabetic retinopathy, demonstrating the feasibility of combining explainable frameworks with deep learning to extract clinically relevant insights [41, 42].

The demonstrator's pipeline within the AI-DAPT adaptive pipeline operates in a continuous feedback loop, where real-world sensor data is validated through expert oversight and explainable AI-driven assessments. This approach allows for the identification of sensor anomalies, such as motion artifacts or environmental interference, and ensures that the AI can dynamically adjust to these factors. Clinicians benefit from intuitive, real-time interpretability tools, which display feature-level explanations of glucose predictions, enabling them to assess whether certain physiological markers are disproportionately influencing results. Such a level of transparency is crucial for ensuring that the model outputs align with established clinical expectations, preventing scenarios where AI-generated recommendations contradict medical best practices.

HITL and XAI: Integration, Benefits, and Challenges

Despite benefits of integrating HITL and XAI, real-world glucose monitoring presents several challenges. One key issue is the trade-off between automation and human oversight. While greater human involvement improves reliability, it also increases cognitive burden on clinicians, requiring them to constantly interpret and validate AI-generated outputs. Additionally, the need for continuous data labelling and validation may introduce workflow inefficiencies, particularly in settings with limited clinical resources. Addressing these challenges requires optimizing the user interface and feedback mechanisms, ensuring that human reviewers can efficiently interact with the system without excessive need for manual intervention.

Another challenge is the scalability of personalized models. The demonstrator relies on individualized patient feedback; the AI's ability to generalize across diverse populations depends on its capacity to learn from heterogeneous data sources. Ensuring that models remain adaptive yet clinically consistent requires a balance between patient-specific fine-tuning and global model standardization. The application of meta-learning techniques may help mitigate these concerns. Models can offer personalized AI by adapting to new patients with minimal retraining while maintaining core physiological consistency.

The Health Demonstrator aims to combine HITL-driven adaptability with XAI-based transparency along with IoT sensor technologies to establish a foundation for clinically trustworthy and patient-friendly diabetes management. This approach presents machine learning as a collaborative tool rather than a black-box decision system.

4.4 The Energy Demo Case: Cross-Vector Residential Demand-Response Through Smart Heating

4.4.1 Real-World Relevance

This section explores how smart heating systems, enhanced by artificial intelligence (AI), can help balance energy supply and demand by being integrated with electricity market prices predicting mechanisms and usage patterns. By combining automated decision-making with *human-in-the-loop* (HITL) input and *explainable AI* (XAI) techniques, the system can potentially illustrate both efficiency and trustworthiness. Users and experts can understand, adjust, and refine the AI's forecasts, improving accuracy and fostering confidence. This approach might not only save energy and costs but can also serve as a model for ethical and transparent AI paradigm.

4.4.2 Brief Description of the Demo Case

The transition to renewable energy sources presents significant challenges in balancing electricity demand and supply, particularly in the residential sector. Smart heating systems equipped with AI-driven demand-response capabilities might offer a solution by optimizing energy consumption in alignment with real-time electricity pricing and grid conditions. The energy demonstrator focuses on cross-vector residential demand-response through smart heating, integrating AI-based predictive modeling, real-time automation, and human-in-the-loop (HITL) methodologies to enhance efficiency, cost savings, and user trust.

In particular, the demonstrator focuses on the seamless integration of smart heat pumps, IoT-enabled sensors, and cloud-based data analytics. By utilizing advanced ML models, the system predicts heating demand based on historical usage patterns, real-time weather data, and electricity market fluctuations [43]. These predictions inform automated heating systems operations, ensuring optimal performance with minimal energy waste. Additionally, smart thermostats and user-friendly control interfaces allow consumers to actively participate in the decision-making process, adjusting preferences and overriding AI-generated recommendations when necessary.

A critical component of the demonstrator is the integration of electricity price and natural gas demand forecasting, which play a vital role in optimizing demand-response strategies. The ability to accurately predict electricity prices and natural gas demand in advance allows for strategic load adjustments that maximize cost savings and improve grid/network flexibility. Advanced AI models analyze historical data, renewable generation [44] forecasts, and grid/network conditions to make informed predictions. HITL approaches ensure that human operators validate these AI-generated forecasts, adjusting parameters to enhance accuracy. Meanwhile, explainable AI (XAI) techniques provide transparency, allowing energy users and utilities to understand the decision-making process and build trust in AI-driven market predictions [45].

Background and Rationale

Traditional heating systems often operate independently of real-time market fluctuations, leading to inefficiencies in energy consumption and unnecessary financial burdens for consumers. Unlike conventional heating models, smart heating systems leverage AI-driven demand-response strategies to dynamically adjust energy usage in response to external factors, such as electricity grid/network load, dynamic pricing signals, and weather conditions.

By accurately predicting energy prices and demand for the following day, heating systems can optimize power usage, reducing costs and alleviating congestion. Demand-response programs benefit significantly from day-ahead forecasting, as consumers can shift loads to low-cost periods, reducing demand during peak hours [46].

HITL plays a key role in refining day-ahead market forecasts by allowing energy analysts to interact with AI models, fine-tuning parameters based on their expertise and market knowledge. This continuous learning process improves forecast accuracy while incorporating real-world insights. XAI further strengthens this integration by ensuring that pricing forecasts remain interpretable, allowing users to validate AI-driven predictions and adjust energy consumption accordingly.

4.4.3 Human-in-the-Loop and Explainable AI

Human-in-the-Loop (HITL) Approaches

HITL methodologies create a bridge between automation and human expertise, enhancing AI-based decision-making in day-ahead forecasting and demand-response optimization. Key potential applications of HITL in this domain include:

1. *Human-Assisted Forecasting Adjustments*: AI-generated electricity price and demand forecasts are reviewed by energy analysts, who make refinements based on external market indicators.
2. *User Preferences and Adaptive Scheduling*: Consumers can manually adjust heating schedules based on their preferences and AI-generated recommendations.
3. *Feedback Loops for Forecast Improvement*: Market experts analyze past forecast deviations and provide corrections, which AI models incorporate into subsequent predictions for better performance.
4. *Risk Mitigation*: Human operators assess forecast uncertainty levels, helping mitigate risks associated with sudden price and/or demand spikes.

By embedding HITL in the above processes, AI models evolve through real-time expert feedback, leading to more robust and adaptable energy predictions.

Explainable AI (XAI) Techniques

Explainable AI (XAI) ensures that AI-driven forecasts and demand-response strategies remain transparent and interpretable for users, operators, and utilities. The energy demonstrator aspires to integrate various XAI techniques to enhance understanding and trust:

1. *Feature Importance Analysis*: Identifies key drivers of forecasting fluctuations, such as weather conditions, demand patterns, and renewable generation forecasts, allowing analysts to validate AI predictions.
2. *Counterfactual Explanations*: Provides alternative scenarios by showing how different inputs (e.g., increased renewable energy generation) would impact day-ahead forecasts, offering insights into energy market behavior.
3. *Interpretable Forecast Visualizations*: Uses graphical representations of AI model outputs to ensure that predictions are easy to interpret and act upon by consumers and utilities.

By integrating XAI into forecasting approaches, the energy demonstrator ensures that AI models remain accountable, enhancing stakeholder confidence in AI-driven energy management decisions.

HITL and XAI: Integration, Benefits, and Challenges

The integration of HITL and XAI into energy forecasting and smart heating demand response might bring numerous **benefits**:

1. *Higher Forecast Accuracy*: Human feedback refines AI-driven price and load predictions, reducing errors and improving market efficiency.
2. *User Trust and Engagement*: Transparent AI decision-making fosters confidence among consumers, enabling greater adoption of AI-driven demand-response solutions.
3. *Energy Cost Savings*: Improved forecast precision allows consumers to align heating consumption with lower-priced electricity periods, reducing overall energy expenses.

However, some **challenges** remain to be addressed:

1. *Balancing Automation with Human Oversight*: Excessive manual interventions can reduce the efficiency of AI systems, while too little oversight can compromise forecast accuracy.
2. *Scalability of HITL Integration*: Expanding human-in-the-loop processes across large-scale energy markets requires efficient data-handling mechanisms to ensure seamless collaboration between AI models and human experts.
3. *Managing Forecast Uncertainty*: Even with AI and human oversight, unexpected events (e.g., extreme weather or policy changes) can disrupt forecast accuracy, requiring robust contingency strategies.

Future developments will focus on enhancing AI adaptability, improving HITL efficiency through automation, and expanding the role of XAI in energy market forecasting and demand-response systems.

5   Conclusions

This chapter has explored the critical intersection of human-in-the-loop (HITL) approaches and explainable AI (XAI) techniques, demonstrating their synergistic potential to build trustworthy, reliable, and ethical AI systems, in line with the ADR Vision 2030, which recommends to make "considerable efforts…to improve human-machine collaboration, ethics and compliance for ADR at the service of society" [44]. The core message of this work is that their true power lies in their combined application. HITL and XAI are not separate endeavors; they are two sides of the same coin, inextricably linked in the pursuit of responsible AI development and deployment. The fundamental premise of HITL is the recognition that human expertise, judgment, and ethical considerations are indispensable in navigating the complexities of AI. AI systems, particularly those based on complex machine learning models, can be prone to biases, errors, and "hallucinations," and therefore human oversight, intervention, and feedback are crucial to mitigating these risks, ensuring that AI systems align with human values and societal norms. HITL provides the mechanism for continuous improvement, allowing humans to refine models, correct errors, and guide the AI toward desired outcomes. However, the effectiveness of HITL is fundamentally limited if the AI system remains a "black box." Without understanding why an AI system makes a particular decision or prediction, human operators are left to rely on intuition or guesswork, potentially overlooking subtle biases or errors. This is where XAI becomes essential. XAI techniques, ranging from feature importance analysis (like SHAP and LIME) to concept-based explanations and the use of large language models for narrative generation, provide the necessary transparency and interpretability. They illuminate the inner workings of AI models, allowing humans to understand the reasoning behind predictions, identify potential biases, and assess the validity of the AI's output. This consideration is paramount to tackle with the concern of loss of human control in the future, as outlined in Sect. 3.5 of the chapter "Roadmaps and Agendas for Research and Innovation in Artificial Intelligence, Data, and Robotics": for instance, the experts might not be considered to be necessary any more to improve the AI diagnostics, leading to their gradual disappearance, thereby fomenting public resistance and disappointment. The relationship between HITL and XAI is therefore not merely additive; it is multiplicative. XAI empowers humans to perform their HITL roles more effectively. The journey toward trustworthy AI is not a destination but a continuous process of refinement and improvement. It requires a commitment to ongoing collaboration between humans and machines, guided by the principles of transparency, accountability, and ethical responsibility. Nevertheless, the combination of HITL and XAI is not a panacea, and challenges remain. However, by embracing this unified approach, we can harness the immense potential of AI while safeguarding human values and ensuring that AI systems serve humanity's best interests.

Acknowledgments This work has been supported by the project "AI-DAPT," which has received funding from the European Union's Horizon Europe Research and Innovation Programme under grant agreement no. 101135826.

References

1. EC, COM (2020) *65 final, White paper on artificial intelligence—A European approach to excellence and trust*
2. Nazer, L.-H., & Zatarah, R. (2023). Bias in artificial intelligence algorithms and recommendations for mitigation. *PLOS Digital Health, 2*, e0000278.
3. HLEG-AI. (2019). Ethics guidelines for trustworthy AI.
4. Mosqueira-Rey, E., Hernández-Pereira, E., Alonso-Ríos, D., et al. (2023). Human-in-the-loop machine learning: A state of the art. *Artificial Intelligence Review, 56*, 3005–3054.
5. EU. (n.d.). Regulation (EU) 2024/1689 of the European Parliament and of the Council of 13 June 2024 laying down harmonised rules on artificial intelligence (Artificial Intelligence Act).
6. Gómez-Carmona, O., & Diego, C.-M. (2024). Human-in-the-loop machine learning: Reconceptualizing the role of the user in interactive approaches. *Internet of Things, 25*, 101048.

7. *Can there be harmony between Human and AI? The key role of Explainable AI and Human in the Loop.* (2024). Retrieved from https://frontiere.io

8. Chromik, M., & Eiband, M. (2021). I think I get your point, AI! The illusion of explanatory depth in explainable AI. In *IUI'21: Proceedings of the 26th International Conference on Intelligent User Interfaces.*

9. Ciravegna, G. (2024). *C-xai: Concept-based explainable AI, Medium.* Retrieved September 16, 2024, from https://medium.com/@gabriele.ciravegna/c-xai-concept-based-explainable-ai-51dece0472f1

10. Schneider, J. (2024). Explainable Generative AI (GenXAI): A survey, conceptualization, and research agenda. *Artificial Intelligence Review, 57,* 289.

11. Zytek, A. P. (2024). LLMs for XAI: Future directions for explaining explanations. arXiv preprint arXiv:2405.06064.

12. Schmidt, P. R. (2018). Introducing WESAD, a multimodal dataset for wearable stress and affect detection. In *Proceedings of the 2018 International Conference on Multimodal Interaction (ICMI'18).*

13. Choi, J. A.-O. (2012). Development and evaluation of an ambulatory stress monitor based on wearable sensors. *IEEE Transactions on Information Technology in Biomedicine, 16*(2), 279–286. https://doi.org/10.1109/TITB.2011.2169804

14. Iqbal, T. S. (2022). Stress monitoring using wearable sensors: A pilot study and stress-predict dataset. *Sensors, 22*(21), 8135. https://doi.org/10.3390/s22218135

15. Najafi, T. A. (2023). Drivers' mental engagement analysis using multi-sensor fusion approaches based on deep convolutional neural networks. *Sensors, 23,* 7346.

16. Kyriakou, K. R. (2019). Detecting moments of stress from measurements of wearable physiological sensors. *Sensors, 19,* 3805.

17. Healey, J. A., Picard, R. W., et al. (2005). Detecting stress during real-world driving tasks using physiological sensors. *IEEE Transactions on Intelligent Transportation Systems, 6*(2), 156–166. https://doi.org/10.1109/TITS.2005.848368

18. Giorgi, A. R. (2021). Wearable technologies for mental workload, stress, and emotional state assessment during working-like tasks: A comparison with laboratory technologies. *Sensors, 21,* 2332.

19. Can, Y. S., et al. (2019). Stress detection in daily life scenarios using smartphones and wearable sensors: A survey. *Journal of Biomedical Informatics, 92,* 103139. https://doi.org/10.1016/j.jbi.2019.103139

20. Giannakakis, G. G., et al. (2022). Review on psychological stress detection using biosignals. *IEEE Transactions on Affective Computing, 13*(1), 440–460. https://doi.org/10.1109/TAFFC.2019.2927337

21. Ucar, A., & Karakose, M. (2024). Artificial intelligence for predictive maintenance applications: Key components, Trustworthiness, and Future Trends. *Applied Science, 14*(2), 898. https://doi.org/10.3390/app14020898

22. Deng, Z., Xuan, X., et al. (2024). A reliable framework for human-in-the-loop anomaly detection in time series. *ACM Transactions on Interactive Intelligent Systems.* https://doi.org/10.1145/3778167

23. Te'eni, D., & Yahav, I. (2023). *Reciprocal human-machine learning: A theory and an instantiation for the case of message classification.* Retrieved from https://pubsonline.informs.org/doi/full/10.1287/mnsc.2022.03518 .

24. Nikitin, A., & Kaski, S. (2022). Human-in-the-loop large-scale predictive maintenance of workstations. In *KDD '22: Proceedings of the 28th ACM SIGKDD Conference on Knowledge Discovery and Data Mining.* https://doi.org/10.1145/3534678.3539196

25. Kumar, S., Datta, S., et al. (2024). *Applications, challenges, and future directions of human-in-the-loop learning.* IEEE.

26. Kuk, M., & Bobek, S. (2023). Feature importances as a tool for root cause analysis in time-series events. In *Computational Science—ICCS 2023.* Springer. https://doi.org/10.1007/978-3-031-36030-5_33

27. Lundberg, S., & Lee, S. (2017). A unified approach to interpreting model predictions. In *NIPS'17: Proceedings of the 31st International Conference on Neural Information Processing Systems*.
28. Fernández-Loría, C., & Provost, F. (2022). Explaining data-driven decisions made by AI systems: The counterfactual approach. *MIS Quarterly, 46*(3), 1635–1660.
29. Cummins, L., & Sommers, A. (2024). Explainable predictive maintenance: A survey of current methods, challenges and opportunities. *IEEE Access.* https://doi.org/10.1109/ACCESS.2024.3391130
30. Kumar, R. S., & Singh, A. R. (2025). Hybrid machine learning framework for predictive maintenance and anomaly detection in lithium-ion batteries using enhanced random forest. *Scientific Reports, 15*, 6243.
31. Tran, K., & Vu, H. (2024). Robust-MBDL: A robust multi-branch deep learning based model for remaining useful life prediction and operational condition identification of rotating machines. *Mathematics, 12*(10), 1569. https://doi.org/10.48550/arXiv.2309.06157
32. Satter, S. T.-H.-D. (2024). EMD-based noninvasive blood glucose estimation from PPG signals using machine learning algorithms. *Applied Sciences, 14*(4), 1406.
33. Manzoni, E. R. (2021). Detection of glucose sensor faults in an artificial pancreas via Whiteness Test on Kalman filter residuals. *IFAC-PapersOnLine, 54*(7), 274–279. https://doi.org/10.1016/j.ifacol.2021.08.371
34. Bequette, B. W. (2021). Human-in-the-loop insulin dosing. *Journal of Diabetes Science and Technology, 15*(3), 699–704. https://doi.org/10.1177/1932296819891177
35. Hettiarachchi, C. M., et al. (2024). A modular reinforcement learning algorithm for glucose control by glucose prediction and planning in Type 1 Diabetes. *Biomedical Signal Processing and Control, 90*, 105839. https://doi.org/10.1016/j.bspc.2023.1
36. Retzlaff, C. O., et al. (2024). Human-in-the-loop reinforcement learning: A survey and position on requirements, challenges, and opportunities. *Journal of Artificial Intelligence Research.* https://doi.org/10.1613/jair.1.15348
37. Mayo Clinic Health System. (2022). *Closing the loop with insulin pumps.* Retrieved from https://www.mayoclinichealthsystem.org/hometown-health/speaking-of-health/closing-the-loop-with-insulin-pumps
38. Ribeiro, M. T., et al. (2016). *Why should I trust you?: Explaining the predictions of any classifier.* ACM.
39. Doshi-Velez, F., & Kim, B. (2017). Towards a rigorous science of inter-pretable machine learning. ArXiv Preprint.
40. Weron, R. (2014). Electricity price forecasting: A review of the state-of-theart with a look into the future. *International Journal of Forecasting, 30*(4), 1030–1081.
41. Kathirgamanathan, A. D., et al. (2021). Data-driven predictive control for unlocking building energy flexibility: A review. *Renewable and Sustainable Energy Reviews, 135*, 110120.
42. Ribeiro, G. C. (2016). Why should I trust you? Explaining the predictions of any classifier. In *Proceedings of the 22nd ACM SIGKDD International Conference on Knowledge Discovery and Data Mining*.
43. A. Papavasiliou, Optimization models in electricity markets. Cambridge University Press., 2024.
44. ADRA. (2025). *Strategic research, innovation and deployment agenda 2025–2027.* ADRA.
45. Wu, X., et al. (2022). A survey on human-in-the-loop for machine learning. *Future Generation Computer Systems, 135*, 364–381.
46. AI-DAPT. (2024). *D1.1 AI-DAPT automated AI pipelines end user needs and scientific technology radar.* AI-DAPT.

Generative Artificial Intelligence to Tackle Visual Data Accessibility Challenges

Lore Goetschalckx, Kaili Wang, Siri Willems, and Tom De Schepper

Abstract Deep neural networks, the cornerstone of the modern AI revolution, had their first breakthroughs in visual tasks (e.g., object recognition). This is not a coincidence given the high interest in such capabilities across industries (e.g., autonomous driving, medical imaging). However, to train, evaluate, and improve vision networks for specific applications, one needs extensive data. Transformative advances in generative AI (GenAI) have recently made it possible to synthesize highly realistic, complex visual data. Harnessing the power of synthetic data can overcome challenges faced with real-world data (e.g., scarcity, cost) and ultimately foster more visually capable machines. Here, we offer an overview of different types of visual GenAI and examine their adoption in real-world applications. We focus on autonomous driving (AD) and medical imaging applications because of their high societal impact and unique data challenges (e.g., privacy concerns for patient data, the long-tail challenge in AD). Finally, we discuss current limitations and promising future directions.

Keywords GenAI · Synthetic data · Vision · Image-to-image translation · Autonomous driving · Medical imaging

1 Introduction

Much of our human intelligence we owe to our ability to process sensory inputs (e.g., light hitting our retina) and infer what is happening in the world around us. With vision often deemed the dominant sense, building artificial intelligence (AI) capable of solving increasingly complex visual tasks has continued to attract research interest in both academic and industrial settings. In 2012, these efforts reached a turning point when AlexNet [1], a deep neural network (DNN),

L. Goetschalckx (✉) · K. Wang · S. Willems · T. De Schepper
imec, Leuven, Belgium
e-mail: Lore.Goetschalckx@imec.be; Kaili.Wang@imec.be; Siri.Willems@imec.be; Tom.DeSchepper@imec.be

© The Author(s) 2026
E. Curry et al. (eds.), *Artificial Intelligence, Data and Robotics*,
https://doi.org/10.1007/978-3-032-10561-5_6

outperformed the competition in the ImageNet Large Scale Visual Recognition Challenge (ILSVRC2012) [2] by a large margin. The impact of deep learning technology has exploded since then. The paper introducing ResNet [3], an AlexNet successor, became one of the most cited AI papers of all time. By 2017, the top-5 error rate on the ILSVRC had dropped to 2.25% [4]. The number of submissions to the annual conference of Neural Information Processing Systems (NeurIPS) has increased tenfold since 2012 [5]. Finally, Geoffrey Hinton and John Hopfield recently winning the Nobel prize in physics for their foundational contributions to deep learning once more emphasized the far-reaching impact of the technology.

Artificial Vision Powered by Data In contrast to traditional methods relying on handcrafted features, DNNs solve vision tasks by coming up with features in a data-driven fashion. Most DNNs essentially model a conditional probability distribution $P(Y|X)$, where X is the multi-dimensional sensory input (i.e., pixel values in an image or video) and Y an outcome variable of interest, such as a class label in an image recognition task or a bounding box in an object detection task (e.g., detecting other cars on the road). The input information is processed by multiple layers (hence "deep") of interconnected neurons (hence "neural network"). The strengths ("weights") of these interconnections constitute the trainable parameters of a DNN, whose values are to be learned from training data. Typically, the first layers of neurons learn to extract relevant image features, whereas the last layers learn how to linearly recombine the feature information into a final output.

The training data generally consists of example images, each provided with a ground truth output label y. One batch of images at a time, randomly initialized parameter values are tweaked by an optimizer algorithm. This algorithm minimizes a loss function $\mathscr{L}$, which quantifies the distance between the predicted labels $\hat{y}$ and the images' ground truth y. This is referred to as supervised learning. By being shown many datapoints with the correct label, the DNN can pick up on image features and their relation to the output labels. As such, they are highly flexible learners, who can cover a relatively broad area of the hypothesis space, which is part of what made them successful. Generally, the bigger the DNN (i.e., number of parameters), the more flexibility. However, this flexibility also comes at a price. Indeed, DNNs do not just have the ability to learn from data but also the *need* to learn from data. The bigger the network, the more "data-hungry" it tends to be. In fact, the 2012 deep learning breakthrough is not merely attributable to advances in algorithms and models but also to the creation of ImageNet [6], a labelled image dataset of unprecedented magnitude, as well as pivotal developments in computing hardware that allowed for efficient training of large networks.

ImageNet was the result of the visionary and relentless efforts of Fei-Fei Li and her team, whose plans to build a larger than ever before labelled image set were initially met with skepticism. However, the success of AlexNet on their ILSVRC opened many eyes when it came to the role of (big) data in AI. Not only did ImageNet fuel many further, major advances in deep learning architectures (e.g., famous vision models such as VGG [7], ResNet [3], EfficientNet [8]), it also kickstarted other initiatives to build large datasets, which in turn gave rise to new artificial vision capabilities. For example, Microsoft's Common Objects in Context dataset (COCO; [9]) enabled the development of object detection and segmentation architectures [10–12]. Similarly, progress in DNNs for video classification and human action recognition happened partly thanks to the release of datasets such as YouTube-8M [13], Kinetics [14], and Moments in Time [15].

Specialized Data for Specialized Applications The enthusiasm surrounding this data-powered AI revolution soon extended beyond the confines of academic publications. Among the first adopters were BigTech companies who applied DNNs to their vision tasks (e.g., Meta's DeepFace [16], Google's InceptionV3 [17]). The broader public got a taste of the power of deep learning through features such as Apple's Face ID on the iPhone X, Google Lens for identifying real-world objects, and traffic-sign recognition on modern cars. Meanwhile, AI vision startups together began to raise millions en route to becoming major players in fields like autonomous driving (AD; e.g., Wayve, Helm.ai, Tractable.ai), medical imaging (e.g., Lunit, Icometrix), and security (e.g., DeepGlint).

Importantly, specialized applications require specialized datasets. ImageNet and other previously mentioned examples constitute broad-domain, publicly available datasets with a wide variety of everyday images (or videos), often collected from Internet sources (e.g., flickr.com, YouTube). However, models trained on these datasets cannot necessarily be taken off the shelf and directly put to work on solving specific real-world problems. At the very least, they require fine-tuning on data that is relevant to the application. In other words, a DNN needs an opportunity to specialize and learn domain-specific features from more tailored data, such as datasets compiled for fashion tasks [18], plant disease detection in agriculture [19], anomaly detection in industrial products [20], sensor fusion in automotive applications [21], lung disease detection in chest X-rays [22], etc. Indeed, every novel application comes with its own data requirements, as well as respective challenges. In many applied settings, relevant images are not readily available on the Web, making data acquisition more cumbersome, not to mention expensive.

This is no less true for autonomous driving (AD) and medical imaging, which we highlight in this chapter as two of the earliest and most influential application areas of deep learning in computer vision. We briefly introduce these areas here, setting the stage for a more focused discussion of their distinct data accessibility challenges in later sections (Sects. 3 and 4).

Autonomous Driving Autonomous driving refers to the technology that uses sensors, AI algorithms and computing power to enable vehicles to navigate and operate with little to no human intervention. It spans different levels of autonomy, from basic driver assistance (Level 0) to full autonomy (Level 5) [23]. Most current systems, such as Tesla's Autopilot [24, 25], operate at Level 2. Companies like Waymo are pioneering fully driverless robotaxi services (Level 4) and are currently live in Phoenix, San Francisco, and Los Angeles [26]. Meanwhile, traditional car manufacturers, such as Volkswagen, are also accelerating their efforts to achieve Level 3 autonomy by collaborating with the AI company Mobileye [27]. Fully autonomous (Level 5) vehicles are still in the experimental phase, and no company has achieved it yet. Higher levels of autonomy require the vehicle to have a more advanced understanding of its surroundings, which begins at the sensor level. However, just as humans rely not only on their eyes but also on their brains to interpret what they see, autonomous driving increasingly depends on AI to interpret sensory inputs and determine appropriate actions. For example, DNNs are trained to detect and classify objects (e.g., recognizing traffic signs), track movement (e.g., of other road users), and segment scenes (e.g., to identify the drivable space). Such training requires large datasets with ground-truth annotations (e.g., bounding boxes, object labels, segmentation maps). Collecting high-quality labelled data is essential for model robustness, reducing false detections, and reliable performance in real-world driving scenarios.

Medical Imaging In the medical field, statistical metrics quantify the benefits of various treatment strategies such that clinicians can make evidence-based recommendations. A shift has occurred from traditional medicine toward personalized medicine, tailoring treatments toward smaller defined subgroups, considering body health and needs. To visualize the inside of the human body, personalized medicine critically relies on various medical imaging technologies. Medical images do not only inform treatment plans but are also used for diagnosis, guiding medical procedures, and monitoring disease progression. They include X-rays, computed tomography (CT), magnetic resonance imaging (MRI), ultrasound (US), and positron emission tomography (PET), with most of them providing a full 3D image of a specific anatomical region. The high data dimensionality in combination with increasingly complex disease-related data patterns calls for powerful analytical methods. For this, the field has turned to AI, and more specifically deep learning. Deep learning has brought advances in image analysis and interpretation, improving accuracy and efficiency [28, 29]. For example, detection and classification of lung nodules in a thoracic CT and microcalcifications in a breast mammography were among the first applications to use DNNs in medical imaging [30, 31].

Synthesizing Data with GenAI A potential solution to data accessibility challenges in AD, medical imaging, and beyond lies in a recent and highly transformative innovation: generative AI (GenAI). In GenAI, the goal is to model the high-dimensional distribution of sensory inputs itself, such that it becomes possible to sample new datapoints from the learned distribution. In the case of image distributions, this allows one to generate synthetic images that look like real data. Some popular applications that capture the imagination include smart photo editing tools, virtual try-on features for online fashion stores, and AI-powered special effects for film. However, synthetic data is also increasingly being explored as a complement to real data in building AI agents or tools that must respond effectively to visual inputs—especially when real-world data is difficult to obtain.

For years, generative modeling in vision remained relatively uncharted territory, explored only by a small community of researchers willing to take on its complexity and uncertainty. This changed drastically though when Ian Goodfellow [32] introduced generative adversarial networks (GANs) in 2014 and presented output images looking much sharper and more realistic than previous methods. The number of publications on generative vision modelling surged after that (e.g., 91K citations for the GAN paper to date). New GAN models were introduced in rapid succession and achieved increasing levels of realism at higher resolutions, with BigGAN [33] (trained on ImageNet) and StyleGAN2 [34] (trained on CelebA [35], a face dataset) two of the most well-known.

However, GANs also had their drawbacks (e.g., unstable training, mode collapse). Around 2022, new-kid-on-the-block diffusion models took over the spotlight, following several key scientific contributions and reported benefits in terms of training stability, output diversity, and quality [36–39]. That year, synthesizing images from a text input ("txt2img") went from a niche research topic to a widely used creative tool, thanks to the release of Dall-E 2 (OpenAI) [40], Imagen (Microsoft) [41], Midjourney (Midjourney, Inc.), and Stable Diffusion (Stability AI) [42], all of which rely on diffusion models. Within months, Dall-E 2 had over one million users, generating over two million images per day [43]. Stable Diffusion, being open source with relatively low computational requirements, democratized AI image generation, allowing independent developers to derive their own applications. These developments not only shook the art world but brought about a paradigm-shift across industries (e.g., creative and design industries, medical imaging, AD, gaming), which was mirrored by concurrent major advances in language models (e.g., release of ChatGPT). However, for the purpose of this chapter, we focus specifically on generative vision models.

Concretely, the rest of the chapter is structured as follows. In Sect. 2, we present an overview of different types of generative vision models and/or tasks along with relevant technical background. Sections 3 and 4 zoom in on data accessibility challenges in AD and medical imaging applications, respectively, as well as explore GenAI-based initiatives to help address them. Finally, we discuss risks and current limitations of synthetic data in Sect. 5 and highlight directions for future research in Sect. 6. Taken together, this chapter relates to the sensing and perception as well as the data enrichment activities of the AI, Data, and Robotics Partnership [44], thus supporting transformation in healthcare and autonomous systems related to mobility.

2 Generative Vision Models

2.1 *From Discriminative to Generative AI*

Given a set of training images from an unknown target distribution $x{\sim}p_{\text{data}}(x)$, can we learn to sample new images such that $x'{\sim}p_\theta(x) \approx p_{\text{data}}(x)$, with θ the parameters of a neural network? For example, starting from a dataset of real face images, can a neural network be trained to generate faces that look real yet are novel? Generative vision models approximate the distribution over the high-dimensional, complex sensory data itself, in contrast to discriminative models that are limited to $p(y|x)$.

Arguably, the former is more challenging. One of the difficulties of explicitly modeling the probability density, e.g.,

$$p_\theta(x) = \frac{f_\theta(x)}{Z_\theta},$$

where f is the unnormalized density function, lies in the (often) intractability of the normalizing constant Z_θ. To deal with this, several approaches have either made assumptions about the form of the (target) distribution or relied on computationally expensive numerical approximation methods. Due to this and other modeling difficulties, as well as qualitatively unrealistic appearing output images, generative efforts did not realize a strong impact until GANs were introduced. GANs essentially bypassed the intractability issues by only modeling the density *implicitly* (see further). In the wake of GANs, the field saw a renewed interest in alternative generative methods, including significant advances in variational auto-encoders (VAEs) [45] and the rise of diffusion models [36, 45]. We now turn to a more technical discussion of these three approaches.

2.2 Types of Generative Vision Models

2.2.1 Generative Adversarial Networks

As mentioned above, GANs do not define the density function $p(x)$ explicitly. Instead, they interact with the modelled distribution only indirectly by sampling from it. Concretely, this is achieved by pitting two DNNs against each other as adversaries, hence the name. One DNN is called the discriminator, D, and has the architecture of a typical image classification network. It models $P(Y = "real"|x) = 1 - P(Y = "fake"|x)$, meaning the discriminator's task is to distinguish between "real" images from the training set and "fake" ones generated by its adversary. Its adversary, the generator G, usually takes in a lower-dimensional noise input z and, through a succession of learned upsampling layers, transforms it into a generated image x'. The noise input z is randomly sampled from a predefined prior distribution, often uniform or Gaussian. The training objective for the generator is to produce outputs that can fool the discriminator into classifying them as "real." Both DNNs are trained simultaneously such that when the discriminator gets better at telling the difference, the generator is encouraged to produce more realistic outputs. Specifically, these are their training objectives respectively:

$$\theta_D^* = \underset{\theta_D}{\operatorname{argmax}}\left(E_{x\sim p_{\text{data}}}\left[\log D_{\theta_D}(x)\right] + E_{z\sim p_z}\left[\log\left(1 - D_{\theta_D}\left(G_{\theta_G}(z)\right)\right)\right]\right)$$

$$\theta_G^* = \underset{\theta_G}{\operatorname{argmin}}\left(-E_{z\sim p_z}\left[\log D_{\theta_D}\left(G_{\theta_G}(z)\right)\right]\right).$$

The former features a binary cross-entropy loss, and the latter expresses that the discriminator ought to predict a high $P\left(Y = "real"|G_{\theta_G}(z)\right)$ (at least from the generator's perspective).

Training GANs requires a delicate balance between D and G, and failures to maintain that balance cause instability. If one adversary gets too far ahead of the other, its loss function will produce values that are too low for the other to make significant parameter updates. Moreover, sometimes, G discovers a single mode that fools D, and sticks to it, rather than learning the full distribution ("mode collapse"). Despite these limitations, GANs have advantages that sustain their relevance today, such as an easy to navigate latent space, fast inference times, and ability to perform unpaired image translations (Sect. 2.4).

2.2.2 Variational Autoencoders

Variational autoencoders (VAE) [45] were introduced before GANs but initially produced images lacking fine details. However, the success of GANs and the interest it drew to generative models promoted further improvements in VAEs and gave rise to hybrid approaches that combined the best of both [46–48].

Standard autoencoders (AE) [49], the basis of VAEs, aim to compress high-dimensional inputs (e.g., images) to a meaningful, lower-dimensional representation, z, with minimal loss of information. That is, z should represent the image well enough to be able to reconstruct it near perfectly from the compressed representation. Common AEs are implemented as a DNN consisting of an encoder and a decoder part. Where the encoder maps the input x to z through learned downsampling layers, the decoder maps z to x' through learned upsampling layers. The training objective is to minimize the reconstruction error between x and x'. Note that standard AEs are not generative and that a given x always maps to the same, single z.

In VAEs, on the other hand, a given x is mapped to a (conditional) distribution, $p(z|x)$. Indeed, VAEs frame the entire model probabilistically, with three key probability distributions involved. The prior distribution $p(z)$, which is typically chosen to be standard Gaussian, represents the distribution of the latent variables before observing any data. The second distribution is the posterior $p(z|x)$, which is approximated by the encoder as $q_\phi(z|x)$, a Gaussian with the mean and variance being a learned function of the input x and ϕ being the parameters of that function. This posterior distribution captures the uncertainty about the latent variable given the observed data. The third distribution is the likelihood $p_\theta(x|z)$, which models how the data is generated from the latent variable z, and is learned by the decoder. Once trained, the idea is to sample a random z from the prior distribution and pass it through the decoder to generate a new image. The training objective is formalized as follows:

$$\left(\theta^*, \phi^*\right) = \underset{\theta,\phi}{\operatorname{argmin}}\left[-E_{q_\phi(z|x)}\left[\log p_\theta(x|z)\right] + \mathrm{KL}\left(q_\phi(z|x)\|p(z)\right)\right],$$

where the first term signifies the reconstruction loss and the second reflects the KL divergence between the approximate posterior and the prior.

VAEs are no longer state of the art when it comes to the quality and realism of the generated output. Nevertheless, they are often computationally more efficient than diffusion models and suffer less from training instabilities than GANs do. Moreover, the setup promotes a well-structured and interpretable latent space, which offers control over the outputs and meaningful continuous image interpolations (e.g., gradual lighting changes). Finally, the learned latent representations could be used in downstream vision tasks, as they encode a lot of the structure in the visual data.

2.2.3 Diffusion Models

Diffusion models have their origin in the field of thermodynamics, where the use cases include modelling how heat diffuses over time due to stochastic noise. When applied in a generative modelling context, an image is gradually destroyed by iteratively adding stochastic (Gaussian) noise over T time steps. A DNN is then tasked with modelling the reverse process [36]. That is, it learns to iteratively recover the denoised image. If trained successfully, novel images can be generated by sampling noise ($t = T$) and denoising it all the way to a clean image ($t = 0$).

Part of why this approach works is because of the favorable properties of Gaussians (e.g., closed-form solutions for many operations) and the fact that the noising iterations constitute Markov chains where each iteration is only dependent on the last. The conditional densities of the reverse (i.e., denoising) process, $p(x_{t-1}|x_t)$, are then approximated by a Gaussian, with the mean, μ, provided by the DNN as a function of x_t and t. However, the authors showed that instead of directly predicting μ, it is easier and more effective to predict the noise ϵ that was added to the data during the forward process [36]. The parameters, θ, of the DNN are then optimized as follows:

$$\theta^* = \underset{\theta}{\operatorname{argmin}} \, E_{x_0, \epsilon, t} \left[\left\| \epsilon - \epsilon_\theta \left(x_t, t \right) \right\|^2 \right]$$

We note here that this denoising framework can also be understood through the lens of score-based generative modeling, as explored by Song et al., but refer to their paper for more details [39]. Moreover, recent developments in flow matching offer an alternative, more general formulation to diffusion models by directly learning smooth, continuous probability paths without discretized diffusion steps. This potentially provides more efficient and flexible generative processes while maintaining a close relation to the original framework [50–52]. Finally, one of the key innovations behind Stable Diffusion was to carry out the diffusion process in the compressed latent space of an AE, rather than in pixel space, which made synthesizing high-resolution images more computationally efficient. This is referred to as a latent diffusion model (LDM).

Diffusion models achieved state-of-the-art output realism in several key generative vision tasks [38, 40, 41] and offer better training stability compared to GANs. In addition, they lend themselves well to txt2img tasks, which provides users with fine-grained control over the outputs in the form of text prompts. A drawback, though, is the need for multiple, sequential denoising steps, which makes the models slow at inference. Speeding up the image generation is an active area of research [53–60].

2.3 *Evaluating Generative Vision Models*

The mathematical goal of GenAI models is to learn $p_\theta(x) \approx p_{\text{data}}(x)$. However, the two distributions cannot directly be compared to validate the model, since $p_{\text{data}}(x)$ is unknown. It is only known through its (high-dimensional) samples. In practice, models are evaluated on their ability to produce realistic and diverse outputs. To measure these variables, one approach is to compute the Inception Score (IS) [61], referring to the image classifier Inception [17], which was trained on ImageNet. The score quantifies the absence of entropy (confidence) of the predicted class distribution per generated image as well as the entropy in predicted classes over all generated images. The former is intended to reflect realism or quality, reasoning that high-quality images should be more confidently classified by Inception. The latter is meant to assess diversity, by means of variance in the kinds (classes) of images generated. The IS has been criticized, though, for not considering real images from the target distribution in its calculation. Therefore, the most widely used metric today determines the Fréchet Distance between the estimated distributions of real and generated image features respectively, where the features are extracted from Inception (Fréchet Inception Distance, FID) [62]. The metric assumes the feature distributions to be multivariate Gaussians. However, this assumption has recently been contested, and an alternative metric has been proposed that is said to better align with human observers' judgements of realism [63]. Indeed, conducting human perceptual experiments is often still considered the golden standard for evaluation, but it is inefficient compared to automated metrics. Taken together, finding the proper ways to evaluate GenAI remains a challenge in the field.

2.4 *Conditional Generation*

Rather than synthesizing samples purely from noise, it is also possible to add a condition, c, to guide the generative process and often produce better results. In this case, the target distribution becomes $p(x|c)$. For example, one might wish to control the class label of the output and train a class-conditional model. BigGAN, which is trained on ImageNet, falls in that category. Generally, the conditioning needs to be present during training already. In diffusion models, however, "classifier guidance" uses a discriminative classification model, capturing $p(c|x)$, to steer the generation post hoc during inference [38], although the results tend to be weaker compared to built-in methods ("classifier-free guidance") [64].

Text-to-image (txt2img) generation goes beyond just a class label and conditions on entire text prompts. Users can then describe in words what they want to see in the output image. As mentioned above, diffusion models established txt2img generation as a prominent paradigm for controllability of generated images. Furthermore, we highlight yet another type of conditioning with a wide range of (vision) applications: image-to-image (img2img) translation. Indeed, translating an image from a source domain A to a target domain B can be seen as conditional generation, where c is a source image. Examples include denoising an image (noisy2clear), colorization (gray2color), super-resolution (low2high resolution), and artistic style transfer (e.g., photo2painting). There is a distinction to be made between paired img2img translation, where the images of A and B form matched pairs, and a harder, unpaired setting where this is not the case. For instance, it is straightforward to create pairs of grayscale and color images for models to learn from, but not to find pairs of paintings and matching photos of the same scene. CycleGAN [65] offered a first solution to the unpaired translation problem. Many VAE and diffusion-based solutions today still borrow principles from the GAN framework when it comes to unpaired translation. Nevertheless, training on paired data generally still yields better results. In the next section, we turn to data accessibility challenges in AD and GenAI-driven ways to tackle them, which include img2img models.

3 Autonomous Driving

As mentioned in the Introduction, a vehicle's perception of its surroundings starts at the sensor level, the first layer in an AD stack. Traditionally, RGB camera sensors have been at the core of the sensor suite in vehicles across all levels of autonomy, from consumer vehicles to robotaxis. Their popularity mostly resides in the relatively low cost and close relation to human perception. RGB cameras offer high-resolution vision, allowing for, e.g., object detection, traffic sign recognition, and lane tracing, but they tend to struggle in low-light and adverse-weather conditions (e.g., snow, fog). Alternative camera modalities being considered include thermal and near-infrared imaging [66, 67]. While some companies, such as Tesla, believe in vision-only solutions [68], most sensor suites today still include radar technology in addition to cameras, which primarily improves performance in low-visibility conditions. Level 4/5 systems (e.g., Waymo's robotaxis) even add a third, more expensive sensor type: LiDAR. LiDAR captures depth information and allows for a more precise 3D mapping of the outside world.

The information captured by the sensors serves as input for the perception layer (understanding the environment) of the AD stack. The stack is completed by a localization layer (determining the precise position of the ego-vehicle on the road), a planning layer (calculating the route the ego-vehicle will take), and a control layer (executing driving actions). However, in this chapter, we focus on the perception layer. The AI models in this layer, often DNNs, infer the position and identity of objects from pixel intensities, track their movement, segment out driveable areas, etc. Next, we turn to the challenges faced in acquiring specialized training and testing data for these DNNs, as well as potential GenAI-driven solutions.

3.1 Data Volume and Diversity

Perception models in an AD stack need to be able to adequately interpret any visual input they might encounter on the road. This requires millions, if not billions of miles worth of training and testing data, which is costly to collect. In addition to volume, diversity also matters. Among other potential biases, existing real-world datasets (e.g., Waymo Open Dataset [69]) often exhibit a "nice-weather bias" [70] and fail to represent the true variety of possible weather conditions. The AD field has therefore turned to GenAI for expanding and augmenting real-world data with synthetic data.

For example, in a 2020 paper [71], Waymo adopted GANs to simulate camera images from a self-driving car. GenDDS [72], on the other hand, builds on Stable Diffusion XL [73] to produce diverse driving images from descriptive text prompts. Furthermore, Zhou et al. [74] compiled DIVA, a dataset consisting of both real videos from the Web and diffusion-generated data. Diffusion models are also used in several multi-view video generation pipelines [75–77]. The challenge with multi-view data (i.e., from multiple cameras on a vehicle) is to generate frames that are visually internally consistent. Finally, Nvidia's DreamDrive [78] leverages a combination of video diffusion and neural rendering to generate 4D spatiotemporal consistent driving videos given a single input image. While it is unclear if they too used diffusion, Helm.ai very recently launched GENSIM-2, an AI-based driving video generator, specifically designed to help train and test AD modules. One of its strengths is the advanced video-editing capability, allowing for flexible modification of the appearance, weather, and illumination of a scene to promote diversity and control.

Apart from imagining novel driving scenes, GenAI can also augment a dataset by creating variations of already included scenes through img2img translations. These include day2night-translations, such as those performed by img2img-turbo [79], a hybrid model combining elements of both Stable Diffusion and CycleGAN. Pizzati et al. [80] even go beyond a discrete setup (day vs. night), to implement a continuous time-of-day transformation. Finally, several studies have also explored weather transformations [70, 81–84], such as dry2rainy, clear2foggy, etc. Taken together, synthetic data augmentations are a valuable approach to promoting generalizability of (perception) modules in the AD stack.

3.2 Long-Tail Scenarios and Edge Cases

A key hurdle to AD development lies in the so-called long-tail problem. It refers to safety-critical scenarios and edge cases (e.g., a large animal crossing the road), which have low occurrences but together form a long tail in the distribution of scenarios. Despite their rarity, they have a major impact on vehicle safety. DNNs often struggle to handle them well because of insufficient representation in real-world training data. Liu and Feng also argued that it is the very high dimensionality of the space of possibilities that condemns interesting events to low probabilities, a phenomenon they named the curse of rarity [85].

To overcome this challenge, many have relied on driving simulators based on graphics and physics engines [86–90]. They are capable of rendering virtual worlds, spanning a variety of scene layouts and agent behaviors. A significant advantage is the level of control that is offered in terms of spatial arrangement. That is, objects and structural elements can be placed such that it closely replicates specific long-tail scenarios. However, these simulations tend to lack the rich visual appearance that is characteristic of the real world, partly because the number of 3D "assets" (e.g., buildings, vegetation, road infrastructure, vehicles) to choose from is limited. Because of these and other discrepancies, AI models trained on data from such simulators do not always generalize well to the real world, a problem known as the Sim2Real gap.

GenAI can help bridge this gap. Specifically, img2img translation can be used to convert the style of simulator images to one that more closely resembles the real world [91–94]. For instance, Richter et al. [95] attempted to enhance photorealism by conditioning on the simulator's graphical buffer data in addition to the simulated image itself. Recently, Zhou et al. [74] introduced a scene generation paradigm called SimGen (Fig. 1), that integrates the controllability of simulators with the photorealism of diffusion models. Using a text prompt and "scene record" (unified description format for simulators [96]) as inputs, SimGen can ultimately generate multiple variants of a given safety-critical scenario (e.g., in different weather, cities, or lighting). Finally, semiconductor giant Qualcomm demonstrated their framework to augment driving scenes with GenAI at NeurIPS 2024 [97]. They report using diffusion models and dynamic Gaussian splatting [98] to realistically reconstruct scenes from the real world and mention how this facilitates investigations into the behavior of perception models in edge-cases.

3.3 Annotations

Precise ground-truth annotations are critical for training high-performing perception models. For instance, object detection models need example driving images with bounding box annotations (i.e., ground truth detections) to learn from. Another type of annotation commonly desired in AD consists of segmentation maps, which indicate for each pixel to which (if any) of a predefined list of classes it belongs (e.g., "road," "car," "tree," "sky"). However, manually annotating every frame from thousands of hours of real-world driving videos is an extremely laborious task. Here too, driving simulators with GenAI-enhanced photorealism (Sect. 3.2) can provide relief. Owing to the controllability of the simulator, it is exactly clear where in the image it will draw which object. As a result, it becomes straightforward to obtain bounding boxes or segmentation maps for such simulated images. Various domain adaptation techniques, leveraging generative modeling, have been explored to aid the generalizability of perception models from simulated to real images [99–101]. In a similar vein, img2img translation can be used to generate driving images from segmentation maps, meaning those generated images already have their own annotations [102–105]. The segmentation inputs themselves could then be created with another GenAI model to further minimize manual effort. In sum, reducing data annotation costs is another reason for the AD field to invest in synthetic data.

3.4 Novel-Sensor Data

Whereas traditional RGB cameras are the most common, the AD sector is also exploring alternative imaging sensors, such as near-infrared (NIR) and thermal imaging. By fusing data from multiple sensor types, weaknesses from one sensor may be compensated by another, and overall uncertainty could be reduced. In other words, novel or additional sensory input types would offer the vehicle a more comprehensive "look" on the outside world. However, because of their more explorative nature compared to RGB cameras' well-established place in the sensor suite, the available data is relatively scarce and does not cover the same variety of scenarios. One potential way to overcome this is to build on the more widely available and

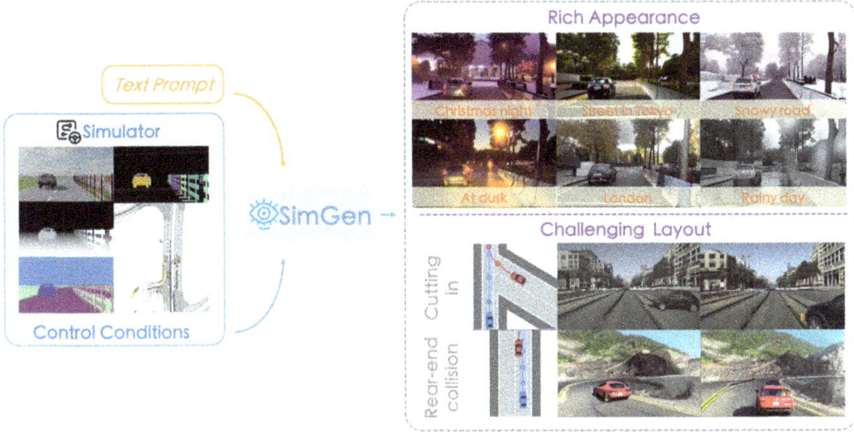

Fig. 1 Illustration of SimGen. SimGen [74] is a simulator-conditioned driving scene generator. Both the layout and the appearance of the outputs can be controlled. (Figure taken from [74] with the authors' permission)

diverse RGB data and map it to the new sensor modality, emulating its unique characteristics. Several studies have considered using img2img translation with generative models to learn such a mapping. Liu et al. [106] addressed the RGB2thermal translation problem with a heatmap-guided model based on CycleGAN. In this approach, heatmaps that highlight key objects and Sobel edge maps are extracted to help the decoder reconstruct important details. Similar CycleGAN-based approaches have been employed to generate synthetic NIR images [107–109]. One major advantage of a CycleGAN-based method is that it does not require *matched* pairs of images representing each modality, unlike the UNet model described in [110]. Emphasizing the fact that one RGB image could have multiple plausible thermal counterparts, Wang et al. [111] recently introduced a novel generative block that incorporates MUNIT [112], to achieve a many-to-many mapping from unpaired data. Finally, diffusion models are considered as well, including an LDM-based RGB2thermal model with additional physics-informed loss functions [113].

3.5 *World Models*

This last part of the section on AD touches upon on a role for generative models that goes much beyond overcoming data accessibility challenges. While most AD solutions today require learning from a massive number of supervised examples (i.e., human experts providing the correct output decision), it takes us humans much less supervision (e.g., limited hours of driving lessons) to master the same tasks. It is hypothesized that we rely on internal models of how our world works, telling us what is (un)likely to happen in it [114–116]. It is believed that we build and update such world models through observation, without the need for external supervision.

Artificial world models incorporate generative modeling to simulate what will happen next. In AD, this can entail imagining plausible actions of other road users, generating possible near-future camera inputs, etc. This then allows for careful action planning. GAIA-1 ("generative AI for autonomy") [76], a world model built by Wayve, combines principles of next-token generation, borrowed from natural language models, and video diffusion models to predict the next frame in a driving sequence. Starting from initial prompts (text description, driving action, and/or starting frame), GAIA-1 can "roll out" a future and produce entire imaginary driving video sequences. By learning to predict the future, the expectation is that the model will embed important world knowledge in its internal representations, which can greatly facilitate downstream driving tasks. Similarly, Waymo's end-to-end multimodal model for AD (EMMA) [77] incorporates world knowledge embedded in a large language model pre-trained to predict the next token in text sequences. EMMA directly generates driving actions from sensor data ("end-to-end"), as opposed to breaking up the problem in multiple, layered modules (e.g., perception, localization, etc.). It achieved competitive motion planning results on popular benchmarks. Finally, Wang et al. [75] introduced DriveDreamer, which generates plausible future video frames along with driving actions using a diffusion backbone.

It takes an initial frame and action as inputs, as well as a text description and structural information about the road. For a more comprehensive overview of world models in AD, we refer to Feng et al. [117].

In sum, while the generative training objective is used here to gather a deeper understanding of the world and arrive at rich internal model representations to serve downstream tasks, we note that the video generation capabilities are themselves already highly advantageous in the light of the data accessibility challenges discussed in this section.

4 Medical Imaging

Medical images, such as X-rays, CT scans, MRI, and ultrasounds, are increasingly being analyzed with the help of AI, and more specifically DNNs. After early success on detection and classification tasks (e.g., detecting microcalcifications in a breast mammography), research expanded rapidly toward segmentation-based approaches for efficient tissue identification, a task that has traditionally been time-consuming in healthcare. Indeed, semi-automatic solutions, which combine AI-generated labels with minimal human corrections, have proven to be more time-efficient and consistent than traditional, manual workflows [118–120]. Today, multiple AI/data-driven tools are being adopted in clinical practice, for example, syngo.via (Siemens), MIM (GE Healthcare), Icobrain (Icometrix), CLARA imaging (Nvidia), or open-source software tools such as MONAI [28, 121]. New trends evolve toward multimodal data analysis that combine multiple imaging modalities as well as electronic health records.

As in other application fields, medical AI models require lots of training data. However, research and innovation in medical imaging face many data-related barriers. Due to legal constraints and privacy issues, open-source healthcare data is scarce, hampering the development and implementation of robust deep learning models across different environments (i.e., different scanners, imaging protocols, etc.). In addition, institutions may have agreements on standardized data capture protocols, but these are not always strictly followed. Small deviations happen based on experience, finance, and manpower, leading to missing data samples over the patient populations. Furthermore, deep learning models tend to underperform on rare diseases, for which only very little data samples are available. Because of these real-world data accessibility challenges, there is an increasing interest in GenAI-driven extensions [122–125], which we discuss here in this section.

4.1   Legal and Ethical Constraints

While datasets such as ImageNet and COCO are publicly accessible online, multiple legal and ethical barriers make sharing healthcare data very difficult. Preserving patients' privacy is key, but perfectly anonymized data does not exist. Even when explicit identifiers (e.g., name and demographics) are removed, sensitive patient information can sometimes still be inferred indirectly from visual features in the images [126]. Apart from privacy, discussions about ownership and responsibility of the data often promote data restrictions over data sharing. This all results in AI methodologies being developed in silos across medical centers. GenAI models, as opposed to data, can be shared more easily and might play a crucial role in breaking these data silos [127]. What is shared then are synthetic representations of a real dataset, but not the data itself. By learning the underlying patterns of the real patient data, GenAI can sample synthetic images from the same distribution. Several deep learning-based approaches trained on a combination of real and synthetic data already showed similar performance compared to those trained on real data only [121, 128–130].

4.2   Unbalanced and Limited Datasets

The quality and size of a dataset has a direct influence on the success of data-driven approaches such as deep learning algorithms. Medical imaging data is generally already scarce (e.g., due to high acquisition costs and required expertise) and prone to class imbalances. That is, certain datasets exhibit a lack of healthy controls, while others underrepresent disease cases. The latter applies especially to rare diseases, which individually cover a small percentage but all together affect millions of lives [126]. Without any intervention during training, the AI models will mainly predict the majority class and misinterpret the underrepresented disease classes. Sampling from learned distributions by GenAI models allows one to generate novel imaging samples in addition to the original training data, maintaining coherence and relevance. This results in data augmentation techniques that can help balance datasets and improve model performance [127–129]. Prusty et al. introduced an LDM framework to augment datasets and showed state-of-the-art performance on four different medical classification tasks [130]. Gupta et al. used a GAN to introduce lesions into healthy bone X-ray images [131]. In addition, Pinaya et al. developed an LDM conditioned on sex, gender, and brain structures for brain image generation. The LDM was used to create a virtual dataset of 100,000 brain images and was made publicly available for research purposes [132]. Furthermore, brainSPADE [133] is a framework combining an LDM and VAE to generate a realistic segmentation map either with or without a pathology present. Subsequently, it uses a GAN for img2img translation of the segmentation map to a realistic medical image. As a result, brainSPADE's generated images come with ground-truth labels for a specific pathology already included, which improves performance on downstream pathology recognition tasks. Next, Kim et al. created SynNodBench, a diffusion-based model to add various sizes and shapes of long nodules to chest X-rays [134], leading to a more balanced dataset for long nodule detection tasks. For rare diseases specifically, Mendez et al. designed a GenAI framework to enrich synthetic ultrasound images and facilitate early detection of hemophilia, a rare bleeding disorder [135]. Finally, GenAI techniques were also used to generate a set of realistic appearing images from patients with Kabuki and Noonan syndromes, hereby improving recognition of these syndromes among pediatric residents [136].

4.3 Missing Modalities

Electronic health records of patients with similar diseases do not always cover the same imaging modalities. For example, slight deviations of protocols might lead to an extra CT image in one patient or a missing MRI scan in another. Because each imaging modality captures different physical properties with unique advantages, filling in the image gaps can be very valuable for creating perfectly paired datasets. Such paired datasets have already proven to provide more information for segmentation tasks [118]. One way to fill in missing data is to convert one modality to another. This is essentially an img2img translation task that can be addressed with GANs or other GenAI models. The goal is to learn a mapping of image appearance across modalities but preserve underlying structural and pathological image information. The ability to generate multimodal data enables clinical flexibility in downstream tasks and ensures that AI models trained on complete data can still be useful if an image modality is missing [137]. For example, Zao et al. used a multimodal masked autoencoder to reconstruct missing modalities. The study showed improved brain tumor segmentation on different rates of missing input modalities [138].

Furthermore, some types of medical imaging are safer for the patient than others. Here too, there is a growing interest in img2img translation solutions, where the safer image is used as the input. A noteworthy example is radiotherapy treatment planning, which requires the intrinsic properties of CT images to carefully define the radiation dose that will be administered to the patient. However, CT itself already exposes patients to radiation, which clinicians want to avoid by shifting to MRI-based planning. Several studies aim to address this concern by generating synthetic CT from MRI [139]. The generated CT images are then evaluated for how well they enable accurate dose planning [140].

Finally, in dynamic and longitudinal imaging, which is used to track temporal changes, missing data can occur in the form of skipped time points (e.g., frame drops in videos). Such time gaps may hamper the development of discriminative AI models. As an example, consider a vision model capturing temporal differences to better understand lung lesions and identify early indicators of future lung cancer. Missing time points could hurt the performance of such model [141]. Studies addressing missing time points include one by Raad et al., who devised a conditional GAN for missing CT images in a longitudinal sequence [142]. Xia et al. furthermore proposed a GAN to fill in missing slices in cardiac MRI videos and avoid the omission of key features for further cardiac monitoring and functional assessment [143]. Finally, sequence-aware diffusion models (SADM) capture longitudinal dependencies and can autoregressively predict next time points, making it a very useful tool not only for generating image sequences but also for disease progression estimations [144].

4.4 Reducing Demographic Dataset Bias

Clinical studies are performed to gather statistical evidence to decide on medical protocols, treatments, etc. [145]. Although many of these studies strive for fair and unbiased evaluations, intrinsic biases can be present in the patient cohort that directly relate to inclusion criteria, ethnicities, geographic regions, and other demographic variables. While different GenAI techniques already proved to be useful in accounting for underrepresented diseases in a dataset (Sect. 4.1), they can also play a crucial role in multi-centric studies covering underrepresented patient populations. Sharing "in silo developed GenAI models," which are intrinsically linked to a specific patient population, means creating a virtual patient cohort. Gathering different virtual patient cohorts over multiple centers should allow researchers to add more heterogeneous data to their study and improve medical insights for various demographic groups [146–149]. For example, Ktena et al. showed that learned data augmentations from diffusion models improve the diagnostic accuracy of discriminative AI models on medical tasks, especially for underrepresented groups and out-of-distribution samples. This increased the fairness of the latter models [150].

4.5 Other GenAI Applications

High-quality medical images are essential for diagnosis and treatment decisions, regardless of whether they are further analyzed by (other) AI models. Another role for GenAI in healthcare, therefore, is to speed up image reconstructions [151] and perform image enhancements [152]. GANs, VAEs, and diffusion models have been used to denoise images or translate low-resolution to high-resolution images [137, 153]. In addition, GenAI models are exploited to create highly realistic versions of phantom simulations, which are used to educate both students and professionals. Three-dimensional replicas of body parts, such as organs, bones, or blood vessels, allow for practicing surgical skills in a controlled environment [154]. Finally, while not the focus of this chapter, non-imaging related applications are gaining attention too. These include synthetic electronic health records to cope with similar data challenges as those mentioned here (e.g., privacy) [155, 156].

5   Risks and Current Limitations

Sections 3 and 4 highlighted the potential of novel solutions to data accessibility challenges enabled by GenAI. Nevertheless, some words of caution are in order, given that the use of GenAI is not without risks.

One important thing to be wary of is the possibility that the generative model might not adequately capture the target distribution even when the output images look convincing to the eye. There is a chance that subtle artefacts will be introduced in the outputs, which downstream models could then overfit to. Such artefacts can include odd textures, unrealistic structural patterns, unnatural edges, etc. For example, Schütte et al. observed artefacts in synthetic chest X-rays that resulted from the model failing to represent medical support devices well (e.g., pacemakers) [157]. Furthermore, sometimes the learned distribution is an oversimplification of the target distribution, not covering all the modes and lacking the intricacies of real-world data. The outputs can even end up appearing too clean in comparison to genuine images. Looking across all output images, one can also wonder about the attenuation or amplification of correlations present in real data. Mikolajczyk et al. trained a GAN on a skin lesion dataset and observed it always added a frame or vignette when generating images of malignant lesions. It thereby amplified a bias already present in the training data [158] and put a downstream classification model more at risk of inappropriately learning that correlation. Taken together, these discrepancies can undermine the ability of models trained on GenAI outputs to generalize to real data. The adverse effects of poor-quality synthetic data become more pronounced as its proportion in the training set increases, which further underlines the importance of quality control in synthetic data generation [159].

Bias amplification by generative models [160–162] implies another risk: when bias involves demographic variables, it raises concerns about fairness. One study reported gender bias in pedestrian detection models at nighttime [163]. To the extent that it is due to dataset bias (e.g., if women are underrepresented in nightly training images), the problem could be exacerbated by training on synthetic data. Friedrich et al. [160] revealed gender-occupation biases in the outputs of Stable Diffusion that were more severe than in its training data (e.g., associating engineering traits more with men). On the other hand, some studies have reported a reduction in bias [164]. Moreover, there is ongoing work to detect and mitigate bias in generative models [160, 165, 166]. Ironically, as discussed in Sect. 4.3, GenAI can also be leveraged to help address known dataset biases. Nevertheless, GenAI must be used with caution to avoid unfair impacts across demographic groups.

Another concern relates to the use of GenAI to avoid privacy limitations faced when dealing with real-world data (Sect. 4.1). Some risk of privacy leakage remains. When GenAI overfits to its training data, for example, it may start to memorize training samples and generate near-identical images at inference time, rather than create novel ones [167]. Especially in medical contexts, this may lead to privacy violations. Furthermore, it is sometimes possible to determine if a given medical image was part of a GenAI model's training set, which increases the risk of re-identification of the patient in the image [168]. Therefore, metrics to quantify a GenAI model's vulnerability to privacy attacks, as well as methods to make them robust against such attacks, are essential to ensuring the safe deployment of GenAI.

Finally, as transformative as GenAI may be, there is no free lunch. Generative vision models themselves also need real-world data to train on. In fact, training them recursively on generated, synthetic data causes irreversible model defects called "model collapse" [169]. Parts of the target distribution will start to erode. While it may appear paradoxical to address data accessibility challenges with yet another data-reliant model, the rationale is not for synthetic data to fully replace real-world data. Rather, generative vision models can help complement or expand existing real-world datasets and get the most out of what is accessible or available. Data augmentation, for example, refers to a technique to increase the size and diversity of a dataset by applying modifications to existing datapoints. Traditionally, these modifications include randomly flipping, cropping, and resizing an image. However, GenAI can achieve much more complex modifications and variations, such as modifying the weather in driving images (Sect. 3.1). Furthermore, generative models can be pretrained on large general-purpose datasets and then fine-tuned on smaller, specialized datasets (so-called transfer learning), which they could then help expand. Finally, by learning the target distribution and interpolating between existing samples, generative vision models can help fill in gaps or overcome imbalances (Sects. 3.2 and 4.1).

6 Concluding Remarks and Future Outlook

"Your AI model is only as good as the data it is trained on" is a phrase commonly heard among AI and data scientists. It is projected that in 2025, the sum of all data created globally will amount to 163 zettabytes [170]. Yet, many companies face a data scarcity paradox, meaning that in spite of this data growth, they struggle to obtain enough high-quality data tailored to their application. Challenges include scarcity per se (e.g., rare diseases in medical imaging), stricter data regulations (e.g., regarding privacy), and high acquisition or annotation costs. In this chapter, we have discussed the promise of synthetic data, produced by GenAI, to tackle such data accessibility challenges in the context of AI for real-world vision applications. While we have focused on AD and medical imaging, synthetic (visual) data has become a transformative force overcoming hurdles in AI development across industries. Other industries include, but are not limited to, agriculture, fashion, security, and e-commerce.

Indeed, some of the biggest breakthroughs in generative modelling happened in the visual domain (e.g., GANs [32]). However, the success extends to other modalities, such as speech and audio generation. In addition, the introduction of the Transformer DNN architecture [171] in 2017 catalyzed a parallel revolution in natural language generation, leading to the release of, e.g., ChatGPT. An important direction for future development will be to integrate different modalities and respective technologies into unified multimodal solutions. For example, in personalized medicine, there is high interest in combining electronic health records, which are text-based, and medical imaging into an overarching AI framework [172].

Moving forward, improving trustworthiness will be key for GenAI to realize its full potential, which will require continued efforts to mitigate the risks outlined above. Take the call for improved quality control. Devising good evaluation metrics has been notoriously difficult in generative vision models (Sect. 2.3). Popular metrics are chosen partly based on how well they align with human realness judgments or preferences [63]. However, in many applications, the consumer of the data will not be human. Indeed, artificial visual systems, such as those in AD, will often be the ones to process the synthetic data. Depending on the goal, it might be possible to evaluate the quality of the synthetic data on its ability to improve performance on the downstream task. For example, does training on synthetic long-tail scenarios improve AD performance in real-world long-tail scenarios [173]? Do synthetic CT images yield accurate dose predictions? In other cases, though, such evaluation will not suffice (e.g., when it is used to test, not just train AD perception models). Thus, finding out the right metrics for the application at hand will be an important domain of further development.

Trust also comes with understanding. Deep neural network models have long been criticized for their black-box nature. Therefore, in recent years, a lot of work has been conducted on explainability and interpretability techniques to gain insight into model decisions [155, 174, 175]. Thus far, most of the efforts have targeted discriminative neural networks. Nevertheless, with the widespread application of generative neural networks, explaining what exactly they have learned about the target domain is moving up the research agenda.

One would want a generative model to learn the right things about the target domain and faithfully represent the physics that gave rise to the training data. Rather than relying on a fully data-driven learning process, hybrid approaches have been put forward as well. In such a hybrid approach, expert domain knowledge is incorporated into the model, either hard-wired or through the choice of loss function [113, 176–178], which promotes trust. Other benefits might include improved robustness and sample efficiency. The latter is particularly helpful in the face of data accessibility challenges. Hence, hybrid approaches to generative modeling make for another promising research direction.

Finally, serious concerns have been raised about the amount of computational resources needed to run AI, especially GenAI, and the massive demands it poses on power and water supplies (e.g., for cooling the hardware) [179, 180]. As an indication, generating one image on average consumes the equivalent energy needed to charge a smartphone to 13%, which is many times more than it takes to classify an image [179]. This raises the pressing question of how GenAI can be made sustainable. In addition to targeting the algorithms themselves, this will entail significant improvements of the computing hardware. There is a huge incentive to invest research efforts into specialized AI chips and efficiency gains (e.g., [181]). However, some have warned that the focus is too narrowly on efficiency, which might ultimately only increase adoption rather than promote sustainability [180].

Taken together, synthetic data generated by generative vision models holds a lot of promise and is already driving innovation across industries. To fulfill its potential, there are significant risks and concerns that must be addressed, but these challenges also present valuable opportunities for ongoing research and development. The present and future of our global data landscape is one that combines real-world data with synthetic data produced by GenAI.

Acknowledgments This work received funding from the Flemish Government under the "Onderzoeksprogramma Artificiële Intelligentie (AI) Vlaanderen" program (Flanders AI Research).

References

1. Krizhevsky, A., Sutskever, I., & Hinton, G. E. (2012). ImageNet classification with deep convolutional neural networks. In *Advances in neural information processing systems (NeurIPS)*. Curran Associates, Inc.
2. Russakovsky, O., Deng, J., Su, H., et al. (2015). ImageNet large scale visual recognition challenge. *International Journal of Computer Vision, 115*, 211–252. https://doi.org/10.1007/s11263-015-0816-y
3. He, K., Zhang, X., Ren, S., & Sun, J. (2016). Deep residual learning for image recognition. In *Proceedings of the IEEE/CVF Conference on Computer Vision and Pattern Recognition (CVPR)* (pp. 770–778).
4. Hu, J., Shen, L., & Sun, G. (2018). Squeeze-and-excitation networks. In *Proceedings of the IEEE/CVF Conference on Computer Vision and Pattern Recognition (CVPR)* (pp. 7132–7141).
5. Yang, J. (2023). *NeurIPS statistics*. Pap. Copilot. Retrieved February 27, 2025, from https://papercopilot.com/statistics/neurips-statistics/
6. Deng, J., Dong, W., Socher, R., et al. (2009). ImageNet: A large-scale hierarchical image database. In *Proceedings of the IEEE/CVF Conference on Computer Vision and Pattern Recognition (CVPR)* (pp. 248–255).
7. Simonyan, K., & Zisserman, A. (2015). Very deep convolutional networks for large-scale image recognition. In Y. Bengio & Y. LeCun (Eds.), *Proceedings of the International Conference on Learning Representations (ICLR)*.
8. Tan, M., & Le, Q. (2019). Efficientnet: Rethinking model scaling for convolutional neural networks. In *Proceedings of the International Conference on Machine Learning (ICML)* (pp. 6105–6114). PMLR.
9. Lin, T.-Y., Maire, M., Belongie, S., et al. (2014). Microsoft COCO: Common objects in context. In *Proceedings of the European Conference on Computer Vision (ECCV)*. Springer.
10. Ren, S., He, K., Girshick, R., & Sun, J. (2015). Faster R-CNN: Towards real-time object detection with region proposal networks. In *Advances in neural information processing systems (NeurIPS)*. Curran Associates, Inc.
11. Redmon, J., Divvala, S., Girshick, R., & Farhadi, A. (2016). You only look once: Unified, real-time object detection. In *Proceedings of the IEEE/CVF Conference on Computer Vision and Pattern Recognition (CVPR)* (pp. 779–788).
12. Carion, N., Massa, F., Synnaeve, G., et al. (2020). End-to-end object detection with transformers. In A. Vedaldi, H. Bischof, T. Brox, & J.-M. Frahm (Eds.), *Proceedings of the European Conference on Computer Vision (ECCV)* (pp. 213–229). Springer.
13. Abu-El-Haija, S., Kothari, N., Lee, J., et al. (2016) *YouTube-8M: A large-scale video classification benchmark.*

14. Kay, W., Carreira, J., Simonyan, K., et al. (2017). *The kinetics human action video dataset.*
15. Monfort, M., Andonian, A., Zhou, B., et al. (2019). Moments in time dataset: One million videos for event understanding. *IEEE Transactions on Pattern Analysis and Machine Intelligence*, 1–8. https://doi.org/10.1109/TPAMI.2019.2901464
16. Taigman, Y., Yang, M., Ranzato, M., & Wolf, L. (2014). DeepFace: Closing the gap to human-level performance in face verification. In *Proceedings of the IEEE/CVF Conference on Computer Vision and Pattern Recognition (CVPR)* (pp. 1701–1708).
17. Szegedy, C., Vanhoucke, V., Ioffe, S., et al. (2016). Rethinking the Inception architecture for computer vision. In *Proceedings of the IEEE/CVF Conference on Computer Vision and Pattern Recognition (CVPR)* (pp. 2818–2826).
18. Liu, Z., Luo, P., Qiu, S., et al. (2016). DeepFashion: Powering robust clothes recognition and retrieval with rich annotations. In *Proceedings of the IEEE/CVF Conference on Computer Vision and Pattern Recognition (CVPR)* (pp. 1096–1104).
19. Geetharamani, G., & Arun Pandian, J. (2019). Identification of plant leaf diseases using a nine-layer deep convolutional neural network. *Computers and Electrical Engineering, 76,* 323–338. https://doi.org/10.1016/j.compeleceng.2019.04.011
20. Bergmann, P., Batzner, K., Fauser, M., et al. (2021). The MVTec anomaly detection dataset: A comprehensive real-world dataset for unsupervised anomaly detection. *International Journal of Computer Vision, 129,* 1038–1059. https://doi.org/10.1007/s11263-020-01400-4
21. Caesar, H., Bankiti, V., Lang, A. H., et al. (2020). NuScenes: A multimodal dataset for autonomous driving. In *Proceedings of the IEEE/CVF Conference on Computer Vision and Pattern Recognition (CVPR)* (pp. 11621–11631).
22. Nguyen, H. Q., Lam, K., Le, L. T., et al. (2022). Vindr-CXR: An open dataset of chest X-rays with radiologist's annotations. *Scientific Data, 9,* 429. https://doi.org/10.1038/s41597-022-01498-w
23. SAE International. (2021). *Taxonomy and definitions for terms related to driving automation systems for on-road motor vehicles.* SAE International.
24. Morris, J. (2021). *Why is Tesla's full self-driving only level 2 autonomous?* Retrieved February 14, 2025, from https://www.forbes.com/sites/jamesmorris/2021/03/13/why-is-teslas-full-self-driving-only-level-2-autonomous/
25. Tesla. (n.d.). *Autopilot and full self-driving (supervised).* Tesla. Retrieved February 14, 2025, from https://www.tesla.com/support/autopilot
26. Waymo. *Ride-hailing app—make the most of your drive—waymo one.* Waymo. Retrieved February 14, 2025, from https://waymo.com/waymo-one/
27. Volkswagen Group. (2024). *Automated driving: Volkswagen Group intensifies collaboration with Mobileye.* Volkswagen Group. Retrieved March 3, 2025, from https://www.volkswagen-group.com/en/press-releases/automated-driving-volkswagen-group-intensifies-collaboration-with-mobileye-18290
28. Cardoso, M. J., Li, W., Brown, R., et al. (2022). *MONAI: An open-source framework for deep learning in healthcare.* Springer Nature.
29. Barragán-Montero, A., Javaid, U., Valdés, G., et al. (2021). Artificial intelligence and machine learning for medical imaging: A technology review. *Physica Medica, 83,* 242–256. https://doi.org/10.1016/j.ejmp.2021.04.016
30. Doi, K., Giger, M. L., Nishikawa, R. M., & Schmidt, R. A. (1997). Computer aided diagnosis of breast cancer on mammograms. *Breast Cancer, 4,* 228–233. https://doi.org/10.1007/BF02966511
31. Lo, S.-C. B., Lou, S.-L. A., Lin, J.-S., et al. (1995). Artificial convolution neural network techniques and applications for lung nodule detection. *IEEE Transactions on Medical Imaging, 14,* 711–718. https://doi.org/10.1109/42.476112
32. Goodfellow, I. J., Pouget-Abadie, J., Mirza, M., et al. (2014). Generative adversarial nets. In *Advances in neural information processing systems (NeurIPS).* Curran Associates, Inc.
33. Brock, A., Donahue, J., & Simonyan, K. (2018). Large scale GAN training for high fidelity natural image synthesis. In *Proceedings of the International Conference on Learning Representations (ICLR).*

34. Karras, T., Laine, S., Aittala, M., et al. (2020). Analyzing and improving the image quality of StyleGAN. In *Proceedings of the IEEE/CVF Conference on Computer Vision and Pattern Recognition (CVPR)* (pp. 8110–8119).
35. Liu, Z., Luo, P., Wang, X., & Tang, X. (2015). Deep learning face attributes in the wild. In *Proceedings of the International Conference on Computer Vision (ICCV)*.
36. Ho, J., Jain, A., & Abbeel, P. (2020). Denoising diffusion probabilistic models. In H. Larochelle, M. Ranzato, R. Hadsell, et al. (Eds.), *Advances in neural information processing systems (NeurIPS)* (pp. 6840–6851). Curran Associates, Inc..
37. Nichol, A. Q., & Dhariwal, P. (2021). Improved denoising diffusion probabilistic models. In M. Meila & T. Zhang (Eds.), *Proceedings of the International Conference on Machine Learning (ICML)* (pp. 8162–8171). PMLR.
38. Dhariwal, P., & Nichol, A. (2021). Diffusion models beat GANs on image synthesis. In *Advances in neural information processing systems (NeurIPS)* (pp. 8780–8794). Curran Associates, Inc..
39. Song, Y., Sohl-Dickstein, J., Kingma, D. P., et al. (2020). Score-based generative modeling through stochastic differential equations. In *Proceedings of the International Conference on Learning Representations (ICLR)*.
40. Ramesh, A., Dhariwal, P., Nichol, A., et al. (2022). *Hierarchical text-conditional image generation with CLIP latents*.
41. Saharia, C., Chan, W., Saxena, S., et al. (2022). Photorealistic text-to-image diffusion models with deep language understanding. In *Advances in neural information processing systems (NeurIPS)* (pp. 36479–36494).
42. Rombach, R., Blattmann, A., Lorenz, D., et al. (2022). High-resolution image synthesis with latent diffusion models. In *Proceedings of the IEEE/CVF Conference on Computer Vision and Pattern Recognition (CVPR)*.
43. OpenAI. (2022). *DALL·E now available without waitlist*. Retrieved March 5, 2025, from https://openai.com/index/dall-e-now-available-without-waitlist/
44. Curry, E., Heintz, F., Irgens, M., et al. (2022). Partnership on AI, data, and robotics. *Communications of the ACM, 65*, 54–55. https://doi.org/10.1145/3513000
45. Kingma, D. P., & Welling, M. (2014). Auto-encoding variational Bayes. In *Proceedings of the International Conference on Learning Representations (ICLR)*.
46. Higgins, I., Matthey, L., Pal, A., et al. (2017). Beta-VAE: Learning basic visual concepts with a constrained variational framework. In *Proceedings of the International Conference on Learning Representations (ICLR)*.
47. Larsen, A. B. L., Sønderby, S. K., Larochelle, H., & Winther, O. (2016). Autoencoding beyond pixels using a learned similarity metric. In *Proceedings of the International Conference on Machine Learning (ICML)* (pp. 1558–1566). PMLR.
48. Mescheder, L., Nowozin, S., & Geiger, A. (2017). Adversarial variational Bayes: Unifying variational autoencoders and generative adversarial networks. In *Proceedings the International Conference on Machine Learning (ICML)* (pp. 2391–2400). PMLR.
49. Hinton, G. E., & Salakhutdinov, R. R. (2006). Reducing the dimensionality of data with neural networks. *Science, 313*, 504–507. https://doi.org/10.1126/science.1127647
50. Lipman, Y., Chen, R. T. Q., Ben-Hamu, H., et al. (2022). Flow matching for generative modeling. In *Proceedings of the International Conference on Learning Representations (ICLR)*.
51. Liu, X., Gong, C., & Liu, Q. (2022). Flow straight and fast: Learning to generate and transfer data with rectified flow. In *Proceedings of the International Conference on Learning Representations (ICLR)*.
52. Liu, G.-H., Vahdat, A., Huang, D.-A., et al. (2023). I^{2}SB: Image-to-image Schrödinger bridge. In *Proceedings of the International Conference on Machine Learning (ICML)* (pp. 22042–22062). PMLR.
53. Song, J., Meng, C., & Ermon, S. (2021). Denoising diffusion implicit models. In *Proceedings of the International Conference on Learning Representations (ICLR)*.
54. Dockhorn, T., Vahdat, A., & Kreis, K. (2022). GENIE: Higher-order denoising diffusion solvers. In *Advances in neural information processing systems (NeurIPS)* (pp. 30150–30166).

55. Zhang, Q., & Chen, Y. (2022). Fast sampling of diffusion models with exponential integrator. In *Proceedings of the International Conference on Learning Representations (ICLR)*.

56. Shaul, N., Perez, J., Chen, R. T. Q., et al. (2024). Bespoke solvers for generative flow models. In *Proceedings of the International Conference on Learning Representations (ICLR)*.

57. Karras, T., Aittala, M., Aila, T., & Laine, S. (2022). Elucidating the design space of diffusion-based generative models. In *Advances in neural information processing systems (NeurIPS)*.

58. Xu, Y., Zhao, Y., Xiao, Z., & Hou, T. (2024). UFOGen: You forward once large scale text-to-image generation via diffusion GANs. In *Proceedings of the IEEE/CVF Conference on Computer Vision and Pattern Recognition (CVPR)* (pp. 8196–8206).

59. Sauer, A., Boesel, F., Dockhorn, T., et al. (2024). Fast high-resolution image synthesis with latent adversarial diffusion distillation. In *Proceedings of the ACM SIGGRAPH Conference on Computer Graphics and Interactive Techniques* (pp. 1–11). Association for Computing Machinery.

60. Sauer, A., Lorenz, D., Blattmann, A., & Rombach, R. (2025). Adversarial diffusion distillation. In A. Leonardis, E. Ricci, S. Roth, et al. (Eds.), *Proceedings of the European Conference on Computer Vision (ECCV)*.

61. Salimans, T., Goodfellow, I., Zaremba, W., et al. (2016). Improved techniques for training GANs. In *Advances in neural information processing systems (NeurIPS)*. Curran Associates, Inc.

62. Kynkäänniemi, T., Karras, T., Laine, S., et al. (2019). Improved precision and recall metric for assessing generative models. In *Advances in neural information processing systems (NeurIPS)*. Curran Associates, Inc.

63. Jayasumana, S., Ramalingam, S., Veit, A., et al. (2024). Rethinking FID: Towards a better evaluation metric for image generation. In *Proceedings of the IEEE/CVF Conference on Computer Vision and Pattern Recognition (CVPR)* (pp. 9307–9315).

64. Ho, J. (2022). Classifier-free diffusion guidance. In *Advances in Neural Information Processing Systems (NeurIPS) Workshop on Deep Generative Models and Downstream Applications*.

65. Zhu, J.-Y., Park, T., Isola, P., & Efros, A. A. (2017). Unpaired image-to-image translation using cycle-consistent adversarial networks. In *Proceedings of the IEEE/CVF International Conference on Computer Vision (ICCV)*.

66. Johan, V., Zürn, J., & Burgard, W. (2020). HeatNet: Bridging the day-night domain gap in semantic segmentation with thermal images. In *Proceedings of the IEEE/RSJ International Conference on Intelligent Robots and Systems (IROS)*.

67. Shin, U., Park, J., & Kweon, I. S. (2023). Deep depth estimation from thermal image. In *Proceedings of the IEEE/CVF Conference on Computer Vision and Pattern Recognition (CVPR)* (pp. 1043–1053).

68. Tesla. (2024). *Tesla vision update: Replacing ultrasonic sensors with tesla vision*. Tesla. Retrieved February 14, 2025, from https://www.tesla.com/support/transitioning-tesla-vision

69. Sun, P., Kretzschmar, H., Dotiwalla, X., et al. (2020). Scalability in perception for autonomous driving: Waymo open dataset. In *Proceedings of the IEEE/CVF Conference on Computer Vision and Pattern Recognition (CVPR)*.

70. Vinod, V., Ram Prabhakar, K., Venkatesh Babu, R., & Chakraborty, A. (2021). Multi-domain conditional image translation: Translating driving datasets from clear-weather to adverse conditions. In *Proceedings of the IEEE/CVF International Conference on Computer Vision (ICCV) Workshops* (pp. 1571–1582). IEEE.

71. Yang, Z., Chai, Y., Anguelov, D., et al. (2020). SurfelGAN: Synthesizing realistic sensor data for autonomous driving. In *Proceedings of the IEEE/CVF Conference on Computer Vision and Pattern Recognition (CVPR)* (pp. 11115–11124). IEEE.

72. Fu, Y., Li, Y., & Di, X. (2024). *GenDDS: Generating diverse driving video scenarios with prompt-to-video generative model*.

73. Podell, D., English, Z., Lacey, K., et al. (2024). SDXL: Improving latent diffusion models for high-resolution image synthesis. In *Proceedings of the International Conference on Learning Representations (ICLR)*.

74. Zhou, Y., Simon, M., Mark, P. Z., et al. (2025). SimGen: Simulator-conditioned driving scene generation. In *Advances in neural information processing systems (NeurIPS)* (pp. 48838–48874).

75. Wang, X., Zhu, Z., Huang, G., et al. (2024). DriveDreamer: Towards real-world-driven world models for autonomous driving. In *Proceedings of the European Conference on Computer Vision (ECCV)*.

76. Hu, A., Russell, L., Yeo, H., et al. (2023). *GAIA-1: A generative world model for autonomous driving*.

77. Hwang, J,-J., Xu, R., Lin, H., et al. (2024). *EMMA: End-to-end multimodal model for autonomous driving*.

78. Mao, J., Li, B., Ivanovic, B., et al. (2025). *DreamDrive: Generative 4D scene modeling from street view images*.

79. Parmar, G., Park, T., Narasimhan, S., & Zhu, J.-Y. (2024). *One-step image translation with text-to-image models*.

80. Pizzati, F., Cerri, P., & de Charette, R. (2021). CoMoGAN: Continuous model-guided image-to-image translation. In *Proceedings of the IEEE/CVF Conference on Computer Vision and Pattern Recognition (CVPR)*.

81. Gong, R., Dai, D., Chen, Y., et al. (2020). Analogical image translation for fog generation. *Proceedings of the AAAI Conference on Artificial Intelligence (AAAI)*.

82. Ye, Y., Chang, Y., Zhou, H., & Yan, L. (2021). Closing the loop: Joint rain generation and removal via disentangled image translation. In *Proceedings of the IEEE/CVF Conference on Computer Vision and Pattern Recognition (CVPR)* (pp. 2053–2062). IEEE.

83. Kim, S., Baek, J., Park, J., et al. (2022). InstaFormer: Instance-aware image-to-image translation with transformer. In *Proceedings of the IEEE/CVF Conference on Computer Vision and Pattern Recognition (CVPR)* (pp. 18321–18331).

84. Zhang, M., Zhang, Y., Zhang, L., et al. (2018). DeepRoad: GAN-based metamorphic testing and input validation framework for autonomous driving systems. In *Proceedings of the IEEE/ACM International Conference on Automated Software Engineering (ASE)* (pp. 132–142).

85. Liu, H. X., & Feng, S. (2024). Curse of rarity for autonomous vehicles. *Nature Communications, 15*, 4808. https://doi.org/10.1038/s41467-024-49194-0

86. Ros, G., Sellart, L., Materzynska, J., et al. (2016). The SYNTHIA dataset: A large collection of synthetic images for semantic segmentation of urban scenes. In *Proceedings of the IEEE/CVF Conference on Computer Vision and Pattern Recognition (CVPR)* (pp. 3234–3243). IEEE.

87. Dosovitskiy, A., Ros, G., Codevilla, F., et al. (2017). CARLA: An open urban driving simulator. In *Proceedings of the Annual Conference on Robot Learning (CoRL)* (pp. 1–16).

88. Weng, X., Man, Y., Cheng, D., et al. (2020). *All-in-one drive: A large-scale comprehensive perception dataset with high-density long-range point clouds*. arXiv Preprint.

89. Richter, S. R., Vineet, V., Roth, S., & Koltun, V. (2016). Playing for data: Ground truth from computer games. In B. Leibe, J. Matas, N. Sebe, & M. Welling (Eds.), *Proceedings of the European Conference on Computer Vision (ECCV)* (pp. 102–118). Springer International Publishing.

90. Shah, S., Dey, D., Lovett, C., & Kapoor, A. (2017). AirSim: High-fidelity visual and physical simulation for autonomous vehicles. In *Field and service robotics*.

91. Hoffman, J., Tzeng, E., Park, T., et al. (2018). CyCADA: Cycle-consistent adversarial domain adaptation. In *Proceedings of the International Conference on Machine Learning (ICML)* (pp. 1989–1998). PMLR.

92. Liu, R., Yang, C., Sun, W., et al. (2020). StereoGAN: Bridging synthetic-to-real domain gap by joint optimization of domain translation and stereo matching. In *Proceedings of the IEEE/CVF Conference on Computer Vision and Pattern Recognition (CVPR)* (pp. 12754–12763). IEEE.

93. Chang, W.-L., Wang, H.-P., Peng, W.-H., & Chiu, W.-C. (2019). All about structure: Adapting structural information across domains for boosting semantic segmentation. In *Proceedings of the IEEE/CVF Conference on Computer Vision and Pattern Recognition (CVPR)* (pp. 1900–1909). IEEE.

94. Zhang, C. Y., & Shrivastava, A. (2023). *AptSim2real: Approximately-paired sim-to-real image translation.*

95. Richter, S. R., AlHaija, H. A., & Koltun, V. (2021). Enhancing photorealism enhancement. arXiv:210504619.

96. Li, Q., Peng, Z., Feng, L., et al. (2023). ScenarioNet: Open-source platform for large-scale traffic scenario simulation and modeling. In *Advances in neural information processing systems (NeurIPS).*

97. *Qualcomm at neurIPS 2024: Our groundbreaking innovations and cutting-edge advancements in AI* | qualcomm. Retrieved 17 February 2025, from https://www.qualcomm.com/news/onq/2024/12/qualcomm-at-neurips-2024-ai-research-demos-papers

98. Kerbl, B., Kopanas, G., Leimkühler, T., & Drettakis, G. (2023). 3D gaussian splatting for real-time radiance field rendering. *ACM Transactions on Graphics, 42.*

99. Kim, M., & Byun, H. (2020). Learning texture invariant representation for domain adaptation of semantic segmentation. In *Proceedings of the IEEE/CVF Conference on Computer Vision and Pattern Recognition (CVPR)* (pp. 12972–12981). IEEE.

100. Pan, F., Shin, I., Rameau, F., et al. (2020). Unsupervised intra-domain adaptation for semantic segmentation through self-supervision. In *Proceedings of the IEEE/CVF Conference on Computer Vision and Pattern Recognition (CVPR).*

101. Chen, Y.-C., Lin, Y.-Y., Yang, M.-H., & Huang, J.-B. (2019). CrDoCo: Pixel-level domain transfer with cross-domain consistency. In *Proceedings of the IEEE/CVF Conference on Computer Vision and Pattern Recognition (CVPR)* (pp. 1791–1800). IEEE.

102. Wang, Y., Qi, L., Chen, Y.-C., et al. (2021). Image synthesis via semantic composition. In *Proceedings of the IEEE/CVF International Conference on Computer Vision (ICCV)* (pp. 13729–13738). IEEE.

103. Berrada, T., Verbeek, J., Couprie, C., & Alahari, K. (2024). Unlocking pre-trained image backbones for semantic image synthesis. In *Proceedings of the IEEE/CVF Conference on Computer Vision and Pattern Recognition (CVPR)* (pp. 7840–7849). IEEE.

104. Wang, W., Bao, J., Zhou, W., et al. (2022). *Semantic image synthesis via diffusion models.*

105. Careil, M., Verbeek, J., & Lathuilière, S. (2023). Few-shot semantic image synthesis with class affinity transfer. In *Proceedings of the IEEE/CVF Conference on Computer Vision and Pattern Recognition (CVPR)* (pp. 23611–23620). IEEE.

106. Liu, T., Liu, Y., Xu, W., et al. (2022). HGGAN: Visible to thermal translation generative adversarial network guided by heatmap. In *Proceedings of the IEEE International Conference on Unmanned Systems (ICUS)* (pp. 171–176).

107. Yan, L., Wang, X., Zhao, M., et al. (2020). A multi-model fusion framework for NIR-to-RGB translation. In *Proceedings of the IEEE International Conference on Visual Communications and Image Processing (VCIP)* (pp. 459–462).

108. Sun, T., Jung, C., Fu, Q., & Han, Q. (2019). NIR to RGB domain translation using asymmetric cycle generative adversarial networks. *IEEE Access, 7,* 112459–112469. https://doi.org/10.1109/ACCESS.2019.2933671

109. Lu, Y., & Lu, G. (2021). Bridging the invisible and visible world: Translation between RGB and IR images through contour cycle GAN. In *Proceedings of the IEEE International Conference on Advanced Video and Signal Based Surveillance (AVSS).*

110. Iwashita, Y., Nakashima, K., Rafol, S., et al. (2019). MU-net: Deep learning-based thermal IR image estimation from RGB image. In *Proceedings of the IEEE/CVF Conference on Computer Vision and Pattern Recognition (CVPR) Workshops* (pp. 1022–1028). IEEE.

111. Wang, K. (2024). Increasing the diversity in RGB-to-thermal image translation for automotive applications. In *Proceedings of the IEEE SENSORS Conference.*

112. Huang, X., Liu, M.-Y., Belongie, S., & Kautz, J. (2018). Multimodal unsupervised image-to-image translation. In *Proceedings of the European Conference on Computer Vision (ECCV).*

113. Mao, F., Mei, J., Lu, S., et al. (2024). *PID: Physics-informed diffusion model for infrared image generation.*

114. LeCun, Y. (2022). A path towards autonomous machine intelligence version 0.9. 2, 2022-06-27. *Open Review, 62,* 1–62.

115. Friston, K., Moran, R. J., Nagai, Y., et al. (2021). World model learning and inference. *Neural Networks, 144,* 573–590. https://doi.org/10.1016/j.neunet.2021.09.011

116. Ha, D., & Schmidhuber, J. (2018). Recurrent world models facilitate policy evolution. In *Advances in neural information processing systems.* Curran Associates, Inc.

117. Feng, T., Wang, W., & Yang, Y. (2025). *A survey of world models for autonomous driving.*

118. Bollen, H., Willems, S., Wegge, M., et al. (2023). Benefits of automated gross tumor volume segmentation in head and neck cancer using multi-modality information. *Radiotherapy and Oncology, 182,* 109574. https://doi.org/10.1016/j.radonc.2023.109574

119. van der Veen, J., Willems, S., Deschuymer, S., et al. (2019). Benefits of deep learning for delineation of organs at risk in head and neck cancer. *Radiotherapy and Oncology, 138,* 68–74. https://doi.org/10.1016/j.radonc.2019.05.010

120. Buelens, P., Willems, S., Vandewinckele, L., et al. (2022). Clinical evaluation of a deep learning model for segmentation of target volumes in breast cancer radiotherapy. *Radiotherapy and Oncology, 171,* 84–90. https://doi.org/10.1016/j.radonc.2022.04.015

121. Pinaya, W. H. L., Graham, M. S., Kerfoot, E., et al. (2023). *Generative AI for medical imaging: Extending the MONAI framework.* arXiv Preprint.

122. Kazerouni, A., Aghdam, E. K., Heidari, M., et al. (2023). Diffusion models in medical imaging: A comprehensive survey. *Medical Image Analysis, 88,* 102846. https://doi.org/10.1016/j.media.2023.102846

123. Chen, Y., Yang, X.-H., Wei, Z., et al. (2022). Generative adversarial networks in medical image augmentation: A review. *Computers in Biology and Medicine, 144,* 105382. https://doi.org/10.1016/j.compbiomed.2022.105382

124. Ghebrehiwet, I., Zaki, N., Damseh, R., & Mohamad, M. S. (2024). Revolutionizing personalized medicine with generative AI: A systematic review. *Artificial Intelligence Review, 57,* 128. https://doi.org/10.1007/s10462-024-10768-5

125. Sakthivel, B., Vanathi, P., Sri, M. R., et al. (2024). Generative AI models and capabilities in cancer medical imaging and applications. In *Proceedings of the International Conference on Sentiment Analysis and Deep Learning (ICSADL)* (pp. 349–355).

126. Visibelli, A., Roncaglia, B., Spiga, O., & Santucci, A. (2023). The impact of artificial intelligence in the odyssey of rare diseases. *Biomedicine, 11,* 887. https://doi.org/10.3390/biomedicines11030887

127. Chatterjee, S., Fruhling, A., Kotiadis, K., & Gartner, D. (2024). Towards new frontiers of healthcare systems research using artificial intelligence and generative AI. *Health Systems, 13,* 263–273. https://doi.org/10.1080/20476965.2024.2402128

128. Hasani, N., Farhadi, F., Morris, M. A., et al. (2022). Artificial intelligence in medical imaging and its impact on the rare disease community: Threats, challenges and opportunities. *PET Clinics, 17,* 13–29. https://doi.org/10.1016/j.cpet.2021.09.009

129. Korkinof, D., Rijken, T., O'Neill, M., et al. (2019). *High-resolution mammogram synthesis using progressive generative adversarial networks.* arXiv Preprint.

130. Prusty, M. R., Sudharsan, R. M., & Anand, P. (2024). Enhancing medical image classification with generative AI using latent denoising diffusion probabilistic model and wiener filtering approach. *Applied Soft Computing, 161,* 111714. https://doi.org/10.1016/j.asoc.2024.111714

131. Gupta, A., Venkatesh, S., Chopra, S., & Ledig, C. (2019). Generative image translation for data augmentation of bone lesion pathology. In *Proceedings of the International Conference on Medical Imaging with Deep Learning* (pp. 225–235). PMLR.

132. Pinaya, W. H. L., Tudosiu, P.-D., Dafflon, J., et al. (2022). Brain imaging generation with latent diffusion models. In A. Mukhopadhyay, I. Oksuz, S. Engelhardt, D. Zhu, & Y. Yuan (Eds.), *Deep generative models. DGM4MICCAI 2022* (Lecture Notes in Computer Science) (Vol. 13609). Springer.

133. Fernandez, V., Pinaya, W. H. L., Borges, P., et al. (2022). Can segmentation models be trained with fully synthetically generated data? In C. Zhao, D. Svoboda, J. M. Wolterink, &

M. Escobar (Eds.), *Simulation and synthesis in medical imaging. SASHIMI 2022* (Lecture Notes in Computer Science) (Vol. 13570, pp. 79–90). Springer.

134. Kim, Y., Jang, S., Kim, S., et al. (2024). *Demo: Harnessing generative AI for comprehensive evaluation of medical imaging AI*. OpenReview.net.

135. Mendez Mora, M. (2023). *Deep generative machine learning to create synthetic 2D ultrasound images for rare disease research*.

136. Waikel, R. L., Othman, A. A., Patel, T., et al. (2024). Recognition of genetic conditions after learning with images created using generative artificial intelligence. *JAMA Network Open, 7*, e242609. https://doi.org/10.1001/jamanetworkopen.2024.2609

137. Sai, S., Gaur, A., Sai, R., et al. (2024). Generative AI for transformative healthcare: A comprehensive study of emerging models, applications, case studies, and limitations. *IEEE Access, 12*, 31078–31106. https://doi.org/10.1109/ACCESS.2024.3367715

138. Zhao, Y., Chen, C., Pang, Q. Y., et al. (2024). *Dealing with all-stage missing modality: Towards a universal model with robust reconstruction and personalization*.

139. Boulanger, M., Nunes, J.-C., Chourak, H., et al. (2021). Deep learning methods to generate synthetic CT from MRI in radiotherapy: A literature review. *Physica Medica, 89*, 265–281. https://doi.org/10.1016/j.ejmp.2021.07.027

140. Kazemifar, S., McGuire, S., Timmerman, R., et al. (2019). MRI-only brain radiotherapy: Assessing the dosimetric accuracy of synthetic CT images generated using a deep learning approach. *Radiotherapy and Oncology, 136*, 56–63. https://doi.org/10.1016/j.radonc.2019.03.026

141. Li, T. Z., Xu, K., Gao, R., et al. (2023). Time-distance vision transformers in lung cancer diagnosis from longitudinal computed tomography. *Proceedings of SPIE, 12464*, 1246412. https://doi.org/10.1117/12.2653911

142. Raad, R., Ray, D., Varghese, B., et al. (2024). Conditional generative learning for medical image imputation. *Scientific Reports, 14*, 171. https://doi.org/10.1038/s41598-023-50566-7

143. Xia, Y., Zhang, L., Ravikumar, N., et al. (2021). Recovering from missing data in population imaging—Cardiac MR image imputation via conditional generative adversarial nets. *Medical Image Analysis, 67*, 101812. https://doi.org/10.1016/j.media.2020.101812

144. Yoon, J. S., Zhang, C., Suk, H.-I., et al. (2023). SADM: Sequence-aware diffusion model for longitudinal medical image generation. In A. Frangi, M. de Bruijne, D. Wassermann, & N. Navab (Eds.), *Information processing in medical imaging. IPMI 2023* (Lecture Notes in Computer Science) (Vol. 13939, pp. 388–400). Springer.

145. Larrazabal, A. J., Nieto, N., Peterson, V., et al. (2020). Gender imbalance in medical imaging datasets produces biased classifiers for computer-aided diagnosis. *Proceedings of the National Academy of Sciences of the United States of America, 117*, 12592–12594. https://doi.org/10.1073/pnas.1919012117

146. Knab, T. D., Clermont, G., & Parker, R. S. (2016). A "virtual patient" cohort and mathematical model of glucose dynamics in critical care. *IFAC-Papers, 49*, 1–7. https://doi.org/10.1016/j.ifacol.2016.12.094

147. Chase, J. G., Preiser, J.-C., Dickson, J. L., et al. (2018). Next-generation, personalised, model-based critical care medicine: A state-of-the art review of in silico virtual patient models, methods, and cohorts, and how to validation them. *Biomedical Engineering Online, 17*, 24. https://doi.org/10.1186/s12938-018-0455-y

148. Galbusera, F., Niemeyer, F., Seyfried, M., et al. (2018). Exploring the potential of generative adversarial networks for synthesizing radiological images of the spine to be used in in silico trials. *Frontiers in Bioengineering and Biotechnology, 6*, 53. https://doi.org/10.3389/fbioe.2018.00053

149. Lim, S. S., Kivitz, A. J., McKinnell, D., et al. (2017). Simulating clinical trial visits yields patient insights into study design and recruitment. *Patient Preference and Adherence, 11*, 1295–1307. https://doi.org/10.2147/PPA.S137416

150. Ktena, I., Wiles, O., Albuquerque, I., et al. (2023). *Generative models improve fairness of medical classifiers under distribution shifts*.

151. Webber, G., & Reader, A. J. (2024). Diffusion models for medical image reconstruction. *BJRArtificial Intelligence, 1*, ubae013. https://doi.org/10.1093/bjrai/ubae013

152. Mall, P. K., Singh, P. K., Srivastav, S., et al. (2023). A comprehensive review of deep neural networks for medical image processing: Recent developments and future opportunities. *Healthcare Analytics, 4*, 100216. https://doi.org/10.1016/j.health.2023.100216

153. Unified multi-modal image synthesis for missing modality imputation | IEEE Journals & Magazine | IEEE xplore. Retrieved March 4, 2025, from https://ieeexplore.ieee.org/document/10589432

154. Martyniak, S., Kaleta, J., Dall'Alba, D., et al. (2024). *SimuScope: Realistic endoscopic synthetic dataset generation through surgical simulation and diffusion models.* IEEE Xplore.

155. Yoon, J., Drumright, L. N., & van der Schaar, M. (2020). Anonymization through data synthesis using generative adversarial networks (ADS-GAN). *IEEE Journal of Biomedical and Health Informatics, 24*, 2378–2388. https://doi.org/10.1109/JBHI.2020.2980262

156. D'Amico, S., Dall'Olio, D., Sala, C., et al. (2023). Synthetic data generation by artificial intelligence to accelerate research and precision medicine in hematology. *JCO Clinical Cancer Informatics, 2023*, e2300021. https://doi.org/10.1200/CCI.23.00021

157. DuMont Schütte, A., Hetzel, J., Gatidis, S., et al. (2021). Overcoming barriers to data sharing with medical image generation: A comprehensive evaluation. *Npj Digital Medicine, 4*, 1–14. https://doi.org/10.1038/s41746-021-00507-3

158. Mikołajczyk, A., Majchrowska, S., & Carrasco Limeros, S. (2022). The (de)biasing effect of GAN-based augmentation methods on skin lesion images. In L. Wang, Q. Dou, P. T. Fletcher, et al. (Eds.), *Medical image computing and computer assisted intervention—MICCAI 2022* (pp. 437–447). Springer Nature Switzerland.

159. Li, T., Zeng, T., Zheng, Y., et al. (2025). *Synthetic poisoning attacks: The impact of poisoned MRI omage on U-Net brain tumor segmentation.* arXiv Preprint.

160. Friedrich, F., Brack, M., Struppek, L., et al. (2024). Auditing and instructing text-to-image generation models on fairness. *AI and Ethics.* https://doi.org/10.1007/s43681-024-00531-5

161. Zhou, M., Abhishek, V., Derdenger, T., et al. (2024). *Bias in generative AI.* arXiv Preprint.

162. Nicoletti, L. (2025). *Equality DBT+. Humans are biased. Generative AI is even worse.* Bloomberg.com.

163. Li, X., Chen, Z., Zhang, J. M., et al. (2025). Bias behind the wheel: Fairness testing of autonomous driving systems. *ACM Transactions on Software Engineering and Methodology, 34*, 82:1–82:24. https://doi.org/10.1145/3702989

164. Chen, T., Hirota, Y., Otani, M., et al. (2024). Would deep generative models amplify bias in future models? In *Proceedings of the IEEE/CVF Conference on Computer Vision and Pattern Recognition (CVPR)* (pp. 10833–10843).

165. Luccioni, S., Akiki, C., Mitchell, M., & Jernite, Y. (2023). Stable bias: Evaluating societal representations in diffusion models. In *Advances in neural information processing systems (NeurIPS).*

166. Kim, E., Kim, S., Rombach, R., et al. (2024). Exploring intrinsic fairness in stable diffusion. In *Advances in Neural Information Processing Systems (NeurIPS) Workshop.*

167. Usman Akbar, M., Wang, W., & Eklund, A. (2025). Beware of diffusion models for synthesizing medical images—A comparison with GANs in terms of memorizing brain MRI and chest x-ray images. *Machine Learning: Science and Technology, 6*, 015022. https://doi.org/10.1088/2632-2153/ad9a3a

168. Bergen, R. V., Rajotte, J.-F., Yousefirizi, F., et al. (2024). Assessing privacy leakage in synthetic 3-D PET imaging using transversal GAN. *Computer Methods and Programs in Biomedicine, 243*, 107910. https://doi.org/10.1016/j.cmpb.2023.107910

169. Shumailov, I., Shumaylov, Z., Zhao, Y., et al. (2024). AI models collapse when trained on recursively generated data. *Nature, 631*, 755–759. https://doi.org/10.1038/s41586-024-07566-y

170. Reinsel, D., Gantz, J., & Rydning, J. (2017). *Data age 2025: The evolution of data to life-critical.* International Data Corporation, sponsored by Seagate.

171. Vaswani, A., Shazeer, N., Parmar, N., et al. (2017). Attention is all you need. In *Advances in neural information processing systems (NeurIPS).* Curran Associates, Inc.

172. Mohsen, F., Ali, H., El Hajj, N., & Shah, Z. (2022). Artificial intelligence-based methods for fusion of electronic health records and imaging data. *Scientific Reports, 12*, 17981. https://doi.org/10.1038/s41598-022-22514-4

173. Jegorova, M., Kaul, C., Mayor, C., et al. (2022). Survey: Leakage and privacy at inference time. *IEEE Transactions on Pattern Analysis and Machine Intelligence*, 1–20. https://doi.org/10.1109/TPAMI.2022.3229593

174. Xu, W., Souly, N., & Brahma, P. P. (2021). Reliability of GAN generated data to train and validate perception systems for autonomous vehicles. In *Proceedings of the IEEE Winter Conference on Applications of Computer Vision Workshops (WACVW)* (pp. 171–180).

175. Fernandez, V., Sanchez, P., Pinaya, W. H. L., et al. (2023). Privacy distillation: Reducing re-identification risk of diffusion models. In *Deep Generative Models: Third MICCAI Workshop, DGM4MICCAI 2023, Held in Conjunction with MICCAI 2023, Vancouver, BC, Canada, October 8, 2023, Proceedings* (pp. 3–13). Springer-Verlag.

176. Brown, E. E., Guy, A. A., Holroyd, N. A., et al. (2024). Physics-informed deep generative learning for quantitative assessment of the retina. *Nature Communications, 15*, 6859. https://doi.org/10.1038/s41467-024-50911-y

177. Zhang, J., Yan, R., Perelli, A., et al. (2024). Phy-diff: Physics-guided hourglass diffusion model for diffusion MRI synthesis. In M. G. Linguraru, Q. Dou, A. Feragen, et al. (Eds.), *Medical image computing and computer assisted intervention (MICCAI)* (pp. 345–355). Springer Nature Switzerland.

178. Lüpke, S., Yeganeh, Y., Adeli, E., et al. (2024). *Physics-informed latent diffusion for multimodal brain MRI synthesis*.

179. Luccioni, S., Jernite, Y., & Strubell, E. (2024). Power hungry processing: Watts driving the cost of AI deployment? In *Proceedings of the 2024 ACM Conference on Fairness, Accountability, and Transparency* (pp. 85–99). Association for Computing Machinery.

180. Bashir, N., Donti, P., Cuff, J., et al. (2024). *The climate and sustainability implications of generative AI*. MIT Explor Gener AI.

181. Imec. (n.d.). *Machine learning accelerators*. imec. Retrieved March 7, 2025, from https://www.imec-int.com/en/expertise/cmos-advanced/compute/accelerators

Hybrid Intelligence: The Fusion of Science-Based and Machine Learning Models

Eleni Lavasa ⓘ, Theodora Chrysoula ⓘ, Vasileios Gkolemis ⓘ,
Charalambos Sinnis ⓘ, Theodore Dalamagas ⓘ, Sotiris Koussouris ⓘ,
Nefeli Bountouni ⓘ, Konstantinos Perakis ⓘ, Viktor Daropoulos ⓘ,
Stratos Keranidis ⓘ, Charalambos Lambri, Giorgos Ioannou ⓘ,
George Pallis ⓘ, Marios Dikaiakos ⓘ, Daniele Crippa, Andre Tabone ⓘ,
Dimitrios Bimpikas ⓘ, Carl Hans, Robert Hellbach ⓘ,
and Dimitris Bouras ⓘ

Abstract Process models and parameter estimation have long been fundamental tools in domains such as manufacturing, robotics, power plants, and biotechnology. Early modeling approaches relied on complex systems of mathematical differential

This project has received funding from the European Union's Horizon Research and Innovation Programme under Grant agreement no. 101135826.

E. Lavasa (✉) · T. Chrysoula · V. Gkolemis · C. Sinnis · T. Dalamagas
Athena Research Center, Marousi, Greece
e-mail: elavasa@athenarc.gr; theodora.chrysoula@athenarc.gr; vgkolemis@athenarc.gr;
charalampos.sinnis@athenarc.gr; dalamag@athenarc.gr

S. Koussouris · N. Bountouni
Suite5 Data Intelligence Solutions Ltd, Limassol, Cyprus
e-mail: sotiris@suite5.eu; nefeli@suite5.eu

K. Perakis · D. Bouras
UBITECH, Chalandri, Greece
e-mail: kperakis@ubitech.eu; dbouras@ubitech.eu

V. Daropoulos · S. Keranidis
domx IoT Technologies, Thessaloniki, Greece
e-mail: viktor@domx.io; stratos@domx.io

C. Lambri · G. Ioannou · G. Pallis · M. Dikaiakos
Department of Computer Science, University of Cyprus, Nicosia, Cyprus
e-mail: lambri.charalambos@ucy.ac.cy; ioannou.george@ucy.ac.cy; pallis.george@ucy.ac.cy;
dikaiakos.marios@ucy.ac.cy

D. Crippa
Consorzio Intellimech, Bergamo, Italy
e-mail: daniele.crippa@intellimech.it

E. Curry et al. (eds.), *Artificial Intelligence, Data and Robotics*,
https://doi.org/10.1007/978-3-032-10561-5_7

A. Tabone
MCS Data Labs, Berlin, Germany
e-mail: andre.tabone@mcs-datalabs.com

D. Bimpikas
Zenith Gas Supply Company, Thessaloniki, Greece
e-mail: d.bimpikas@zenith.gr

C. Hans
OHS Engineering GmbH, Bremen, Germany
e-mail: carl.hans@ohs-engineering.de

R. Hellbach
BIBA—Bremer Institut für Produktion und Logistik GmbH, Bremen, Germany
e-mail: hel@biba.uni-bremen.de

and algebraic equations, known as science-based models, to capture process knowledge and scientific expertise. These models leverage physical and chemical properties, static and dynamic behaviors, and causal relationships among observed quantities to support predictive control and operational optimization. As white-box models, they provide transparency by uncovering the inner logic and decision-making steps of the process. However, the advent of Industry 4.0 has brought a surge in available data from industrial processes, driving the rapid growth of machine learning (ML) models. Those models excel at discovering patterns and nonlinear relationships in data, but often they are black-box models, i.e., they lack interpretability. To combine the strengths of science-based and ML models, hybrid models have emerged as a powerful solution. By integrating the transparency and domain knowledge of science-based approaches with the adaptability and predictive capabilities of ML, hybrid models enhance accuracy, robustness, and scalability. This chapter explores the foundations of hybrid models, their development, and applications, providing a comprehensive perspective on their transformative potential across various scientific and engineering domains. This work is done in the context of the AI-DAPT EU project.

Keywords Science-guided ML · ML-assisted science · Physics-informed ML · Knowledge-guided ML (KGML) · Domain-informed ML

1 Introduction

The pursuit of understanding, predicting, and optimizing complex systems has historically relied on science-based (or first-principles) models, which encode domain expertise through differential equations, conservation laws, and causal relationships [1, 2]. These models provide transparency and interpretability, making them indispensable in fields like chemical engineering, robotics, and aerospace. However, their reliance on idealized assumptions or imperfect knowledge limits their ability to capture highly nonlinear, dynamic systems, while high computational costs and parameter calibration challenges further constrain scalability and operational applicability [3, 4].

The rise of Industry 4.0 and its effect on increasing data availability have driven the adoption of machine learning (ML) models, which excel at uncovering hidden patterns, adapting to evolving conditions, and approximating high-dimensional relationships [3, 5]. Applications of those models have transformed several scientific domains, e.g., materials science, healthcare, and climate modeling, to name a few, yet their black-box nature raises concerns about interpretability, generalizability, and compliance with physical constraints [5, 6].

To exploit the advantages and address the limitations of purely data-driven or purely science-based models, science-guided (or theory-guided, knowledge-guided, physics-informed, scientific) ML emerges as a new paradigm of hybrid intelligence that fuses theoretical understanding with machine learning [1–7]. Hybrid science-ML models may incorporate physics-based constraints in traditional ML algorithms to ensure predictions remain physically plausible, or ML components may be employed to augment science-based frameworks by learning residual behaviors, refining parameters in real time, and approximating computationally intractable phenomena [3, 5, 7]. This synergy mitigates ML's interpretability and causality limitations, as well as its reliance on extensive (labeled) training data, which may be scarce or sensitive, while addressing the computational cost and calibration challenges inherent in traditional first-principles models [3, 7, 8]. Hybrid models are proving essential in applications that span multiple scales such as bioprocessing and chemical engineering [8], multi-scale environments operating under large uncertainty such as climate modeling [9], and real-time decision-making systems as in biomedical devices and autonomous control [1].

This chapter explores the principles, methodologies, and applications of hybrid intelligence. Section 2 introduces science-based and data-driven modeling paradigms, outlining their advantages and limitations. Section 3 presents hybrid modeling strategies, in which we identify three broad families: science-guided ML refers to modeling approaches that integrate scientific principles and domain knowledge into ML models; ML-assisted science includes approaches that utilize ML to enhance the effectiveness of science-based models; and fully integrated hybrid systems involve strategies to incorporate both ML and science-based components into a concrete framework. Section 4 discusses real-world applications in energy, robotics, manufacturing, and health, demonstrating the benefits of hybrid approaches. Finally, Sect. 5 addresses challenges and future directions, focusing on scalable, interpretable, and domain-adjustable hybrid frameworks, and concluding remarks are drawn. The chapter aligns with the strategic objectives of the AI, Data, and Robotics Partnership, which emphasizes trustworthy, human-centric, and scientifically grounded AI innovation in Europe [10]. By integrating theory-driven knowledge with data-driven adaptability, hybrid intelligence unlocks new frontiers in predictive accuracy, robustness, and knowledge discovery across scientific and engineering domains.

2 Foundations of Modeling Paradigms

2.1 Science-Based Models

2.1.1 Scientific Simulators

Scientific simulators are computational methods designed to simulate physical processes or phenomena by integrating fundamental principles of real-world systems, such as physical laws, chemical reactions, or biological mechanisms. These models employ numerical methods to approximate solutions to complex problems, requiring input parameters that define initial conditions and interactions between system components.

Advancements in computational power and numerical algorithms have led to the development of sophisticated simulators across various scientific domains. Several well-known examples include computational fluid dynamics (CFD) models, which predict the behavior of liquid and gases based on the governing equations of mass, momentum, and energy [11, 12]. In climate modeling and weather forecasting, numerical weather prediction (NWP) models simulate atmospheric processes through systems of partial differential equations [13]. A simpler yet widely used example is the General Lake Model (GLM), a one-dimensional hydrodynamic model that simulates the water balance and vertical stratification of lakes [14].

A key advantage of scientific simulators is their ability to provide a transparent and interpretable representation of cause-effect relationships between input and output variables, ensuring that system behavior is well understood rather than inferred solely from data. Beyond relying on existing knowledge, simulators help advance scientific understanding by enabling researchers to refine hypotheses and fill gaps in understanding through the alignment of model outputs with real-world observations. Moreover, because simulators are built upon well-established mathematical equations, they can replace costly and time-consuming physical trials, allowing systematic exploration of different scenarios. This capability significantly contributes to the generalizability and trustworthiness of models, making them valuable tools for scientific exploration and engineering applications.

Despite their strength, scientific simulators face challenges that limit their real-world effectiveness. Governing equations may not fully capture physical processes due to incomplete knowledge or inherent stochasticity, leading to simplifying assumptions that introduce bias and uncertainty into predictions, particularly in extreme or previously unobserved conditions. Additionally, parameter estimation and calibration often require solving inverse problems, a highly computational process due to the need for numerous simulations. Poorly constrained parameters can further introduce bias, affecting the reliability of the model. These limitations highlight the potential for integrating data-driven approaches, paving the way for hybrid models that combine scientific knowledge with the flexibility of machine learning.

2.1.2   Process Modeling

Process modeling is a fundamental approach used across industries for describing, analyzing, optimizing, and controlling complex systems. Unlike scientific simulators, which aim for high-fidelity representations of physical and chemical phenomena, process models develop computationally efficient mathematical representations to enable monitoring, control, and decision-making in industrial applications. These models integrate scientific laws alongside empirical correlations and statistical methods to balance accuracy and computational efficiency. Reduced order equations, linearization, and steady-state approximations are often applied in the encoding of scientific principles and domain knowledge in order to reduce the computational cost of exact solutions, since the goal is not scientific research and discovery but operational optimization and control.

Several seminal works have shaped the field. Structured process systems analysis was introduced by [15], laying the groundwork for studying process dynamics and feedback control. This was expanded with real-time control, multivariable interactions, and industrial automation by [16], while [17] emphasized practical automation strategies and empirical parameter tuning. In recent years, the field has evolved to address the increasing complexity of industrial processes and the need for real-time optimization and process synthesis in large-scale systems [18].

Process modeling, from its early emphasis on dynamic system representation and control to the current focus on integrating advanced optimization techniques, is employed to address complex industrial challenges in various domains. In chemical engineering, process models improve catalytic reactor efficiency, such as in the Haber-Bosch process for ammonia production and methanol synthesis, by integrating reaction kinetics (Arrhenius law), mass and energy balances, and Computational Fluid Dynamics (CFD) [19]. In bio-industry, fermentation process models integrate metabolic network models, kinetic equations, and oxygen/pH control systems to optimize biopharmaceutical production, improving monoclonal antibody and insulin yields [20]. In aerospace, process models enhance gas turbine efficiency through thermodynamic cycle models (Brayton cycle), combustion simulations, and structural heat transfer analysis, reducing fuel consumption, emissions, and enabling predictive maintenance [21]. Across these fields, process models enhance efficiency, product quality, and operational stability, ensuring reliable monitoring and control.

Despite their advantages, traditional process models have inherent limitations. Many rely on simplifications and assumptions, such as linear approximations and steady-state conditions, which may fail to capture nonlinear, dynamic behaviors. Extensive experimental calibration is often required for parameter estimation, and these models struggle with scalability and robustness in complex, multi-scale systems [22]. These challenges motivate the shift toward hybrid modeling techniques that integrate science-based and ML-driven methods.

2.2 Data-Driven Models

2.2.1 Machine Learning and Deep Learning

Machine learning (ML) and deep learning (DL) are subsets of artificial intelligence (AI) that focus on developing algorithms and systems that can learn and infer from data. The main approaches of ML are [7, 23]:

Supervised learning is the process of training models on labelled data, where the proper outputs are known during training. The model learns to map inputs to outputs by minimizing errors, based on the features used for training. Common techniques include regression, which predicts continuous values, and classification, which categorizes data into discrete labels.

Unsupervised learning is the process of analyzing unlabeled data to identify patterns and relationships without predefined outcomes. Techniques such as clustering, association rule learning, density estimation, and dimensionality reduction help uncover underlying structures in the data.

Reinforcement learning is the process of training models (e.g., agents) to learn how to interact with an environment and create a series of choices to achieve a goal. Reinforcement learning is commonly used in robotics and autonomous systems.

Some of the algorithms commonly used in supervised settings include decision tree (DT), tree-based ensembles, and support vector machine (SVM). DTs are hierarchical models where data is split into branches based on feature values, forming an interpretable treelike structure. They recursively partition data into homogeneous subsets under a set of sequential if-then-else decision rules [24]. Tree-based ensembles train multiple DTs on different parts of the data using bootstrapping techniques, such as the random forest algorithm [25], or gradient boosting methods, as in XGBoost [26], with both approaches finding great success in various tasks due to their ability to handle complex relations while reducing overfitting. SVMs find the optimal hyperplane that maximizes the margin between different classes in a high-dimensional space [27]. By using kernel functions, SVMs can efficiently handle non-linearly separable data.

Neural networks are a foundational concept of ML, consisting of interconnected layers of nodes that process information through weighted connections. For solving complicated tasks, such as image classification or natural language processing (NLP), multiple layers are inserted to uncover complex relationships from the data. When multiple layers are included for understanding highly abstract patterns, the subset of AI is called DL.

The most common DL architectures include [3, 7, 28]:

- Artificial neural networks (ANNs), which are made up of multiple layers that transform input data through weighted connections and activation functions.

- Convolutional neural networks (CNNs) are used to process pictures and other data with a known gridlike architecture. They are essential for tasks like image classification and object detection.
- Recurrent neural networks (RNNs) are used to predict sequential data in applications such as time-series analysis.
- Transformers use approaches like attentiveness to determine the relative importance of various words or image components in a sequence. They have revolutionized fields like NLP (e.g., GPT).

ML and DL have contributed to the advancement and transformation of many industries [23, 29], offering solutions to complex problems. Their efficiency and scalability enable them to handle huge datasets while improving predictions as data volumes increase, making them invaluable in sectors like finance and healthcare. Also, their flexibility and adaptability help in applying the same models across various domains from medical diagnostics to autonomous vehicles while maintaining effectiveness. Other advantages include their performance and accuracy, particularly in DL, which allows for high precision in tasks like image recognition and NLP, often surpassing human-level performance.

However, they come not only with advantages but also with various challenges and limitations [2, 3, 23]. One major challenge is data dependency and bias, as these models rely on high-quality, diverse datasets. If training data is sparse, biased, or incomplete, the models may produce inaccurate or unfair results, making them unreliable for high-risk decisions. Additionally, ensuring that training data accurately reflects future data remains a challenge. Another limitation is the lack of interpretability and transparency, particularly in DL models, which function as "black boxes" with complex internal representations that make understanding their findings challenging. This raises concerns in critical areas such as legal systems, where explainability is essential. Furthermore, energy consumption and computational costs cannot be ignored. Training and deploying ML and DL models require significant computing power, leading to high operational costs and environmental impacts.

2.2.2 Probabilistic and Statistical Models

Probabilistic and statistical models play a crucial role in quantifying uncertainty and uncovering patterns in data. These models use probability distributions to represent relationships between variables, ranging from simpler models like linear regression (for modeling relationships between continuous variables) and logistic regression (for binary outcomes) to more complex approaches such as hidden Markov models (for sequential data), Gaussian mixture models (for data from multiple distributions), and Bayesian networks and probabilistic graphical models. These models are widely used in fields like economics, biology, and machine learning for tasks like prediction, classification, and inference. For further exploration, see [30–34].

Bayesian inference can be seen as the probabilistic equivalent of the training procedure in traditional machine learning. Initially, we have a parametric model but don't know the parameter values that best fit the data. After fitting the model, only specific parameter values align with the data. These values form the posterior distribution, which represents updated beliefs about the model parameters after observing real-world data.

The main advantage of statistical models, compared to traditional machine learning models, is their ability to predict with uncertainty. Bayesian inference allows for a principled way to incorporate uncertainty in the model parameters and make predictions that reflect this uncertainty. There are several methods for Bayesian inference. Below, we categorize them into three broad classes: exact inference methods, sampling methods, and advanced deep learning methods.

Exact inference methods compute the posterior distribution without introducing approximations. They work well in simpler settings with manageable computational complexity, such as Gaussian models and linear regression. However, they become computationally intractable as the model's complexity or dimensionality increases. Examples include variable elimination and sum-product networks [35].

Sampling methods like Markov Chain Monte Carlo (MCMC) and Importance Sampling are used when exact inference is impractical, especially in high-dimensional or complex models. MCMC generates samples from the posterior by constructing a Markov chain, with algorithms like Gibbs sampling and Metropolis-Hastings [30]. Importance sampling uses a proposal distribution to sample and reweight the samples to approximate the desired distribution. These methods scale better than exact inference in high-dimensional settings but still face challenges.

Advanced deep learning methods, such as Variational Inference (VI), Variational Autoencoders (VAEs), and Normalizing Flows, combine deep learning and probabilistic models for efficient inference in high-dimensional settings. VI reformulates Bayesian inference as an optimization problem by approximating the posterior with a simpler distribution, making it faster and scalable for high-dimensional problems [36]. VAEs, introduced by [37], use neural networks to learn latent variable models for efficient probabilistic generative modeling while maintaining uncertainty quantification. Normalizing Flows [38] enhance this by transforming simple distributions into complex ones through invertible functions, allowing for more expressive posterior approximations. These methods are particularly useful in tasks like anomaly detection, semi-supervised learning, and data compression. Their ability to model complex data distributions has revolutionized areas like natural language processing, computer vision, and bioinformatics [36, 37].

3 Hybrid Models: Bridging Science and Data

3.1 Integrating Domain Knowledge with ML Systems

Hybrid scientific ML has emerged as a powerful paradigm that combines the strengths of science-based modeling and data-driven ML, leveraging both domain expertise and statistical learning to enhance predictive accuracy, interpretability, and adaptability. Various strategies have been proposed to integrate domain knowledge into ML systems, which can be classified into three main categories (Fig. 1):

Science-guided ML—ML models remain the primary component but are augmented with scientific constraints, such as physics-based loss functions, regularization techniques, or embedded governing equations, ensuring physically consistent predictions.

ML-assisted science—Science-based models serve as the primary component, with ML modules assisting in parameter calibration, uncertainty quantification, or acceleration of numerical solvers, improving efficiency and adaptability.

Fully integrated hybrid systems—These approaches combine both science-based and data-driven components in a tightly coupled manner, where ML and mechanistic models operate in an orchestrated framework, exchanging information dynamically.

This section explores key methodologies and applications within each of these hybrid modeling approaches, illustrating how the fusion of theoretical principles and data-driven learning is shaping the future of intelligent systems.

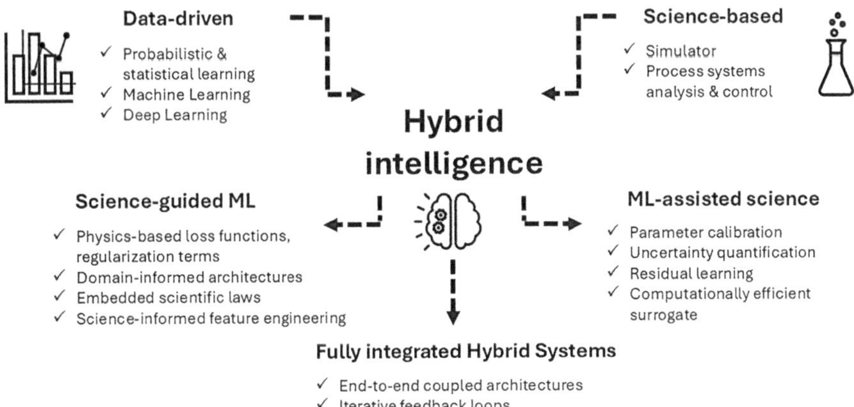

Fig. 1 Conceptual landscape of hybrid intelligence as the fusion of domain knowledge and data-driven learning, summarizing the three hybrid modeling categories—science-guided ML, ML-assisted science, and fully integrated hybrid systems

3.2   Science-Guided ML

Machine learning algorithms have been employed across a wide array of domains, including bioinformatics, cheminformatics, computer networks, DNA sequence classification, economics, and banking, among others [39]. This widespread application indicates that these models are utilized in areas where substantial prior knowledge has already been established. Consequently, incorporating domain-specific knowledge, often referred to as prior knowledge, into machine learning models has the potential to enhance their overall effectiveness for tasks relevant to each specific field of expertise.

Incorporating domain knowledge into ML models offers significant advantages, particularly in scenarios where data is scarce, costly to obtain, or of poor quality, as it enhances model efficiency [40]. This approach typically reduces the volume of required data compared to standard methods that rely on non-specific optimization while also improving generalization ability by limiting overall variability [41]. Another important result is the increased interpretability of the models, which is crucial for domains where the justification of the decision-making process is important such as the healthcare industry. Furthermore, few-shot learning algorithms can swiftly generalize to new tasks with only a few supervised samples when prior knowledge is integrated, enhancing their performance in data-limited contexts [42].

The incorporation of prior knowledge into ML models can be achieved through various methods, with three principal integration strategies identified in the literature [41, 43]. The first type of approach emphasizes data, wherein an informed feature extraction process may be employed based on the task at hand and any additional knowledge available about the dataset. The second methodology focuses on the incorporation of prior knowledge through the regularization of loss functions. Finally, domain-specific model architectures can be leveraged to embed existing knowledge directly into the models.

3.2.1   Science-Informed Feature Engineering

Feature engineering is a critical step in ML, where raw data is transformed into meaningful representations that enhance model performance. In traditional ML, feature selection and extraction are often data driven, relying on statistical methods or deep learning architectures to discover relevant patterns [44]. However, in scientific and engineering applications, integrating domain knowledge into feature generation significantly enhances model interpretability, generalizability, and physical consistency [2, 5]. By embedding physical, biological, or clinical constraints into feature representations, science-guided feature engineering addresses common limitations of purely data-driven approaches, such as the lack of explainability, and non-physical extrapolations. This strategy has proven particularly effective across diverse domains:

Protein engineering and bioinformatics: The use of domain-guided feature generation has advanced protein engineering, where ML models predict protein function and stability based on sequence and structural properties. A machine learning-guided protein engineering approach leverages biophysical knowledge, incorporating thermodynamic stability, hydrophobicity indices, and active-site constraints to optimize biocatalyst design and drug discovery [45]. Similarly, BioAutoML automates feature extraction from biological sequence data, embedding functional genomics knowledge to optimize predictions in bioinformatics and drug discovery [46].

Space weather prediction and solar physics: Science-driven feature extraction has been used to improve solar flare forecasting with ML in Space Weather operations. A set of magnetic field properties, derived from vector magnetograms by spatial averaging and integration, have been employed for the prediction task [47]. These physics-based features have been combined with novel shape-based descriptors, derived from topological data analysis of solar active regions, to improve prediction accuracy. By incorporating physics-informed feature sets, the model not only improved flare forecasts but also provided more interpretable insights into solar dynamics [48].

Industrial metrology and quality control: Physics-based feature engineering has been applied to point cloud measurements produced by a 3D laser scanner, for the prediction of the point-wise accuracy of the measurement, for a given surface and laser orientation. Driven by basic optics and geometrical principles, informative features were generated to encode the incident and reflection angles of the laser light onto the different parts of a surface. The use of physics-based features has enhanced both prediction accuracy and interpretability, enhancing the efficiency of precision measurement and industrial quality control [49].

These examples illustrate how science-guided feature engineering bridges the gap between domain knowledge and ML adaptability, improving interpretability and predictive power.

3.2.2   Domain-Specific Regularization and Embedding of Physical Laws

The regularization of the loss function facilitates the incorporation of prior knowledge from various domains and tasks by introducing terms that explicitly favor or penalize solutions in alignment with this knowledge [41]. These additional regularization terms can encompass tasks related to logical operations [43] or constraints expressed as algebraic or differential equations, probabilistic relations, or knowledge graphs [50]. This allows for the integration of domain expertise, including physical models, mathematical constraints, scientific laws, and relationships among features [40].

A notable example is physics-informed neural networks (PINNs), which embed differential equations representing physical laws into the loss function, serving as constraining factors for the solutions generated by the neural network. PINNs were introduced in 2017 [51, 52] as a way of using neural networks to solve problems incorporating nonlinear partial differential equations. The use of PINNs, their variants and applicability, has exploded since the time they were first introduced.

PINNs were introduced as a way to solve and discover differential equations, employing neural networks and automatic differentiation [53], by enforcing the governing equation that describes the problem into the loss function. Specifically, given a differential equation, the network's weights are optimized such that the model's output approximates the solution of the ODE or PDE. This is achieved by crafting a cost function consisting of different terms, each one trying to guide the network toward physically consistent solutions, specifically, the residual term that ensures that the network's output satisfies the governing equation, the initial and boundary conditions terms, as well as the data term that depends on the error with observed data, if available. Each term in the cost function is weighted to control its contribution to the total loss. The estimation of the weights used in the cost function has been the research topic of many works, such as the one of [54] where the authors employed a Neural Tangent Kernel for weight calibration. Most solutions using PINN have utilized feedforward neural networks for function approximation; however, some researchers have also explored the use of convolutional neural networks (CNNs), recurrent neural networks (RNNs), and generative adversarial networks (GANs) [55].

The range of use cases in which PINNs have been evaluated is quite broad. For instance, in the foundational works of [51, 52], the PINNs were tested on the Burgers', Schrödinger, Allen-Cahn, and Navier-Stokes equations for forward and inverse problem-solving. The first case corresponds to approximating the solution of the differential equation given known initial/boundary conditions along with the parameters describing the system, whereas the latter case falls in the category of system identification where data is used to infer the parameters governing the physical process. Since then, PINNs have also been applied for stochastic PDEs [56] and fractional PDEs [57] and have also been incorporated into domain decomposition settings [58, 59]. Moreover, advancements have been made in optimizing PINNs since often, the terms within the loss function lead to conflicting updates. The ConFig [60] framework tackles this issue by ensuring a positive dot product between the final update and each loss-specific gradient. Although PINNs have achieved significant success, they continue to face issues and challenges. For instance, the research conducted in [61] demonstrated that the framework still underperforms when compared to the finite element method (FEM). Challenges still remain in a number of categories related to optimization [62], scalability, generalization and predictive ability, as well as in scenarios with scarce and noisy data [63].

3.2.3 Domain-Informed Architectures

The integration of domain knowledge directly into the architecture of hybrid ML models, which combine science-based models and machine learning, is not merely beneficial but essential for achieving highly accurate and performant pipelines. This is because domain knowledge provides crucial context, constraints, and interpretability that are often missing when relying solely on data-driven approaches [64]. Domain knowledge serves as a bridge between the abstract world of data and the concrete reality of the problem being addressed [65]. It encompasses established theories, empirical relationships, known constraints, and expert intuition about the system's behavior. By explicitly incorporating this knowledge into the model's structure, we can guide the learning process, reduce the reliance on vast amounts of training data, and improve the model's ability to generalize to unseen situations. This is particularly important in domains where data is scarce, expensive to acquire, or subject to significant noise and uncertainty. Furthermore, domain knowledge enhances the interpretability of hybrid AI models. Instead of being a "black box" that simply produces predictions, such as most AI models, hybrid models, when structured wisely by incorporating domain knowledge, can offer insights into why a particular outcome is expected, providing domain experts with valuable explanations grounded in scientific principles, which they can understand and follow.

The "one-size-fits-all" principle in the architectural tasks, often employed in traditional machine learning, is demonstrably inapplicable when constructing hybrid ML models. Instead, such an architecture must be carefully tailored to the specific problem at hand, reflecting the unique characteristics of the domain and the underlying scientific principles involved. This necessitates a deep understanding of both the scientific domain and the capabilities and limitations of various machine learning techniques. For example, Physics-informed neural networks (PINNs) embed prior knowledge and physics laws directly into their architecture, so that the network can by design respect the physics constraints [55]. This can be achieved by incorporating symmetry and invariance principles (such as symmetry constraints or equivariance), utilizing domain-specific network layers (e.g., RNNs for time-dependent problems or GNNs for geometric problems), and incorporating physical constraints (such as physical variables) into the network structures.

Other approaches involve constraining the model's search space or guiding the selection of appropriate ML algorithms. Ultimately, the choice of architecture depends on the specific problem and the available domain knowledge, while as seen before, there is always the chance of architecture refinements, as the changing conditions in which a model operates might bring this need to the surface. For example, DeepMind's AlphaFold was initially based on a deep learning model but later integrated structural biology principles into its architecture to improve protein-folding predictions [66].

3.3 ML-Assisted Science

Scientific simulators are indispensable tools in studying and predicting complex physical and chemical systems; however, their usually high computational complexity and incorporation of imperfect scientific knowledge limit operational use. ML methods can be applied to address such shortcomings and improve the performance and efficiency of science-based models. We identify four main directions to that end. First, surrogate ML models can be trained to mimic the behavior of scientific simulators while operating in a small fraction of the simulator's computational cost and execution time. Another strategy is to refine the science-based model using an ML module that enables the dynamic update of model parameters, learns from the residual errors in model predictions, or discovers more accurate forms of the governing equations from data. A third direction involves the employment of ML models to infer free parameter distributions in scientific simulators, addressing thus a major issue, which is parameter calibration. Finally, the use of ML models enables uncertainty estimation in the predictions of scientific simulators, a property highly desired in several scientific and engineering domains that operate under large uncertainties.

3.3.1 Surrogate Models

Surrogate models, also known as metamodels, response surfaces, or emulators, are simplified approximations of more complex, higher-order models [67]. These models are designed to emulate the input-output relationships of the original system when the actual relationship between the two is unknown or computationally expensive to evaluate, thus significantly reducing computational costs [68]. Surrogate models have emerged as a pivotal tool in computational science and engineering and are particularly useful in scenarios where direct measurement or computation of an outcome is challenging or where repeated evaluations of the original model are required, such as optimization, uncertainty quantification, and sensitivity analysis. Surrogate models can also be constructed for use in surrogate-based optimization when a closed analytical form of the relationship between input data and output data does not exist or is not conducive for use in traditional gradient based optimization methods [69].

Surrogate models used in scientific research range from simple, commonly used techniques to advanced and less commonly employed methods. The most commonly used surrogate models include, yet are not limited to, (1) polynomial regression (response surface models), (2) radial basis functions (RBF), (3) Gaussian process regression (Kriging), (4) artificial neural networks (ANNs), (5) deep neural networks (DNNs), (6) support vector machines (SVMs), and (7) Bayesian neural networks (BNNs) [70].

The utilization of surrogate models demonstrates numerous advantages over traditional simulation methods. Simpler surrogate models, e.g., linear or low-order polynomial regressors, decision trees, etc., are, in their nature, intuitive and straightforward and are relatively easy to implement and explain, making them accessible to individuals without extensive data science or ML backgrounds while approximating predictions from both complex "white-box" models as well as from "black-box" models [71]. Surrogate models in general also demonstrate increased flexibility in choosing statistical frameworks that better suit the modelers or better address specific business needs, ranging from decision trees and linear regression models to more complex ones, allowing for adaptability in various scenarios [72]. In addition, surrogate models demonstrate increased scalability and can be applied to problems with high-dimensional input spaces, while recent advancements in ML, such as deep neural networks and Gaussian processes (GPs), have further enhanced their scalability potential and have empowered them to handle complex, high-dimensional problems [73]. Moreover, surrogate models seamlessly integrate with hybrid intelligence systems, where they complement physics-based models with data-driven insights, enhancing predictive accuracy and robustness [74].

Nevertheless, indisputably, the primary advantage of surrogate models is their ability to drastically reduce computational costs in terms of time and resources consumed and thus drastically improve computational efficiency [75], acting as low-cost proxies for high-fidelity models. By mimicking the behavior of simulation models at a fraction of the computational cost, decoupling computational intensity from the number of evaluations required, they enable routine tasks that would otherwise be impossible, making them ideal for iterative processes such as multi-objective optimization and uncertainty quantification [76]. For example, by replacing expensive high-fidelity simulations with fast approximations, they enable the exploration of large parameter spaces and the execution of iterative processes that would otherwise be infeasible [67].

Despite their significant advantages, surrogate models come with their limitations, including dependence on training data, difficulty in capturing complex behaviors, and lack of interpretability, which must be carefully managed to ensure reliable performance. A surrogate model is only as good as the model it is trained on, as it doesn't have access to actual observations, which means that the surrogate models inherit the limitations of the original models. In addition, surrogate models draw conclusions about the original model and not about the underlying data, while regarding their accuracy, it is highly dependent on the quality and quantity of the training data, which means that poorly sampled or insufficient data can lead to inaccurate predictions. Thus, surrogate models may struggle to capture complex, non-linear behaviors present in the original systems [77], which is particularly challenging in systems with discontinuities or sharp gradients [78]. Additionally, the construction of surrogate models often requires significant upfront computational effort to generate training data, which can offset some of their efficiency gains especially in cases where the original model is extremely expensive to evaluate [79]. Surrogate models are typically domain specific and may not generalize well to new problems without retraining, which limits their applicability in interdisciplinary research [80]. Moreover, surrogate models, particularly those based on ML, are prone to overfitting, especially when the training data is limited. They also may not adequately quantify uncertainty [81, 82], while many surrogate models, especially those based on deep learning, lack interpretability, making it difficult to understand the underlying physical processes [83].

Nevertheless, because of their significant benefits, and despite their limitations, surrogate models have found extensive applications across a wide range of scientific and engineering disciplines, enabling researchers to tackle complex problems with reduced computational burden and bridging the gap between science-based models and data-driven approaches [84]. In example, among many others, in aerospace engineering, surrogate models are used to solve the optimal power flow (OPF) problem [85], while in mechanical engineering, surrogate models have been used to model compressor characteristics for predicting compressor efficiency and pressure ratio [86]. In civil engineering, surrogate models have been used to assess the structural integrity of buildings and bridges under extreme conditions, including for real-time seismic risk assessment [87], while in energy systems, surrogate models have been used to optimize the layout of wind farms [88]. In climate and environmental science, surrogate models have been used to emulate earth system models, enabling researchers to conduct large-scale climate simulations with reduced computational costs [89], while in biomedical research, surrogate models have been used for the parameter identification problem in probabilistic cellular automaton epidemic models [90].

In summary, surrogate models enhance computational efficiency by providing rapid evaluations, reducing costs, offering flexibility in modeling approaches, facilitating design exploration, supporting iterative improvements, and enabling multi-objective optimization. These advantages make them a powerful tool in engineering design and analysis compared to traditional simulation methods.

3.3.2 Data-Driven Refinements of Science-Based Models

First-principle or science-based models provide reliable foundations based on established scientific knowledge, while ML components can adapt to complex patterns and uncertainties not captured by theoretical frameworks [91]. This section explores strategies for refining science-based models through ML, with particular focus on approaches that enable dynamic parameter updates, residual error learning, and equation discovery.

The residual modeling approach represents one of the most effective strategies for enhancing first-principle models. This hybrid technique first employs an established model, such as an empirical correlation or physics-based equation, to compute an initial estimate of the target output. The ML component then focuses on predicting the residual (the difference between the base model's prediction and actual experimental values) [3]. In a study on critical heat flux (CHF) prediction in nuclear reactors, researchers follow this approach by first using physical models to generate initial predictions. Afterward, they trained various ML models to predict the residuals between these model outputs and experimental values. The final prediction was obtained by adding the ML-predicted residual to the base model's output [92]. The residual modeling approach essentially provides a correction mechanism that preserves valuable insights from domain knowledge while leveraging data to improve precision [8].

ML models can dynamically update the parameters of science-based models in real time. This continuous learning from incoming data ensures that the models remain accurate and relevant, even as conditions change [93]. In a study on tool wear prediction in manufacturing processes, researchers implemented a dynamic parameter update approach by first using established physics-based models to generate initial predictions of tool wear. Subsequently, they employed ML models to continuously update the parameters of the physics-based models based on real-time sensor data [94].

By employing these techniques, science-based models not only become more accurate but also more robust and capable of handling the complexities of real-world applications.

3.3.3 Simulation-Based Inference and Parameter Estimation

Simulation-based inference (SBI) is a fundamental approach for testing scientific theories and studying complex natural phenomena. Central to this approach is the scientific simulator—a computer program that models a natural system by taking a set of input parameters, applying a sequence of deterministic or probabilistic steps, and producing an observable output.

Due to the probabilistic nature of the simulator, for a particular set of input parameters, the output is not deterministic; it is a sample from a distribution of possible outputs. SBI focuses on the reverse question: given a specific output, what set of parameters could have produced it? By applying SBI to real-world observational data, researchers can identify the posterior distribution, i.e., the plausible parameter values that explain the observed phenomena [95].

Unlike opaque ML models, scientific simulators offer transparency. This happens because they encode the natural processes into a series of transparent steps. After identifying the posterior distribution, researchers can validate or challenge scientific theories [96].

Two principal challenges characterize this domain. The first is the development of accurate simulators. Crafting a simulator is difficult as it requires deep domain expertise to translate intricate natural phenomena into a computational model. Furthermore, this process does not scale, as this modelling part should be repeated from scratch for each application [97].

The second challenge in simulation-based inference is the estimation of model parameters. Estimating the parameters that produce simulated outputs closely matching observed data can be difficult due to the computational expense of simulators and the often-intractable likelihood function.

Recently, significant research has focused on addressing this challenge [98], developing both methodologies [99–102] and software tools [103–105]. Lately, many neural likelihood approaches have been introduced, which have shown promising improvements in both the speed and accuracy of posterior estimation [106, 107].

In summary, simulation-based inference and parameter estimation represent a dynamic and evolving field. By integrating detailed parametric models with innovative inference techniques, researchers are now better equipped to unravel the complexities of natural phenomena and enhance the predictive power of scientific theories.

3.3.4 Uncertainty Quantification

Hybrid modeling, which integrates physical principles with data-driven adaptability, has emerged as a powerful approach for addressing complex systems. However, a critical challenge lies in uncertainty quantification (UQ), which assesses the reliability and confidence of predictions. Traditional numerical simulations often rely on deterministic assumptions, failing to capture the stochastic nature of real-world systems. UQ addresses this limitation by quantifying uncertainties from various sources, including uncertain input parameters, model approximations, and probabilistic input output relationships, in both forward and inverse problems.

Uncertainty arises from two primary sources:

- Aleatoric uncertainty (data uncertainty) stems from inherent noise or limitations in the data. While it cannot be reduced by collecting more data, it can often be explicitly modeled.
- Epistemic uncertainty (model uncertainty) results from limitations in the model, such as insufficient training data or errors in approximating physical laws. This type of uncertainty can be reduced by improving the model or acquiring more data.

To address these uncertainties, several advanced techniques have been developed. Bayesian neural networks (BNNs) model parameter uncertainty by placing probability distributions over weights, enabling uncertainty-aware predictions [108]. Variational inference (VI) offers a scalable alternative to exact Bayesian inference, approximating posterior distributions to reduce computational costs [109]. Recent advancements, such as Bayesian physics-informed neural networks (B-PINNs), combine BNNs with equations (PDEs) while quantifying aleatoric uncertainty, noise in data [110]. In addition to deep learning-based approaches, Gaussian processes (GPs) offer a non-parametric Bayesian framework for uncertainty estimation, particularly useful in low-data regimes. However, their scalability remains a challenge for high-dimensional hybrid models [111, 112].

Generative adversarial networks (GANs) [113] are a powerful approach to solve stochastic PDEs in high dimensions; by generating multiple outputs for the same input, GANs enable variability analysis, which reflects the model's confidence in its predictions. In [114], physics-informed generative adversarial networks (PI-GANs) embed stochastic PDEs into GAN architectures, addressing both forward and inverse stochastic problems while overcoming the curse of dimensionality.

For computationally efficient UQ, Monte Carlo Dropout (MC Dropout) approximates Bayesian inference by randomly deactivating neurons during training and inference, generating a distribution of outcomes [115]. Ensemble learning, which trains multiple neural networks independently, estimates epistemic uncertainty by analyzing prediction variance [116]. This is achieved through random initialization or perturbing initial conditions. Common applications include climate modeling, weather forecasting, and engineering, where ensemble methods quantify epistemic uncertainty in forward problems [117–119].

Incorporating UQ into hybrid models enhances the reliability of predictions by balancing physics-based constraints with data-driven learning. This integration improves interpretability, robustness, and trust in ML-assisted scientific discovery and decision-making.

3.4 Fully Integrated Hybrid Systems

3.4.1 Coupled Models and End-to-End Hybrid Pipelines

The interplay between ML and scientific disciplines has been discussed in the previous sections, highlighting instances where each enhances the other's capabilities and outputs. Recently, the concept of fully integrated hybrid systems (FIHS) has emerged as a prominent research area. FIHS moves beyond treating ML and science-based models as distinct entities, instead fostering deep integration where scientific reasoning and AI methodologies are interconnected at a fundamental level. This synergistic approach capitalizes on the strengths of both paradigms, leading to improvements in efficiency, interpretability, adaptability, and robustness across diverse domains. Realizing the full potential of FIHS necessitates a comprehensive understanding of the logical underpinnings and requirements of both model types by data scientists. Furthermore, careful orchestration of their execution is crucial to achieve the anticipated performance gains.

Instead of employing sequential or loosely coupled modules, hybrid pipelines should be architected to enable seamless collaboration between models, facilitating the delivery of required outputs for various tasks. Techniques within FIHS are interwoven, promoting real-time information sharing and collaboration. This departure from traditional sequential processing enables complex problem-solving and a more holistic approach to intelligence provision. Consequently, these pipelines can be conceptualized as complex directed acyclic graphs (DAGs) rather than simple workflows.

While the fundamental principles for constructing FIHS pipelines align with those of mainstream ML pipelines, encompassing data acquisition, preprocessing, model integration, testing, and deployment, the key distinction lies in the model engineering and integration phase. This phase requires careful consideration of how scientific and ML models will interact, ultimately leading to the creation of a hybrid model that drives the pipeline's intelligence generation.

The applicability of FIHS spans a wide range of domains. In personalized medicine, ML systems can analyze patient data and medical history to suggest optimal treatment options while providing explanations grounded in scientific principles derived from fields like medicine, biology, and chemistry. In advanced robotics, FIHS empowers machines to interact with the world in a more humanlike manner by integrating perception, planning, and reasoning capabilities. Similarly, in autonomous systems, such as unmanned aerial vehicles (drones), FIHS facilitates navigation in complex environments and real-time decision-making by combining AI for object recognition with knowledge of complex fluid dynamics (e.g., aerodynamics, thermodynamics) for precise maneuvering.

3.4.2 Iterative Feedback Loops for Continuous Model Refinement

The optimization and refinement of integrated ML and science-based models within FIHS can be significantly enhanced through the implementation of iterative feedback loops. These loops, embedded within the FIHS pipeline logic, facilitate continuous monitoring of system performance and enable both automated adjustments and human-guided refinements to achieve optimal outcomes. Functionally, these loops contribute to error correction, parameter refinement, and uncertainty reduction by systematically identifying and reconciling discrepancies between the predictions of the ML model, the science-based model, and real-world observations. This reconciliation process often involves statistical methods to quantify the significance of discrepancies and guide adjustments. Furthermore, these feedback mechanisms enable adaptive learning, allowing the FIHS to dynamically adjust to changing operational conditions and assimilate new data streams, thereby maintaining predictive accuracy and relevance over time. Critically, these loops also enhance the explainability of the hybrid model's output by analyzing the complex interactions between the ML and science-based components, revealing how each model influences the other and contributing to a deeper understanding of the underlying phenomena. This enhanced explainability is crucial for building trust in the FIHS and for extracting valuable scientific insights from its operation.

Contrary to the misconception that feedback loops primarily benefit ML models, they are equally crucial for optimizing science-based models within the FIHS framework. In the context of ML model refinement, optimization strategies typically involve hyperparameter fine-tuning, feature engineering, and potential architectural modifications guided by feedback indicating the potential for superior performance with alternative model structures. For science-based models, feedback loops facilitate calibration of model parameters to better align with the specific conditions of the operational environment. This calibration may involve adjusting parameters related to boundary conditions, material properties, or other factors influencing the model's behavior. Additionally, feedback can inform modifications to the model's structure or governing equations that describe the underlying phenomena, ensuring the model accurately captures the relevant physics or chemistry. Moreover, feedback loops play a vital role in identifying the inherent limitations of the science-based model, defining the boundaries beyond which its equations cannot provide trustworthy or reliable results. This understanding of model limitations is essential for ensuring the responsible application of the FIHS and for avoiding erroneous conclusions based on extrapolations beyond the model's domain of validity.

As previously discussed, these feedback loops can trigger automated actions to enhance model performance. However, they can also generate notifications to human operators, suggesting specific actions for manual intervention. Even in cases where automated actions are implemented, human oversight and pre-approval are essential. This is particularly critical in AI applications categorized as high-risk under regulations such as the EU AI Act, where regulatory constraints mandate human control over automated decision-making processes. Furthermore, even in minimal-risk applications where human oversight is not strictly required, automated actions should be governed by pre-defined thresholds and limits established by human experts. This precautionary approach mitigates the risk of model hallucination, where the AI component generates erroneous or nonsensical outputs, potentially leading to flawed decisions and resource wastage. By maintaining human oversight of the feedback loop and automated refinement processes, FIHS can be effectively managed to ensure both accuracy and responsible application across a range of risk categories. This human-in-the-loop approach allows for the combination of the computational power of AI with the expertise and judgment of human scientists and engineers.

4 Applications

4.1 Energy

The efficient management of energy consumption in residential buildings requires accurate modeling of their thermal dynamics. A widely used physics-based approach is the resistance-capacitance (RC) thermal model, which represents heat transfer and storage through electrical circuit analogies. Thermal resistance characterizes how materials impede heat transfer (e.g., insulation properties of walls and windows), while thermal capacitance describes a material's ability to store thermal energy. These models are typically formulated using state-space equations or ordinary differential equations (ODEs) and serve as the foundation for energy control systems. However, they often struggle with capturing nonlinear thermal dynamics, transient effects, and varying environmental conditions. To address these challenges, hybrid models integrate ML components to improve predictive accuracy. One approach is black-box residual learning, where an ML model corrects the errors of an RC-based predictor, refining its estimates. Another method leverages PINNs, which enforce the governing ODEs of RC models as constraints within an ML framework. For instance, [120] introduced an RC-PINN framework for predicting room temperature, ensuring physically consistent and data-driven predictions. Similarly, a physics-informed ARMAX (Autoregressive-Moving-Average with Exogenous Inputs) model has been shown to outperform purely ML-based models in terms of both prediction accuracy and computational efficiency [121]. In the work of [122], a PINN was used to predict the water heating load of a heat pump, outperforming traditional neural networks.

Model Predictive Control (MPC) further enhances the efficiency of hybrid thermal models by continuously adjusting heating and cooling operations based on real-time predictions. Unlike conventional rule-based control, MPC anticipates future states by solving an optimization problem over a finite time horizon, selecting the most efficient energy usage strategy while maintaining comfort constraints. When combined with hybrid thermal models, MPC can dynamically adapt to changing weather conditions, occupancy patterns, and insulation performance, improving energy efficiency in real-world applications. A novel hybrid approach integrates stability and dissipativity constraints into the learned system dynamics by restricting the eigenvalues of the learned matrices describing the system dynamics, ensuring that AI-augmented models remain stable and physically realistic [123]. Additionally, ML-enhanced MPC frameworks incorporate self-learning mechanisms, allowing control strategies to evolve based on historical performance and environmental changes.

Future directions in hybrid intelligence for building energy management include extending these models to integrate energy market price predictions. Inspired by financial models, hybrid AI can analyze daily fluctuations in electricity prices, linking them to historical market trends, demand forecasts, and supply variations [124]. Advanced ML techniques can be used to capture nonlinear interactions between energy consumption, weather patterns, and energy stock market dynamics. By integrating these market-aware forecasts into MPC-driven energy control systems, buildings can optimize heating and cooling operations not only based on physical constraints but also on economic efficiency, scheduling energy-intensive activities during cost-effective periods. This hybrid approach could lead to more responsive, cost-efficient, and grid-supportive demand-side management solutions.

4.2 Robotics

Applied cognitive ergonomics in robotics is an interdisciplinary field focused on optimizing human-robot interaction (HRI) by considering users' mental processes, such as perception, memory, attention, and decision-making. This approach is essential for designing robotic systems that not only assist with physical tasks but also interact with humans intuitively and naturally, reducing cognitive load and enhancing operational efficiency. In industrial settings, collaborative robotics ("cobotics") enables direct human-robot interaction, overcoming traditional labor separation. Cobots share workspaces with humans, enhancing production flexibility and cognitive ergonomics by reducing stress and mental workload [125].

A key research area within cognitive ergonomics is the analysis of operators' mental states, which requires a multidisciplinary approach, integrating engineering, psychology, and cognitive sciences. Various studies demonstrate that physiological signals can be leveraged to estimate mental states, with different applications depending on the context [126–134]. Correlations have been identified between stress levels and physiological indicators, such as heart activity (ECG, BVP), electrodermal activity (EDA), respiration, and body temperature. On the other hand, high cognitive load and inattention are often linked to brain signals (EEG) and ocular activity (EOG). External factors such as temperature, humidity, and noise levels in industrial environments can significantly impact cognitive load, making robust feature extraction and interpretation essential.

Addressing mental state estimation using ML requires careful consideration of the complexity of physiological signals, as these signals can also indicate unrelated medical conditions. This limits the applicability of unsupervised learning strategies, such as anomaly detection, for mental state identification. As a result, literature primarily proposes supervised classification-based approaches, which require proper data labeling [126–134], often conducted via subjective questionnaires administered to operators. Traditional approaches employ ML techniques with signal processing-based feature extraction to maximize the relevance of specific physiological markers. Depending on the target phenomenon, different preprocessing transformations are applied, such as frequency filtering for power band extraction (e.g., alpha waves from EEG), decomposition techniques to remove noise (e.g., SLC/SCR processing of EDA), or spectrogram-based imaging for multivariate signal representation (e.g., multichannel ECG) [126, 127]. Alternative methods use ML models directly on raw signals without predefined assumptions about how mental state information is encoded, reducing the need for extensive feature engineering but requiring larger datasets and longer training times [133].

To improve efficiency and interpretability, hybrid ML models integrate domain knowledge into deep learning architectures, refining both feature extraction and classification. A simple approach involves informed data labeling, where instead of relying solely on subjective questionnaires, expert-driven assessments provide additional training labels. More structured models apply sequential feature extraction before ML-based classification, leveraging domain knowledge to simplify learning and improve pattern recognition [126]. Advanced architectures embed traditional feature extraction directly within deep learning models, combining CNN-based image processing for EEG/ECG with parallel 1D-convolutional or recurrent layers (LSTMs) for temporal feature modeling [133]. More comprehensive approaches first apply science-guided preprocessing to filter and transform signals, followed by hybrid CNN-LSTM architectures, which refine predictions by incorporating expert-informed constraints and task-specific priors [134].

Future advancements in hybrid ML applications in cognitive ergonomics will focus on scalable, multimodal systems capable of integrating heterogeneous sensor data, analyzing the combined effects of cognitive and physical factors, and adapting to task-specific demands in real-world industrial settings. The challenges of signal artifacts caused by physical movement, sensor placement, and minimizing the number of wearable devices while ensuring high-quality data acquisition require further exploration.

4.3 Manufacturing

Hybrid ML approaches combine physical/scientific models with machine learning methods and are applied in production and logistics from different perspectives. In most cases, such projects, initiatives, or products are found in the context of "Condition Monitoring," "Predictive Maintenance," or "Digital Twins."

Although not exclusively related to the manufacturing sector, the most prominent application can be found in the aviation sector, particularly in the monitoring of jet engines. Rolls-Royce launched their "IntelligentEngine" project in 2018 [135]. It aims to transform aircraft engine development by incorporating AI and ML. The focus is on real-time data collection from engines and related systems to enhance predictive maintenance, leading to cost savings and increased uptime. Engines can self-optimize based on operational conditions, automatically adjusting performance for efficiency. AI analyzes data to drive decision-making and optimize engine operations. Real-time monitoring and remote diagnostics allow for immediate action in case of performance anomalies. The program also integrates AI-driven inspection tools, such as the Intelligent Borescope, which detects issues during engine inspections.

Virtual commissioning is another interesting field for the application of simulation, digital twins, and hybrid intelligence in the manufacturing sector. It combines physical models with machine learning to optimize system design and virtual commissioning [136]. By using system simulations based on physics, the behavior of complex systems can be simulated under various conditions. These simulations help predict how systems will perform in the real world. Siemens enhances this approach by integrating machine learning algorithms, which analyze real-time data to detect patterns, predict failures, and optimize system performance. The combination of physical simulations with AI allows for more accurate predictions and efficient decision-making. This hybrid approach helps reduce development time, improve system reliability, and optimize performance by identifying potential issues early in the design phase. The virtual commissioning aspect of this model ensures that systems are thoroughly tested in a simulated environment before being deployed in the real world, improving overall efficiency and reducing costs.

Based on the abovementioned examples of industrial applications of hybrid intelligence, further utilization of hybrid intelligence appears to be possible and meaningful but depends largely on the focus. In this context, the availability of correct physical, scientific, or simulation models must be considered as a major challenge.

With such models available, potential application areas for hybrid intelligence could include data quality improvement through ML-based anomaly detection for the automatic identification of missing, inconsistent, or incorrect data entries, flagging anomalies, and suggesting corrections.

More visionary application areas—especially in the maintenance sector—could address dynamic pricing models for maintenance service providers, enabling them to improve customer attractiveness by offering dynamic, usage-based, or performance-based pricing to their customers. This would reduce maintenance costs while ensuring optimal service quality, instead of relying on fixed-price contracts. During the offering phase of long-term service contracts, business risks could be reduced while simultaneously more accurate pricing models could be calculated, supported by AI-driven predictive contracting and smart service-level agreements (SLAs) that are based on reliable physical, scientific, or simulation models. Contracts could become data driven and adaptive, with SLAs adjusting based on real-time equipment performance rather than static, time-based agreements.

4.4 Health

The management of chronic illnesses like diabetes requires continuous patient engagement in daily self-care activities such as blood sugar monitoring and treatment adherence, which significantly reduce complications [137]. However, barriers such as complex regimens and limited healthcare access make self-management challenging. Innovative technologies provide personalized feedback, real-time monitoring, and enhanced patient-provider communication, encouraging active health management [137].

Hybrid intelligence in healthcare has gained traction with the availability of bio-medical data and improved ML models. Science-based mechanistic models and data-driven approaches complement each other, balancing physiological insight with data richness for more accurate and interpretable models [138, 139]. Explainability and consistency are critical in clinical AI applications. While CNNs and gated recurrent units (GRUs) forecast personalized glucose levels using real-time IoT data, sensor fusion integrates continuous glucose monitor (CGM) signals, insulin pump data, and activity trackers for metabolic insights. A stacked LSTM with Kalman smoothing enhances CGM signal quality [140]. However, these remain data-driven models, capturing complex patterns but lacking physiological integration, making them black-box approaches. Conversely, purely mechanistic models like the Hovorka Model for T1D [141] and the Bergman Minimal Model for T1D and T2D [142] offer interpretability but struggle with metabolic variability [143].

To bridge this gap, hybrid frameworks integrate ML with physiological knowledge. Grammatical evolution aids in simulating insulin–glucose dynamics [139], while PINNs embed metabolic constraints into deep learning [144]. Multi-scale models further integrate cellular and whole-body glucose metabolism, capturing fast and slow regulatory processes [145, 146].

Feature engineering plays a crucial role in AI-driven healthcare models. Traditional methods such as statistical filtering, wrapper approaches, and deep learning-based extraction often prioritize data correlations over clinical interpretability. Science-informed feature engineering incorporates clinical expertise, enhancing both transparency and predictive accuracy. Authors in [147] showed that clinician-guided feature selection can yield simpler, more interpretable models without compromising performance.

Hybrid intelligence enables both invasive and noninvasive glucose monitoring. The UVA/Padova T1D simulator generates synthetic training data for hybrid models, improving glycemic response predictions [139]. Noninvasive methods leverage breath biomarkers [148] and photoplethysmography (PPG) signals [149], processed via wavelet transforms (DWT, CWT) for fingerprick-free glucose estimation, enhancing long-term adherence and monitoring efficiency.

Hybrid intelligence adoption is driven by limitations in standalone models. Mechanistic models provide biological realism but struggle with high-dimensional metabolic variability [141, 150], whereas data-driven models excel at pattern recognition but lack interpretability for clinical use. Hybrid models [137, 139, 140, 144, 145] address these issues by embedding mechanistic constraints into ML architectures, ensuring physiological plausibility alongside adaptability. These advances have improved glucose monitoring, cardiovascular care, tumor growth prediction, and neurological disorder analysis. However, successful deployment requires seamless integration into healthcare IT systems. Data from CGM devices, PPG sensors, and electronic health records (EHRs) are critical, yet clinician adoption remains a challenge due to the complexity of AI technologies. Promising research directions include the development of hybrid AI pipelines that merge diverse data streams while ensuring interpretability and clinical usability.

5 Conclusions, Challenges, and Future Directions

The nascent field of hybrid intelligence presents a promising avenue for advancing scientific discovery and technological innovation. The landscape of these hybrid methods—alongside purely data-driven and purely science-based approaches—has been explored throughout this chapter, and Table 1 summarizes their respective advantages and limitations. However, realizing the full potential of HI requires addressing a range of significant challenges, encompassing both technical and methodological dimensions, akin to those encountered in other pioneering research domains.

From a methodological standpoint, the paramount challenge lies in the formation and effective operation of multidisciplinary teams. Constructing robust HI models necessitates a confluence of expertise, including domain-specific scientific knowledge, theoretical modelling proficiency, applied research experience, and technical ML implementation skills. Establishing effective communication channels and fostering genuine collaboration among individuals with diverse backgrounds and perspectives are crucial for successful project execution. Bridging the potential gaps in understanding and terminology between these disciplines is essential to ensure a cohesive and synergistic research effort. This necessitates developing shared conceptual frameworks and promoting mutual learning among team members.

Table 1 Summary table listing the main advantages and limitations of the different families of methods discussed in this chapter

Family	Methods	Advantages	Limitations
Science based			
Simulators	CFD and NWP models, GLM, FEM numerical models	• Theory-driven causality and explainability • Extrapolate safely within known physics regime • No large training data required	• High computational cost • Unknown physics introduce bias in extreme or unobserved conditions • Parameter calibration often requires computationally heavy inverse problem-solving
Process (low-order) models	RC thermal networks, fermentation process models, Brayton-cycle models, catalytic reactor models	• Fast execution enables real-time optimization and control • Still interpretable • Easy to embed in MPC systems	• Linear/steady-state assumptions may fail in highly non-linear and dynamic systems • Require empirical tuning • Scalability issues in large plants
Data driven			
Machine learning and deep learning	SVM, ANN, random forest, XGBoost, DNN, CNN, transformers	• Excel at complex, high-dimensional pattern recognition and discovery • Improve performance with larger data volumes • Flexible and adaptable to different domains	• Opaque "black-box" reasoning • Heavy data, labelling and computational demands (in training) • Data dependency, bias in case of incomplete, unrepresentative/biased training data
Probabilistic and statistical learning	Gaussian mixture models, BNNs, VI, VAEs, normalizing flows, MCMC	• Native uncertainty quantification • Often data-efficient in low-sample settings • Incorporation of prior beliefs	• Exact inference can be intractable in high complexity/dimensionality settings • Sampling methods (e.g., MCMC) still have high computational cost
Hybrid: science and data driven			
Science-guided ML	PINNs, science-informed architectures	• Physically consistent outputs with less data • Improved generalization outside training set • Partial interpretability retained	• Optimization and scalability challenges in sparse/noisy data scenarios • No "one-size-fits-all" architectures, deep domain and algorithmic knowledge required

(continued)

Table 1 (continued)

Family	Methods	Advantages	Limitations
ML-assisted science	Surrogate models	• Orders-of-magnitude faster than simulators • Enable simulator optimization, UQ, sensitivity analysis • Enable real-time what-if exploration	• Surrogates inherit simulator bias • Reliability drops when extrapolating far from training region
	Residual learning	• Preserves known physics while ML provides data-driven corrections to boost performance, esp. when complex nonlinear terms are unknown • Enables real-time parameter updates to simulators based on incoming data	• Requires high-quality paired data (simulator and observations) over the relevant operating domain • Risk of overfitting residuals, poor generalization outside training domain
	Parameter calibration, SBI, neural likelihood methods	• Enables the estimation of the posterior distribution of simulator free/unknown parameters that produce the observed data • Allows researched to validate or challenge scientific theories	• Accuracy and scalability challenges due to computational cost of simulator and/or intractability of posterior distribution • Requires an accurate simulator—if the physics is incomplete, no parameter set will yield good fit
	UQ with BNN, B-PINN, VI, GP, PI-GAN, MC Dropout	• Enable UQ for science-based simulators (deterministic) taking into account stochastic processes involved	• Scalability challenges, esp. in high-dimensional systems
Fully integrated hybrid systems	Coupled multi-scale pipelines, hybrid MPC, iterative feedback loops	• Enable end-to-end optimization while retaining physical consistency • Enable real-time adaptive control • Can achieve high performance and interpretability/causal reasoning	• Conceptualization and design require deep understanding of domain knowledge and ML/DL model mechanics • Most complex hybrid systems to orchestrate, require heavy integration and maintenance effort

Furthermore, the selection of appropriate models and architectures constitutes a significant challenge. Researchers must carefully evaluate the capabilities and limitations of both the ML and science-based components, ensuring in parallel compatibility and seamless integration between these disparate models. This involves not only technical considerations, such as data formats and software interfaces, but also conceptual alignment, ensuring that the outputs and interpretations of the different models are mutually consistent and physically meaningful. The selection process must also account for the specific research question being addressed, the availability and the quality of the data, and the computational resources available.

Another key challenge is the inherent uncertainty associated with both ML and science-based models. ML models are often trained on incomplete or noisy data, leading to uncertainties in their predictions. Similarly, science-based models, while grounded in physical laws or established theories, may involve simplifications and assumptions that introduce uncertainties. When these models are combined, the uncertainties can be amplified, potentially leading to unreliable or misleading results. Developing robust methodologies for quantifying, managing, and mitigating uncertainty in such models is crucial for ensuring their credibility and practical applicability.

Overcoming the aforementioned technical and methodological obstacles is crucial for realizing the transformative potential of hybrid intelligence, which can significantly advance scientific frontiers and drive impactful technological progress. Potential applications span diverse areas, from accelerating scientific discovery across various domains by generating novel insights from complex datasets and enabling the formulation and testing of hypotheses at unprecedented scales to enhancing decision-making processes and facilitating the development of innovative technologies, such as next-generation nanomaterials with tailored properties or human-cognizant robots capable of sophisticated interaction and collaboration.

Appendix

Acronyms used in the chapter are defined in Table 2.

Table 2 Acronym definitions

Acronym	Definition
AI	Artificial intelligence
ANN	Artificial neural network
ARMAX	Autoregressive-moving-average with exogenous inputs
BNN	Bayesian neural network
B-PINN	Bayesian physics-informed neural network
BVP	Blood volume pulse
CFD	Computational fluid dynamics
CGM	Continuous glucose monitor
CHF	Critical heat flux
CNN	Convolutional neural network
CWT	Continuous wavelet transform
DAG	Directed acyclic graph
DL	Deep learning
DNA	Deoxyribonucleic acid

(continued)

Table 2 (continued)

Acronym	Definition
DNN	Deep neural network
DT	Decision tree
DWT	Discrete wavelet transform
ECG	Electrocardiogram
EDA	Electrodermal activity
EEG	Electroencephalogram
EHR	Electronic health records
EOG	Electrooculogram
EU	European Union
FEM	Finite element method
FIHS	Fully integrated hybrid systems
GAN	Generative adversarial network
GLM	General Lake model
GNN	Graph neural network
GP	Gaussian process
GPT	Generative pre-trained transformer
HRI	Human-robot interaction
IoT	Internet of Things
LSTM	Long-short-term memory
MC	Monte Carlo
MCMC	Markov chain Monte Carlo
ML	Machine learning
MPC	Model predictive control
NLP	Natural language processing
NWP	Numerical weather prediction
ODE	Ordinary differential equation
PDE	Partial differential equation
PI-GAN	Physics-informed generative adversarial networks
PINN	Physics-informed neural network
PPG	Photoplethysmography
RBF	Radial basis function
RC	Resistance-capacitance
RNN	Recurrent neural network
SBI	Simulation-based inference
SCR	Skin conductance response
SLA	Service-level agreement
SLC	Sidelobe cancellation
SVM	Support vector machine
T1D	Type 1 diabetes
T2D	Type 2 diabetes
UQ	Uncertainty quantification
VAE	Variational autoencoders
VI	Variational inference

References

1. Willard, J. D., et al. (2020). *Integrating physics-based modeling with machine learning: A survey*. ArXiv abs/2003.04919.
2. Karpatne, A., et al. (2016). Theory-guided data science: A new paradigm for scientific discovery from data. *IEEE Transactions on Knowledge and Data Engineering, 29*, 2318–2331.
3. Karpatne, A. et al. (2024). *Knowledge-guided machine learning: Current trends and future prospects*. ArXiv abs/2403.15989.
4. Schmidt, M. D., & Lipson, H. (2009). Distilling free-form natural laws from experimental data. *Science, 324*, 81–85.
5. Freiesleben, T., & Molnar, C. (2024). *Supervised Machine Learning for Science: How to stop worrying and love your black box*. Retrieved from https://ml-science-book.com/
6. Karniadakis, G. E., Kevrekidis, I. G., Lu, L., et al. (2021). Physics-informed machine learning. *Nature Reviews Physics, 3*, 422–440. https://doi.org/10.1038/s42254-021-00314-5
7. Quarteroni, A. M. et al. (2025). *Combining physics-based and data-driven models: Advancing the frontiers of research with Scientific Machine Learning*. ArXiv abs/2501.18708.
8. Sharma, N., & Liu, Y. A. (2021). A hybrid science-guided machine learning approach for modeling chemical processes: A review. *AIChE Journal*.
9. Slater, L. J., et al. (2023). Hybrid forecasting: Blending climate predictions with AI models. *Hydrology and Earth System Sciences, 27*, 1865–1889. https://doi.org/10.5194/hess-27-1865-2023
10. Curry, E., Heintz, F., Irgens, M., Smeulders, A. W., & Stramigioli, S. (2022). Partnership on AI, data, and robotics. *Communications of the ACM, 65*(4), 54–55.
11. Anderson, J. D. (1995). *Computational fluid dynamics: The basics with applications*. McGraw-Hill.
12. Bender, E. (1981). Numerical heat transfer and fluid flow. Von S. V. Patankar. Hemisphere Publishing Corporation, Washington—New York—London. McGraw Hill Book Company, New York 1980. 1. Aufl., 197 S., 76 Abb., geb., DM 71,90. *Chemie Ingenieur Technik, 53*, 225–225.
13. Bauer, P., et al. (2015). The quiet revolution of numerical weather prediction. *Nature, 525*, 47–55.
14. Hipsey, M. R., et al. (2017). A General Lake Model (GLM 2.4) for linking with high-frequency sensor data from the Global Lake Ecological Observatory Network (GLEON). *Geoscientific Model Development Discussions*, 1–60.
15. Coughanowr, D. R., & Koppel, L. B. (1965). *Process systems analysis and control*. McGraw-Hill.
16. Seborg, D. E., et al. (1989). *Process dynamics and control*. John Wiley & Sons, Inc..
17. Smith, C. A., & Corripio, A. B. (1985). *Principles and practice of automatic process control*. John Wiley & Sons, Inc..
18. Grossmann, I. E. (2021). *Advanced optimization for process systems engineering* (Cambridge Series in Chemical Engineering). Cambridge University Press.
19. Froment, G. F., Bischoff, K. B., & De Wilde, J. (2011). *Chemical reactor-analysis and design* (3rd ed.). John Wiley & Sons, Inc..
20. Shuler, M. L., Kargı, F., & DeLisa, M. (2017). *Bioprocess engineering: Basic concepts*. Prentice Hall.
21. Saravanamuttoo, H. I. H., Rogers, G. F. C., & Cohen, H. (2001). *Gas turbine theory*. Pearson Education.
22. Pistikopoulos, E. N., & Tian, Y. (2024). Advanced modeling and optimization strategies for process synthesis. *Annual Review of Chemical and Biomolecular Engineering, 15*, 81.
23. Srivastava, A., & Jha, S. (2022). Data-driven machine learning: A new approach to process and utilize biomedical data. In *Predictive modeling in biomedical data mining and analysis* (pp. 225–252). Academic Press.
24. Breiman, L., et al. (1984). *Classification and regression trees*. CRC Press.

25. Breiman, L. (2001). Random forests. *Machine Learning, 45*, 5–32.
26. Chen, T., & Guestrin, C. (2016). Xgboost: A scalable tree boosting system. In *Proceedings of the 22nd ACM SIGKDD International Conference on Knowledge Discovery and Data Mining*.
27. Cortes, C., & Vapnik, V. (1995). Support-vector networks. *Machine Learning, 20*, 273–297.
28. Uddin, S., Ong, S., & Lu, H. (2022). Machine learning in project analytics: A data-driven framework and case study. *Scientific Reports, 12*(1), 15252.
29. Raissi, M., Perdikaris, P., & Karniadakis, G. E. (2019). Physics-informed neural networks: A deep learning framework for solving forward and inverse problems involving nonlinear partial differential equations. *Journal of Computational Physics, 378*, 686–707.
30. Gelman, A., et al. (2013). *Bayesian data analysis* (3rd ed.). CRC Press.
31. Bishop, C. M. (2006). *Pattern recognition and machine learning*. Springer.
32. Neal, R. M. (1996). *Bayesian learning for neural networks*. PhD thesis, University of Toronto.
33. Murphy, K. P. (2012). *Machine learning: A probabilistic perspective*. MIT Press.
34. Koller, D., & Friedman, N. (2009). *Probabilistic graphical models: Principles and techniques*. MIT Press.
35. Friedman, N., Geiger, D., & Goldszmidt, M. (2003). Bayesian network classifiers. *Machine Learning, 29*(2), 131–163.
36. Blei, D. M., Kucukelbir, A., & Jordan, M. I. (2017). Variational inference: A review for statisticians. *Journal of the American Statistical Association, 112*(518), 859–877.
37. Kingma, D. P., & Welling, M. (2014). Auto-encoding variational Bayes. In *Proceedings of the 2nd International Conference on Learning Representations (ICLR)*.
38. Rezende, D. J., & Mohamed, S. (2015). Variational inference with normalizing flows. In *Proceedings of the 32nd International Conference on Machine Learning (ICML)*.
39. Sarker, I. H. (2021). Machine learning: Algorithms, Real-World applications and research directions. *SN Computer Science, 2*(3), 160.
40. Gupta, H. D., & Sheng, V. S. (2020). A roadmap to domain knowledge integration in machine learning. In *2020 IEEE International Conference on Knowledge Graph (ICKG)* (pp. 145–151).
41. Kaden, M., Saralajew, S., & Villmann, T. (2024). Domain knowledge integration in machine learning systems—An introduction. In *ESANN 2024—Proceedings* (pp. 405–412).
42. Wang, Y., et al. (2020). Generalizing from a few examples: A survey on few-shot learning. *ACM Computing Surveys, 53*, 1.
43. Dash, T., et al. (2022). A review of some techniques for inclusion of domain-knowledge into deep neural networks. *Scientific Reports, 12*(1), 1040.
44. Chatzimparmpas, A., et al. (2022). FeatureEnVi: Visual analytics for feature engineering using stepwise selection and semi-automatic extraction approaches. *IEEE Transactions on Visualization and Computer Graphics, 28*(4), 1773–1791.
45. Kouba, P., et al. (2023). Machine learning-guided protein engineering. *ACS Catalysis, 13*(21), 13863–13895. https://doi.org/10.1021/acscatal.3c02743
46. Bonidia, R. P., et al. (2022). BioAutoML: Automated feature engineering and metalearning to predict noncoding RNAs in bacteria. *Briefings in Bioinformatics, 23*.
47. Camporeale, E. (2019). The challenge of machine learning in space weather: Nowcasting and forecasting. *Space Weather, 17*, 1166–1207.
48. Sun, H., Manchester, W., IV, & Chen, Y. (2021). Improved and interpretable solar flare predictions with spatial and topological features of the polarity inversion line masked magnetograms. *Space Weather, 19*, e2021SW002837. https://doi.org/10.1029/2021SW002837
49. Lavasa, E., et al. (2024). Toward explainable metrology 4.0: Utilizing explainable AI to predict the pointwise accuracy of laser scanning devices in industrial manufacturing. In J. Soldatos (Ed.), *Artificial intelligence in manufacturing*. Springer.
50. von Rueden, L., et al. (2023). Informed machine learning—A taxonomy and survey of integrating prior knowledge into learning systems. *IEEE Transactions on Knowledge and Data Engineering, 35*(1), 614–633.

51. Raissi, M., Perdikaris, P., & Karniadakis, G. E. (2017). *Physics informed deep learning (Part I): Data-driven solutions of nonlinear partial differential equations.* arXiv preprint arXiv:1711.10561.

52. Raissi, M., Perdikaris, P., & Karniadakis, G. E. (2017). *Physics informed deep learning (Part II): Data-driven discovery of nonlinear partial differential equations.* arXiv preprint arXiv:1711.10566.

53. Baydin, A. G., et al. (2018). Automatic differentiation in machine learning: A survey. *Journal of Machine Learning Research, 18*(153).

54. Wang, S., Yu, X., & Perdikaris, P. (2022). When and why PINNs fail to train: A neural tangent kernel perspective. *Journal of Computational Physics, 449.* https://doi.org/10.1016/j.jcp.2021.110768

55. Cuomo, S., et al. (2022). Scientific machine learning through physics–informed neural networks: Where we are and what's next. *Journal of Scientific Computing, 92*(3), 88.

56. Zhang, D., et al. (2019). Quantifying total uncertainty in physics-informed neural networks for solving forward and inverse stochastic problems. *Journal of Computational Physics, 397,* 108850.

57. Pang, G., Lu, L., & Karniadakis, G. E. (2019). fPINNs: Fractional physics-informed neural networks. *SIAM Journal on Scientific Computing, 41*(4), A2603–A2626.

58. Jagtap, A. D., & Karniadakis, G. E. (2020). Extended physics-informed neural networks (XPINNs): A generalized space-time domain decomposition based deep learning framework for nonlinear partial differential equations. *Communications in Computational Physics, 28,* 5.

59. Dolean, V., et al. (2022). Finite basis physics-informed neural networks as a Schwarz domain decomposition method. In *International Conference on Domain Decomposition Methods* (pp. 165–172). Springer.

60. Liu, Q., et al. (2024). *ConFIG: Towards conflict-free training of physics informed neural networks.* arXiv abs/2408.11104.

61. Grossmann, T. G., et al. (2024). Can physics-informed neural networks beat the finite element method? *IMA Journal of Applied Mathematics, 89,* 143.

62. Bonfanti, A., Bruno, G., & Cipriani, C. (2024). *The challenges of the nonlinear regime for physics-informed neural networks.* arXiv preprint arXiv:2402.03864.

63. Philipp, P., & Niklas, W. (2024). Physics-informed neural networks with unknown measurement noise. In *6th Annual Learning for Dynamics & Control Conference* (pp. 235–247). PMLR.

64. Rudolph, M., Kurz, S., & Rakitsch, B. (2024). Hybrid modeling design patterns. *Journal of Mathematics in Industry, 14,* 3. https://doi.org/10.1186/s13362-024-00141-0

65. Yin, H., et al. (2020). *1—The importance of domain knowledge.* Machine Learning Blog. ML@CMU. Retrieved from https://blog.ml.cmu.edu/2020/08/31/1-domain-knowledge/

66. Abramson, J., Adler, J., Dunger, J., et al. (2024). Accurate structure prediction of biomolecular interactions with AlphaFold 3. *Nature, 630,* 493–500. https://doi.org/10.1038/s41586-024-07487-w

67. Wang, C., et al. (2014). An evaluation of adaptive surrogate modeling based optimization with two benchmark problems. *Environmental Modelling & Software, 60,* 167–179.

68. Han, Z., & Zhang, K. (2012). Surrogate-based optimization. In *Real-world applications of genetic algorithms.* InTech Europe.

69. Williams, B. A., & Cremaschi, S. (2021). Surrogate model selection for design space approximation and surrogate based optimization. *Computer Aided Chemical Engineering.*

70. Semenova, E. (2025). *Case for a unified surrogate modelling framework in the age of AI.* arXiv:2502.06753v1.

71. Wortmann, T., Costa, A., Nannicini, G., & Schroepfer, T. (2015). Advantages of surrogate models for architectural design optimization. *Artifcial Intelligence for Engineering Design, Analysis and Manufacturing, 29,* 471–481. https://doi.org/10.1017/S0890060415000451

72. Barragán-Montero, A., et al. (2021). Artificial intelligence and machine learning for medical imaging: A technology review. *Physica Medica, 83,* 242–256. https://doi.org/10.1016/j.ejmp.2021.04.016

73. Raul, V., & Leifsson, L. (2021). Surrogate-based aerodynamic shape optimization for delaying airfoil dynamic stall using Kriging regression and infill criteria. *Aerospace Science and Technology, 111*, 106555.
74. Brunton, S. L., Noack, B. R., & Koumoutsakos, P. (2020). Machine learning for fluid mechanics. *Annual Review of Fluid Mechanics, 52*, 477–508.
75. Esteghamati, M., & Flint, M. (2023). Do all roads lead to Rome? A comparison of knowledge-based, data-driven, and physics-based surrogate models for performance-based early design. *Engineering Structures, 286*(13). https://doi.org/10.1016/j.engstruct.2023.116098
76. Dushatskiy, A., et al. (2023). *Multi-objective population based training.* arXiv:2306.01436v1.
77. Razavi, S., Toplson, B. A., & Burn, D. H. (2012). Review of surrogate modeling in water resources. *Water Resources Research, 48*(7), W07401.
78. Zhao, Y., et al. (2022). Surrogate modeling of nonlinear dynamic systems: A comparative study. *Journal of Computing and Information Science in Engineering, 23*(1), 1–48.
79. Gorissen, D., et al. (2010). A surrogate modeling and adaptive sampling toolbox for computer-based design. *Journal of Machine Learning Research, 11*, 2051–2055.
80. Bassetti, S., et al. (2023). *DiffESM: Conditional emulation of earth system models with diffusion models.* Physics>arXiv:2304.11699.
81. Schneider, T., et al. (2022). Machine learning based surrogate models for the thermal behavior of multi-plate clutches. *Applied System Innovation, 5*(5), 97. https://doi.org/10.3390/asi5050097
82. Vuillod, B., et al. (2024). Handling noise and overfitting in surrogate models based on non-uniform rational basis spline entities. *Computer Methods in Applied Mechanics and Engineering, 425*, 116913.
83. Franco, N. R., et al. (2023). *Deep Learning-based surrogate models for parametrized PDEs: Handling geometric variability through graph neural networks.* math>arXiv:2308.01602.
84. Chen, W. (2024). The application of different surrogate models in engineering predictions. *Applied and Computational Engineering, 80*(1), 87–93.
85. Mohammadi, S., Bui, V.-H., Su, W., & Wang, B. (2024). Surrogate modeling for solving OPF: A review. *Sustainability, 16*(22), 9851. https://doi.org/10.3390/su16229851
86. Yang, Q., et al. (2024). Research on surrogate models and optimization algorithms of compressor characteristic based on digital twins. *Journal of Engineering Research.* https://doi.org/10.1016/j.jer.2024.01.025
87. Li, S., Li, C., Huang, Y., & Zhai, C. (2024). Real-time seismic damage simulation for urban building portfolio based on basic building information and machine learning. *International Journal of Disaster Risk Reduction, 111*, 104687.
88. Chen, Y., Wang, L., & Huang, H. (2023). An effective surrogate model assisted algorithm for multi-objective optimization: Application to wind farm layout design. *Frontiers in Energy Research, 11*. https://doi.org/10.3389/fenrg.2023.1239332
89. Bassetti, S., Hutchinson, B., Tebaldi, C., & Kravitz, B. (2023). *DiffESM: Conditional emulation of earth system models with diffusion models.* Physics>arXiv:2304.11699.
90. Pereira, F. H., Schimit, P. H. T., & Bezerra, F. E. (2021). A deep learning based surrogate model for the parameter identification problem in probabilistic cellular automaton epidemic models. *Computer Methods and Programs in Biomedicine, 205*, 106078.
91. Schweidtmann, A. M., Zhang, D., & von Stosch, M. (2024). A review and perspective on hybrid modeling methodologies. *Digital Chemical Engineering, 10*, 100136.
92. Mao, C., & Jin, Y. (2024). Uncertainty quantification study of the physics-informed machine learning models for critical heat flux prediction. *Progress in Nuclear Energy, 170*, 105097.
93. Willard, J., et al. (2022). Integrating scientific knowledge with machine learning for engineering and environmental systems. *ACM Computing Surveys, 55*(4), 1–37.
94. Hanachi, H., et al. (2019). *Hybrid data-driven physics-based model fusion framework for tool wear prediction.*
95. Cranmer, K., Brehmer, J., & Louppe, G. (2020). The frontier of simulation-based inference. *Proceedings of the National Academy of Sciences, 117*(48), 30055–30062.

96. Beaumont, M. A. (2010). Approximate Bayesian computation in evolution and ecology. *Annual Review of Ecology, Evolution, and Systematics, 41*, 379–406.

97. Marin, J.-M., et al. (2012). Approximate Bayesian computational methods. *Statistics and Computing, 22*(6), 1167–1180.

98. Lintusaari, J., et al. (2017). Fundamentals and recent developments in approximate Bayesian computation. *Systematic Biology, 66*(1), e66.

99. Radev, S. T., et al. (2020). BayesFlow: Learning complex stochastic models with invertible neural networks. *IEEE Transactions on Neural Networks and Learning Systems, 33*(4), 1452–1466.

100. Ikonomov, B., & Gutmann, M. U. (2020). Robust optimisation Monte Carlo. In *International Conference on Artificial Intelligence and Statistics* (pp. 2819–2829). PMLR.

101. Gutmann, M. U., & Cor, J. (2016). Bayesian optimization for likelihood-free inference of simulator-based statistical models. *Journal of Machine Learning Research, 17*(125), 1–47.

102. Thomas, O., Dutta, R., Corander, J., Kaski, S., & Gutmann, M. U. (2022). Likelihood-free inference by ratio estimation. *Bayesian Analysis, 17*(1), 1–31.

103. Tejero-Cantero, A., et al. (2020). *SBI—A toolkit for simulation-based inference.* arXiv preprint arXiv:2007.09114.

104. Gkolemis, V., Gutmann, M., & Pesonen, H. (2024). An extendable Python implementation of robust optimization Monte Carlo. *Journal of Statistical Software, 110*, 1–26.

105. Lintusaari, J., et al. (2018). Elfi: Engine for likelihood-free inference. *Journal of Machine Learning Research, 19*(16), 1–7.

106. Ramesh, P., et al. (2022). *GATSBI: Generative adversarial training for simulation-based inference.* arXiv preprint arXiv:2203.06481.

107. Papamakarios, G., Sterratt, D., & Murray, I. (2019). Sequential neural likelihood: Fast likelihood-free inference with autoregressive flows. In *Proceedings of the 36th International Conference on Machine Learning (ICML).*

108. Neal, R. M. (1992). *Bayesian training for backpropagation networks by the hybrid Monte Carlo method.*

109. Graves, A. (2011). Practical variational inference for neural networks. In *Advances in neural information processing systems.*

110. Yang, L., Meng, X., & Karniadakis, G. E. (2021). B-PINNs: Bayesian physics-informed neural networks for forward and inverse problems with noisy data. *Journal of Computational Physics.*

111. Gilboa, E., Saatci, Y., & Cunningham, J. (2013). Scaling multidimensional Gaussian processes using projected additive approximations. In *International Conference on Machine Learning.*

112. He, W., & Jiang, Z. (2023). *A survey on uncertainty quantification methods for deep learning.*

113. Goodfellow, I. J., et al. (2014). *Generative adversarial nets.* Neural Information Processing Systems.

114. Yang, L., et al. (2018). Physics-informed generative adversarial networks for stochastic differential equations. *SIAM Journal on Scientific Computing, 42*, A292–A317.

115. Yarin Gal, Z. G. (2016). Dropout as a Bayesian approximation: Representing model uncertainty in deep learning. In *International Conference on Machine Learning.*

116. Lakshminarayanan, B., et al. (2016). *Simple and scalable predictive uncertainty estimation using deep ensembles.* Neural Information Processing Systems.

117. Horat, N., et al. (2024). *Uncertainty quantification for data-driven weather models.* EGU General Assembly.

118. Kurth, T., et al. (2023). FourCastNet: Accelerating global high-resolution weather forecasting using adaptive Fourier neural operators. In *Proceedings of the Platform for Advanced Scientific Computing Conference (PASC '23), Article 13* (pp. 1–11). Association for Computing Machinery. https://doi.org/10.1145/3592979.3593412

119. Chu, M., & Qian, W. (2024). *A deep learning approach for epistemic uncertainty quantification of turbulent flow simulations.*

120. Gokhale, G., Claessens, B., & Develder, C. (2022). Physics-informed neural networks for control-oriented thermal modeling of buildings. *Applied Energy, 314*, 118852.
121. Bünning, F., et al. (2022). Physics-informed linear regression is competitive with two machine learning methods in residential building MPC. *Applied Energy, 310*, 118491. https://doi.org/10.1016/j.apenergy.2021.118491
122. Chifu, V. R., et al. (2024). Physics-informed neural networks for heat pump load prediction. *Energies, 18*(1), 8.
123. Drgoňa, J., et al. (2021). Physics-constrained deep learning of multi-zone building thermal dynamics. *Energy and Buildings, 243*, 110992. https://doi.org/10.1016/j.enbuild.2021.110992
124. Wu, L., & Shahidehpour, M. (2010). A hybrid model for day-ahead price forecasting. *IEEE Transactions on Power Systems, 25*(3).
125. Lagomarsino, M., Lorenzini, M., De Momi, E., & Ajoudani, A. (2022). Robot trajectory adaptation to optimise the trade-off between human cognitive ergonomics and workplace productivity in collaborative tasks. In *IEEE/RSJ International Conference on Intelligent Robots and Systems (IROS)* (pp. 663–669). https://doi.org/10.1109/IROS47612.2022.9981424
126. Schmidt, P., et al. (2018). Introducing WESAD, a multimodal dataset for wearable stress and affect detection. In *Proceedings of the 2018 International Conference on Multimodal Interaction (ICMI '18) Boulder, CO, USA*. ACM.
127. Giannakakis, G., et al. (2022). Review on psychological stress detection using biosignals. *IEEE Transactions on Affective Computing, 13*(1), 440–460.
128. Choi, J., Ahmed, B., & Gutierrez-Osuna, R. (2012). Development and evaluation of an ambulatory stress monitor based on wearable sensors. *IEEE Transactions on Information Technology in Biomedicine, 16*(2), 279–286. https://doi.org/10.1109/TITB.2011.2169804
129. Iqbal, T., et al. (2022). Stress monitoring using wearable sensors: A pilot study and stress-predict dataset. *Sensors, 22*(21), 8135. https://doi.org/10.3390/s22218135
130. Kyriakou, K., et al. (2019). Detecting moments of stress from measurements of wearable physiological sensors. *Sensors, 19*, 3805.
131. Giorgi, A., et al. (2021). Wearable technologies for mental workload, stress, and emotional state assessment during working-like tasks: A comparison with laboratory technologies. *Sensors (Basel), 21*, 2332.
132. Ishaque, S., Khan, N., & Krishnan, S. (2023). Physiological signal analysis and stress classification from VR simulations using decision tree methods. *Bioengineering, 10*, 766. https://doi.org/10.3390/bioengineering10070766
133. Ahmad, Z., Rabbani, S., Zafar, M. R., Ishaque, S., Krishnan, S., & Khan, N. (2021). Multilevel stress assessment from ECG in a virtual reality environment using multimodal fusion. *IEEE Sensors Journal, 23*, 29559–29570. https://doi.org/10.1109/JSEN.2023.332329
134. Rashid, N., et al. (2021). Feature augmented hybrid CNN for stress recognition using wrist-based photoplethysmography sensor. In *43rd Annual International Conference of the IEEE Engineering in Medicine & Biology Society (EMBC)* (pp. 2374–2377). https://doi.org/10.1109/EMBC46164.2021.9630576
135. https://www.rolls-royce.com/media/press-releases/2018/05-02-2018-rr-launches-intelligentengine.aspx
136. *System simulation and machine learning for concept optimization and virtual commissioning.* (2022). Siemens.
137. Samal, L., et al. (2021). Health information technology to improve care for people with multiple chronic conditions. *Health Services Research, 56*(S1), 1006–1036. https://doi.org/10.1111/1475-6773.13860
138. Alkanhel, R. I., et al. (2024). Hybrid CNN-GRU model for real-time blood glucose forecasting: Enhancing IoT-based diabetes management with AI. *Sensors (Basel), 24*(23), 7670. https://doi.org/10.3390/s24237670
139. Contreras, I., Oviedo, S., Vettoretti, M., Visentin, R., & Vehí, J. (2017). Personalized blood glucose prediction: A hybrid approach using grammatical evolution and physiological models. *PLoS One, 12*(11), e0187754. https://doi.org/10.1371/journal.pone.0187754

140. Rabby, M. F., Tu, Y., Hossen, M. I., Lee, I., Maida, A. S., & Hei, X. (2021). Stacked LSTM based deep recurrent neural network with Kalman smoothing for blood glucose prediction. *BMC Medical Informatics and Decision Making, 21*(1), 101. https://doi.org/10.1186/s12911-021-01462-5

141. Hovorka, R., et al. (2002). Partitioning glucose distribution/transport, disposal, and endogenous production during IVGTT. *American Journal of Physiology-Endocrinology and Metabolism, 282*(5), E992–E1007. https://doi.org/10.1152/ajpendo.00304.2001

142. Bergman, R. N., Phillips, L. S., & Cobelli, C. (1981). Physiologic evaluation of factors controlling glucose tolerance in man: Measurement of insulin sensitivity and beta-cell glucose sensitivity from the response to intravenous glucose. *Journal of Clinical Investigation, 68*(6), 1456–1467. https://doi.org/10.1172/JCI110398

143. Bequette, B. W. (2007). Analysis of algorithms for intensive care unit blood glucose control. *Journal of Diabetes Science and Technology, 1*(6), 813–824. https://doi.org/10.1177/193229680700100604

144. Schliess, F., et al. (2019). Artificial pancreas systems for people with type 2 diabetes: Conception and design of the European CLOSE Project. *Journal of Diabetes Science and Technology, 13*(2), 261–267. https://doi.org/10.1177/1932296818803588

145. Borri, A., et al. (2017). Luenberger-like observers for nonlinear time-delay systems with application to the artificial pancreas: The attainment of good performance. *IEEE Control Systems, 37*(4), 33–49. https://doi.org/10.1109/MCS.2017.2696759

146. Cobelli, C., Renard, E., & Kovatchev, B. (2011). Artificial pancreas: Past, present, future. *Diabetes, 60*(11), 2672–2682. https://doi.org/10.2337/db11-0654

147. Roe, K. D., et al. (2020). Feature engineering with clinical expert knowledge: A case study assessment of machine learning model complexity and performance. *PLoS One, 15*(4), e0231300. https://doi.org/10.1371/journal.pone.0231300

148. Gade, A., Vijaya Baskar, V., & Panneerselvam, J. (2024). Hybrid model with optimal features for non-invasive blood glucose monitoring from breath biomarkers. *Biomedical Signal Processing and Control, 88*, 105036. https://doi.org/10.1016/j.bspc.2023.105036

149. Munoz-Organero, M. (2020). Deep physiological model for blood glucose prediction in T1DM patients. *Sensors, 20*(14), 3896. https://doi.org/10.3390/s20143896

150. Bergman, R. N. (2021). Origins and history of the minimal model of glucose regulation. *Frontiers in Endocrinology, 11*. https://doi.org/10.3389/fendo.2020.583016

Frugal Machine Learning
for Energy-Efficient and Resource-Aware
Artificial Intelligence

John Violos, Konstantina-Christina Diamanti, Ioannis Kompatsiaris,
and Symeon Papadopoulos

Abstract Frugal machine learning (FML) refers to the practice of designing machine learning (ML) models that are efficient, cost-effective, and mindful of resource constraints. This field aims to achieve acceptable performance while minimizing the use of computational resources, time, energy, and data for both training and inference. FML strategies can be broadly categorized into input frugality, learning process frugality, and model frugality, each focusing on reducing resource consumption at different stages of the ML pipeline. This chapter explores recent advancements, applications, and open challenges in FML, emphasizing its importance for smart environments that incorporate edge computing and IoT devices, which often face strict limitations in bandwidth, energy, or latency. Technological enablers such as model compression, energy-efficient hardware, and data-efficient learning techniques are discussed, along with adaptive methods including parameter regularization, knowledge distillation, and dynamic architecture design that enable incremental model updates without full retraining. Furthermore, it provides a comprehensive taxonomy of frugal methods, discusses case studies across diverse domains, and identifies future research directions to drive innovation in this evolving field.

Keywords Frugal machine learning · Pruning · Knowledge distillation · Quantization · Feature selection · Data sampling · Hardware accelerators

J. Violos (✉) · I. Kompatsiaris · S. Papadopoulos
Information Technologies Institute, Centre for Research and Technology Hellas,
Thessaloniki, Greece
e-mail: violos@iti.gr; ikom@iti.gr; papadop@iti.gr

K.-C. Diamanti
Department of Informatics & Telematics, Harokopio University, Athens, Greece
e-mail: it218116@hua.gr

© The Author(s) 2026

E. Curry et al. (eds.), *Artificial Intelligence, Data and Robotics*,
https://doi.org/10.1007/978-3-032-10561-5_8

1   Introduction

FML aims to make AI more accessible and sustainable, particularly in resource constrained environments such as smart environments, IoT systems, and embedded platforms [1]. Unlike conventional deep learning approaches that assume abundant resources, FML focuses on creating models that perform effectively on devices with limited computational power, data, memory, and energy [2]. This is increasingly important as more applications move toward edge computing, which requires real-time processing on low-power devices [3].

FML also contributes to sustainability by reducing energy consumption, making AI development more environmentally friendly, while democratizing access to powerful models in settings with limited infrastructure [4]. It enables scalable and cost-effective deployment across industries such as IoT and edge computing [5], wearable and mobile AI [6], autonomous systems [7], and cybersecurity [8], benefiting regions with constrained resources. The objectives of FML are summarized in Table 1.

While closely related to tiny machine learning (TinyML), FML and TinyML differ in focus. FML emphasizes reducing costs associated with data, computation, and time while maintaining model performance across various ML tasks. TinyML, on the other hand, is specifically designed for ultralow-power devices like microcontrollers and edge devices, optimizing models for minimal memory, processing power, and energy consumption. FML is broader and cost driven, while TinyML is specialized for small, resource-limited hardware platforms.

Key technological advancements have fueled the rise of FML, including the shift from centralized cloud computing to edge computing. This transition has driven the demand for lightweight models that can operate on devices with limited processing power. Techniques such as model pruning [9], quantization [10], and knowledge distillation [11] have enabled significant reductions in model size with minimal

Table 1 Objectives of FML

Objective	Description
Data size	Minimize the amount of (labelled) data for the training process
Computation	Reduce the computational load by building lightweight models
Energy	Minimize energy consumption during training and inference
Scalability	Scale across various devices in constrained environments with different levels of resource availability
Monetary cost	Optimize computational, memory, and data resources to reduce the overall expense of deploying and maintaining ML systems
Model size	Decrease the number of parameters (weights and biases) in ANNs
Compression	Reduce the time and resources required to compress, fine-tune, and validate a compressed ANN
Trade-offs	Balance the above objectives with performance, ensuring efficient operation without compromising model accuracy

performance loss, making them suitable for embedded systems. Energy-efficient hardware, such as specialized AI accelerators like GPUs [4] and TPUs [12], also play a crucial role in optimizing ML algorithms for low-power devices. These developments support frugal algorithms by reducing the energy consumption required for AI computations. Additionally, techniques such as feature selection [13] and data selection [14] help retain the most informative data points, reducing the amount of training data required without compromising the accuracy of ML models.

Key insights from the literature reveal that techniques like pruning and quantization can reduce model size and inference latency by up to 90% while preserving accuracy, making large neural networks viable for edge devices [15]. Knowledge distillation enables small models to inherit the performance of larger ones through soft-label learning [5], while low-rank approximations and neural architecture search help streamline models structurally [16]. Furthermore, input frugality methods employ data-efficient approaches, enabling ML models to be trained and operate with significantly smaller amounts of data, thereby reducing the need for computational resources [14].

The chapter relates to the efficiency and sustainability of the AI, Data, and Robotics Partnership [17]. Its main contribution is an accessible survey of FML, with the aim to help readers easily understand and engage with the topic. Section 2 proposes a taxonomy of most prominent FML methods, models, and approaches. Section 3 presents various use cases and applications of FML. Section 4 offers insights into the underlying technologies and methodologies and casts a critical eye in FML discussing limitations and research gaps and providing suggestions for future work. Finally, we conclude the article in Sect. 5.

2 A Taxonomy of FML Methods

Frugality in machine learning models can be achieved through various approaches, as illustrated in Fig. 1. The most common approach involves compressing large artificial neural network (ANNs) into smaller ones (Sect. 2.1). Another approach leverages specialized hardware accelerators such as GPUs, TPUs, FPGAs, and NPUs to enhance computational efficiency while lowering energy consumption (Sect. 2.6). Algorithm optimization focuses on selecting lightweight ML techniques or refined versions of existing models to improve efficiency (Sect. 2.2). Feature selection streamlines input data by eliminating non-essential features that have minimal impact on model performance (Sect. 2.3). Similarly, data sampling reduces training data by discarding entire samples that contribute little to the learning process (Sect. 2.4). Furthermore, inverse modelling techniques facilitate the rapid determination of hyperparameter values that influence ML model performance (Sect. 2.5).

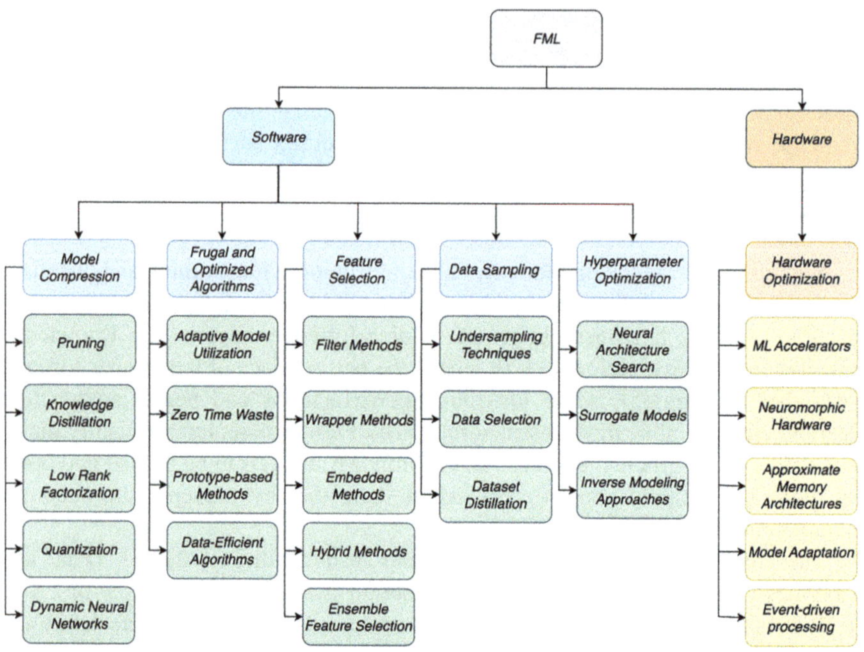

Fig. 1 Frugal machine learning methods

2.1 Model Compression

Model compression in FML refers to techniques that reduce the size and computational complexity of ANNs [15, 18]. This can be achieved through methods such as pruning, quantization, knowledge distillation, and low-rank factorization or dynamic neural networks that dynamically shrink the size of the ANN based on available computational resources.

Pruning is a widely used FML technique that reduces the size and complexity of ANNs by eliminating redundant or insignificant parameters while maintaining overall performance [9]. This approach is particularly valuable for deploying deep learning models in environments with limited computational and memory resources, such as mobile devices and embedded systems. Pruning typically follows a two-step process: first, an over-parameterized network is trained to achieve high accuracy; then, unnecessary weights, neurons, or filters are identified and pruned [19]. To mitigate the potential accuracy loss from pruning, a fine-tuning phase is often incorporated to restore the model's performance.

Pruning can be structured and unstructured. Structured pruning involves removing entire structures such as neurons, filters, channels, or layers, thereby maintaining a regular architecture and facilitating efficient hardware implementation [20]. This approach contrasts with unstructured pruning, which eliminates individual weights, leading to irregular sparsity patterns that are hard to leverage using standard hardware [21]. An example of structured pruning is filter pruning in CNNs, where entire filters are removed based on specific criteria, for instance, removing filters with small L1-norm values to compress CNNs [22]. A third approach is dynamic pruning, which refers to methods that adjust the network's structure during training or inference, allowing the model to adaptively prune and regrow connections based on certain criteria [23].

An important pruning technique is magnitude-based pruning, which assesses the importance of parameters based on their magnitude, with smaller weights typically considered less critical [24]. Sensitivity-based pruning goes further by analyzing the effect of removing specific weights or neurons on the model's loss function, often using techniques like Taylor series approximations for precise pruning decisions [25]. Additionally, clustering and similarity-based methods identify and group highly similar or duplicate parameters to reduce redundancy while retaining the network's representational power [26].

Recent advancements in pruning methodologies include optimization-driven approaches such as filter pruning and dynamic network surgery [27]. These methods adaptively prune filters or feature maps during training to ensure sustained performance. The introduction of concepts like the Lottery Ticket Hypothesis has further improved pruning strategies by uncovering high-performing subnetworks within large trained models [28]. Identified through iterative pruning and fine-tuning, these sub-networks have achieved remarkable reductions in model size while maintaining accuracy in tasks such as image recognition and natural language processing.

Knowledge Distillation (KD) is a model compression technique designed to transfer knowledge from a large teacher model to a smaller lightweight student model [11]. This approach enables the student model to mimic the behavior of the teacher while achieving a comparable level of performance. KD is especially advantageous in scenarios with constrained computational resources, such as IoT and edge computing devices [5]. The process involves the teacher model generating "soft targets," which are class probabilities that encapsulate relational information among different classes. These soft targets guide the student model during training, allowing it to achieve high performance with significantly fewer parameters and reduced computational complexity. Recent advancements, such as self-distillation and multi-teacher frameworks, have further extended KD's utility, enabling efficient optimization and scalability for diverse tasks [29].

The KD process comprises three key components: knowledge, distillation algorithm, and teacher-student architecture [30]. Knowledge can be derived from multiple facets of the teacher model, including the final output layer (response-based knowledge), intermediate feature representations (feature-based knowledge), or the relationships among data points (relation-based knowledge) [31]. The distillation algorithm typically uses loss functions, such as Kullback-Leibler divergence, to align the outputs of the student model with those of the teacher model [32]. A teacher-student architecture should maintain a balanced capacity between the two models to facilitate effective knowledge transfer [11].

Low-Rank Factorization is a technique used in ANN compression, where large weight matrices are approximated by smaller, low-rank matrices [33]. This approach leverages the fact that many weight matrices in neural networks contain redundancies and can be efficiently represented by decomposing them into lower-dimensional forms. One common method of low-rank factorization is Singular Value Decomposition (SVD), which decomposes a matrix into three smaller matrices, and the rank is reduced by retaining only the largest singular values [34]. Another method is Matrix Decomposition, which includes techniques like CUR decomposition and Alternating Least Squares (ALS) [35]. These techniques factorize the original matrix into smaller components, further reducing its dimensionality and improving computational efficiency.

Quantization reduces model size and computational requirements by lowering the number of bits needed to represent each parameter. This process can reduce the precision of model weights and activations from 32-bit floating-point to 8-bit integers [36]. Although post-training quantization may lead to performance loss due to reduced precision, quantization-aware training (QAT) mitigates this by addressing quantization effects during training, allowing models to adjust to the process and maintain higher accuracy [37].

These compression techniques, though varied in terms of approach, can be combined to maximize efficiency. For instance, pruning and quantization can be applied together to simultaneously reduce model size and inference time [38]. Quantized distillation leverages distillation during the training process, by incorporating distillation loss, expressed with respect to the teacher, into the training of a student network whose weights are quantized to a limited set of levels [39]. Quantization also has been combined with low-rank factorization by performing Canonical Polyadic decomposition with factor matrices constrained to a lower precision format (e.g., INT8/INT6/INT4) [10].

Dynamic Neural Networks. Typically, ANNs are designed as static models with fixed architecture and computation patterns. Dynamic neural networks (DyNNs) have been proposed to introduce flexibility by allowing adjustable parameters such as depth, width, and activation pathways, which can change dynamically during inference or training [40]. These possess several key features that enhance their efficiency and adaptability. They utilize adaptive computation to adjust the number of layers, neurons, or operations based on input complexity. Conditional execution enables selective activation of network components, such as skipping layers or neurons, based on learned policies [41]. Additionally, input dependent processing ensures efficient resource allocation by dedicating fewer computations to simpler inputs while allocating more resources for complex ones [42]. As a result, DyNNs offer energy efficiency, reducing unnecessary computations and optimizing power consumption—making them particularly beneficial for real-time applications and deployment on resource-constrained edge devices.

DyNNs can be categorized into different types based on how they adapt their structure and computation. Spatially adaptive networks modify their architecture at different spatial locations in an image, such as using dynamic convolution to process regions of varying complexity more efficiently [43]. Temporally adaptive networks adjust the number of timesteps in recurrent architectures like RNNs or transformers, allowing them to allocate computational resources based on the temporal dependencies of the input sequence [44]. Routing-based networks leverage mechanisms like gating or reinforcement learning to selectively activate sub-networks, as seen in Mixture of Experts models, which dynamically route inputs to the most relevant network components [45]. Last, self-pruning networks enhance efficiency by learning to remove redundant weights or neurons during training or inference, reducing computational cost while maintaining performance [46]. These dynamic architectures enable efficient and flexible deep learning models suited for a variety of tasks.

One specific category of DyNNs are slimmable neural networks (SNNs). SNNs are designed to dynamically adjust their architecture based on available computational resources, offering flexible configurations in terms of depth (number of layers) and width (number of neurons per layer) [47]. During inference, SNNs can select different model configurations depending on input complexity and hardware constraints, making them adaptable for a variety of environments [48]. This scalable computation allows the network to use lightweight architectures for simpler tasks and scale up to deeper or wider configurations for more complex tasks. SNNs are trained to support all possible configurations, with mechanisms that enable parts of the network to be turned on or off as needed.

In addition to depth and width slimming, channel slimming reduces the number of channels (filters) in each convolutional layer of CNNs [49]. Similarly, block slimming applied in modular networks enables the dynamic selection of blocks, such as residual or attention blocks, based on the task requirements or available resources [48]. Last, Skip Connections and Residual Slimming adjusts the connections between layers, skipping unnecessary ones to further reduce computation [50]. Each type of slimming contributes to enhancing model flexibility and efficiency, depending on the computational demands and input complexity. Similar to SNNs, adaptive neural networks adjust computational complexity based on resource constraints, but while SNNs switch between predefined network widths, adaptive neural networks dynamically modify their structure per input using mechanisms like early exiting or conditional activation [51].

2.2 *Frugal and Optimized Algorithms*

This category encompasses methods that dynamically adjust their computational demands based on task complexity. It includes approaches that minimize computational waste in early-exit strategies; methods inherently designed to be lightweight, such as prototype-based techniques; and those specifically developed to operate effectively with limited data.

Adaptive Model Utilization (AMU) focuses on using computationally intensive DNNs only when necessary while relying on lightweight DNNs otherwise. In computer vision, one proposed approach activates DNNs selectively to detect new objects or re-identify those that undergo significant appearance changes [52]. During intervals between DNN executions, lightweight methods track detected objects efficiently without compromising performance. A more efficient AMU approach employs a dual-network architecture that dynamically alternates between a lightweight CNN for simple tasks and a larger CNN for more complex tasks, depending on whether the confidence score of the smaller CNN drops below a predefined threshold [53]. Furthermore, instead of using a lightweight and a larger CNN, another AMU method proposes the synergy of two complementary lightweight CNNs and a memory component [2], where each network compensates for the other's weaknesses in predictions. If the first CNN has low confidence in its output, the second CNN is activated to improve accuracy. This method also uses a memory component that stores past classifications to avoid redundant computations.

Zero Time Waste (ZTW) is an optimization method for early exit neural networks that minimizes computational waste by reusing information from previous internal classifiers (ICs) rather than discarding their predictions [54]. Conventional early exit methods allow certain ICs to return predictions early for easy-to-classify samples, reducing overall inference time [55]. However, when an IC does not meet the confidence threshold to exit, its computations are wasted. ZTW addresses this inefficiency by introducing cascade connections, which pass the outputs of earlier ICs to later ones, allowing each classifier to refine previous predictions rather than starting from scratch.

Prototype-Based Methods have been characterized as frugal, because they rely on a small set of representative prototypes rather than optimizing over large datasets. They work by comparing data points with the representative examples of each class called prototypes [56]. Incremental neural networks are prototype-based neural classifiers that dynamically adjust and expand their structure by incorporating new prototypes as data is progressively introduced [57]. In this way, they enable efficient learning without the need for full retraining, even when integrating data from different data distributions. Adaptive resonance theory models also belong to the prototype-based methods by comparing the input data to existing patterns [58]. If the new input is similar to a stored pattern, then the existing pattern is updated. Otherwise, a new category is created.

Data-Efficient Algorithms are designed to optimize learning processes by reducing the need for large amounts of labelled data [59]. These employ various strategies to enhance learning efficiency, making them particularly suitable for data-scarce environments. Semi-supervised learning is a key method, combining labelled and unlabeled data to improve learning without extensive labeling [60]. Techniques such as self-training, co-training, and graph-based methods are employed to iteratively label and train on unlabeled data [61]. Additionally, generative models and transformation-equivalent representations are utilized to learn robust features, further reducing the need for large datasets.

2.3 *Feature Selection*

Feature selection reduces data dimensionality by identifying the most relevant features. Techniques such as filter, wrapper, embedded, hybrid, and ensemble methods decrease computational and memory demands. By eliminating redundant or irrelevant features, they enhance input efficiency in traditional ML models. While feature selection methods may introduce additional computational overhead before or during training, the resulting models built on fewer and more informative features are typically more efficient. This trade-off often justifies the extra effort during training because it leads to faster inference times, lower resource consumption, and, in many cases, more accurate and compact models that are well suited for deployment.

Filter Methods select relevant features independently of any ML model by using statistical measures to assess their importance. These operate as a preprocessing step, ranking features based on their correlation with the target variable. For example, symmetrical uncertainty ranks features based on the association between each feature and the target variable using entropy calculations [62]. RELIEF evaluates feature importance by examining the values of the nearest instances of the same and different classes, capturing feature dependencies effectively [63].

Wrapper Methods evaluate feature subsets by training and testing a specific ML model on different combinations of features. These use model performance as the criterion for selecting the best subset of features. A widely used wrapper method is forward selection, which begins with an empty feature set and adds the most significant feature at each step until performance reaches its peak [64]. In contrast, backward elimination starts with all available features and removes the least important ones at each step, guided by the model's performance [65].

Embedded Methods. Unlike filter methods, which evaluate features independently of the model, and wrapper methods, which rely on separate evaluation processes, embedded methods integrate feature selection directly into model training. These have been applied to traditional ML models such as SVMs and ANNs. In SVMs, feature selection has been formulated as a min-max optimization problem that tries to balance model complexity and classification accuracy [66]. By leveraging duality theory, this method reformulates the problem into a solvable nonlinear optimization framework. In ANNs, an embedded method has been introduced, incorporating a binary mask layer that dynamically selects the most relevant features during training using gradient-based optimization [13]. By applying regularization constraints to enforce sparsity, the embedded method forces the selection of a predefined maximum number of features, ensuring computational efficiency while maintaining accuracy.

Hybrid Methods combine the strengths of filter and wrapper methods to balance computational efficiency and model performance. The hybrid approach combines the speed of filter methods with the relevance search capabilities of wrapper methods [8]. In the first phase, two filter methods, information gain and random forest, are utilized to reduce the feature subset search space efficiently. In the second phase, a wrapper method applies recursive feature elimination, further refining the feature set while preserving the most relevant features and considering feature similarities.

Ensemble Methods entail combining multiple feature selection methods and aggregating their results, often using weighted voting [67]. Homogeneous ensemble methods apply the same selection algorithm to different training data subsets, whereas heterogeneous ensemble methods integrate various selection algorithms on the same dataset. This enhances stability by reducing the risk of selecting inconsistent feature subsets and ensures more reliable feature ranking.

2.4 Data Sampling

Data sampling, initially used to address data imbalance, gains attention in FML due to its resource efficiency. Techniques such as under-sampling significantly reduce the size of training data, resulting in shorter training times, lower computational costs, and decreased resource consumption. By selecting the most informative data samples, data sampling aims to preserve performance, ensuring that models remain effective even with a more compact dataset.

Under-sampling Techniques. Random under-sampling in the majority class is a straightforward approach but may lead to the loss of valuable information. Edited Nearest Neighbors, eliminates majority class samples that are misclassified by their nearest neighbors, thereby enhancing the classifier's ability to distinguish between classes [68]. A different approach identifies and removes samples in the majority class that are close to minority class samples [69].

Data Selection Methods use ranking criteria to assess and prioritize samples for training an ANN or for knowledge distillation. Manifold learning-based data selection works by projecting high-dimensional data into a lower-dimensional plane. Then, clustering techniques group the data points into distinct classes, from which the training dataset is sampled [70]. An entropy-based data selection method has also been proposed that identifies smaller subsets from the original dataset by focusing on instances that retain the most informational value [14]. This method calculates entropy using the logit vectors of data samples and selects those with the highest entropy.

Dataset Distillation compresses knowledge from a large dataset into a much smaller synthetic dataset by optimizing it so that models trained on it perform comparably to those trained on the original data [71]. The process involves initializing a small synthetic dataset, training a neural network on it, and optimizing the data using the following three objectives [72]: (a) performance matching, which ensures models trained on synthetic data achieve similar accuracy as those trained on real data; (b) parameter matching, which matches the internal parameters of networks trained on real and synthetic data; and (c) distribution matching, which ensures the synthetic data maintains statistical properties similar to the original dataset. The synthetic dataset is then refined through iterative steps to maximize its informativeness.

2.5 *Hyperparameter Optimization*

ANNs and traditional ML models involve numerous hyperparameters related to both their architecture and training process. Identifying the optimal hyperparameters is often done through trial and error, requiring practitioners to test multiple combinations to achieve the most accurate model. Hyperparameter optimization aims to streamline this process by efficiently exploring the hyperparameter space, reducing the number of experiments needed to achieve the FML objectives. Although hyperparameter optimization can be computationally intensive due to the large number of tested parameter combinations, the resulting model can be highly efficient and lightweight. By optimizing conflicting objectives such as model accuracy versus size, inference time, or resource usage, this process enables the development of models that strike an optimal balance between performance and compactness, aligning well with the goals of FML.

Neural Architecture Search is an automated method for designing optimal neural network architectures that consists of three key components: (a) search space, which defines the range of possible architectures, including factors like the number of layers, filter sizes, and connections [16]; (b) search strategy, which determines how different architectures are explored, using methods such as reinforcement learning, genetic algorithms, or gradient-based optimization [73]; and (c) evaluation method, which measures the performance of candidate architectures based on FML criteria like accuracy, efficiency, or computational cost.

Surrogate Models are simplified, computationally efficient approximations of complex and expensive-to-evaluate functions, commonly used in ML to replace costly simulations [74]. The process begins with data collection, where a set of sample points is gathered from the original function or system. Using these samples, a surrogate model is constructed to approximate the function, providing a more efficient way to evaluate potential solutions [75]. This is then used for optimization and exploration, identifying promising solutions while minimizing computational costs. To enhance accuracy, new samples from the original function can be iteratively incorporated, refining the surrogate model over time.

Inverse Modeling Approaches are computational methods used to estimate unknown system parameters like ANN architectures by analyzing observed outputs [76]. These techniques work backward from the results, making them valuable for optimization and inference. The process begins by defining a forward model that captures the relationship between inputs and outputs. Observational data, such as real-world measurements, serve as the known outputs, and optimization or statistical inference techniques are then applied to estimate the unknown inputs by minimizing the discrepancy between the model's predictions and actual observations. The estimated parameters are validated using independent data, and the model is refined iteratively to improve accuracy.

Hyperparameter Optimization Beyond Neural Networks. Hyperparameter optimization is not limited to ANN architectures but is also widely applied in classical ML models. For instance, grid search has been used to fine-tune the hyperparameters of models such as SVMs, random forests, and K-nearest neighbors. This involves exhaustively evaluating all possible combinations of parameters, such as the kernel type in SVMs, the number of estimators in random forest, and the number of neighbors in KNN, in order to enhance classification performance [77]. In contrast, random search explores hyperparameter combinations by sampling from predefined distributions. Although it does not guarantee evaluation of every combination, it enables a wider and more diverse search of the hyperparameter space, which is particularly useful when only a few parameters substantially affect model performance [78]. Building on this idea, Bayesian optimization constructs a probabilistic model of the objective function, often using Gaussian processes, and chooses the next set of hyperparameters to evaluate aiming to efficiently converge to close to optimal configurations [79].

2.6 Hardware Optimization

Hardware optimization involves the use of processing units designed for ML tasks, neuromorphic hardware, and efficient memory architectures. It also includes techniques that adapt ANNs to the deployed devices and event-driven processing methods, where ANN models operate only when necessary.

ML Accelerators are specialized processing units engineered to support the objectives of FML. Graphics processing units (GPUs), initially developed for graphics rendering, have become essential for ML due to their parallel processing capabilities, significantly minimizing energy consumption [4] while accelerating neural network training and inference [80]. Additionally, tensor processing units (TPUs) are specifically optimized for deep learning workloads, enabling low-computational matrix multiplications while consuming less power than GPUs and CPUs [12]. Field-programmable gate arrays (FPGAs) offer customizable ML acceleration, enabling optimized energy-efficient computations for embedded AI systems [81]. Neural processing units (NPUs), such as the Apple Neural Engine and Qualcomm Hexagon NPU, are dedicated AI accelerators designed for efficient deep learning inference on edge devices and mobile platforms [6].

Neuromorphic Hardware consists of computing systems engineered to replicate the structure and function of the human brain by leveraging spiking neural networks (SNNs) and specialized circuits that function similarly to biological neurons and synapses [82]. Unlike conventional von Neumann architectures, which distinctly separate memory and processing, neuromorphic computing systems both store and process data in individual neurons, resulting in lower latency and data transfer [83]. These systems process information using discrete spikes rather than continuous signals, reducing unnecessary computations. Additionally, asynchronous processing ensures computations occur only when necessary, minimizing energy consumption. By incorporating in-memory computing, neuromorphic hardware eliminates data transfer bottlenecks, improving speed and efficiency.

Approximate Memory Architectures are memory systems designed to improve energy efficiency and performance by relaxing accuracy constraints in data storage and retrieval [84]. An approximate memory architecture uses soft approximation through row-level refresh control and hard approximation via data truncation, adjusting DRAM refresh rates and data precision based on significance.

Model Adaptation Through Progressive Shrinking and Once-for-All Networks. Progressive shrinking is a training strategy that optimizes large neural networks by gradually training smaller sub-networks, reducing interference between different architectures while preserving accuracy. This approach enables efficient adaptation to diverse hardware constraints, allowing models to scale seamlessly across different devices [7]. Once-for-All (OFA) Networks extend this concept by training a single, flexible model that can be specialized for different hardware setups without retraining. By decoupling training from neural architecture search and leveraging progressive shrinking, OFA Networks enable highly efficient deep learning deployment, particularly for resource-constrained environments like mobile and edge devices [85].

Event-Driven Processing is a computing approach where operations are executed only when specific events occur such as sensor inputs, user interactions, or data changes, rather than continuously processing data in a clock-driven manner [86]. This is particularly useful in hardware optimization because it reduces unnecessary computations, leading to lower energy consumption and improved efficiency.

3 Use Cases and Applications of Frugal Machine Learning

FML is particularly useful in scenarios requiring efficient AI with minimal computational resources, energy consumption, and data availability. Some key applications and use cases are briefly described in the following subsections.

3.1 IoT and Edge AI

FML enables lightweight AI models to run efficiently on low-power sensors, smart home devices, and IoT systems, reducing reliance on cloud computing [2]. By enabling on-device inference, these models improve real-time decision-making with local data, enhancing both privacy and control over the users' data [5]. Furthermore, FML facilitates decentralized intelligence, allowing IoT and Edge AI systems to operate collaboratively through federated learning and peer-to-peer communication, enabling more robust, adaptive, and resilient AI applications in dynamic environments [3].

3.2 Wearable and Mobile AI

FML powers fitness trackers, activity recognition, and smartwatches, optimizing energy consumption for prolonged battery life [6]. By using compressed ANNs and efficient inference strategies, wearable AI can track heart rate, sleep patterns, and physical activity without frequent cloud interactions [64]. This enables real-time health alerts and personalized recommendations while ensuring that devices last longer on a single charge. Advanced techniques of FML and on-device learning further enhance the capabilities of wearables, allowing them to adapt and improve over time with minimal data exchange [47].

3.3 Autonomous Systems and Robotics

In drones, self-driving cars, and warehouse robots, FML enhances efficiency by enabling real-time, low-power inference for navigation, obstacle detection, and decision-making [57]. Autonomous agents demonstrate robust adaptability across diverse environments, leveraging frugal learning to efficiently process and learn from sensor data. Applications include autonomous delivery robots, smart transportation systems, and industrial automation, where energy-efficient AI ensures prolonged operation, better scalability, and reduced computational overhead while maintaining accuracy in complex environments [7].

3.4 Healthcare and Medical AI

FML supports low-resource diagnostics, remote health monitoring, and medical imaging in areas with limited infrastructure [64]. By deploying lightweight AI models on portable medical devices and mobile health apps, doctors and caregivers can perform real-time patient monitoring, early disease detection, and predictive analytics even in remote locations. For example, compact AI models can analyze X-rays, detect anomalies in electrocardiogram readings, or provide AI-driven telemedicine support, improving accessibility to healthcare without requiring expensive computing resources [51].

3.5   Ambient Intelligence

FML enhances ambient intelligence in smart homes, smart cities, smart buildings, and smart agriculture, by enabling low-power AI models to process data locally on edge devices [1]. In smart homes and buildings, energy-efficient AI optimizes heating, ventilation, and air conditioning, security, and appliance automation. Smart cities benefit from real-time traffic management, waste monitoring, and public safety enhancements. In smart agriculture, FML supports precision farming, pest detection, and irrigation control using low-power remote sensors.

3.6   Bandwidth Constrained Systems

FML facilitates AI applications in remote and rural areas by reducing data transmission needs and preserving bandwidth [72]. By enabling on-device data processing and compression, AI models can extract key insights before sending only essential information to centralized servers. This is particularly useful in satellite-based IoT, remote health diagnostics, and off-grid communication networks, where connectivity is expensive or intermittent [36]. FML ensures efficient data utilization, making AI-powered services accessible even in bandwidth-constrained environments.

3.7   Embedded AI for Consumer Electronics

FML powers smart cameras, voice assistants, and AR/VR devices with optimized processing for real-time responsiveness [52]. By enabling low-latency AI inference directly on devices, users experience faster interactions with technologies like speech recognition, facial authentication, and augmented reality overlays [7]. Techniques such as quantized neural networks and hardware-aware optimizations allow complex AI tasks to be performed efficiently on compact, battery-operated devices, enhancing usability while extending battery life.

3.8 Cybersecurity

FML provides efficient threat detection in networked systems without excessive computational overhead [8]. Lightweight AI models deployed in firewalls, intrusion detection systems, and endpoint security solutions can analyze network traffic, identify anomalies, and prevent cyberattacks in real time without straining system resources [60]. Applications include fraud detection in financial transactions, malware classification on mobile devices, and anomaly detection in industrial networks, ensuring strong security while data requirements and computational load remain low.

4 Discussion

4.1 Observations

To achieve frugality, three main approaches exist: input frugality, learning process frugality, and model frugality [14]. Input frugality emphasizes minimizing the cost associated with data acquisition and feature utilization by using fewer training data or fewer features, driven by data availability, resource, or privacy constraints. Learning process frugality emphasizes minimizing the computational and memory resources required for training a model, often resulting in a less accurate but more resource-efficient model, driven by constraints such as limited computational power and training time. Model frugality emphasizes minimizing the memory, computational resources, and energy required to store and use a ML model, often at the expense of optimal prediction quality.

A key requirement for many ML models is the ability to adapt continuously to new tasks. Instead of retraining an entire ANN from scratch, a process that requires substantial computational power and energy, models can be updated incrementally, preserving learned representations in a more resource-efficient manner. Three core strategies support this approach: parameter regularization, knowledge distillation, and dynamic architecture design [40]. Parameter regularization limits changes to critical weights, ensuring previously acquired knowledge is retained. Knowledge distillation allows newer models to inherit insights from earlier versions without storing entire datasets. Meanwhile, dynamic architectures enable networks to expand selectively, efficiently incorporating new tasks.

Last, FML often aligns with efforts to reduce the environmental impact of AI, sometimes called "green computing." Cutting back on resource use, including energy consumption, can help lower the overall carbon footprint of ML. However, standard methods to measure these gains and compare them across different projects are still not standardized. As research in FML continues, the field is expected to bring more practical, cost-friendly, and environmentally responsible ways to deploy AI in real-world applications.

4.2   Open Challenges of Frugal Machine Learning

Despite significant progress in FML, several challenges must be overcome for broader adoption by ML researchers and practitioners. A key issue stems from the resource-constrained environments where FML is typically deployed, such as IoT and edge computing. Smart environments often handle noisy, redundant, or rapidly changing data, making efficient processing difficult. While techniques like undersampling, feature selection, and data selection help reduce input size and select the most appropriate data samples, they fail to fully compensate for limited processing resources. Additionally, the risk of catastrophic forgetting further complicates the task of ensuring stable and continuous learning, making long-term deployment even more challenging.

Another key challenge lies in balancing model interpretability and fairness with aggressive optimization. Compression techniques, while improving efficiency, often make it harder to understand and explain how a model arrives at its decisions. Additionally, these optimizations can unintentionally introduce bias, particularly when data distributions shift or certain subgroups become underrepresented. Maintaining transparency and fairness is crucial, especially in high-stakes applications like healthcare and finance. Striking the right balance between efficiency and trustworthiness remains an ongoing challenge in the adoption of FML.

One more open challenge is the lack of standardized benchmarks and evaluation metrics capable of capturing various dimensions of resource constraints, from memory and inference latency to overall energy consumption and carbon footprint. Existing studies often focus on either accuracy or speed, employing different datasets and methodologies, which makes fair comparisons and reproducibility difficult. Establishing robust benchmarks that reflect real-world conditions, such as limited connectivity on edge devices or dynamic data streams, would facilitate direct comparisons of new techniques and accelerate progress.

Finally, there is need for better guidance on when and how to deploy various FML strategies across different stages of the ML life cycle, from hyperparameter search to model deployment and maintenance. The sequence in which techniques like pruning and quantization are applied can significantly impact final performance, yet there are no standardized best practices or unified pipelines to streamline these processes. Future advancements could benefit from end-to-end tool chains that seamlessly integrate hyperparameter optimization, model compression, and hardware-aware deployment into a cohesive framework. Additionally, incorporating life cycle cost assessments will be essential to ensure that reductions in model size or training time lead to substantial sustainability gains rather than incremental optimizations.

5 Conclusions

FML offers a resource-aware approach to AI development by employing smaller models, efficient data use, and optimized algorithms. This enables strong performance even under resource constraints like limited processing power, scarce data, or low energy availability. Techniques such as feature selection, model compression, data sampling, and hardware optimization demonstrate FML's adaptability across diverse settings, from mobile devices and IoT sensors to large-scale cloud systems. By promoting efficient resource use, FML supports AI deployment in low-bandwidth environments while advancing sustainability. As the field evolves, it is poised to drive innovation in accessible, efficient, and responsible ML solutions.

Acknowledgments This work was funded by the European Union's Horizon Europe Research and Innovation Programme under grant agreement No. 101121447 Covert and Advanced multimodal Sensor Systems for tArget acquisition and reconnAissance (CASSATA) and innovation program under grant agreement No. 101120237 European Lighthouse of AI for Sustainability (ELIAS).

References

1. Bimpas, A., Violos, J., Leivadeas, A., & Varlamis, I. (2024). Leveraging pervasive computing for ambient intelligence: A survey on recent advancements, applications and open challenges. *Computer Networks, 239*, 110156. https://doi.org/10.1016/j.comnet.2023.110156. Retrieved from https://www.sciencedirect.com/science/article/pii/S1389128623006011
2. Kinnas, M., Violos, J., Kompatsiaris, I., & Papadopoulos, S. (2025). Reducing inference energy consumption using dual complementary CNNs. *Future Generation Computer Systems, 165*, 107606. https://doi.org/10.1016/j.future.2024.107606. Retrieved from https://www.sciencedirect.com/science/article/pii/S0167739X24005703
3. Llisterri Giménez, N., Monfort Grau, M., Pueyo Centelles, R., & Freitag, F. (2022). On-device training of machine learning models on microcontrollers with federated learning. *Electronics, 11*(4), 573. https://doi.org/10.3390/electronics11040573. Retrieved from https://www.mdpi.com/2079-9292/11/4/573
4. Li, D., Chen, X., Becchi, M., & Zong, Z. (2016). Evaluating the energy efficiency of deep convolutional neural networks on CPUs and GPUs. In *2016 IEEE International Conferences on Big Data and Cloud Computing (BDCloud), Social Computing and Networking (SocialCom), Sustainable Computing and Communications (SustainCom) (BDCloud-SocialCom-SustainCom)* (pp. 477–484). https://doi.org/10.1109/BDCloudSocialCom-SustainCom.2016.76. Retrieved from https://ieeexplore.ieee.org/document/7723730/\\?arnumber=7723730
5. Violos, J., Papadopoulos, S., & Kompatsiaris, I. (2024). Towards optimal trade-offs in knowledge distillation for CNNs and vision transformers at the edge. In *2024 32nd European Signal Processing Conference (EUSIPCO)* (pp. 1896–1900). https://doi.org/10.23919/EUSIPCO63174.2024.10715301. Retrieved from https://ieeexplore.ieee.org/abstract/document/10715301
6. Tan, T., & Cao, G. (2023). Deep learning on mobile devices with neural processing units. *Computer, 56*(8), 48–57. https://doi.org/10.1109/MC.2022.3215780. Retrieved from https://ieeexplore.ieee.org/abstract/document/10207072

7. Capra, M., Bussolino, B., Marchisio, A., Masera, G., Martina, M., & Shafique, M. (2020). Hardware and software optimizations for accelerating deep neural networks: Survey of current trends, challenges, and the road ahead. *IEEE Access, 8*, 225134–225180. https://doi. org/10.1109/ACCESS.2020.3039858. Retrieved from https://ieeexplore.ieee.org/documen t/9269334/\\?arnumber=9269334

8. Yin, Y., Jang-Jaccard, J., Xu, W., Singh, A., Zhu, J., Sabrina, F., & Kwak, J. (2023). IGRF-RFE: A hybrid feature selection method for MLP-based network intrusion detection on UNSW-NB15 dataset. *Journal of Big Data, 10*(1), 15. https://doi.org/10.1186/ s40537023-00694-8

9. Cheng, H., Zhang, M., & Shi, J. Q. (2024). A survey on deep neural network pruning: Taxonomy, comparison, analysis, and recommendations. *IEEE Transactions on Pattern Analysis and Machine Intelligence, 46*(12), 10558–10578. https://doi.org/10.1109/TPAMI.2024.3447085. Retrieved from https://ieeexplore.ieee.org/abstract/document/10643325

10. Cherniuk, D., Abukhovich, S., Phan, A.-H., Oseledets, I., Cichocki, A., & Gusak, J. (2024). Quantization aware factorization for deep neural network compression. *Journal of Artificial Intelligence Research, 81*, 973–988. https://doi.org/10.1613/jair.1.16167. Retrieved from https://www.jair.org/index.php/jair/article/view/16167

11. Gou, J., Yu, B., Maybank, S. J., & Tao, D. (2006). Knowledge distillation: A survey. *arXiv:2006.05525 [cs], 129*(6), 1789–1819. https://doi.org/10.1007/s11263-021-01453-z. Retrieved from http://arxiv.org/abs/2006.05525

12. Pandey, P., Basu, P., Chakraborty, K., & Roy, S. (2019). GreenTPU: Improving timing error resilience of a near-threshold tensor processing unit. In *Proceedings of the 56th Annual Design Automation Conference 2019, DAC '19* (pp. 1–6). Association for Computing Machinery. https://doi.org/10.1145/3316781.3317835

13. Cancela, B., Bolón-Canedo, V., & Alonso-Betanzos, A. (2023). E2E-FS: An end-to-end feature selection method for neural networks. *IEEE Transactions on Pattern Analysis and Machine Intelligence, 45*(7), 8311–8323. https://doi.org/10.1109/TPAMI.2022.3228824. Retrieved from https://ieeexplore.ieee.org/abstract/document/9983480

14. Kinnas, M., Violos, J., Karapiperis, N. I., & Kompatsiaris, I. (2025). Selecting images with entropy for frugal knowledge distillation. *IEEE Access, 13*, 28189–28203. https://doi. org/10.1109/ACCESS.2025.3540384. Retrieved from https://ieeexplore.ieee.org/abstract/ document/10878975

15. Karathanasis, A., Violos, J., Kompatsiaris, I., & Papadopoulos, S. (2025). A brief review for compression and transfer learning techniques in DeepFake detection. *arXiv:2504.21066 [cs].* https://doi.org/10.48550/arXiv.2504.21066

16. Salmani Pour Avval, S., Eskue, N. D., Groves, R. M., & Yaghoubi, V. (2025). Systematic review on neural architecture search. *Artificial Intelligence Review, 58*(3), 73. https://doi. org/10.1007/s10462-024-11058-w

17. Curry, E., Heintz, F., Irgens, M., Smeulders, A. W. M., & Stramigioli, S. (2022). Partnership on AI, data, and robotics. *Communications of the ACM, 65*(4), 54–55. https://doi. org/10.1145/3513000

18. Karathanasis, A., Violos, J., & Kompatsiaris, I. (2025). A comparative analysis of compression and transfer learning techniques in DeepFake detection models. *Mathematics, 13*(5), 887. https://doi.org/10.3390/math13050887. Retrieved from https://www.mdpi. com/2227-7390/13/5/887

19. Liebenwein, L., Baykal, C., Carter, B., Gifford, D., & Rus, D. (2021). Lost in pruning: The effects of pruning neural networks beyond test accuracy. *arXiv:2103.03014 [cs].* https://doi. org/10.48550/arXiv.2103.03014

20. He, Y., & Xiao, L. (2024). Structured pruning for deep convolutional neural networks: A survey. *IEEE Transactions on Pattern Analysis and Machine Intelligence, 46*(5), 2900–2919. https://doi.org/10.1109/TPAMI.2023.3334614. Retrieved from https://ieeexplore.ieee.org/ abstract/document/10330640

21. Zou, J., Rui, T., Zhou, Y., Yang, C., & Zhang, S. (2018). Convolutional neural network simplification via feature map pruning. *Computers & Electrical Engineering, 70*, 950–958. https://doi.org/10.1016/j.compeleceng.2018.01.036. Retrieved from https://linkinghub.elsevier.com/retrieve/pii/\\S0045790617326393

22. Li, H., Kadav, A., Durdanovic, I., Samet, H., & Graf, H. P. (2017). Pruning filters for efficient ConvNets. *arXiv:1608.08710 [cs]*. https://doi.org/10.48550/arXiv.1608.08710

23. Chen, C., Li, K., Zou, X., & Li, Y. (2021). DyGNN: Algorithm and architecture support of dynamic pruning for graph neural networks. In *2021 58th ACM/IEEE Design Automation Conference (DAC)* (pp. 1201–1206). https://doi.org/10.1109/DAC18074.2021.9586298. Retrieved from https://ieeexplore.ieee.org/document/9586298

24. Li, G., Yang, P., Qian, C., Hong, R., & Tang, K. (2024). Stage-wise magnitude-based pruning for recurrent neural networks. *IEEE Transactions on Neural Networks and Learning Systems, 35*(2), 1666–1680. https://doi.org/10.1109/TNNLS.2022.3184730. Retrieved from https://ieeexplore.ieee.org/document/9808105

25. Molchanov, P., Tyree, S., Karras, T., Aila, T., & Kautz, J. (2017). Pruning convolutional neural networks for resource efficient inference. *arXiv:1611.06440 [cs]*. https://doi.org/10.48550/arXiv.1611.06440

26. Song, K., Yao, W., & Zhu, X. (2023). Filter pruning via similarity clustering for deep convolutional neural networks. In *Neural Information Processing: 29th International Conference, ICONIP 2022, Virtual Event, November 22–26, 2022, Proceedings, Part I* (pp. 88–99). Springer. https://doi.org/10.1007/978-3-031-30105-6_8

27. Guo, Y., Yao, A., & Chen, Y. (2016). Dynamic network surgery for efficient DNNs. In *Advances in neural information processing systems* (Vol. 29). Curran Associates, Inc. Retrieved from https://proceedings.neurips.cc/paper/2016/hash/2823f4797102ce1a1aec05359cc16dd9-Abstract.html

28. Frankle, J., & Carbin, M. (2019). The lottery ticket hypothesis: Finding sparse, trainable neural networks. *arXiv:1803.03635 [cs]*. https://doi.org/10.48550/arXiv.1803.03635

29. Pham, C., Hoang, T., & Do, T.-T. (2023). Collaborative multi-teacher knowledge distillation for learning low bit-width deep neural networks. *arXiv*, 6435–6443. Retrieved from https://openaccess.thecvf.com/content/WACV2023/html/Pham_Collaborative_Multi-Teacher_Knowledge_Distillation_for_Learning_Low_Bit-Width_Deep_Neural_WACV_2023_paper.html

30. Tsanakas, S., Hameed, A., Violos, J., & Leivadeas, A. (2024). A light-weight edge-enabled knowledge distillation technique for next location prediction of multitude transportation means. *Future Generation Computer Systems, 154*, 45–58. https://doi.org/10.1016/j.future.2023.12.025. Retrieved from https://www.sciencedirect.com/science/article/pii/S0167739X23004867

31. Yang, C., Yu, X., An, Z., & Xu, Y. (2023). Categories of response-based, feature-based, and relation-based knowledge distillation. In W. Pedrycz & S.-M. Chen (Eds.), *Advancements in knowledge distillation: Towards new horizons of intelligent systems* (pp. 1–32). Springer International Publishing. https://doi.org/10.1007/978-3-031-32095-8_1

32. Wu, T., Tao, C., Wang, J., Yang, R., Zhao, Z., & Wong, N. (2025). Rethinking Kullback-Leibler divergence in knowledge distillation for large language models. In O. Rambow, L. Wanner, M. Apidianaki, H. Al-Khalifa, B. D. Eugenio, & S. Schockaert (Eds.), *Proceedings of the 31st International Conference on Computational Linguistics* (pp. 5737–5755). Association for Computational Linguistics. Retrieved from https://aclanthology.org/2025.coling-main.383/

33. Cai, G., Li, J., Liu, X., Chen, Z., & Zhang, H. (2023). Learning and compressing: Low-rank matrix factorization for deep neural network compression. *Applied Sciences, 13*(4), 2704. https://doi.org/10.3390/app13042704. Retrieved from https://www.mdpi.com/2076-3417/13/4/2704

34. Yang, H., Tang, M., Wen, W., Yan, F., Hu, D., Li, A., Li, H., & Chen, Y. (2020). *Learning low-rank deep neural networks via singular vector orthogonality regularization and singular value sparsification* (pp. 678–679). IEEE Xplore. Retrieved from https://openaccess.thecvf.com/content_CVPRW_2020/html/w40/Yang_Learning_Low-Rank_Deep_Neural_Networks_via_Singular_Vector_Orthogonality_Regularization_CVPRW_2020_paper.html

35. Park, S., & Moon, S.-M. (2025). CURing large models: Compression via CUR decomposition. *arXiv:2501.04211 [cs]*. https://doi.org/10.48550/arXiv.2501.04211
36. Nalepa, J., Antoniak, M., Myller, M., Ribalta Lorenzo, P., & Marcinkiewicz, M. (2020). Towards resource-frugal deep convolutional neural networks for hyperspectral image segmentation. *Microprocessors and Microsystems, 73*, 102994. https://doi.org/10.1016/j.micpro.2020.102994. Retrieved from https://linkinghub.elsevier.com/retrieve/\\pii/S0141933119302844
37. Menghani, G. (2021). Efficient deep learning: A survey on making deep learning models smaller, faster, and better. *ACM Computing Surveys, 55*(12), 1–37. https://doi.org/10.1145/3578938
38. Tabani, H., Balasubramaniam, A., Marzban, S., Arani, E., & Zonooz, B. (2021). Improving the efficiency of transformers for resource-constrained devices. *arXiv:2106.16006 [cs]*. Retrieved from http://arxiv.org/abs/2106.16006
39. Polino, A., Pascanu, R., & Alistarh, D. (2018). Model compression via distillation and quantization. *arXiv:1802.05668 [cs]*. https://doi.org/10.48550/arXiv.1802.05668
40. Liu, H., Zhou, Y., Liu, B., Zhao, J., Yao, R., & Shao, Z. (2023). Incremental learning with neural networks for computer vision: a survey. *Artificial Intelligence Review, 56*(5), 4557–4589. https://doi.org/10.1007/s10462-02210294-2
41. Wang, X., Yu, F., Dou, Z.-Y., Darrell, T., & Gonzalez, J. E. (2018). SkipNet: Learning dynamic routing in convolutional networks. In *Proceedings of the European Conference on Computer Vision (ECCV)* (pp. 409–424) Retrieved from https://openaccess.thecvf.com/content_ECCV_2018/html/Xin_Wang\ _SkipNet_Learning_Dynamic_ECCV_2018_paper.html
42. Liu, L., & Deng, J. (2018). Dynamic deep neural networks: Optimizing accuracy-efficiency trade-offs by selective execution. In *Proceedings of the AAAI Conference on Artificial Intelligence* (Vol. 32(1)). https://doi.org/10.1609/aaai.v32i1.11630. Retrieved from https://ojs.aaai.org/index.php/AAAI/article/view/11630
43. Verelst, T., & Tuytelaars, T. (2020). *Dynamic convolutions: Exploiting spatial sparsity for faster inference* (pp. 2320–2329). CVF Open Access. Retrieved from https://openaccess.thecvf.com/content_CVPR_2020/html/Verelst_Dynamic_Convolutions_Exploiting_Spatial_Sparsity_for_Faster_Inference_CVPR_2020_paper.html
44. Varghese, A. J., Bora, A., Xu, M., & Karniadakis, G. E. (2024). TransformerG2G: Adaptive time-stepping for learning temporal graph embeddings using transformers. *Neural Networks, 172*, 106086. https://doi.org/10.1016/j.neunet.2023.12.040. Retrieved from https://www.sciencedirect.com/science/article/pii/S0893608023007475
45. Zhou, Y., Lei, T., Liu, H., Du, N., Huang, Y., Zhao, V., Dai, A. M., Chen, Z., Le, Q. V., & Laudon, J. (2022). Mixture-of-experts with expert choice routing. *Advances in Neural Information Processing Systems, 35*, 7103–7114. Retrieved from https://proceedings.neurips.cc/paper_files/paper/2022/hash/2f00ecd787b432c1d36f3de9800728eb-Abstract-Conference.html
46. Elkerdawy, S., Elhoushi, M., Zhang, H., & Ray, N. (2022). Fire together wire together: A dynamic pruning approach with self-supervised mask prediction. In *Proceedings of the IEEE/CVF Conference on Computer Vision and Pattern Recognition (CVPR)* (pp. 12454–12463) Retrieved from https://openaccess.thecvf.com/content/CVPR2022/html/Elkerdawy_Fire_Together_Wire_Together_A_Dynamic_Pruning_Approach_With_Self-Supervised_CVPR_2022_paper.html
47. Yu, J., Yang, L., Xu, N., Yang, J., & Huang, T. (2018). Slimmable neural networks. *arXiv:1812.08928 [cs]*. https://doi.org/10.48550/arXiv.1812.08928
48. Li, C., Wang, G., Wang, B., Liang, X., Li, Z., & Chang, X. (2021). *Dynamic slimmable network* (pp. 8607–8617). CVF Open Access. Retrieved from https://openaccess.thecvf.com/content/CVPR2021/html/Li_Dynamic_Slimmable_Network_CVPR_2021_paper.html
49. Qiu, J., Chen, C., Liu, S., Zhang, H.-Y., & Zeng, B. (2021). SlimConv: Reducing channel redundancy in convolutional neural networks by features recombining. *IEEE Transactions on Image Processing, 30*, 6434–6445. https://doi.org/10.1109/TIP.2021.3093795. Retrieved from https://ieeexplore.ieee.org/abstract/document/9477103

50. Xu, G., Wang, X., Wu, X., Leng, X., & Xu, Y. (2024). Development of skip connection in deep neural networks for computer vision and medical image analysis: A survey. *arXiv:2405.01725 [eess]*. https://doi.org/10.48550/arXiv.2405.01725

51. Lee, K. H., & Verma, N. (2013). A low-power processor with configurable embedded machine learning accelerators for high-order and adaptive analysis of medical-sensor signals. *IEEE Journal of Solid-State Circuits, 48*(7), 1625–1637. https://doi.org/10.1109/JSSC.2013.2253226. Retrieved from https://ieeexplore.ieee.org/document/6493458/\\?arnumber=6493458

52. Apicharttrisorn, K., Ran, X., Chen, J., Krishnamurthy, S. V., & Roy-Chowdhury, A. K. (2019). Frugal following: Power thrifty object detection and tracking for mobile augmented reality. In *Proceedings of the 17th Conference on Embedded Networked Sensor Systems* (pp. 96–109). ACM. https://doi.org/10.1145/3356250.3360044

53. Park, E., Kim, D., Kim, S., Kim, Y.-D., Kim, G., Yoon, S., & Yoo, S. (2015). Big/little deep neural network for ultra low power inference. In *2015 International Conference on Hardware/Software Codesign and System Synthesis (CODES+ISSS)* (pp. 124–132). https://doi.org/10.1109/CODESISSS.2015.7331375. Retrieved from https://ieeexplore.ieee.org/document/7331375/?arnumber=7331375

54. Wojcik, B., Przewiezlikowski, M., Szatkowski, F., Wolczyk, M., Balazy, K., Krzepkowski, B., Podolak, I., Tabor, J., Smieja, M., & Trzcinski, T. (2023). Zero time waste in pre-trained early exit neural networks. *Neural Networks, 168*, 580–601. https://doi.org/10.1016/j.neunet.2023.10.003. Retrieved from https://www.sciencedirect.com/science/article/pii/S0893608023005555

55. Rahmath, H. P., Srivastava, V., Chaurasia, K., Pacheco, R. G., & Couto, R. S. (2024). Early-exit deep neural network—A comprehensive survey. *ACM Computing Surveys, 57*(3), 1–37. https://doi.org/10.1145/3698767

56. Cholet, S., & Biabiany, E. (2023). A framework for frugal supervised learning with incremental neural networks. *Applied Sciences, 13*(9), 5489. https://doi.org/10.3390/app13095489

57. Kawewong, A., Tangruamsub, S., Kankuekul, P., & Hasegawa, O. (2011). Fast online incremental transfer learning for unseen object classification using self-organizing incremental neural networks. In *The 2011 International Joint Conference on Neural Networks* (pp. 749–756). https://doi.org/10.1109/IJCNN.2011.6033296. Retrieved from https://ieeexplore.ieee.org/abstract/document/\\6033296

58. Grossberg, S. (1988). Nonlinear neural networks: Principles, mechanisms, and architectures. *Neural Networks, 1*(1), 17–61. https://doi.org/10.1016/0893-6080(88)90021-4. Retrieved from https://www.sciencedirect.com/science/article/pii/\\0893608088900214

59. Adadi, A. (2021). A survey on data-efficient algorithms in big data era. *Journal of Big Data, 8*(1), 24. https://doi.org/10.1186/s40537-021-00419-9

60. Sarantos, P., Violos, J., & Leivadeas, A. (2025). Enabling semi-supervised learning in intrusion detection systems. *Journal of Parallel and Distributed Computing, 196*, 105010. https://doi.org/10.1016/j.jpdc.2024.105010. Retrieved from https://www.sciencedirect.com/science/article/pii/S0743731524001746

61. Tu, H., & Menzies, T. (2021). FRUGAL: Unlocking semi-supervised learning for software analytics. In *2021 36th IEEE/ACM International Conference on Automated Software Engineering (ASE)* (pp. 394–406). https://doi.org/10.1109/ASE51524.2021.9678617. Retrieved from https://ieeexplore.ieee.org/abstract/\\document/9678617

62. Senthamarai Kannan, S., & Ramaraj, N. (2010). A novel hybrid feature selection via Symmetrical Uncertainty ranking based local memetic search algorithm. *Knowledge-Based Systems, 23*(6), 580–585. https://doi.org/10.1016/j.knosys.2010.03.016. Retrieved from https://www.sciencedirect.com/science/article/pii/S0950705110000602

63. Urbanowicz, R. J., Meeker, M., La Cava, W., Olson, R. S., & Moore, J. H. (2018). Relief-based feature selection: Introduction and review. *Journal of Biomedical Informatics, 85*, 189–203. https://doi.org/10.1016/j.jbi.2018.07.014. Retrieved from https://www.sciencedirect.com/science/article/pii/S1532046418301400

64. Saha, P., Patikar, S., & Neogy, S. (2020). A correlation—Sequential forward selection based feature selection method for healthcare data analysis. In *2020 IEEE International Conference on Computing, Power and Communication Technologies (GUCON)* (pp. 69–72). https://doi.org/10.1109/GUCON48875.2020.9231205. Retrieved from https://ieeexplore.ieee.org/abstract/document/9231205

65. Nguyen, H. B., Xue, B., Liu, I., & Zhang, M. (2014). Filter based backward elimination in wrapper based PSO for feature selection in classification. In *2014 IEEE Congress on Evolutionary Computation (CEC)* (pp. 3111–3118). https://doi.org/10.1109/CEC.2014.6900657. Retrieved from https://ieeexplore.ieee.org/abstract/document/6900657

66. Jiménez-Cordero, A., Morales, J. M., & Pineda, S. (2021). A novel embedded min-max approach for feature selection in nonlinear Support Vector Machine classification. *European Journal of Operational Research, 293*(1), 24–35. https://doi.org/10.1016/j.ejor.2020.12.009. Retrieved from https://www.sciencedirect.com/science/article/pii/S0377221720310195

67. Seijo-Pardo, B., Porto-Díaz, I., Bolón-Canedo, V., & Alonso-Betanzos, A. (2017). Ensemble feature selection: Homogeneous and heterogeneous approaches. *Knowledge-Based Systems, 118*, 124–139. https://doi.org/10.1016/j.knosys.2016.11.017. Retrieved from https://www.sciencedirect.com/science/article/pii/S0950705116304749

68. Mohammed, R., Rawashdeh, J., & Abdullah, M. (2020). Machine learning with oversampling and undersampling techniques: Overview study and experimental results. In *2020 11th International Conference on Information and Communication Systems (ICICS)* (pp. 243–248). https://doi.org/10.1109/ICICS49469.2020.239556. Retrieved from https://ieeexplore.ieee.org/document/9078901/

69. Werner de Vargas, V., Schneider Aranda, J. A., Dos Santos Costa, R., Da Silva Pereira, P. R., & Victória Barbosa, J. L. (2023). Imbalanced data preprocessing techniques for machine learning: A systematic mapping study. *Knowledge and Information Systems, 65*(1), 31–57. https://doi.org/10.1007/s10115-022-01772-8

70. Chen, S., Dorn, S., Lell, M., Kachelrieß, M., & Maier, A. (2018). Manifold learning-based data sampling for model training. In A. Maier, T. M. Deserno, H. Handels, K. H. Maier-Hein, C. Palm, & T. Tolxdorff (Eds.), *Bildverarbeitung für die Medizin 2018* (pp. 269–274). Springer. https://doi.org/10.1007/978-3-662-56537-7_70

71. Wang, T., Zhu, J.-Y., Torralba, A., & Efros, A. A. (2020). Dataset distillation. *arXiv:1811.10959 [cs]*. https://doi.org/10.48550/arXiv.1811.10959

72. Yu, R., Liu, S., & Wang, X. (2023). Dataset distillation: A comprehensive review. *arXiv:2301.07014 [cs]*. Retrieved from http://arxiv.org/abs/2301.07014

73. Tsanakas, S., Hameed, A., Violos, J., & Leivadeas, A. (2021). An innovative neuro-genetic algorithm and geometric loss function for mobility prediction. In *Proceedings of the 19th ACM International Symposium on Mobility Management and Wireless Access, MobiWac '21* (pp. 25–32). Association for Computing Machinery. https://doi.org/10.1145/3479241.3486706

74. Cho, H., Kim, Y., Lee, E., Choi, D., Lee, Y., & Rhee, W. (2020). Basic enhancement strategies when using Bayesian optimization for hyperparameter tuning of deep neural networks. *IEEE Access, 8*, 52588–52608. https://doi.org/10.1109/ACCESS.2020.2981072

75. Diaw, A., McKerns, M., Sagert, I., Stanton, L. G., & Murillo, M. S. (2024). Efficient learning of accurate surrogates for simulations of complex systems. *Nature Machine Intelligence, 6*(5), 568–577. https://doi.org/10.1038/s42256-024-00839-1. Retrieved from https://www.nature.com/articles/s42256-024-00839-1

76. Chan, T. C. Y., Mahmood, R., & Zhu, I. Y. (2023). *Inverse optimization: Theory and applications*. Operations Research Publisher: INFORMS. https://doi.org/10.1287/opre.2022.0382. Retrieved from https://pubsonline.informs.org/doi/full/10.1287/opre.2022.0382

77. Shetty, A. M., Aljunid, M. F., Manjaiah, D. H., & Shaik Afzal, A. M. S. (2024). Hyperparameter optimization of machine learning models using grid search for amazon review sentiment analysis. In S. J. Nanda, R. P. Yadav, A. H. Gandomi, & M. Saraswat (Eds.), *Data science and applications* (pp. 451–474). Springer Nature. https://doi.org/10.1007/978-981-99-7814-4_36

78. Bergstra, J., & Bengio, Y. (2012). Random search for hyper-parameter optimization. *Journal of Machine Learning Research, 13*(10), 281–305. Retrieved from http://jmlr.org/papers/v13/bergstra12a.html

79. Bischl, B., Binder, M., Lang, M., Pielok, T., Richter, J., Coors, S., Thomas, J., Ullmann, T., Becker, M., Boulesteix, A.-L., Deng, D., & Lindauer, M. (2023). Hyperparameter optimization: Foundations, algorithms, best practices, and open challenges. *Wiley Interdisciplinary Reviews. Data Mining and Knowledge Discovery, 13*(2), e1484. https://doi.org/10.1002/widm.1484

80. Parnell, T., Duenner, C., Atasu, K., Sifalakis, M., & Pozidis, H. (2017). Large-scale stochastic learning using GPUs. In *2017 IEEE International Parallel and Distributed Processing Symposium Workshops (IPDPSW)* (pp. 419–428). https://doi.org/10.1109/IPDPSW.2017.140. Retrieved from https://ieeexplore.ieee.org/abstract/document/7965076

81. Yan, F., Koch, A., & Sinnen, O. (2024). A survey on FPGA-based accelerator for ML models. *arXiv:2412.15666 [cs]*. https://doi.org/10.48550/arXiv.2412.15666

82. Balaji, A., Das, A., Wu, Y., Huynh, K., Dell'Anna, F. G., Indiveri, G., Krichmar, J. L., Dutt, N. D., Schaafsma, S., & Catthoor, F. (2020). Mapping spiking neural networks to neuromorphic hardware. *IEEE Transactions on Very Large Scale Integration (VLSI) Systems, 28*(1), 76–86. https://doi.org/10.1109/TVLSI.2019.2951493. Retrieved from https://ieeexplore.ieee.org/document/8913677

83. Schuman, C. D., Potok, T. E., Patton, R. M., Birdwell, J. D., Dean, M. E., Rose, G. S., & Plank, J. S. (2017). A survey of neuromorphic computing and neural networks in hardware. *arXiv:1705.06963 [cs]*. https://doi.org/10.48550/arXiv.1705.06963

84. Nguyen, D. T., Hung, N. H., Kim, H., & Lee, H.-J. (2019). An approximate memory architecture for energy saving in deep learning applications. *IEEE Transactions on Circuits and Systems I, Regular Papers, 67*(5), 1588–1601. https://doi.org/10.1109/TCSI.2019.2962516. Retrieved from https://ieeexplore.ieee.org/document/8957353/

85. Cai, H., Gan, C., Wang, T., Zhang, Z., & Han, S. (2019). Once-for-all: Train one network and specialize it for efficient deployment. *arXiv:1908.09791 [cs, stat]*. Retrieved from http://arxiv.org/abs/1908.09791

86. Kuhrt, M., Glombiewski, N., Körber, M., Morgen, A., Brandenstein, D., & Seeger, B. (2025). Efficient event processing on modern hardware. In K.-U. Sattler, A. Kemper, T. Neumann, & J. Teubner (Eds.), *Scalable data management for future hardware* (pp. 65–89). Springer Nature Switzerland. https://doi.org/10.1007/978-3-031-74097-8_3

Transformations

The Role of Digital Transformation in Manufacturing Under the Light of AI Integration

John Angelopoulos and Kosmas Alexopoulos

Abstract Digital transformation in manufacturing represents a strategic evolution beyond digitization and digitalization, aiming to redefine industrial processes through the integration of intelligent technologies, such as artificial intelligence (AI). This chapter explores how the integration of AI facilitates this transformation within the frameworks of Industry 4.0 and Industry 5.0 (I5.0). While the main theme of Industry 4.0 is the promotion of automation and smart manufacturing, in I5.0, emphasis is focused on human centricity, resilience, and sustainability, emerging new challenges over traditional implementations. Through critical literature review and the presentation of real-life industrial case studies, the organizational, operational, and technological aspects of digital transformation are examined, with special attention to challenges faced by small and medium-sized enterprises (SMEs). The analysis is framed within the Technology Organization Environment (TOE) model, addressing the need for adaptable and inclusive transition strategies. Furthermore, the contribution of this chapter extends to the provision of a roadmap for AI-enabled digital maturity, promoting resilient, low-carbon, and human-aligned manufacturing systems. The discussion aligns with the priorities of the AI, Data, and Robotics Partnership and supports the strategic objectives of Europe's digital and green transitions.

Keywords Digital transformation · Artificial intelligence (AI) · Industry 5.0 · Sustainability · Servitization

J. Angelopoulos
Laboratory for Manufacturing Systems and Automation, Department of Mechanical Engineering and Aeronautics, University of Patras, Rio Patras, Greece
e-mail: angelopoulos@lms.mech.upatras.gr

K. Alexopoulos (✉)
Laboratory for Manufacturing Systems and Automation, Department of Mechanical Engineering and Aeronautics, University of Patras, Rio Patras, Greece

Department of Digital Industry Technologies, National and Kapodistrian University of Athens, Psachna, Greece
e-mail: alexokos@lms.mech.upatras.gr; kalexopoulos@dind.uoa.gr

E. Curry et al. (eds.), *Artificial Intelligence, Data and Robotics*,
https://doi.org/10.1007/978-3-032-10561-5_9

1 Introduction

The evolution of the manufacturing sector has been shaped by successive industrial revolutions. Industry 1.0 introduced mechanical production powered by water and steam in the late eighteenth century. Industry 2.0 followed in the late nineteenth century, integrating electricity and introducing the mass production paradigm. The third revolution, Industry 3.0, in the 1970s, introduced automation through electronics and information technologies (IT).

The current phase, Industry 4.0 (I4.0), is characterized by the integration of technologies such as the Internet of Things (IoT), cloud computing, AI, and digital twins. These technologies enable the creation of smart cyber-physical systems (CPS), which tightly couple the digital and physical realms, enabling real-time optimization and advanced decision-making. Despite these advances, I4.0 has primarily focused on efficiency, flexibility, and digital connectivity, often at the expense of sustainability and human well-being.

In response to these limitations, I5.0 has emerged as a complementary paradigm that emphasizes human-centricity, resilience, and sustainability. While I4.0 remains ongoing, I5.0 repositions technology to serve broader societal goals, promoting a vision aligned with Europe's digital and green transitions and related to the broader concept of Society 5.0. To that end, I5.0 is foreseen to touch upon key areas, such as sustainability of operations (e.g., energy efficiency, waste reduction, social responsibility etc.), resilience as a medium to mitigate the effect disruptions have in manufacturing and production systems (e.g., supply chain disruptions, labor shortages etc.), and emerging paradigms including among others circular economy and green manufacturing, which promote reuse and resource optimization.

To contextualize the broader strategic transformation, it is essential to distinguish related terms, like "digitization" and "digitalization." Digitization refers to the conversion of analog information into digital form, while digitalization involves the utilization of digital technologies for improving business processes [1]. Digital transformation extends these concepts by representing a holistic strategic transformation, which integrates digital technologies across all aspects of an organization. Concretely, it includes changes in culture, operations, and value delivery, enabling organizations to continuously adapt and innovate.

Digital transformation refers to the comprehensive integration of digital technologies into all areas of a business, resulting in fundamental changes to how organizations operate and deliver value to customers. Unlike digitization (conversion of analog to digital data) or digitalization (use of digital tools to improve processes), digital transformation is a strategic and cultural shift that redefines business models, operations, and customer engagement [2]. IBM frames this shift as a business strategy initiative that incorporates digital technologies across all areas of an organization [3]. Therefore, processes, operations, products, services, and technologies have to be carefully evaluated, in order to enable continuous adaptation and rapid and customer-driven innovation. McKinsey defines digital transformation as a more vital component for business sustainability [4]. Specifically, it is defined as a complete rewiring of an organization, aiming at the continuous integration of technology at a scale, mandatory for a company to survive as well as to maintain their competitive edge. In Fig. 1, digital transformation has been divided into four key dimensions, i.e., (1) technology, (2) management, (3) change, and (4) outcomes, clarifying the actions involved as well as the expected outcomes.

According to recent market research insights, the global digital transformation market was valued at $2.27 trillion during 2023 and is projected to grow to $12.35 trillion by 2032, exhibiting a compound annual growth rate (CAGR) of 20.9% during the forecast period. This exponential growth underscores the increasing importance of digital transformation as a strategic imperative for manufacturing industries worldwide. However, the uneven adoption of digital solutions globally highlights that many regions and sectors are still lagging in their transition toward Industry 4.0 maturity. North America dominated the market with a 44.49% share in 2023, and the US digital transformation market is anticipated to reach $2.39 trillion by 2032, driven by rising adoption and investment in digital technologies (e.g., cloud infrastructure, IoT, and 5G technologies) [5], positioning it as a global leader in advanced manufacturing digitalization.

Among the digital technologies driving this transformation, AI is one of the driver technologies. AI enables predictive analytics, autonomous systems, intelligent decision-making, and real-time optimization, all of which are fundamental to the shift toward truly data-driven and adaptable operations. As such, AI is much more than a single component of digital transformation. It can be considered as one of its primary enablers, particularly in modern manufacturing environments where responsiveness, precision, and resilience are critical objectives.

Although the concept of AI was introduced decades ago, its integration into industrial systems became feasible with the recent advances in data availability, cloud infrastructure, and computing power [6]. Specifically, within the context of I4.0, data became an additional by-product of the manufacturing and production processes, necessitating the development and integration of complex smart algorithms (i.e., AI algorithms), which provide added value to the stakeholders, by

Fig. 1 The four dimensions of digital transformation. (Created by the authors)

analyzing in near-real-time vast amounts of data, and in most cases predicting certain outcomes [7–9]. As of late, market analysis indicates that by 2030, the market size of AI will reach \$15.7 trillion, which corresponds to a 14% increase in global GDP [10]. Therefore, AI takes the leading position in the concurrent market landscape. Delving more into the market analysis, AI is expected to deliver productivity gains, which are estimated at \$6.6 trillion, and increased consumer demand (AI-driven personalization and quality improvements), which is estimated at \$9.1 trillion.

This chapter aims to analyze the current progress of digitalization at a global level, identify key barriers to the transition from I4.0 to I5.0, and evaluate how AI enables sustainable and human-centric manufacturing transformation. Therefore, AI, as a central enabler of digital transformation, is a cornerstone technology not only in data analysis but also in driving automation, adaptability, and innovation.

The remainder of the chapter is organized as follows. Section 2 presents a detailed overview of digital transformation in manufacturing. Then, Sect. 3 investigates servitization along with green manufacturing paradigms, highlighting how digital transformation facilitates the shift from product-based models to sustainable serviced-based models. In Sect. 4, industrial case studies are provided to illustrate the role of AI in key areas of manufacturing (scheduling, maintenance, predictive analytics). Finally, Sect. 5 concludes the chapter and outlines future research directions.

2   Overview Digital Transformation in Manufacturing

2.1   Transition from Industry 4.0 to Industry 5.0

Industrial revolutions have been well documented in the last 10 years, especially after the buzz created from I4.0 and the digital transformation of everything. Furthermore, technological inequality between developed and developing countries has been a key discussion point, creating further confusion, especially after the introduction of the I5.0 term. The key question at this point is how less developed countries and companies close the gap. Although this is an intricate point, the solution might be easier. Concretely, I4.0 and I5.0 should be regarded as mutually complementary rather than mutually exclusive. In Fig. 2, the authors have attempted to identify the key differences between I4.0 and I5.0.

Sarkar et al. [11] discuss the transition from I4.0 to I5.0 and the most influential factors shaping this shift in smart manufacturing. The article highlights that while I4.0 ushered in automation, digitalization, and smart manufacturing, it lacked sustainability, resilience, and human-centric approaches. Conversely, I5.0 aims to integrate high technology with human intelligence, sustainability, and flexibility to deliver a more comprehensive and future-proof industrial system. The transition toward I5.0 is being driven by rapid development in information and

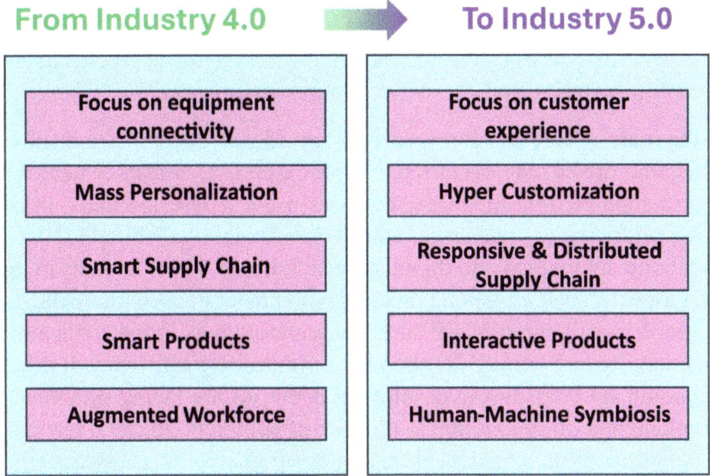

Fig. 2 The leap from Industry 4.0 to Industry 5.0. (Created by the authors)

communication technologies (ICTs) such as big data analytics, AI, the IoT, CPS, robotics, augmented and virtual reality (AR/VR), and cloud computing. These technologies have an essential function in making the manufacturing processes more intelligent, efficient, and responsive to market as well as supply chain modifications. Among the most important findings of the study is the fact that I5.0 is necessary to resolve the deficiencies of I4.0. The COVID-19 pandemic and global supply chain disruptions, among other incidents, demonstrated the weakness of highly automated systems with zero flexibility and human involvement. I5.0 tries to bridge this gap by introducing human intelligence again into production to ensure resilience and sustainability. Considering the above mentioned, there are several critical success factors for the transition from I4.0 to I5.0, which are summarized in the following seven points below:

1. Companies must invest in the adoption of advanced digital technologies (e.g., AI, IoT, real-time data analytics) in order to create added value in the context of I5.0 [12].
2. The smart manufacturing paradigm dictates that high-level data analysis capabilities are required in order to optimize operations as well as to increase operational efficiency [13].
3. Digital transformation and the subsequent transition to I4.0 and I5.0 requires significant investment in technology and training [11].
4. The transition from I4.0 to I5.0 is a continuous adaptation trip. Therefore, companies need to integrate I5.0 philosophies into their organizational long-term strategy [14].
5. Digital transformation, despite its benefits for companies, comes with risks related to cybersecurity. Therefore, data security measures are essential to maintain sensitive data safe [15].

6. Employees require upskilling and reskilling in order to handle new technology and establish harmony between human and machine interactions [16].
7. I5.0 in general focuses on the adoption of the paradigms of green manufacturing, low carbon footprint, and circular economy principles [17].

One of the main differences between I4.0 and I5.0 is the shift from fully autonomous, machine-driven manufacturing to a more adaptive, human-centered approach. This is also supported by the three keywords used to describe I5.0—in particular, (1) human centricity, (2) resilience, and (3) sustainability. While I4.0 was focused on maximizing automation and digital connectivity, I5.0 puts humans in the center, in several aspects, such as, among others, involvement in decision-making, creativity, and problem-solving, to make sure that technology is in the service of humans and sustainability and not just for the sake of improving efficiency. It is a common phenomenon that companies in developing countries are facing significant hurdles to the implementation of I5.0 due to the lack of sufficient financial resources, scarcity of skilled workforce, and weak regulatory frameworks. Consequently, policymakers must support businesses with tax incentives, creation of digital infrastructure, and skill-development programs to enable smooth transition.

Transition from I4.0 to I5.0, especially in the manufacturing sector, necessitates the design, development, and implementation of a systematic roadmap for digital transformation, since the latter constitutes a multidimensional and strategic change rather than a technological upgrade [18]. Although efforts under I4.0 have been focused on automation, smart manufacturing, and data-driven decision-making through the corresponding digital technologies (i.e., AI, IoT, cloud computing, CPS, etc.), the shift toward I5.0 requires a human-centric, resilient, and sustainable approach, which is also its main theme.

The shift becomes imperative by the fact that fully autonomous systems leave very little space for adaptability and flexibility during times of crisis (e.g., the COVID-19 pandemic) that highlighted vulnerabilities in supply chain and manufacturing processes. In fact, the global pandemic has served as a practical example for verifying that currently, systems suffer from low/no adaptability to sudden changes, leading to the ripple effect evidenced on a global scale [19].

Another dimension of digital transformation, as a strategic change, extends to cultural transformation, business model transformation, and process re-engineering, indicating the need for further exploration of suitable strategies, which will facilitate companies to complete their digitalization process. Beyond that, companies must balance technological integration and human touch, ensuring that digital tools focus on enhancing decision-making and not replacing human intervention (e.g., knowledge, experience, intuition etc.)

Another priority area highlighted in the research is how digital and collaborative industrial ecosystems can aid in the transition to I5.0 [20]. Businesses need to adopt collaborative platforms [21], data-sharing networks [22], and cross-industry partnerships to establish resilience and efficiency [23]. Cybersecurity and data privacy are also pressing concerns that should be addressed in an attempt to ensure a seamless and secure digital transformation [15].

Small and medium-sized enterprises (SMEs) comprise the majority of companies, particularly within the European Union (EU) [24]. Therefore, it is of great importance to identify the challenges for the digital transformation of SMEs, since they lack financial resources or technical expertise to adopt I5.0 technologies. The role of government policies, fiscal incentives, and skills training programs is emphasized as required to enable a smooth transition. A clear roadmap, strategic planning, employees' training, and integration of digital and sustainable business practices are indispensable in order to succeed in digital transformation [25]. The companies that embrace the transition will be better prepared to handle future industrial challenges and remain competitive in an increasingly digitized world. According to the AI, Data, and Robotics Partnership, substantial developments in skills, trustworthy data spaces, and innovation ecosystems are critical enablers for a smooth transition to human-centric manufacturing systems [26].

This approach for the digital transformation of SMEs is further elaborated by the authors in [27]. Concretely, it is pointed out that the adoption of sustainable practices in tandem with economic, environmental, and social dimensions is required. However, existing roadmaps in the literature for the digital transformation of SMEs are not comprehensively touching upon sustainability [28].

Based on the above mentioned, a high-level roadmap for the sustainable digital transformation of companies is presented in Fig. 3, highlighting key actions to be undertaken by the stakeholders. Although the roadmap offers benefits, there are still certain challenges that need to be addressed by SMEs when they are undergoing digital transformation. Some of these challenges include limited resources, limited know-how, and change resistance. Government policies and fiscal incentives, the research suggests, can prove to be powerful in helping SMEs through this change.

In total, the study emphasizes that sustainable digital transformation is not only a trend but a requirement for SMEs to be competitive and resilient. Through embracing a structured roadmap that incorporates technology, sustainability, and business strategy, SMEs can spur innovation while reducing environmental footprint and promoting long-term growth [29].

2.2 Sustainability of Operations

This section explores how sustainability is tackled through digital transformation in manufacturing. Among its key enablers, AI supports data-driven optimization, which facilitates the reduction of environmental impact while improving operational efficiency. As discussed in the previous paragraphs, I4.0 focuses on technological advancement of operation, disregarding environmental and sustainability issues. In order to address the sustainability challenges in the field of operations, the Triple Bottom Line (TBL) approach, which balances economic, social, and environmental factors has been adopted [30]. I5.0 introduces a shift from a purely technology-centric approach to one that considers resource efficiency, waste reduction, and circular economy principles. Furthermore, digital transformation practices,

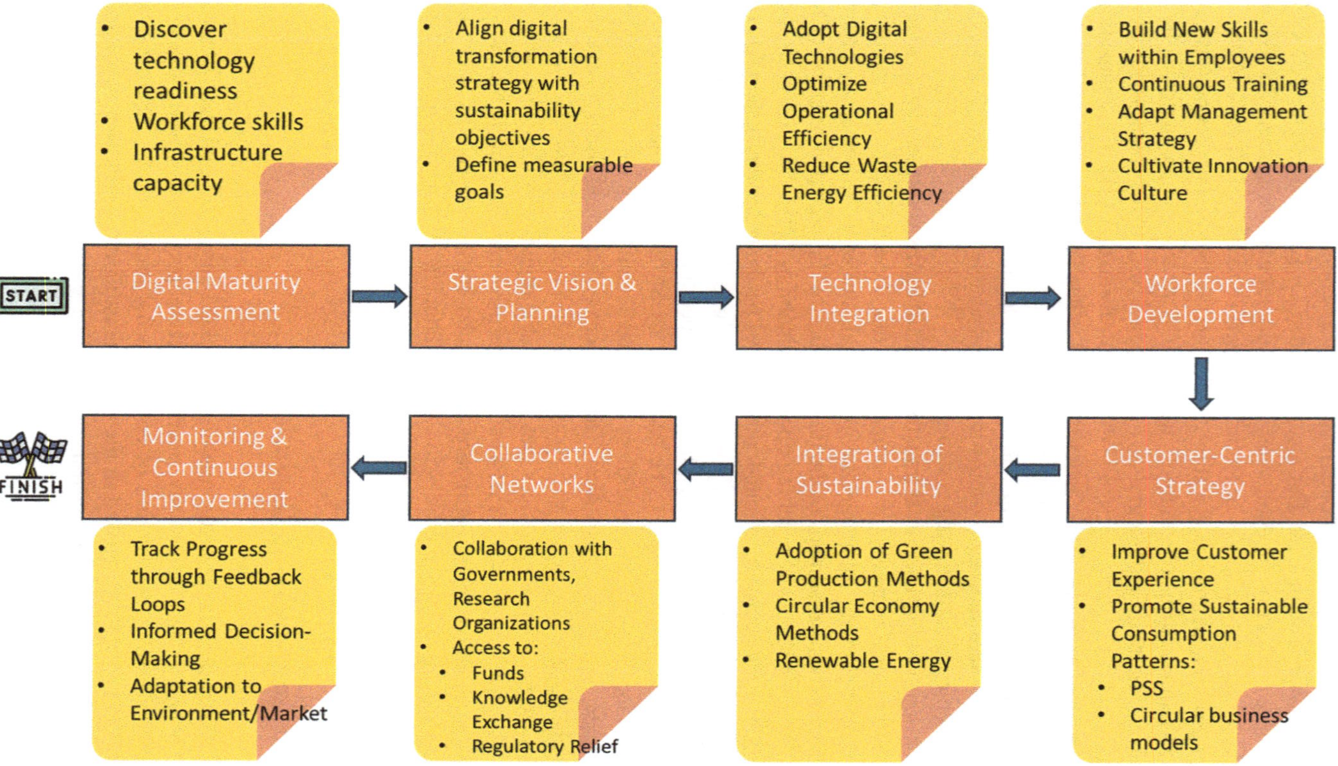

Fig. 3 Roadmap for sustainable digital transformation in the context of Industry 4.0 and Industry 5.0. (Created by the authors)

particularly those integrating AI, play a crucial role in optimizing supply chains, and with the provision of valuable insights, reduced carbon footprints and improved energy management are feasible. Other aspects of AI, like predictive maintenance and process automation among others, can lead to even lower material waste and enhanced sustainability in manufacturing [31–33]. Within the digital transformation context, AI-powered analytics and machine learning enhance decision-making processes by identifying inefficiencies and proposing environmentally friendly solutions [34, 35]. Further to that, AI can enable real-time monitoring of resource consumption and utilization [36], helping companies align their production processes with sustainable development goals (SDGs) [37]. However, the contribution of digital twins in combination with AI-based simulations extends to the reduction of environmental impact and improving industrial resilience [38]. To achieve sustainability in manufacturing, organizations must adopt a structured approach that will be based on the integration of digital technologies following sustainable practices. Figure 4 outlines the most significant steps that companies should follow in order to embed sustainability in digitalized factories, ensuring both environmental and operational benefits.

The adoption of new and improved manufacturing paradigms, such as lean, green, and smart manufacturing, can enhance sustainability and efficiency of manufacturing operations. This is imperative for companies aiming toward the reduction of waste, improvement of resource utilization, and adoption of intelligent systems for further optimizing their shop floor operations [39]. Sustainability on its own poses a great challenge for companies. However, it can be addressed through the green manufacturing approach [40]. Key aspects to be considered when addressing sustainability of operations include, among others, (1) reducing environmental waste, (2) improving workplace conditions, and (3) minimizing carbon footprints by optimizing operational processes. On the contrary, engineers are obliged to face issues such as (1) high carbon emissions, (2) inefficient waste disposal, (3) poor air

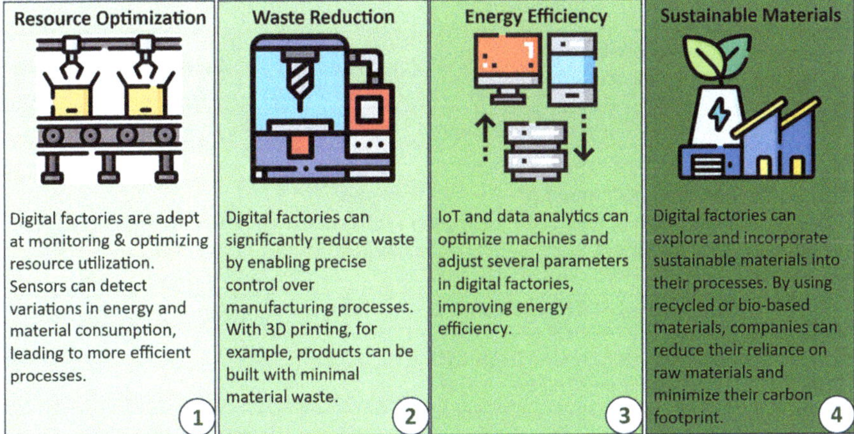

Fig. 4 Steps for sustainability in digitalized factories. (Created by the authors)

quality, and (4) excessive energy consumption. All the above mentioned contribute to unsustainable industrial practices. However, with the integration of better ergonomics practices and through continuous environmental life cycle assessment, it is feasible to create a cleaner and more efficient working environment [41, 42]. To that end, AI-driven solutions facilitate real-time monitoring of manufacturing and production processes, predictive maintenance, and data-driven decision-making. By incorporating such intelligent tools along with IoT-based monitoring and big data analytics, process automation can be improved. By extension, this reduces defects and optimizes resource allocation. The framework by Tripathi et al. [39] has been tested in a mining machinery manufacturing industry, where it delivered remarkable results. It led to an 85% improvement in productivity, a 21% increase in machinery utilization, a 65% reduction in production time, a 96% decrease in manufacturing defects, and a 75% reduction in production costs. These findings suggest that integrating AI-driven smart manufacturing with lean and green principles can create a sustainable, efficient, and resilient industrial ecosystem.

2.3 Systems Resilience

The transition to resilient manufacturing systems is a strategic priority within digital transformation. AI contributes significantly to this goal, enabling adaptive operations, predictive planning, and recovery strategies that support long-term sustainability. Resilience can be shortly defined as the ability of a system or organization to anticipate, adapt to, and recover from disruptions, challenges, or unforeseen changes while maintaining essential functions and stability [43]. Therefore, for manufacturing companies, resilience becomes a multifaceted challenge. Resilient manufacturing systems are essential to mitigate supply chain disruptions, adapt to market volatility, and ensure operational stability. In Fig. 5, key challenges for

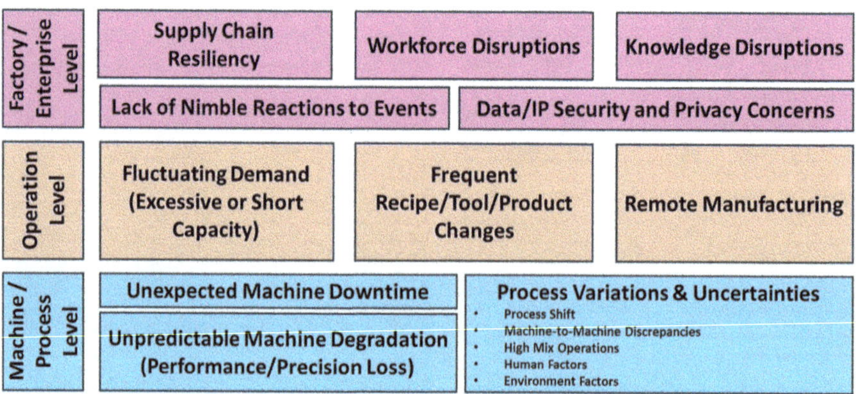

Fig. 5 Challenges and needs for resilient manufacturing systems. (Adopted from [44])

manufacturing companies in achieving resilience have been compiled, providing information on the technological and strategic approaches needed to overcome them.

Resilience in digital transformation is increasingly enabled by AI-driven predictive analytics, which allow manufacturers to anticipate and adapt to disruptions. Some of the most common domains include infrastructure, disaster management, climate resilience, healthcare, business continuity, cybersecurity, and social resilience. The role of AI is indeed pivotal, so that adaptability, robustness, and long-term sustainability in an increasingly complex and highly volatile landscape are ensured. Modern AI-driven systems are capable of forecasting potential failures in critical infrastructure like roads, bridges, and power grids, allowing proactive intervention before catastrophic breakdowns occur. For example, the integration of Building Information Modeling (BIM) and AR with AI facilitates the optimization of construction progress, ensuring resilience early from the design phase onward.

In disaster response, AI facilitates real-time situational awareness through data analytics, machine learning, and predictive modeling. AI-driven algorithms analyze information from satellite imagery, sensor networks, and social media to predict disasters, assess damage, and optimize emergency resource allocation. The study highlights the use of AI-powered drones and robots for searching and rescuing missions in hazardous environments, minimizing human risk. Furthermore, AI-enhanced natural language processing (NLP) systems improve emergency communication and coordination during crises.

For climate resilience, AI helps predict extreme weather events and models long-term climate change scenarios. AI-driven climate modeling tools use satellite data and environmental metrics to identify vulnerabilities and recommend adaptation strategies. The study also discusses AI applications in smart grids and energy optimization, reducing greenhouse gas emissions by optimizing power distribution and integrating renewable energy sources.

AI's impact on healthcare resilience is evident in disease diagnostics, outbreak prediction, and personalized treatment plans. Machine Learning models analyze large-scale medical data, enabling early detection of diseases like cancer and cardiovascular conditions. AI-powered predictive analytics enhance pandemic preparedness by identifying potential outbreaks and optimizing hospital resource allocation. Additionally, AI-powered telemedicine and remote patient monitoring improve healthcare accessibility and continuity.

In business continuity and crisis management, AI-driven risk assessment models predict supply chain disruptions, economic downturns, and cybersecurity threats. AI enhances supply chain resilience by forecasting demand fluctuations, optimizing inventory levels, and improving logistics efficiency. The study also highlights AI's role in automating cybersecurity defenses through real-time threat detection, predictive analytics, and automated incident response systems.

AI contributes to urban resilience through smart city technologies, enhancing transportation efficiency, optimizing public services, and improving air and water quality. AI-driven digital twins create real-time virtual models of cities, allowing for predictive infrastructure maintenance and disaster response planning.

3 Servitization and Green Manufacturing

Before delving into the digital enablers, it is important to define the concepts of servitization and green manufacturing. Servitization refers to the transformation of manufacturing firms from solely product providers to solution providers, offering value-added services such as maintenance, upgrades, and life cycle support alongside physical products. Green manufacturing, on the other hand, focuses on minimizing environmental impacts during production by enhancing energy efficiency, reducing waste, and adopting environmentally friendly materials and processes. These two paradigms are increasingly linked with sustainability efforts in I5.0 and are strongly supported by digital technologies, especially AI.

Digitalization extends beyond the context of manufacturing systems to manufacturing networks and the supply chain in general. As discussed in [45], digitalization is imperative for the servitization of manufacturing companies. However, management of supply chain is more complicated than the integration of digital technologies. Concretely, supply chain can be defined by the concentration of customers and suppliers. Sun and Xi have come to the conclusion that digitalization and servitization as entities have direct impact on the negative effects of poor customer and supplier concentrations. Therefore, companies have to carefully design their digitalization strategy so that operational risks can be mitigated and servitization is facilitated in the case of concentrated customer pool. Similarly, in the case of concentrated supplier pool, digitalization requires even more careful planning, in order to minimize exposure risks, by managing efficiently supplier relationships.

Insights from a micro-survey done in China also reveal the significance of digitalization in supporting breakthrough service innovation and incremental innovation, enabling manufacturing companies to develop more advanced service offerings [46]. Digitalization benefits several aspects that contribute to servitization, such as, among others, customer engagement, process optimization, and data-driven service innovations. This research work also confirms that digitalization and servitization are mutually reinforcing, leading to improved company performance, enhanced sustainability, and greater competitiveness. The key challenge for companies is to balance both elements strategically to maximize efficiency and innovation while contributing to carbon reduction goals [47].

Rabetino et al. [48] highlight how digital transformation and servitization contribute to sustainable manufacturing. Servitization, if designed in a sustainable way, can facilitate companies to achieve more efficient resource utilization, reduce waste, and extend life cycle for products by implementing common good practices (e.g., reuse, remanufacturing, and optimization). Nevertheless, not all forms of servitization automatically lead to sustainability, since they could also generate environmental unsustainability if managed adequately. Digitalization plays a key role in enabling sustainable servitization by integrating cutting-edge digital technologies (e.g., IoT, big data, AI, and blockchain). Industry 4.0–related digital technologies ensure predictive maintenance, optimize asset utilization, and provide real-time monitoring, all of which lead to reduced emissions and energy consumption. Other

Industry 4.0 technologies, such as additive manufacturing (3D printing) and digital twins also assist in minimizing waste and promoting material circularity, and by extension resulting in cleaner and more resource-efficient production processes. The integration of servitization with circular economy principles is another key aspect in promoting sustainability [49]. Circular business models, where companies sell the use of products rather than physical products, encourage reduced material consumption and more effective end-of-life product management. Digital technologies allow companies to track product usage, organize repair and remanufacturing operations, and optimize reverse logistics to ensure that materials are recycled and not discarded. Use-based or outcome-based business models, rather than ownership-based models, have been shown to promote more sustainable outcomes for businesses. Despite these advantages, there are also certain dilemmas in achieving green servitization. Not all servitization models will be supportive of sustainability by default, and some servitization may actually increase overall consumption of resources. Supply chain stickiness and supplier dependencies would also render the uptake of circular practices challenging for businesses. Consumer acceptance of servitized offerings, for instance, leasing instead of outright ownership, remains a limitation in the majority of industries. Regulation and policy interventions by the government are equally responsible for deciding the fate of sustainable servitization efforts.

The authors in [50] talk about how digitalization and servitization are being applied in green manufacturing. Their research focuses on how manufacturing firms can apply green servitization and ISO 14001 to improve their environmental and financial performance and tackle sustainability-related issues [50]. Green servitization expands the traditional definition of servitization by integrating environmentally friendly services into production processes. This approach reorders firms from product-oriented business models to service-based strategies that promote cleaner production, waste reduction, and energy efficiency. Through offering services alongside products, firms are able to optimize the use of resources, reduce emissions, and generate extra revenue streams while maintaining environmental stewardship. Digitalization plays a significant role in facilitating green servitization. Digital technologies such as IoT, big data, and AI enable predictive maintenance, real-time monitoring, and process enhancement. These digital technologies help businesses manage their environmental impacts through reduced energy consumption, reduced waste, and improved overall performance. Digitalization also facilitates the circular economy by enabling material traceability, product life extension, and remanufacturing process improvement.

ISO 14001, the widely accepted international standard environmental management system, serves as the facilitator of green servitization to sustainable performance. It provides an orderly procedure via which businesses may follow environmental legislation, raise the level of corporate social responsibility (CSR), and develop stakeholder confidence. Research finds that organizations that implement ISO 14001 alongside digitalization and servitization receive substantial improvement in their sustainability performance. By incorporating green servitization into ISO 14001 standards, businesses are able to attain long-term resilience, mitigate environmental risks, and enhance business competitiveness.

While such benefits, studies pose several challenges to the adoption of green servitization. The majority of manufacturing firms remain hesitant due to the perceived cost and operational risks of embracing sustainability initiatives. Additionally, firms typically find it difficult to transition from traditional production paradigms to service-based green business models. Policy incentives and regulatory systems play a crucial role in promoting firms' adoption of green servitization, ensuring compliance with environmental standards while guaranteeing economic survival.

Green servitization is the development of traditional product-based business models into service-based models focused on sustainability [51]. This means that product manufacturers are not just offering products but also services such as maintenance, repair, and leasing, which extend the life cycles of products and reduce the use of resources. The study confirms that green servitization makes a direct contribution to circular economy practices, which strive for reuse, recycling, and reduction of waste for improved overall sustainability. Digitalization is a major facilitator for enhancing green servitization operations through more efficient use of resources, optimized production processes, and easier monitoring of the environmental impact [52]. IoT, AI, and data analytics allow firms to track product usage, predict maintenance, and implement more energy-efficient processes. The study highlights that green digital technology is a main enabler of sustainable performance, but its impact depends on the degree to which companies integrate it into their servitization strategies. The study recognizes that circular economy practices (CEPs) act as an interface between green servitization and sustainability. Companies that adopt CEPs, such as remanufacturing, waste minimization, and sustainable supply chain management, achieve better environmental and economic performance. This means that green servitization alone is not enough; companies must integrate circular economy principles to fully realize the benefits of sustainability. Despite the benefits, many firms struggle with adopting green servitization due to high initial costs, lack of digital capabilities, and resistance to change. The study also points out that some firms prioritize short-term financial gains over long-term sustainability, making it difficult to shift toward servitization. Additionally, policy and regulatory encouragement is important in encouraging firms to adopt environmentally sustainable service models. The paper suggests that policymakers implement incentives for firms to adopt green servitization and circular economy models. These include financial incentives, tax breaks, and regulatory policies that encourage firms to invest in environmentally friendly business practices. On the business level, companies should develop internal green competencies, invest in digital technologies, and pursue customer-centric sustainable solutions to remain competitive in a changing market.

To sum up, for manufacturing companies aiming to transition from product-based to service-oriented business models, digitalization can be a powerful tool. However, its impact depends heavily on supply chain structure. While it can enhance customer engagement and service efficiency, it may also increase supplier dependency risks if not managed carefully. Looking ahead, firms should focus on developing business models that integrate servitization and sustainability objectives.

More empirical research is required to measure the actual impact of servitization on environmental performance. There is a growing need to examine how digitalization can further support the circular economy through advanced service models. It imperative to stress out that green servitization, digitalization, and circular economy practices are intertwined and essential for achieving sustainable manufacturing. While digitalization enhances servitization by improving efficiency and reducing waste, circular economy principles ensure that sustainability goals are met. However, firms need strategic planning, strong policy support, and investment in digital infrastructure to fully benefit from green servitization.

4 Industrial Case Studies

The following section is dedicated to the presentation and discussion of real-life pilot cases derived from the manufacturing sector. Through these pilot cases, it becomes evident that the integration of AI in manufacturing in combination with the digital transformation of the companies can provide valuable outcomes both for the academia/research organizations as well as for the companies in terms of overall operational efficiency.

4.1 Case Study 1: Dynamic Scheduling in Manufacturing

One of the most transformative applications of digital transformation in manufacturing is the integration of AI with multi-agent systems (MAS) and digital twins for dynamic production scheduling. The research presents an advanced AI-powered scheduling framework, specifically designed for the bicycle manufacturing industry, where complex production environments require real-time adaptability to optimize operations. The implementation of a multi-agent system (Fig. 6), in combination with deep reinforcement learning (DRL) algorithms, has enabled manufacturers to enhance scheduling efficiency, minimize bottlenecks, and improve overall production performance [53].

The proposed framework leverages AI agents to autonomously manage scheduling decisions across different production phases, including wheel assembly, painting lines, and bike assembly operations. The integration of digital twin technology allows for continuous monitoring of the production process, enabling predictive scheduling adjustments to mitigate disruptions before they impact production flow. A key advantage of this approach is its ability to process large volumes of real-time data, continuously refining scheduling decisions and improving production rates while reducing delays.

The framework was implemented in an Original Equipment Manufacturer (OEM) specializing in bicycle production, where dynamic scheduling challenges often arise due to fluctuating customer demands and variable component

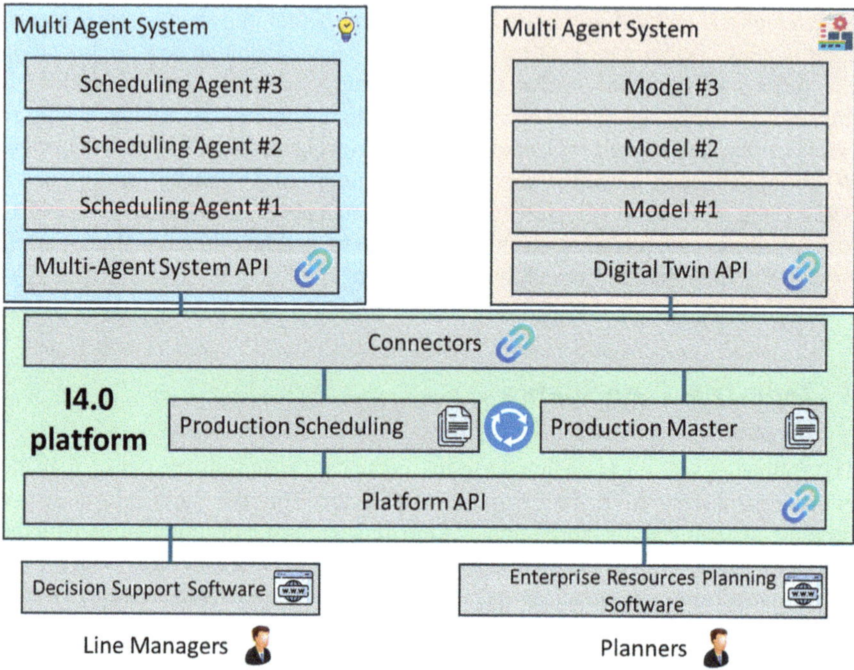

Fig. 6 Multi-agent framework for dynamic scheduling. (Adapted from [53], licensed under CC-BY 4.0)

availability. The three primary production stages—wheel assembly, painting, and final assembly—were individually optimized using AI-powered scheduling agents.

The wheel assembly stage is characterized by manual operations and dedicated assembly lines for different wheel types. The AI agent employed Deep Q-Networks (DQN) to determine the optimal sequence of operations, ensuring that resources were allocated efficiently. The DRL-based scheduling approach minimized production delays by dynamically adjusting task sequences in real time, effectively responding to variations in production demand.

The painting stage introduced an additional challenge: sequence-dependent setup delays when switching between different colors. The scheduling problem was modeled as an optimization problem, where AI-driven heuristics and mathematical programming methods were used to minimize the total flow time while preventing buffer overflows. This allowed for more efficient utilization of painting lines and reduced material waste.

The final assembly stage was modeled as a flow-shop scheduling problem with blocking, where production efficiency was constrained by the sequential nature of operations. The AI-driven heuristic optimized product sequencing, ensuring continuous material flow and minimizing idle time between different assembly steps. This significantly improved cycle time and overall production throughput.

While the multi-agent AI scheduling system demonstrated significant improvements in efficiency and scalability, the study also identified key challenges:

- Computational complexity: Training deep reinforcement learning (DRL) models requires extensive computational resources. To address this, a digital twin environment was used to simulate production scenarios and accelerate AI training.
- Real-time adaptability: The need for rapid reconfiguration in case of production disturbances was addressed by integrating digital twin simulations, enabling predictive adjustments in scheduling.
- Interoperability with existing systems: The Asset Administration Shell (AAS) standard was implemented to enhance communication between AI agents, ERP systems, and production machinery, ensuring seamless integration with Industry 4.0 architectures.

Despite these challenges, the results demonstrated that AI-powered dynamic scheduling significantly outperforms traditional manual scheduling methods, offering:

- Higher production efficiency by reducing idle times and increasing throughput
- Lower operational costs through optimized resource utilization
- Enhanced resilience to production disruptions by enabling real-time, adaptive scheduling adjustments

This case study provides a concrete example of how AI, multi-agent systems, and digital twins can revolutionize dynamic scheduling in manufacturing. By implementing AI-driven decision-making, manufacturers can optimize production scheduling, improve response times to demand fluctuations, and ensure seamless coordination between different production phases.

The findings underscore the critical role of AI in digital transformation, particularly in complex, multi-stage production environments. Future research is expected to explore hybrid AI models, integrating machine learning, heuristic optimization, and real-time IoT data to further enhance scheduling precision and adaptability (Fig. 7).

This AI-powered scheduling approach not only improves operational efficiency but also supports the goals of I5.0 by reducing resource waste, minimizing downtime, and enabling resilient, human-in-the-loop decision-making in complex production environments. By leveraging real-time optimization, it helps manufacturers reduce energy consumption and material loss, thereby contributing to sustainable production practices.

4.2 Case Study 2: Integration of LLMs in Industrial Maintenance

One of the most compelling examples of digital transformation in manufacturing is the integration of AI and AR for industrial maintenance. A recent study explores how large language models (LLMs) can be leveraged to assist maintenance

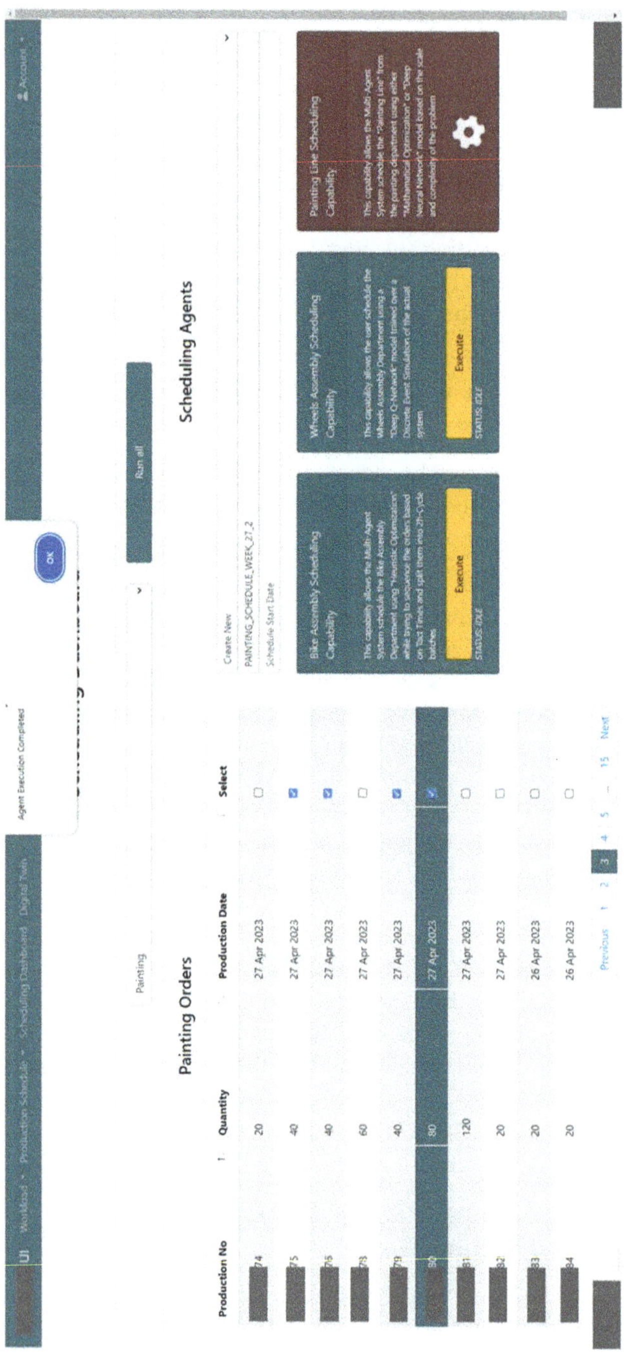

Fig. 7 GUI for dynamic scheduling of multi-agent system. (Adopted from [36], licensed under CC-BY 4.0)

personnel in maintenance and repair operations (MRO) by providing actionable insights and reducing downtime through intuitive, humanlike interactions. This case study exemplifies the shift from traditional, labor-intensive maintenance procedures to AI-powered predictive and prescriptive maintenance models, demonstrating the transformative impact of digitalization in the manufacturing sector [54, 55].

The research focuses on fine-tuning an LLM to enhance maintenance decision-making, particularly in environments where unforeseen events, such as equipment breakdowns, require rapid response and accurate troubleshooting. Unlike general-purpose AI models, this domain-specific LLM is trained using historical maintenance records, equipment manuals, and past troubleshooting cases, ensuring it provides tailored and context-aware maintenance recommendations. The framework also incorporates AR visualizations, allowing shop-floor technicians to interact with AI-driven instructions in real time through head-mounted displays (HMDs), tablets, or smartphones. This dual integration of AI-driven guidance and immersive AR interactions ensures that technicians receive step-by-step assistance while working on complex machinery, ultimately reducing human error and improving operational efficiency.

A real-world implementation of this framework has been carried out in a dairy manufacturing plant, where maintenance is critical to ensuring seamless production and quality control. The facility relies on intricate machinery for processing and packaging milk-based products, requiring frequent maintenance due to high operational loads. The case study tested the AI-AR framework on the replacement of an LCD module in a Programmable Logic Controller (PLC), an essential component for industrial automation.

The LLM was fine-tuned using historical maintenance logs, technical manuals, and past failure reports to ensure the model could generate precise troubleshooting steps. When a technician initiated a maintenance request, the system provided a customized set of instructions, including detailed procedural steps for safely replacing the module. The AR module further enhanced this interaction by overlaying visual markers on the physical equipment, highlighting the exact components that needed to be addressed. Compared to traditional methods—where technicians relied on printed manuals or expert guidance, this AI-enhanced workflow led to faster issue resolution, reduced downtime, and improved knowledge retention among maintenance staff. The generalized system architecture is illustrated in Fig. 8.

While the integration of LLMs and AR in manufacturing maintenance has shown promising results, several technical challenges remain. The research highlights that LLM responses, though highly relevant, sometimes lack technical depth, particularly regarding specific hardware details such as bolt sizes or component specifications. This underscores the need for continuous fine-tuning and domain-specific training using a more extensive dataset that includes manufacturer-specific MRO guidelines. Additionally, multimodal AI models, capable of processing both text and images, could further enhance the framework by allowing AI systems to interpret real-time sensor data and visual inspections.

Future developments aim to improve model adaptability by integrating self-learning mechanisms, where the AI refines its recommendations based on

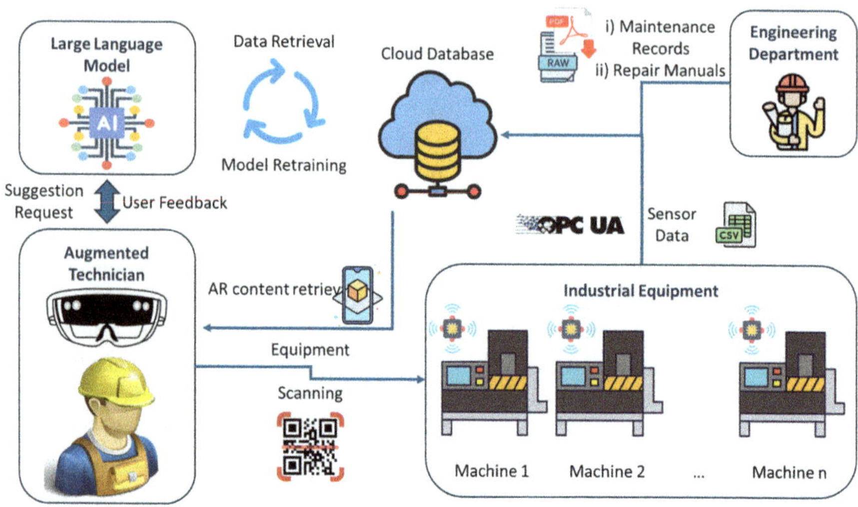

Fig. 8 Framework integrating LLM and AR for digital transformation. (Adopted from [54], licensed under CC-BY 4.0)

real-world feedback from technicians. Moreover, the incorporation of cloud-based LLM architecture will allow decentralized manufacturing facilities to access real-time AI insights, further enhancing digital transformation in industrial maintenance.

This case study highlights how AI and AR can significantly enhance industrial maintenance practices, making them more proactive, efficient, and cost-effective. By leveraging LLM-driven insights and AR-based visualization tools, manufacturing facilities can transition from reactive to predictive maintenance, aligning with Industry 4.0 and 5.0 paradigms. As AI models continue to evolve, their role in industrial automation, knowledge augmentation, and real-time decision-making will be pivotal in achieving higher digital maturity in manufacturing.

Beyond improving response time and technician support, the integration of LLMs and AR into industrial maintenance contributes to sustainability by reducing unplanned downtimes, extending equipment life, and preventing wasteful over-maintenance. These outcomes align with I5.0's emphasis on resilience, workforce empowerment, and more efficient use of physical and informational resources.

4.3   Case Study 3: Predictive Maintenance Based on AI

The integration of AI for condition monitoring and predictive maintenance (PdM) has revolutionized modern manufacturing, significantly improving asset utilization, failure prevention, and operational efficiency. A recent study proposes a deep learning–based PdM framework, leveraging Long Short-Term Memory (LSTM)-Autoencoders and Transformer encoders, to predict machine failures and estimate

Remaining Useful Life (RUL). This approach provides a more proactive alternative to traditional preventive maintenance, helping manufacturers minimize unplanned downtime, reduce maintenance costs, and enhance overall production sustainability [56].

This research introduces a two-stage AI model designed to detect faults, classify machine health conditions, and predict RUL. First, LSTM-Autoencoders are used for feature extraction and anomaly detection, determining whether a machine is in a healthy or deteriorating state. If deterioration is detected, the Transformer encoder is activated to further analyze time-series sensor data and predict the RUL of the asset. The combined model enhances accuracy in predicting machine degradation patterns, enabling manufacturers to schedule maintenance before failures occur.

The proposed AI-based PdM system was implemented in a steel production facility, focusing on a hot rolling mill machine responsible for producing metal bars. The machine's coated rolling segments, which experience wear and tear due to high force and heat, typically require replacement every 16 days. Unexpected failures of these components lead to production halts and costly maintenance interventions.

The rolling mill machine was equipped with 27 industrial sensors that monitored critical parameters, including surface temperature and hydraulic force on the rolling cylinders. Sensor data was collected at 5-ms intervals and stored in a PLC database. Before analysis, the dataset was preprocessed to eliminate noise, inconsistencies, and missing values, ensuring data integrity and model reliability.

For the identification of potential failures, two separate LSTM-Autoencoder networks were trained using historical maintenance records. These models were designed to classify the health status of the machine as either in a "Good State" or a "Bad State." When incoming sensor data showed an abnormal degradation pattern, it triggered the activation of the Transformer encoder for a more detailed assessment of the asset's condition.

Once engaged, the Transformer model was tasked with predicting the Remaining Useful Life (RUL) of the machine components. It categorized the RUL into three classes: Class 0 indicated 3–4 days remaining before failure; Class 1 corresponded to 2–3 days; and Class 2 signaled only 1 day remaining. This classification framework enabled the maintenance team to proactively schedule component replacements, thereby minimizing the likelihood of emergency shutdowns and production losses.

Although the LSTM-Transformer framework delivered significant improvements in failure prediction and maintenance planning, the implementation posed several challenges. First, the computational demands for training and deploying deep learning models in real-time environments were substantial, potentially limiting adoption among small and medium-sized enterprises (SMEs). Second, managing and processing large volumes of multivariate time-series sensor data introduced complexities related to data storage, processing latency, and overall model efficiency. Lastly, while the framework was highly effective for the specific case of hot rolling mills, its direct application to other industrial assets would require adaptation. This issue reflects a broader challenge faced by many AI-driven systems that rely on conventional, non-transferable architectures.

The application of predictive maintenance directly supports the sustainability objectives of I5.0 by optimizing maintenance cycles, reducing unnecessary part replacements, and preventing energy-intensive breakdown scenarios. Furthermore, the early identification of component wear reduces material waste and extends equipment longevity, fostering more sustainable and resilient industrial operations.

5 Conclusions and Outlook

There can be no doubt that the application of AI in manufacturing continues to redefine the industrial landscape, making possible hitherto unprecedented efficiency, resilience, and sustainability. While the evolution from I4.0 to I5.0 marks a transition from completely automated systems to human-centered, resilient, and sustainable industrial ecosystems, it also makes clear the complementarity of the two revolutions. While AI-driven digital transformation offers numerous benefits, e.g., enhanced operational efficiency, predictive maintenance, and optimization of resources, numerous challenges remain unresolved as new ones emerge, e.g., cybersecurity risks, workforce upskilling, and budget constraints, particularly for SMEs, which constitute the vast majority of firms, at least at the European level.

Throughout the literature review of this chapter, the necessity of a structured roadmap for digital transformation is highlighted. Apart from this, it is imperative that policymakers and stakeholders realize that a one-size-fits-all approach is infeasible and ineffective. Companies must strategically embrace AI, IoT, and cloud-based technologies to enhance decision-making and long-term resilience. Apart from this, the use of green manufacturing and servitization models, enabled through digitalization, plays a crucial role in achieving sustainable industrial processes. Servitization, if blended with circular economy concepts, can reduce waste, increase resource efficiency, and provide long-term economic sustainability.

The proposed strategic roadmap and discussions in this chapter are in good alignment with the strategic vision outlined in the Strategic Research, Innovation and Deployment Agenda (SRIDA) 2025–2027 by the AI, Data and Robotics Association (ADRA) [57]. As Europe aims to establish technological sovereignty as well as to accelerate the digital and green transitions, the SRIDA emphasizes the need for trustworthy AI, sustainable robotics, and robust data ecosystems. Notably, SRIDA highlights the critical role of human-centric, interoperable, and regulatory-compliant AI systems as enablers of next-generation manufacturing innovation. The integration of these principles into future industrial strategies will be essential for ensuring both economic competitiveness and societal well-being.

Industrial case studies illustrate the actual impact of large language model-powered AI solutions on production scheduling, maintenance, and predictive analytics. Dynamic scheduling facilitated by AI fosters manufacturing agility, while LLMs coupled with AR aid in enhanced maintenance operations. Similarly, predictive maintenance based on deep learning models reduces downtime as well as operational costs. However, concerns of computational demands, scalability, and data management complexities are indicative of the need for continuous improvement of AI models.

Looking ahead, the manufacturing industry must take an integrated approach that harmonizes technological innovation and human capabilities. Future research must focus on hybrid AI models integrating heuristic optimization, machine learning, and real-time IoT data for decision-making enhancement. Governments and policymaking bodies must also provide enabling policies, such as tax rebates and worker retraining schemes, to chart a course toward I5.0 in a seamless fashion.

Generative AI presents an enormous potential for manufacturing in that it enables the creation of optimized designs, process simulation automation, and product innovation acceleration. Through the ability to design new prototypes, materials, and manufacturing pathways, generative AI can accelerate research and development cycles while reducing material waste and production costs. Looking ahead, the integration of generative AI with real-time production data will allow self-improving systems to respond in real time to changing conditions, further driving efficiency and innovation in smart factories.

Agentic AI, which emphasizes autonomous decision-making and proactive system management, is expected to be a primary driver of industrial automation. Unlike traditional AI systems that rely on pre-programmed inputs, agentic AI systems can interpret complex situations independently, coordinate machine actions, and streamline workflows with minimal human intervention. Future multi-agent AI architectures will achieve fully adaptive manufacturing environments, where intelligent agents communicate, negotiate, and adapt in real time to optimize efficiency, reduce downtime, and improve resilience to disruptions.

Lastly, the successful digitalization of manufacturing will require a collective push from industry stakeholders, researchers, and policymakers. By the strategic and sustainable application of AI and digitalization, the manufacturing sector can gain the dividends of higher productivity, reduced environmental impact, and enhanced resilience, resulting in a more adaptive and future-ready industrial ecosystem.

References

1. Chryssolouris, G., Alexopoulos, K., & Arkouli, Z. (2023). Artificial intelligence in manufacturing systems. In *A perspective on artificial intelligence in manufacturing* (Studies in systems, decision and control) (Vol. 436). Springer. https://doi.org/10.1007/978-3-031-21828-6_4
2. Pellicelli, M. (2023). Introduction. In *Elsevier eBooks* (pp. ix–xv). Elsevier. https://doi.org/10.1016/b978-0-323-85532-7.02001-6
3. Ibm. (2025). *Digital transformation. What is digital transformation?* Retrieved February 2, 2025, from https://www.ibm.com/think/topics/digital-transformation
4. *What is digital transformation?* (2024). McKinsey & Company. Retrieved February 15, 2025, from https://www.mckinsey.com/featured-insights/mckinsey-explainers/what-is-digital-transformation#/
5. *Digital transformation market size, share, growth, trends [2032].* (n.d.). Retrieved March 3, 2025, from https://www.fortunebusinessinsights.com/digital-transformation-market-104878

6. Rashid, A. B., & Kausik, M. A. K. (2024). AI revolutionizing industries worldwide: A comprehensive overview of its diverse applications. *Hybrid Advances, 7*, 100277. https://doi.org/10.1016/j.hybadv.2024.100277

7. Mourtzis, D., Angelopoulos, J., & Panopoulos, N. (2023). Blockchain integration in the era of industrial metaverse. *Applied Sciences, 13*(3), 1353. https://doi.org/10.3390/app13031353

8. Aivaliotis, P., Arkouli, Z., Georgoulias, K., & Makris, S. (2023). Methodology for enabling dynamic digital twins and virtual model evolution in industrial robotics—A predictive maintenance application. *International Journal of Computer Integrated Manufacturing, 36*(7), 947–965. https://doi.org/10.1080/0951192x.2022.2162591

9. Stavropoulos, P., Papacharalampopoulos, A., & Sabatakakis, K. (2023). Data attributes in quality monitoring of manufacturing processes: The welding case. *Applied Sciences, 13*(19), 10580. https://doi.org/10.3390/app131910580

10. PricewaterhouseCoopers. (n.d.). *PwC's Global Artificial Intelligence Study: Sizing the prize.* PwC. Retrieved February 20, 2025, from https://www.pwc.com/gx/en/issues/artificial-intelligence/publications/artificial-intelligence-study.html

11. Sarkar, B. D., Shardeo, V., Dwivedi, A., & Pamucar, D. (2024). Digital transition from Industry 4.0 to Industry 5.0 in smart manufacturing: A framework for sustainable future. *Technology in Society, 78*, 102649. https://doi.org/10.1016/j.techsoc.2024.102649

12. Wu, H., Li, G., & Ivanov, D. (2025). The transformative power of generative AI for supply chain management: Theoretical framework and agenda. *Frontiers of Engineering Management.* https://doi.org/10.1007/s42524-025-4240-x

13. Hu, Y., Jia, Q., Yao, Y., Lee, Y., Lee, M., Wang, C., Zhou, X., Xie, R., & Yu, F. R. (2024). Industrial Internet of Things intelligence empowering smart manufacturing: A literature review. *IEEE Internet of Things Journal, 11*(11), 19143–19167. https://doi.org/10.1109/jiot.2024.3367692

14. Caiado, R. G. G., Machado, E., Santos, R. S., Thomé, A. M. T., & Scavarda, L. F. (2024). Sustainable I4.0 integration and transition to I5.0 in traditional and digital technological organisations. *Technological Forecasting and Social Change, 207*, 123582. https://doi.org/10.1016/j.techfore.2024.123582

15. Saeed, S., Altamimi, S. A., Alkayyal, N. A., Alshehri, E., & Alabbad, D. A. (2023). Digital transformation and cybersecurity challenges for businesses resilience: Issues and recommendations. *Sensors, 23*(15), 6666. https://doi.org/10.3390/s23156666

16. Krishnaveni, N., Swathi, N., Schwartz, E., & Sangeetha, N. (2024). *The role of Human-Centric Solutions in tackling challenges and unlocking opportunities in Industry 4.0* (pp. 365–375). Auerbach Publications eBooks. https://doi.org/10.1201/9781032694375-20

17. Narkhede, G. B., Pasi, B. N., Rajhans, N., & Kulkarni, A. (2024). Industry 5.0 and sustainable manufacturing: A systematic literature review. *Benchmarking an International Journal.* https://doi.org/10.1108/bij-03-2023-0196

18. Zaoui, F., & Souissi, N. (2020). Roadmap for digital transformation: A literature review. *Procedia Computer Science, 175*, 621–628. https://doi.org/10.1016/j.procs.2020.07.090

19. Dolgui, A., Ivanov, D., & Sokolov, B. (2017). Ripple effect in the supply chain: An analysis and recent literature. *International Journal of Production Research, 56*(1–2), 414–430. https://doi.org/10.1080/00207543.2017.1387680

20. Schallmo, D., Williams, C. A., & Boardman, L. (2017). Digital transformation of business models—Best practice, enablers, and roadmap. *International Journal of Innovation Management, 21*(08), 1740014. https://doi.org/10.1142/s136391961740014x

21. Belkadi, F., Boli, N., Usatorre, L., Maleki, E., Alexopoulos, K., Bernard, A., & Mourtzis, D. (2018). A knowledge-based collaborative platform for PSS design and production. *CIRP Journal of Manufacturing Science and Technology, 29*, 220–231. https://doi.org/10.1016/j.cirpj.2018.08.004

22. Catti, P., et al. (2024). Data analytics and AI for quality assurance in manufacturing: Challenges and opportunities. In S. Thiede & E. Lutters (Eds.), *Learning factories of the future. CFL 2024* (Lecture notes in networks and systems) (Vol. 1059). Springer. https://doi.org/10.1007/978-3-031-65411-4_25

23. Koilo, V. (2025). Driving the circular economy through Digital servitization: Sustainable business models in the maritime sector. *Businesses*. https://doi.org/10.3390/businesses5010012
24. SMEs. (n.d.). *Internal market, industry, entrepreneurship and SMEs*. Retrieved March 1, 2025, from https://single-market-economy.ec.europa.eu/smes_en
25. Issa, A., Hatiboglu, B., Bildstein, A., & Bauernhansl, T. (2018). Industrie 4.0 roadmap: Framework for digital transformation based on the concepts of capability maturity and alignment. *Procedia CIRP, 72*, 973–978. https://doi.org/10.1016/j.procir.2018.03.151
26. Curry, E., Heintz, F., Irgens, M., Smeulders, A. W. M., & Stramigioli, S. (2022). Partnership on AI, data, and robotics. *Communications of the ACM, 65*(4), 54–55. https://doi.org/10.1145/3513000
27. Mick, M. M. A. P., Kovaleski, J. L., & De Genaro Chiroli, D. M. (2024). Sustainable digital transformation roadmaps for SMEs: A systematic literature review. *Sustainability, 16*(19), 8551. https://doi.org/10.3390/su16198551
28. Nachit, M., & Okar, C. (2020). Digital transformation of human resources management: A roadmap. In *2020 IEEE International Conference on Technology Management, Operations and Decisions (ICTMOD)* (pp. 1–6). https://doi.org/10.1109/ictmod49425.2020.9380608
29. Madhavan, M., Sharafuddin, M. A., & Wangtueai, S. (2024). Measuring the Industry 5.0-Readiness level of SMEs using industry 1.0–5.0 practices: The case of the seafood processing industry. *Sustainability, 16*(5), 2205. https://doi.org/10.3390/su16052205
30. Garrido, S., Muniz, J., & Ribeiro, V. B. (2024). Operations management, sustainability & Industry 5.0: A critical analysis and future agenda. *Cleaner Logistics and Supply Chain, 10*, 100141. https://doi.org/10.1016/j.clscn.2024.100141
31. Zhang, Y., Ren, S., Liu, Y., & Si, S. (2016). A big data analytics architecture for cleaner manufacturing and maintenance processes of complex products. *Journal of Cleaner Production, 142*, 626–641. https://doi.org/10.1016/j.jclepro.2016.07.123
32. Javaid, M., Haleem, A., Singh, R. P., Suman, R., & Rab, S. (2021). Role of additive manufacturing applications towards environmental sustainability. *Advanced Industrial and Engineering Polymer Research, 4*(4), 312–322. https://doi.org/10.1016/j.aiepr.2021.07.005
33. Javaid, M., Haleem, A., Singh, R. P., Suman, R., & Gonzalez, E. S. (2022). Understanding the adoption of Industry 4.0 technologies in improving environmental sustainability. *Sustainable Operations and Computers, 3*, 203–217. https://doi.org/10.1016/j.susoc.2022.01.008
34. Minghai, Y., Saliu, F., & Khan, W. A. (2024). *AI-powered decision-making applications for sustainable development* (pp. 233–251). CRC Press. https://doi.org/10.1201/9781032635170-18
35. Alijoyo, F. A. (2024). AI-powered deep learning for sustainable Industry 4.0 and Internet of Things: Enhancing energy management in smart buildings. *Alexandria Engineering Journal, 104*, 409–422. https://doi.org/10.1016/j.aej.2024.07.110
36. Siatras, V., Bakopoulos, E., Mavrothalassitis, P., Nikolakis, N., & Alexopoulos, K. (2024). Production scheduling based on a multi-agent system and digital twin: A bicycle industry case. *Information, 15*(6), 337. https://doi.org/10.3390/info15060337
37. United Nation. (n.d.) *THE 17 GOALS. Sustainable development*. Retrieved February 25, 2025, from https://sdgs.un.org/goals
38. Bibri, S. E., Huang, J., Jagatheesaperumal, S. K., & Krogstie, J. (2024). The synergistic interplay of artificial intelligence and digital twin in environmentally planning sustainable smart cities: A comprehensive systematic review. *Environmental Science and Ecotechnology, 20*, 100433. https://doi.org/10.1016/j.ese.2024.100433
39. Tripathi, V., Chattopadhyaya, S., Mukhopadhyay, A. K., Sharma, S., Kumar, V., Li, C., & Singh, S. (2023). Lean, green, and smart manufacturing: An ingenious framework for enhancing the sustainability of operations management on the shop floor in Industry 4.0. *Proceedings of the Institution of Mechanical Engineers Part E Journal of Process Mechanical Engineering, 238*(4), 1976–1990. https://doi.org/10.1177/09544089231159834
40. Abualfaraa, W., Salonitis, K., Al-Ashaab, A., & Ala'raj, M. (2020). Lean-green manufacturing practices and their link with sustainability: A critical review. *Sustainability, 12*(3), 981. https://doi.org/10.3390/su12030981

41. Lin, C. J., Belis, T. T., & Kuo, T. C. (2019). Ergonomics-based factors or criteria for the evaluation of sustainable product manufacturing. *Sustainability, 11*(18), 4955. https://doi.org/10.3390/su11184955

42. Chaloulos, A., Catti, P., Nikolakis, N., & Alexopoulos, K. (2024). A process-level LCA for evaluating the contribution of digitalization in the greening of a manufacturing system. *Procedia CIRP, 126*, 904–908. https://doi.org/10.1016/j.procir.2024.08.346

43. Caputo, A. C., Pelagagge, P. M., & Salini, P. (2019). A methodology to estimate resilience of manufacturing plants. *IFAC-Papers OnLine, 52*(13), 808–813. https://doi.org/10.1016/j.ifacol.2019.11.229

44. Lee, J., Siahpour, S., Jia, X., & Brown, P. (2022). Introduction to resilient manufacturing systems. *Manufacturing Letters, 32*, 24–27. https://doi.org/10.1016/j.mfglet.2022.02.002

45. Sun, B., & Xi, Y. (2024). Supply chain concentration, digitalization and servitization of manufacturing firms. *Journal of Manufacturing Technology Management, 36*(1), 112–133. https://doi.org/10.1108/jmtm-03-2024-0114

46. Song, X., & Yang, J. (2023). Assessing the impact of digitization and servitization of manufacturing firms in the context of carbon emission reduction: Evidence from a microsurvey in China. *Energy & Environment, 35*(7), 3340–3385. https://doi.org/10.1177/0958305x231167470

47. Stavropoulos, P., & Panagiotopoulou, V. C. (2022). Carbon footprint of manufacturing processes: Conventional vs. non-conventional. *Processes, 10*(9), 1858. https://doi.org/10.3390/pr10091858

48. Rabetino, R., Kohtamäki, M., Parida, V., & Vendrell-Herrero, F. (2024). Sustainable servitization for cleaner and resource-wise production and consumption: Past, present, and future. *Journal of Cleaner Production, 469*, 143179. https://doi.org/10.1016/j.jclepro.2024.143179

49. Nikolakis, N., Catti, P., Chaloulos, A., van de Kamp, W., Coy, M. P., & Alexopoulos, K. (2024). A methodology to assess circular economy strategies for sustainable manufacturing using process eco-efficiency. *Journal of Cleaner Production, 445*. https://doi.org/10.1016/j.jclepro.2024.141289

50. Oyelakin, I. O., Ting, D. H., Yusuf, A. H., Arbak, S., & Dhar, B. K. (2025). Building resource capabilities through green servitization and ISO 14001 for sustainable performance: Perspectives from manufacturing firms. *Corporate Social Responsibility and Environmental Management.* https://doi.org/10.1002/csr.3150

51. Johl, S. K., Ali, K., Shirahada, K., & Oyewale, O. I. (2024). Green servitization, circular economy, and sustainability a winning combination analysis through hybrid SEM-ANN approach. *Business Strategy and the Environment.* https://doi.org/10.1002/bse.3950

52. Alexopoulos, K., Nikolakis, N., & Xanthakis, E. (2022). Digital transformation of production planning and control in manufacturing SMEs—The mold shop case. *Applied Sciences, 12*(21), 10788. https://doi.org/10.3390/app122110788

53. Bakopoulos, E., Siatras, V., Mavrothalassitis, P., Nikolakis, N., & Alexopoulos, K. (2024). Digital-twin-enabled framework for training and deploying AI agents for production scheduling. In J. Soldatos (Ed.), *Artificial intelligence in manufacturing*. Springer. https://doi.org/10.1007/978-3-031-46452-2_9

54. Angelopoulos, J., Manettas, C., & Alexopoulos, K. (2025). Industrial maintenance optimization based on the integration of Large Language Models (LLM) and Augmented Reality (AR). In *Lecture notes in mechanical engineering* (pp. 197–205). https://doi.org/10.1007/978-3-031-86489-6_20

55. Angelopoulos, J., Stavropoulos, P., Manettas, C., & Alexopoulos, K. (2025). LLM and XR remote maintenance tool for empowering operator 5.0. *Procedia CIRP, 134*. https://doi.org/10.1016/j.procir.2025.03.070

56. Bampoula, X., Nikolakis, N., & Alexopoulos, K. (2024). Condition monitoring and predictive maintenance of assets in manufacturing using LSTM-autoencoders and transformer encoders. *Sensors, 24*(10), 3215. https://doi.org/10.3390/s24103215

57. Heintz, F., Belbachir, N., & Curry, E. (2024). *Strategic research, innovation, and deployment agenda 2025–2027.* ADRA. Retrieved from https://adr-association.eu

AI-Based Management Approaches for Production Processes: Identifying Requirements for Frameworks and Architectures

Anna Sumereder, Robert Woitsch, Damiano Falcioni, and Marlene Mayr

Abstract Among the most prevalent challenges for organizations is the artificial intelligence (AI) integration including process transformation and organizational complexity reflecting management obstacles. Processes exist in various formats and formalizations such as textual documentation, graphical instructions, or the like. Here, appropriate combinations of AI can be effective to increase trustworthiness, provide reliable information, and reduce users' effort. We propose an incremental approach considering early-stage experiments, prototypes, industrial scenario demonstrations, and exploitable results to introduce AI into industry. This is showcased by the European research project FAIRWork that applies hybrid AI-based management approaches to (a) improve information access supporting maintenance and information reliability, (b) validate document compliance, and (c) assist decision-making for worker allocation and production planning. Requirements for AI frameworks and architectures are identified, and the outlook presents a recap on how hybrid AI-based management approaches may entail a wider impact on AI-supported decision-making across domains.

Keywords Artificial intelligence · Reliable information · Industry · Frameworks and architectures · Production processes

1 Introduction

The chapter begins by presenting the underlying motivation, followed by the identified research challenges and interests, highlighting the proposed contributions aimed at introducing AI into the production industry.

A. Sumereder (✉) · R. Woitsch · D. Falcioni · M. Mayr
Research Group, BOC Products & Services AG, Vienna, Austria
e-mail: anna.sumereder@boc-group.com; robert.woitsch@boc-group.com;
damiano.falcioni@boc-group.com; marlene.mayr@boc-group.com; http://www.boc-group.com/en/

E. Curry et al. (eds.), *Artificial Intelligence, Data and Robotics*,
https://doi.org/10.1007/978-3-032-10561-5_10

1.1 Motivation

Gartner [1] predicts that the domain-specific generative artificial intelligence (AI) models will significantly increase in the next years. More than 50% of the generative AI models used by enterprises are projected to have a targeted industry or business function by 2027, in contrast to approximately 1% in 2023. Compared to general purpose models, domain models have the benefit of being smaller, computationally more efficient, and less prone to hallucination. However, this means that organizations require strategies to integrate off the shelf, customize existing, deploy own, and manage (multiple) AI models for heterogeneous use cases. ChatGPT—an AI-powered natural language processing model answering given prompts—is expected to revolutionize Industry 4.0 supporting automation, digitalization, and real-time connectivity of systems to raise productivity, quality assurance, and efficiency [2]. Exemplary applications in manufacturing are task automation, the creation of technical documentation, predictive maintenance improvement, quality control, or the optimization of maintenance schedules to name only a few. However, it can be stated that effective knowledge management is critical to deal with the upcoming technologies and related complexity of data as well as for enabling innovation in dynamic environments [3]. Knowledge management in organizations acts as an organizer and curator of knowledge, providing access to relevant resources and preserving information agility, where AI models can augment traditional knowledge management processes. Although ChatGPT provides responses resembling human answers, it comes with drawbacks such as missing in-depth expertise, concerns on intellectual property rights and data privacy, intransparency resulting in a lack of explainability, unintentional biases reducing accuracy and reliability, and the like.

1.2 Research Focus and Contributions

Hence, among the most prevalent challenges for organizations is the integration of AI in order to keep up with the efficiency of competitors while making sure that the AI-based information is reliable, which requires process transformation and the reflection of obstacles ranging from organizational complexity to technical issues. Based on this observation, two main AI-focused research interests have been identified for this work. Those are (1) *how information technology (IT) governance can support AI applications*, for instance, in terms of infrastructure selection, deployment decisions, resource efficiency, performance increase, or the like, and (2) *how the reliability of information can be enhanced* by ensuring the trustworthiness of solutions applying AI-based services.

This work relates to the fundamentals of the "AI, Data and Robotics Partnership" [4] by showcasing requirements and best practices to introduce AI into the production industry in a trustworthy way. For this, patterns—describing AI interactions on a meta-level—are anticipated to act as a facilitator for identified conceptual and technical requirements. So far, patterns in an AI context often refer to identifying common structures for optimization, maintenance, or forecasting services (e.g., recognizing recurrency in workflows, detecting machine failures, analyzing historical data, etc.). In this work, a general-purpose AI is applied on selected industrial samples to identify and validate the requirements heading for generalizing them in the form of patterns for a broader applicability.

A European project in the production domain is utilized as a showcase. The FAIRWork [5] project aims at bringing humans, AI, and robots together by designing fair decision models to transform decision-making and underlying current production processes toward a cooperative decision environment. Human decision-makers are supported in the design and execution of processes. Currently, the decision-making follows a hierarchical structure requiring humans to be aware of all relevant information. The idea of FAIRWork (Fig. 1) is to democratize the decision-making approach, where different actors—considered as agents—negotiate to propose optimized decision recommendations to the human decision-maker. Complex decisions can be made by considering multiple parameters configured to optimize production processes, an improved perception of trust and fairness, as well as an increased flexibility.

This work is structured to provide a comprehensive exploration of AI adoption in industrial production environments, emphasizing reliable information access and integration. The first section introduces the motivation, the relevancy of AI for industry, and the related challenges. Section 2 reviews related work, covering key aspects such as production process modelling, AI services, and microservices for the realization. The next section delves into reliable information, highlighting the applied method and trustworthiness. The fourth section presents the AI adoption in industry using samples from FAIRWork detailing the machine maintenance demonstrator, including problem, process, realization, and feedback descriptions. Building on these insights, the fifth section illustrates requirements and derived patterns that aim at frameworks and architectures to guide future AI developments. Finally, a recap of the findings and an outlook on future research directions are provided.

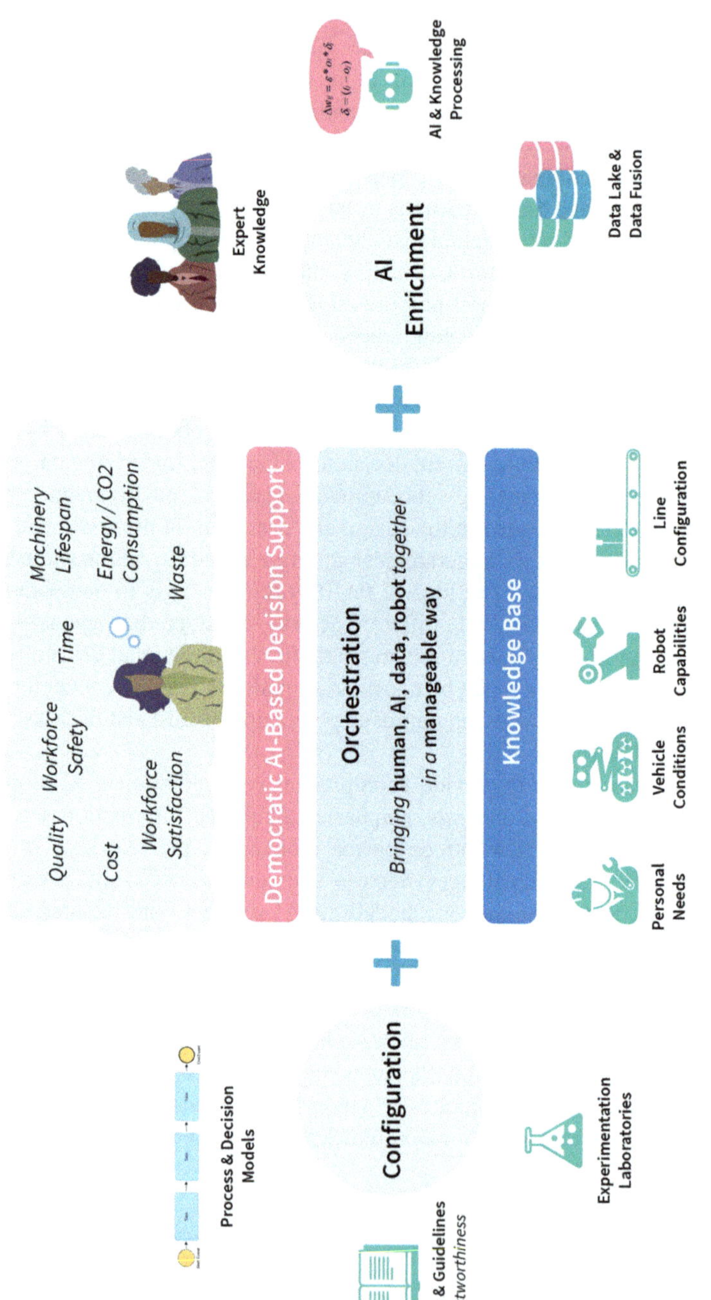

Fig. 1 FAIRWork project idea [5]

2   Related Work

As a foundation for the AI-based management approaches, the following subsections summarize a selection of related work on modelling, AI, and microservices.

2.1   Modelling Production Processes

Manufacturing can be complex, being both predictable yet influenced by multiple factors and unpredictable due to unforeseen events. A symbiosis of human and machine interpretable approaches can address both complexity of socio-technical systems [6] and unpredictability. Concept models are essential components in management approaches used for process modelling (e.g., using Business Process Model and Notation, BPMN[1]), IT governance (e.g., using ArchiMate[2]), and compliance and risk management. Such models can be represented graphically in a semi-formal manner to facilitate human interpretation, as well as in a formalized format suitable for machine processing. Concept models are defined by meta-models using meta-modelling platforms [7] like ADOxx[3] that structure the expressiveness of concepts and their relations. Patterns [8] such as references, graph-rewriting, merging, or transfer concepts, support the integration of different conceptual representations into a holistic model. Depending on the used (meta-)modelling platform, there may be support for (a) combining relevant algorithms and mapping underlying concepts; (b) interpreting corresponding structures, models, and methods using annotation or vectorization; (c) accessing and integrating data; and (d) orchestrating processes offering decision support to the user. For years, process models have been considered a foundation for simulation [9] and optimization [10] in production domains. In recent years, conceptual models also support the digitization of business ecosystems [11] and integrate data [12] aiming at digital twins.

In today's agile academic and industrial environments, business process management knowledge and techniques can be considered as both a necessity and an opportunity for managing dynamic business processes [13]. The authors differentiate two dimensions—namely, closed- vs. open-ended and operational vs. contextual. These can be combined, resulting in four approaches for managing dynamics:

1. Closed ended and operational, where processes are quite stable, such as for configuration, monitoring, prediction, or recommendation.

---

[1] The standardized Business Process Model and Notation (BPMN) is used to model and understand business processes (https://www.bpmn.org/, last accessed May 9, 2025).

[2] The standardized ArchiMate Modeling Language is used to model and analyze enterprise architecture (https://www.opengroup.org/archimate-forum/archimate-overview, last accessed May 9, 2025).

[3] The ADOxx meta modelling platform is used for implementing domain-specific modelling tools (https://www.adoxx.org/index.html, last accessed May 9, 2025).

2. Open ended and operational, requiring some freedom such as for ad hoc planning, template models having various concrete instantiations, or improvisation.
3. Closed ended and contextual, such as exploitative or scenario analysis.
4. Open ended and contextual, dealing with unknown dynamics such as trend scouting, ramping up, or scaling.

In general, the operational perspective refers to focusing on the business processes, in contrast to the contextual perspective dealing with surroundings setting the frame for performing the business processes. In closed-ended environments, future dynamics can be anticipated quite well; however, uncertainty increases when it comes to an open-ended future. For this work, it is assumed that both contextual and operational perspectives have a relevance for supporting the IT governance as well as organizational challenges using AI. Due to the characteristics of AI approaches, symbolic AI may be more appropriate for closed-ended dynamics in terms of explicit models, while sub-symbolic AI may rather support open-ended scenarios where implicit knowledge is required.

2.2   Artificial Intelligence Enrichment

Symbolic approaches using explicit models such as rules or ontologies and sub-symbolic approaches using implicit models such as neural networks, machine learning, or large language models (LLMs) can be differentiated [14]. In recent years, there is also a tendency to hybrid "in-between" methods, bridging the gap and exploiting benefits of both methods [15]. Enabling AI-based data interpretation, aforementioned human-approved graphical models can be converted into knowledge graphs (e.g., using SPARQL,[4] GraphQL,[5] or Neo4j[6] for formal representations and querying). When representing elements such as rules, fuzzy rules, workflows, or semantic models, via concept models, particularly symbolic AI is utilized. Here, various approaches for knowledge representation (e.g., using Entity Relationship Models [16], Unified Modeling Language,[7] or Resource Description Framework and Web Ontology Language[4]) can be applied.

---

[4] A set of Semantic Web Standards, published by W3C, supports the handling of data. This includes, for instance, the Resource Description Framework (RDF) for data exchange, the Web Ontology Language (OWL) for representing knowledge about things, and SPARQL Query Language for RDF (https://www.w3.org/2001/sw/wiki/Main_Page, last accessed May 4, 2025).

[5] GraphQL is a data query language developed by the GraphQL Foundation (https://graphql.org/, last accessed May 4, 2025).

[6] Neo4j is a graph database developed by Neo4j Inc. (https://neo4j.com/, last accessed May 4, 2025).

[7] The standardized Unified Modeling Language (UML) supports system architects and software engineers/developers in analyzing, designing, and implementing software-based systems (https://www.uml.org/, last accessed May 4, 2025).

Foundational models are considered a logical progression in the evolution of machine learning [17]. Representative LLMs include, for instance, GPT-4,[8] Llama,[9] Gemini,[10] DeepSeek,[11] and Mistral,[12] all of which offer broad applicability, natural language processing, and interaction capabilities. Some LLMs incorporate initial built-in ethical safeguards, while fine-tuning enables the development of context-specific models. Key challenges associated with LLMs include (a) hallucinations, where outputs may be inaccurate, and (b) lack of robustness, as identical inputs can yield different results [18]. A major concern of global LLMs is their ability to learn from user interactions, which may lead to potential unreliability and confidentiality risks. In contrast, locally deployed models mitigate data privacy concerns but at the cost of reduced learning capabilities. This raises significant issues related to privacy, robustness, and ethical considerations that surpass the safeguards currently embedded in these models. Especially, LLM-based AI struggles with transparency, compared to more explicit symbolic approaches. Enhancing LLMs with domain-specific knowledge bases through retrieval-augmented generation (RAG) improves robustness by enabling a more controlled knowledge utilization. A RAG-enhanced LLM can retrieve relevant information from structured sources such as databases, documents, sensors, or the Internet. Recent research [19] explores the extent to which RAG, combined with fine-tuning using domain-specific data, can mitigate hallucinations and enhance model reliability. Also structured "chain-of-thought" reasoning methods improve LLMs' performance [20] using a zero-shot strategy in dynamic environments [21].

Also, AI significantly advances software engineering [22], for instance, by AI-driven code generation for automating repetitive tasks or the identification of bugs. However, clean and structured data is required to successfully apply AI, and explainability is important, where especially rule-based systems offer improvement potential. Still, numerous technical (e.g., compatibility, integration, etc.) and organizational (e.g., resistance to change, upskilling requirement, etc.) challenges need to be addressed. Promising in software engineering are self-healing and adaptive systems that may be relevant also for other domains such as manufacturing.

---

[8] ChatGPT and related LLMs such as GPT-4 and GPT-4o are developed by OpenAI, an American organization working on AI technologies (https://openai.com/, last accessed May 9, 2025).

[9] Llama is a set of large language models developed by Meta (https://www.llama.com/, last accessed May 9, 2025).

[10] Gemini is a set of large language models developed by the British-American organization Google DeepMind (https://deepmind.google/technologies/gemini/, last accessed May 4, 2025).

[11] DeepSeek is a set of large language models developed by the Chinese organization DeepSeek (https://www.deepseek.com/en, last accessed May 4, 2025).

[12] Mistral is a set of large language models developed by the French organization Mistral AI. Le Chat can be used for interaction with the models (https://mistral.ai/, last accessed May 4, 2025).

2.3 Microservices as Realization Approach

In recent years, traditional monolithic software applications have been increasingly supplanted by lightweight microservice and cloud-oriented solutions. The implementation and operation of microservice-based environments, however, present significant challenges due to the complexity inherent in distributed services and data management. Cloud service deployment and management necessitate considerable manual effort, as users are required to define deployable models that specify infrastructural details and functionalities. While existing tools (e.g., Cloudify[13] or Kubernetes[14]) facilitate application deployment, management, and monitoring, there remains a pressing need for suitable configuration solutions. For instance, in software engineering, the challenge of automatically deriving feasible configurations that are in line with specified requirements has been addressed through the software product lines approach [23]. Research has demonstrated its utility in modeling variability requirements within microservices, as evidenced by the works of [24] and [25]. Furthermore, [26] employed a metaheuristic optimization approach to extract feature models from interdependent microservices.

Pattern-based deployment strategies [27] and constraint-based algorithms [28] have been proposed to streamline cloud application deployment. Tools (e.g., Aeolus Blender or Metis) automate the synthesis and deployment of cloud applications, utilizing a configuration optimizer (e.g., Zephyrus) [29–31]. Generative programming and machine learning–based solutions are addressing the need for automated cloud configuration, orchestration, and resource elasticity in heterogeneous and multi-cloud environments [32, 33]. Standards (e.g., TOSCA and CAMP[15]) have been developed to support cloud orchestration and management. Meanwhile, tools to integrate software development and operation (e.g., Ansible,[16] Puppet,[17] or Chef[18]) often result in vendor lock-in. Domain-specific languages and uniform Application Programming Interfaces (API) aim to abstract the complexity of integrating development and operation, although they lack automation support for developers [34,

---

[13]Cloudify is an open-source cloud orchestration framework (https://docs.cloudify.co/, last accessed May 4, 2025).

[14]Kubernetes is an open-source container orchestration system (https://kubernetes.io/, last accessed May 4, 2025).

[15]Topology and Orchestration Specification for Cloud Applications (TOSCA) and Cloud Application Management for Platforms (CAMP) are standards related to cloud applications by OASIS (https://www.oasis-open.org/standards/, last accessed May 9, 2025).

[16]Ansible is a unified automation solution for strategic automation by Red Hat (https://www.red-hat.com/en/technologies/management/ansible, last accessed May 9, 2025).

[17]Puppet is an infrastructure automation and operation platform by Perforce (https://www.puppet.com/, last accessed May 9, 2025).

[18]Chef is a configuration, deployment, and management solution for application infrastructure by Progress Software Corporation (https://www.chef.io/, last accessed May 9, 2025).

35]. In addition, solutions for automated mapping are emerging, driven by standards or building on new domain-specific languages [36, 37].

For Internet-of-Things applications, data-driven decision-making services are increasingly supported by AI, with microservices providing scalable, robust, and real-time processing capabilities for decentralized (edge) devices [38]. Microservices are also being leveraged in communication domains through cloud and AI technologies, necessitating configurations in environments like Kubernetes to enhance performance. Generative AI models and modular approaches with API gateways enable microservices to efficiently handle concurrent calls and operate across various contexts, thereby improving latency and bandwidth [39]. Summarizing, the integration of AI into enterprise applications has fundamentally transformed business operations by facilitating automation and enhanced decision-making, with AI microservices playing a pivotal role in modular and efficient deployment [40]. Generative AI and LLMs are rapidly reshaping software system architectures by enabling automation, personalization, and advanced data processing, although they introduce challenges related to complexity, ethics, and accountability [41].

3 Reliable Information and Artificial Intelligence for Industry

Starting with, the methodology for introducing AI for industry is presented and followed by a reflection on trustworthiness.

3.1 Introducing Artificial Intelligence to Industry

The challenges related to introducing AI into industrial organizations are tackled in line with the design science research approach [42]. Focusing on the needs of relevant production scenarios, a set of experiments, prototypes, demonstrators, and solutions have been worked out at different maturity levels (e.g., differing in technology readiness, scope, closeness to industry, etc.) and in appropriate environments. Figure 2 illustrates how several generic AI experiments are concretized in prototypes, introduced in industrial demonstrators, and condensed to well-chosen and domain expert aligned items for industry.

The worldwide OMiLAB (Open Models Initiative Laboratory) community provides, develops, and research features and tools for transforming conceptual models in different formal expressiveness for different purposes. In this context, the OMiLAB Innovation Environment [43] shown in Fig. 3 assists organizations by offering physical and virtual infrastructure in order to facilitate the AI-oriented transformation. Design thinking is applied and combined with co-creative innovation techniques, collective intelligence mechanisms strengthen competitiveness, and haptic modelling enables flexible holistic business scenario modelling in order

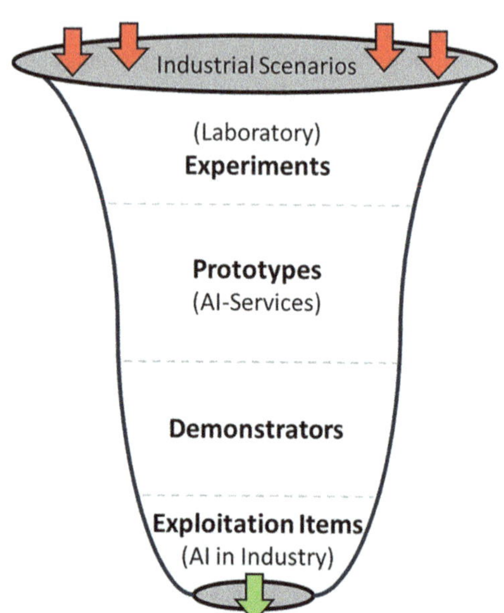

Fig. 2 Condensation of AI approaches enhancing industrial scenarios

Fig. 3 OMiLAB Innovation Corner at BOC Vienna ([43], Fig. 1)

to support the transformation toward AI. Process modelling techniques are utilized to gather, structure, and enrich information in order to enable valuable analysis, simulation, and decision-making, as well as workflows for configuring and executing AI algorithms or production robots. A microservice environment is used to support AI enrichment enabling the creation of decision support systems facilitating reliable information while considering data and AI services from different sources.

3.2 Reflecting Trustworthiness in Artificial Intelligence

The regulation of AI is evolving rapidly worldwide, with the EU leading the way through the AI Act [44]. Early AI systems faced significant challenges, particularly regarding bias and privacy concerns [45]. This has driven increasing efforts toward trustworthy AI [46]. Issues related to bias, privacy, explainability, and ethics present risks not only at the individual user level but also for the society as a whole [47]. Existing trustworthy AI frameworks [48] emphasize that AI must be ethical, lawful, and robust. Seven key requirements have been identified, which are (1) human agency and oversight; (2) technical robustness and safety; (3) privacy and data governance; (4) transparency; (5) diversity, non-discrimination, and fairness; (6) societal and environmental well-being; and (7) accountability [49]. These principles aim to ensure the responsible development and deployment of AI systems while mitigating potential risks.

AI governance [50] requires at least (a) the synthesis of currently fragmented literature on AI governance, (b) a contextual understanding of how AI ethics principles are transferred to practice by organizations, (c) practical AI governance tools and frameworks including accessibility for design science research, and (d) AI auditing. Here, key elements for IT audits are data as insufficient data may result in bias, quality metrics, and concrete measurable actions based on legal and ethical principles. The operationalization of the generic principles for the domain-specific application of LLMs is needed to increase the transparency; to align legal obligations with aspects such as reliability, explainability, and robustness; as well as to incorporate domain-specific requirements for AI services such as different transparency levels for distinct user groups [51] or various application domains [52].

4 Machine Maintenance Support: FAIRWork Use Case

Starting from bottom up, the above-described condensation method with proof-of-concepts at various maturity levels is used to showcase AI scenarios in the production domain. Identified insights anticipate an applicability in a broader context. A combination of several experiments and prototypes contributed to the FAIRWork use case described in the following subsections. Starting from the AI adoption in the production industry (Sect. 4.1), FAIRWork applies hybrid AI-based management approaches in three demonstrators to (a) improve information access supporting maintenance and information reliability, (b) validate document compliance, and (c) assist decision-making for worker allocation and production planning. One of those demonstrators—the machine maintenance case of FLEX[19] focusing on decision support at the production lines—is outlined in the following subsections, including

---

[19] Flextronics (FLEX) is an advanced end-to-end manufacturing partner and one of the use case companies in the FAIRWork project (https://flex.com/, last accessed May 9, 2025).

use case challenges, maintenance processes, demonstration, and users' feedback (comprehensive information can be found in FAIRWork [5, 53–60]).

4.1 Artificial Intelligence Adoption in the Production Industry

Following subsection elaborates on the proof-of-concept types (compare Fig. 2) and presents samples in the context of FAIRWork for each type. An overview is illustrated in Fig. 4. By condensing suitable AI approaches, the relevancy for industry, such as the production domain, increases from generic experiments to targeted solution items. An exemplary selection of experiments, prototypes, and demonstrators is outlined in more detail below. Taken together, concretized and refined, the aforementioned proof-of-concepts can be condensed to a solution—respectively an exploitation item in the context of the FAIRWork project—that can be utilized in an industrial environment.

4.1.1 Sample Experiment: Design Thinking on Machine Maintenance

Based on on-site organization visits and design thinking workshops with domain experts from FLEX, industrial scenarios were extracted in the form of customized Scene2Model Scenes [53, 61, 62]. Those scenes provide the foundation for identifying where introducing AI is a benefit (Fig. 5).

4.1.2 Sample Prototype: Document Transformation

AI is utilized to transform unstructured documents into structured BPMN models. Organizations, such as FLEX, often document instructions in unstructured formats, where AI-driven transformation significantly reduces time and manual effort. The AI-driven conversion process encompasses three key functions. First, LLMs analyze existing documents to interpret content, identify relationships, and incorporate domain-specific knowledge when applicable. Second, AI processes various media types within documents, including plain text, titles, section headings, numerical data, lists, and images. Third, LLMs assist in structuring and positioning elements within the BPMN model. The output of these text-to-model AI services is a structured process model that accurately represents the document's content. However, human experts remain essential for reviewing and approving the AI-generated processes. The models are created using the ADONIS[20] process modelling tool, which is also employed for modelling maintenance and decision-making processes. Here,

---

[20] The ADONIS modelling environment by BOC is used to model business processes (https://www.boc-group.com/en/adonis/, last accessed May 9, 2025).

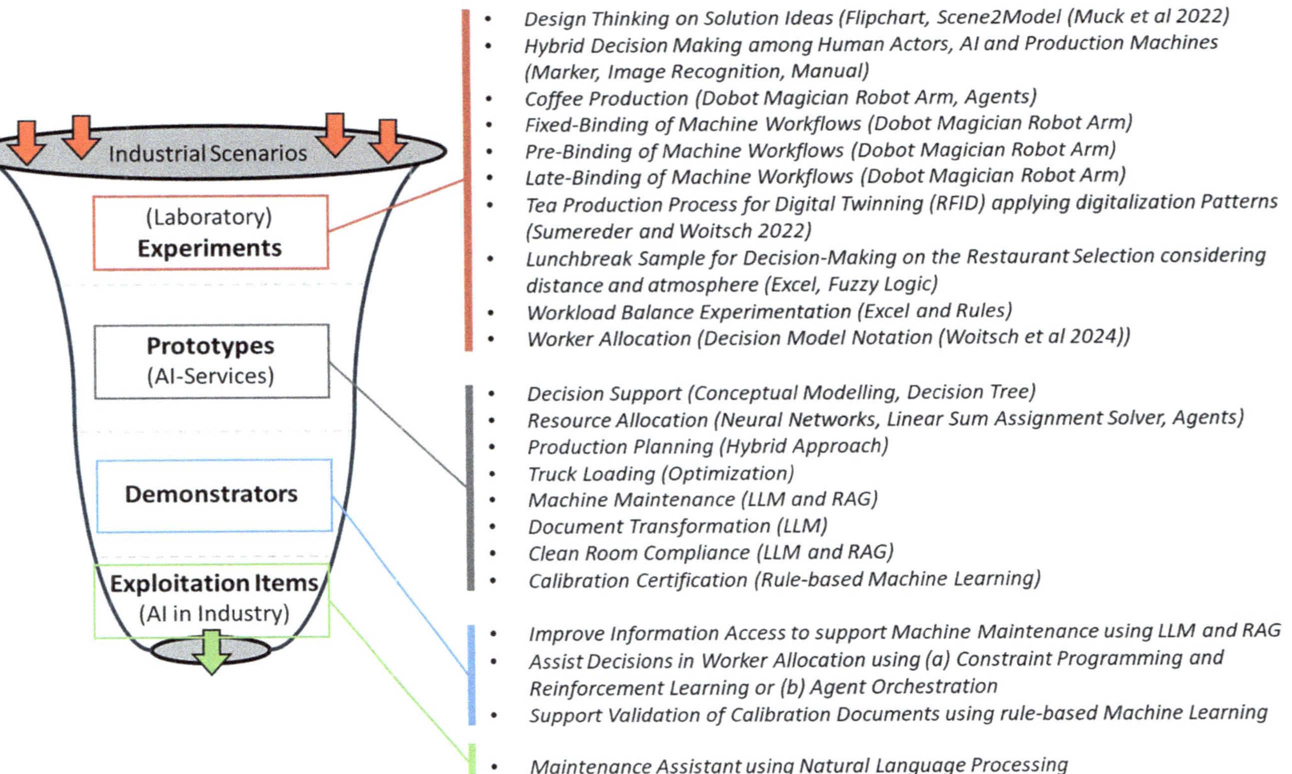

Fig. 4 Selected experiments, prototypes, and demonstrators from FAIRWork

Fig. 5 Workshop on the machine maintenance after breakdown scenario ([53], Fig. 4)

OpenAI's general-purpose LLM "GPT-4o" is used to interpret documents, generate a corresponding model file in an interchangeable format, and arrange the model elements. Other LLMs may be configured to optimize the results. The OLIVE framework [63] orchestrates the entire transformation process by managing document uploads, model creation, and AI model selection. Additionally, OLIVE enhances prompt optimization and facilitates the development of a user interface for the system.

4.1.3 Sample Prototype: Support Compliance for Clean Room

For compliance with FLEX's clean room standards, they seek to compare entirely different documents or various versions of a single document to provide employees with an overview of the latest relevant compliance regulations. For instance, when new versions of regulations take effect in the European Union or its member states, an AI-powered solution can generate a high-level summary of the changes. This allows organizations to quickly assess the relevance of updates without reviewing the entire document. The prototype follows four main steps and applies the RAG approach. First, documents are divided into smaller chunks and stored as structured information in a vector database. Documents are split into chapters, which are further segmented into paragraphs. Each paragraph is processed to generate a semantic summary using the LLM, and then those are aggregated into chapter summaries, with both summaries stored in the vector database. The second step involves identifying semantically corresponding content between two documents using the vector database. Starting from chapters and then moving to paragraph-level comparison, the LLM examines smaller text segments for similar content and flags any paragraphs present in one document but absent in another. In the final step, the LLM compiles a comprehensive list of identified differences and categorizes them. A key advantage of this approach, which extracts the semantic context of documents and converts it into vector representations, is its capability to compare regulatory content against other data sources, such as repository data. This enhances compliance monitoring by enabling broader contextual analysis beyond just document-to-document comparison.

4.1.4 Sample Prototype: Assist Decisions in Resource Allocation (The Watchdog Realization)

An ethical watchdog agent is developed to alert the decision-maker on any pre-established and previously analyzed factor in the process that is insulting on ethical issues. The aim of this watchdog is to set an alarm every time the result of a process infringes an ethical rule observed by the watchdog. In the FAIRWork context, a worker allocation process is monitored with the ethical watchdog. The negotiation for allocating the most suitable workers to a given production line takes place taking into account data such as the resilience to carry out the given task on the line, the worker's preference, gender balance at a production line, the workers' age average, and parameters for conflict resolution. Whenever the score of a worker to be assigned ties, an experience parameter and job rotation aspects are taken into account to resolve the conflict situation. After an initial worker allocation, the watchdog analyzes whether any of the relevant monitoring parameters have been over threshold. The watchdog agent raises the alarm and informs the decision-maker which parameter exceeded the expected value for the given ethical issue. If the watchdog sets off an alarm, the decision-maker is responsible for triggering the renegotiation of the worker allocation, considering the results of the observation. The system recommends a solution considering the initially exceeded parameter as well as the other aspects to provide a result that simultaneously pays attention to the parameters of the first negotiation that are desired for production line allocation. Several renegotiation rounds, in which the order of considering the observed parameters is refined, may be needed. The ethical watchdog agent supports decision-makers not only in the context of production specific aspects but also analysis ethical factors and individual interests related to the recommendations.

4.1.5 Sample Demonstrator: Support Validation of Certification Documents

At FLEX, the verification of calibration certificates is a manual, repetitive, time-consuming, and error-prone process. Around 3000 certificates are processed annually. The certificates are checked for missing or incorrect information related to critical data points such as instrument information (e.g., serial number/ID of the instrument being calibrated, manufacturer, calibration dates, and results) or formal aspects. Overlooked errors can have serious consequences, for example, if faulty measuring equipment remains in use and affects the quality of products or services. The format of calibration certificates may vary depending on the laboratory that conducted the instrument calibration and issued the certification document. To optimize the manual verification process at FLEX, it is replaced with an automated solution using a rule-based approach and applying Python libraries. At the beginning, the service enables the end user to configure and modify paths for data, configuration, and output folders. Through a layout configurator, the document layout structure information (individual calibration laboratories have different document structures) is provided. A PDF Parser is applied to extract tables, plain text, and images from the provided calibration certificate and cleans the data by removing unnecessary whitespaces, adjusting capitalization, and refining table cells utilizing Python libraries. Finally, the extracted information is compared against the guidelines (e.g., results of calibration tests "passed") defined by FLEX, and the results are reported in a user-friendly overview of certificates that are not compliant with these guidelines.

4.2   Use Case Specification

Machine errors vary from simple operator failures to complex technical issues. The time technicians need to fix machine malfunctions can range from minutes to hours. The relevant information for solving machine errors or maintenance issues is distributed across multiple data sources. Among those sources are:

(a) A maintenance database collecting historical data on issues including timestamps, machine types, duration to solve the issue, error source, maintenance instructions, and the like.
(b) An internal collaborative platform for maintenance engineers collecting information on equipment and procedure issues as well as graphical descriptions on previously working solutions.
(c) A specific database for official documents from the equipment and machine manufacturers such as operating instructions, installation manuals, specifications, and handbooks for product descriptions or safety requirements.

The decision-makers have to identify solution procedures in order to solve complex problems. The abovementioned sources can be tapped by bidirectional communication either in written form or via speech interaction.

Taken together, the key challenge for the FAIRWork use case is to *improve the information access supporting maintenance and information reliability*.

4.3   Maintenance Processes in Use Case

The use case–specific maintenance process for machines used in production lines was worked out in informal workshops and refined using BPMN. On top of the processes, suitable AI services are applied that fit the different types of decision challenges.

Figure 6 visualizes the maintenance process after a machine breaks down that can be roughly summarized as follows. When a machine malfunctions, a ticket is created in the ticket system, prompting a technician to assess whether the issue is due to an operator error or a technical failure. The former can be resolved fairly quickly. The latter requires checking the failure database to determine if the issue is already known. Known problems—either previously encountered or identifiable via an error code—are addressed by initiating the corresponding solution process. However, more complex technical issues, such as those requiring spare parts or extensive repairs, may exceed a critical time window and disrupt production. If the malfunction is a new or unidentified issue, all available technicians from different teams (e.g., engineering, maintenance) collaborate to diagnose and resolve it. Since the resolution time may be uncertain, production disruptions become more likely. To mitigate delays, countermeasures such as reallocating production to other machines, rescheduling tasks, and adjusting workforce assignments must be considered. The revised production plan should optimize resource usage, maintain product quality, and ensure timely order fulfillment. Running simulations can provide insights into the impact of different alternatives, assisting decision-makers in selecting the most effective course of action.

The decision process for finding a good strategy after a machine breakdown is illustrated in Fig. 7. The process starts with the ticket creation. Then the first aspect that the maintenance technician must check is whether an unknown error occurred. If this is the case, the error type is directly associated with an unknown and uncertain maintenance time. When the error is known, it must be determined if it is relatively simple, such as an operator or software error, or if it is coming from a different source of error. The latter requires analyzing the specific error type. It can be either a simple or a rather severe error. If the simple error occurs, it is attributed to category A, meaning that it should be resolved soon. The production is paused until maintenance is finished. If a severe failure is identified, it must be considered if multiple errors caused the defect. If only one error occurs, category B for single severe errors is determined. For category B errors, where the maintenance could take some time, it must be checked whether production on a backup line is possible. If it is feasible, then the order is allocated to another line. Otherwise, the production must be stopped till maintenance is finished. If multiple errors arise, they are attributed to category C, resulting in a longer resolution time compared to category A or B. Likewise, for the unknown error category D, the orders are rescheduled to other lines.

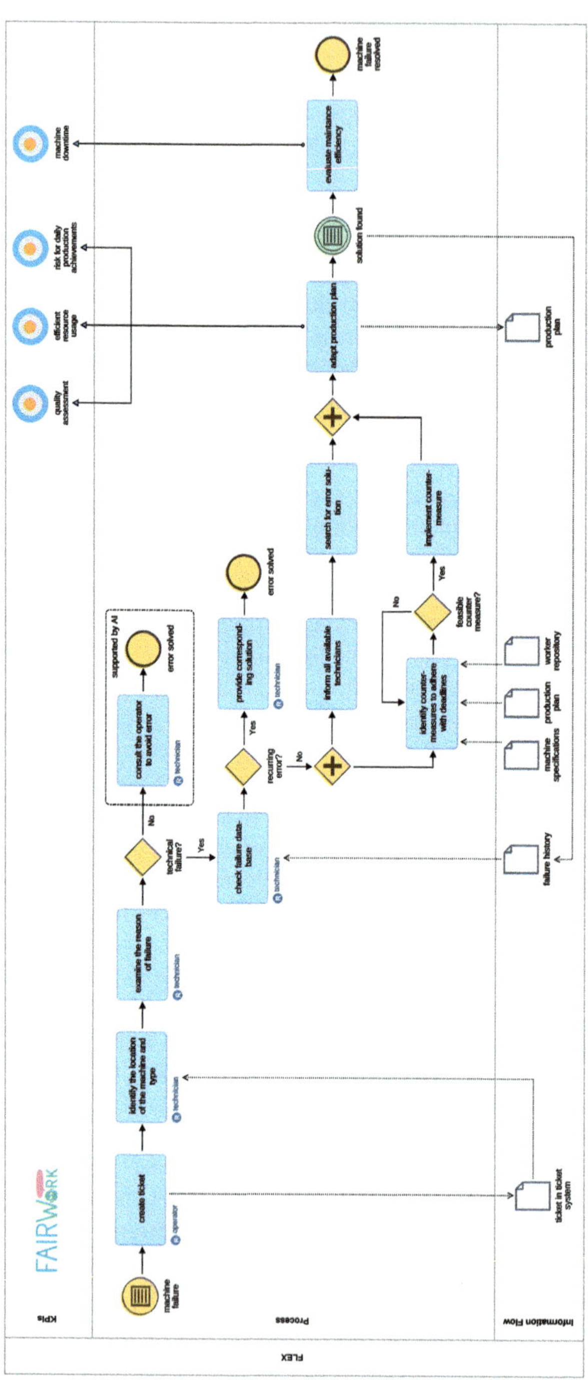

Fig. 6 Machine maintenance after breakdown at FLEX ([56], Fig. 14)

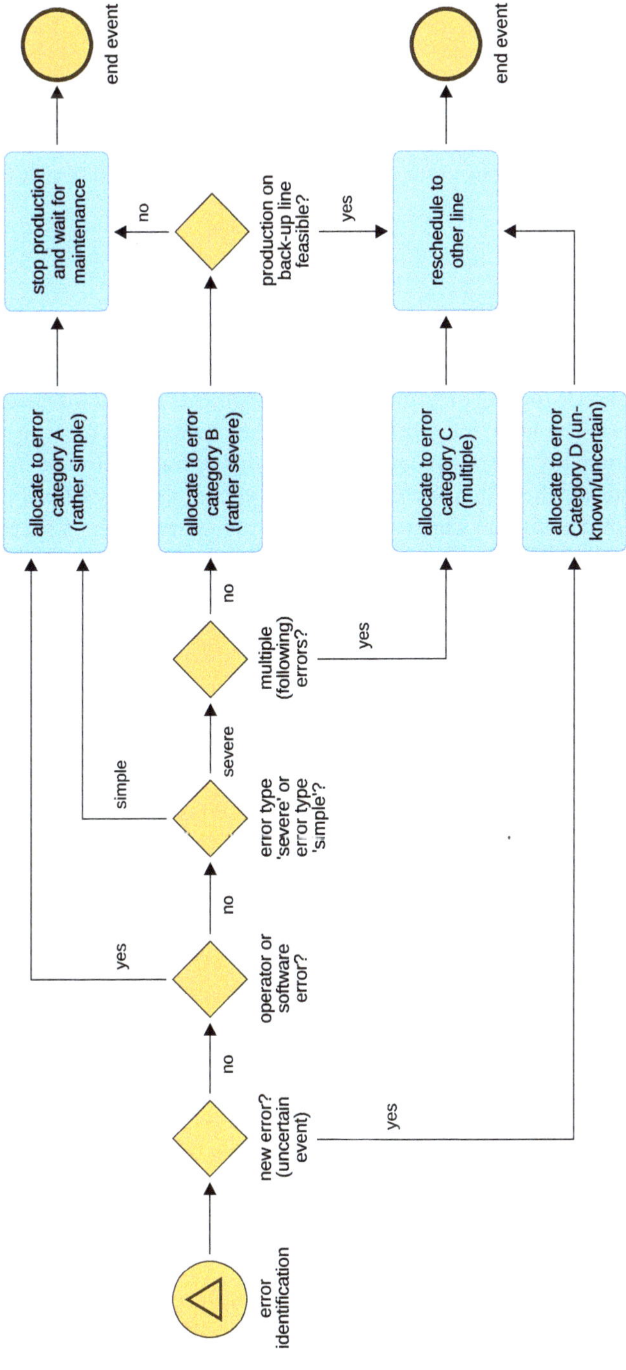

Fig. 7 Decision model of machine maintenance after breakdown at FLEX ([56], Fig. 15)

4.4 Demonstrator

As described in the use case specification, the aim is to reduce the corrective maintenance and machine breakdowns by receiving fast and correct AI-based support. Thus, the requirement to support the decision-makers with suggestions is to create a service that considers information from heterogeneous sources. Key building blocks to demonstrate the machine maintenance use case are presented in the following subsections.

4.4.1 User Interface (UI)

Two UIs are used to improve the information access to support the maintenance. First, there is the UI for indexing the different documents containing machine-specific procedures. The interface allows to upload documents for preprocessing and provides functionality to trigger an AI service that extracts relevant information chunks and indexes them in a vector database. Second, the query UI (Fig. 8, left) can be asked in natural language (including functionalities such as speech-to-text for the user query and text-to-speech for reading the results) how to solve a specific issue in a machine as soon as the indexing is finalized. An LLM-based AI processes the request by checking the information stored in the vector database. The best solution for the machine issue is presented, and the original documents are referenced for further lookup.

4.4.2 AI Services

A chatbot is used to support the machine maintenance by enabling targeted workers to query relevant information via mobile devices using natural language. For the demonstrator, the following aspects are considered particularly relevant (Fig. 8, middle):

Structured and Unstructured Information Structured information on the different databases for maintenance information and underlying (enterprise) architecture concepts are collected via ADOIT.[21] Queries on structured information are relatively straightforward. However, also unstructured information, like maintenance history or relevant notes stored in documents including text and images, must be handled. Here, AI is used to interpret these documents, identify patterns, establish a structure, and get chunks in the form of domain or use case–specific information into the LLM.

---

[21] The ADOIT modelling environment by BOC is used to model IT infrastructure (https://www. boc-group.com/en/adoit/, last accessed May 9, 2025).

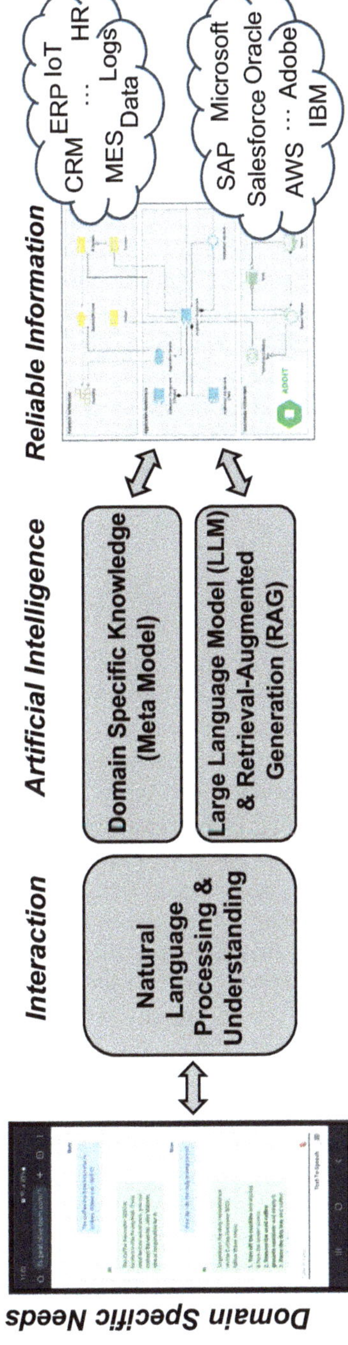

Fig. 8 Exemplary solution for machine maintenance using AI

Interaction with AI Natural language querying is used to improve the interaction of the maintenance workers with the AI by using mobile or desktop devices to ask queries on specific machines and to receive maintenance status information. Both directions—human actor asking the AI and AI providing answer to human actor—are supported via voice (speech, audio) or written format (text) interaction. Interactions may be customized with respect to the used device, for instance, comprehensive answers including links to the corresponding documents, and figures can be provided via desktop interfaces, while mobile devices provide condensed results.

The demonstrator follows the RAG pattern to improve the quality of the LLM's response by including external knowledge stored and retrieved from vector databases. The knowledge about maintenance is either stored in documents and manuals potentially including information from models such as definitions of the data sources or infrastructure aspects or is available on the Internet. Therefore, RAG is used to forward the LLM to retrieve the relevant information from predefined knowledge sources that are not used for the LLMs training. In the FAIRWork case, RAG guides the LLM to use, in addition to its trained knowledge, the information of the FLEX maintenance documents so that answers are based on more precise domain knowledge.

4.4.3   Configuration

The configuration covers all relevant information and definitions such as client components or client functions and defines the wiring to handle endpoints, connectors, and flows contributing to reliable information. For instance, the wiring manages the transformation of uploaded documents into suitable input for the vector database by interpreting different document types, splitting the text into chunks, extracting text from images, adding preprocessed chunks to the database, and the like. The retrieval of information is based on RAG using the Agentic Knowledge Retrieval approach with LLM-based agents. For this, an agent pattern is created in LangChain[22] that aims to fulfill the goal of answering maintenance questions from the user. The agent can access its (a) short-term memory enabling in-context learning and awareness of past interactions and (b) long-term memory composed of information stored over extended periods, mostly in the form of vector stores. In the maintenance use case, an agent queries tools (Fig. 8, right top) such as the maintenance database, ADOIT, the Internet, or the knowledge of an LLM by invoking external APIs. The general-purpose LLM model "GPT-4o" from OpenAI is used to differentiate services by using an agent approach and to initialize data into the vector database. The OLIVE framework connects the individual endpoints such as the uploading of documents and a UI. It is used to improve the prompts and select suitable LLM models for individual endpoints.

---

[22]Langchain is used to build AI apps (https://www.langchain.com/langchain, last accessed May 9, 2025).

4.4.4 Deployment

The deployment (Fig. 8, right bottom) is twofold: (1) the OLIVE server deploys instances of the modelling toolkits (e.g., ADONIS for the decision processes) and (2) an Amazon Web Services[23] (AWS) environment that deploys the implementation of the concrete logic for the use case relevant services, the UI, and the configuration components. An OpenAI endpoint can be called from the applications in the AWS environment. Both the configuration environment and the workflow engine are deployed in AWS as a combination of serverless AWS Lambda functions and AWS EC2 instances relying on AWS DynamoDB for file storage and AWS API Gateway for the exposure of the APIs. The UI components are deployed in an AWS Bucket and distributed through the AWS CloudFront content delivery network. Local deployment options, where trustworthiness aspects can be followed easier (e.g., building on Docker[24]), may serve as an alternative to the AWS environment. Specific deployment scripts are created to automate the complete deployment process of the configuration environment and integrate it into a continuous integration and deployment pipeline. In addition, the configuration environment allows to register different UI components that follow a specific format. By providing the UI component name, the category, a label, and the React[25]-based UI template source code (zip package), the system automatically builds, deploys, and assigns the component to a new entry in the UI builder part of the configuration environment.

4.5 Users' Feedback

Currently, the demonstrator provides correct answers and references the right documents for data that is stored in the vector database. However, the AI scope is limited to certain very specific use cases of which the relevant data was preprocessed and is covered by the vector database. RAG is used to improve the quality of the LLM response including the usage of external sources of knowledge. Nevertheless, the questions must be very precise so that the chatbot provides the correct answer. Here, advancements may include additional parameter fields in the UI to improve the answers. Also, different LLMs may be configured to receive better results for targeted requests. Additionally, a general improvement in the natural language

---

[23] Amazon Web Services environment is used for the deployment and offers a variety of services (e.g., Lambda functions, EC2, DynamoDB, CloudFront, etc.) for computing resources, databases, storage, Web interfaces, etc. to build flexible and scalable applications (https://aws.amazon.com/, last accessed May 9, 2025).

[24] The Docker environment is used to build, share, and run containerized applications and microservices (https://www.docker.com/, last accessed May 9, 2025).

[25] React is used to build user interfaces out of individual pieces called components (https://react.dev/, last accessed May 9, 2025).

understanding should be made. It already provides good answers for English or German input but with background noise or dialect the chatbot faces difficulties.

5 Identified Requirements for Frameworks and Architectures

In recent years, reference frameworks are emerging throughout Europe such as the RAMI 4.0 reference framework for digitalization [64] or the IDSA reference architecture for designing and implementing data spaces [65]. Those reference frameworks provide technical and operational guidance considering relevant key components, management strategies, interaction mechanisms, as well as governance principles. For instance, such a framework can offer procedures on how different models can cooperate, for instance, statistics for calculation, rules to identify optima, qualitative resources interpreted by humans, or (semi-)automatic workflows. The models coming with complexity and heterogeneity create a need for conceptual and technical integration. Microservices can be used to establish technical integration (e.g., for accessing heterogenous data sources or accessing third-party algorithms) supported via low-code platforms like the microservice platform OLIVE. In addition, security frameworks may be integrated. Hybrid approaches separating the users' prompts in individual questions may result in different processing strategies, call various framework components, and combine most appropriate algorithms.

Heading for an appropriate architecture incorporating AI, the FAIRWork architecture (Fig. 9), and the AI interaction workflows (Fig. 10) presented below are

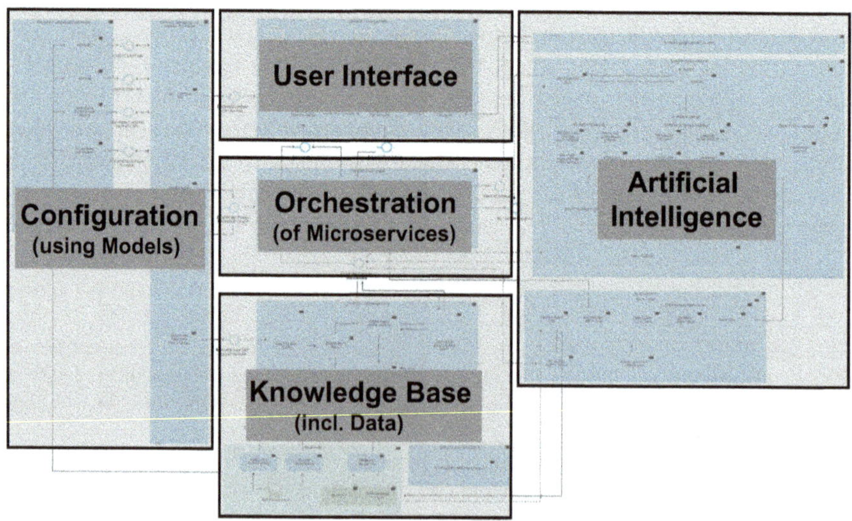

Fig. 9 Architecture for the democratic AI-based decision support system (based on [58], Fig. 1)

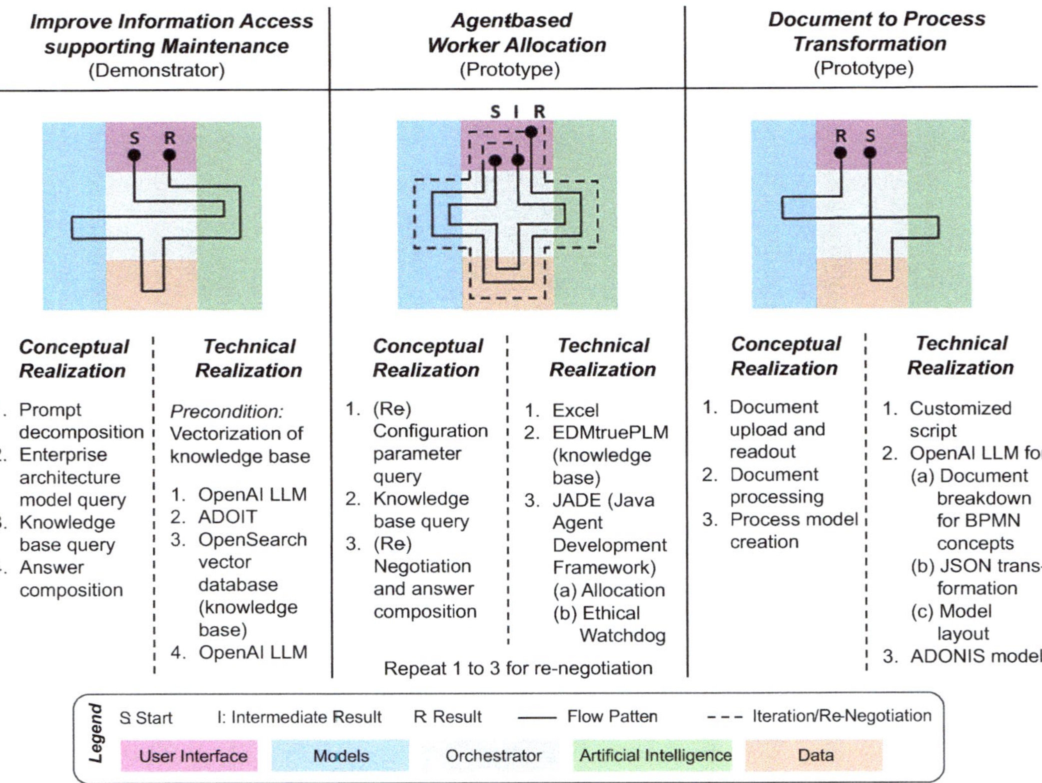

Fig. 10 Selected AI interaction workflows as basis for architecture and framework requirements

considered as a starting point. For depicting the orchestration (Fig. 9, center) of heterogenous services and components, the so-called AI interaction workflows are envisioned. In FAIRWork, a model-based management approach is applied to realize the conceptual integration. Specifically, concept models are utilized, having the capacity to mediate between and integrate different models and components with varying formal expressiveness (e.g., non-formal models like text) and providing mechanisms to overcome semantic gaps.

Three exemplary AI interaction workflows (also referred to as AI patterns or AI interaction paths) from FAIRWork are showcased in Fig. 10— from both, a conceptual as well as from a technical realization perspective. These constitute basic requirements on the interaction procedure that are valid for related AI frameworks and architectures.

The first workflow reflects the scenario "Improve Information Access supporting Maintenance" (details in Sect. 4 "Machine Maintenance Support: FAIRWork Use Case"), where a prompt is requested to the user through the UI, decomposed by the AI in its main topics, and used for the query in the enterprise architecture model that contain instructions to query the data in the knowledge base. The final answer is then composed by the AI and reported to the user through the same UI. These conceptual steps have been implemented using OpenAI LLM as AI, ADOIT as enterprise architecture modelling environment, OpenSearch Vector Database[26] as knowledge base, and OLIVE as orchestrator.

The second workflow realizes the scenario "Agent-based Worker Allocation" (details in Sect. 4.1.4, "Sample Prototype: Assist Decisions in Resource Allocation"), where the user is asked to provide some configuration parameters that will be used to query a knowledge base, and the resulting data will be used by the AI to negotiate iteratively with the user the best worker allocation, taking into account certain parameters. The technical implementation relies on an uploaded Excel in the UI for depicting the configuration parameters, EDMtruePLM[27] as knowledge base, and JADE[28] for the custom AI solution realizing both the initial worker allocation and the ethical compliance negotiation.

Finally, the third workflow depicts the scenario "Document to Process Transformation" (details in Sect. 4.1.2, "Sample Prototype: Document Transformation"), where a document containing the description of a specific business process is uploaded by the user via an UI in a storage space, processed by an AI that extracts the process information, converting them in a structured format ready for the usage by a process modelling environment that provides a diagrammatic representation of the process for the user. The technical realization is based on

---

[26] The vector database by OpenSearch is open source and supports building flexible, scalable, and future-proof AI applications (https://opensearch.org/platform/os-search/vector-database/, last accessed May 9, 2025).

[27] The EDMtruePLM by Jotne consists of a set of tools for managing the entire product life cycle, from concept design to product retirement (https://jotneconnect.com/about-jotne/, last accessed May 9, 2025).

[28] The Java Agent DEvelopment Framework (JADE) is a software framework simplifying the implementation of multi-agent systems (https://jade.tilab.com/, last accessed May 9, 2025).

custom scripts for user interaction, using OpenAI LLM for document analysis, transformation and model layouting, as well as on ADONIS as a process modelling environment.

Before testing the applicability for a reference infrastructure, the patterns are applied to manufacturing use cases in the MODAPTO project, specifically focusing on process management aspects. MODAPTO project [66] envisions modular industrial systems that are enhanced by distributed intelligence via interoperable digital twins to increase efficiency in production line design, reconfiguration, and decision support. In MODAPTO, the previously proposed decision support architecture—the underlying structure for the patterns—is mapped as follows: a smart services catalogue provides access to all domain-specific smart services, enabling their instantiation, orchestration, or assignment to a digital twin in the targeted production environment; data comes directly from the digital twin of the physical asset or from internal knowledge bases; a model based environment is used for the configuration of the different digital twins in standard Asset Administration Shell (AAS) format and for their involvement in different production schemas represented in standard BPMN format; finally, the user interacts with the system by targeted UIs specifically for different evaluations and decision support. In this context, the "Sample Prototype: Document Transformation" described above (see Sect. 4.1) has been applied and tested for the automatic generation of production schemas in BPMN model format starting from manufacturing process documentation—referring to the pattern right pattern in Fig. 10 realized with a different technology suitable for digital twinning.

6   Recap and Outlook

Summarizing, this chapter explores the implementation of AI experiments, prototypes, demonstrations, and solutions as a foundation for understanding how AI can be effectively integrated into real-world industrial applications such as the presented production scenarios. A key focus is on enhancing reliable information by trustworthy AI, emphasizing the development of domain-specific and technical concept models for configuring systems such as the ethical watchdog to ensure compliance with ethical and regulatory standards. Furthermore, the research contributes to more generic AI frameworks and architectures, particularly through the FAIRWork architecture and prototype implementations. Additionally, AI patterns—interaction workflows—identified in this study provide valuable insights for shaping standardized approaches within such a European AI ecosystem. While the conceptual realizations contribute to a reference framework, the technical realizations are expected to contribute to a reference architecture in the long run. The FAIRWork project was used to collect requirements for both and explore the pattern-based approach in the production industry, while the MODAPTO project is used to test the applicability of the patterns in a modular manufacturing and digital twinning context.

Acknowledgments This work was supported by the European Horizon Europe projects FAIRWork, with grant agreement ID 101069499 (fairwork-project.eu), and MODAPTO, with grant agreement ID 101091996 (modapto.eu). We thank the FAIRWork consortium, in particular Roland Sitar (FLEX), for sharing his domain expertise on the use case aspects; Gustavo Vieira (MORE CoLAB), for his implementation of the agent-based ethical watchdog; and Herwig Zeiner (JR), for his work on validating the certification documents.

References

1. Gartner. (2024). *3 bold and actionable predictions for the future of GenAI*. Retrieved February 25, 2025, from https://www.gartner.com/en/articles/3-bold-and-actionable-predictions-for-the-future-of-genai
2. Javaid, M., Haleem, A., & Singh, R. P. (2023). A study on ChatGPT for Industry 4.0: Background, potentials, challenges, and eventualities. *Journal of Economy and Technology, 1*, 127–143. https://doi.org/10.1016/j.ject.2023.08.001
3. Sherif, A., Salloum, S. A., & Shaalan, K. (2024). Systematic review for knowledge management in Industry 4.0 and ChatGPT applicability as a tool. In A. Al-Marzouqi, S. A. Salloum, M. Al-Saidat, A. Aburayya, & B. Gupta (Eds.), *Artificial intelligence in education: The power and dangers of ChatGPT in the classroom* (Studies in big data) (Vol. 144). Springer. https://doi.org/10.1007/978-3-031-52280-2_19
4. Curry, E., Heintz, F., Irgens, M., Smeulders, A. W., & Stramigioli, S. (2022). Partnership on AI, data, and robotics. *Communications of the ACM, 65*(4), 54–55.
5. FAIRWork Consortium. (2025). *Flexibilization of complex ecosystems using democratic AI based decision support and recommendation systems at work*. Retrieved May 9, 2025, from https://fairwork-project.eu/
6. Woitsch, R., Hrgovcic, V., & Buchmann, R. (2012). Knowledge product modelling for industry: The PROMOTE approach. 14th IFAC Symposium on Information Control Problems in Manufacturing. *IFAC Proceedings Volumes, 45*, 1208.
7. Karagiannis, D., & Kühn, H. (2002). Metamodelling platforms. In K. Bauknecht, A. Min Tjoa, & G. Quirchmayer (Eds.), *Proceedings of the Third International Conference EC-Web 2002—Dexa 2002, Aix-en-Provence, France, September 2–6* (LNCS) (Vol. 2455, p. 182). Springer-Verlag.
8. Kühn, H., Bayer, F., Junginger, S., & Karagiannis, D. (2003). Enterprise model integration. In K. Bauknecht, A. M. Tjoa, & G. Quirchmayr (Eds.), *E-Commerce and web technologies. EC-Web 2003* (Lecture notes in computer science) (Vol. 2738). Springer. https://doi.org/10.1007/978-3-540-45229-4_37
9. Ryan, J., & Heavy, C. (2006). Process modeling for simulation. *Computers in Industry, 57*(5), 437–450. https://doi.org/10.1016/j.compind.2006.02.002
10. Göthe-Lundgren, M., Lundgren, J. T., & Persson, J. A. (2002). An optimization model for refinery production scheduling. *International Journal of Production Economics, 78*(3), 255–270. https://doi.org/10.1016/S0925-5273(00)00162-6
11. Sumereder, A., & Dokken, T. (2022). Model-based guide toward digitization in digital business ecosystems. In *Domain-specific conceptual modeling* (pp. 411–433). Springer International Publishing.
12. Woitsch, R., Sumereder, A., & Falcioni, D. (2022). Model-based data integration along the product & service life cycle supported by digital twinning. *Computers in Industry, 140*, 103648.
13. Grisold, T., Janiesch, C., Roeglinger, M., & Wynn, M. (2024). Managing dynamics in and around business processes. *Business & Information Systems Engineering, 66*. https://doi.org/10.1007/s12599-024-00895-2

14. Lieberman, H. (2016). Symbolic vs. subsymbolic AI. Retrieved May 9, 2025, from https://courses.media.mit.edu/2016spring/mass63/wp-content/uploads/sites/40/2016/02/Symbolic-vs.-Subsymbolic.pptx_.pdf

15. Ilkou, E., & Koutraki, M. (2020). *Symbolic vs sub-symbolic AI methods: Friends or enemies?* Retrieved from https://ceur-ws.org/Vol-2699/paper06.pdf

16. Chen, P. (1976). The entity-relationship model—Toward a unified view of data. *ACM Transactions on Database Systems, 1*(1), 9–36. https://doi.org/10.1145/320434.320440

17. Schneider, J., et al. (2024). Foundation models. A new paradigm for artificial intelligence. *Business & Information Systems Engineering, 66*(2), 221–231.

18. Szczuko, P. (2024). Dos and don'ts of LLMs: How to survive past the trough for disillusionment? *Keynote ISD, 2024.*

19. Yang, L., et al. (2024). Give us the facts: Enhancing large language models with knowledge graphs for fact-aware language modeling. *IEEE Transactions on Knowledge and Data Engineering, 36*, 3091.

20. Wei, J., Wang, X., Schuurmans, D., Bosma, M., Ichter, B., Xia, F., Chi, E. H., Le, Q. V., & Zhou, D. (2022). Chain-of-thought prompting elicits reasoning in large language models. In *Proceedings of the 36th International Conference on Neural Information Processing Systems (NIPS '22)*. Curran Associates Inc..

21. Kojima, T., Gu, S. S., Reid, M., Matsuo, Y., & Iwasawa, Y. (2022). Large language models are zero-shot reasoners. In *Proceedings of the 36th International Conference on Neural Information Processing Systems (NIPS '22)*. Curran Associates Inc..

22. Müller, M. (2024). Advancements in software engineering through artificial intelligence. *International Journal of Scientific Research and Engineering Trends.*

23. Pohl, K., Böckle, G., & van der Linden, F. (2005). *Software product line engineering: Foundations, principles, and techniques* (1st ed.). Springer.

24. Naily Moh, A., Setyautami, M. R. A., Muschevici, R., & Azurat, A. (2018). A framework for modelling variable microservices as software product lines. In A. Cerone & M. Roveri (Eds.), *Software engineering and formal methods* (pp. 246–261). Springer International Publishing. https://doi.org/10.1007/978-3-319-74781-1_18

25. Setyautami, M. R. A., Fadhlillah, H. S., Adianto, D., Affan, I., & Azurat, A. (2020). Variability management: Re-engineering microservices with delta-oriented software product lines. In *Proceedings of the 24th ACM Conference on Systems and Software Product Line: Volume A* (pp. 1–6). https://doi.org/10.1145/3382025.3414981

26. Mendonça W. D. F., Assunção W. K. G., Estanislau L. V., Vergilio S. R. & Garcia A. (2020). Towards a Microservices-Based Product Line with Multi-Objective Evolutionary Algorithms. IEEE Congress on Evolutionary Computation (CEC), Glasgow, UK, pp. 1–8. https://doi.org/10.1109/CEC48606.2020.9185776

27. Lu H., Shtern M., Simmons B., Smit M. & Litoiu M. (2013). Pattern-Based Deployment Service for Next Generation Clouds. *2013 IEEE Ninth World Congress on Services*, Santa Clara, CA, USA, pp. 464–471. https://doi.org/10.1109/SERVICES.2013.54

28. Fischer J., Majumdar R. & Esmaeilsabzali S. (2012). Engage: a deployment management system. *33rd ACM SIGPLAN Conference on Programming Language Design and Implementation (PLDI '12)*. Association for Computing Machinery, New York, NY, USA, 263–274. https://doi.org/10.1145/2345156.2254096

29. Di Cosmo, R., Mauro, J., Zacchiroli, S., & Zavattaro, G. (2014). Aeolus: A component model for the cloud. *Information and Computation, 239*, 100–121.

30. Di Cosmo, R., Eiche, A., Mauro, J., Zacchiroli, S., Zavattaro, G., & Zwolakowski, J. (2015). Automatic deployment of services in the cloud with aeolus blender. In *International Conference on Service-Oriented Computing* (pp. 397–411). Springer Berlin Heidelberg.

31. Lascu, T. A., Mauro, J., & Zavattaro, G. (2015). Automatic deployment of component-based applications. *Science of Computer Programming, 113*, 261–284.

32. Brabra, H. (2020). *Supporting management and orchestration of cloud resources in a multi-cloud environment* (Doctoral dissertation, Institut Polytechnique de Paris; Université de Sfax (Tunisie). Faculté des Sciences économiques et de gestion).
33. Kiss, T., et al. (2019). Micado—Microservice-based cloud application-level dynamic orchestrator. *Future Generation Computer Systems, 94*, 937–946.
34. Wurster, M., Breitenbücher, U., Brogi, A., Falazi, G., Harzenetter, L., Leymann, F., Soldani, J., & Yussupov, V. (2019). The EDMM modeling and transformation system. In *Service-Oriented Computing—ICSOC 2019 Workshops, Toulouse, France* (pp. 294–298). Springer-Verlag. https://doi.org/10.1007/978-3-030-45989-5_26
35. Kovács J. & Kacsuk P. (2018). Occopus: a Multi-Cloud Orchestrator to Deploy and Manage Complex Scientific Infrastructures. *J. Grid Comput. 16*(1), 19–37. https://doi.org/10.1007/s10723-017-9421-3
36. Bhattacharjee, A., Barve, Y., Gokhale, A., & Kuroda, T. (2019). *Cloudcamp: Automating cloud services deployment and management.* arXiv.
37. Sandobalin, J., Insfran, E., & Abrahao, S. (2017). *End-to-end automation in cloud infrastructure provisioning.* AIS eLibrary.
38. MyoungLee, G., Um, T. W., & Choi, J. (2018). AI as a microservice (AIMS) over 5G networks. In *ITU kaleidoscope: Machine learning for a 5G future (ITU K), Santa Fe, Argentina.*
39. Jhingran, S., Bansal, N., Chaturvedi, R., Singh, A., & Arora, Y. (2024). Decentralized generative AI model deployment using microservices. In *2024 International Conference on Artificial Intelligence and Quantum Computation-Based Sensor Application (ICAIQSA), Nagpur, India* (pp. 1–5). https://doi.org/10.1109/ICAIQSA64000.2024.10882424
40. Bablu, T. A. (2025). AI microservices in enterprise applications: A comprehensive review of use cases and implementation frameworks. *Journal of Computational Social Dynamics.*
41. Sarma, W., Tiwari, S., & Dey, S. (2023). Architecting next-generation software systems with generative AI and large language models: Challenges, opportunities, and best practices. *International Journal of Innovative Research in Computer and Communication Engineering.* Retrieved from https://ijircce.com/admin/main/storage/app/pdf/mIloN4e3CnJ2XUpjwhM-KadKjnT3lWqzeACN8qhrM.pdf
42. Hevner, A. R., March, S. T., Park, J., & Ram, S. (2004). Design science in information systems research. *Management Information Systems Quarterly, 28*, 75–105.
43. Woitsch, R. (2020). Industrial digital environments in action: The OMiLAB innovation corner. In J. Grabis & D. Bork (Eds.), *The practice of enterprise modeling. PoEM 2020* (Lecture notes in business information processing) (Vol. 400). Springer. https://doi.org/10.1007/978-3-030-63479-7_2
44. European Commission. (2025). *AI Act.* Retrieved May 9, 2025, from https://digital-strategy.ec.europa.eu/en/policies/regulatory-framework-ai
45. Li, B., et al. (2023). Trustworthy AI: From principles to practices. *ACM Computing Surveys, 55*(9), 1–46. https://doi.org/10.1145/3555803
46. Mittelstadt, B. (2019). Principles alone cannot guarantee ethical AI. *Nature Machine Intelligence, 1*(11), 501–507. https://doi.org/10.1038/s42256-019-0114-4
47. Siau, K., & Wang, W. (2020). Artificial intelligence (AI) ethics: Ethics of AI and ethical AI. *Journal of Database Management, 31*(2), 74–87. https://doi.org/10.4018/JDM.2020040105
48. European Commission. (2019). *Ethics guidelines for trustworthy AI.* AI HLEG. Retrieved May 9, 2025, from https://ec.europa.eu/digital-single-market/en/news/ethics-guidelines-trustworthy-ai
49. European Commission. (2020). *The Assessment List for Trustworthy Artificial Intelligence (ALTAI).* Retrieved May 9, 2025, from https://digital-strategy.ec.europa.eu/en/library/assessment-list-trustworthy-artificial-intelligence-altai-self-assessment
50. Mäntymäki, M., Minkkinen, M., Birkstedt, T., & Viljanen, M. (2022). Defining organizational AI governance. *AI Ethics, 2*, 603–609. https://doi.org/10.1007/s43681-022-00143-x

51. van Nuenen, T., Ferrer, X., Such, J. M., & Cote, M. (2020). Transparency for whom? Assessing discriminatory artificial intelligence. *Computer, 53*(11), 36–44. https://doi.org/10.1109/MC.2020.3002181

52. Werz, J., et al. (2024). Explainability as a means for transparency? Lay users requirements towards transparent AI. In *Cognitive Computing and Internet of Things, Bd.124.AHFE Open Access, 2024.* https://doi.org/10.54941/ahfe1004712

53. FAIRWork Consortium. (2023). *D2.1 Specification of FAIRWork use case and DAI-DSS prototype report.* Retrieved May 9, 2025, from https://fairwork-project.eu/deliverables/D2.1_Specification%20of%20FAIRWork-v1.0a-preliminary.pdf

54. FAIRWork Consortium. (2023). *D3.1 DAI-DSS research specification.* Retrieved May 9, 2025, from https://fairwork-project.eu/deliverables/D3.1_DAI-DSS%20Research%20Specification-preliminary.pdf

55. FAIRWork Consortium. (2023). *D4.2 Initial DAI-DSS prototype.* Retrieved May 9, 2025, from https://fairwork-project.eu/deliverables/D4.2_Initial%20DAI-DSS%20Prototype%20v1.0a-preliminary.pdf

56. FAIRWork Consortium. (2023). *D5.1 DAI-DSS FAIRWork knowledge base at use case site.* Retrieved May 9, 2025, from https://fairwork-project.eu/deliverables/d5-1/D5.1_DAI-DSS%20FAIRWork%20knowledge%20base_v1.0-preliminary.pdf

57. FAIRWork Consortium. (2024). *D3.2 First DAI-DSS research collection.* Retrieved May 9, 2025, from https://fairwork-project.eu/deliverables/D3.2_First-DAI-DSS-ResearchCollection_V1.0_preliminary.pdf

58. FAIRWork Consortium. (2024). *D4.1 DAI-DSS Architecture and initial documentation and test report.* Retrieved May 9, 2025, from https://fairwork-project.eu/deliverables/D4.1.1_DAI-DSS_ArchitectureAndInitialDocumentationAndTestReport_V1.0_preliminary.pdf

59. FAIRWork Consortium. (2025). *D4.3 final DAI-DSS prototype, documentation and test report.* Retrieved May 9, 2025, from https://fairwork-project.eu/deliverables/D4.3%20Final%20DAI-DSS%20Prototype%20v1.0-preliminary.pdf

60. FAIRWork Consortium. (2025). *D3.3 final DAI-DSS research collection.* Retrieved May 9, 2025, from https://fairwork-project.eu/deliverables/D3.3%20DAI-DSS%20Research%20Collection%20v1.0-preliminary.pdf

61. Muck, C., & Palkovits-Rauter, S. (2022). Conceptualizing design thinking artefacts: The Scene2ModelStoryboard approach. In D. Karagiannis, M. Lee, K. Hinkelmann, & W. Utz (Eds.), *Domain-specific conceptual modeling: Concepts, methods and ADOxxTools* (pp. 567–587). Springer. https://doi.org/10.1007/978-3-030-93547-4_25

62. Woitsch, R., Muck, C., Utz, W., & Zeiner, H. (2024). Enable flexibilisation in FAIRWork's democratic AI-based decision support system by applying conceptual models using ADOxx. *Complex Systems Informatics and Modeling Quarterly, 38,* 27–53. https://doi.org/10.7250/csimq.2024-38.02

63. Falcioni, D., & Woitsch, R. (2021). OLIVE, a model-aware microservice framework. In E. Serral, J. Stirna, J. Ralyté, & J. Grabis (Eds.), *The practice of enterprise modeling. PoEM 2021* (Lecture notes in business information processing) (Vol. 432). Springer. https://doi.org/10.1007/978-3-030-91279-6_7

64. Platform Industrie 4.0. (2018). *Reference architectural model Industrie 4.0 (RAMI4.0)—An introduction.* Retrieved May 9, 2025, from https://www.plattform-i40.de/IP/Redaktion/EN/Downloads/Publikation/rami40-an-introduction.html

65. IDSA. (2025). *International Data Spaces Association.* Retrieved May 9, 2025, from https://internationaldataspaces.org/

66. MODAPTO Consortium. (2025). *Modular manufacturing and distributed control via interoperable digital twins.* Retrieved May 9, 2025, from https://modapto.eu/

Data, AI, Robotics Transformative Power in Industry 5.0

New Applications and Pilot Cases for a More Sustainable and Circular Manufacturing

Sergio Gusmeroli, Davide Dalle Carbonare, Anibal Reñones, Gema Antequera García, Carmen Polcaro, Cinzia Rubattino, Gabriella Monteleone, Pietro Pittaro, Juanan Arreta, Marielena Marquez Barreiro, Santiago Muños-Landin, Ignacio Lacalle, and Jesus Alonso

Abstract Industry 5.0 (I5.0) is an evolution of the fourth Industrial Revolution (Industrie 4.0) aiming at a more human-centric, resilient, and sustainable manufacturing. This proposed chapter aims at illustrating the decisive contribution of ADR (AI Data Robotics) technologies in the development, implementation, and demonstration of Industry 5.0 principles and in particular of its sustainable and

S. Gusmeroli (✉) · G. Monteleone
Politecnico di Milano, Milan, Italy
e-mail: sergio.gusmeroli@polimi.it; gabriella.monteleone@polimi.it

D. D. Carbonare · C. Rubattino
Engineering Ingegneria Informatica, Rome, Italy
e-mail: davide.dallecarbonare@eng.it; cinzia.rubattino@eng.lt

A. Reñones
CARTIF Technology Center, Valladolid, Spain
e-mail: aniren@cartif.es

G. A. García
CTAG Automotive Technology Center of Galicia, Pontevedra, Vigo, Spain
e-mail: gema.antequera@ctag.com

C. Polcaro · J. Alonso
Innovalia Association, Bilbao, Spain
e-mail: cpolcaro@innovalia.org; jalonso@innovalia.org

P. Pittaro
Prima Industrie, Collegno, Torino, Italy
e-mail: pietro.pittaro@primaindustrie.com

J. Arreta
IDEKO, Elgoibar, Spain
e-mail: jarrieta@ideko.es

E. Curry et al. (eds.), *Artificial Intelligence, Data and Robotics*,
https://doi.org/10.1007/978-3-032-10561-5_11

M. M. Barreiro
Gradiant Technology Center, Pontevedra, Vigo, Spain
e-mail: mmarquez@gradiant.org

S. Muños-Landin
AIMEN Technology Centre, Pontevedra, Vigo, Spain
e-mail: santiago.muinos@aimen.es

I. Lacalle
Universitat Politècnica de València, València, Spain
e-mail: iglaub@upv.es

circular manufacturing pillar. In the domain of EC-funded projects (Horizon and Digital Europe programs, HEP and DEP), several initiatives have been recently addressing the I5.0 sustainability/circularity challenge from different viewpoints: the technology perspective in HEP ADRA partnership; the application perspective in HEP, Made in Europe (MiE) partnership; and the deployment perspective in DEP initiatives such as data spaces, AI TEFs (Testing Experimentation Facilities), and EDIH (European Digital Innovation Hubs). The analysis has been conducted along three main dimensions: a **Technology** dimension, where data, AI, and robotics convergence is enabling unprecedented innovations regarding sustainability and circularity for EU manufacturing industries; a **Project** dimension, to demonstrate the fundamental role of EU-funded research-innovation-deployment actions to create, nurture, and diffuse an Industry 5.0 mindset among researchers; a **Pilot** dimension, where significant industrial players are showing the improvement of their global competitiveness, thanks to Industry 5.0. Main conclusions regard (1) the fundamental role in Industry 5.0 of the robotics and automation physical world and the need for edge-to-cloud data and AI continuum; (2) the confidentiality and high strategic value of the data exchanged in Industry 5.0 value networks and the need for open, secure, and interoperable data infrastructures (data spaces); and (3) the competitiveness and resilience of manufacturing supply chains as a top priority for Europe and the need for integrated ADR technologies with methods and tools for environmental impact assessment and materials-components-products circularity promotion. The **BDVA (Big Data Value Association) Smart Manufacturing Industry (SMI)** group is a business-technology community that studies, analyzes, and promotes the impact of data spaces, AI, and digital twins to the manufacturing industry. It encompasses more than 130 experts representing more than 60 organizations involved in dozens of EC-funded projects. The content of this chapter originates from the last years of workshops, discussions, and webinars inside the SMI group, where experts, projects, and industrial pilots actively participated in the definition of the role of ADR technologies in the several different manufacturing business processes according to Industry 5.0 principles.

Keywords Smart manufacturing Industry 5.0 · Data technologies · Artificial intelligence · Robotics · Human-centric manufacturing · Resilience · Sustainability

1 Introduction

Industry 5.0 is an evolution of Industry 4.0 with a specific focus on human-centric, resilient, and sustainable manufacturing. In complement to Industry 4.0 where efficiency and productiveness focus through the integration of digital technologies (cyber-physical production systems), Industry 5.0 proposes a more decisive and important role of humans (empowerment) and environmental compliance topics along the different phases of the product life cycle, implementing the so-called Twin Transition (Digital-Green) from a human perspective. At **pre-production** time, Industry 5.0 is promoting support to ESPR (Ecodesign for Sustainable Products Regulation) directive via data-AI-driven collaborative environments such as the Industrial Metaverse, and a flexible and resilient virtual commissioning environment for engineers to re-design and configure supply chains according to resilience and sustainability-circularity criteria. **During production**, human-robot and human-machine interaction collaborative intelligence patterns are enabled by ADR technologies in order to monitor workers' behavior, well-being, and ergonomics, also in a waste-resource management perspective. In the **post-production** steps, during operations, smart products are enabled by ADR technology to sense the environment, reason, and take decisions also regarding environmental impact; during maintenance, new business Models-as-a-Service can be activated and provide additional revenue opportunities to the manufacturer; during End-of-Life, de-manufacturing, re-manufacturing, re-purposing, and recycling techniques are all supported by the combination of robotics and AI technologies, implementing circular economy virtuous loops. In the ADR technological convergence, **data** technologies and data spaces in particular not only guarantee the collaboration among the different stakeholders in the value chain but also assure and certify data availability, data quality, data transparency, and data traceability through Digital Product Passports. **AI** technologies, both generative AI and machine learning techniques, are essential for generating product options and simulate their behaviors in virtual worlds, fundamental for quality and environmental impact assessment through machine vision techniques, necessary for predictions and optimization of decisions. Thanks to Industry 5.0, **robotic** technologies are not just cyber-physical systems, but their interaction and collaboration with humans are also studied and exploited.

As a final remark, the *Draghi Report* on EU competitiveness is harmonizing I4.0 efficiency and optimization with I5.0 sustainability and resilience aspects, suggesting a new model of industrialization where humans are empowered and become protagonist of the changes and the socioeconomic impact of ADR technologies.

In this document, Sect. 2 is devoted to a technological review, which analyses state of the art, identified opportunities, and challenges and envisages a future pathway toward full adoption of ADR in Industry 5.0. In particular, for the data dimensions, recent discussions held during BDVA events (EBDVF and DATAWEEK) will be reported especially regarding data spaces for AI and data spaces for DPP and Industry 5.0. Section 3 reports on a series of recent EU projects (Horizon and Digital Europe) aiming at the integration of ADR technologies in the key Industry 5.0

twin-transition business processes, including a new role for humans in terms of co-creation, participation, and inclusiveness. Section 4 reports real-life industrial pilots where not just technologies are demonstrated but also business and social impact measured and assessed. Section 5 reports on some conclusions and future outlook of the full adoption of ADR technologies in Industry 5.0 business scenarios. This chapter aligns with the objectives of the AI, Data, and Robotics Partnership by demonstrating how European-funded initiatives and pilot implementations are advancing the integration of trustworthy ADR technologies in Industry 5.0 contexts [1].

2 ADR in Industry 5.0: The Technology Perspective

This chapter aims at depicting how ADR technologies convergence could address I5.0 main challenges (human centricity, resilience, sustainability-circularity). Three main paragraphs are included for data (Sect. 2.1), AI (Sect. 2.2), and robotics (Sect. 2.3) technologies, respectively, coming from our discussions in BDVA SMI group.

2.1 Data Technologies for Industry 5.0

2.1.1 State of the Art

In the context of Industry 5.0, data technologies are pivotal in advancing the integration of human-centric manufacturing with industrial processes. A key component of this evolution is the digital twin, adopted to facilitate enhanced decision-making and operational efficiency by providing a comprehensive digital representation of physical assets. Data sharing is a complementary critical aspect of Industry 5.0, promoting transparency and collaboration across the value chain.

The landscape of technologies dealing with data, at various levels, is broad and includes, for example, cloud computing, edge computing, and blockchain that are instrumental in the digitalization process and to provide the necessary infrastructure for real-time data processing, secure data sharing, and decentralized decision-making.

Data technologies such as Big Data and the Internet of Things (IoT) are relevant to Industry 5.0 scenarios to enable real-time data collection, analysis, and decision-making, which are crucial for creating responsive and adaptive industrial systems. In addition to that, seamless data exchange between different stakeholders, including suppliers, manufacturers, and customers, is essential for optimizing resource allocation, circular economy paradigms, and improving decision-making processes.

2.1.2 Opportunities and Challenges

Opportunities in the context of Industry 5.0 are multiple, and their relevance is strongly related to every specific case. Some of them are summarized below:

- **Optimization of manufacturing processes** by utilizing digital twins. Businesses can create real-time virtual replicas of their physical assets, allowing for continuous monitoring and analysis. This enables companies to identify inefficiencies, predict maintenance needs, and optimize production schedules, leading to enhanced operational efficiency and cost savings.
- **Advancing Environmental, Social, and Governance** (ESG) initiatives. Data sharing across the supply chain allows for better transparency and accountability. By leveraging, for example, on blockchain technology alongside digital twins, companies can ensure the integrity and security of shared ESG data. This transparency helps in tracking sustainability metrics, verifying compliance, and improving overall ESG performance, thereby enhancing brand reputation and meeting regulatory requirements.
- **Improvements in supply chain management**. Digital twins provide end-to-end visibility and traceability, allowing businesses to monitor and optimize their supply chain operations in real-time. This capability helps in identifying potential disruptions, optimizing inventory levels, and improving coordination among stakeholders. As a result, companies can achieve greater supply chain resilience, reduce costs, and enhance overall efficiency.

Below are only some of the reported challenges, from a technical perspective, that are mostly affecting Industry 5.0 scenarios:

- With the integration of digital twins, businesses must manage data from diverse sources and formats, and then, one of the most significant technical challenges is ensuring **data interoperability**. Achieving seamless data exchange and integration across different systems and platforms is crucial for accurate real-time simulations and decision-making.
- Maintaining **data security and privacy**. As data sharing becomes more prevalent, protecting sensitive information from cyber threats and unauthorized access is paramount. Implementing robust encryption, access controls, and compliance with data protection regulations are essential to safeguard data integrity and build trust among stakeholders.
- **Scalability of data infrastructure**. The vast amount of data generated by digital twins and shared across networks requires scalable storage and processing capabilities. Businesses must invest in advanced data management solutions and cloud technologies to handle the increasing data volume and ensure efficient data processing and analysis.

2.1.3 Technology Roadmap

As a significant case for describing the technology roadmap for data technologies toward Industry 5.0, we took inspiration from two sessions held in Luxembourg on December 10 during the DATAWEEK 2024 Part 3.

The first session, entitled "Manufacturing Data Spaces, as the Data Infrastructure for advanced AI applications," focused on the key role of data availability, suitability, and quality in the training phase of advanced AI models. In this perspective, the EC recently published the AI OFFICE ecosystem framework, where data spaces are put in relation with other seven initiatives. As an outcome of the DATAWEEK session, we could interpret the framework for the manufacturing sector and Industry 5.0 in particular. The technology roadmap for I5.0 oriented data spaces in the AI OFFICE ecosystem includes:

1. **DS and AI TEF** (Testing and Experimentation Facilities). The AI TEF for Manufacturing (**AI MATTERS**) services aim at providing AI suppliers with diverse scenarios, business cases, and datasets for enhancing the TRL of their solutions. Data spaces are essential for that.
2. **DS and EDIHs** (European Digital Innovation Hubs). The SME orientation of both initiatives needs to be improved: Data spaces are usually dominated by large manufacturers, and their federation should allow inter-DS and even cross-sectoral interoperability. EDIHs offer D BEST (Data Business Ecosystem Skills Test before invest) services for improving SMEs preparedness to new digital technologies such as data spaces.
3. **DS and the AIoD** (AI on Demand) platform. In the old AIoD manufacturing platform (AI4Europe), we had 44 projects registered and just 6 datasets. In the new version, we have now a flat list of more than 600k datasets and 141 contain manufacturing (RAIL). The main challenge here is to transform datasets into data products (FAIR) and to incentivize the population of the platform. Data and AI pipelines are closely connected indeed.
4. **DS and the EDIC** (European Digital Infrastructure Consortium). EDIC is an instrument made available to Member States under the Digital Decade Policy Programme 2030 to speed up and simplify the setup and implementation of multi-country projects. ALT EDIC (Language Technologies) and EUROPEUM--EDIC (Blockchain) are less relevant to I5.0, while CitiVerse could instead offer very interesting scenarios for urban manufacturing and social-technical-business integration of human centric, resilient, sustainable factories in an urban environment.
5. **DS and the AI Skills Academy** (to be developed). Training and education are of key relevance for Industry 5.0 adoption. Beyond the role of EDIHs and their services, we envisage the birth of hands-on I5.0 didactic factories where I5.0 principles could be taught but also concretely experimented (see below).
6. **DS and AI Regulatory Sandboxes**. Regulatory sandboxes are the ideal environment where to assess compliance of data and AI solutions to the current EU regulations. A first example can be found in the AIREDGIO5.0 project and its

TERESAs (TEchnology REgulatory SAndboxes) where workspaces are assessed according to WISE principles (Wellbeing, Inclusiveness, Safety, and Ergonomics).

7. **DS and AI FACTORIES** (HPC). Data spaces and AI FACTORIES are expected to go hand in hand in several sectors, including manufacturing. The seven chosen EU AI FACTORIES, possibly linked with BDVA iSPaces will represent an unvaluable opportunity to join data spaces and AI.

A second session was instead dedicated to "Manufacturing Data Spaces, as the Data Infrastructure for Digital Product Passports," where the following pathways have been identified:

1. **DS for Circular Manufacturing**, as an extension of existing linear manufacturing data spaces (OEM and its Customers/Suppliers) to new circular business processes including waste collection/sorting, re-use/re-purposing, de- and re-manufacturing, and, finally, recycling of components and materials.

   - **New Stakeholders** are usually not included in B2B linear value chains and often represent multi-sector, multi-product, and multi-material entities in the EPR (Extended Producer Responsibility) landscape. How to qualify and certify them? Are they suitable to join existing domain-specific data spaces (e.g., the Battery Data Space, the WEEE Data Space) in order to get data from the manufacturing value chain, or shall they remain separated in another data space?
   - **New ADR Technologies** are now in place for TR[3] purposes (TRaceability TRansparency TRust) in a circular data space and especially devoted to implement Digital Product Passports (see below) for those sectors requiring it; also for other sectors, ADR for circularity technologies (identification, storage, elaboration, optimization) embedded in the product or at the far edge. Edge-to-cloud continuum challenges will be very relevant in the future not just for performance reasons but also for energy consumption and environmental impact motivations.
   - **New Business Models** are maybe the most challenging aspect for a future roadmap. To which extent are circular business models (secondhand market, re-purposing, Circularity-as-a-Service) of interest of manufacturers, and to which extent will they be of interest for third parties and service providers? The car industry is usually paving the way for a full servitization of the business, including the end of life of products, components, and materials.

2. **DS for Digital Product Passports**, as a technology enabler for supporting the whole DPP life cycle (first generation, authorized update, re-use, cancellation) at disposal of regulatory authorities, citizens, and companies.

   - **New Stakeholders** who are driving the DPP revolution (governmental agencies like Customs Anti-Fraud Market-Regulation authorities) are to be interlinked with the long-lasting B2B value network of manufacturing, characterized by their own standards, data models, and, above all, industrial trusted agreements. Is a remanufacturer to be certified by a national government authority in order to have access to a B2B DS (batteries)?

- **New Industrial Sectors** beyond the DPP traditional promoters (batteries, WEEE, textile) are now discovering the usefulness for tagging, identifying products, and getting life cycle information from them, especially in the domain of CO_2 and environmental footprint (e.g., furniture, consumer goods, construction, machinery). Irrespective of the entry-into-force date of a regulatory DPP (if any), such sectors could benefit from the developed ADR technologies and improve their business processes.
- **New Business Models**, again DPPs can enable aaS business models also by newcomers and even citizens do-it-by-yourself in a democratization process that seems unstoppable. So, not just B2B models but also more collaborative, participative, and incentivized business are envisaged for the near future.

2.2 Artificial Intelligence Technologies for Industry 5.0

2.2.1 State of the Art

The integration of artificial intelligence (AI) in Industry 5.0 is revolutionizing manufacturing by combining human-centric approaches with intelligent automation. AI-powered digital twins and generative AI are key enablers, providing real-time simulations and adaptive learning capabilities that enhance efficiency and decision-making processes.

In manufacturing, AI-driven digital twins enable real-time monitoring and predictive maintenance, reducing downtime and optimizing resource allocation. Generative AI enhances design processes by automating the creation of complex structures, improving production speed, and reducing waste. Additionally, AI in logistics supports demand forecasting, route optimization, and automated inventory management.

Despite these advancements, challenges such as data interoperability, cybersecurity risks, and the need for skilled personnel persist. Addressing these issues is crucial for widespread AI adoption in Industry 5.0.

2.2.2 Opportunities and Challenges

The following opportunities can be expected in this field:

- Enhanced customization and flexibility: AI-powered digital twins allow manufacturers to simulate and test different production scenarios, enabling mass customization and reducing lead times.
- Sustainable and resilient manufacturing: AI optimizes energy consumption and waste reduction, supporting green manufacturing initiatives and circular economy principles.

- Human-AI collaboration: Industry 5.0 emphasizes human-centric automation, where AI assists workers rather than replacing them, improving workplace safety and operational efficiency.
- Supply chain resilience and optimization: AI enhances visibility across the supply chain, enabling predictive analytics to mitigate disruptions and improve logistics.

The following are the challenges:

- AI-data integration convergence and interoperability: Standardizing data across different manufacturing systems remains a significant hurdle.
- AI-robotics convergence and integration.
- AI cybersecurity risks: Increased connectivity and reliance on AI expose manufacturing environments to potential cyber threats.
- AI workforce upskilling: The shift toward AI-driven manufacturing necessitates new skills, requiring investment in training and workforce development.
- AI regulatory compliance and ethical considerations: The use of AI in decision-making raises concerns regarding transparency, bias, and regulatory alignment.

2.2.3 Technology Roadmap

Short Term (1–3 Years)

- Deployment of AI-driven predictive maintenance solutions to reduce equipment downtime
- Implementation of AI-enhanced digital twins for real-time monitoring and process optimization
- Development of generative AI models to support automated design and prototyping
- Standardization of data formats and integration protocols for improved interoperability

Medium Term (3–5 Years)

- Expansion of AI capabilities in robotics for adaptive and collaborative manufacturing
- Strengthening cybersecurity frameworks for AI-driven manufacturing environments
- Enhancement of supply chain intelligence using AI-driven analytics and blockchain integration
- Workforce training programs focused on AI literacy and human-AI collaboration

Long Term (5+ Years)

- Full-scale implementation of autonomous AI-driven factories with self-optimizing systems

- Integration of AI-powered digital twins at a systemic level for end-to-end production life cycle management
- Development of regulatory frameworks and ethical AI standards to guide responsible AI adoption
- Advancements in quantum computing for enhanced AI-driven manufacturing solutions

2.3 Robotic Technologies for Industry 5.0

2.3.1 State of the Art

Industry 5.0 envisions a paradigm shift where robotics not only enhances automation but also prioritizes human centricity, resilience, and sustainability. Human centricity is key in Industry 5.0, ensuring that robotic systems are designed to support and enhance human capabilities rather than replace them. Robotic technologies play a crucial role in advancing these principles through enhanced human-robot collaboration (HRC), adaptive automation, and intelligent robotic systems. Advancements in these areas are needed through research projects that foster innovation in robotics solutions.

Current robotic technologies for Industry 5.0 focus on enhancing flexibility, interaction, and safety in human-robot collaboration. The evolution from traditional automation to collaborative robots (cobots) enables seamless integration of robotics into industrial workflows, allowing human workers and robots to share tasks efficiently.

Recent advancements include multimodal perception systems, improved dexterity, and AI-driven decision-making, which enhance robots' ability to understand and adapt to dynamic industrial settings. These innovations contribute to reducing downtime, improving efficiency, and ensuring safer working environments. Additionally, fostering small and medium-sized enterprises (SMEs) in the field of robotic solutions development and integration is crucial to overcoming technical complexity and ensuring a diverse and competitive market. Furthermore, ergonomics and worker well-being are increasingly being integrated into robotic system designs to ensure that interactions between humans and robots promote comfort, safety, and long-term physical health.

2.3.2 Opportunities and Challenges

The following opportunities can be expected in this field:

- **Human-Centric Robotics**: The integration of advanced sensing and AI enables robots to adapt to human behavior, enhancing safety and productivity in shared workspaces.

- **Sustainable Manufacturing**: Robotics supports circular economy initiatives by improving resource efficiency, enabling automated disassembly, and facilitating remanufacturing.
- **Interoperability and Modularity**: Open-source middleware and standardized interfaces facilitate seamless integration of collaborative robotics solutions not only as isolated solutions but also within the operational technology (OT) and information technology (IT) infrastructures of manufacturing companies.
- **AI and Machine Learning Integration**: Continuous learning, self-optimization, and the huge potential of generative AI enable robots to enhance precision, efficiency, and adaptability over time, opening new avenues for autonomous decision-making and interaction.
- **Enhanced Ergonomics and Well-Being**: Robotics is increasingly being leveraged to improve workplace ergonomics by assisting workers in physically demanding tasks, reducing strain and injury risk. Adaptive robotic systems can support load handling, precision work, and repetitive tasks, enhancing worker well-being and efficiency.

With respect to the challenges of integrating robotic technologies to help Industry 5.0 paradigm become a reality, it can be said that they belong to the following categories:

- **Trust and Acceptance**: Human workers need to feel comfortable and safe working alongside robots, requiring intuitive and transparent interaction models.
- **Technical Complexity**: Developing interoperable and adaptable robotic systems demands advancements in software architecture, real-time processing, and robotic skills applications such as computer vision for environment perception and interpretation. Additionally, targeted support is needed to help SMEs adopt and integrate these complex robotic solutions effectively.
- **Economic Viability**: While robotic solutions offer long-term benefits, initial investments and operational costs remain a barrier to widespread adoption.

2.3.3 Technology Roadmap

The expected advancements in robotics need to flourish along a collaborative, resilient, and sustainable manufacturing ecosystem aligned with the Industry 5.0 paradigm. For that to happen, it is needed to define and execute a solid roadmap involving several key developments:

1. **Creation of Innovative Solutions**: Development of novel solutions such as open-source middleware, robotic skills applications, and other software components that enhance integration and adaptability
2. **Initial Validation in Relevant Scenarios**: Testing and experimentation in controlled environments such as Testing and Experimentation Facilities (TEFs) to validate human-centric robotic applications

3. **Maturity and Expansion in Industrial Settings**: Further refinement of solutions in diverse industrial scenarios to address initial challenges and barriers while generating additional open-source components to enhance the ecosystem
4. **Creation of a Robotics Community**: Establishing a network to provide sustainable support for new players and drive progress in Industry 5.0
5. **Advancing Generative AI and Robotics Integration**: Exploring new applications of generative AI in robotics to develop the next generation of general-purpose robotic solutions
6. **Safety and Ethical Considerations**: Ensuring that ethical and safety guidelines are embedded throughout the development and deployment of robotic solutions

3 Research, Innovation, and Deployment Challenges in EU Projects

Several technology-oriented EU projects are gravitating around the BDVA SMI Group, in the data, AI, and robotics perspective. In this chapter, we present a selection of some representative projects demonstrating the value for data, AI, and robotics convergence.

3.1 Data-Driven Projects for Industry 5.0

3.1.1 aerOS Project: Autonomous, scalablE, tRustworthy, Intelligent European Meta Operating System

The aerOS project responds to the Horizon Europe CL4-DATA-01-05-2021 call, aiming to create a platform agnostic meta operating system for the IoT-Edge-Cloud Continuum, with a strong focus on **trusted data sharing, orchestration, and AI** across a wide range of industries [2]. Data is the main pillar in order to achieve real-time intelligence at the edge, translating into a reduction of bandwidth use, latency times, and cloud dependencies, objectives critical for enabling the vision of Industry 5.0 of human-centric and sustainable manufacturing ecosystems [3].

Key Data-Driven Innovations

aerOS improves data governance by introducing tools for ownership and interoperability across heterogeneous devices and platforms, enabling secure and traceable dataflows within most industrial ecosystems. It supports federated data sharing so that multiple stakeholders can access data in a controlled and secure manner, fostering circular economy business models.

Collection, processing, storage, and distribution of actionable data are at the core behind the concept of IoT edge-cloud continuum [4]. Through open-source tools and customized developments, aerOS supports needs of data producers/consumers with mechanisms such as:

1. Data ownership, control, and governance enhanced with (cryptographic) non-repudiation and traceability assurance
2. Timely delivery of data to (ML) processors via just-in-time predictive projections
3. Semantic and syntactic data interoperability mediation to ease integration and widen possible range of cooperating entities, services, and stakeholders
4. ML-powered data aggregation and fusion that benefits from semantic interoperability
5. Facilitation of advanced services and data sharing topologies, e.g., data lakes (or puddles), and ad hoc (or persistent) local clouds
6. Rich semantic metadata annotation for deep querying and reasoning

aerOS orchestration redistributes computing capabilities based on demands, ensuring that critical tasks operate close to the data source.Furthermore, aerOS uses Frugal AI techniques—models that require the minimal amount of data—at the edge to improve response speed and minimize energy expenditure of data transfers to the cloud, essential for lowering both costs and environmental footprint.

Relevance to Industry 5.0

By shifting repetitive tasks and heavy data processing to lower layers of the continuum (local devices), aerOS promotes the principle of human centric by putting people at the center [5], making easier and more cost-effective to give them real-time insights without replacing their expertise. Also, reinforcing this friendliness and automation capacities, aerOS management tools reduce complexity and improve configuration capabilities, providing cross-cutting coordination between core and supporting components. The portal is usable not only by humans (e.g., to initially deploy and, later, to manage services and data) but also embeds some (semi-)autonomous actions undertaken by the ecosystem (e.g., self-adaptation based on self-observation) [6].

aerOS federated design increases resilience by enabling uninterrupted dataflows to continue running if certain node fails and avoiding to shut down the entire system.

Finally, processing data at the edge saves energy and supports circular economy like reusing, recycling, or refurbishing products or materials [7].

Overall aerOS data-driven solution enables human-centric, resilient, and environmentally conscious manufacturing processes, making the project a perfect example of Industry 5.0.

3.1.2 SM4RTENANCE Project: European Deployment of Smart Manufacturing Asset 4.0 MultilateRal DaTa Sharing SpacEs for an AutoNomous Operation of CollAborative MainteNance and Circular SErvices

The SM4RTENANCE project is dedicated to streamlining industrial maintenance and asset management through data-space-mediated data sharing [8]. The project seeks to enhance operational efficiency, optimize asset performance, and support collaborative maintenance [9]. Seamless data sharing between stakeholders will increase predictive analytics, therefore reducing unplanned downtime.

The project seeks to expand cross-data-space and cross-sectoral interoperability by creating a federation, where stakeholders can exchange data with their counterparts from a different data space. Within the initiative, open-source components and open industrial standards and protocols are favored, as are existing certification schemes, in order to accelerate the deployment of the federation ecosystem.

Deployment Challenges Considerations

Data space adoption requires addressing a number of concerns and difficulties:

- **Integration with Legacy Systems**: Harmonizing new data-space technologies with pre-existing industrial infrastructure remains a significant effort, requiring compatibility middleware solutions.
- **Data Standardization**: As data sharing takes place, its structure has to become searchable and interpretable by all the involved parties. Standards for data models should be followed whenever possible.
- **Governance and Scalability**: Beyond the few-participant early stages, data exchange across diverse stakeholders requires clearly defined participation rules and responsibilities.
- **SME Adoption Barriers**: Solutions have to be SME-friendly with regard to cost-effectiveness and initial operational overhead and time-to-market; modular solutions would be favored to facilitate adoption.
- **Trusted Exchange of Confidential Data**: Industry stakeholders are oftentimes in possession of valuable trade secrets, and they could be naturally reluctant to partake in the data-sharing economy. The data sovereignty-by-design enables them to precisely define the extent to which the data is shared and to whom [10].
- **Lack of General Awareness** about data space technology: as a still-novel and maturing field, a significant effort is necessary to popularize what a data space offers to a manufacturers' data operations.

With the maturation of the field, and technologies becoming more widely standardized, data-space adoption is envisioned to become a simple logical step for any business willing to partake in the data economy.

Pilot Implementations

SM4RTENANCE is in the process of demonstrating the benefits of data spaces through nine pilots in sectors like e-vehicles, industrial machines, textiles, and metrology [11]. These trials will show how the data space can improve asset resilience, optimize equipment effectiveness, extend asset lifetimes, and overall improve the value offering of businesses:

- Metrology Value Chain Digital Twins: This trial involves the development of digital twins for both Coordinate Measuring Machines (CMM) and manufactured products. The data exchange between parties will flow through a company-owned data space.
- Energy Consumption Monitoring for Industrial Robots: data spaces and software applications integrate with the robot manufacturer's platform to collect energy consumption data from customer-owned industrial robots and facilitate predictive maintenance.
- Fabric Manufacturing Data Space for Digital Product Passport Compliance: Preemptive compliance with the upcoming 2027 EU Digital Product Passport regulations motivates this pilot, enabling data sharing among fabric manufacturing facilities via a data space to support traceability requirements.
- Supply Chain Optimization for Weaving Machines: Component-level condition monitoring optimizes predictive maintenance and operational efficiency for weaving machines operated by customers.
- Predictive Maintenance Integration for Milling Machines and Metrology Data: This trial correlates milling-machine-operation conditions with metrology data on finished parts commissioned by OEM or Tier-1 clients to improve predictive maintenance strategies and diagnose machine component wear or miscalibrations.
- Failure-Prognostics Model for Electric Vehicles: Software as-a-Service failure-prognostics model generator for electric vehicles, which will be funneled through the consolidated Catena-X data space.
- Collaborative Asset Monitoring and Maintenance via data spaces: Asset information from a variety of sources, including production and maintenance instances, will be collected under a unified digital twin, with cross-data space data sharing.
- Smart Predictive Maintenance for 2D Laser Cutting Systems: AI analysis of 2D laser cutting systems will enable predictive maintenance. Continuous cloud collection of laser-cutter status data from customer-owned machines will support anomaly detection. Having maintenance sessions scheduled in advance reduces inventory stocks by relying more on just-in-time replacement parts delivery by suppliers.
- Milling Machine Usage Data Sharing: Milling machine monitoring data from will be shared sharing between the OEM, the machine owners/users, and parts suppliers. The OEM will be able to offer pre-scheduled maintenance operations. Based on part-failure data, parts manufacturers will be able to improve their

designs for durability, and end users will benefit from usage analytics and minimized unplanned downtimes.

SM4RTENANCE seeks to bolster intelligent asset management and predictive maintenance through the implementation of data space technology. The project advances data interoperability and compliance across multiple industries, driving efficiency, sustainability, and resilience in industrial maintenance and operations. Federation efforts push the current boundaries of data space technology and seek to even further tear down data-silos walls and maximize the value of industrial data.

3.1.3 RE4DY Project: European Data as a PRoduct Value Ecosystems for Resilient Factory 4.0 Product and ProDuction ContinuitY and Sustainability

The RE4DY project is not only creating a groundbreaking platform for industrial decision-making but also providing businesses with the tools they need to implement it. Grounded in the Manufacturing Data Networks vision, RE4DY prioritizes "Data as a Product" and empowers businesses to develop, share, and utilize data securely and efficiently [12]. The key to this platform is a digital continuum that seamlessly connects every stage of the manufacturing process, from design and production to maintenance and end-of-life management. This comprehensive approach, combined with AI and advanced data exploitation, empowers businesses to optimize operations, enhance competitiveness, and embrace the core principles of Industry 5.0.

RE4DY recognizes the urgent need for data-driven digital manufacturing processes to incorporate innovative and active resiliency strategies at both production and supply chain levels [13]. This is crucial for European manufacturers to maintain their sovereignty and competitiveness while upholding European digital values such as excellence, privacy, and trust. The project aims to demonstrate that the European industry can jointly build unique data-driven digital value networks 4.0 to sustain competitive advantages through digital continuity and sovereign data spaces.

By establishing a "Data as a Product" core concept, RE4DY facilitates the implementation of digital continuity across digital threads, data spaces, digital twin workflows, and AI/ML/Data pipelines. This concept leverages resiliency on top of advanced manufacturing digital processes and value ecosystems, supporting the development and implementation of digital continuity.

Key Innovations

- **Establishing the Digital Continuum**: RE4DY envisions a holistic digital representation of the entire manufacturing value chain, providing a unified view of operations for real-time monitoring and analysis [14].

- **AI-Powered Decision Support**: The project integrates AI algorithms that analyze data from across the digital continuum, providing actionable insights and recommendations for process optimization, predictive maintenance, and resource allocation.
- **Data Interoperability and Exploitation**: RE4DY emphasizes seamless data exchange and integration through a data space, ensuring that information from various sources can be effectively utilized for informed decision-making while preserving data sovereignty.
- **Developing the Digital Continuum Toolkit**: RE4DY is creating a comprehensive toolkit that will provide businesses with the resources they need to build and manage their own digital continuums, tailored to their specific needs and industry requirements.
- **Data Sovereignty and Trusted AI**: RE4DY prioritizes data sovereignty and trusted AI by establishing secure data spaces and ensuring responsible AI development and deployment.
- **Executable Cognitive Digital Twins**: RE4DY develops executable digital twins that can be used to simulate and optimize manufacturing processes, improving efficiency and resilience.

Relevance to Industry 5.0

RE4DY significantly contributes to the advancement of Industry 5.0's core principles. It places humans at the center of its operations by implementing an AI-powered decision support system. This system is designed to empower workers by providing them with data-driven insights and recommendations, ultimately enhancing their decision-making capabilities. The goal is to augment human skills and expertise, not to replace them, ensuring that workers remain valuable assets in the evolving industrial landscape.

Furthermore, RE4DY demonstrates a strong commitment to sustainability. By leveraging data-driven insights and analytics, it enables manufacturers to optimize their resource utilization and minimize waste. This leads to more environmentally responsible practices and helps reduce the ecological footprint of industrial operations. The project also focuses on increasing resilience within industrial processes. Through real-time monitoring and predictive capabilities, RE4DY can proactively identify potential disruptions or issues. This allows manufacturers to take preemptive action and mitigate risks, ensuring smoother and more reliable operations.

Data sovereignty is another key aspect that RE4DY addresses. The project ensures that businesses retain control and ownership of their data, allowing them to manage and share data securely and reliably. This fosters trust and confidence in data-sharing practices while also promoting responsible data governance. Lastly, RE4DY is committed to the responsible development and deployment of AI technologies. It emphasizes transparency and trust in AI-powered decision-making processes, ensuring that AI systems are used ethically and for the benefit of all stakeholders.

3.2 AI-Driven Projects for Industry 5.0

3.2.1 s-X-AIPI Project: Advancing Self-X AI in Process Industries

The s-X-AIPI project, funded under Horizon Europe, is pioneering self-X AI technologies to transform the European process industry, enhancing efficiency, sustainability, and resilience. The project introduces the **self-X paradigm**, where AI systems autonomously manage their configuration, optimization, healing, and protection. At the core of this approach is the **Autonomic Manager (AM)**, an AI-driven orchestration system that continuously monitors, evaluates, and adapts models to ensure sustained performance with minimal human intervention [15]. The AM integrates human-in-the-loop (HITL) mechanisms, reinforcing explainability and trust in AI decision-making.

Asphalt Pilot Case: AI-Driven Process Optimization

The asphalt industry faces challenges in quality control, energy efficiency, and predictive maintenance. The asphalt pilot within s-X-AIPI addresses these challenges by integrating AI-powered solutions, incorporating self-adaptive AI capabilities through an Autonomic Manager. This AI-driven framework ensures continuous monitoring, model retraining, and human-in-the-loop (HITL) mechanisms to enhance explainability and trust.

The AI-powered dashboards track mix composition deviations, ensuring compliance with quality standards [16]. Human-in-the-loop interaction allows operators to refine AI predictions and deviations. The AM ensures model accuracy and initiates retraining when necessary [17].

AI models analyze burner performance, predicting failures in critical components like the baghouse filter exhauster. Energy consumption data is used to recommend efficiency improvements, reducing unplanned downtime and extending equipment lifespan.

AI predicts laying and compacting temperatures based on real-time sensor data and weather conditions. A Web-based interface provides plant operators with GPS-tracked paving insights for real-time decision-making.

AI links laboratory test results with production data, enabling predictive monitoring of volumetric and mechanical properties. Anomaly detection models identify deviations early, with operators providing feedback to refine AI learning models [18].

Deployment Challenge Considerations

- **Integration with Legacy Systems**: AI models must align with existing infrastructure, requiring robust data harmonization and middleware solutions.

- **Adoption of Human-in-the-Loop (HITL) Mechanisms**: Ensuring operator engagement in model refinement and decision validation is critical to fostering acceptance.
- **Modular AI Deployment**: Developing AI solutions with modular architectures to facilitate adaptation across different process industries.
- **Interoperability and Standardization**: Ensuring seamless integration of AI models within industrial workflows through adherence to industry standards and protocols.

The s-X-AIPI asphalt pilot exemplifies the transformative potential of AI-driven automation and self-X technologies in process industries. By focusing on continuous learning, human-AI collaboration, and real-world validation, the project sets a strong foundation for future Industry 5.0 advancements.

3.2.2 Circular TwAIn Project: AI Platform for Integrated, Sustainable, and Circular Manufacturing

Circular TwAIn, funded under Horizon Europe, researches, develops, and validates an innovative **AI platform** designed to enhance circular manufacturing value chains. This platform facilitates the creation of **interoperable circular twins**, enabling end-to-end **sustainability across the manufacturing life cycle**.

By leveraging trustworthy AI techniques, Circular TwAIn promotes human-centric, sustainable manufacturing, accelerating the transition to Industry 5.0. It integrates and harmonizes diverse data sources, unlocking the benefits of seamless data sharing within trusted and efficient manufacturing data spaces while ensuring a strong focus on sustainability.

The project's impact is demonstrated through three pilot cases spanning both discrete manufacturing and process industries, where circularity and sustainability are tested and refined through trusted AI technologies:

- De- and re-manufacturing Li-ion battery packs in e-mobility by COBAT, RAEEMAN, and POLIMI
- De- and re-manufacturing consumer WEEE by RECYCLIA, REVERTIA, and AIMEN
- Energy optimization in petrol-chemical production plants by SOCAR and TEKNOPAR

In-depth analyses have been conducted on stakeholders, KPIs, data availability (repositories, semantics, ontologies), AI requirements, data sovereignty, and data space needs. The role of common data spaces for circular value chains has been assessed, leading to the definition of a FAIR dataset with domain-specific circular data models and ontologies.

An industrial data platform has been specified, detailing IDS connectors, an App Store, and a Clearing House, alongside AI-driven confidentiality and trust models for enforcing data sovereignty agreements. Furthermore, the AI-based circular digital twin ecosystem has been designed to enable collaborative intelligence through a swarm of digital twins, supporting AI applications across circular manufacturing. To this end, three highly relevant sources of data and information to train and feed the human-AI applications have been identified, (de-)manufacturing processes, products to be de-manufactured, and human operators, reinforcing the transformative power of data, AI, and robotics in Industry 5.0.

3.2.3 AIREDGIO5.0 Project: Regions and (E)DIHs Alliance for AI-at-the-Edge Adoption by European Industry 5.0 Manufacturing SMEs

AI REDGIO 5.0 aims at renovating and extending the AI REGIO alliance between Vanguard EU regions and DIHs [19] for a competitive AI-at-the-Edge Digital Transformation of Industry 5.0 Manufacturing SMEs.

1. The project enhances and combines existing reference models toward an architecture for end-to-end Industry 5.0. The objective is to develop a secure computation continuum between edge and cloud and to implement data pipelines for data quality and Industry 5.0 and data spaces at different levels with standard data models and ontologies. Then, the project develops, validates, and uses a multidisciplinary methodology, which follows a structured digital transformation approach toward AI adoption in manufacturing 5.0 scenarios.
2. The project develops an Industrial 5.0 **data space** (connected to existing manufacturing data spaces) with specific focus on sustainability and resilience aspects as well as human-AI interaction (collaborative intelligence).
3. The project develops an **interoperability framework** between regional DIHs resources and the AI-on-demand platform and contributes to the creation of a manufacturing industry vertical in the AI4Europe portal and experimentation platform.
4. The project manages the transition from regional DIHs to a network of EDihs. An EDIH hubs supports SMEs in building, deploying, and fully leveraging novel edge AI systems for manufacturing use cases, through access to tools, provision of training and consulting services, as well as focused AI integration services.
5. The AI REGIO Network of **didactic factories** is extended especially in Eastern EU countries and regions.
6. The project liaises with existing platforms, projects, and communities to develop a strong and vibrant ecosystem of relevant stakeholders around AI driven Industry 5.0.

Deployment Challenge Considerations

The evolution of digital manufacturing platforms and technologies is posing new challenges to EU Regions and their SMEs:

The manufacturing industry is now aiming at harmonizing Industry 4.0 principles with the new Industry 5.0 challenges of human centricity, sustainability, and resilience. There is a need for an Industry 5.0 transition of EU manufacturing industry.

- Digital technologies and especially AI is progressively migrating from the cloud to the edge, creating a data sovereignty computational continuum between industrial IoT and cloud, supported by initiatives like GAIA-X and the Data Space Business Alliance. Manufacturers (especially SMEs) could greatly benefit from the ongoing shift of data from cloud to edge, as this provides opportunities for applications that reside closer to the field and feature low-latency and real-time performance while boosting industrial data protection. Nevertheless, there is a still a need for enabling manufacturers to develop, deploy, and validate edge AI application through lowering the efforts and costs for data collection for edge AI, AI model building, and model deployment. Overall, there is a need for data spaces to enable trusted exchanges of data across different edge/cloud infrastructures within factories and across the supply chain. Such data spaces can provide a sound basis for training and deploying distributed AI at edge as well.
- The AI4Europe on-demand platform and its ecosystem of satellite projects have become an extraordinary tangible example of the EU-way to AI, where all the projects addressing AI could find a unique access point to trustworthy and ethical AI-driven digital transformation. There is a need for a pan-EU AI-on-demand platform dedicated to the manufacturing industry to support emerging edge AI paradigms.
- The transition from DIHs to EDIHs is promoting inter-regional collaborations (corridors) with new roles and opportunities for EU avant-garde regions. But there is a need for new methods and tools to govern the transition from regional DIHs to EDIH-to-EDIH cross-border collaboration.

Pilot Implementations

In order to show and exploit the benefits of applying artificial intelligence in manufacturing, three types of experiments are conducted:

1. **(TERESA) experiments in the Didactic Factories** of 14 regions. **Didactic Factories (DF)** offer training and education and perform test activities and experimentations, creating awareness and disseminating beneficial effects of the application of innovative digital technologies (i.e., artificial intelligence) in manufacturing applications. In the project, the Didactic Factory experiments focus on TEchnology and REgulatory SAndboxes (TERESA). A selection of innovative AI applications/tools/services for human-machine interaction are tested/experimented on a limited scale and in a secure and controlled way, according to the "test before invest" paradigm (technical sandbox).
2. SME-driven experiments in selected seven regions to demonstrate the development and implementation of an **AI-based solution in a real industrial**

environment. The beneficial impact of the use of AI is evaluated through the monitoring of specific key performance indicators (KPIs).

3. **Twenty additional SMEs-driven experiments** as outcome of two waves of Open Calls.

The seven SME-driven experiments are applications of innovative AI-driven solutions in real industrial environments:

1. **Real-time monitoring for control and detection of production, Lombardy (IT)**: The use case focuses on the application of monitoring technologies for process self-adaptivity to a sheet metal pressing line as any variation in processes' parameters could lead to non-conformity of the final product. The plant consists of five presses in series that have the task of gradually deforming the sheet metal by applying a predefined load. The line is currently equipped with a control software that allows the setting of process parameters while monitoring is limited to specific technical parameters. The integration of the line control software with edge technologies allows the data to be accessed and processed, enabling the quantitative characterization of the line's operating conditions, the assessment of process quality, and the determination of any deviations in an objective manner. Moreover, through edge computing technologies, it is possible to apply AI methodologies to operating data to carry out knowledge extraction, allowing to pass from quantitative data to structured information

2. **AI and digital twins for agility in mold making, Auvergne Rhone Alpes (FR)**: the Experiment aims to demonstrate the impact of artificial intelligence in shopfloor regarding productivity and agility. AI is used as a decision-making tool to organize the manufacturing sequences in a shopfloor where resources are shared for different kinds of business unit. It has to be link to company MES system to analyze the current status of every task and predict if some delays (or advance) are expected. In case of delays (or advance), alerts are launch to inform the users, and some advices to improve the planning are sent to the users. Advices are defined to limit the impact according to the company decision rules. The final call is always done by the user, a human, with ability to not follow the tools advices.

3. **AI-based autonomous machine for safer faster agriculture, Trentino (IT)**: The aim of the experiment is to create an autonomous agricultural machine to make human work more secure. There are in fact some situations where the agricultural land is very steep and the rows of trees are very narrow. In these conditions, the maneuver of the machine might be dangerous for the driver and, on some occasions, even fatal. The objective is to create a prototype of a semiautonomous agricultural machine that is able to follow the path between the tree lines. To enable this function, a pool of sensors robust to the weather condition and the environment is used, able to recognize road trees, obstacles, and measure distances. The hardware is selected in order to process images and, through AI, evaluate movements depending on the objects around. The system interfaces with the existing driving equipment and interact with the machine control unit.

4. **Predictive maintenance and zero-defect production of thermoplastic products, (SI)**: This use case is a small-scale demonstration (for one production cell/process) of predictive maintenance scenario in POLYCOM. The scenario consists of data infrastructure, edge device, and algorithms that detect and measure the condition of the molding tools. Demonstrated system enables operators to continuously monitor the condition of machines and tools and general process operations (changes/events on daily base). Predictive maintenance is based on active monitoring of the production process, where several data sources are gathered from the production setup (material, production parameters and signals, end-quality measurements). Deviations in the performance of specific asset (tool and machine) are monitored using AI technologies deployed on the local edge device.

5. **AI-enabled digital twins for virtual commissioning, Galicia (ES)**: For this use case, digital twin technologies based on the RAMI AAS are applied as a basis that enables a virtual representation of the plant, focusing not only on process variables obtained from PLCs or MES systems but also considering virtual carbon footprints and waste related parameters (water and electricity consumption, raw materials waste kg, etc.). This digital twin is created as a virtual representation at the edge, where models of the different involved machines are continuously created and improved when drifting is detected. Moreover, approaches such as federated and incremental learning algorithms are used to better create models and extract insights from the process. The main objective is to apply analytic technologies at the edge digital twins to prescribe the best process parameters that minimize waste while keeping quality, efficiency, and productivity at its best. xAI (explainable AI) methodologies are applied to provide valid feedback to humans regarding the proposed decisions by the algorithms to increase trustworthiness. Security is considered from the beginning of the use case, thus applying security by design.

6. **Intelligent contextualized visual system for error reduction, Wales (UK)**: The experiment integrates different production systems (inc. Industreweb™, PLCs, and other control devices), embedded technologies (AI, CPS, IoT), and human operator knowledge, into a single unified digital visualization platform. Using the AI REDGIO 5.0 architecture, all relevant systems being developed are connected to create added value within the production process, enabling a variety of outcomes, including the access to specialized knowledge in the production system and to production data for process improvement and efficiency and the transfer of human-to-machine and machine-to-human knowledge.

7. **Quality assurance of clothing production, (RO)**. This use case consists in testing the product defect detection system prototype integrating AI-at-the-edge technology developed during the project. System installation with local network connectivity is set up on the QA production zone of KAF to capture product images with low occlusion and good, consistent lighting. Several relevant types/styles of clothing products are used for the experiment. The images captured include examples of both defective and non-defective products to investigate the possibility of defining the non-defective products, by using fuzzy system.

Furthermore, the possibility of triggering an automated model building workflow to include a sequence of steps including data augmentation, model training, and postprocessing is evaluated.

3.3   Robotic-Driven Projects for Industry 5.0

3.3.1   EMPYREAN: Trustworthy, Cognitive, and AI-Driven Collaborative Associations of IoT Devices and Edge Resources for Data Processing

EMPYREAN EU project proposes a new hyper-distributed computing paradigm encompassing heterogeneous IoT devices and computing, storage, and connectivity resources that may belong to different providers at different segments of the continuum. EMPYREAN will build federations of collaborative resources, to be referred to as IoT-Edge Associations or simply Associations, that will be created and be operated autonomously and seamlessly using the EMPYREAN AI-enabled management scheme. This Association-based continuum will utilize distributed, cognitive, and dynamic AI-enabled decision-making to balance computing tasks and data locally inside an Association and between federated Associations in a decentralized, multi-agent manner as well as across central computing environments. It will optimize resources and provide scalability, resiliency, energy efficiency, and quality of service [20].

An Association will constitute a secure execution environment where the trust between data-generating and data-processing entities will be continuously validated using distributed trust services, while identity and data access management will assure controlled access and confidentiality of data among different organizations [21]. Automated tools and mechanisms for seamless interconnection, efficient data processing of ML workloads, and secure distributed edge storage will empower the EMPYREAN platform. EMPYREAN will also develop technologies for Associations-native application development and deployment that contribute to the entire application life cycle and interoperability for AI-driven value extraction of high-volume and dynamic IoT data generated at the edge of the network from multiple sources.

EMPYREAN EU project has the following objectives:

- Objective 1: Enable collaborative autonomy in the IoT-edge-cloud continuum
- Objective 2: Cognitive associations, speculative resource orchestration, and AI-driven adaptability
- Objective 3: Ensure security, privacy, and multi-party trust in the IoT-edge-cloud continuum
- Objective 4: Build mechanisms for efficient IoT data processing of ML-workloads and hyper-distributed applications
- Objective 5: Workflow-based, AI-augmented application development and seamless deployment on the edge-cloud continuum

- Objective 6: Demonstrate the capabilities of the EMPYREAN platform in supporting hyper-distributed, highly demanding, and dynamic applications

EMPYREAN will contribute to key factors such as intelligence, automation, trustworthiness, security, heterogeneous IoT edge mesh connectivity, interoperability, elasticity, and energy efficiency that are required to shift in practice the paradigm from cloud-native toward continuum-native applications that will fully leverage the highly anticipated X-continuum paradigm.

3.3.2 ARISE Project: Advancing Industry 5.0 Through Human-Centric Robotics

ARISE project plays a pivotal role in driving robotic innovation within the Industry 5.0 paradigm, emphasizing human-centric, adaptable, and sustainable robotics solutions. By leveraging open-source middleware, dedicated Testing and Experimentation Facilities (TEFs), and a structured Financial Support to Third Parties (FSTP) program, ARISE fosters the development and deployment of next-generation collaborative robotic systems in industrial environments.

Open-Source Middleware for Seamless Integration

A core technological achievement of ARISE is its **real-time enabled all-in-one middleware**, which facilitates seamless communication between **field data sources** and **digital data products**. The middleware operates across two key planes. The **Operational Plane** collects real-time data streams from machinery, robots, and field applications, integrating them with collaborative robotics applications to enhance industrial workflows. The **Analytical Plane** utilizes AI models, simulations, and operational dashboards to process and optimize industrial performance. It also provides API protocols for IT system integration, ensuring smooth interaction with enterprise IT systems (ERP, CRM, MES, etc.). Built on **ROS2 and FIWARE**, ARISE's middleware enhances modularity, adaptability, and scalability, enabling seamless deployment of collaborative robotic solutions within manufacturing environments. This approach ensures that robotics collaborative applications are not isolated but deeply integrated with broader IT/OT infrastructures, fostering more intelligent, data-driven decision-making processes.

Testing and Experimentation Facilities (TEFs)

To validate and refine robotic-driven solutions, ARISE operates four **Testing and Experimentation Facilities** (TEFs), each dedicated to distinct industrial challenges. These TEFs serve as real-world environments where robotic applications can be tested, optimized, and validated before large-scale deployment. The TEFs focus on collaborative robotics for industrial assembly, enhancing human-robot synergy in manufacturing tasks [22]. They also address ergonomic and well-being support robotics, developing solutions that reduce physical strain on human workers. AI-driven autonomous robotic systems enable adaptive decision-making and automation in dynamic industrial settings. Additionally, flexible robotics for smart factories support agile and reconfigurable production processes.

Deployment Challenges and Considerations

While ARISE presents a robust framework for advancing robotics in Industry 5.0, several deployment challenges must be addressed to ensure successful adoption. Scalability and adaptability remain critical, ensuring robotic solutions can be easily scaled and adapted to various industrial applications without extensive customization. Economic and investment barriers, including high upfront costs, need to be addressed by demonstrating clear return on investment (ROI) to encourage broader industry adoption. Regulatory and safety compliance is essential, navigating complex regulations to ensure robotic systems meet stringent safety and ethical standards for human-robot collaboration. Workforce training and acceptance must be prioritized by supporting worker upskilling and fostering trust in robotic systems to enhance adoption and maximize productivity.

4 ADR Adoption in Industrial Pilots and Demonstrators

After the ADR technological analysis and the EC research and innovation, the BDVA SMI Group has selected here some representative industrial cases to show Industry 5.0 as the DNA for EU manufacturing industry and a competitive advantage for Europe in the global landscape.

4.1 PRIMA INDUSTRIE Pilot in SM4RTENANCE: Machine Fleet Management and Predictive Maintenance

The PRIMA INDUSTRIE Pilot in SM4RTENANCE implements the Predictive Maintenance functionality supported by artificial intelligence for the global and integrated management of the maintenance service through the analysis of machine data, components, and service technicians' data [23]. The continuous collection of data in the PRIMA INDUSTRIE Fleet Management in the cloud (more than 8500 systems) is thus valorized to provide customers with a maintenance and consumables management service that can plan the necessary actions in advance for needs based on the real condition of the individual system. The sharing of data between different warehouses and internal and external service providers enables optimized and automated management of the service process.

The new functionality from the SM4RTENANCE project to orchestrate maintenance activities based on the analysis and sharing of data of the various stakeholders will bring a significant advantage in terms of efficiency and cost reduction. The centralized planning of the activities of the approximately 400 service engineers and 35 external service providers and the reduction in inventories possible through predictive analysis and data sharing with suppliers will be an advantage for PRIMA INDUSTRIE and its customers alike [24].

PRIMA INDUSTRIE partners (customers and suppliers) need to be supported by a secure and trustworthy system for enabling and stimulating data sharing under data sovereignty principles: a data space has been designed and implemented. Through data space technology, data providers can define the data products that can be shared under certain conditions (data access and usage policies). The definition of data models boosts the data exchange by enabling the data products' common understanding (between the data space counterparts).

The data products are described in the data space catalogue to support the data discovery and data space federation. In this way, any potential data consumer can discover and start a contract negotiation phase (with the data provider) for the data product of interest.

The continuous analysis of machine data will also allow the realization of customized ad hoc company service contracts for each customer based on the specific characteristics of system use. The information on the entire life of the system will allow, from a circular economy perspective, the analysis of the overall state and individual significant components to direct reuse or recycling choices of all or parts of the system.

The PRIMA INDUSTRIE Pilot, supported by data space technology, can scale easily in terms of participants, data products, and supported business scenarios paving the way for new and extended use cases (i.e., new company services, integration of third-party service providers, supply chain improvements in terms of network and efficiency, etc.).

4.2 SOCAR Pilot in Circular TwAIn: Advancing Energy Optimization in Petrochemical Production

The Circular TwAIn project is pioneering sustainability in the petrochemical industry through its PETRO pilot, operated by SOCAR and hosted at the PETKIM Ethylene Oxide/Ethylene Glycol (EO/EG) plant in Turkey and supported by TEKNOPAR. The pilot showcases the transformative potential of **artificial intelligence** and **hybrid circular twin** technologies. The offered solution drives substantial improvements in energy efficiency, process optimization, and environmental sustainability. The PETRO Pilot integrates a multi-layered technological architecture to enhance the petrochemical production processes' efficiency, sustainability, and circularity. This advanced architecture facilitates seamless integration between physical systems, AI frameworks, and digital twins.

4.2.1 Optimizing Production for a Greener Future

The PETRO Pilot focuses on enhancing the performance of ethylene oxide and ethylene glycol production processes. It aims to:

- Optimize steam consumption to improve energy efficiency
- Reduce CO_2 emissions and minimize environmental impact
- Utilize waste streams to support circular manufacturing goals

By implementing AI-driven frameworks, the pilot is introducing innovative solutions to long-standing industrial challenges, reinforcing sustainability through data-driven decision-making.

4.2.2 Key Technological Components

The PETRO Pilot integrates cutting-edge AI and automation technologies to streamline operations and maximize efficiency. Its architecture consists of the following core components:

1. **Data Acquisition and Integration**
   Process data is continuously collected from sensors and control systems across the EO/EG plants. Critical parameters, such as temperature, pressure, and flow rates, are monitored and stored in the Process History Database (PHD) for real-time analysis.
2. **AutoML Framework**
   A powerful AutoML tool autonomously processes data and trains predictive models for energy optimization, anomaly detection, and proactive maintenance scheduling
3. **Digital Twin Integration**
   A Hybrid Circular Twin (HCT) serves as a real-time virtual representation of physical plant operations. It enables dynamic process simulation, workflow optimization, and data-driven decision-making for improved operational efficiency.
4. **Circular Resource Utilization**
   The PETRO Pilot actively supports circularity by integrating waste management processes, maximizing resource recovery, and promoting sustainable production practices.

4.2.3   Impact and Measurable Outcomes

The PETRO Pilot demonstrates significant environmental and operational benefits, including:

- A **5% reduction in steam consumption** leads to lower natural gas usage and decreased operational costs. Through predictive analytics and advanced process control, we achieved a 5% reduction in steam consumption. This directly led to decreased natural gas usage, resulting in lower operational costs and a significant improvement in energy efficiency.
- **Reduced CO_2 emissions**, aligning with global sustainability goals and improving the plant's carbon footprint. By optimizing steam use and implementing real-time monitoring, we reduced CO_2 emissions, directly contributing to global sustainability goals and improving the plant's carbon footprint. This aligns with the European Green Deal and SOCAR's environmental KPIs, demonstrating a commitment to sustainable practices.
- **Enhanced automation and reliability** through predictive analytics, reducing downtime and maintenance requirements. Predictive analytics for equipment maintenance and system performance reduced downtime and maintenance requirements, leading to higher system reliability. This resulted in a more efficient operation, directly reducing costs related to unplanned shutdowns and increasing overall productivity.

These outcomes provide a scalable framework that can be implemented across industries to achieve similar sustainability goals.

4.2.4   Driving the Future of Sustainable Petrochemical Manufacturing

Through AI-powered analytics, real-time monitoring, and digital twin technology, the PETRO Pilot is setting new benchmarks for energy-efficient and circular petrochemical production. By demonstrating scalable solutions, the Circular TwAIn project paves the way for a more sustainable and resilient industrial future.

4.3   *QUESCREM Pilot in AIREDGIO5.0 Project: AI at the Edge for Zero Defect Food Industry and Sustainability Gain*

The Galician region's industrial landscape in northwest Spain features a significant weight of mid-sized manufacturing industries based on bio-inputs coming from its important bio- and primary sector (agrifood, fish, wood, and others). The *SustGAIN* SME-driven experiment is being carried out in this region, led by Quescrem and Gradiant, within the framework of the AI REDGIO 5.0 project.

Quescrem is a cream cheese manufacturing company in the dairy industry that has always been heavily committed to innovation, by creating high-quality products for the food industry and the professional cooking and bakery, as well as for consumers. Thus, the industrial processes of the company are heavily oriented toward achieving the highest level of quality in its products, as well as reducing variability in the organoleptic properties, such as creaminess. Also, one of the main objectives of the company to comply with sustainability goals is to reduce waste generation and optimize the use of raw materials, water, and energy in manufacturing activities. Meanwhile, Gradiant takes on the role of technology provider for the experiment, being a Spanish ICT technology center with more than 15 years of experience in the fields of connectivity, intelligence, and security. For this experiment in particular, Gradiant provides the technological knowledge and expertise regarding edge computing and AI technologies required for designing and developing the pilot's system that aims at helping Quescrem achieve its objectives.

Considering this context, this AI REDGIO 5.0 experiment focuses precisely on the application of AI and edge technologies to improve the quality of Quescrem's end product and to reduce waste, in order to improve sustainability. The idea is to use sensorized process data (temperatures, pressures, nutritional characteristics of the ingredients, etc.) stored in Quescrem's systems, as well as real-time data streams collected from its production lines, in order to prescribe the optimal parameters that ensure the provision of a high-quality output (e.g., homogenized cream cheese texture) and improve efficiency as well. To this end, Gradiant is developing real-time analysis of manufacturing process data through AI algorithms at edge level, which will enable the discovery of complex patterns and the provision of accurate predictions and prescriptions that support decision-making.

In traditional edge computing, the model is trained offline at the cloud with all the data at once before its deployment at edge devices. Processing and storing increasingly large datasets to train and update AI models at central servers is extremely inefficient and produces high energy-consuming pipelines and processes. Moreover, this is not the best solution considering non-stationary scenarios. Robust AI approaches, able to both (1) process data with extremely low latency and (2) incrementally train models at the edge adapting to evolving patterns within data streams, are challenging for the development of low-energy and data-efficient autonomous agents and real-time decision-making systems at the edge for industry. One of the main purposes of this experiment is the development of AI algorithms that address these issues and provide a solution for them. The specific model that is being proposed and implemented is an AI-edge continual learning model, which provides an online solution with effective performance that enable AI models to adapt and improve over time as they learn incrementally from new tasks and dynamic data streams while being robust against the forgetting of old knowledge. To accomplish this, the model integrates the combination of a prototype model based on ILVQ (XuILVQ [25]) and Deep Neural Decision Forests (DNDF) [26], which, by being partially updatable, favor its online use and its good performance, especially in tabular data. The resulting model is able to incrementally learn as new data becomes available, efficiently adapting to unseen patterns and concept-drift issues without significant performance losses.

The implementation of this model, in the context of the experiment, allows the monitoring of the cream cheese pasteurization process, which is executed prior to final product packaging. The model forecasts the expected hardness of the final product, whether it is expected to be out of specification or within the corresponding quality thresholds, this indicator being one of the main quality KPIs for Quescrem's products. In case an out-of-specification final hardness is being predicted, the optimizer module also developed for the experiment comes into play. The algorithm of this module recommends adjustments of certain variables, which can be modified on the fly during the pasteurization process, to maximize the probability of obtaining a correct final product. In this way, the entire system makes it possible to correct the manufacturing process in real time, avoiding having to wait until the end of the process, the laboratory analysis of a sample, and, in case of a bad quality result, having to reprocess the product again, with all that this implies.

The architecture of the experiment system is shown in the diagram below. As can be seen, the prediction module allows future incremental retrainings of the model. Following the human-in-the-loop (HITL) paradigm, based on introducing humans in the retraining of the AI system, the operator can provide new labeled data once the production process is finalized, comparing the actual value of the KPI obtained with the prediction of the model, and triggering a retraining of the model. It must be stressed that the prediction and prescription algorithms will only provide valuable information and support to the human operator during the decision-making process, but the user will remain in charge of the final decision. The AI system result will only be informative; it will not take any automated action, hence following the "Machine Assist Humans" approach promoted by the Industry 5.0 paradigm. The feedback loop is handled through the usage of the Collaborative Intelligence Platform, one of the components of the AI toolkit developed in the AI REDGIO 5.0 project.

4.4 Three Industrial Pilots for aerOS Edge-to-Cloud Meta Operating System

Pilot d.1: Green manufacturing and CO_2 footprint monitoring (Switzerland Innovation Park Biel)
The Swiss Smart Factory in Biel/Bienne (Switzerland) hosts a 1000 m² demonstration platform for advanced Industry 4.0 applications, with a particular focus on green manufacturing. In this facility, SIPBB manages a production line for quadcopter and hexacopter drones that operates with extreme customization (batch size 1).

In relation to green manufacturing, energy consumption, and CO_2 footprint, this pilot has integrated SMC's Air Management System and is using **analytics**

tools (like Node-red), which helps pinpoint energy inefficiencies and develop smarter, **more sustainable workflows**.

Among the already proven benefits of aerOS in this Industry, three can be highlighted: (1) the integration of disparate data under homogenized, federated NGSI-LD representation, (2) the monitoring of different machinery (e.g., robot arm, SETAGO work station, smart conveyor, THT assembly or AGVs), and (3) implementation of a Digital Product Passport (DPP) and raise customer awareness of their specific

Eventually, when aerOS is finally deployed on the site, it will support **real-time data sharing** between the continuum formed by machines, the factory network, and the cloud, making it easier for engineers, managers, and other stakeholders to **collaborate** and refine processes. This not only advances the **sustainability** aspect of Industry 5.0 but also keeps workers engaged in meaningful tasks, reinforcing the **human-centric** pillar.

Pilot d.2: Zero ramp-up safe PLC reconfiguration for lot-size-1 production (Siemens)

In Nuremberg, Germany, SIEMENS runs a pilot on their Tech Hall laboratory to demonstrate a modular, adaptive production system driven by automated guided vehicles (AGVs) and robotic arms. These elements communicate through different communication protocols (e.g., OPCUA, ROS2, MQTT), forming a heterogeneous cyber-physical system that can quickly adapt to changes in production demands.

Levering on **decentralized intelligence**, Siemens orchestrates all these machines in a flexible, **resilient**, and transparent way, responding quickly to changing demands or sudden issues, besides releasing mental load on the personnel.

This means reassigning tasks between AGVs and robotic arms when order requirements shift or coordinating the AGVs to deliver parts exactly when and where they are needed. And on top on that, the human remains central by data-driven insights (e.g., **real-time** machine feedback), which leads to a better and faster decision-making, ensuring **resilient** and **sustainable** operations rather than a purely automated one. aerOS supports the distribution of such tasks among the AGVs, thanks to the reinforcement-learning powered orchestration engine and the decentralized deployment of low-code tools. All of the previous seamlessly integrated with Siemens' own software.

Pilot d.3: AGV Zero breakdown Logistics (MADE & Politecnico di Milano)

In Milan, Italy, the MADE Competence Center and Politecnico di Milano (POLIMI) conducts a pilot simulating a mechanical valve production chain for the oil and gas sector. The line is divided into a mechanical processing station, a manual assembly station, and an automated testing station, all connected by transport lines and AGVs.

By deploying aerOS in this environment, the pilot aims to move a significant part of the computations toward the edge, reducing data transfer resources and reducing the risks of potential overloads. In addition, synchronization among

AGVs and pathfinding between the required production steps is being achieved, thanks to the federated availability of NGSI-LD data and the ubiquitous deployment of loads between cloud or edge domains that can belong to different tenants.

Furthermore, aerOS **frugal AI** methods avoids sending massive datasets to unnecessary cloud servers, enhancing response times and reducing network load. In this pilot as well, the **human remains in the loop**, acting as core and guiding the process changes, responding to **data-driven** inputs and ensuring **safe** operations. The pilot covers **Industry 5.0** goals of **resilience** and **sustainability**.

Finally, **data interoperability** features enable heterogenous group of components to communicate seamlessly, **optimizing** AGV travel, **cutting times**, and enhancing overall **efficiency**.

4.5 *Machine Manufacturer Pilot in EMPYREAN, Process Monitoring, and Anomaly Detection in Robotic Machining Cells*

Short Description: This pilot aims at developing a system able to perform process monitoring in robotic machining cells, composed of robots performing drilling and milling operations that are managed by specialized personnel. Process monitoring involves real-time monitoring and alerting of abnormal machining operations. High-precision sensors mounted on robots are used to measure various variables, such as temperature, vibration, cutting force, and tool wear, in order to create a unique fingerprint for each process and use this for future comparison with similar operations. Methods for fingerprinting include statistical, machine learning, time-series analysis, and other data analytic techniques. The system will detect anomalies based on a previous process fingerprint analysis and will alert the operator in real time so as to take corrective actions.

The Need: The companies that manufacture extremely expensive, high added-value parts are very demanding in terms of machine availability and quality assurance. Predictive maintenance, remaining lifetime assessment, and diagnosis of critical machine elements are state-of-the-art practices. However, those techniques focus primarily on the machine rather than the process itself. Monitoring the process for early detection of anomalies can make the difference between a final piece with the demanded quality or scrap. In production lines where manufacturing a single piece requires many hours of work, effective monitoring of the manufacturing process is even more critical. Producing a part with quality problems can disrupt the automated workflow since the machines (and the robotic cell) have to be stopped and the problem corrected, causing delays in production and in meeting delivery deadlines, while incurring repair and replacement costs and even affecting the company's profitability. In particular, the following challenges can be identified:

1. Data that is generated from a variety of sources (e.g., robots with sensors) with high frequency need to be efficiently transferred, stored, and processed.
2. High variability of operations, with drilling and milling operations all along the machining part, warrants the system to find the best-fit variables for process fingerprinting in real time.
3. The extensive list of operations necessitates a system with significant computing power in order to generate normality patterns and perform timely comparisons for each operation.

EMPYREAN Solutions: This pilot focuses on addressing the aforementioned challenges through the EMPYREAN'S Associations-based continuum. An Association will provide a secure and trusted execution environment for the interactions between robots and edge resources. Additionally, resource orchestration and autoscaling mechanisms will efficiently and dynamically handle the increased demands inside or across multiple Associations (e.g., remote plants of the same organization) for processing and storage in a multi-agent manner, supporting latency sensitive (transaction) processing operations of the pilot. Finally, EMPYREAN will provide efficient and secure storage for generic and time-series data from various sensors of the robotic cells. These developments will demonstrate EMPYREAN's platform as a key enabler for the ongoing industrial revolution.

4.6 The WEEE Pilot in Circular TwAIn: Robotic-Driven De-manufacturing and Re-manufacturing of PCs

Within the Circular TwAIn Project, collaborative robotics is a key enabling technology to enhance the circularity of manufacturing frameworks. In the WEEE (Waste of Electrical and Electronic Equipment) Pilot, one of the main industrial challenges is the standardization of disassembly processes to optimize remanufacturing and recycling. Given the high variability of electronic products—influenced by different manufacturers, models, and modifications over their life cycle—disassembly requires significant manual labor, making it difficult to achieve efficiency and consistency. This variability highlights the urgent need for adaptive and automated solutions that address the demands of dismantlers and Extended Producer Responsibility (EPR) organizations.

To meet this need, the integration of AI-based object detection and collaborative robotics is being explored to automate disassembly processes, ensuring efficiency, flexibility, and precision. In this approach, AI-driven computer vision—powered by deep learning models—accurately identifies and localizes key PC components (e.g., motherboards, processors, screws) using 2D and 3D imaging. These precise coordinates are then utilized by a collaborative robot, which autonomously performs component unmounting and screw unfastening, significantly reducing human effort, thereby allowing operators to focus on higher-value tasks.

A crucial aspect of this system is the dynamic human-robot collaboration, where operators supervise and fine-tune robotic actions in real time. Through *freedrive* mode, human workers can manually adjust robot trajectories, ensuring adaptability and safety in handling complex and delicate disassembly tasks. Additionally, AI-powered dynamic task planning assigns tasks based on real-time data, optimizing workflow efficiency.

This synergy between AI and collaborative robotics exemplifies the principles of Industry 5.0, where automation is not just about efficiency but also about sustainability and human-machine collaboration. The benefits of this approach include:

- Standardized and efficient disassembly—Reduces manual effort and variability across different product types.
- Enhanced resource recovery—Improves circular economy practices by automating recycling and remanufacturing workflows.
- Increased adaptability—AI-driven robots adjust to different product designs and conditions without requiring reprogramming.
- Improved safety and ergonomics—Minimizes human exposure to hazardous disassembly tasks while allowing operator intervention when needed.

One key finding from the project is the performance dependency of this methodology on the product manufacturer. This insight highlights the opportunity to develop new business models where circularity indicators become the primary metrics for success, while data and AI-driven models serve as the core value drivers.

By leveraging collaborative robotics and AI, the Circular TwAIn Project is pioneering a smarter, more sustainable approach to WEEE disassembly, aligning with the broader vision of Industry 5.0—where technology and human expertise work hand in hand to build more efficient, circular, and intelligent manufacturing ecosystems.

4.7   Implementing Manufacturing Data Networks: Four Pilots of the RE4DY Project

The RE4DY project involves four large-scale pilot projects, each focusing on different aspects of industrial manufacturing and supply chains. These pilots are designed to test and validate the RE4DY framework and its ability to enhance resilience and sustainability in real-world industrial settings.

4.7.1 Logistics of the Future: Connected Resilient Logistics Design and Planning Pilot

The pilot focuses on the automotive industry, specifically the logistics of the future for connected and resilient logistics design and planning. The pilot aims to optimize the logistics process by using AI and simulation to improve planning, shop floor implementation, and resource optimization. The key business processes involved in this pilot include:

- **Autonomous/Automatic Planning**: This process involves using AI to automate and optimize the planning process, improving efficiency and reducing lead times.
- **Shop Floor Implementation**: This process focuses on using real-time data and AI to optimize the shop floor operations, improving productivity and reducing waste.
- **Resource Optimization**: This process aims to optimize the allocation of resources, such as equipment and personnel, to improve overall efficiency and reduce costs.

RE4DY's Logistics of the Future pilot utilizes data from various sources, including ERP, MES, and PLC data, to train AI algorithms and develop digital twins to support decision-making and process optimization.

4.7.2 Collaborative Ecosystem Resilient Product/Production System Engineering for Electric Battery Pilot

This pilot focuses on the manufacturing of battery systems for e-mobility. The pilot aims to improve the sustainability, customization, and productivity of the battery system manufacturing process by using AI and data analytics to optimize various aspects of the production process. The key business processes involved in this pilot include:

- **Sustainability**: This process focuses on using data and AI to improve the sustainability of the battery system manufacturing process, reducing environmental impact and resource consumption.
- **Customization**: This process aims to enable the customization of battery systems to meet specific customer requirements, improving flexibility and responsiveness.
- **Productivity OEE**: This process involves using data and AI to optimize the overall equipment effectiveness (OEE) of the battery system manufacturing process, improving productivity and reducing costs.

RE4DY's Electric Battery pilot utilizes data from various sources, including production data, quality data, and energy consumption data, to train AI algorithms and develop digital twins to support decision-making and process optimization.

4.7.3 Collaborative Ecosystem Integrated Machine Tool Performance Self-Optimization Pilot

This pilot focuses on the tool manufacturing industry, specifically the production of milling cutters. The pilot aims to improve the efficiency and sustainability of the milling cutter production process by optimizing tool selection, process management, machine maintenance, and adaptative digital manufacturing. The key business processes involved in this pilot include:

- **Defining Milling Strategy and Tool Selection**: This process involves selecting the best tool for a specific milling operation based on available data and recommendations.
- **Process Management, Tool Refurbishment, and Recycling**: This process focuses on managing the lifetime of tools, including refurbishment and recycling, to minimize waste and environmental impact.
- **Machine Maintenance**: This process aims to optimize machine maintenance schedules and procedures to reduce downtime and improve overall equipment effectiveness.
- **Adaptative Digital Manufacturing**: This process involves using real-time data and AI to adapt the manufacturing process to changing conditions and optimize performance.

RE4DY's Machine Tool pilot utilizes data from various sources, including CAD data, training data, PLC data, sensor data, and ERP data. The data is used to train AI algorithms and develop digital twins to support decision-making and process optimization.

4.7.4 Cooperative Multi-plant Turbine Production with Predictable Quality Chains

The pilot focuses on the aeronautical manufacturing industry, specifically the production of jet engine components. The pilot aims to improve the quality and efficiency of the visual inspection process by using AI and machine learning to support operators in identifying defects. The key business processes involved in this pilot include:

- **Mitigating the Impact of Human Factor**: This process involves using AI to assist operators in the visual inspection process, reducing the risk of human error and improving defect detection rates.
- **Enhancing Quality Process Stability**: This process focuses on using data and AI to stabilize the quality process and reduce variability, leading to more consistent product quality.

RE4DY's Turbine Production pilot utilizes image data and quality notifications to train AI algorithms for defect detection. The pilot also leverages the concept of

Asset Administration Shell (AAS) to enable interoperability and data sharing between different systems.

5 Recommendations and Future Outlook Discussions in BDVA SMI Group

Industry 5.0 is the way Europe interprets the role of manufacturing industry in our society: human centric, resilient, sustainable, and circular. Inside the BDVA SMI group, we opened a discussion on the impact and future outlook of Industry 5.0 in Europe, and some recommendations have been formulated on how to improve EU competitiveness leveraging on Industry 5.0 principles.

5.1 Data Technologies for Industry 5.0

5.1.1 Key Recommendations

To effectively guide the evolution of data technologies within Industry 5.0, policy-makers and public funding bodies should prioritize the following recommendations:

- **Secure, interoperable data ecosystems**. This necessitates funding the development of standardized data governance frameworks and investing in infrastructure that facilitates seamless data exchange across diverse industrial sectors. This will foster collaborative innovation and unlock the potential of data-driven solutions.
- **Advanced data analytics capabilities**. This involves investments in real-time data processing, predictive analytics, and data visualization tools, as well as supporting initiatives that promote the adoption of these technologies by small and medium enterprises. This will enable businesses to obtain actionable insights from their data and optimize operational efficiency.
- **Robust digital twin platforms**. This requires the development of standardized digital twin architectures, supporting the integration of diverse data sources and promoting the use of digital twins for process simulation and optimization. This will empower businesses to enhance decision-making and drive innovation.
- **Data security and privacy infrastructure**. Actions should be directed toward the development of advanced cybersecurity solutions, the establishment of clear data privacy regulations, and the promotion of ethical data handling practices. This will build trust in data-driven technologies and ensure the responsible use of data.
- **Workforce development** initiatives focused on data literacy and data management skills. Activities should be dedicated to educational programs that train data analysts, data engineers, and data scientists, as well as initiatives that pro-

vide workers with the skills needed to follow the evolutions brought by Industry 5.0. This will ensure a skilled workforce capable of leveraging the opportunities presented by these technologies.

5.1.2 Future Outlook

The evolving landscape of Industry 5.0 represents a promising future for data and its associated technologies, producing a substantial impact for business operations. Digital twins, powered by increasingly sophisticated data collection and analysis, will become necessary for predictive maintenance, process optimization, and the creation of highly personalized products. The setup and running of secure, interoperable data-sharing platforms will foster collaborative ecosystems, enabling businesses to unlock new value through cross-industry partnerships and data monetization.

The emphasis on human centricity within Industry 5.0 will require the development of data-driven solutions that prioritize individual needs and preferences, creating new markets for tailored products and services. Businesses that strategically leverage these data technologies will gain a significant competitive advantage, driving innovation and fostering sustainable growth in collaborative and intelligent manufacturing.

The use of data, and data technologies, in Industry 5.0 scenarios represents a significant shift toward a more human-centric and sustainable industrial future. This approach not only augments human potential but also ensures that technological advancements contribute positively to society and the environment.

5.2 AI Technologies for Industry 5.0

The future of AI in Industry 5.0 is set to redefine manufacturing processes through a synergy of automation, intelligent systems, and human collaboration. The incorporation of AI-driven digital twins will enable factories to achieve unprecedented levels of efficiency, flexibility, and resilience.

5.2.1 Key Recommendations

- Adoption of AI-driven digital twins: Industries should invest in digital twin technology to create accurate virtual representations of physical assets, enabling predictive maintenance, process optimization, and real-time decision-making.
- Strengthening AI integration in manufacturing: AI-powered solutions such as machine learning algorithms and computer vision should be leveraged to enhance quality control, defect detection, and process automation.

- Human-AI collaboration frameworks: AI should be implemented in a way that supports human workers, enhancing safety, reducing cognitive load, and improving operational efficiency.
- Enhanced data management and interoperability: A standardized data infrastructure must be developed to ensure seamless communication between AI-driven systems and existing industrial platforms.
- Investment in AI talent development: Workforce training initiatives should focus on equipping employees with AI literacy and the necessary skills to interact with AI-driven systems effectively.
- Sustainability-driven AI applications: AI should be used to optimize resource usage, reduce waste, and improve energy efficiency to align with global sustainability goals.
- AI-powered supply chain resilience: Leveraging AI analytics to enhance demand forecasting, supply chain visibility, and risk mitigation strategies.

5.2.2 Future Outlook

In the next decade, AI will continue to drive innovation in Industry 5.0 by enabling smart, self-optimizing manufacturing ecosystems. Digital twins will evolve into more advanced predictive models, incorporating real-time data streams from IoT-enabled devices to create a fully autonomous and adaptive industrial environment.

Furthermore, the convergence of AI with emerging technologies such as 5G, blockchain, and quantum computing will unlock new possibilities in manufacturing, making industrial processes more resilient, transparent, and efficient.

By embracing AI-driven technologies, manufacturing industries can transition toward a more intelligent, sustainable, and human-centric future, ensuring long-term competitiveness in an increasingly digitalized global market.

5.3 Robotic Technologies for Industry 5.0

As robotics plays an increasingly crucial role in Industry 5.0, several deployment challenges must be addressed to ensure successful adoption. Scalability, adaptability, and modularity remain critical, ensuring robotic solutions can be easily scaled and adapted to various industrial applications without extensive and costly customization, which prevents mass adoption. Interoperability issues present technical barriers in integrating robots with existing industrial infrastructures, particularly respect legacy systems. Economic and investment barriers, including high upfront costs, need to be addressed by demonstrating clear return on investment (ROI) to encourage broader industry adoption. Regulatory and safety compliance is essential, navigating complex regulations to ensure robotic systems meet stringent safety and ethical standards for human-robot collaboration. Workforce training and acceptance must be prioritized by supporting worker upskilling and fostering trust in robotic systems to enhance adoption and maximize productivity.

Robotic-driven projects under the Industry 5.0 paradigm are shaping the transformation of modern industrial environments. By fostering collaboration between technology providers and industry stakeholders, robotics is expected to drive the future of a more sustainable, adaptable, and human-focused industrial landscape. The continued development of open-source middleware, dedicated testing environments, and financial support programs will be instrumental in overcoming existing challenges and ensuring the widespread adoption of robotics in industrial applications specially respect SMEs technical capacities for developing or acquiring such advanced solutions.

Acknowledgments This work has been supported by the following projects, which have received funding from the European Union's Horizon Europe Research and Innovation and Digital Europe Deployment Programmes:

1. "Regions and (E)DIHs alliance for AI-at-the-Edge adoption by European Industry 5.0 Manufacturing SMEs" (AI REDGIO 5.0), GA No. 101092069
2. "Agile, human-centric, and Real-tIme enabled open SourcE technologies advancing industrial HRI in Europe" (ARISE), GA No. 101135784
3. "European Lighthouse to Manifest Trustworthy and Green AI" (ENFIELD), GA No. 101120657
4. "Autonomous, scalablE, tRustworthy, intelligent European meta Operating System for the IoT edge-cloud continuum" (aerOS), GA No. 101069732
5. "European Deployment of Smart Manufacturing Asset 4.0 MultilateRal DaTa Sharing SpacEs for an AutoNomous Operation of CollAborative MainteNance and Circular SErvices" (SM4RTENANCE), GA No. 101123490
6. "European Data as a PRoduct Value Ecosystems for Resilient Factory 4.0 Product and ProDuction ContinuitY and Sustainability" (RE4DY), GA No. 101058384
7. "self-X Artificial Intelligence for European Process Industry digital transformation" (s-X-AIPI), GA No. 101058715
8. "AI Platform for Integrated Sustainable and Circular Manufacturing" (Circular TwAIn), GA No. 101058585
9. "Trustworthy, Cognitive And AI-Driven Collaborative Associations Of Iot Devices And Edge Resources For Data Processing" (EMPYREAN), GA No. 101136024.

References

1. Curry, E., Heintz, F., Irgens, M., Smeulders, A. W. M., & Stramigioli, S. (2022). Partnership on AI, data, and robotics. *Communications of the ACM, 65*(4), 54–55. https://doi.org/10.1145/3513000
2. Kong, X., Wu, Y., Wang, H., & Xia, F. (2022). Edge computing for internet of everything: A survey. *IEEE Internet of Things Journal, 9*(23), 23472–23485. https://doi.org/10.1109/JIOT.2022.3200431
3. Dauda, A., Flauzac, O., & Nolot, F. (2024). A survey on IoT application architectures. *Sensors (Basel), 24*(16), 16. https://doi.org/10.3390/s24165320
4. Vaño, R., Lacalle, I., Sowiński, P., S-Julián, R., & Palau, C. E. (2023). Cloud-native workload orchestration at the edge: A deployment review and future directions. *Sensors (Basel), 23*(4), 4. https://doi.org/10.3390/s23042215

5. Syed, D. A., Quadrini, W., Rahmani Choubeh, N., Pinzone, M., & Gusmeroli, S. (2025). Approaching interoperability and data-related processing issues in a human-centric industrial scenario. In M. Presser, A. Skarmeta, S. Krco, & A. González Vidal (Eds.), *Global Internet of Things and Edge Computing Summit* (pp. 21–34). Springer Nature Switzerland. https://doi.org/10.1007/978-3-031-78572-6_2

6. S-Julián, R., Lacalle, I., Vaño, R., Boronat, F., & Palau, C. E. (2023). Self-capabilities of cloud-edge nodes: A research review. *Sensors (Basel), 23*(6), 6. https://doi.org/10.3390/s23062931

7. Maenhaut, P.-J., Volckaert, B., Ongenae, V., & De Turck, F. (2020). Resource management in a containerized cloud: Status and challenges. *Journal of Network and Systems Management, 28*(2), 197–246. https://doi.org/10.1007/s10922-019-09504-0

8. Bacco, M., Kocian, A., Chessa, S., Crivello, A., & Barsocchi, P. (2024). What are data spaces? Systematic survey and future outlook. *Data in Brief, 57,* 110969. https://doi.org/10.1016/j.dib.2024.110969

9. Jan, Z., et al. (2023). Artificial intelligence for industry 4.0: Systematic review of applications, challenges, and opportunities. *Expert Systems with Applications, 216,* 119456. https://doi.org/10.1016/j.eswa.2022.119456

10. Huber, M., Wessel, S., Brost, G., & Menz, N. (2022). Building trust in data spaces. In B. Otto, M. ten Hompel, & S. Wrobel (Eds.), *Designing data spaces: The ecosystem approach to competitive advantage* (pp. 147–164). Springer International Publishing. https://doi.org/10.1007/978-3-030-93975-5_9

11. Xu, X., & Hua, Q. (2017). Industrial big data analysis in smart factory: Current status and research strategies. *IEEE Access, 5,* 17543–17551. https://doi.org/10.1109/ACCESS.2017.2741105

12. Dognini, A., et al. (2024). Blueprint of the common European energy data space. *Zenodo.* https://doi.org/10.5281/zenodo.10964387

13. Redeker, M., Weskamp, J. N., Rössl, B., & Pethig, F. (2022). A digital twin platform for Industrie 4.0. In E. Curry, S. Scerri, & T. Tuikka (Eds.), *Data spaces: Design, deployment and future directions* (pp. 173–200). Springer International Publishing. https://doi.org/10.1007/978-3-030-98636-0_9

14. Schneider, T., & Šimkus, M. (2020). Ontologies and data management: A brief survey. *Künstliche Intelligenz, 34*(3), 329–353. https://doi.org/10.1007/s13218-020-00686-3

15. Quadrini, W., et al. (2024). A reference architecture to implement Self-X capability in an industrial software architecture. *Procedia Computer Science, 232,* 446–455. https://doi.org/10.1016/j.procs.2024.01.044

16. Sanz, L. (2025). Anomaly detection model demo for asphalt process. *Zenodo.* https://doi.org/10.5281/zenodo.15189591

17. Galende, M., Corral, A., & Reñones, A. (2024). selfX: An open-source R package with autonomic abilities for data …. *Open Research Europe | Open Access Publishing Platform, 4*(156). https://doi.org/10.12688/openreseurope.18070.1

18. de la Rosa, M. (2025). s-X-AIPI Asphalt automatic training for ML models. *Zenodo.* https://doi.org/10.5281/zenodo.15111509

19. Gavkalova, N., et al. (2024). Digital Innovation Hubs and portfolio of their services across European economies. *Oeconomia Copernicana, 15*(1), 1. https://doi.org/10.24136/oc.2757

20. Kretsis, A., et al. (2024). EMPYREAN: Trustworthy, cognitive and AI-driven collaborative associations of IoT devices and edge resources for data processing. In *Proceedings of the 33rd International Symposium on High-Performance Parallel and Distributed Computing, in HPDC '24* (pp. 385–388). Association for Computing Machinery. https://doi.org/10.1145/3625549.3658814

21. Allegretta, M., Siracusano, G., González, R., Gramaglia, M., & Caballero, J. (2025). Web of shadows: Investigating malware abuse of internet services. *Computers & Security, 149,* 104182. https://doi.org/10.1016/j.cose.2024.104182

22. Rahmani Choubeh, N., Zarei, M., Quadrini, W., Gusmeroli, S., & Fumagalli, L. (2025). Safety-driven electronic components disassembly through human-robot collaboration framework. *Procedia Computer Science, 253,* 1103–1112. https://doi.org/10.1016/j.procs.2025.01.172

23. Chryssolouris, G., Alexopoulos, K., & Arkouli, Z. (2023). Artificial intelligence in manufacturing equipment, automation, and robots. In G. Chryssolouris, K. Alexopoulos, & Z. Arkouli (Eds.), *A perspective on artificial intelligence in manufacturing* (pp. 41–78). Springer International Publishing. https://doi.org/10.1007/978-3-031-21828-6_3

24. Cannizzaro, D., et al. (2021). In-situ monitoring of additive manufacturing. In T. Cerquitelli, N. Nikolakis, N. O'Mahony, E. Macii, M. Ippolito, & S. Makris (Eds.), *Predictive maintenance in smart factories: Architectures, methodologies, and use-cases* (pp. 207–228). Springer. https://doi.org/10.1007/978-981-16-2940-2_10

25. González Soto, M., Fernández Castro, B., Díaz Redondo, R. P., & Fernández Veiga, M. (2022). XuILVQ: A river implementation of the incremental learning vector quantization for IoT. In *Proceedings of the 19th ACM International Symposium on Performance Evaluation of Wireless Ad Hoc, Sensor, & Ubiquitous Networks, in PE-WASUN '22* (pp. 1–8). Association for Computing Machinery. https://doi.org/10.1145/3551663.3558676

26. Kontschieder, P., Fiterau, M., Criminisi, A., & Bulò, S. R. (2015). Deep neural decision forests. In *2015 IEEE International Conference on Computer Vision (ICCV)* (pp. 1467–1475). https://doi.org/10.1109/ICCV.2015.172

Toward the Irish Mobility Data Space: Challenges, Opportunities, and Requirements

Rafiqul Haque, Diarmuid Ó. Conchubhair, Muhammad Asif Razzak, Majjed Al-Qatf, Fatemeh Ahmadi Zeleti, Wassim Derguech, and Edward Curry

Abstract The rapid evolution of digital technologies is transforming the transportation sector in Ireland, with Mobility-as-a-Service (MaaS) emerging as a promising solution to enhance mobility efficiency and sustainability. However, the successful implementation of MaaS in Ireland faces significant barriers, primarily due to the lack of standardized, interoperable, and secure data-sharing frameworks. The mobility data spaces (MDSs) are potential solutions that provide a trusted and secure data sharing among stakeholders, an essential foundation for integrated mobility services. Yet the development of such data spaces in the Irish context is constrained by fragmented data ecosystems, a lack of coherent governance and technical infrastructure, and limited trust among stakeholders, particularly regarding data control and usage.

This chapter presents and discusses the findings from a study conducted within the MaaS4IRL (A Trusted and Interoperable Mobility-as-a-Service Ecosystem for Ireland) project (https://universityofgalway.ie/dsi/maas4irl/), funded by the Sustainable Energy Authority of Ireland (https://www.seai.ie/). The study explores the challenges, opportunities, and requirements associated with mobility data sharing in the Irish context. Drawing on the structured workshops with key public and private stakeholders, it identifies critical technical, legal, organizational, and trust-related barriers to effective mobility data exchange. Based on these findings, the study proposes the establishment of the Irish Mobility Data Space (IMDS) as a framework to facilitate secure, standardized, and trustworthy data exchange among public and private mobility stakeholders. The proposed IMDS is intended to support

R. Haque (✉) · M. A. Razzak · M. Al-Qatf · F. A. Zeleti · E. Curry
Insight Research Ireland Centre for Data Analytics, Data Science Institute, University of Galway, Galway, Ireland
e-mail: rafiqul.haque@insight-centre.org; m.asif.razzaq@insight-centre.org; majjed.alqatf@insight-centre.org; fatemeh.ahmadizeleti@insight-centre.org; edward.curry@insight-centre.org

D. Ó. Conchubhair · W. Derguech
Future Mobility Campus Ireland, Shannon, Clare, Ireland
e-mail: diarmuid@futuremobilityireland.ie; wassim@futuremobilityireland.ie

E. Curry et al. (eds.), *Artificial Intelligence, Data and Robotics*,
https://doi.org/10.1007/978-3-032-10561-5_12

critical areas such as MaaS, EV charging, and infrastructure planning, contributing to a more integrated and resilient mobility ecosystem in Ireland.

Keywords Mobility data space · Data spaces · Mobility data as a service · Smart mobility · Data governance · Data interoperability

1   Introduction

The digital transformation of the transportation sector offers unprecedented opportunities to advance efficiency, sustainability, and user - centricity. However, it also presents complex challenges, particularly in the domain of data sharing. In the context of Ireland, where mobility demands are diverse and infrastructural disparities are significant, the absence of a standardized, trusted, and regulation-compliant data-sharing framework remains a critical barrier to realizing the full potential of an integrated and intelligent transport system.

The MaaS4IRL project, funded by the SEAI, undertakes a foundational investigation into the structural and systematic barriers impeding effective mobility data sharing. This study lies within the broader discourse on Mobility-as-a-Service (MaaS) and the Common European Mobility Data Space (CEDS), to assess Ireland's readiness to establish the Irish Mobility Data Space (IMDS). The section provides an overview of Ireland's mobility landscape, reviews existing data sharing practices in Ireland, introduces the scope of the MaaS4IRL study, and outlines the key contributions of this chapter.

1.1   *Ireland's Mobility Landscape*

The transport system in Ireland presents a complex interplay of infrastructural constraints, demographic distribution, and modal inefficiencies. The modal split remains heavily skewed toward private vehicle use: approximately 70% of all journeys are undertaken by private car, while public transport modes—including buses, trams, and rail—account for 18–20% of trips.[1] Active travel modes such as walking and cycling represent 10–12% of journeys, with uptake concentrated in urban centers such as Dublin, Cork, and Galway.

The distribution of transport infrastructure is uneven. Dublin, for example, is served by a multi-modal network including buses, the Luas light rail system, and suburban rail lines.[2] In contrast, regional towns and rural communities remain dependent on infrequent bus services with limited integration or real-time

---

[1] Central Statistics Office: https://www.cso.ie/en/releasesandpublications/

[2] National Transport Authority (NTA): https://www.nationaltransport.ie/

information.[3] These geographic disparities reinforce transport inequality and restrict access to mobility services outside urban cores.

Ireland's mobility network currently includes over 200 licensed public and private transport operators, a fleet of approximately 2500 public buses and 300 trains,[4] and more than 1000 licensed taxis. Shared mobility initiatives such as bike sharing and electric scooter pilots have been deployed in selected urban areas, while the national electric vehicle (EV) fleet exceeded 100,000 in 2024.[5] Despite these developments, modal integration remains limited, particularly in rural areas where approximately 35.5% of the population resides.[6]

There are various constraints on the Irish transportation ecosystem. A major constraint is the overburdening of existing transport infrastructure, particularly in urban areas. Dublin's severe traffic congestion highlights the urgent need for innovative mobility solutions in Ireland. According to the 2023 TomTom Traffic Index, Dublin ranks as the second slowest city in the world, with a 10-km journey taking an average of 29 min and 30 s [1]. Only London fares worse, with the same journey taking over 37 min [2]. Additionally, Dublin's congestion level stands at 66%, the highest among all surveyed cities [3]. These highlight the increasing inefficiency of conventional transport systems and emphasize the necessity of adopting innovative, data-driven mobility solutions to reduce the pressure on urban infrastructure.

In recent years, the Irish government has taken significant strides toward integrating diverse transport services under the Mobility-as-a-Service (MaaS) framework. A collaborative study by Smart Dublin, Urban Foresight, and the Department of Transport highlights the critical need for MaaS in Ireland, identifying models best suited to the country's transport landscape [4]. MaaS initiatives aim to enhance urban mobility, promote sustainable travel behaviors, and support healthier communities. They align closely with Ireland's strategic objectives and the broader goals of the European Union, offering tangible social, environmental, and economic benefits by fostering a multimodal and user-centric approach to transportation. Key MaaS initiatives include:

- *Journey Planner*: It enables end users to plan journeys using different modes of transport—connected seamlessly from origin to destination.
- *Next Generation Ticketing (NGT)*: This program, part of a project by the National Transport Authority (NTA), is aimed at upgrading ticketing infrastructure through account-based payment.
- *Automatic Vehicle Location*: This NTA project involved a technology upgrade to provide real-time information, helping passengers access accurate and timely live updates about bus locations.
- *Leap Card Top-Up*: The Transport for Ireland (TFI) Leap Card is a contactless smart card operated for the automated collection of fares on various transit services.

---

[3] Department of Transport: https://www.gov.ie/en/publication/

[4] Transport Infrastructure Ireland: https://www.tii.ie/

[5] Sustainable Energy Authority of Ireland: https://www.seai.ie/

[6] Central Statistics Office: https://www.cso.ie/

- *DublinBikes*: Developed by Transport for Ireland, the TFI Live App allows users to access live departure information and plan journeys across the entire TFI network, informing them about how they can travel from A to B.
- *Bleeperbike*: It offers a dockless bike hiring system, which is an automated solution enabled via smartphone apps and GPS technology that allows users to rent a bike from anywhere in Dublin.

However, while urban centers like Dublin, Galway, and Cork benefit from MaaS applications, significant challenges remain for rural areas, where nearly 35.53% of the population faces limited access due to inadequate connectivity infrastructure [5]. This disparity highlights the need for an equitable transport strategy that includes rural populations.

1.2 Mobility Data-Sharing Practices in Ireland

A core enabler of MaaS—and more broadly of integrated and intelligent mobility services—is the capacity to share and reuse mobility data across public and private stakeholders. MaaS operators rely heavily on data for operational optimization and enhancing the customer experience. Hence, data-sharing practices are vital for MaaS. Beyond supporting MaaS operations, data sharing can unlock broader opportunities for the transport sector in Ireland. It is crucial for infrastructure planning, emissions monitoring, enhancing rural connectivity, and implementing demand-responsive transit systems. The successful implementation of an interoperable and secure data-sharing framework is essential in realizing a smart and sustainable transport ecosystem for Ireland.

A robust framework for mobility data sharing requires multiple essential components. According to [6], establishing data exchange agreements between key stakeholders is fundamental to enabling the efficient operation of shared mobility hubs. The agreement defines terms, responsibilities, and technical requirements for data exchange, fostering greater coordination and interoperability among different mobility service providers. Furthermore, a standardized mobility data model is critical for achieving interoperability. One widely recognized standard is the Mobility Data Specification [7], which provides a structured format for data exchange between mobility operations and other types of stakeholders such as regulators and MaaS application developers. Compliance with national and European regulatory frameworks is imperative to ensure lawful and secure data-sharing practices.

The mobility data ecosystem in Ireland is evolving, driven by the increasing demand for real-time information, urban sustainability goals, and alignment with EU-wide data governance initiatives. While several public and private actors are generating and disseminating mobility data, the country currently lacks a unified framework for interoperable and trusted data sharing across the mobility sector.

Ireland maintains a national mobility data hub through the Open Data Portal (Data.Gov.IE[7]). The National Transportation Authority plays a central role in enabling open access to public transport data via Transport for Ireland (TFI).[8] Available datasets include General Transit Feed Specification (GTFS) schedules, real-time vehicle locations, and service alerts for buses, LUAS, and other public modes. In addition, several government bodies—including the Central Statistics Office (CSO), Department of Transport, Dublin City Council,[9] Road Management Office (RMO),[10] and County Council—contribute mobility data to the portal.

The Irish mobility data hub plays a significant role in enabling data-driven insights across various areas of mobility, including public transport and infrastructure planning. It allows regional and city authorities to better understand demand and patterns and allocate resources more efficiently to optimize transport operations. Moreover, it supports responsive management of dynamic conditions—for example, adjusting public bus operations during poor weather conditions or coordinating traffic control in response to sudden congestion. Furthermore, platforms such as Smart Dublin are fostering innovation by making urban mobility datasets, such as pedestrian footfall, cycling data, and traffic flow, publicly accessible to data consumers such as researchers, developers, and policymakers. However, the lack of a coherent data sharing framework limits their integration into the broader data ecosystem.

1.3   Scope of the Maa4IRL Study

The premise of the MaaS4IRL study is that effective data sharing is indispensable for the realization of a smart, inclusive, and sustainable mobility ecosystem. The current lack of a coherent data sharing framework impedes the implementation of intelligent mobility solutions capable of addressing pressing challenges such as congestion, emissions, and rural-urban mobility disparities in Ireland.

This study seeks to assess the feasibility of establishing an Irish Mobility Data Space (IMDS), inspired by the principles of the European Common Data Spaces. The research adopts a multidimensional approach involving technical, legal, and governance analyses, with the following objectives.

Data sharing is the key to building a smart and sustainable mobility ecosystem. The intelligent applications in the mobility sector, such as forecasting traffic and route optimization, need data. However, the Irish mobility ecosystem lacks a data-sharing framework and is therefore missing many opportunities to implement smart mobility that can solve ongoing challenges. This study aims to identify what are the

---

[7] https://data.gov.ie/

[8] Transport for Ireland: https://www.transportforireland.ie/

[9] https://www.dublincity.ie/residential

[10] https://www.rmo.ie

current challenges by analyzing the data from the stakeholders. Inspired by the European Common Data Spaces, MaaS4IRL project investigates the opportunities of the Irish Mobility Data Space (IMDS) and what is required to establish a mobility data space in Ireland. The purpose is to identify whether the Irish Mobility Data Space can effectively create adequate opportunities to address the data-sharing challenges and identify requirements to set up the Irish Mobility Data Space. Given below is the summary of the study:

- *Stakeholder Engagement and Challenge Identification*: Structured workshops with government agencies, service providers, and regulators to identify systemic barriers to data sharing.
- *Opportunity Mapping*: Identification of value propositions and strategic benefits that an IMDS could unlock for stakeholders across the mobility ecosystem.
- *Requirements Elicitation*: Extraction of technical, regulatory, and organizational prerequisites necessary for developing a trusted and sustainable data space.
- *Barriers to Implementation of IMDS*: Identification of constraints and risks to building and operating the IMDS, informed by stakeholder input.

1.4 Key Contributions of the Study

Data spaces constitute a cornerstone of the European Data Strategy, facilitating sovereign, secure, and interoperable data exchange across sectors. Several pioneering initiatives—such as deployEMDS, EONA-X, and the German Mobility Data Space (GMDS)—have been successfully implemented in countries including Finland, Spain, France, and Germany. Despite these advancements, countries like Ireland have yet to establish a dedicated data space within the mobility domain.

This chapter presents findings from the MaaS4IRL study, which serves as a foundational effort to conceptualize an Irish Mobility Data Space (IMDS). The insights generated aim not only to inform national strategy but also to provide a transferable knowledge base for other countries at a similar stage of development. This chapter delivers two principal contributions of the study:

- *Comprehensive Analysis of Data Collected from Stakeholders*: The study presents an in-depth analysis based on the structured workshops with key public and private stakeholders, including policymakers, transport operators, and industry leaders in Ireland. These workshops employ qualitative thematic analysis to identify major challenges in mobility data sharing and opportunities for a mobility data space in Ireland. The findings highlight high-priority areas where the mobility data space can improve operational efficiency, support policymaking, and advance sustainable urban mobility. The analysis spans multiple dimensions, including technological efficiency, data standardization, interoperability, governance, and legal compliance. Additionally, we emphasize the role of stakeholder collaboration in fostering a cohesive and trusted mobility data-sharing ecosystem.

- *Irish Mobility Data Space*: Based on the research findings, this study recommends the establishment of an Irish Mobility Data Space (IMDS) to enable secure, standardized, and interoperable mobility data sharing across Ireland. This chapter presents the technical foundations of the IMDS, including its conceptual architecture. The proposed architecture is informed by an extensive review of established reference models—most notably deployEMDS and the International Data Spaces Association (IDSA) Reference Architecture [3]—to identify critical gaps and develop a design tailored to the specific needs of Ireland's mobility ecosystem. The IMDS architecture integrates international best practices, privacy-preserving mechanisms, and advanced technologies (e.g., secure data exchange protocols) essential for trustworthy implementation. By introducing a structured and governance-driven framework, this research lays the foundation for an adaptive, secure, and innovation-oriented data-sharing environment that aligns with European regulatory standards and supports a wide range of applications, including MaaS, EV charging coordination, and infrastructure planning in the Irish transport sector.

1.5 Organization of the Chapter

The remainder of this chapter is organized as follows: Sect. 2 details the methodology, focusing on stakeholder workshops, data collection techniques, and analysis methods. Section 3 presents the key findings from the stakeholder engagement, highlighting major challenges and opportunities in mobility data sharing. Section 4 introduces the IMDS, detailing its architecture, governance framework, and technological components for implementation. Finally, Sect. 5 concludes with a summary of findings and potential future directions to advance mobility data sharing in Ireland.

2 Methodology

This section presents the methodology employed to identify the challenges, opportunities, and requirements for building a mobility data space in Ireland. The methodology involves three workshops, MaaS4IRL-FMCI, MaaS4IRL-DSAI, and ITS Ireland-MaaSIRL, which are discussed in detail in this section.

2.1 Data Collection

2.1.1 Participant Selection

As a first step, the potential stakeholders to participate in the workshops were identified. The aim was to get every stakeholder's point of view regarding MaaS, mobility data sharing, and the challenges and opportunities of mobility data sharing in Ireland to consolidate a complete set of needs/concerns that can be prioritized and transformed into the set of requirements, optimal approaches to enable data-sharing solutions, requirements for enabling interoperability in mobility data sharing, and the potential governance structure for IMDS.

Given that the identification of the proper stakeholders is a key point for understanding data-sharing ecosystem requirements, the list of stakeholders was created, taking into consideration the following criteria: (1) domain (private/public) of the stakeholders and (2) the organization's wider influence in the transport sector.

Stakeholders from different domains, such as transport operators, consultants, data providers, representatives from national, regional or city public authorities, etc., were identified to achieve a balanced representation of key actors that are considered part of the data-sharing ecosystem. Some of the key participants are listed in Table 1.

Table 1 Description of selected participants

Participants	Information
Consultancies and technology vendors	A mix of consultancies, technology vendors (e.g., KAPSCH[a], Commsignia[b])
Academic Institutes	Academic institutions like University College Dublin and Civic representation highlight the involvement of educational and community-based organizations
Department of Transport	A department of the Government of Ireland that is responsible for transport policy and overseeing transport services and infrastructure
National Transport Authority	The transport authority in Greater Dublin. It also operates TFI
Transport Infrastructure Ireland	A state agency in Ireland, dealing with road and public transport infrastructure
City Council	The local authorities in Ireland. Three city councils participated in stakeholder workshops, including Dublin City Council[c], Meath County Council[d], and Cork City Council[e]
Arup	A design, engineering, and sustainability expert company

[a] https://www.kapsch.net/en
[b] https://www.commsignia.com/
[c] https://www.dublincity.ie/residential
[d] https://www.meath.ie/
[e] https://www.corkcity.ie/en/

2.1.2 Questionnaire

We developed two types of materials for data collection during the workshops: presentations and questionnaires. The questionnaires were designed to gather insights from participants, focusing on the following topics:

- Benefits of mobility data sharing for various stakeholders, including public transport authorities, transport operators, end users, data providers, etc.
- Governance mechanisms for sharing data focus on policies and regulatory compliance.
- Barriers to implementing the mobility data sharing platforms in Ireland include institutional, operational, and legal challenges.

Table 2 presents three questions asked of participants during the workshop. The primary focus of our research is to gain insights into the challenges, opportunities, and requirements for an efficient mobility data-sharing platform in the transportation sector in Ireland.

Attendees responded to the questionnaire via Slido[11] individually, after which a discussion followed among the table participants to create a consolidated version, if agreement was achievable.

2.1.3 Guiding Framework for the Design of the Protocol

Roadmapping is a practical and widely used foresight approach and a flexible and structured method for exploring and communicating the relationships between societal and market drivers, enabling technologies and solutions needed over time [8]. We employed the *learning by foresighting and evaluating* (LIFE) method to facilitate dialogue between MaaS stakeholders. The idea of the LIFE method is to establish a collaborative and multi-voiced learning and innovation process within and across organizational settings.

Table 2 Questionnaire used in workshops for data collection

Questions
Q1. What are the main challenges in the Irish mobility ecosystem?
Q2. What are the key opportunities of data sharing in Ireland?
Q3. What are the primary requirements for setting up a mobility data space in Ireland?
Q4. What are the key barriers to implementing a mobility data space in Ireland?

[11] https://www.slido.com/

LIFE has been adopted as a roadmap methodology for the development of Roadmap 2025 for MaaS in Europe [9]. We used LIFE to design the protocol for the stakeholder engagement to serve as a guiding tool for exchanging information, contributions, and interactions during the feedback sessions, as well as the analysis of the data that emerged. LIFE is useful in gathering data around the four major areas specified in this roadmap model: (1) the need for change; (2) impact assessment; (3) creating a new model; and (4) implementation and consolidation of services.

2.1.4   Data Analysis

The roadmapping framework discussed in the earlier section was used to analyze the data collected during the workshops. The pull for service coming from society and markets (why) was mapped along with the push coming from technological development, legislation, and other areas (how). Finally, we brainstormed about how to meet the current and future demands of service emerging from why and how levels of the roadmap (what).

2.1.5   Procedure

Each workshop starts with a keynote speech followed by the main workshop session. The workshop session starts with a short presentation on the structure of the session, participating stakeholders, process, and expectations from the participants. Topics discussed in each workshop session concern the objectives of the workshop.

The presentation includes the session topic, with subsequent interactive round table discussions among the participants. After the completion of the session, the workshop's closing remarks were presented, and the next workshop date and venue were announced.

3   Results

This section presents the findings from stakeholder workshops, structured into three categories: challenges, opportunities, and requirements. *Challenges* synthesize responses related to barriers within the Irish mobility ecosystem (Q1) and the challenges associated with implementing the Irish Mobility Data Space (Q4). *Opportunities* capture insights on key opportunities for data sharing in Ireland (Q2), and *Requirements* outline the primary requirements for establishing mobility data spaces in Ireland, based on the stakeholder responses (Q3). In addition, we discuss the critical factors influencing the development of data spaces in Ireland.

3.1 Challenges

Q1. What are the main challenges in the Irish mobility ecosystem?

Findings from the MaaS4IRL-FMCI, MaaS4IRL-DSAI, and ITS Ireland-MaaS4IRL workshops highlight several specific challenges. Participants identified data sharing and data standardization as the two most critical challenges. They emphasized the absence of well-defined data-sharing approaches, the lack of integration between mobility services, and inadequate broadband connectivity, all of which hinder seamless data exchange. They also identified the absence of standardized frameworks as a major obstacle, contributing to fragmentation and limited interoperability within the ecosystem. Furthermore, cybersecurity risks emerged as a key concern, emphasizing the need for robust data governance measures. Figure 1 presents the percentage of participant responses across 15 types of challenges.

- *Lack of Trust*: Ninety-five percent of participants (based on the *Mean* value of all responses from the three workshops) pointed out that trust is the key to practicing data sharing between different actors. The transport sector in Ireland consists of public and private operators of different modalities, such as rail, road, shared mobility, etc. Some of the operators, such as MaaS operators, are foreign companies operating in Ireland. There is no initiative in Ireland for establishing trust among these operators so that they can share data without concern of regulatory violation. Participants expressed concern that mobility operators are often unwilling to share data due to the absence of a robust framework that guarantees the responsible use of data and protects against regulatory breaches.
- *Lack of Standardization*: The session with the participants of public service organizations, including TFI, National Transport Authority (NTA),[12] and Transport Infrastructure Ireland (TII),[13] touched on standardization of mobility data. According to 93% of participants, currently, transport operators in Ireland do not use any standard data format, creating significant barriers to data sharing. Participants acknowledged that while certain standard data models are practiced within the mobility ecosystem, their adoption remains inconsistent, further complicating interoperability and integration efforts.
- *Data Sharing*: Ninety-two percent of participants identified data sharing as a significant challenge, underscoring critical concerns about data management and inter-organizational collaboration.
- *Fragmentation and Interoperability*: A significant challenge is the fragmented nature of data management across sectors. The absence of standardized data-sharing protocols and governance structures has resulted in inefficiencies and the underutilization of valuable data resources. Eighty-six percent of participants highlighted that this fragmentation limits Ireland's ability to fully leverage

---

[12] https://www.nationaltransport.ie/
[13] https://www.tii.ie/

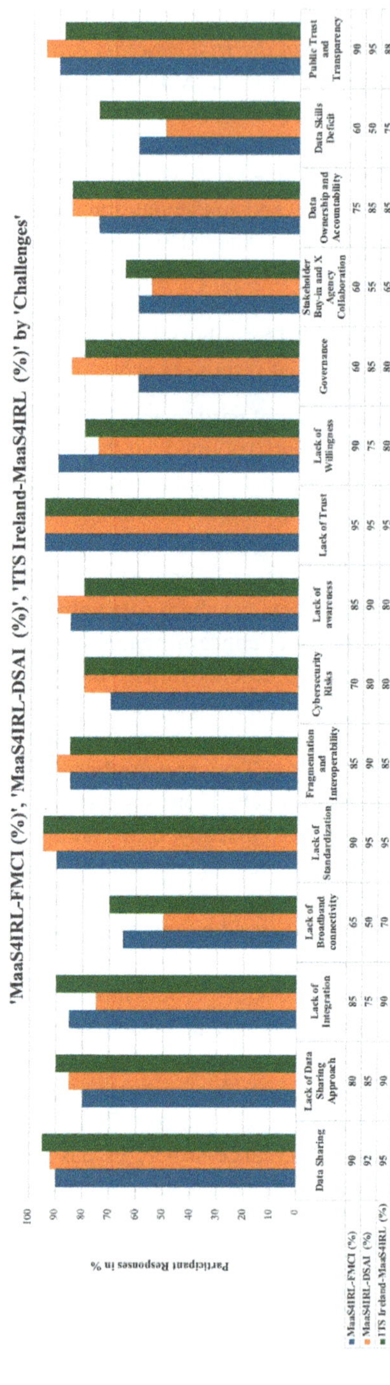

Fig. 1 Responses to mobility data-sharing challenge questions

data to optimize public service delivery and foster innovation in sectors such as healthcare, education, and transport.

- *Lack of Data-Sharing Approaches*: Participants explained the current data sharing approaches. Eighty-five percent of the participants thought that there is no standard framework for sharing transport data within the organizations in the transport sectors, such as NTA, TII, and TFI. They share data on demand using standard channels such as email or File Transfer Protocol or other file-sharing applications.
- *Lack of Integration*: Identified as a significant hurdle, the lack of integration presents major challenges in unifying diverse systems and services. Eighty-three percent of participants highlighted difficulties arising from fragmented infrastructure, incompatible data formats, and the absence of standardized interoperability frameworks.
- *Lack of Broadband Connectivity*: Sixty-one percent of participants pointed out the need for improved broadband in inner cities to support real-time information on transport availability. Enhancing digital infrastructure will be critical to ensure data-sharing platforms can offer reliable and up-to-date services, especially for underserved populations.
- *Lack of Willingness*: Eighty-one percent of participants emphasized that operators who are designated data custodians are not willing to share data primarily because of a lack of trust among these operators. The mistrust stems from concerns over the potential misuse of shared data and the absence of clear, enforceable agreements on data usage. Additionally, according to some participants, operators may fear the risk of competitive disadvantage.
- *Cybersecurity Risks*: Slightly more than 75% identified cybersecurity as a critical concern, particularly in relation to potential vulnerabilities in data. They believe that sharing data will expose sensitive information to various threats, such as data breaches.
- *Data Ownership and Accountability*: Eighty-one percent of participants emphasized that there is a lack of clarity regarding data ownership and accountability, particularly in sectors where public and private data overlap. Without a clear delineation of roles, the potential for data misuse or mismanagement increases, risking public trust and regulatory breaches.
- *Lack of Awareness*: Eighty-five percent of participants highlighted that the data space topic still lacks awareness in the Irish transport sector.
- *Stakeholder Buy-In and X Agency Collaboration*: Sixty percent of participants suggested that coordination among different stakeholders and agencies is crucial yet challenging.
- *Data Skills Deficit*: Sixty-one percent of participants from both the public and private sectors face significant skill shortages in data management, analytics, and governance. As the demand for data expertise grows, Ireland must ensure that its workforce is equipped with the necessary skills to manage, analyze, and protect data effectively.
- *Transparency*: While Ireland has made progress in promoting open data, there is still work to be done in building public trust in how data is collected, managed,

and used. Ninety-one percent of participants believe that the success of any data governance framework hinges on the public's confidence that their data is handled transparently and securely.

Some participants highlighted the complexity of establishing clear authority and operational frameworks. Furthermore, in a one-to-one consultation with the participants from Smart Dublin, the participants identified that there is a lack of a unified governance framework that hinders data-sharing implementation and scalability.

These limitations outline the need for an advanced data-sharing framework that provides a standardized, interoperable, and secure platform to overcome fragmentation and enable real-time data sharing, aligning with EU regulations like GDPR [10] and the Data Governance Act [11].

3.2 Opportunities

Q2. What are the key opportunities for mobility data spaces in Ireland?

This was an open-ended question presented to the participants, prompting the identification of key opportunities across two distinct steps of the discussion. In step 1, participants identified critical areas where the Irish Mobility Data Space is essential for advancing mobility solutions. These areas encompass Mobility-as-a-Service, the EV Charging Ecosystem, First-Mile-Last Mile Freight Delivery Solutions, Mobility Hubs, and Demand-Responsive Transit. These areas are not only pivotal for the development of IMDS but also align strategically with the Government's net-zero emission targets and the overarching Climate Action Plan.

In step 2, participants further explored the specific opportunities, in the form of use cases, within each of these identified areas. A detailed explanation of these opportunities is provided below.

3.2.1 Mobility-as-a-Service

Ireland's transport network has remained largely unchanged for an extended period; however, this is rapidly evolving due to the increasing demand for integrated, sustainable, and customer-centric mobility solutions. Establishing a connection between IMDS and the MaaS platform for multimodal trip planning and booking presents a significant opportunity. This integration would facilitate secure and interoperable data sharing between public and private transport services, enhancing accessibility and efficiency.

Moreover, the IMDS enables real-time data exchange among transportation operators, city planners, and service providers, fostering a more dynamic, responsive, and seamless mobility experience. This real-time capability enhances journey planning and booking by improving accuracy, efficiency, and adaptability to changing conditions.

The key opportunities summarized by the participants include the following:

- *Facilitating Uninterrupted Transport Mode Data Interoperability*: IMDS will encourage the standardizing data formats and protocols in order to ensure seamless interoperability across modes, including the Dublin Bus, Iarnród Éireann (Irish Rail), bike-sharing operators, e-scooter providers, and others. It will provide access to real-time data across these services, allowing for journey planning, scheduling, and ticketing to be integrated without delay so that users can have reliable multimodal choices.
- *Enabling Data-Driven, Real-Time Decision-Making for Users and Transit Agencies*: The transit authorities will be able to share information (availability, real-time schedules, travel disruption) via IMDS. If a train line is delayed, the platform can notify users immediately and provide alternative transport options. Public authorities can also use this data for demand forecasting, adjusting the frequency of buses or trams based on travel peaks, and identifying where additional services are needed.
- *Encouraging Cooperation Between the Public and Private Sector for a Sustainable Mobility Ecosystem*: IMDS will provide a common platform for public transit authorities, bike-sharing programs, e-scooter companies, and ride-hailing providers to collaborate while safeguarding regional interests. It will create a transparent data-sharing partnership mechanism, which brings mutual value to service providers who can then collaborate in co-developing solutions aligned with Ireland's climate and sustainability targets by exchanging insights.

3.2.2 EV Charging Ecosystem

Ireland's commitment to reducing greenhouse gas emissions and fostering sustainable mobility aligns with the growth of EVs across the country. The IMDS has high potential to help enable secure, standardized data exchange between the stakeholders participating in it, such as municipalities, EV charging operators, energy providers, and/or transit authorities when planning for EV infrastructure. IMDS may provide these participants with both real-time and historical mobility data that they can leverage to make informed decisions about siting, sizing, and operating EV charging stations across Ireland.

- *Data-Driven Site Selection for Charging Stations*: With data on mobility trends, human behavior regarding high-demand routes and bus stops, IMDS will provide an alternative for planners to strategically place EV charging stations for maximum utility.
- *Real-Time Demand Monitoring and Capacity Management*: Using IMDS, municipalities and energy providers will gain real-time access to charging station data such as occupancy rates, wait times, and frequency of use for the different charging stations.
- *Integration with Multimodal Mobility Hubs*: IMDS has high potential to enable seamless integration of EV charging points within multimodal mobility hubs,

providing EV drivers with easy access to public transit, bike-sharing, or e-scooter options.

3.2.3 Demand Response Transit

Ireland's commitment to sustainable mobility and reducing greenhouse gas emissions aligns closely with the implementation of demand-responsive transit (DRT) solutions. The participants identified a list of potential use cases within DRT.

- *Dynamic Route Optimization*: IMDS will provide access to mobility trends and user demand data, enabling transit operators to design routes dynamically based on real-time needs.
- *Capacity Management and Resource Allocation*: With real-time data from IMDS, transit agencies will be able to monitor vehicle occupancy, trip requests, and service frequency, allowing for dynamic adjustments to fleet size and deployment.
- *Integration with Multimodal Transit Options*: IMDS will facilitate the integration of DRT services within Ireland's broader multimodal transit network. This includes seamless connections to buses, trains, and active mobility options like bike sharing and e-scooters.

3.2.4 Infrastructure Planning

Ireland has seen urbanization, population growth, and changing mobility demand, making it evident that its transportation infrastructure needs adaptation. The IMDS has high potential to create a path for urban planners and transportation authorities to receive standardized, anonymized, real-time mobility data from public and private mobility providers by enabling secure data sharing. The participants outlined a list of potential use cases:

- *Access to Real-Time Mobility Data for Informed Decision-Making*: IMDS enables access to high-quality and secure datasets from anonymized sources, from these different mobility providers of ridesharing, public transit, bike sharing, and e-scooters.
- *Enhanced Ability to Identify Infrastructure Gaps and High-Demand Areas*: IMDS will enable urban planners to pinpoint where infrastructure improvements are a priority.
- *Long-Term Planning with Historical Mobility Data*: With IMDS, organizations can access various data sources on the evolution of mobility patterns and can discover seasonal demand trends, impacts of population growth, and changes in travel behavior over time.
- *Data-Driven Sustainability Planning and Environmental Impact Reduction:* IMDS will enable data from low-carbon vehicle adoption to be shared in a way that municipalities can use to rank projects on their environmen-

tal mitigation potential. Using information from public transportation, bike sharing, and e-scooter providers, planners can also recognize where expanding sustainable infrastructure—like Electric Vehicle charging stations, bicycle lanes, and pedestrian zones—is an opportunity. This complements the Climate Action Plan with respect to diminished car dependence and also greening mobility in Ireland.

3.3 Requirements

Q3. What are the key requirements for setting up a mobility data space in Ireland?
Participants from all workshops identified a set of essential requirements for establishing a trusted and interoperable mobility data space in Ireland. We grouped these requirements into *governance frameworks*, *technology*, *data skills*, *awareness*, and *security*. This section explores these key requirements in detail, highlighting their significance in shaping a seamless, data-driven transportation ecosystem.

- *Governance and Regulatory Frameworks*

  - Government and Industry Collaboration: Stakeholders emphasized the need for stronger collaboration between government, industry, and academic institutions to unlock new opportunities for innovation. By fostering partnerships with industry players and learning from international best practices, Ireland can accelerate its transition to sustainable and flexible MaaS solutions. Additionally, stakeholders called for government incentives to encourage greater private-sector engagement and help shift public perception toward the benefits of multimodal transport systems.
  - Preferred Governance Model for MaaS in Ireland: Dominant Preference— The majority of participants supported a public-led back-end model, where MaaS providers operate under government regulations. This reflects strong stakeholder support for public oversight, ensuring uniform standards and alignment with national transport goals.
  - Governance and Regulatory Framework: A structured governance approach with significant public oversight is preferred, promoting trust, equity, and system reliability. Stakeholders recommended establishing a centralized data governance body or coordinating mechanism to oversee and harmonize data management practices across different sectors. This would ensure consistent implementation of data protection, sharing, and security protocols.
  - Policy and Standardization: Ireland must develop a comprehensive regulatory framework to standardize data sharing, user privacy protection, and private operator integration into MaaS. Stakeholders emphasized the need to align with European standards to improve data accessibility, interoperability, and compliance across all transport modes.
  - Legal and Data Protection Frameworks: Strict adherence to GDPR is essential for MaaS platforms managing user data, ensuring privacy, security, and legal

compliance. Additionally, clear contractual agreements should be established to govern partnerships between transport providers and MaaS operators, ensuring transparency, accountability, and fair collaboration.

- National Data Governance Strategy: Ireland should develop a unified, cross-sectoral data governance strategy that clarifies data ownership, streamlines sharing protocols, and defines roles across public and private sector stakeholders.

- *Technologies*

  - Technological Integration: Develop interoperable platforms that support multimodal payments and seamlessly integrate all transport operators into a unified system.
  - User-Centric Design and Seamless Experience: The success of MaaS adoption will depend on intuitive, user-friendly platforms that simplify mobility and enhance accessibility, convenience, and efficiency for all users.
  - Provenance and Traceability: Ensure data provenance and traceability to enhance transparency, security, and trust within the MaaS ecosystem.

- *Data Skills*

  - Strengthen Data Skills: Establish and fund training programs and capacity-building initiatives to develop a workforce skilled in data analytics, governance, and cybersecurity. This should be achieved through strategic partnerships with academic institutions, industry, and government to ensure long-term sustainability and alignment with evolving technological needs.

- *Awareness Campaigns*

  - Public Awareness Campaigns: Enhance public trust by implementing clear and transparent communication of data governance practices. Conduct outreach initiatives to educate citizens on how their data is used and protected, as well as their rights under GDPR, ensuring greater awareness and engagement.

- *Security*

  - Ensure Cybersecurity Alignment: Ensure data governance alignment with cybersecurity policies to safeguard sensitive data and critical infrastructure from breaches and cyberattacks. Implement robust security frameworks that address emerging threats and enhance resilience across the mobility ecosystem.

3.4   Barriers to Implementation of IMDS

Q4. What are the key barriers to implementing a mobility data space in Ireland?

The development of a mobility data space in Ireland presents several complex challenges. Based on extensive discussions with workshop participants, we identified five critical challenges: *stakeholder coordination, data ownership and sovereignty, liability and accountability, trust among stakeholders, and public engagement and awareness.* These challenges are discussed in the analysis below.

3.4.1 Stakeholder Coordination

In the IMDS, stakeholder coordination is a critical yet challenging aspect of establishing a unified and effective governance framework. The IMDS encompasses a wide array of stakeholders, including public authorities like the NTA, private mobility providers such as ride-sharing and public transport operators, and intermediaries such as MaaS operators and other service providers. Each of these entities has unique objectives, interests, and operational priorities, which can create barriers to seamless collaboration.

Public authorities, for example, may prioritize compliance with national policies and European Union (EU) regulations, focusing on accessibility, sustainability, and public interest goals. On the other hand, private companies may be driven by profitability, competition, and data monetization opportunities. Similarly, citizen groups are likely to prioritize privacy, transparency, and equitable access to mobility services. Balancing these diverse priorities under a unified governance model requires significant effort.

Conflicting interests can slow down decision-making and adoption within the IMDS. For instance, private companies may hesitate to share proprietary data due to competitive concerns, while public authorities may enforce stringent regulations to safeguard public interest, delaying the approval of operational frameworks. Additionally, citizen groups may challenge initiatives they perceive as invasive to privacy or inequitable, creating further resistance.

3.4.2 Data Ownership and Sovereignty

In the context of the IMDS, clarifying data ownership rights is a critical challenge. Mobility data often originates from a variety of sources, including public transit systems, private operators, and citizen-generated data through apps and IoT devices. Determining who owns the data—whether it is the provider, the processor, or the individual generating it—can lead to disputes, especially when private stakeholders are involved.

Additionally, the IMDS must ensure that data sharing adheres to sovereignty principles, particularly within the framework of EU regulations like the General Data Protection Regulation (GDPR) [10] and the Data Governance Act [11]. Data sovereignty dictates that data remains under the jurisdiction of the originating country or organization, complicating international data sharing. For example, Irish mobility data shared with EU-wide systems must maintain compliance with Ireland's national laws while aligning with EU-level standards, raising potential conflicts over governance and control.

3.4.3 Liability and Accountability

The question of who is liable for issues such as data misuse, breaches, or inaccuracies presents a major challenge in the IMDS. Data flows across multiple entities, including providers, processors, and consumers, making it difficult to assign accountability. For example, if an inaccurate dataset from a mobility app leads to operational disruptions, determining whether the responsibility lies with the app developer, the data space operator, or the end user can be contentious.

Additionally, as mobility data is shared and processed within the IMDS, liability frameworks must address potential risks such as data breaches, privacy violations, and misuse of proprietary information. Without clear accountability mechanisms, stakeholders may hesitate to share data, fearing financial or reputational damage in the event of legal disputes or breaches.

3.4.4 Trust Among Stakeholders

Trust is a cornerstone for the success of the IMDS, where collaboration among public authorities, private companies, small and medium-scale enterprises (SMEs), and citizens is essential. However, building trust among these diverse stakeholders is inherently challenging. Concerns about data misuse, unfair competition, and lack of transparency often create barriers to participation. For instance, private companies may hesitate to share proprietary data, fearing it could be misused by competitors or leveraged against them in the marketplace. Similarly, public authorities may worry about potential breaches of public trust if shared data is mishandled, while citizen advocacy groups may raise concerns about privacy and ethical use.

3.4.5 Public Engagement and Awareness

Public engagement is equally critical for the adoption and success of the IMDS. Citizens, as both contributors to and beneficiaries of the data space, must understand its value and feel confident in how their data is handled. Misunderstandings or mistrust about data privacy or potential misuse can significantly limit public participation, thereby undermining the effectiveness of the IMDS.

3.5 Discussion

The stakeholder workshops provide critical insights into the feasibility and requirements for establishing the IMDS. We used LIFE to design the protocol for the stakeholder engagement to serve as a guiding tool for exchanging information, contributions, and interactions during the feedback sessions, as well as the analysis of the data that emerged. LIFE is useful in gathering data around the four major areas specified in this roadmap model: (1) the need for change; (2) impact assessment; (3) creating a new model; and (4) implementation and consolidation of services.

The identified challenges—particularly the lack of trust (95% of the participants), standardization (93%), and fragmentation (86%)—reflect systematic issues in Ireland's mobility ecosystem, compounded by the limitations of existing data-sharing initiatives such as data.gov.ie and Smart Dublin. These challenges suggest that IMDS implementation will require significant coordination to align with diverse stakeholders, including public authorities (e.g., NTA, TFI), private operators, and citizens.

The opportunities identified, such as enabling real-time MaaS interoperability and data-driven planning of EV charging infrastructure, align closely with Ireland's Climate Action Plan and net-zero emissions targets. However, realizing these benefits hinges on overcoming barriers like high implementation costs, potential stakeholder resistance, and the need for robust cybersecurity frameworks. For instance, while IMDS could streamline multimodal trip planning, the absence of standardized data formats (e.g., GTFS adoption) could delay integration, requiring investment in technical solutions like interoperable platforms.

To address the challenges related to trust, a public-led governance model, as supported by stakeholders, is essential to ensure accountability and trust. Key themes emerged across the workshop sessions, with governance being a dominant focus and stakeholders overwhelmingly supporting a public-led back-end governance model to ensure accountability, uniformity, and alignment with public interests. There was consensus that while private-sector innovation is valuable, it must operate within a structured and regulated framework. Legislative gaps were identified as significant barriers, particularly in relation to data governance, private operator integration, and autonomous transport systems. Data privacy and trust were highlighted as essential to user adoption, with calls for transparent policies and robust protections.

Mechanisms such as blockchain for data provenance or standardized protocols like GTFS could mitigate trust and interoperability issues. However, the financial and logistical complexities of scaling IMDS, particularly in rural areas with limited broadband connectivity (61% of participants), pose risks that necessitate phased implementation and government incentives.

Public engagement is another critical factor. The lack of awareness (85% of participants) highlights the need for transparent communication to build trust and encourage adoption. Without public buy-in, IMDS risks underutilization, limiting its impact on sustainable mobility. Furthermore, the skill deficit in data management (61% of participants) highlights the need for workforce training programs, potentially through partnerships with academic institutions.

In conclusion, while IMDS offers transformative potential, its success depends on addressing systemic challenges through robust governance, standardized technologies, and public trust. By building on the strengths of Ireland's existing ecosystem and learning from EU-wide initiatives, IMDS can position Ireland as a leader in data-driven mobility solutions, provided implementation is carefully managed to balance innovation with practical constraints.

4 Irish Mobility Data Space

As outlined in the EU's Data Strategy, the Mobility Data Space is one of several common European data spaces designed to foster secure and interoperable data sharing across sectors. Following an extensive analysis of the data gathered from stakeholders, the study identifies data spaces as one of the most promising technological approaches to address current challenges in mobility data sharing in Ireland. Consequently, we strongly recommend the establishment of an Irish Mobility Data Space. This recommendation is further supported by the recognition of the notion of data spaces at the European level and their strategic alignment with the EU's broader objectives—namely, promoting innovation, sustainability, and regulatory compliance, particularly with frameworks such as the GDPR and the Data Governance Act. The section provides technical details of how to implement the Irish Mobility Data Space.

4.1 Irish Mobility Data Space

We propose the IMDS as a trusted, interoperable, compliant, inclusive, and user-centric data-sharing ecosystem for the mobility stakeholders in Ireland. It serves as a single platform for regulatory authorities, government agencies related to transportation, and private operators for exchanging data. For the government, it facilitates the availability of data, allowing for smarter policymaking. It allows mobility operators to maximize resource usage and improve user experience.

The core concept of IMDS was adopted from the definition of data spaces provided in the Data Spaces Support Centre (DSSC) project [12]. Figure 2 presents the concept of IMDS refined from the conceptual model of DSSC.

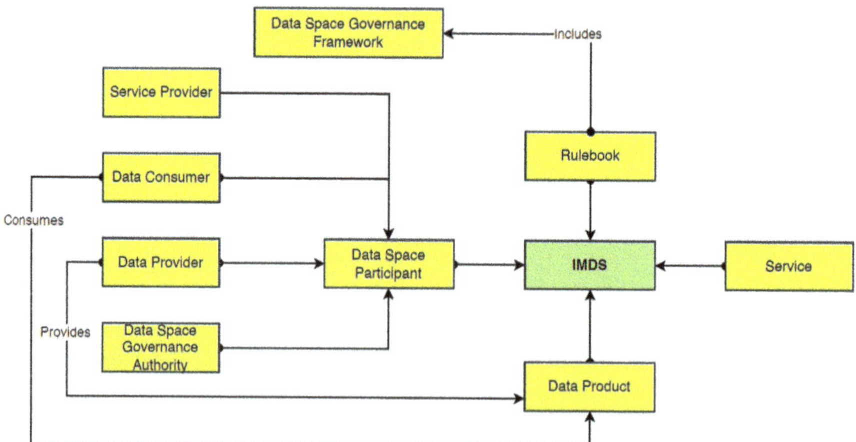

Fig. 2 The conceptual model of IMDS

The key concepts of the IMDS include Data Space Participant, Data Products, Services, Governance Frameworks, and Rulebook. The IMDS ecosystem consists of various types of participant roles, including data provider, data consumer, service provider, and data governance authorities. Data providers offer data products that are structured solutions that encompass datasets, access policies, delivery models, pricing and billing information, license terms, and other metadata.

To facilitate seamless data exchange, IMDS incorporates a range of services essential for its operation, including federation services, value creation services, and participant agent services. The governance of the IMDS is defined by the data governance frameworks codified within the Rulebook. A rulebook assists data providers and data consumers in assessing any requirements imposed by applicable legislation and contracts appropriately, in addition to guiding them in adopting practices that promote the use of data and management of risks [13]. The data space governance authority is responsible for managing and maintaining the Rulebook, ensuring compliance with regulatory standards and interoperability requirements.

Designed in compliance with European data regulations and governance frameworks, IMDS provides a privacy-preserving and collaborative environment that aligns with European Union sustainability objectives. This makes IMDS well suited for Ireland, where regulatory compliance and environmental sustainability are key national priorities. The European data strategy emphasizes the development of common, interoperable data spaces in strategic sectors, aiming to eliminate technical and legal barriers to data sharing while unlocking the transformative potential of data-driven innovation. Mobility is a key focus area within this strategy, with countries like Spain, France, and Finland establishing data spaces that contribute to the broader European Mobility Data Spaces (EMDS) [14].

To remain competitive and fully capitalize on collaborative European mobility innovations, Irish mobility policymakers must take proactive steps to initiate the creation of a national mobility data space. Aligning with the EMDS is crucial to ensuring Ireland's strategic integration into the EU's vision for sustainable mobility, enabling the country to leverage cutting-edge solutions and drive long-term transport efficiency.

This section introduces key characteristics of IMDS, the blueprint, and the best-practice technologies for implementing IMDS.

4.2   Characteristics of IMDS

IMDS has a diverse set of characteristics, spanning quality, interoperability and standardization, security, compliance, and sustainability, all of which are explored in this section.

4.2.1 Provide High-Quality Mobility Data

Artificial intelligence/machine learning (AI/ML) models often face significant challenges due to the scarcity of high-quality industrial data, particularly in the mobility sector. Accessing reliable mobility data is inherently difficult, as companies hesitate to share their datasets due to concerns over data privacy, competitive advantage, and a lack of trust in data-sharing frameworks. This limitation restricts the development of data-driven AI solutions for MaaS operators or data analytics for government agencies. The IMDS addresses this challenge by facilitating high-quality data exchange between participants (e.g., data product consumers). It enables data product consumers (e.g., MaaS application developers) to seamlessly discover and access high-quality data offered by data product providers (e.g., public transport operators, mobility service providers). Through its quality assurance mechanisms, IMDS ensures that only high-fidelity mobility datasets are available, preventing the circulation of low-quality and unreliable data.

4.2.2 Strengthens the Collaboration of Different Stakeholders

IMDS facilitates collaboration among public transit agencies, private mobility providers, urban planners, and researchers while creating a safe data infrastructure to solve complex issues of urban mobility. The IMDS model's collaborative framework allows multiple stakeholders to come together and collaborate with each other, sharing the resources and insights to address common problems like congestion, reduction in emissions, or last-mile connectivity. Projects such as Gaia-X [15] demonstrate how collaborative governance approaches can facilitate top-down methods to achieve efficient, resilient mobility systems across Europe in which multi-actor participation enhances co-creation to link together various urban transport networks and nodes.

4.2.3 Data Interoperability and Standardization

One of the great features of IMDS is that it embraces interoperability by using standardized formats and open protocols such as those behind NGSI-LD [16] (starting from FIWARE [17]). It enables disparate mobility systems to share data seamlessly without the risk of a data silo. Standardized data, for instance, allows integration across multiple transportation modes—buses, bike-sharing, and ride-hailing services—which is critical for effective urban mobility planning. According to the European Union Urban Mobility Observatory, applying these standards allows for more seamless inter-agency coordination and promotes a system that minimizes inefficiencies and supports a coherent network driven by user needs.

4.2.4   Data Sovereignty and Trust

IMDS is closely aligned with GDPR, placing data sovereignty front and center. This ensures that data providers retain control over their data while complying with privacy-focused regulations and building trust. Unlike centralized models, IMDS allows each participant to manage their data independently, deciding when and how it is shared. Data sovereignty is crucial in the context of European regulations, which mandate exceptionally high privacy standards—a point recently underscored by ENISA, the European Union Agency for Cybersecurity. ENISA [18] emphasizes that "data ownership is one of the key enablers contributing to compliant and, therefore, secure data-sharing solutions in mobility ecosystems."

4.2.5   Governance and Regulatory Alignment

Data spaces are only as effective as the governance and regulatory alignment supporting them. IMDS is designed to conform to regulatory frameworks at both national and EU levels, incorporating controls that ensure compliance with the Data Act, Data Governance Act, and similar frameworks. IMDS promotes clear protocols for governing data sharing, defining access provisions and use rights, and specifying data quality standards. By doing so, IMDS encourages participants to share their data while operating within a comprehensive governance framework aligned with both national policies and EU regulatory instruments, all underpinned by a decentralized model. The successful convergence of local rule setting with European-level regulatory requirements, as seen in initiatives like Gaia-X, exemplifies a Data-Access/Security strategy that enables borderless data use without compromising sovereignty.

4.2.6   Sustainability Focus

A mobility data space is well-known for its attention to regulatory compliance, collaborative governance, and sustainable mobility solutions. By leveraging data-fueled insights on traffic flows, congestion patterns, and public transit demand, IMDS enables environmentally sustainable choices for urban planning. According to the European Commission's (EC's) Green Mobility program, cities that have adopted data-driven methodologies to monitor and reduce traffic or vehicle routes showed substantial reductions in carbon emissions, improved air quality, as well as an increase in sustainable transportation solutions. This is particularly crucial as cities across Europe strive to meet demanding climate targets while reducing reliance on traditional travel methods to foster healthier and more sustainable urban environments.

4.3   Building Blocks of IMDS

We adopt DSSC building blocks to design the IMDS. The building blocks are categorized into business and governance building blocks and technical building blocks.

There are two categories of building blocks: business and organizational building blocks and technical building blocks.

4.3.1   Business and Organizational Building Blocks

Building a mobility data space is not just about the technology. The business and governance building blocks provide the essential foundation for IMDS, covering business, governance, and legal aspects.

- Business Building Blocks include business model, use case development, data space offering, and intermediaries and operators.

  - *Business Model*: A good incentive mechanism is one of the cornerstones to unlock the IMDS, ensuring that a broad spectrum of stakeholders is compensated and incentivized to actively participate in the data space from public transport operators, private mobility providers, and technology companies alike. Such an approach could include monetary and non-monetary incentives, as they have been effective for individual motivation, whereby the two components would synergize to promote active participation by those involved in the data-sharing and data-exchanging processes.
  - *Use Case Development*: Mobility use cases demonstrate the intrinsic value of data spaces in this sector. These are particularly relevant in scenarios in the Irish mobility sector where two or more participants, e.g., a MaaS operator and government organizations such as TII, exchange data with the intent of creating business, social, or environmental benefits. Such use cases serve as pivotal elements of the data space, extending its functionality and enhancing its overall value.
  - *Data Space Offering*: IMDS will offer data products that serve as fundamental units of data sharing, incorporating resources (such as data and/or data services) alongside metadata that delineates licensing terms, resource attributes, and other vital details in a machine-readable format. These data products are essential for fostering seamless data exchange, enabling participants to discover and utilize data in a self-service manner.
  - *Data Intermediaries and Operators*: These are entities that facilitate the exchange of data between providers and consumers by providing enablement and intermediation services. While not directly involved in data transactions, they are critical stakeholders of the data space. Consequently, participation management must make sure that intermediaries and operators conform to interoperability standards, resulting in easy data exchange.

- Governance Building Blocks include organization forms and governance authority and participation management. The governance framework of a data space is an internal framework of rules and policies that guide all participants in the data space. Such a framework guarantees consistency in practices, compliance, and shared accountability while creating a trustworthy and secure ground for data sharing and collaboration. Among data space governance elements, participation management addresses the processes that ensure participants make use of the data space in a regulated manner. This covers everything from how participants will be identified and onboarded, offboarded, and the governance rules related to data transactions, as well as options for enablement and intermediation services.
- Legal Building Blocks include regulatory compliance and a contractual framework.

  - *Regulatory Compliance*: Regulatory compliance is a fundamental pillar of the IMDS, ensuring that all its operations adhere to the relevant legal frameworks at both national and EU levels. This encompasses a wide range of activities, including interpreting and implementing regulatory requirements, embedding compliance into every aspect of the data space, and facilitating lawful data sharing and collaboration among participants. Given the complexity and diversity of data handled within the IMDS, compliance spans personal data protection, intellectual property rights, trade secret safeguards, and non-personal data regulations. The governance authority plays a pivotal role in overseeing these efforts, identifying relevant laws, and establishing mechanisms to ensure full compliance.
  - *Contractual Framework*: The Governance Authority and the participants of IMDS must establish a contractual framework. The contractual framework consists of legally binding agreements establishing the relationships between data space participants and transaction participants. This framework falls into two types of agreements, namely, data space agreements and data transaction agreements, as defined in the DSSC blueprint. The two differentiating components between these agreement types are (1) the event that makes agreements binding (e.g., joining a data space or executing a specific transaction) and (2) the parties covered by the agreement (all data space participants versus only those participating in a specific transaction).

4.3.2 Technical Building Blocks

The technical building blocks of IMDS cover data interoperability, data sovereignty, and trust and data value creation enablers.

- Data Interoperability: Data interoperability is composed of data model, data exchange, and provenance and traceability.

  - *Data Model*: Data models are essential to guarantee the coherent interpretation of data when shared within a data space. These models carry metadata that conveys semantic information, enabling participants to understand the data they are consuming. Data models are especially relevant for interactions between two participants in a data space who wish to exchange data. By adopting a common data model, participants can achieve semantic interoperability, facilitating seamless data exchange across the data space ecosystem. To achieve semantic interoperability, data space participants must describe their data (or any products offered) following a harmonized data model as defined by the proprietary governance framework of the data space. These standardized data models are provided by data model providers under the supervision of the data space governance authority. When these models are published, they are stored in a vocabulary service—essentially a repository used for finding and referencing data models throughout the data space. During the actual data exchange, all participants consult their respective vocabulary services and refer to the same data model.

- *Data Exchange*: Data exchange requirements within a data space must be defined based on the specific data transfer characteristics of the intended use cases. These characteristics directly determine how data exchange is implemented. The implementation of data exchange within a data space is fundamentally tied to the Data Plane, which is inherently interconnected with the Control Plane. The Control Plane facilitates generic processes and protocols, such as the catalogue protocol and the contract negotiation protocol. The Data Plane, in contrast, is domain - specific and directly handles the transmission and access of data between participants.

- *Data Provenance and Traceability*: Provenance and traceability are essential for building trust, ensuring compliance, and maintaining transparency within the IMDS. By leveraging existing standards and guidelines, rather than creating new frameworks, the IMDS can efficiently meet legal, regulatory, and operational requirements. Provenance and traceability help stakeholders understand the origin, handling, and usage of data, supporting secure and ethical data sharing across the mobility ecosystem.

- Data Sovereignty and Trust: Data sovereignty is an essential principle underpinning the management, access, and control of data in the Irish Mobility Data Space, according to relevant legislation (EU based). This means that personal and organizational data that could be stored or processed in a country like Ireland is under the laws and regulations of data privacy for that jurisdiction. This is the key to building trust, enabling cooperation, and ensuring alignment with EU priorities on data protection, security, and strategic autonomy.

- Data Value Creation Enabler: This pillar is composed of data and services, offering descriptions, publication and discovery, and value creation services.

  - In the IMDS, providers utilize metadata to describe their organizations, their data product offerings, and the resources composing these offerings. These metadata records, referred to as Self-Descriptions, are designed to be comprehensible not only to human data consumers but also to automated machine agents that support data processing and decision-making. This universal comprehensibility is achieved by adhering to established semantic Web and linked data standards.

- *Publication and Discovery*: It serves to publish and discover data product offerings in a data space. Data product offerings descriptions are created and maintained by their data providers. These descriptions are stored in a catalogue, and it is the data provider's responsibility to manage them throughout their life cycle—from publication to being retired or removed from availability. Once data product offerings are published, data consumers can query the catalogue to find and evaluate existing offerings, identifying those that best suit their needs. A robust Data Product Discovery mechanism ensures that data products are Findable, thus adhering to the FAIR (Findability, Accessibility, Interoperability, and Reusability) principle.

- *Value Creation Services*: It addresses the technical aspects related to the definition of mobility services, such as value-added MaaS applications aiming at creating value out of the data shared in the data space, and sets up the conditions to ensure the appropriate provision, delivery, and utilization of those services.

4.4   Architecture of IMDS

Various reference architectures have been proposed for building data spaces, among which the International Data Spaces Association (IDSA) is widely adopted by data space initiatives. The Common European Data Spaces (CEDS) rely on the architecture proposed in deployEMDS [19]. In this work, we propose an architecture[14] for the IMDS. The architecture of IMDS has been adapted from the IDSA reference architecture and the architectural components of deployEMDS. Figure 3 shows the architecture of IMDS.

- *Data Product Pipelines*: The data product pipelines are a set of building blocks that enable data providers and data owners to perform data processing operations such as Extract, Transform, and Load (ETL) to produce high-quality data. These pipelines manage the data life cycle, including change management and data product packaging. Data providers can also use data applications (data apps) available in an app store provided by service providers.

- *Data Space Connector*: The Data Space Connector is the key component of a data space. It is an integrated suite of components that every organization participating in a data space needs to deploy to "connect" to a data space. A connector integrates the information model, the catalogue of data products offered by the data providers, and usage control.

---

[14]The architecture is development is an ongoing work. This is the first version of the IMDS architecture.

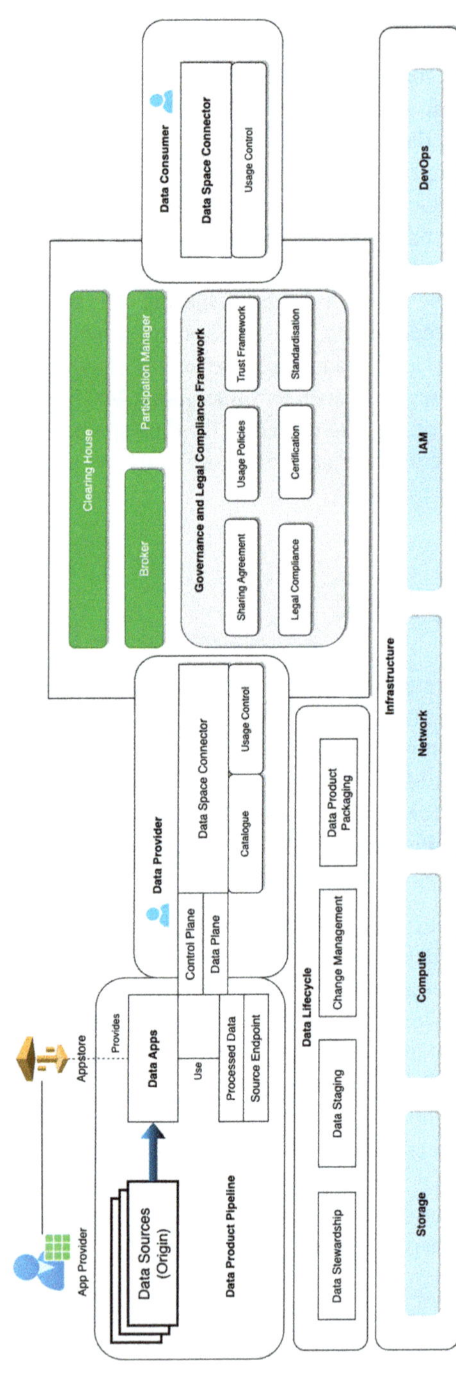

Fig. 3 The high-level architecture of Irish Mobility Data Space

- *Core Components*: The core components of IMDS are used to automate the fundamental operations of data spaces.
  - *Clearing House*: The Clearing House is a broker that provides approval and settlement services for all financial and data exchange transactions.
  - *Broker*: A data space depends on a broker to distribute its resource inventory and associated metadata, including details such as source, name, location, size, creation date, owner, and more. Each connector can discover data assets within the data space by querying the broker. This intermediary service specializes in search functionalities and facilitates the sharing of relevant information—such as status updates and newly available data—with subscribed connectors. By following the PUB/SUB paradigm, the broker ensures that metadata describing data assets remains accessible within the ecosystem.
  - *Participation Manager*: It manages the onboarding and offboarding of data space participants, including data providers and data consumers. It manages registering participants and updating their information.
  - *Infrastructure*: The IMDS is deployed on infrastructure comprising storage, compute, and network, which are essential for deploying the mobility data space. It integrates an Identity and Access Management (IAM) framework to control user access to data space resources. Additionally, the infrastructure includes DevOps capabilities to streamline the development and deployment of IMDS.

4.5 Tools and Technologies for Implementing IMDS

A wide range of technologies can be used for implementing the IMDS architecture (shown in Fig. 3). In this section, we present some of the best-practice technologies commonly used for building mobility data spaces. We focus on key components such as connectors, clearing houses, data interoperability, compliance mechanisms, and brokers. Below, we provide an overview of these technologies.

4.5.1 Connectors

- Eclipse Data Connector (EDC) [20]: EDC is a comprehensive framework (concept, architecture, code, samples) providing a basic set of features (functional and non-functional) that data space implementations can re-use and customize by leveraging the framework's defined APIs and ensure interoperability by design.
- Sovity [21]: It provides easy access to data spaces with industry-proven software. Maintain full control while we empower your company to share data with others effortlessly in a community. Trust via interoperable and secure usage control.
- EGI Datahub Connector [22]: EGI Datahub Connector is a high-performance data management solution that offers unified data access across globally distrib-

uted environments and multiple types of underlying storage. It allows researchers to share, collaborate, and perform computations on the stored data easily.

4.5.2 Platform/Infrastructure

- iShare [23]: An integrated set of agreements for identification, authentication, and authorization facilitates seamless data access between organizations. Originally pioneered by the Dutch transport and logistics sector, this framework is now expanding its reach into new sectors and even crossing international borders.
- FIWARE: It is an open-source platform engineered to drive the development of intelligent solutions and applications for the digital economy. Its core mission is to provide developers, companies, and cities with a comprehensive framework and a suite of standardized technologies. This empowers them to efficiently build and deploy innovative digital solutions while reducing costs.
- Gaia-X: It stands as a European initiative committed to constructing a federated and highly secure data infrastructure, with the primary objectives of advancing data sovereignty, interoperability, and innovation. At its core, Gaia-X aims to establish a robust and transparent digital ecosystem, empowering individuals, enterprises, and governmental bodies to securely and meticulously share and harness data resources.
- EONA-X [24]: EONA-X, the global data space for Mobility, Transport, and Tourism, is designed to facilitate data exchange, ultimately enhancing the safety, planning, sustainability, comfort, resilience, accessibility, and enjoyment of all related activities. This initiative opens opportunities for economic growth and improved service quality, both nationally and internationally. When it comes to tourism, it is worth noting that the sector consists of many players, particularly SMEs, who often have limited capacity for digital investment.
- Catena-X [25]: Catena-X is the first open, collaborative, and decentralized data ecosystem for the automotive industry, built on common standards and shared principles. The Catena-X data ecosystem consists of three areas: the Catena-X Automotive Network, e.g., the development environment, and the operating environment.
- IDSA Metadata Broker [26]: The IDS Metadata Broker is an IDS Connector, which contains an endpoint for the registration, publication, maintenance, and query of Self-Descriptions. A participant can interact with an IDS Metadata Broker by using the processes defined on the Process Layer, the descriptions defined on the Information Layer, and the descriptions defined on the System Layer.

4.5.3 Data Model

- Mobility data are distributed across multiple public and private sector organizations and are stored in heterogeneous data formats, including relational databases, graph models, XML files, and flat files. Consequently, well-defined data models are essential to ensure semantic consistency, interoperability, coherent interpretation of data assets across the mobility data space. To create a robust and effective data model for the IMDS, the following standards may be used for modeling data: General Transit Feed Specification (GTFS) [27], Service Interface for Real-Time Information (SIRI) [28], NeTEx (Network Exchange) [29], DATEX II [30], General Bikeshare Feed Specification (GBFS) [31], MDS, Transport Infrastructure Data Exchange System (TIDES) [32], Next Generation Service Interface-Linked Data (NGSI-LD), and Transport Network Intelligent Transport Systems (TN-ITS).

4.5.4 Trust and Sovereignty

- *Gaia-X Trust Framework*: The Gaia-X Trust Framework establishes the fundamental set of rules required to participate in the Gaia-X Ecosystem. These rules ensure a unified governance structure and a foundational level of interoperability across various ecosystems while granting users full control over their decisions.
- *Access Control and Authentication*: Various technologies can be used for access control and authentication. Role-Based Access Control (RBAC) [33] and Multi-Factor Authentication (MFA) [34] are widely used to restrict data access. Identity verification standards, such as OpenID and OAuth, are also effective in securing access.
- *Verifiable Credentials (VC)* [35]: Gaia-X uses verifiable credentials; however, it is an open technology that can be integrated with any custom system. The Verifiable Credentials model enables verifiers to trust the data without relying on the trustworthiness of its source while also facilitating the easy identification of credential holders.
- *Decentralized Identifier (DID)* [36]: It is a novel type of identifier defined by the World Wide Web Consortium (W3C)[15] that enables verifiable digital identities. Unlike traditional identifiers, DIDs are entirely controlled by their owners and operate independently of centralized registries, identity providers, or certificate authorities.

4.5.5 Compliance

- Gaia-X Compliance Framework: The Gaia-X Compliance Framework consists of two primary subsystems. The first is the Trust Framework, which is described earlier. The second subsystem is the Labeling Framework, which is optional and

---

[15] https://www.w3.org/

allows for the verification of adherence to specific rule sets that meet market requirements.

4.5.6 Data, Services, and Offerings Descriptions

- The IDSA Dataspace Protocol [37] comprises a series of specifications to stream-line interoperable data sharing among entities. Governed by usage control and grounded in Web technologies, these specifications delineate the schemas and protocols necessary for entities to publish data, negotiate usage agreements, and access data within a federation of technical systems, collectively termed a dataspace.
- mobilityDCAT-AP [38]: It provides a structured, interoperable and harmonized approach to describing and exchanging metadata about datasets related to mobil-ity and in particular related to Intelligent Transport Systems (ITS).
- Data Quality Vocabulary (DQV) [39]: It is a (meta) data model implemented as a Resource Description Framework (RDF) vocabulary, which extends the Data Catalogue Vocabulary (DCAT) [40] with properties and classes suitable for expressing the quality of datasets and their distributions. DQV has been con-ceived as a high-level, interoperable framework that must accommodate various views over data quality.
- Open Digital Rights Language (ODRL) [41]: The Open Digital Rights Language (ODRL) is a policy expression language that offers a flexible and interoperable information model, along with a standardized vocabulary and encoding mecha-nisms, to define and manage usage policies for content and services.

4.5.7 Catalogue

The IDSA Data Space Protocol facilitates the implementation of a catalogue by defining a structured mechanism for representing datasets and their associated offers, which are advertised by provider participants within the data space. It enables standardized metadata exchange and discovery within a federated data ecosystem. CKAN [42], an open-source data management system, provides an extensible archi-tecture for cataloguing, storing, and accessing datasets. It features a robust front-end, a fully RESTful API supporting both data and metadata operations, advanced visualization capabilities, and modular plugin support for enhanced interoperability and integration within complex data infrastructures.

4.5.8 Data Product Pipelines

Data product pipelines can be implemented using an integrated solution available in the App Store, comprising a suite of tools for data ingestion, processing, and quality enhancement. Industry-standard technologies such as Apache Kafka [43] and

Apache NiFi [44] are leveraged for scalable, real-time, and batch data ingestion, ensuring efficient data flow management. For advanced data processing, Apache Spark [45] provides a distributed computing framework capable of executing complex transformations, handling missing values, enhancing data accuracy, and facilitating seamless integration across heterogeneous data sources. These best-practice technologies collectively enable a high-performance, scalable, and fault-tolerant data pipeline architecture. However, IMDS initiatives may choose other open-source technologies.

5   Conclusion

In an era of rapid technological advancement, mobility data has emerged as a cornerstone for innovation, efficiency, and the development of intelligent, user-centric transport solutions. Ireland—renowned for its dynamic technology sector and host to numerous global tech enterprises—is uniquely positioned to leverage this data-driven potential to advance a sustainable and digitally enabled transport ecosystem.

This study, grounded in extensive stakeholder engagement through structured workshops, investigated the transformative role of mobility data across multiple domains, including MaaS, infrastructure planning, EV charging, and mobility hub deployment. The findings emphasize the considerable opportunities that enhanced data sharing offers, particularly in optimizing multimodal coordination, improving operational efficiency, enabling demand-responsive transit, and fostering inclusive access to services. Nevertheless, the research also revealed persistent challenges: the absence of a cohesive data-sharing framework, insufficient mechanisms to ensure compliance with regulatory and governance standards, and the lack of standardized, interoperable data models across stakeholders. Furthermore, the study concludes that effective data sharing is not merely a technical necessity but a foundational pillar for fostering innovation, operational efficiency, and equitable access to mobility solutions.

To address these challenges and capitalize on the opportunities identified, the study proposes the establishment of the IMDS. The IMDS is a transformative initiative designed to redefine Ireland's mobility ecosystem by providing an interoperable and scalable framework for integrating heterogeneous mobility datasets. This framework will foster collaboration between public and private mobility operators, enhancing data analytics capabilities, supporting strategic EV charging network deployment, optimizing demand-responsive transit, and improving infrastructure planning. Furthermore, the proposed IMDS aligns with broader European initiatives such as the European Strategy for Data and is consistent with regulatory frameworks like the GDPR and the Data Governance Act (DGA). As such, the architectural components, stakeholder engagement methodology, and governance recommendations put forward in this study may be adapted and replicated in other national or regional contexts pursuing similar policy objectives.

By implementing the IMDS, Ireland has a unique opportunity to lead in transportation excellence, creating a pioneering mobility ecosystem that embodies innovation, efficiency, and environmental stewardship. This bold initiative offers a clear pathway to a sustainable, data-driven future, positioning Ireland as a global leader in smart mobility solutions.

It is worth noting that although this investigation is rooted in the Irish context, the issues it addresses—such as data fragmentation, limited interoperability, and stakeholder mistrust—are common across many European mobility ecosystems. As such, the insights gained from this study contribute to a reference model for other EU countries aiming to modernize their transportation infrastructures through data-centric strategies. Effective data sharing, as demonstrated here, is not merely a technical imperative but a critical enabler of system-wide innovation, equity, and resilience, particularly in addressing regional disparities between urban and rural mobility services.

Acknowledgments The MaaS4IRL study has been supported with financial contributions from the Sustainable Energy Authority of Ireland and the Department of Transport under the SEAI Research, Development and Demonstration Funding Programme 2023, Grant number 23/RDD/907.

References

1. Boland, L. Retrieved March 19, 2025, from https://www.thejournal.ie/what-are-the-traffic-transport-road-changes-dublin-city-6469975-Aug2024/
2. Gallagher, F. Retrieved March 27, 2025, from https://www.irishtimes.com/transport/2025/01/06/dublin-ranks-as-europes-third-most-congested-city-after-london-and-paris/
3. Ireland's Town Centre First. Retrieved March 15, 2025, from https://www.urban-initiative.eu/news/irelands-town-centre-first
4. Smart Dublin. *Rethinking Mobility in Ireland—The case for MaaS*. Smart Dublin. Retrieved March 27, 2025, from https://smartdublin.ie/rethinking-mobility-in-ireland-the-case-for-maas/
5. O'Cearbhaill, M. Retrieved March 15, 2025, from https://www.thejournal.ie/dublin-second-slowest-city-6268171-Jan2024/
6. Department of Transport. Shared mobility hubs issues, challenges and opportunities. Retrieved March 27, 2025, from https://www.gov.ie/en/consultation/7e2fc-public-consultation-shared-mobility-hubs-issues-paper/
7. Open Mobility Foundation. Retrieved March 27, 2025, from https://www.openmobilityfoundation.org/about-mds/
8. Halonen, M., Kallio, K., & Saari, E. (2010). Towards co-creation of service research projects: A method for learning in networks. *International Journal of Quality and Service Sciences, 2*(1), 128–145.
9. Eckhardt, J., Aapaoja, A., Nykänen, L., Sochor, J., Karlsson, M., & König, D. (2018). The European roadmap 2025 for mobility as a service. In *Proceedings of the 7th Transport Research Arena TRA*.
10. General Data Protection Regulation (GDPR)—Legal text. Retrieved March 12, 2025, from https://gdpr-info.eu/
11. European Data Governance Act. Retrieved March 19, 2025, from https://digital-strategy.ec.europa.eu/en/policies/data-governance-act

12. Data Spaces Support Centre. Retrieved March 26, 2025, from https://dssc.eu/
13. Pitkänen, O., Turpeinen, M., & Lähteenoja, V. (Eds.). (2025). *Rulebook model for a fair data economy part 2: Templates (version 3.0)*. 1001 Lakes Oy, Funded by Sitra.
14. Creating a common European mobility data space. Retrieved March 21, 2025, from https://transport.ec.europa.eu/transport-themes/smart-mobility/creating-common-european-mobility-data-space_en
15. Gaia-X European Association for Data and Cloud AISBL Home. Retrieved March 19, 2025, from https://gaia-x.eu/
16. NGSI-LD. Retrieved March 20, 2025, from https://fiware-datamodels.readthedocs.io/en/stable/ngsi-ld_howto/
17. De Panfilis, G. *FIWARE—Open APIs for open minds*. FIWARE. Retrieved March 2, 2025, from https://www.fiware.org/
18. ENISA. Retrieved March 26, 2025, from https://www.enisa.europa.eu/
19. DeployEMDS. Retrieved March 27, 2025, from https://deployemds.eu/
20. Delgado, M. T. Retrieved March 26, 2025, from https://projects.eclipse.org/projects/technology.edc
21. Sovity. *Sovereign data exchange with your partners in Data Spaces*. Sovity. Retrieved March 26, 2025, from https://sovity.de/en/sovity-en/
22. DataHub. Retrieved March 26, 2025, from https://www.egi.eu/service/datahub//
23. iShare. Retrieved March 23, 2025, from https://ishare.eu/
24. EONA-X. Retrieved March 27, 2025, from https://eona-x.eu/
25. Catena-X. Retrieved March 27, 2025, from https://catena-x.net/en/1
26. Metadata Broker. Retrieved March 27, 2025, from https://docs.internationaldataspaces.org/ids-knowledgebase/ids-ram-4/layers-of-the-reference-architecture-model/3-layers-of-the-reference-architecture-model/3_5_0_system_layer/3_5_4_metadata_broker
27. General Transit Feed Specification. Retrieved March 27, 2025, from https://gtfs.org/
28. Service Interface for Real time Information (SIRI). Retrieved March 27, 2025, from https://www.vdv.de/siri.aspx
29. NeTEx. Retrieved March 27, 2025, from https://netex.ie/overview.htm
30. DATEX II. Retrieved March 26, 2025, from https://datex2.eu/
31. General Bikeshare Feed Specification. Retrieved March 27, 2025, from https://staging.gbfs.org/
32. Transit ITS Data Exchange (TIDES). Retrieved March 27, 2025, from https://tides-transit.org/main/
33. Ferraiolo, D., Cugini, J., & Kuhn, D. R. (1995). Role-based access control (RBAC): Features and motivations. In *Proceedings of the 11th Annual Computer Security Application Conference* (pp. 241–248).
34. Ometov, A., Bezzateev, S., Mäkitalo, N., Andreev, S., Mikkonen, T., & Koucheryavy, Y. (2018). Multi-factor authentication: A survey. *Cryptography, 2*(1), 1.
35. Verifiable Credentials Data Model v2.0. Retrieved March 27, 2025, from https://www.w3.org/TR/vc-data-model-2.0/
36. Decentralized Identifiers (DIDs) v1.0. Retrieved March 27, 2025, from https://www.w3.org/TR/did-1.0/
37. Dataspace Protocol. Retrieved March 25, 2025, from https://docs.internationaldataspaces.org/ids-knowledgebase/dataspace-protocol
38. MobilityDCAT-AP, GitHub. Retrieved March 27, 2025, from https://github.com/mobilityDCAT-AP
39. Data Quality Vocabulary. (2016, December 15). Retrieved March 26, 2025, from https://www.w3.org/TR/vocab-dqv/
40. Data Catalogue Vocabulary (DCAT). Retrieved March 27, 2025, from https://www.w3.org/TR/vocab-dcat-3/
41. Open Digital Rights Language (ODRL) ontology. Retrieved March 27, 2025, from https://www.w3.org/ns/odrl/2/ODRL20.html

42. CKAN—The open source data management system. Retrieved March 27, 2025, from https://ckan.org/
43. Apache Kafka. Retrieved March 27, 2025, from https://kafka.apache.org/
44. Apache NiFi. Retrieved March 27, 2025, from https://nifi.apache.org/
45. Apache Spark. Retrieved March 27, 2025, from https://spark.apache.org/

Toward a Holistic Framework for Human-AI Collaboration in Safety-Critical Systems

Ricardo J. Bessa, Milad Leyli-Abadi, Mouadh Yagoubi, Daniel Boos,
Clark Borst, Alberto Castagna, Ricardo Chavarriaga, Duarte Dias,
Adrian Egli, Andrina Eisenegger, Joost Ellerbroek, Anna Fedorova,
Cristina Felix, Anton Fuxjäger, Joaquim Geraldes, Samira Hamouche,
Mohamed Hassouna, Sjoerd Kop, Bruno Lemetayer, Giulia Leto,
Roman Liessner, Jonas Lundberg, Antoine Marot, Maroua Meddeb,
Manuel Meyer, Hélio Sales, Viola Schiaffonati, Manuel Schneider,
Irene Sturm, Julia Usher, Herke Van Hoof, Jan Viebahn, Toni Wäfler,
and Giacomo Zanotti

Abstract The integration of artificial intelligence (AI) into safety-critical systems, where human operators remain central to decision-making, introduces various challenges that existing AI frameworks struggle to address comprehensively. Key concerns involve designing a socio-technical system that balances AI transparency, trust, and explainability with the imperative for robust and reliable decision-making.

R. J. Bessa (✉) · D. Dias
INESC TEC, Porto, Portugal
e-mail: ricardo.j.bessa@inesctec.pt; duarte.f.dias@inesctec.pt

M. Leyli-Abadi · M. Yagoubi · M. Meddeb
IRT SystemX, Paris, France
e-mail: milad.leyli-abadi@irt-systemx.fr; mouadh.yagoubi@irt-systemx.fr; maroua.meddeb@
irt-systemx.fr

D. Boos · A. Egli
SBB Swiss Federal Railways, Bern, Switzerland
e-mail: daniel.boos@sbb.ch; adrian.egli@sbb.ch

C. Borst · J. Ellerbroek · G. Leto
Delft University of Technology, Delft, The Netherlands
e-mail: c.borst@tudelft.nl; j.ellerbroek@tudelft.nl; G.Leto@tudelft.nl

A. Castagna · A. Fuxjäger
EnliteAI, Wien, Austria
e-mail: a.castagna@enlite.ai; a.fuxjaeger@enlite.ai

R. Chavarriaga · A. Fedorova
Zurich University of Applied Sciences, Zurich, Switzerland
e-mail: char@zhaw.ch; fedo@zhaw.ch

E. Curry et al. (eds.), *Artificial Intelligence, Data and Robotics*,
https://doi.org/10.1007/978-3-032-10561-5_13

A. Eisenegger · S. Hamouche · J. Usher · T. Wäfler
University of Applied Sciences and Arts Northwestern Switzerland, Olten, Switzerland
e-mail: andrina.eisenegger@fhnw.ch; samira.hamouche@fhnw.ch; julia.usher@fhnw.ch; toni.waefler@fhnw.ch

C. Felix · J. Geraldes · H. Sales
NAV Portugal, Lisbon, Portugal
e-mail: Cristina.Felix@nav.pt; Joaquim.Geraldes@nav.pt; Helio.Sales@nav.pt

M. Hassouna
Fraunhofer IEE, Kassel, Germany
e-mail: mohamed.hassouna@iee.fraunhofer.de

S. Kop · J. Viebahn
TenneT, Arnhem, The Netherlands
e-mail: sjoerd.kop@tennet.eu; jan.viebahn@tennet.eu

B. Lemetayer · A. Marot
Réseau de Transport d'Electricité, Paris, France
e-mail: bruno.lemetayer@rte-france.com; antoine.marot@rte-france.com

R. Liessner · I. Sturm
Deutsche Bahn, Berlin, Germany
e-mail: roman.liessner@deutschebahn.com; Irene.Sturm@deutschebahn.com

J. Lundberg
Linköping University, Norrköping, Sweden
e-mail: jonas.lundberg@liu.se

M. Meyer · M. Schneider
Flatland Association, Bern, Switzerland
e-mail: manuel.meyer@flatland-association.org; manuel.schneider@flatland-association.org

V. Schiaffonati · G. Zanotti
Politecnico di Milano, Milan, Italy
e-mail: viola.schiaffonati@polimi.it; giacomo.zanotti@polimi.it

H. Van Hoof
University of Amsterdam, Amsterdam, The Netherlands
e-mail: h.c.vanhoof@uva.nl

Presently, while numerous sector-specific solutions exist, a holistic framework that effectively integrates human expertise with AI capabilities remains absent, leaving critical gaps in system design, deployment, and oversight. This chapter proposes a multidisciplinary conceptual framework to enhance human-AI collaboration in critical infrastructures such as power grids, railways, and air traffic management. The different design steps were guided by the requirements of these industrial domains. The framework combines key design principles that support human cognition, leveraging insights from decision theory, mathematics, and specialized engineering domains to optimize AI-assisted decision-making. Furthermore, it embeds trustworthiness and risk assessment methodologies, using tools such as the Assessment List for Trustworthy Artificial Intelligence (ALTAI) tool to ensure compliance with ethical and regulatory requirements.

Keywords Safety-critical systems · Framework · Human centric · Power grid · Railway · Air traffic management

Acronyms

AI	Artificial intelligence
ALTAI	The Assessment List for Trustworthy Artificial Intelligence
ATCOs	Air traffic controllers
ATM	Air traffic management
CAB	Cockpit and bidirectional assistant
CSE	Cognitive systems engineering
EID	Ecological interface design
HMI	Human-machine interface
JCF	Joint Control Framework
KPIs	Key performance indicators
MBSE	Model-based system engineering
RL	Reinforcement learning
TAI	Trustworthy AI
UQ	Uncertainty quantification
UTM	Unmanned aircraft system traffic management
XAI	Explainable AI

1 Introduction

In the context of emerging climate change impacts and growing demand for energy and mobility networks, complemented with the implementation of digitalization and decarbonization roadmaps, artificial intelligence (AI) can be a transformative technology in the management and operation of critical infrastructures. These sectors are classified as high risk in the European Union AI Act. Therefore, safety, transparency, and adherence to ethical principles are fundamental to mitigating risks across the technical, social, and environmental dimensions. Since humans play a pivotal role in operational decision-making within these infrastructures (e.g., train redispatch, power grid congestion management), achieving seamless human-AI collaboration remains a critical challenge. This requires integrating human expertise into AI-driven processes, promoting mutual adaptation and shared decision-making while preventing individual decision-making. Various methodologies can be employed, including co-learning paradigms where AI and human operators iteratively enhance their knowledge and hybrid systems that enable AI to function autonomously while maintaining human oversight for critical decisions.

This chapter describes a conceptual framework tailored for the operation of critical infrastructures aided by AI-based systems and relates to the *cross-sectional reasoning and decision-making* areas of the AI, Data, and Robotics Partnership [1]. The framework incorporates AI algorithms, AI-friendly digital environments, a hypervision-based human-AI interface, and socio-technical design principles to enhance seamless human-machine collaboration. It aims to enhance real-time and predictive operational efficiency while offering structured methodologies for managing uncertainty and mitigating operational risks. Operators of critical infrastructures and academia designed this framework for the electrical power grid, railway network, and air traffic management.

1.1   AI Frameworks for Power Grid Operations

Marot et al. introduced a framework for AI-driven agents to assist in real time human operators in solving technical problems, such as power line overloads in power grids. There, the AI agent proactively alerts the operator when the confidence level in its suggested remedial actions is low, thereby preventing a human-out-of-the-loop scenario [2]. To reduce excessive alarms, an attention budget mechanism was implemented. The framework combines three key principles: *credibility*, achieved via transparency and clear explanations of AI-driven actions; *reliability*, maintained via consistent performance and generalization across different scenarios; and *intimacy*, where detailed explanations for incorrect actions are provided to build trust between human operators and AI. A framework based on naturalistic decision-making is presented in Greitzer and Podmore [3], combining concepts such as recognition-primed decision-making, recognition/meta-recognition, and situational awareness. It models how power grid operators process information, interact with environmental cues, and make decisions in real time. Two internal simulation loops allow for rapid meta-cognitive evaluation and prediction of the potential effects of control actions, thereby supporting decision-making in grid operations. Fan et al. introduce a human-machine hybrid-augmented intelligence framework for grid dispatching and control [4]. It integrates AI-driven decision aid, a digital twin, and human expertise to enhance operational efficiency in dynamic and complex grid management tasks. Key elements include bidirectional data exchange, knowledge co-evolution between AI and humans, and flexible task redistribution between operators and AI systems to optimize decision-making. Hilliard et al. propose the Work Domain Analysis framework that structures power grid operations into five hierarchical levels, linking physical grid components to overarching objectives such as reliability and safety [5]. The framework employs means-ends relationships, hierarchical decomposition, and cause-effect modeling to enhance operator decision support, improve the visual representation of grid states, and facilitate structured decision-making.

1.2   AI for Railway Traffic Management

Traditionally, traffic management in railway networks is performed by human operators, nowadays supported by traffic management systems (TMS) with various degrees of automation, such as the Swiss Federal Railway's Rail Control System [6], which provides real-time traffic monitoring and dynamic routing options to the human dispatcher. While such TMS leverage the long history of operations research in the railway domain, experimental systems like the automatic dispatching assistant developed by Deutsche Bahn [7, 8] use machine learning to enhance the human decision support. In recent years, there has been growing interest in machine learning approaches, such as deep reinforcement learning (DRL) and multi-agent reinforcement learning (MARL) in particular, for planning and (re-)scheduling tasks in railway [9], and fundamental features of DRL for vehicle rescheduling problems [10] in railway were demonstrated. For example, the international community of researchers and railway enthusiasts around the Flatland digital environment [11] illustrated basic generalization and scalability capabilities of DRL through open machine learning challenges [12]. In addition, researchers showed the effectiveness of MARL for real-time rescheduling [13], how advanced machine learning model architectures enhance coordination between trains [14], and how communication between MARL agents can improve solution quality [15]. Researchers further demonstrated that MARL-based rescheduling can save energy in train operations [16]. While DRL-based traffic management remains an active area of research, a recent study showcased the application of MARL for real-world scenarios [17], and novel open-source tools such as the Open Source Railway Designer [18] support the translation of basic research results to real-world settings.

1.3   Framework for Air Traffic Management (ATM) Systems

Currently, ATM remains predominantly human centric, with responsibility for operational safety resting on human operators, including strategic flight planners and tactical Air Traffic Controllers (ATCOs). Nunes et al. [19] developed an AI-based support system tailored for tactical ATCOs, aiming to align conflict detection and resolution recommendations (CD&R) with human decision-making strategies and expectations. The system uses images as inputs for a Reinforcement Learning (RL) agent, allowing it to generate a human-interpretable state representation for learning. Project MAHALO [20] studied the role of transparent and conformal AI in CD&R, showing that automation designed to mimic human strategies closely—referred to as conformal automation—is generally more accepted than purely transparent AI. Despite the benefits of such tools in reducing ATCOs workload, human-in-the-loop experiments have highlighted challenges such as decreased situational awareness and issues related to trust and acceptance. Notably, greater transparency did not always resolve these concerns, as an increased flow of information sometimes resulted in delayed responses, which is particularly problematic in time-sensitive operations. In Unmanned Aircraft System Traffic Management (UTM), where automation plays a dominant role due to the absence of human pilots, AI-driven multi-agent path planning has demonstrated significant potential in optimizing flight routes and resolving conflicts [21, 22]. Additionally, some studies have explored implementing adaptive sector structures, aiming to enhance workload distribution among controllers using heuristic and learning-based optimization techniques [23, 24].

1.4   Socio-Technical Systems and Human-AI Interaction

Frameworks addressing the socio-technical dimensions of decision-making are becoming increasingly relevant. The Joint Control Framework (JCF), for example, emphasizes collaborative decision-making between humans and AI systems [25]. It provides methodologies for adaptive control and co-learning, ensuring that AI-driven decisions incorporate human expertise, especially in high-stakes situations where full automation is impractical.

1.5   Trustworthy AI Frameworks

Trust plays a fundamental role in decision-making frameworks for critical infrastructures. The Confiance.AI framework [26, 27] and the Human AI Ethical Framework [28] highlight the necessity of developing AI systems that uphold transparency, ethical standards, and human values. These frameworks establish principles to ensure that AI-driven decision-making systems remain reliable, interpretable, and aligned with societal expectations.

1.6   Risk Management and Regulations

The Assessment List for Trustworthy Artificial Intelligence (ALTAI) tool, the European Union's AI Act, and ISO/IEC 42001 AI management provide a structured methodology for evaluating and mitigating the risks associated with AI deployment in critical infrastructures [29]. These regulatory frameworks facilitate the assessment of both ethical considerations and technical functionalities, ensuring that AI-driven decision processes adhere to stringent safety, accountability, and compliance standards.

All these frameworks offer valuable and different perspectives on integrating AI into decision-making processes within critical infrastructures. However, their domain-specific focus often results in fragmented approaches, addressing isolated challenges rather than providing a unified solution. There is an increasing demand for a comprehensive conceptual framework that accommodates the diverse requirements of critical infrastructures, promoting a holistic decision-making strategy that integrates technical, ethical, and human-centered considerations. The proposed framework results from the European project AI4REALNET (*AI for REAL-world NETwork operation*) and focuses on developing AI-driven decision systems that are not only technically sound but also ethically aligned and context aware, improving both social and technical dimensions of system performance. Furthermore, it covers three distinct levels of interaction between humans and AI: (a) full human control with AI assistance, (b) co-learning where AI and human operators continuously refine their collaboration, and (c) autonomous AI with human supervision.

The design and implementation of this framework are guided by industry-relevant use cases that address key operational challenges network operators face. These use cases, described in the Appendix section, are grounded in real-world operational settings where human oversight plays a crucial role. This framework is intended for use by various stakeholders, including AI developers, innovation managers, network operation managers, regulatory authorities, and standardization organizations.

An interdisciplinary approach is adopted to develop the proposed holistic conceptual framework for decision-making in critical infrastructures by integrating traditionally distinct fields such as psychology and cognitive engineering. The framework also drew on mathematics, decision theory, computer science, and specialized engineering domains, particularly energy and mobility. A high-level overview of the proposed conceptual framework is illustrated in Fig. 1, which also outlines the structure of this chapter.

The systems engineering and theories are adapted for trustworthy AI integration in designing the conceptual framework's operational, functional, and logical

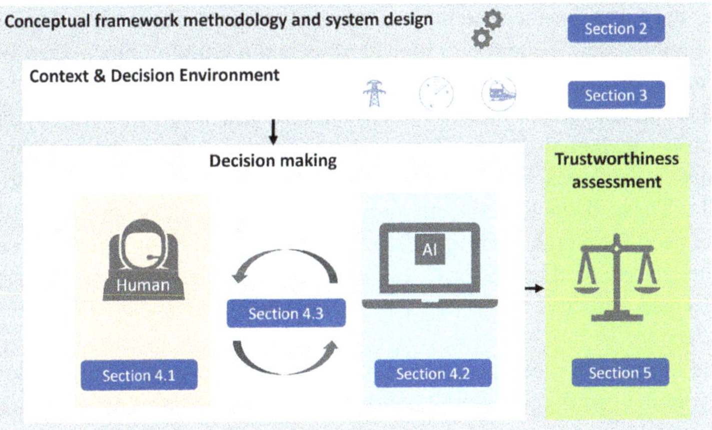

Fig. 1 The proposed conceptual framework, comprising three main components governed by system design principles and their corresponding sections

architectures to meet both functional and non-functional requirements of critical infrastructures (Sect. 2). Based on the observed context and decision environment in a specific industrial domain (Sect. 3), the decision-making process (Sect. 4) is responsible for ensuring the stability and resilience in the corresponding critical infrastructure when an event is triggered. Decisions may be the result of collaboration between human operators and AI-based assistants through a provided interface. As such, the human cognitive process could be increased by AI-based assistants through different perspectives. Furthermore, the decisions must comply with established trustworthy key performance indicators (KPIs) and undergo validation by a regulatory authority (Sect. 5). The subsequent sections provide a detailed explanation of each component of the framework.

2 Conceptual Framework Methodology and System Design

The system design level focuses on the description of the technical system while incorporating the perspectives of stakeholders of the system and the environment in which the system is intended to operate. Recent advances in artificial intelligence enable the development of systems capable of performing increasingly complex tasks, opening new opportunities to assist industries in various sectors while raising important ethical and social issues; the design of these systems is crucial to ensure their effectiveness and acceptance [30, 31]. To cope with the complex environment and tasks the system will operate in and execute, multiple viewpoints (of stakeholders) on the envisioned system are taken to derive both functional and non-functional requirements. It also guides the development process of the proposed framework's submodules and their interactions. Specifically, the system design level described in the following focuses on three distinct views. The operational view captures characteristics of the intended use of the system in a real-world setting. A functional view analyzes the functions the system should be able to provide. Lastly, a logical architecture segments the system into logical units and a building block view outlining the technical structure of the human-AI system.

Further, the AI-based operation of critical infrastructure—i.e., employing a system with a substantial degree of automation operating in an environment where high-impact decisions must be taken in real-time—imposes particular quality demands. These demands concern functional suitability (the provided functions meet the need for a high degree), reliability (a specified performance level is maintained under specified conditions), and operability (understandable, learnable, and usable by and attractive to the user), as well as quality characteristics according to *ISO 25010*.

These initial views, definitions, and considerations outlined in this section play a crucial role in ensuring the overall coherence and alignment of the system between the different aspects during development. Figure 2 summarizes the adopted

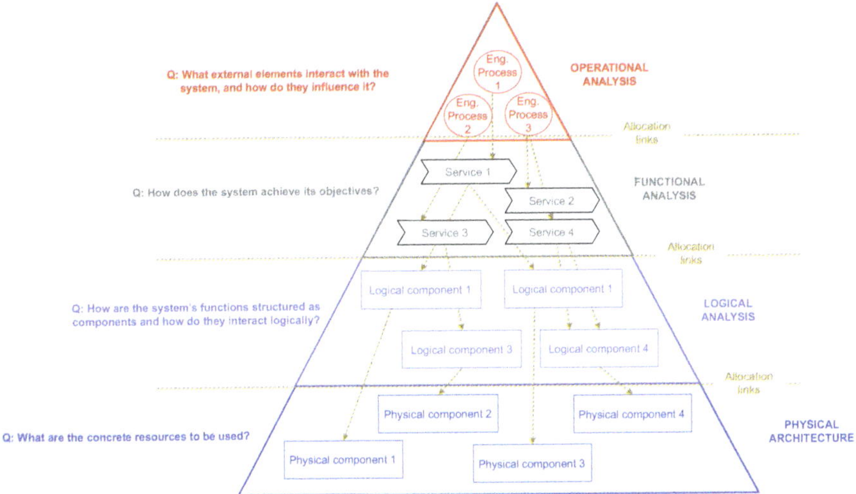

Fig. 2 The Arcadia modeling pyramid used for system design of the conceptual framework

Model-Based System Engineering (MBSE) approach using the ARCADIA method [32]. With the aim of proposing a conceptual framework, the physical architecture at the lowest level of the pyramid is not considered. Each of the operational, functional, and logical analyses with corresponding design processes is described in the following.

2.1 Operational View

The operational view focuses on how the system functions in real-world scenarios. It addresses the operational aspects of the system, including the interactions between users, systems, and external entities, and uses various diagrams to visually represent them. It also allows us to identify the operational, functional, and non-functional requirements.

Using this analysis, various stakeholders involved in the three previously mentioned critical infrastructures are primarily identified. These stakeholders may interact with the proposed frameworks during the different phases of the decision-making process and could be attributed to different roles. For example, an operator could be responsible for implementing a decision and also providing the AI-based system with some feedback at the same time. In the following, these stakeholders and their interactions with the conceptual framework are analyzed. It is followed by the identification of operational requirements and those based on cooperation with AI-based assistance, which may increase productivity and human capacity.

2.1.1 Operational Environment Diagram

An operational environment diagram in system design visually represents the external entities (e.g., stakeholders), interactions, and contexts in which a system operates, helping define its boundaries and interfaces. Figure 3 illustrates these interactions, depicting the system as a black box with data flows between various stakeholders, including human operators, AI-based decision systems, and regulatory agents. This diagram is designed based on the analysis performed on three critical infrastructures.

The *environment* provides real-world context and data, which are utilized by different actors—*data profiles*, for instance, are responsible for training AI decision systems supporting human operators. *Operators* and *supervisors* interact with both the *environment* and *simulators* via the conceptual framework interface, to assess or monitor the impact of their actions before execution in real-world settings. *Secondary actors* as stakeholders in critical infrastructures are those who are not directly involved in the operations but still play significant roles in supporting, influencing, or being affected by these infrastructures. Additionally, operators engage directly with the framework throughout the decision-support process, requesting AI assistance in different operational scenarios. To ensure security and compliance, *regulatory agents* analyze system decisions to verify adherence to established guidelines and *standards*.

2.1.2 Operational Requirements of Three Critical Infrastructures

Operational requirements define the conditions and performance criteria a system must meet to function effectively in its intended environment. These requirements encompass functionality, performance, reliability, security, and user interactions and are specified for each industrial domain and its corresponding scenarios. They are categorized into functional (F) vs. non-functional (N) and generic (G) vs. specific (S) requirements. Generic requirements are directly integrated into the functional view of the framework, while specific requirements are generalized to ensure broader applicability across use cases. Table 1 presents an extract of this classification, aiding in the structured design of the framework's functional architecture.

2.1.3 Human-in-the-Loop and Oversight Requirements

In addition to the operational requirements intrinsic to the domain presented earlier, a set of functional requirements has been identified to enable human-in-the-loop (HITL) decision-making under risk and uncertainty. Table 2 lists these requirements, providing a short description and a categorization based on the part of the system and the actor, when possible, that must adhere to these requirements. The identified actors are *Operator*, the human interacting with the system; *Agent*, the AI-powered side of the decision-making process; and *Environment*, either the true

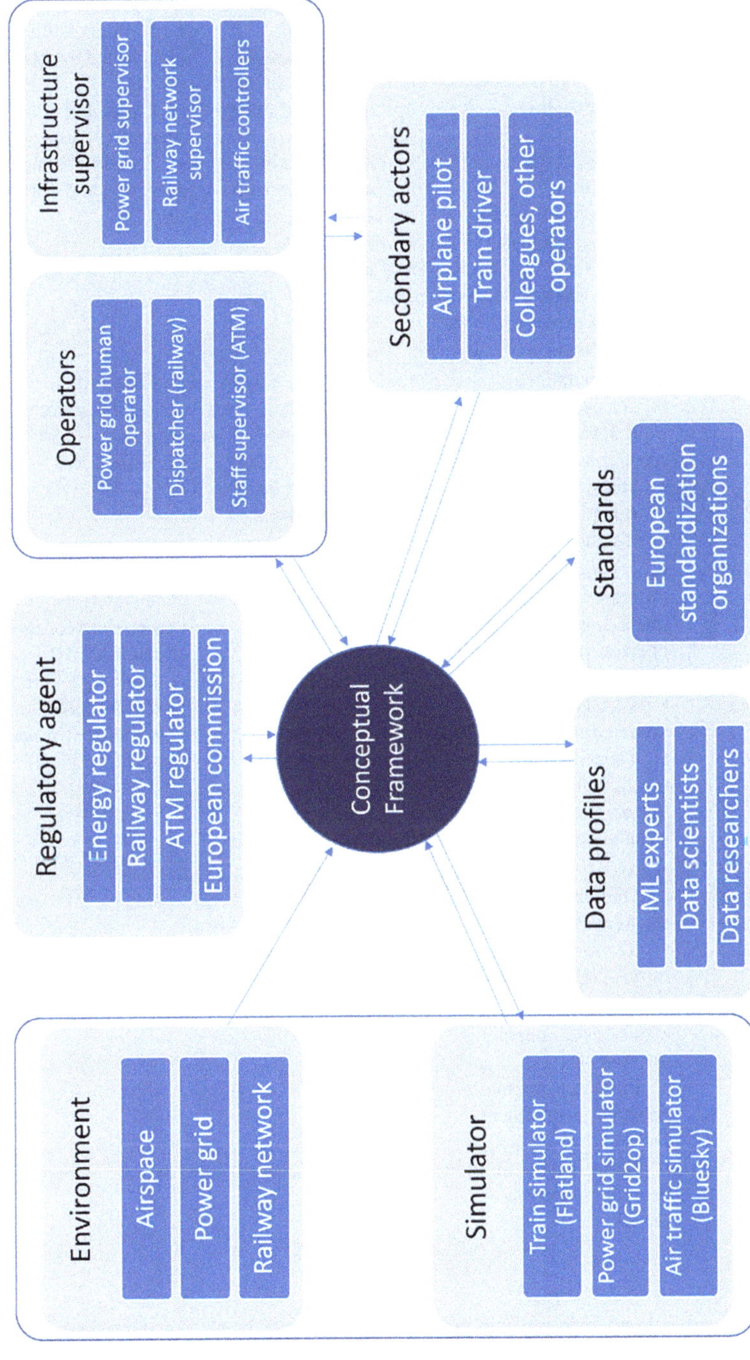

Fig. 3 Operational environment diagram illustrating the external entities, interactions, and contexts in which a system operates

Table 1 Operational requirements for three industrial domains

Category	Power grid	Railway	Air traffic
Robustness	• AI informs the human operator about its confidence in the output recommendation (self-awareness) (F, G) • Reproducibility and traceability of recommendations for *postmortem* analysis (F, G) • Robustness to attacks targeting model space and reward function (N, G)	• Reasonable recommendations in new situations (not seen during model training) (F, G) • Good performance in operating scenarios with high variability (N, G) • Retrospective quality control (N, G)	• System resilience to unexpected events (N, G) • Cyber and data security (N, G) • System's reliable operation and decisions (N, G)
Efficiency	• Computational efficiency (N, G) • Relevance of the recommendations (N, G) • Scalability (N, G) • Adequate training environment (N, G)	• Capacity to handle operating scenarios with high complexity (N, G) • Scalability (N, G) • Generalization to different scenarios (F, G)	• Capability to optimize resources and operations (F, S) • Scalability (N, G)
Interpretability	• Adaptability to different levels of interaction and human operator preferences (F, G) • Transparency during system training (F, G) • Capacity to explain recommendation(s) to the human operator (and other stakeholders) (F, G)	• Interpretability of suggestions (F, G)	• Provide clear, understandable explanations for its decisions (F, G) • Usability of the system from the human and other stakeholders' perspective (N, G)
Non-discrimination and fairness	• Avoid creating or reinforcing unfair bias in the AI system (F, G) • Regular monitoring of fairness (F, S)	• Distribution of delays (N, G)	×
Human agency and oversight	• Additional training about AI for human operators (N, G) • Mitigate addictive behavior from humans (N, G) • Mitigate de-skilling in the human operators (N, G)	×	×

Table 1 (continued)

Category	Power grid	Railway	Air traffic
Regulatory and legal	• Compliance with existing operational policies (N, G) • European AI Act (F, G) • Transparency to humans in terms of interaction with an AI system (N, G)	• Compliance with legal standards and regulations (N, G) • RUOM favoritism (N, S)	• Compliance with legal standards and regulations (N, G)
Data governance	• Processing of human operator data (N, G)	×	×
Accountability	• Allow audits for the AI recommendations and human operator actions (N, S) • Reporting of potential vulnerabilities, risks, or biases (F, G)	×	×
Other	×	• Maintainability (N, G) • Environmental sustainability (N, G)	• Maintainability (N, G) • Environmental sustainability (N, G)

one or the cloned copy accessed by the agent for forecasting and assessment. The identification of these requirements and their association with different actor classes facilitates the decision-making process in the context of the proposed interactive framework.

2.1.4   Abstract Base User Story

The user story allows us to clearly define a system's functionality from the end user's perspective, ensuring that development aligns with user needs and expectations. The user stories were derived during a cross-domain workshop to identify commonalities across analyzed critical infrastructures and guide the development of the AI-decision system, particularly in relation to the human-machine interface.

As can be seen in Fig. 4, the abstract base user story has three manifestations depending on the time horizon: *planning* (preparing for foreseen events), *near real time* (preventing predicted events), and *real time* (correcting events that have occurred). Each manifestation follows a structured sequence: within a given context, a trigger initiates three key actions. First, the situation is observed, either through real-time monitoring, simulation of potential futures, or identification of foreseen events. If a deviation or potential issue is detected, possible measures are explored. Finally, an intervention takes place, where the human operator, using the assistance provided by the AI-based system (depending on the considered

Table 2 Human-in-the-loop and oversight requirements

Categories ID	Category name for requirement	Category description	Actor
R-01	Alarm triggering human	An operator can trigger the alarm to interrupt an execution and step into the decision-making	Operator
R-02	Inspect status	An operator can inspect the system's undergoing situation to observe what has happened and what caused their intervention	Operator
R-03	Provide action	An operator can take action if the suggestions given by the agent are not exhaustive	Operator
R-04	Inspect feedback	An operator can analyze the feedback provided by the autonomous agent to decide whether a recommendation should be followed	Operator
R-05	Estimate epistemic uncertainty	The agent can estimate the epistemic uncertainty of its decision model to establish its level of confidence within an observed state	Agent
R-06	Alarm triggering agent	The agent can raise an alarm to draw human attention. Consequently, the human operator will need to provide some input	Agent
R-07	Simulate remedial plan	An agent can interact with a copy of an environment to simulate a remedial plan and/or provide feedback	Agent
R-08	Provide visual human readable state	An environment should provide a graphical depiction of the ongoing situation to allow human intervention	Environment
R-09	Estimate aleatoric uncertainty	An environment should support the estimation of aleatoric uncertainty derived from an external source, such as weather conditions	Environment
R-10	Raise alarm	The system should have an alarm accessible by a different range of actors to enable a human to intervene	Environment
R-11	Communicate alarm context	On the triggering, the system should provide extensive information about the causes that triggered the alarm	Other

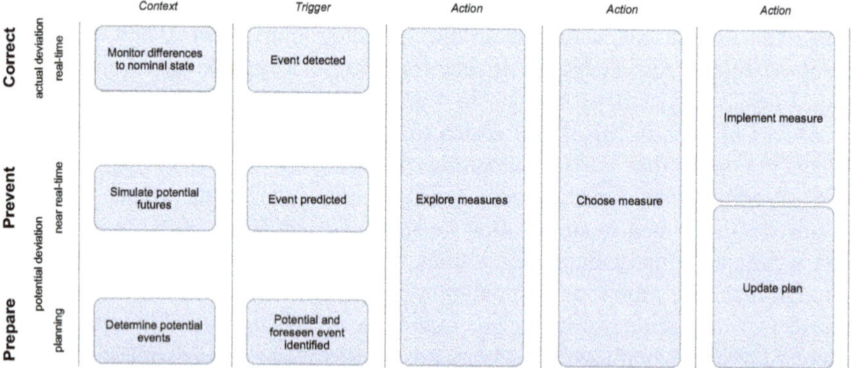

Fig. 4 Abstract base user story

interaction mode), implements a corrective or preventive action or integrates the measure into the operational plan for future execution. This structured approach ensures adaptability across different domains.

2.2   *Functional View*

The functional view answers the following question: "How does the system achieve its objectives?" Thereby, it defines the system functions and shows how system components process inputs, produce outputs, and interact. Based on the functional requirements identified in the operational analysis, the functional view contributes to the definition of system functions that satisfy those requirements through a functional decomposition approach. Next, it establishes how the identified functions interact with each other and with stakeholders while also outlining the associated data flow.

2.2.1   Functional Decomposition Diagram

Based on the identified functional requirements, the functional view outlines eight key functions as shown in Fig. 5. Each of these functions is broken down into subfunctions to ensure effective management of critical infrastructures and is identified based on the three studied industrial domains. These core functions include *interacting with the operator* to facilitate communication and support, *determining realtime context* by analyzing internal and external data, *selecting human-AI interaction*

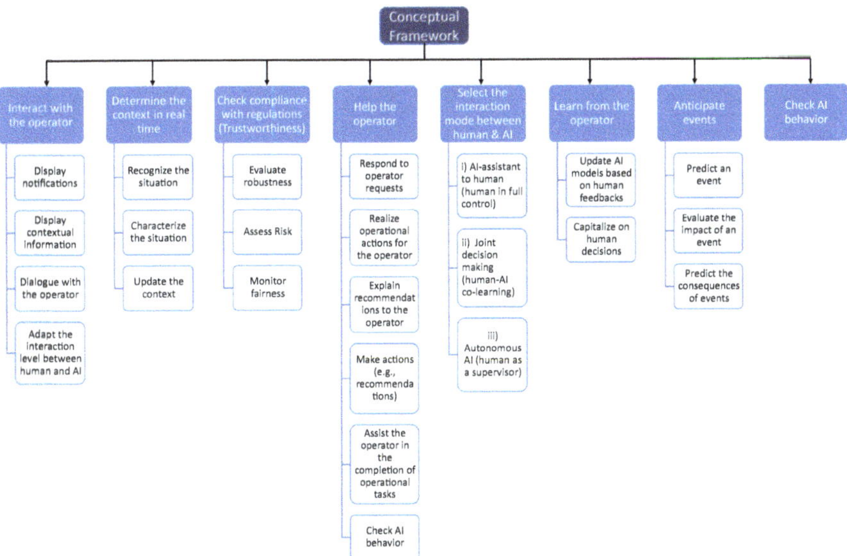

Fig. 5 Functional decomposition diagram

modes to customize collaboration, *assisting the operator* with decision-making, *learning from the operator*'s actions to enhance the system's knowledge base, *anticipating future events* to proactively address potential issues, ensuring *compliance with regulations* to maintain trustworthiness, and *continuously monitoring AI* behavior to detect and respond to anomalies. The next section illustrates the interaction and data flow between these functions and also with the external entities in the context of the human-AI decision-making process.

2.2.2 Functional Interaction Diagram

The functional interaction diagram (Fig. 6) illustrates the interaction and data flow between the identified system functions in the previous section, starting from high-level functions down to elementary ones that directly interact with stakeholders. It integrates the eight main functionalities while also incorporating interactions with the operator, simulator, and regulatory agent. As shown in the diagram, the operator can engage directly with the platform for assistance, select an interaction mode, and provide feedback on recommendations, which is then stored as new knowledge. The operator could simulate the impact of their actions using the simulators. When AI-based assistance is provided to increase the human cognitive process, their behavior is continuously monitored through a set of technical KPIs. A regulatory agent is also responsible for verifying the decision process and its compliance with standards and regulations.

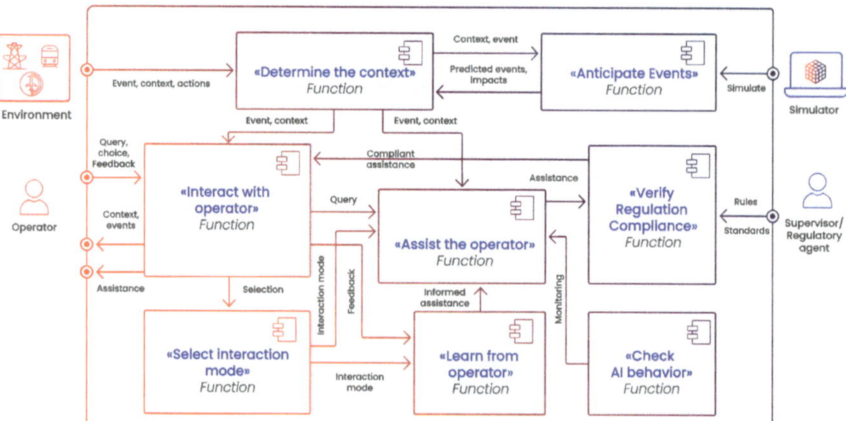

Fig. 6 Functional interaction diagram representing the relationships, dependencies, and information flow between system functions to illustrate how components interact to achieve overall system objectives and allow for determining the data flow among the eight identified key functions

2.3 Logical View

The logical view provides a blueprint for integrating the identified core functions, focusing on seamless interaction among subsystems (representing multiple functions performing a specific task) and stakeholders (internal or external). It also focuses on the abstract structure of the conceptual framework, defining the logical components and their interactions to meet system requirements. This view breaks down the system into three interaction modes: human in full control, joint human-AI decision-making (co-learning), and autonomous AI with human supervision. Each mode emphasizes varying levels of human involvement in decision-making processes, from full authority to collaborative learning and oversight of autonomous AI operations. This decomposition aids in understanding the system's functionality and supports stakeholders by presenting a clear, conceptual model. However, for the purpose of consistency, these three different interaction modes and their corresponding logical architecture are described in the decision-making section and under the human-AI interaction in Sect. 4.3.

In the following, the process view illustrating the generic architecture, including the subsystems and stakeholders, and their interactions, is presented. This generic process is then instantiated for each of the abovementioned interaction modes.

2.3.1 Process View

The generic process is presented in Fig. 7 as an overview of the high-level interaction between different subsystems that encompass one or more functions, identified during the functional analysis, to perform a task (e.g., AI-based agent including *assist the operator* and *learn from operator* functions). As shown in this process view, the *environment*, consisting of both the real-world setting and simulators, provides contextual information to the *human operator* and *AI-based* agent. The

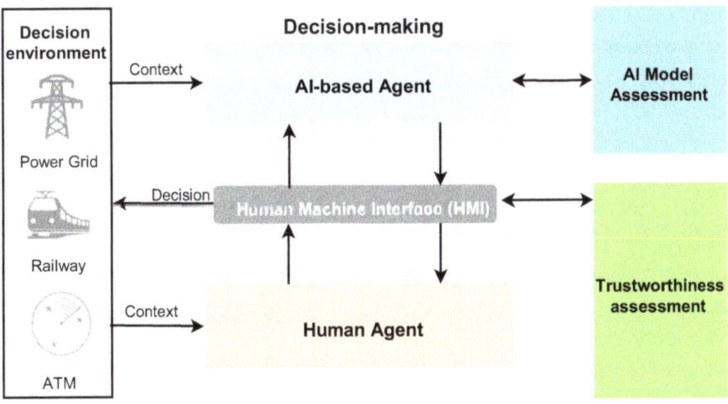

Fig. 7 Process view as an overview of the high-level interaction between different subsystems

AI-based agent assists the human operator via the *human-machine interface* (*HMI*) by offering recommendations or increasing his/her cognitive process, while its behavior is continuously monitored by the *AI Model Assessment* subsystem to ensure robustness and reliability. In return, the human operator can provide feedback, allowing the AI-based agent to refine its performance for future assistance. The interaction should also be conducted in a way to avoid the cognitive load of the human operator (see hypervision definition in Sect. 4.3.3). Finally, the human-AI decision-making process is evaluated through *trustworthiness assessment* KPIs and verified by a regulatory agent for regulatory compliance before being implemented in the decision environment.

2.3.2   Building Block View

As presented in the previous section, the conceptual framework is structured into various layers. At the system design level for human-AI interaction, the focus shifts to translating these layers into practical applications. To enhance the connection between research questions and real-world applications, a high-level conceptual prototype is developed, which allows for testing and refining ideas, ensuring research outcomes meet practical needs. It aims to evolve and serve as initial design guidelines for future applications. This offers a hierarchical representation of the system from a technical perspective. Figure 8 shows the scope, context, and high-level view of the AI-based (conceptual) system.

The system's *context* includes neighboring systems to provide real-time operational information (production information system) and implement decisions taken within the system in live operations (production dispatching system). Further, users, such as operators, supervisors, and regulatory agents, are also part of the context and interact with the system.

In *Level 1*, the system is organized into modules based on function. The Human-Machine-Interaction module manages how AI interacts with humans, providing notifications, contextual information, and assisting with tasks. The Adaptation module recognizes situations, adjusts human-AI interaction, and updates AI models based on feedback. The Prediction module forecasts events, assesses their impact, and stores important data. The Recommendation module suggests actions and explanations to operators, while the Execution module implements operational actions. Lastly, the Assessment module evaluates AI behavior, robustness, and fairness.

In *Level 2*, the Prediction module, central to the system, includes an evaluation submodule, a simulation engine, AI agents, and a digital environment. It receives current system data and requests simulations to predict events and consequences. The simulation results are evaluated and sent to the recommendation module, with all relevant data stored for future use.

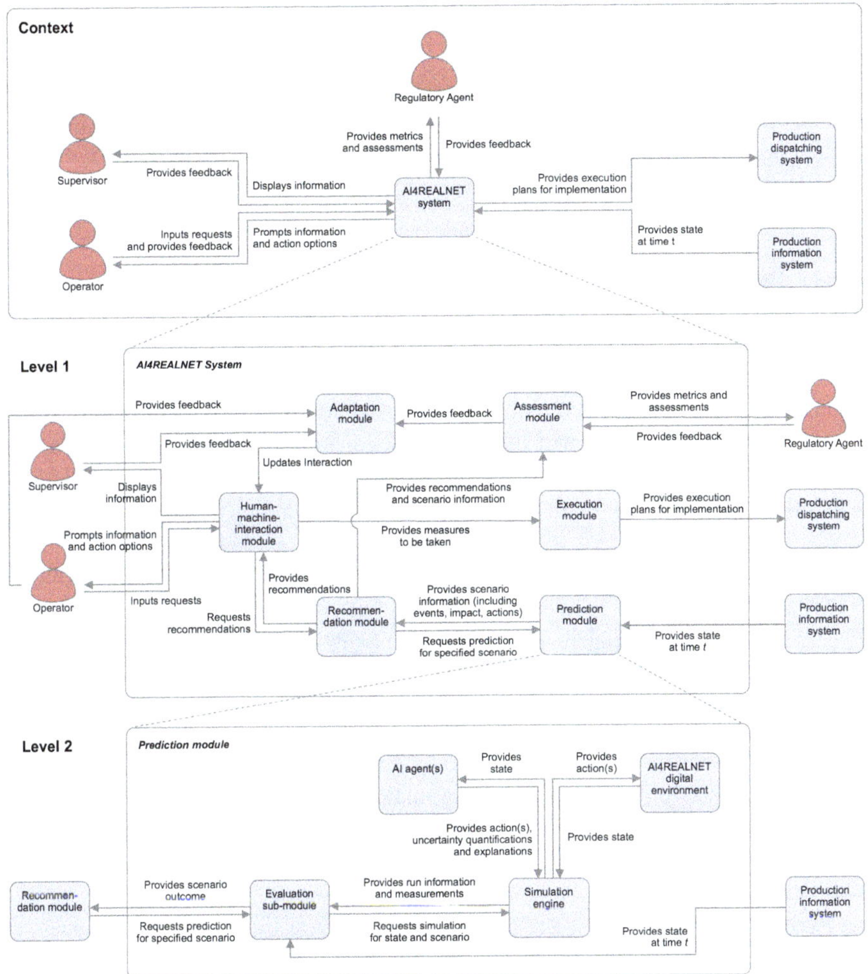

Fig. 8 Hierarchical representation of the systems' building blocks and context

3   Context and Decision Environment in Critical Infrastructures

This section starts with a description of various use cases related to the three considered critical infrastructures, namely, the power grids, the railway network, and air traffic management. Then comes the analysis of context and decision environment related to these critical infrastructures.

3.1   Use Cases Selected by Industry

This section presents the industry-relevant and domain-specific use cases that (a) are focused on critical challenges and tasks of network operators, considering strategic long-term goals, and (b) reproduce real operating scenarios with human operators.

3.1.1   Power Grid

Business Problem   Power grids are evolving due to decarbonization and digitalization, integrating clean energy, electrified demand, and new technologies. These changes increase complexity, requiring enhanced control room supervision. Aging infrastructure, automation, and limited developments further challenge operators, who face cognitive overload due to fragmented multiscreen applications. AI can support decision-making by managing complexity, uncertainty, and workload.

Today's Operations   Power system engineers are highly specialized, requiring thorough studies, accurate planning, and complex decision-making rather than merely following established protocols. They depend significantly on simulation tools, using both real-time and forecast data. However, they have limited access to decision-support tools like automated assistants. When faced with a problem, they manually explore solutions and verify their decisions using simulation tools. They can adjust line connectivity on the grid to redirect power flows, modify (re-dispatch) generation levels, limit consumption by a small percentage, or use battery storage to change power flows in the electrical grid. Despite the range of options, their process relies heavily on experience and manual simulation to determine appropriate remedial measures.

Use Case 1: AI-Assisted Congestion Management

Objectives   The AI assistant aids TSOs in managing electricity transmission while ensuring (a) safe grid operation and overload mitigation; (b) maximized renewable energy use, reducing thermal dispatch; (c) lower workload for human operators; and (d) enhanced explainability and trust in AI recommendations.

Description   The AI assistant monitors the transmission grid using SCADA data and EMS tools, identifying issues and categorizing them for operator intervention. It anticipates congestion, sending alerts with confidence levels while avoiding unnecessary notifications. Action recommendations include topological changes, redispatching, and renewable energy curtailment. Operators review suggestions, explore alternatives, and make final decisions. The AI assistant verifies the decision via load flow calculations and logs interactions for continuous learning.

Human-AI Interaction   Human-AI interaction follows a hybrid approach where the operator retains decision-making authority. The AI assistant continuously learns from operator feedback and refines recommendations.

Use Case 2: Transferring AI from Simulation to Real Grid Operations

Objectives   Evaluates AI deployment in real-world operations, where actual conditions may differ from training environments. The objectives include (a) ensuring AI effectiveness beyond simulated conditions; (b) strengthening human trust in AI-assisted operations; and (c) enabling iterative refinements through operator feedback.

Description   Two operational pathways are explored:

1. Adaptive AI monitoring: The AI assistant continuously assesses the grid, raises alerts, and suggests actions while accounting for uncertainties from noisy or missing data. Operators validate and refine AI recommendations.
2. Human-enhanced AI: When data is incomplete, the AI requests additional input from the operator. Human expertise fills knowledge gaps, refining AI learning and improving recommendations.

In cases where data limitations restrict AI autonomy, it alerts the operator, prompting manual input. The operator's decisions, informed by simulation tools, guide AI forecasting and recommendation assessment.

3.1.2   Railway Network

Business Problem   Growing environmental concerns and evolving mobility policies are driving increased demand for railway network capacity, leading to denser traffic and a greater need for efficiency and resilience in railway traffic management. Addressing these challenges requires either significant infrastructure investments or advanced dispatching technologies. AI-based support systems can enhance dispatchers' capabilities by automating certain decision-making processes and assisting human operators in managing complex scenarios.

Today's Operations   In railway operations, the already densely planned schedules are disturbed by unexpected events, such as delays, infrastructure defects, or short-term maintenance. The execution of the planned timetable can only be achieved by acting on these events with frequent adaptation and re-scheduling of the planned train runs. Today, maintaining smoothly running operations requires that in operational centers, highly skilled personnel monitor the flow of traffic day and night and quickly make re-scheduling decisions. Re-scheduling measures include changing a train's speed, path, or platform. In a densely utilized railway network, local re-scheduling decisions potentially affect the entire flow of traffic, and their effect can propagate far into the future. This means that the re-scheduling task is a complex decision-making task that must integrate much context information under time constraints.

Use Case 1: Automated AI-Based Re-scheduling

Objectives The system aims to fully automate railway re-scheduling, ensuring all services are fulfilled while minimizing passenger delays.

Description AI-driven systems dynamically adapt schedules in response to disruptions, such as infrastructure failures or delays. The system evaluates multiple interventions, including modifying train speeds, reordering train sequences, rerouting trains, or adjusting station platforms. This AI assistant continuously optimizes railway schedules in real time, ensuring efficient network use while reducing delays. Human operators oversee the system, adjusting configurations and identifying the need for updates.

Human-AI Interaction The AI system autonomously performs re-scheduling, monitoring real-time train and track states. It identifies required interventions, implements changes, and evaluates outcomes. Since fully automated railway rescheduling is a novel concept, the AI system is introduced as a supervised tool. Human experts evaluate its performance, ensuring reliability and effectiveness. Operators monitor key indicators, including:

- Traffic conditions (e.g., train counts, network bottlenecks)
- Key performance indicators (e.g., current delay statistics)
- AI confidence levels in its recommendations
- The extent of AI-driven interventions (e.g., platform changes)

Based on these insights, human supervisors decide whether to override AI recommendations, reconfigure system parameters, or adjust the AI model.

Use Case 2: AI-Assisted Human Re-scheduling

Objectives This approach leverages AI-based tools to assist human dispatchers in re-scheduling, ensuring all services are maintained with minimal delays.

Description The AI assistant continuously analyzes real-time train and track data, identifying optimal re-scheduling options in response to disruptions. These recommendations are presented to dispatchers in near real time, allowing them to make informed decisions. The system also anticipates future disruptions, assessing their impact on network stability. The AI assistant enhances human decision-making by providing predictive insights, enabling faster and more efficient responses to deviations from planned schedules.

Human-AI Interaction The AI assistant generates re-scheduling options based on real-time data. Dispatchers can:

- Select an AI-recommended action
- Request alternative solutions
- Override AI suggestions and implement their own decisions

Additionally, the AI system supports human learning by providing contextual evidence for or against specific re-scheduling hypotheses. Supervisors continuously assess the AI's performance, refining decision criteria and system parameters as needed.

3.1.3 Air Traffic Management

Business Problem Air traffic density in European airspaces is rising, driven by economic and environmental concerns that push for time- and trajectory-based operations. Increased traffic and operational complexities may lead to excessive tactical air traffic controller (ATCO) workload, threatening ATM system safety and efficiency. Military airspace activations in regions like the Lisbon FIR further restrict upper airspace usage, requiring frequent traffic deviations.

Today's Operations Today, sectorization is the sole responsibility of the ATC supervisor, who exclusively decides when and how to split and merge sectors best, warranted by situational demands and available ATCO personnel. Only scattered information is available on different platforms to aid ATC supervisors in this task. Still, there is no traffic pre-analysis tool and/or integrated decision-support system to assist in, or even fully automate, the structuring of sectors with trajectory efficient routes (e.g., flight time and fuel burn) and sectorizations to keep the workload of the ATCO within acceptable thresholds, i.e., without exceeding sector capacity limits.

Use Case 1: AI-Assisted Airspace Sectorization

Objectives Automate sectorization decisions to assist or replace ATC supervisors in balancing ATCO workload while ensuring safe and efficient traffic flows.

Description ATC supervisors manually determine how to split and merge airspace sectors based on real-time demands. Sectorization involves horizontal (2D) and/or vertical (altitude) adjustments, often using predefined configurations. However, unexpected events—such as adverse weather, flight emergencies, or ATCO shortages—require non-standard sectorizations. An AI assistant provides recommendations or automates decisions, optimizing horizontal and vertical sector configurations to balance workload while ensuring safety.

Human-AI Interaction The AI system continuously monitors real-time air traffic data, predicts optimal sectorization adjustments, and presents recommendations. Human operators can:

- Evaluate AI-generated recommendations
- Request additional information or explanations
- Accept, reject, or manually adjust AI advisories

Automation levels range from advisory support (where AI suggests actions) to full automation, where AI executes decisions unless overridden by human operators. Logged decisions and interactions enable continuous AI learning and adaptation.

Use Case 2: AI-Assisted Flow and Airspace Management

Objectives Optimize sector capacity and environmental efficiency when military airspace is activated by recommending deviations with minimal impact on sector workload and flight efficiency.

Description When military airspace is activated, ATCOs must reroute flights to avoid restricted areas, impacting sector capacity and fuel efficiency. An AI assistant supports ATC and Flow Management Position (FMP) supervisors by analyzing airspace configurations and suggesting optimal traffic flow adjustments. The system dynamically optimizes routing decisions to minimize en route inefficiencies while maintaining adherence to sector workload limits.

Human-AI Interaction The AI system observes real-time airspace data, forecasts traffic flow disruptions, and recommends airspace and routing adjustments. Human operators can:

- Review AI-generated routing and sectorization recommendations
- Provide feedback and nudge AI-generated decisions
- Modify AI-based configurations based on operational constraints

At lower automation levels, AI assists supervisors by providing recommendations, while higher automation levels allow AI to autonomously execute sectorization adjustments unless overridden. The AI system continuously learns from human feedback to refine its recommendations.

3.2   Decision Environment

Decisions in critical network infrastructure operations are at the heart of operational processes. They can be described in three main points (see Fig. 9). Firstly, they are made to manage constraints on a network capacity that can stem from external events (operational disruptions or emergencies). These events are detected through

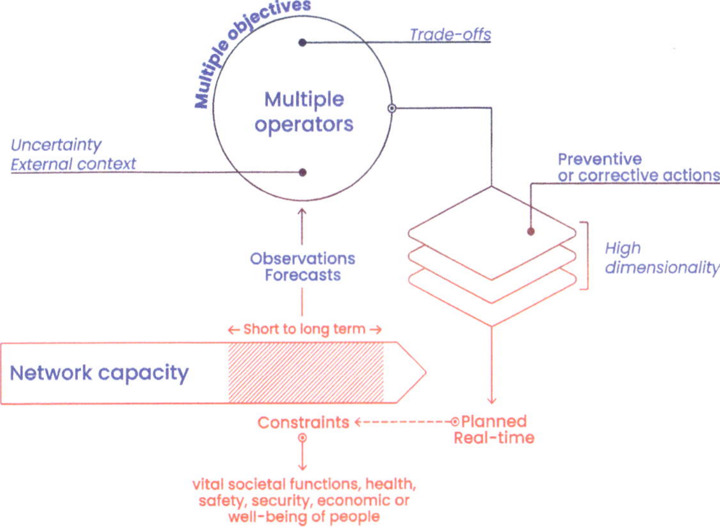

Fig. 9 Decisions in critical network infrastructure operations

observations or forecasts of the state of the infrastructure that include a certain level of uncertainty and an external context. Secondly, they involve multiple operators or stakeholders from short to long-term horizons. Lastly, they are made under time constraints and trade-offs between multiple and conflicting objectives and lead to both preventive and corrective actions that are chosen within a large action space and are planned or implemented in real time, respectively.

The criticality of the decisions is directly linked to the critical nature of the underlying infrastructure for ensuring vital societal functions, health, safety, security, economic, or well-being of people, namely:

> *European critical infrastructure means an asset, system or part thereof located on EU territory, which is essential for the maintenance of vital societal functions, health, safety, security, economic or well-being of people, and the disruption or destruction of which would have a significant impact on at least two Member States, as result of the failure to maintain those functions. The significance of the impact is assessed against distinct cross-cutting criteria, which encompass casualties and economic, environmental, and public effects.*

Three industrial domains are considered in this study to extract generic aspects of decisions across critical infrastructures: power grid management, air traffic management, and railway network management. To provide a better description of the decision process from a business perspective, an analysis was performed on data collected using a detailed questionnaire for each domain. This questionnaire is structured into four main topics: context, characteristics, impacts, and evaluation of decisions. Based on all data collected, a similarity score was computed across pairs of domains to give an idea of the commonalities between the three domains. The results are shown in Table 3.

Table 3 Decision analysis comparison across domains

Decision analysis	Air Traffic-Electricity	Electricity-Railway	Railway-Air Traffic
Context	13%	13%	12%
Characteristics	40%	50%	50%
Impacts	0%	4%	0%
Evaluation (KPIs)	23%	38%	46%

Even if the decision context is different for each domain (which can be explained by the fact that each domain remains intrinsically different), it could be observed that the characteristics of the decision have a higher degree of similarity, which is of the same level of magnitude across the different pairs of domains. This illustrates the interest in performing multi-domain work for the development of a holistic conceptual framework.

Then, the second highest similarity is obtained on the evaluation part, which defines a set of KPIs for each domain. The following KPIs were similar across all domains: assistance, relevance, and trust in the AI system. Within the multidomain work, this shows the interest in the evaluation that will be carried out in the future. On the other hand, the level of similarity across domains is almost zero for the "impact of a decision" topic: this can be explained by the domain-specific impacts of each decision. In line with the similarity scores obtained for the "impact of a decision" topic, there are no similar words across all domains for this topic. Finally, as could be observed, the two most similar pairs of domains are "Railway-Air Traffic" (highest) and "Electricity-Railway." The following section introduces the decision-making process, which is mainly based on observations and analyses of the context.

4 Decision-Making Process in Critical Infrastructures

This section presents the core concept of the conceptual framework, which is the decision-making process in critical systems. It involves a dynamic interplay between human expertise, AI-based decision-making capabilities, and their collaborative interaction. Based on the decision context introduced in the previous section, three key dimensions of decision-making process are explored in the following: human decision-making, which leverages domain knowledge and intuition; AI-based decision-making, which provides data-driven insights and scalability; and human-AI interaction, which integrates these strengths to optimize decisions under complex and uncertain conditions.

4.1   Human Agent and Decision-Making

This section presents the conceptual framework from a socio-technical systems perspective, emphasizing that work systems consist of interacting social and technical subsystems, requiring joint optimization for overall performance. Optimizing one subsystem in isolation may reduce effectiveness due to its interdependence. This leads to two key conclusions: AI design must consider human factors, and the social subsystem—including skills, task design, and organizational culture—must be adapted to effectively utilize AI. For instance, if AI provides recommendations, human decision-makers need both the skills and the autonomy to evaluate them without fear of blame, which could otherwise lead to blind acceptance. While socio-technical design principles are well established [33], further research is needed on AI integration [34–36].

Against this background, three aspects of AI integration in socio-technical systems are examined in the following sections, namely, different design approaches and their effects on human behavior, normative aspects of AI design, and descriptive aspects of AI design. Most of the use cases described in Sect. 3.1 involve interaction between a human operator and an AI agent, making the analysis of human decision-making behavior essential to effectively addressing their objectives.

4.1.1   Different Design Approaches and Their Effects on Human Behavior

The integration of AI into socio-technical systems, particularly in critical infrastructures such as air traffic management (ATM), requires careful design to balance automation and human involvement. Two primary approaches involve *automate*, which replaces human effort, and *informate*, which enhances human capabilities. Full automation is impractical for complex tasks, often leading to unintended negative consequences such as operator fatigue, loss of skills, and overreliance on technology. To maintain system resilience, AI must support human flexibility and decision-making by enhancing their cognitive process rather than relegating humans to passive oversight roles. Research emphasizes the need for AI-human collaboration, where AI design fosters human engagement, autonomy, and motivation. Ensuring that AI systems empower rather than undermine human involvement is crucial for effective performance and safety.

4.1.2   Normative Aspects

Miller [37] outlines and compares five types of explainable AI (XAI) frameworks that support decision-making in human-AI collaboration, ranging from simple AI recommendations to more interactive models where AI explains, interprets, or evaluates human decisions (Evaluative AI). These types influence *human decision-making, motivation, learning*, and *trust* in AI. In the following, their impact is examined through the scenarios considered for interaction between humans and AI: AI assisting humans, joint human-AI decision-making, and autonomous AI with human supervision.

Human Decision-Making

AI-based decision-support systems primarily rely on recommendations, but research indicates that recommendations alone—even when supplemented with explanations—are often insufficient for effective decision-making [37–39]. Instead, joint human-AI decision-making, which leverages the complementary strengths of both, is necessary [34]. Effective decision-making requires an understanding of human cognitive processes, including biases like the anchoring effect and confirmation bias [38, 40]. Human decision-making is not merely a selection of predefined options but a complex cognitive process that involves sense-making, anticipation, and attention management [41, 42]. To support these cognitive functions, AI should facilitate knowledge of operational processes, process monitoring, situation awareness, and bias mitigation. However, current AI systems often fail to sufficiently support these elements, impacting human performance and decision quality [41, 43].

A simplified representation of human decision-making is illustrated in Figure 10, where the human cognitive processes are summarized as monitoring (observation of the current context or situation); building situation awareness as suggested by Endsley [44], which is to be able to understand the current situation and project to a

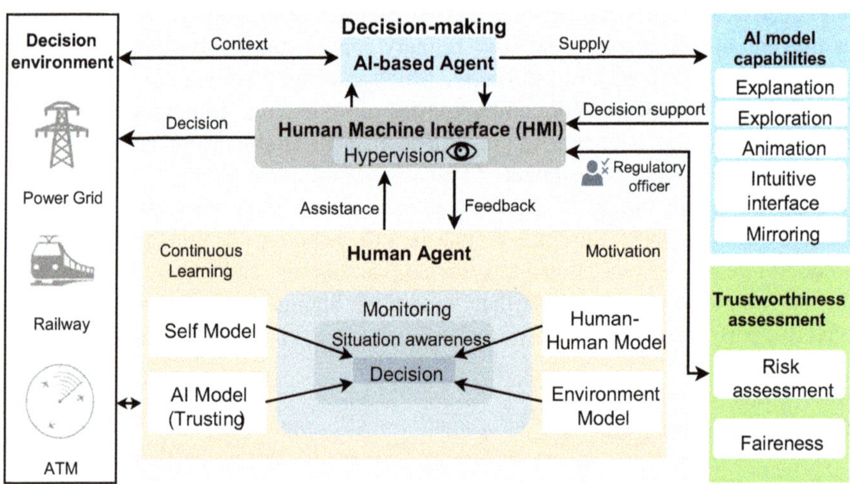

Fig. 10 Decision-making from a human perspective in the proposed conceptual framework involving monitoring, situation awareness, and the decision itself. Human cognitive processes required for decision-making may be supported by AI through explanation, exploration, animation, and mirroring

future state; and finally the actual decision. Beyond the transparency required in AI recommendations (interpretability and explainability), AI can support human cognitive decision-making processes by allowing humans to explore and test their assumptions as well as the implications of their decisions. AI can also support monitoring by requiring humans to animate in detected unusual patterns and mirror the individualized patterns in human behavior to make humans aware of their own biases and variabilities in decision-making. These observations show that AI should support humans in various cognitive processes rather than just making simple recommendations. These are mainly required for the AI-assisted human decision-making use cases, such as AI-assisted congestion management in power grids, AI-assisted human rescheduling for railway networks, and AI-assisted airspace sectorization for ATM.

Human Motivation

Intrinsic motivation is crucial for the effective use of AI, as algorithm aversion and reluctance to adopt IT tools are common [45–47]. AI has an impact on humans' tasks and therefore on their motivation [43]. Task design significantly impacts motivation, which should support the following three critical mental states necessary for intrinsic motivation: meaningfulness, autonomy, and feedback [48]. AI should be designed to enhance these aspects by providing clear explanations, enabling genuine choices, and offering constructive feedback to human, inciting both good work and their usage. In decision-making scenarios, AI recommendations without explanations reduce user engagement, whereas interpretable models and explained recommendations improve transparency and task-related motivation. More advanced AI approaches, such as cognitive forcing and evaluative AI [37], further enhance motivation by increasing user involvement and responsibility in decision-making.

The proposed conceptual framework enables the exploration of different AI-human collaboration models, each affecting intrinsic motivation differently. In an AI-assistant scenario, where AI provides recommendations with or without explanations, user motivation is often low due to a lack of deep understanding, limited autonomy, and missing feedback. Joint human-AI decision-making, including cognitive forcing and evaluative AI, fosters intrinsic motivation by involving users in the decision process, enhancing their sense of responsibility and effectiveness. Conversely, fully autonomous AI, where humans act only as supervisors, poses challenges for motivation and effectiveness due to limited involvement, aligning with [49] irony of automation. In general, AI must be designed to support meaningfulness, autonomy, and feedback to enhance human motivation and performance in AI-assisted tasks.

Human Learning

In addition to the information that could be observed by monitoring and enhanced using situation awareness for decision-making, the human knowledge that may be gained through experience is more enduring. This knowledge could be used as a basis for monitoring and decision-making. Polanyi [50] proposes the distinction between explicit (transferable to others) and tacit (implicit and not transferable) knowledge. This distinction is important because tacit knowledge is the main source of decisions made by domain experts. It may also be interdependent to other individuals' knowledge, forming a community [51]. In this sense, it is important that AI supports humans to earn tacit knowledge through experience.

Four different domains of human knowledge are depicted in Fig. 10 as human mental models, which are the environment model (knowledge about decision context), human-human model (interrelation between the different individuals' work), AI model (knowledge about their capabilities and limitations which could induce the human trust), and self-model (human knowledge about themselves). Human learning is a multifaceted process encompassing psychological, physical, and social dimensions, shaping how individuals perceive and interact with the world [52]. Based on Kolb's [53] Experiential Learning Theory, describing learning as a cyclical process involving four stages (i.e., concrete experience, reflective observation, abstract conceptualization, and active experimentation), AI should also provide the opportunity for the human to learn through exploration, animation, and mirroring. This could be helpful for all the use cases of critical infrastructures (see the previous section), where the human intervention as a supervisor is enhanced using the AI assistance.

Human Trust

Human trust in AI is a dynamic relationship influenced by beliefs, knowledge, emotions, and experiences with the AI and with their mental AI model as shown in Fig. 10 [54, 55]. Unlike trustworthiness, which is a static attribute of a particular AI, trust evolves with repeated interactions [56]. As outlined by Parasuraman and Riley [57], appropriate trust requires users to understand AI's capabilities and limitations to avoid both overreliance (where human blindly approves the AI recommendation) and under-utilization (not having the opportunity to experience and develop trust). On the other hand, the transparency of AI models should not only provide explanations but also allow users to explore and test AI functionality through direct experience and develop the appropriate trust accordingly [58].

The trust is examined across three previously mentioned human-AI collaboration scenarios considered in the conceptual framework. The AI-assistant to human model provides recommendations but limits user involvement, often leading to overreliance or rejection of AI decisions [41, 49]. Joint human-AI decision-making, which integrates cognitive forcing and Evaluative AI [37], fosters trust by increasing transparency, interaction, and feedback opportunities. Autonomous AI, where humans act as supervisors, presents the greatest challenge for trust development due to minimal human engagement. Ultimately, fostering appropriate trust requires AI systems to prioritize transparency, interactive exploration, and feedback mechanisms, ensuring that trust is informed by hands-on experience and a clear understanding of AI's strengths and limitations.

Acceptance

Initial acceptance of new technology is crucial for its successful deployment, as rejection can occur even before use, creating a paradox where trust develops through experience but may not form without initial engagement. Research from sociology, psychology, and information systems highlights human-technology compatibility as a key factor in overcoming this hurdle, with acceptance increasing when a system aligns with users' values, experiences, and needs. The compatibility exists at various cognitive levels, from basic handling to decision-making strategies. Studies in air traffic control (ATC) show that strategic conformance—where AI solutions resemble human decision-making—enhances acceptance, as seen in higher approval of personalized recommendation systems over generalized ones. However, while strategic conformity aids initial adoption, its long-term benefits may diminish with prolonged human-AI interaction [59].

4.2 AI-Based Decision-Making

AI-based decision-making is increasingly transforming the landscape of human-AI collaboration, offering unprecedented capabilities in processing complex data, identifying patterns, and generating insights that surpass human cognitive limits. In the context of human-AI decision-making, AI systems can augment human judgment by providing data-driven recommendations, enhancing efficiency, and reducing bias in critical decisions. However, this synergy also brings challenges, including the need for transparency, trust, and ethical considerations to ensure that AI supports, rather than undermines, human autonomy and values. Balancing the strengths of AI with human intuition and expertise is essential to harness the full potential of AI-based decision-making responsibly and effectively.

Figure 11 presents the logical view of human-AI decision-making, emphasizing the role of the AI-based agent. Both AI and human agents observe the context and decision environment, but the AI-based agent excels in processing and analyzing complex contextual information in real-time—capabilities beyond human capacity. Leveraging various strategies and methods, such as knowledge-assisted AI, meta-awareness for AI assistants, human-AI co-learning, and multi-objective learning, the AI-based agent anticipates events and provides recommendations to the human agent. For effective integration into the proposed conceptual framework, AI-based models must exhibit specific characteristics that facilitate seamless interactions

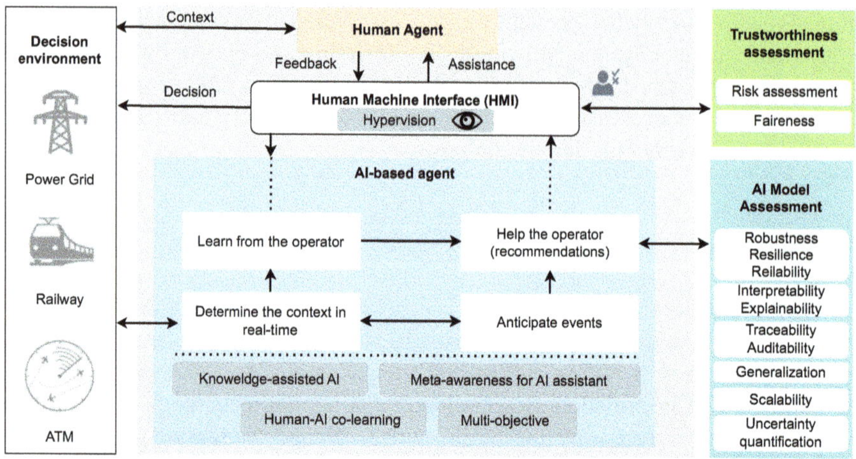

Fig. 11 Decision-making from an AI perspective in the proposed conceptual framework

between AI and human decision-makers across different scenarios and interaction modes. Additionally, the AI-based agent should continuously learn from human feedback and incorporate it into future recommendations. Ultimately, any decision must align with predefined trustworthiness KPIs to ensure reliable implementation in the decision environment. The following sections provide a detailed discussion of the AI-based agent's characteristics, along with the methodological and algorithmic aspects of AI-based models.

4.2.1 Characteristics of AI-Based Decision-Making

In the context of human-AI decision-making process of a proposed conceptual framework, AI systems can augment human judgment by providing data-driven recommendations or by increasing their cognitive process. This helps enhance efficiency and reduce bias in critical decisions. However, this synergy also brings challenges for which AI-based assistance should adhere to a set of properties (see Fig. 11).

They should be robust and resilient to perturbations and maintain reliability with respect to what they have learned during the training phase. Their assistance must be explainable to human operators, with an architecture that is easily interpretable. Additionally, generalization and scalability are essential for deploying these solutions in real-world conditions. Lastly, their assistance should be accompanied by clear indications of uncertainty levels, increasing the trust of the human operator. These properties are discussed in greater detail in the following sections.

Robustness, Reliability, and Resilience

Robustness in AI can be examined from two perspectives: technical robustness and social robustness. *Technical robustness* refers to a system's ability to sustain its performance despite natural or adversarial perturbations (ISO/IEC 24029-2). It can be assessed locally (for specific inputs) or globally (across all inputs). Evaluation methods include measuring sensitivity—how output changes with input variations—or introducing adversarial perturbations to test system stability [60]. Metrics like output variance or reward function deterioration can quantify robustness. Furthermore, different learning strategies may impact the robustness, e.g., online learning impacts robustness, requiring continuous test-time monitoring. *Social robustness*, on the other hand, ensures that the AI system duly considers the context and environment in which it operates, guided by frameworks like ALTAI. Digital environments associated with each critical infrastructure may help assess social impacts, such as carbon emissions reduction.

Reliability in AI refers to its ability to perform as expected, even with novel inputs (EU-U.S. Terminology and Taxonomy for AI). Unlike robustness, which considers AI performance under disruptions, reliability focuses on stable performance within a consistent data distribution [61]. It relates closely to generalization, ensuring AI models perform similarly across different test datasets. Estimating epistemic uncertainty can also help assess reliability by correlating performance with uncertainty levels. Reliable AI systems should demonstrate consistent results across standard operational conditions, independent of external disturbances.

Resilience is the AI system's ability to prepare for, withstand, and recover from unexpected perturbations or attacks (EU-US Terminology and Taxonomy for AI). Its quantification is related to the magnitude and/or duration of reward/loss function performance degradation compared to an unperturbed system for the same context. Figure 12 depicts a conceptual definition of resilience quantification for a reward function. In this scheme, resilience can be quantified by (a) the gray area between the reward curves of the unperturbed and perturbed AI system, (b) the minimum reward in the degradation state and the maximum reward in the restorative state, and (c) the duration of the degradation and restorative stages. These metrics should be

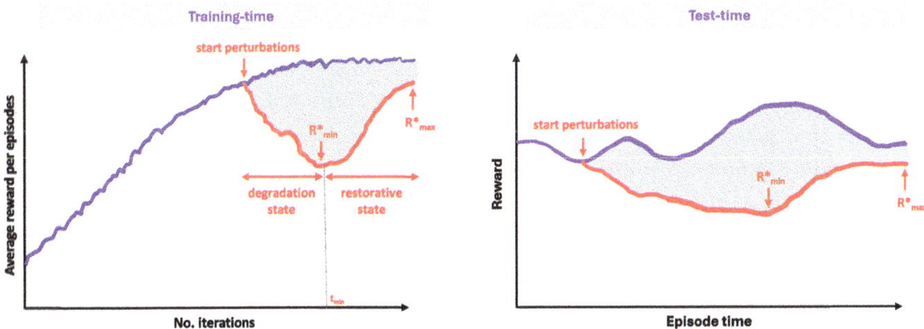

Fig. 12 Concept of resilience quantification in training-time and test-time phase

computed for different probability levels of the perturbation or by defining a maximum number of perturbations or a perturbation budget.

Interpretability and Explainability

Explainability measures the capability of a human user to understand how models make predictions or decisions, where the model's transparency (interpretability) is a way to support explainability [37]. According to Molnar [62], explainability is contextualized by a specific input, and it often requires additional information, which the decision model does not generally generate. XAI techniques refer to the set of methods that aim to generate local explanations for black-box models, e.g., artificial neural networks predictions.

To ensure a thorough evaluation, Vouros [63] proposes a set of additional human-related metrics evaluating the interaction between a human and the explanation. These metrics focus on the effectiveness of the explanation from a user's perspective. The metrics relevant to explainability include:

- Comprehensibility: assessing the capacity of a human to understand an explanation
- Preferability: estimating the relevance of an explanation given to the user
- Cognitive load: estimating the cognitive effort required by a human to appreciate and comprehend the provided explanation
- Actionability: assessing the utility of an explanation by capturing how well an explanation enables end users to make informed decisions

Considering these metrics from an early stage of development allows for the design of a human-friendly AI, where humans can be in control as it enables users to understand the system's decision.

Traceability and Auditability

Traceability refers to the process of establishing a clear and direct connection between the stakeholders' requirements and the product developed in the context of software development. When applied to AI systems, this connection covers the design elements, code implementation, test scenarios, and data used to train the system. To ensure continuous traceability of RL in human-centered AI, the human-machine interaction, along with the corresponding context, must be logged to trace the human influence in future decisions.

Auditability consists of a thorough analysis of data, algorithms, and design processes to ensure alignment with the desired objectives, standards, and legal and technical requirements, such as those outlined by the European Commission [64]. This concept is pivotal in building human trust in AI systems. One might argue that, when deploying AI into real-world systems, auditability is as important as model performance.

In the proposed conceptual framework, auditability and traceability serve three main purposes. First, they ensure human control and safe operation by detecting changes in RL elements, maintaining task consistency, and preventing external interference. Second, they support regulatory compliance, monitoring, and inspection of AI and human inputs. Third, they enable tracking of both AI recommendations and human influence on decisions. Additionally, they help identify performance issues, improving system maintainability.

Generalization

Generalization, in the context of AI, is the ability of a trained model to perform well on previously unseen data that is derived from the same distribution as the data explored during the training phase. For RL, Nichol et al. [65] consider an agent to generalize well when it can adapt to previously unencountered situations drawn from the same task explored during the training phase, such as varying difficulty levels in a simulation or different configurations in an operational environment. As in RL, the policy is optimized based on a reward signal, which is often formulated ad hoc for specific scenarios. The generalization capability may be limited by the overfitting problem [66], where an agent is trained to maximize the cumulative reward in a fixed scenario. The trade-off between generalization and overfitting is still an open challenge [67]. The problem of generalization in RL should be addressed on three different levels: domain diversity (exploration of heterogeneous environment configurations), exploration-exploitation trade-off (find a balance to optimize on various experiences to prevent overfitting) [68], and finally reward engineering (ensuring that reward formulation is not overly tailored to a specific scenario).

In the context of the proposed conceptual framework, it is required to ensure that the AI model is robust and capable of generalizing across diverse scenarios. Where possible, the trained model should maintain its performance in unseen and out-of-distribution scenarios, such as areas of observation and action space not visited during training. If this is not feasible, the AI agent must alert the operator about its uncertainty. The generalization capability of a model could be measured by observing the change in the reward/loss of the model when visiting novel data.

Scalability

In AI, scalability is the ability of a model to adapt to different workloads, similar to the algorithm's scalability as outlined by Paliouras [69] and Ulanov et al. [70], where they assess the scalability of an AI distributed model by measuring the empirical speedup obtained from a system while increasing the computational resources.

The conceptual framework describes an RL-powered AI agent supporting human operators working in control rooms to maintain a critical infrastructure operational. When a disruption occurs, operators usually have a tight time frame to analyze the situation and devise a remedial plan to return to safe conditions. Consequently, the computational overhead of the AI agent, as well as the additional workload resulting from its suggestions to the operator, must be kept minimal regardless of the scale and complexity of the problem.

As a result, the scalability challenge is twofold. On one hand, from an engineering perspective, an AI decision-making model should scale based on hardware availability. On the other hand, from a theoretical RL perspective, the system's effectiveness and performance should not be compromised by the complexity resulting from the combinatorial nature of Multi-Agent RL (MARL) with an arbitrary number of agents [71]. The combinatorial complexity can be mitigated by factorizing the learning process across multiple agents, enabling each agent to observe and make decisions simultaneously within a shared environment.

Uncertainty Quantification

Uncertainty quantification (UQ) systematically characterizes and manages uncertainties inherent in both AI models and real-world data, and it is crucial when an AI agent supports the decision-making of a human operator for critical infrastructures [72, 73]. In human-AI interactions, UQ ensures decisions are reliable and robust by addressing different aspects of uncertainty. Epistemic uncertainty arises from insufficient training data, leading to gaps in the model's knowledge and limiting its ability to generalize to unseen scenarios. Quantifying epistemic uncertainty helps humans interacting with an AI system better understand AI limitations, which is crucial for the conceptual framework integrating a human-in-the-loop strategy. Aleatoric uncertainty is intrinsic to the data, as it stems from noise, incompleteness, or inaccuracies in input data. Since the proposed conceptual framework is designed for critical infrastructures, environments should support aleatoric uncertainty estimation, particularly from external sources unobservable by an AI agent, such as weather conditions, to ensure more reliable AI-assisted decision-making.

Beyond these technical dimensions, UQ plays a crucial role in decision-making under uncertainty, offering probabilistic frameworks to evaluate risks and outcomes for critical infrastructures. This helps human operators make informed decisions despite uncertainty. Human-AI collaboration is also enhanced, as transparent uncertainty metrics allow humans to interpret AI predictions with greater awareness of their limitations. For instance, AI-generated decisions are augmented with confidence levels in power grid use cases, ensuring that human operators understand the system's constraints. Finally, UQ contributes to resilience and reliability by ensuring AI systems are not only accurate but also aware of their limitations. This awareness leads to more robust designs and operational strategies, improving the overall dependability of critical infrastructures.

4.2.2 AI-Based Methods and Strategies

This section explores four key methodologies and strategies employed in the design of AI-based models. These models play a crucial role within the proposed conceptual framework, ensuring the fulfillment of objectives related to the analyzed critical infrastructures.

Knowledge-Assisted AI

Knowledge-assisted AI integrates preexisting knowledge with data-driven methods, often resulting in hybrid or neuro-symbolic approaches [74, 75]. This knowledge can come from heuristics, symbolic rules, or physics equations, enhancing generalization, particularly in data-scarce environments. While such knowledge improves transparency and interpretability, it may not cover all possible scenarios, necessitating data-driven elements to fill gaps, resulting in knowledge-assisted AI approaches. In the closely related area of "informed machine learning," key classification factors of existing approaches include the source, representation, and integration of knowledge [75].

The primary goal of knowledge-assisted AI is to enhance system performance rather than add new functionality, making AI systems more robust, interpretable, and generalizable. These are of particular importance in the considered critical infrastructures, and evaluation could specifically target these aspects by, for example, testing the system under abnormal conditions. Knowledge structured in an accessible format helps establish cross-domain applicability, whereas more exotic formats could be useful in domain-specific applications. The type of knowledge that could be considered varies from one domain to another. For example, in power grids, the physics constraints or expert knowledge could be exploited by the models to provide a more robust and compliant set of actions.

While various approaches to knowledge-assisted AI systems have been explored, fewer studies have directly examined their impact on decision-making. Decision-making in AI is primarily studied in reinforcement learning (RL), which can be categorized into model-based and model-free methods. Model-based approaches learn an explicit model of the environment, typically using supervised learning techniques, and could, therefore, benefit directly from neuro-symbolic and informed machine learning methods surveyed by van Harmelen and Teije [74] and Von Rueden et al. [75].

In contrast, model-free methods do not rely on an explicit model but can still incorporate different types of knowledge. For example, prior works have leveraged logical rules for high-level transitions [76–79], spatial invariances [80], knowledge graphs [81], differential equations [82], and human feedback [83–85].

Meta-Awareness for AI Assistants

Meta-awareness in AI assistants is essential for effective human-AI teaming, where AI systems complement human operators by handling large data volumes while humans manage unforeseen and edge-case situations [41]. Unlike ML-based systems that lack causal reasoning, AI assistants must anticipate events, manage uncertainty, and flag anomalous situations [86, 87]. To enhance reliability, AI systems should quickly learn from failures, alert users to high-uncertainty states, and ensure meaningful human control [34], emphasizing the need for AI-based systems to have a level of meta-awareness. This awareness enables them to recognize situations that exceed their capabilities and prompts them to seek human assistance (e.g., send alarms to the operator when the proposed actions are of low confidence).

In the context of the proposed conceptual framework, the meta-awareness framework considers the following phases: (1) monitoring infrastructure with contextual knowledge extraction [88], (2) predicting system behavior and anomalies using uncertainty quantification, and (3) transferring control to humans when uncertainty is high [89]. Deferral mechanisms, such as those in Bondi et al. [90], enable AI to decide when to defer to human judgment based on uncertainty levels and operational context, ensuring reliable AI-assisted decision-making.

Figure 13 illustrates a prototype of a deferral mechanism that, following the nomenclature proposed by Bondi et al. [90], learns to defer decision-making from the AI model to a human. This mechanism considers aleatoric and epistemic uncertainty, as well as the network context, and the rule-based system can also include a constraint related to the deferral rate (i.e., an acceptable level of human effort or the

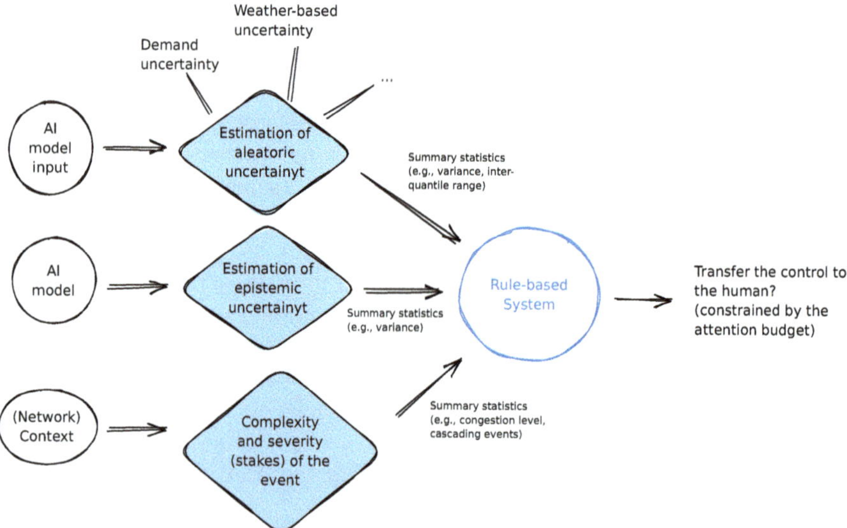

Fig. 13 Prototype schematic of a deferral mechanism that learns to defer decision-making from the AI model to a human

attention budget). This can be evaluated in real time (i.e., for the current operating scenario) or predicted for the next lead time, where aleatoric uncertainty needs to be considered in the model.

Human-AI Co-learning

Human-AI co-learning, often referred to as human-AI teaming or human-machine collaboration, aims to establish a continuous and mutual learning process between humans and AI systems. Unlike traditional AI-human interactions, co-learning focuses on leveraging the strengths of both agents while mitigating their respective weaknesses, leading to superior team performance.

A preliminary design for a co-learning AI agent is inspired by the work of van den Bosch et al. [91], which outlines six essential models that an AI system must develop and refine to enable effective collaboration. These include the taxonomy model (shared language), team model (work agreements and hierarchy), task model (knowledge about tasks and strategies), self-model (AI's internal state), Theory-of-Mind model (understanding human agents' inner states), and communication model (exchange of information based on shared understanding).

These models collectively enable the AI agent to align with human cognition, adapt dynamically, and foster productive interactions within the team.

As illustrated in Fig. 14, these models interact through structured information flows, ensuring that AI agents can process human communication, infer behavioral cues, and communicate their internal states transparently. This approach provides a conceptual foundation for human-AI co-learning, offering insights into necessary functionalities without yet specifying technical implementations.

An example of co-learning is AI-assisted human rescheduling in railway operations, where human and AI capabilities could be improved in parallel. Human feedback could be used to improve future AI recommendations, and AI systems support human learning by providing contextual evidence for or against specific rescheduling hypotheses.

Multi-Objective Reinforcement Learning

Designing multi-objective AI agents requires addressing both training and operational phases while integrating human preferences. In training, the AI optimizes a total reward function, R_{tot}, which aggregates individual objectives through a scalarization function (R_1, R_2, ..., R_n) [92]. Without human input, this function must be predefined, though a more flexible but complex alternative would be training an agent to handle multiple reward combinations dynamically.

During operation, AI-generated solutions are ranked by R_{tot} and presented to human operators, who may override the AI's ranking based on their expertise, implying an implicit preference function different from the predefined function U. To refine AI decision-making, a feedback loop is introduced, capturing

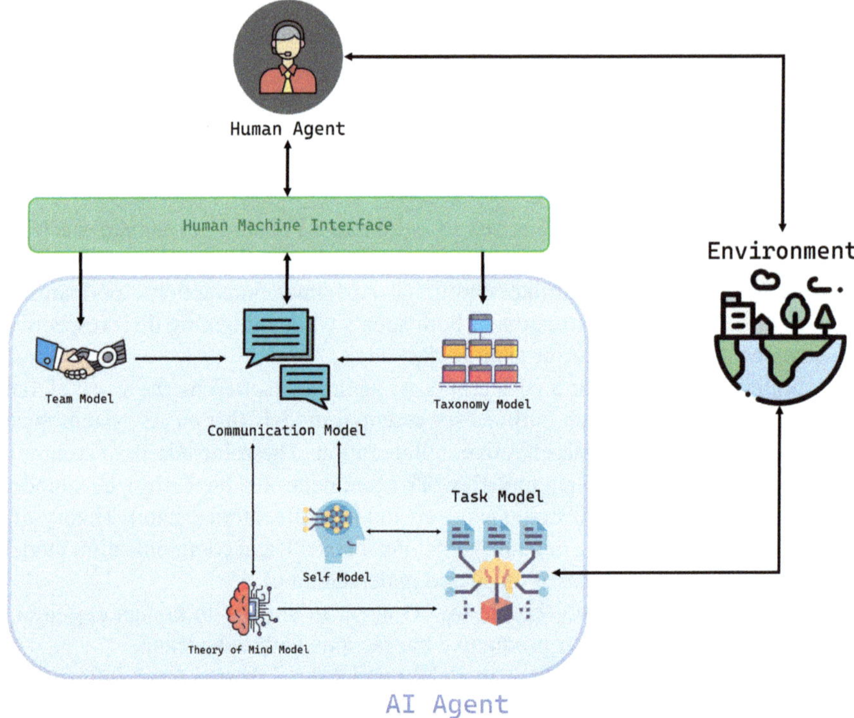

Fig. 14 Descriptive schematic of a co-learning AI agent

discrepancies between AI recommendations and human choices. Ideally, this data would be used to adjust the AI's reward function to better reflect expert preferences, although perfect alignment may be unattainable. Instead, heuristic methods developed in collaboration with experts can help approximate a suitable U-function.

A roadmap for developing multi-objective AI agents emphasizes a progressive approach, beginning with heuristic models and visualization tools (e.g., spider charts) to enhance explainability and transparency before integrating human feedback. The final step involves refining the utility function through iterative learning. Figure 15 illustrates a multi-objective visualization in a power grid congestion management use case, where operators select among AI-generated actions based on multiple conflicting objectives. Recording operator choices inform future AI adjustments, fostering alignment between human and AI decision-making.

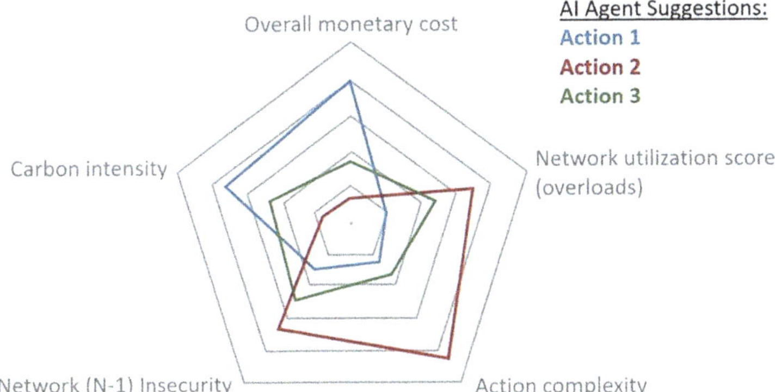

Fig. 15 Example of multi-objective visualization

4.3 Human-AI Interaction in Critical Infrastructures

Within the conceptual framework and in the context of human-AI interaction, three human-AI teamwork configurations are considered: (1) AI-assisted human control (human in control), (2) joint human-AI decision-making (including co-learning), and (3) autonomous AI (human as supervisor). In cognitive engineering, these scenarios are embedded in the notion of "stages and levels of automation" (see Fig. 20). At each stage, the levels of automation consider the division of roles and responsibilities between humans and machines and the delegation between the two of both autonomy (i.e., how independently the system is permitted to initiate system changes) and authority (i.e., the level of automation capability available to the system).

The following sections describe in greater detail the interaction scenarios, the design steps based on existing frameworks, and the hypervision tool, which provides the human operator with real-time insights, system diagnostics, and performance analytics, enabling better oversight and informed decision-making.

4.3.1 Interaction Scenarios

This section describes the three interaction levels between humans and AI that can be leveraged within the proposed conceptual framework based on the observed context.

Human in Full Control

This refers to scenarios where humans retain ultimate authority over decisions and actions influenced or assisted by AI systems. This concept emphasizes that while AI can provide insights, recommendations, or even perform tasks, the final

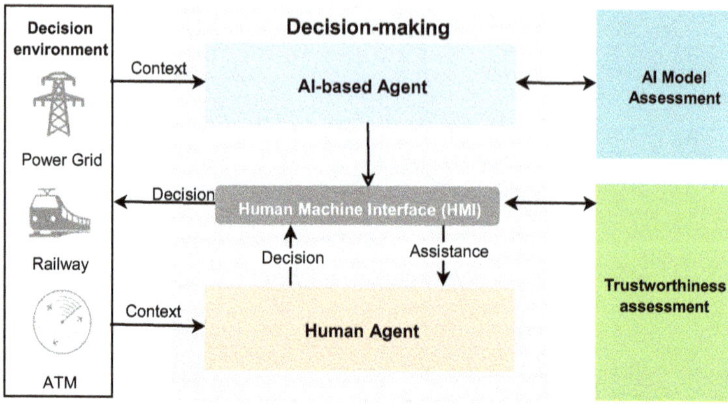

Fig. 16 Human in full control scenario

decision-making power rests with humans. Overall, maintaining human control in AI interactions ensures that technology serves to augment human capabilities (cognitive processes) while safeguarding against unintended consequences or misuse. The logical view corresponding to this mode of interaction is shown in Fig. 16. In addition to the observed context, the digital environments provide us with a set of tools to simulate real scenarios, enabling the assessment of the decision's impact before its application in a real-world context. When an event occurs, human operators should take some actions (decisions) to keep the environment in a stable state. They could interact with AI assistance to enhance their capabilities (exploration) at the decision-making step. The AI assistant may also provide explanations to guide human operators in the selection of recommendations. Once a candidate's decision is made by the human operator, the regulatory agent can verify the trustworthiness of the decision through various KPIs. This mode of interaction is required for both use cases of the power grid domain and the airspace sectorization assistant in ATM.

Human-AI Co-learning

Human-AI co-learning in the context of critical infrastructure involves a synergistic partnership where humans and AI systems continuously learn from each other to enhance the efficiency, reliability, and resilience of essential services. This collaboration is crucial for managing infrastructure such as power grids, water supply systems, transportation networks, and cybersecurity frameworks. In this co-learning process (see Fig. 17), AI systems can analyze vast amounts of data in real time, identify patterns, and predict potential issues before they occur. For example, in a power grid, AI can monitor the network and detect anomalies that might indicate a fault. Human operators, on the other hand, bring contextual understanding and decision-making capabilities that AI lacks. They can interpret AI-generated insights within the broader context of socioeconomic and environmental factors, make

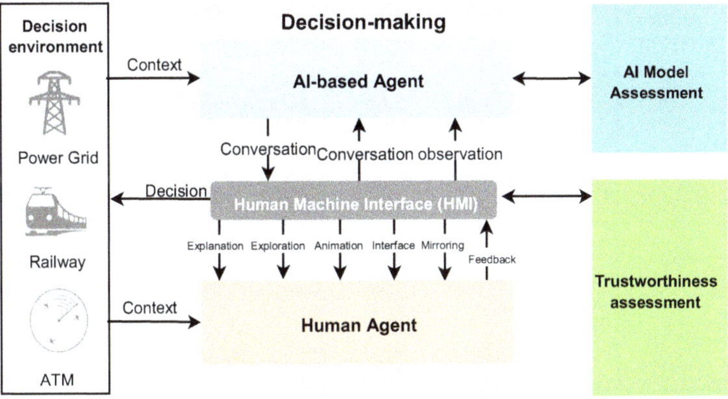

Fig. 17 Human-AI co-learning scenario

nuanced decisions, and adapt strategies as needed. In co-learning, the human learning process is explicitly supported by AI to increase human decision-making skills and cognitive processes. The overarching goal is to continuously improve human mental models about the environment, the AI, the self, and the cooperation with other people. AI can support these learning processes in different ways (e.g., by checking human assumptions or by mirroring his/her decision-making patterns). It is crucial that the collaboration between humans and AI is deliberately designed in such a way that it supports the human learning processes. Moreover, humans can provide feedback to AI systems, refining their algorithms and improving their accuracy over time. This feedback loop ensures that AI systems are not static but evolve based on real-world experiences and expert knowledge. In critical infrastructure, this means that AI can help anticipate and mitigate risks more effectively, to improve human-AI joint decision-making. This mode of interaction is required by the AI-assisted human rescheduling in railway operation and flow and airspace management assistant in ATM.

Human as Supervisor

Autonomous AI systems with human supervision (see Fig. 18) in the context of critical infrastructure refer to AI technologies that operate independently to manage and control essential networks like the power grid, railway, air traffic sectors, and information and communication networks. These AI systems use advanced algorithms and ML to monitor, analyze, and make decisions to optimize performance, detect anomalies, and respond to emergencies. However, given the high stakes and potential risks associated with critical infrastructure, human supervision remains crucial. This supervisory role involves overseeing the AI's decisions, intervening in complex or unforeseen situations, and ensuring that the AI operates within ethical and regulatory boundaries. Humans provide the necessary oversight to manage the

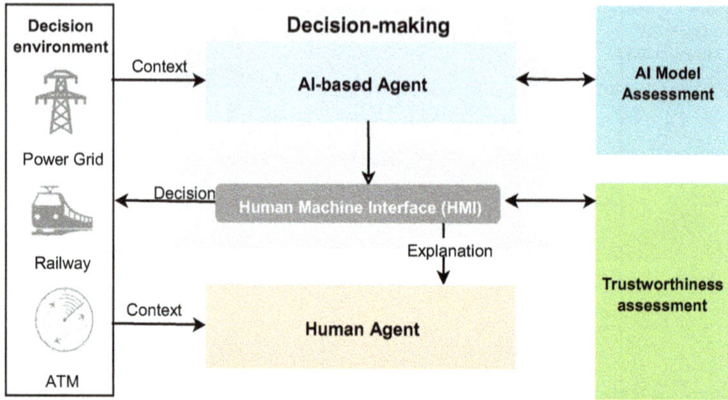

Fig. 18 Human as supervisor or autonomous AI scenario

AI's limitations, address biases, and make judgment calls that require human intuition and experience. This is an extremely demanding task for humans and, therefore, requires appropriate automation transparency as well as targeted leverage points for interventions. In summary, while autonomous AI can significantly enhance the efficiency and reliability of critical infrastructure, the human supervisor ensures safety, accountability, and compliance, creating a balanced and effective system. This is required by the automated rescheduling in railways operations, where a human as supervisor ensures the reliability and effectiveness of AI-based decisions and could override AI recommendations.

4.3.2 Describing and Designing Human-AI Interaction

For describing and designing human-AI interactions, cognitive engineering offers insights from human-automation teamwork, emphasizing shared control, autonomy, and transparency. However, a universal design framework is lacking. The proposed conceptual framework explores integrating two complementary frameworks: JCF [25, 94], which focuses on planning and executing activities among agents, and Ecological Interface Design (EID) [95], which enhances transparency by visualizing constraints. Both are rooted in Cognitive Systems Engineering (CSE), which prioritizes designing interactions based on the work environment rather than specific agents. Figure 19a illustrates this triadic approach, where JCF structures activities, while EID ensures clarity in decision-making by depicting system constraints. Combining these frameworks allows for a structured approach to human-AI collaboration, where EID dictates what information to display and JCF determines when and how it should be presented to support decision-making. Figure 19b represents this integration, showing how AI can enhance perception and action leverage at different abstraction levels, ensuring alignment between what is seen, decided, and executed.

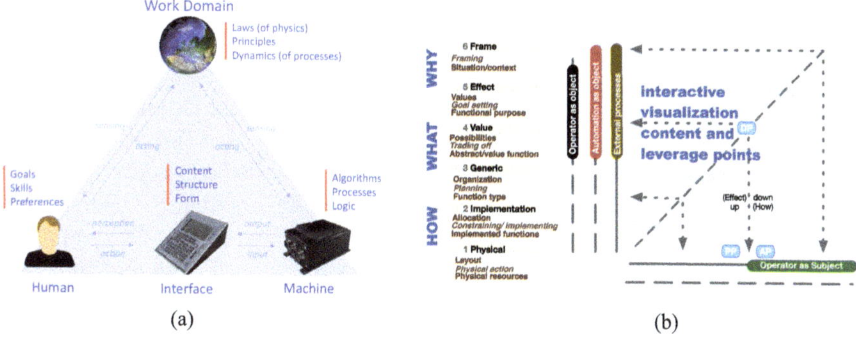

Fig. 19 Designing human-AI interaction: toward a common framework. (**a**) Triadic approach to human-AI interaction. (**b**) Merger of JCF and EID on a functional level

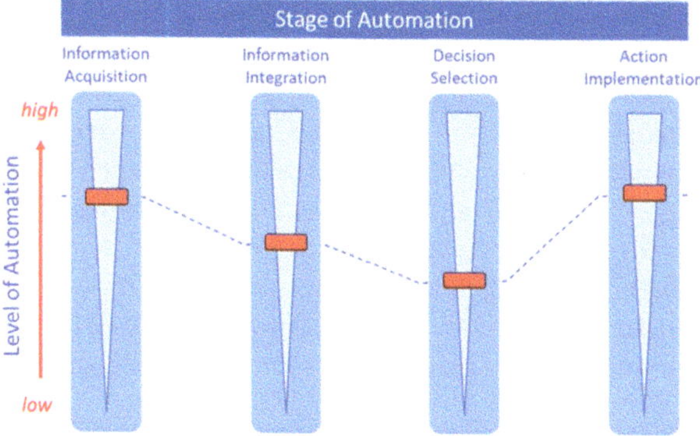

Fig. 20 Stages and levels of automation modeled after human information processing steps [93]

The proposed conceptual framework considers three human-AI teamwork configurations: AI-assisted human control (human in control), joint human-AI decision-making (co-learning), and autonomous AI (human as supervisor). These scenarios align with the established "stages and levels of automation" model (Fig. 20), which determines how autonomy and authority are distributed between humans and AI across four stages: information acquisition, integration, decision selection, and action implementation. In AI-assisted control, AI supports human decision-making but does not execute actions. In joint decision-making, AI and humans work in parallel, learning from each other's actions. In autonomous AI, the system operates independently, with human intervention reserved for system failures. The appropriate level of automation depends on operational context and system capabilities, meaning a universal model does not exist and must be tailored to specific applications. Figure 20 links these automation levels to the interface design from Fig. 19b, showing how AI's role in decision-making evolves based on its autonomy level.

4.3.3 Hypervision

Today's supervision tooling is inherited from successive waves of IT implementation over the last decades: operator supervision over many screens and applications leaves the user the cognitive load to prioritize, organize, and link disparate displayed information and alarms before considering any decision or action.

More variable and complex infrastructure dynamics—driven, for example, by energy transition on electric transmission systems—tend to increase the complexity of tooling: in such a context, supervision becomes impractical, with numerous and complex information to process and non-integrated applications under heterogeneous formats. It contributes to the problem of information overload, which dilutes the operator's attention. To be effective at continuous decision-making, it is often important to focus on the highest priority task at a time, using only the most relevant information. The sub-optimal design of human-machine interfaces and interactions has even been identified as a risk factor for human error in operations [96].

Hypervision aims to deliver the right information to the right person at the right time while tracking user progress for each task [97]. It provides a unified interface that synthesizes key information and centralizes real-time business events to support decision-making and task prioritization. Operators use Hypervision to understand the operational context, diagnose alerts, and implement solutions efficiently.

By enhancing event prioritization and syncretization, hypervision goes beyond real-time monitoring to anticipate future tasks through forecasting. This shift moves the focus from alarm management to proactive task completion (see Fig. 21a).

Hypervision is structured into four layers as shown in Fig. 21b: *Synthesis*, which integrates data from various online tools; *Formatting*, which determines the best

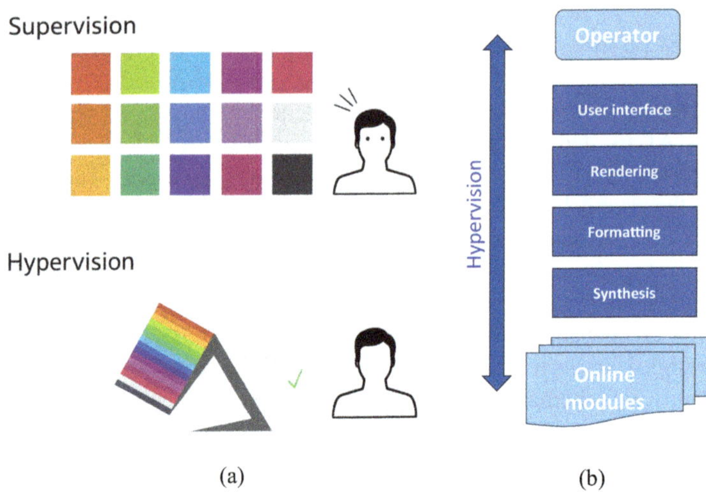

Fig. 21 Hypervision, providing unified interface that synthesizes key information and centralizes real-time business events to support decision-making and task prioritization. (**a**) From supervision to hypervision. (**b**) Hypervision implementation

way to present information (text, tables, graphs, etc.); *Rendering*, which connects formatted data to context (e.g., visualizing line overloads on a map); *User Interface*, which supports synthesis, prioritization, human-machine interaction, and collaborative decision-making.

As an example of hypervision, Amokrane-Ferka et al. [98] propose a virtual bidirectional assistant to support augmented decision-making in complex steering systems (Cockpit and Bidirectional Assistant (CAB) project[1]). This assistant continuously learns from real-time data flows and human decisions, enabling seamless interaction between human experts and AI. As can be seen in Fig. 22, the interface follows a hypervision-based framework with multiple panels: a context panel for real-time environmental visualization, a timeline panel for tracking past events, an alerts panel for notifying operators of risks and system changes, and a recommendations panel that provides AI-driven suggestions. Operators can choose to follow these recommendations based on their expertise and the complexity of the situation.

A major focus of this approach is enhancing the explainability of AI-generated recommendations to support human decision-making. The assistant dynamically assesses the operator's profile and cognitive workload to tailor the flow of information, ensuring optimal handling of complex or atypical situations. The system aims to improve situation awareness, reduce cognitive overload, and enhance overall efficiency by adapting recommendations and interactions to the operator's needs.

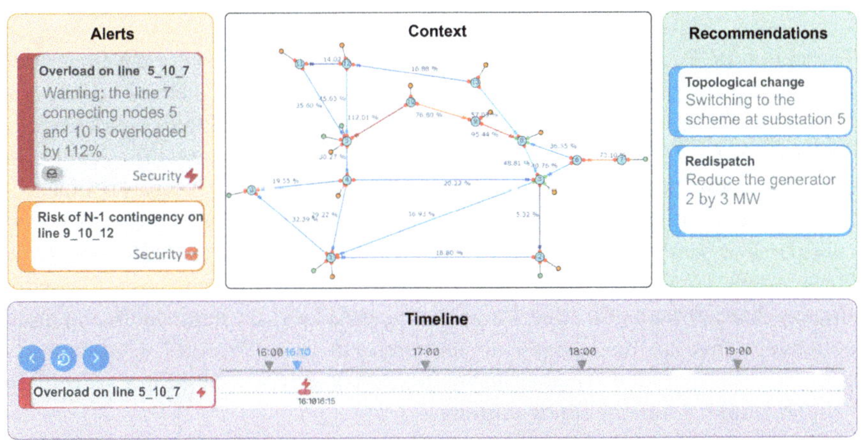

Fig. 22 Example of hypervision interface (CAB project)

[1] https://www.irt-systemx.fr/en/projets/cab/

5   Epistemological and Philosophical Foundations of Trustworthy AI and Their Adaptation for the Conceptual Framework

This subsection investigates the epistemological and normative foundations of the notion of Trustworthy AI (TAI) and analyses the different components of risk and their application to AI with a particular focus on safety-critical systems. The goal is to lay the foundation, from an epistemological and philosophical perspective, for a non-calculative approach to AI risk assessment. The starting point is the assessment list for TAI (ALTAI) elaborated by the high-level expert group appointed by the European Commission. The endpoint is a revised and improved ALTAI that focuses on key requirements for safety-critical systems and takes into consideration, when needed, the three main components of risk (hazard, exposure, vulnerability). Overall, this part of the conceptual framework aims at devising a theoretical approach capable of dealing with risk and uncertainty that is difficult to quantify, suggesting that some problems must be addressed with methods that have a philosophical—or at least non-purely quantitative—nature.

5.1   The Epistemological and Normative Grounds of the Notion of TAI

The notion of TAI plays a central role in ensuring AI systems are developed and deployed responsibly, particularly within the European Union's ethics-based regulatory framework. According to the European Commission's Ethics Guidelines for Trustworthy AI, TAI must be lawful, ethical, and robust throughout the system's life cycle. However, there is ongoing debate about what constitutes trustworthiness in AI and whether it is feasible for all systems. Explainability, for example, is widely considered essential for TAI but remains challenging in many AI applications. Furthermore, some critics argue that the concept of TAI is more of a marketing strategy than a meaningful ethical framework, as AI lacks the motivations and moral obligations that are fundamental to interpersonal trust [99, 100]. Consequently, some scholars question whether attributing trustworthiness to AI is conceptually valid or simply a form of ethics washing.

Despite these criticisms, in the design of the conceptual framework, the value of applying the notion of trustworthiness to AI is substantial, as it integrates both technical reliability and ethical acceptability. This aligns with the European Ethics Guidelines for Trustworthy AI and the ALTAI framework, which the conceptual framework uses to assess risks, establish requirements, and validate AI systems. However, to avoid conceptual errors, a distinction is made between trustworthiness in human-human (H-H) and human-AI (H-AI) interactions [101]. While both H-H and H-AI trust involve reliability and ethics, they differ in how ethical considerations manifest. Human trust relies on goodwill and moral obligations, whereas AI trustworthiness is based on compliance with ethical requirements. For example, in a use case where an AI assistant is transferred from simulation to real-world operation, AI must be designed to prevent human manipulation, such as misleading feedback or misuse of the AI learning process. By acknowledging these differences, it is ensured that TAI remains a practical and meaningful framework for developing ethical and technically robust AI systems.

5.2 AI-Related Risk and Uncertainty

The notion of TAI not only integrates crucial aspects of AI system design, deployment, and assessment but also plays a key role in addressing AI-related risk. Traditionally, trust and trustworthiness have been associated with situations of risk and vulnerability [102], making risk assessment a central concern for the conception of the framework. However, risk itself is a complex concept with no universally accepted definition. The classic definition by the Royal Society [103] emphasizes probability, while modern interpretations define risk as the combination of an event's probability and its consequences [104]. This perspective is also reflected in the AI Act, which considers risk as the likelihood and severity of harm (Art. 3, 2). However, the AI Act does not further articulate how risk should be assessed or mitigated, making it necessary to refine the methodological foundations of AI-related risk assessment.

5.2.1 The Components of Risk

A structured approach to risk assessment, often used in disaster risk management, breaks risk into three key components: hazard, exposure, and vulnerability. Hazard refers to the source of potential harm, such as system malfunctions, and is typically assessed through probabilistic estimates. Exposure considers the entities—people or material assets—that could be affected by the hazard. Vulnerability refers to the factors determining how susceptible these entities are to harm. Risk arises from the interaction of these three components, meaning even a low-hazard system can pose significant risks if exposure or vulnerability is high. Conversely, high hazard levels do not automatically translate into high risks if exposure and vulnerability are minimal. This multi-component analysis, commonly used in natural risk management, is also applicable to technological risks, including AI-related risks. As illustrated in Fig. 23, different AI systems may present risks due to varying levels of these components, enabling targeted mitigation strategies such as restricting access to AI-based services or reducing user vulnerability.

Hazard	**Exposure**	**Vulnerability**
It refers to the source of potential harm (both natural and non-natural hazards)	It refers to what could be armed, critical infrastructure like hospitals, and industrial plants relying on uninterrupted electricity	It refers to circumstance and measures making people or assets more or less prone to being damaged
Examples — Malfunction in a predictive maintenance algorithm leading to failure in detecting a transformer fault — Error in rescheduling trains following a disruption, leading to cascading delays across the network — Misjudgement in-flight conflict detection, causing delayed response to potential mid-air collisions	**Examples** — Power supply to households, critical infrastructure like hospitals, and industrial plants relying on uninterrupted electricity — Passengers, freight deliveries, and train operators across affected routes — Aircraft, passengers, crew members, and airport infrastructure in densely trafficked airspace	**Examples** — Dependence on AI models without human oversight, outdated backup systems, and lack of redundant infrastructure — Lack of fallback manual systems, dependence on real-time AI decision-making in complex rail networks, and outdated sensor data feeding the AI model — Limited human intervention due to over-reliance on AI systems, insufficient AI model training data for rare events

Fig. 23 AI-related risk and its components

5.2.2 Uncertainty

Risk assessment often assumes that potential outcomes can be assigned precise probabilities, yet this is rarely realistic, especially in AI-related scenarios [105]. While probabilistic models, including second-order probabilities and probabilistic intervals, can sometimes quantify uncertainties, large-scale deployments of innovative technologies pose additional challenges. In many cases, historical data is insufficient to inform probability estimates, leading to deep uncertainty where even approximate probabilities are difficult to assign [106]. To address this, AI risk assessment must go beyond the design phase and involve continuous evaluation in real-world contexts. The tool provided in the following pages is designed for ongoing assessments, ensuring that AI systems remain aligned with safety and ethical considerations throughout their life cycle.

5.3 A Non-calculative Tool for Risk Assessment in Safety-Critical Systems

In the design of the conceptual framework, the multi-component analysis of risk serves as the foundation for applying the ALTAI framework to a specific context. ALTAI is structured around seven key requirements central to TAI: Human Agency and Oversight, Technical Robustness and Safety, Privacy and Data Governance, Transparency, Diversity and Fairness, Societal and Environmental Well-being, and Accountability. However, in the case of considered safety-critical systems, four of these requirements—Human Agency and Oversight, Technical Robustness and Safety, Societal and Environmental Well-Being, and Accountability—are particularly relevant. These requirements align well with the multi-component risk analysis and help address key challenges, such as the risk of overreliance on AI systems. For example, Requirement #2 (General Safety) from ALTAI, which asks whether threats to an AI system have been identified, can be refined by considering hazard (the probability and impact of threats), exposure (the system's susceptibility to threats), and vulnerability (measures taken to reduce susceptibility). Similarly, the requirement related to Societal and Environmental Well-Being is refined by assessing de-skilling risks in terms of affected skills, workforce, and vulnerability.

In addition to adapting ALTAI's questions, a requirement on risk acceptability is proposed, recognizing that acceptable levels of risk depend on context and available alternatives. Unlike ALTAI, which primarily functions as a tool for ex post self-assessment, the proposed approach reconceives ALTAI's questions as positive requirements to be applied during the design phase. This shift encourages proactive responsibility in AI development, ensuring that safety-critical systems are assessed for risk mitigation from the outset rather than as an afterthought. The key requirements, tailored for the conceptual framework's specific needs, are summarized in Table 4.

Table 4 Summary of the key requirements derived from the ALTAI framework and adapted for conceptual framework's safety-critical systems

Relevant ALTAI requirement	Relevant ALTAI sub-requirement	Conceptual framework's requirements
#1 Human agency and oversight	Human agency and autonomy	• Make sure that users are adequately informed about (1) the fact that they are interacting with an AI system and (2) the kind of inferential mechanism behind the system's output • Establish mechanisms for (1) preventing overreliance on the system and (2) monitoring the actual use of the system to constantly check for overreliance dynamics, especially in those scenarios with scarce data • Assess the risks stemming from overreliance by considering these risks in terms of hazard (the potential harming consequences of overreliance), exposure (people and assets exposed to such harm), and vulnerability • Make sure that humans maintain meaningful control over the system and that their autonomy is not limited by a loss of competence due to their regularly outsourcing decisions—e.g., by blindly following recommendations—to the AI system (cf. [107])
	Human oversight	• Besides giving human operators specific training on how to exercise oversight, make sure that they are provided information on the basic working principles of RL as well as on its risks

(continued)

Table 4 (continued)

Relevant ALTAI requirement	Relevant ALTAI sub-requirement	Conceptual framework's requirements
#2 Technical robustness and safety	Resilience to attacks and security	• Assess the risks stemming from potential hazards related to technical faults, outages, attacks, as well as inappropriate and malicious use • Identify the people and material assets exposed to the potential harm resulting from such hazards • Implement strategies to reduce the vulnerability to such hazards of (1) the system and (2) the exposed people and assets • Plan regular monitoring to continuously assess the involved risks and collect information on the system's real-world deployment
	General safety	• Identify possible threats by considering both their probability of occurrence and their magnitude/impact on the system • Identify the system's levels of exposure to such threats, both in terms of quantity and duration • Implement sufficient measures to make the system less vulnerable to such threats
	Accuracy	• Identify risks stemming from low levels of accuracy of the system by identifying possible hazards, the related levels of exposure, and the vulnerability of exposed people and assets, as well as measures to reduce such vulnerability
	Reliability, fallback plans, and reproducibility	• Since the deployment of AI systems is often characterized by elements of uncertainty, make sure that the introduction of the system occurs in different steps so that it is possible to evaluate risks in progressively broader controlled contexts
#6 Societal and environmental well-being	Environmental well-being	• Identify the potential environmental impact of the system by considering both the training and the deployment phases
	Impact on work and skills	• Assess whether and how the systematic deployment of the system might cause human de-skilling by identifying: – The affected skills and the magnitude of the phenomenon – The affected workforce – The contexts and features that make humans more or less prone to de-skilling, taking measures to mitigate de-skilling risks and providing training and material to enable re- and up-skilling

Table 4 (continued)

Relevant ALTAI requirement	Relevant ALTAI sub-requirement	Conceptual framework's requirements
#7 Accountability	Risk management	• Organize risk training to ensure that all three components of risk are considered • Put in place by-design mechanisms in case of applications that can adversely affect individuals in terms not only of hazard but also exposure and vulnerability
Additional requirements on risk acceptability		• Given a certain system and the involved risks, make sure that there are no alternative options (with or without the use of AI) reasonably involving lower levels of risk in view of comparable positive outcomes

6 Conclusions

This chapter presented a conceptual and technology-agnostic framework designed to integrate AI into decision-making processes in cross-sector critical infrastructures while maintaining an appropriate balance between automation and human oversight. The framework fosters collaboration between human operators and AI systems. Through an iterative co-learning approach, human operators engage with AI, refining system behavior and enhancing decision quality over time. Furthermore, the framework supports real-time operations by incorporating integrated information and predictive insights, allowing for corrective and preventive actions at different levels of automation.

More precisely, the conceptual framework was shaped by systems engineering principles, which supported the identification of requirements, functionalities, and interactions among various components. It was further grounded in three critical infrastructures—power grids, railways, and air traffic management—along with their respective use cases. Insights from human cognition and decision theory were applied to inform the decision-making model and to establish a foundation for human-AI interaction. Established procedures for requirements definition and risk assessment, such as IEC 62559-2, ISO/IEC TR 24030, and the ALTAI assessment tool, were adopted to ensure a structured and standardized approach. When integrated into the proposed framework, it provides a foundation for developers and end users to incorporate essential requirements from the initial stages of AI-based system design, fostering seamless integration and compliance with evolving regulatory landscapes.

Finally, this work highlights that integrating AI into high-risk applications and sectors extends beyond the AI component or software itself—it requires a holistic perspective that considers the entire socio-technical system in which the AI solution operates. Achieving this demands an interdisciplinary approach to shift from individual decision-making toward collaborative human-AI decision-making, drawing from fields such as philosophy and cognitive engineering. In other words, the successful adoption of emerging technologies demands a comprehensive understanding of the broader operational and social context in which they are deployed.

Acknowledgments The research leading to this work is part of the AI4REALNET (*AI for REAL-world NETwork operation*) project, which received funding from the European Union's Horizon Europe Research and Innovation Programme under the Grant Agreement No 101119527 and from the Swiss State Secretariat for Education, Research and Innovation (SERI). This project is funded by the European Union and SERI. Views and opinions expressed are however those of the author(s) only and do not necessarily reflect those of the European Union and SERI. Neither the European Union nor the granting authority can be held responsible for them.

References

1. Curry, E., Heintz, F., Irgens, M., Smeulders, A. W. M., & Stramigioli, S. (2022). Partnership on AI, data, and robotics. *Communications of the ACM, 65*(4), 54–55.
2. Marot, A., Donnot, B., Chaouache, K., Kelly, A., Huang, Q., Hossain, R.-R., & Cremer, J. L. (2022). Learning to run a power network with trust. *Electric Power Systems Research, 212*, 108487.
3. Greitzer, F. L., & Podmore, R. (2008). *Naturalistic decision making in power grid operations: Implications for dispatcher training and usability testing*. Tech. rep. PNNL-18040. Pacific Northwest National Laboratory. Retrieved from https://www.pnnl.gov/main/publications/external/technical_reports/PNNL-18040.pdf
4. Fan, S., Guo, J., Ma, S., Li, L., Wang, G., Haotian, X., Yang, J., & Zhao, Z. (2024). Framework and key technologies of human machine hybrid-augmented intelligence system for large-scale power grid dispatching and control. *CSEE Journal of Power and Energy Systems, 10*(1), 1–12.
5. Hilliard, A., Brath, R., & Jamieson, G. A. (2024). Work domain analysis of electric transmission networks and operation. *IEEE Systems Journal, 18*(1), 474–484.
6. SBB. (2020). *Rail control system*. German. Retrieved April 3, 2025, from https://bahninfrastruktur.sbb.ch/de/produkte-dienstleistungen/bahninformatiksysteme/verkehrssteuerung/rcs.html
7. Rittner, M., Richta, H. N., & Große, S. (2022). *Automatische Dispositionsunterstützung mit ADA-PMB*. Retrieved October 3, 2023, from https://www.system-bahn.net/aktuell/automatische-dispositionsunterstuetzung-mit-ada-pmb/
8. Wälter, J., Mehta, F. D., & Rao, X. (2020). Aiding vehicle scheduling and rescheduling using machine learning. *International Journal of Transport Development and Integration, 4*(4), 308–320. https://doi.org/10.2495/TDI-V4-N4-308320. Retrieved March 7, 2025, from http://www.witpress.com/doi/journals/TDI-V4-N4-308-320
9. Parvez Farazi, N., Zou, B., Ahamed, T., & Barua, L. (2021). Deep reinforcement learning in transportation research: A review. *Transportation Research Interdisciplinary Perspectives, 11*, 100425. https://doi.org/10.1016/j.trip.2021.100425. Retrieved March 7, 2025, from https://linkinghub.elsevier.com/retrieve/pii/S2590198221001317
10. Li, J.-Q., Mirchandani, P. B., & Borenstein, D. (2007). The vehicle rescheduling problem: Model and algorithms. *Networks, 50*(3), 211–229. https://doi.org/10.1002/net.20199. Retrieved March 4, 2025, from https://onlinelibrary.wiley.com/doi/10.1002/net.20199
11. Mohanty, S., Nygren, E., Laurent, F., Schneider, M., Scheller, C., Bhattacharya, N., Watson, J., Egli, A., Eichenberger, C., Baumberger, C., Vienken, G., Sturm, I., Sartoretti, G., & Spigler, G. (2020). Flatland-RL: Multi-agent reinforcement learning on trains. Version number 2. https://doi.org/10.48550/ARXIV.2012.05893. Retrieved March 4, 2025, from https://arxiv.org/abs/2012.05893
12. Laurent, F., Schneider, M., Scheller, C., Watson, J., Li, J., Chen, Z., Zheng, Y., Chan, S.-H., Makhnev, K., Svidchenko, O., Egorov, V., Ivanov, D., Shpilman, A., Spirovska, E., Tanevski, O., Nikov, A., Grunder, R., Galevski, D., Mitrovski, J., Sartoretti, G., Luo, Z., Damani,

M., Bhattacharya, N., Agarwal, S., Egli, A., Nygren, E., & Mohanty, S. (2021). Flatland competition 2020: MAPF and MARL for efficient train coordination on a grid world. In H. J. Escalante & K. Hofmann (Eds.), *Proceedings of the NeurIPS 2020 Competition and Demonstration Track* (Proceedings of Machine Learning Research. PMLR) (Vol. 133, pp. 275–301). Retrieved from https://proceedings.mlr.press/v133/laurent21a.html

13. Lövétei, I., Kővári, B., Bécsi, T., & Aradi, S. (2022). Environment representations of railway infrastructure for reinforcement learning-based traffic control. *Applied Sciences, 12*(9), 4465. https://doi.org/10.3390/app12094465. Retrieved March 7, 2025, from https://www.mdpi.com/2076-3417/12/9/4465

14. Jiang, Y., Zhang, K., Li, Q., Chen, J., & Zhu, X. (2022). Multiagent path finding via tree LSTM. Version number 2. https://doi.org/10.48550/ARXIV.2210.12933. Retrieved March 7, 2025, from https://arxiv.org/abs/2210.12933

15. Roost, D., Meier, R., Huschauer, S., Nygren, E., Egli, A., Weiler, A., & Stadelmann, T. (2020). Improving sample efficiency and multi-agent communication in RL-based train rescheduling. In *2020 7th Swiss Conference on Data Science (SDS)* (pp. 63–64). IEEE. https://doi.org/10.1109/SDS49233.2020.00024. Retrieved March 7, 2025, from https://ieeexplore.ieee.org/document/9145010/

16. Shang, M., Zhou, Y., Mei, Y., Zhao, J., & Fujita, H. (2023). Energy-saving train operation synergy based on multi-agent deep reinforcement learning on spark cloud. *IEEE Transactions on Vehicular Technology, 72*(1), 214–226. https://doi.org/10.1109/TVT.2022.3205379. Retrieved March 7, 2025, from https://ieeexplore.ieee.org/document/9882357/

17. Schneider, S., Ramesh, A., Roets, A., Stirbu, C., Safaei, F., Ghriss, F., Wülfing, J., Güral, M., Siboni, N., Gentry, R., Liessner, R., Hustache, T., Lecat, T., Deekshith, U., Markin, V., Le, V., Bejjani, W., Küpper, M., & Sturm, I. (2024). Intelligent railway capacity and traffic management using multi-agent deep reinforcement learning. In *2024 IEEE 27th International Conference on Intelligent Transportation Systems (ITSC), Edmonton, Canada.*

18. SNCF. (2022). Open Source Railway Designer. Retrieved April 3, 2025, from https://github.com/OpenRailAssociation/osrd

19. Nunes, T. M. M., Borst, C., van Kampen, E.-J., Hilburn, B., & Westin, C. (2021). Human-interpretable input for machine learning in tactical air traffic control. In *SESAR Innovation Days 2021.*

20. Westin, C., Hilburn, B., Borst, C., Van Kampen, E.-J., & Bång, M. (2020). Building transparent and personalized AI support in air traffic control. In *2020 AIAA/IEEE 39th Digital Avionics Systems Conference (DASC)* (pp. 1–8).

21. Yilmaz, E., Sanni, O., Kotwicz Herniczek, M. T., & German, B. (2021). Deep reinforcement learning approach to air traffic optimization using the MuZero algorithm. In *AIAA aviation 2021 forum* (p. 2377).

22. Yutong, C., Minghua, H., Yan, X., & Lei, Y. (2023). Locally generalised multi-agent reinforcement learning for demand and capacity balancing with customised neural networks. *Chinese Journal of Aeronautics, 36*(4), 338–353.

23. Ahrenhold, N., Gerdes, I., Mühlhausen, T., & Temme, A. (2023). Validating dynamic sectorization for air traffic control due to climate sensitive areas: Designing effective air traffic control strategies. *Aerospace, 10*(5), 405.

24. Lui, G. N., Lulli, G., Lema-Esposto, M. F., & Martinez, R. L. (2024). Airspace sector design: An optimization approach.

25. Lundberg, J., & Johansson, B. J. E. (2021). A framework for describing interaction between human operators and autonomous, automated, and manual control systems. *Cognition, Technology & Work, 23*, 381–401.

26. Braunschweig, B., Gelin, R., & Terrier, F. (2022). The wall of safety for AI: Approaches in the confiance.ai program. In *Workshop on Artificial Intelligence Safety (SAFEAI).*

27. Gelin, R. (2024). Confiance.ai program software engineering for a trustworthy AI. In *Producing artificial intelligent systems: The roles of benchmarking, standardisation and certification* (pp. 11–29). Springer Nature Switzerland.

28. Dignum, Virginia. 2019. *Humane AI ethical framework. HumanE AI Deliverable 1.3*. Tech. rep. Retrieved from https://www.humane-ai.eu/wp-content/uploads/2019/11/D13-HumaneAIframework-report.pdf

29. Golpayegani, D., Pandit, H. J., & Lewis, D. (2022). Comparison and analysis of 3 key AI documents: EU's proposed AI Act, assessment list for trustworthy AI (ALTAI), and ISO/IEC 42001 AI management system. In *Artificial Intelligence and Cognitive Science (AICS 2022)* (pp. 189–200).

30. Amokrane, K., Rousseaux, V., Dussartre, M., Zouinar, M., & Renoir, N. (2024). Combining user centered design and system engineering to the design of a generic AI-based assistant. In *3rd INCOSE International Conference on Human Systems Integration (HSI), Jeju Island, South Korea*. Retrieved from https://hal.science/hal-04621899

31. Zouinar, M., Amokrane-Ferka, K., & Rousseaux, V. (2024). A user centered approach for the design of a generic AI based assistant system for work activities. In *22nd Triennial Congress of the International Ergonomics Association (IEA), Jeju Island, South Korea*. Retrieved from https://hal.science/hal-04685856

32. Roques, P. (2016). MBSE with the arcadia method and the Capella tool. In *8th European Congress on Embedded Real Time Software and Systems (ERTS 2016)*.

33. Clegg, C. W. (2000). Sociotechnical principles for system design. *Applied Ergonomics, 31*, 463–477.

34. Endsley, M. R. (2023). Supporting human-AI teams: Transparency, explainability, and situation awareness. *Computers in Human Behavior, 140*, 107574.

35. Naikar, N., Brady, A., Moy, G., & Kwok, H.-W. (2023). Designing human-AI systems for complex settings: Ideas from distributed, joint, and self-organising perspectives of sociotechnical systems and cognitive work analysis. *Ergonomics, 66*(11), 1669–1694.

36. National Academies of Sciences, Engineering, and Medicine. (2022). *Human-AI teaming: State-of-the-art and research needs*. The National Academies Press.

37. Miller, T. (2023). Explainable AI is dead, long live explainable AI! Hypothesis driven decision support using evaluative AI. In *Proceedings of the 2023 ACM Conference on Fairness, Accountability, and Transparency (FAccT'23)* (pp. 333–342). Association for Computing Machinery. https://doi.org/10.1145/3593013.3594001

38. Eisbach, S., Langer, M., & Hertel, G. (2023). Optimizing human-AI collaboration: Effects of motivation and accuracy information in AI-supported decision-making. *Computers in Human Behavior: Artificial Humans, 1*(2), 100015.

39. Ngo, T., & Krämer, N. (2022). I humanize, therefore I understand? Effects of explanations and humanization of intelligent systems on perceived and objective user understanding. PsyArXiv [Preprint]. https://doi.org/10.31234/osf.io/6az2h

40. Ha, T. W., & Kim, S. (2023). Improving trust in AI with mitigating confirmation bias: Effects of explanation type and debiasing strategy for decisionmaking with explainable AI. *International Journal of Human-Computer Interaction*, 1–12. https://doi.org/10.1080/10447318.2023.2285640

41. Endsley, M. R. (2023). Ironies of artificial intelligence. *Ergonomics, 66*(11), 1656–1668.

42. Klein, G. (2018). Macrocognitive measures for evaluating cognitive work. In E. S. Patterson & J. E. Miller (Eds.), *Macrocognition metrics and scenarios: Design and evaluation for real-world teams* (1st ed.). CRC Press. https://doi.org/10.1201/9781315593173

43. Parker, S. K., & Grote, G. (2022). Automation, algorithms, and beyond: Why work design matters more than ever in a digital world. *Applied Psychology, 71*(4), 1171–1204. https://doi.org/10.1111/apps.12241

44. Endsley, M. R. (2000). Situation models: An avenue to the modeling of mental models. *Proceedings of the Human Factors and Ergonomics Society Annual Meeting, 44*(1), 61–64. https://doi.org/10.1177/154193120004400117

45. Fildes, R., Goodwin, P., Lawrence, M., & Nikolopoulos, K. (2009). Effective forecasting and judgmental adjustments: An empirical evaluation and strategies for improvement in supply-

chain planning. *International Journal of Forecasting, 25*(1), 3–23. https://doi.org/10.1016/j. ijforecast.2008.11.010

46. Niehaus, S., Hartwig, M., Rosen, P. H., & Wischniewski, S. (2022). An occupational safety and health perspective on human in control and AI. *Frontiers in Artificial Intelligence, 5*, 868382. https://doi.org/10.3389/frai.2022.868382

47. Schaap, G., Bosse, T., & Vettehen, P. H. (2023). The ABC of algorithmic aversion: Not agent, but benefits and control determine the acceptance of automated decision-making. *AI & Society.* https://doi.org/10.1007/s00146-023-01649-6

48. Hackman, J. R., & Oldham, G. R. (1976). Motivation through the design of work: Test of a theory. *Organizational Behavior and Human Performance, 16*, 250–279. https://doi. org/10.1016/0030-5073(76)90016-7

49. Bainbridge, L. (1983). Ironies of automation. *Proceedings of IFAC, 19*(6), 775–779.

50. Polanyi, M. (2012). *Personal knowledge*. Routledge.

51. Wäfler, T., & Rack, O. (2021). Kooperation und künstliche intelligenz. In *Kooperation in der digitalen Arbeitswelt: Verlässliche Führung in Zeiten virtueller Kommunikation* (pp. 77–88).

52. Alexander, P. A., Schallert, D. L., & Reynolds, V. E. (2009). What is learning anyway? A topographical perspective considered. *Educational Psychologist, 44*(3), 176–192.

53. Kolb, D. A. (1984). *Experiential learning: Experience as the source of learning and development*. Prentice-Hall.

54. Jacovi, A., Marasović, A., Miller, T., & Goldberg, Y. (2021). Formalizing trust in artificial intelligence: Prerequisites, causes and goals of human trust in AI. arXiv preprint arXiv:2010.07487. Retrieved from http://arxiv.org/abs/2010.07487

55. Lee, J. D., & See, K. A. (2004). Trust in automation: Designing for appropriate reliance. *Human Factors, 46*, 50.

56. Hoffman, R. (2017). A taxonomy of emergent trusting in the human–machine relationship. In P. J. Smith (Ed.), *Cognitive systems engineering: The future for a changing world*. CRC Press. https://doi.org/10.1201/9781315572529

57. Parasuraman, R., & Riley, V. (1997). Humans and automation: Use, misuse, disuse, abuse. *Human Factors: The Journal of the Human Factors and Ergonomics Society, 39*(2), 230–253. https://doi.org/10.1518/001872097778543886

58. Koopman, P., & Hoffman, R. R. (2003). Work-arounds, make-work, and kludges. *IEEE Intelligent Systems, 18*(6), 70–75.

59. Westin, C., Borst, C., & Hilburn, B. (2016). Strategic conformance: Overcoming acceptance issues of decision aiding automation? *IEEE Transactions on Human Machine Systems, 46*(1), 41–52. https://doi.org/10.1109/THMS.2015.2482480

60. Behzadan, V., & Munir, A. (2017). Whatever does not kill deep reinforcement learning, makes it stronger. arXiv preprint arXiv:1712.09344.

61. Zissis, G. (2019). The r3 concept: Reliability, robustness, and resilience [president's message]. *IEEE Industry Applications Magazine, 25*(4), 5–6.

62. Molnar, C. (2020). *Interpretable machine learning*. Lulu.com. Retrieved from https://christophmolnar.com/books/interpretable-machine-learning/

63. Vouros, G. A. (2022). Explainable deep reinforcement learning: State of the art and challenges. *ACM Computing Surveys, 55*, 1–39.

64. European Commission (EC). (2024). *Ethics guidelines for trustworthy AI*. Tech. rep. Retrieved from https://digital-strategy.ec.europa.eu/en/library/ethics-guidelines-trustworthy-ai

65. Nichol, A., Pfau, V., & Hesse, C. (2018). Gotta learn fast: A new benchmark for generalization in RL. arXiv preprint arXiv:1804.03720. https://arxiv.org/abs/1804.03720

66. Irpan, A. (2018). *Deep reinforcement learning doesn't work yet*. Retrieved February 15, 2025, from https://www. alexirpan.com/2018/02/14/rl-hard.html

67. Cobbe, K., Klimov, O., & Hesse, C. (2019). Quantifying generalization in reinforcement learning. In *International Conference on Machine Learning* (pp. 1282–1289).

68. Olteanu, A., Castillo, C., & Diaz, F. (2019). Social data: Biases, methodological pitfalls, and ethical boundaries. *Frontiers in Big Data, 2*, 13. https://doi.org/10.3389/fdata.2019.00013

69. Paliouras, G. (1993). *Scalability of machine learning algorithms*. Doctoral dissertation, University of Manchester.
70. Ulanov, A., Simanovsky, A., & Marwah, M. (2017). Modeling scalability of distributed machine learning. In *2017 IEEE 33rd International Conference on Data Engineering (ICDE)* (pp. 1249–1254).
71. Hernandez-Leal, P., Kartal, B., & Taylor, M. E. (2019). A survey and critique of multiagent deep reinforcement learning. *Autonomous Agents and Multi-Agent Systems, 33*, 750–797. https://doi.org/10.1007/s10458-019-09421-1
72. Cobb, A. D., Jalaian, B., Bastian, N. D., & Russell, S. (2021). Toward safe decision-making via uncertainty quantification in machine learning. In *Systems engineering and artificial intelligence* (pp. 379–399).
73. Nemani, V., Biggio, L., Huan, X., Hu, Z., Fink, O., Tran, A., Wang, Y., Zhang, X., & Hu, C. (2023). Uncertainty quantification in machine learning for engineering design and health prognostics: A tutorial. *Mechanical Systems and Signal Processing, 205*, 110796.
74. van Harmelen, F., & Ten Teije, A. (2019). A boxology of design patterns for hybrid learning and reasoning systems. *Journal of Web Engineering, 18*, 97–123.
75. Von Rueden, L., Mayer, S., Beckh, K., Georgiev, B., Giesselbach, S., Heese, R., Kirsch, B., Pfrommer, J., Pick, A., Ramamurthy, R., et al. (2021). Informed machine learning—A taxonomy and survey of integrating prior knowledge into learning systems. *IEEE Transactions on Knowledge and Data Engineering, 35*(1), 614–633.
76. Araki, B., Li, X., Vodrahalli, K., DeCastro, J., Fry, M., & Rus, D. (2021). The logical options framework. In *International Conference on Machine Learning* (pp. 307–317).
77. Lyu, D., Yang, F., Liu, B., & Gustafson, S. (2019). SDRL: Interpretable and data-efficient deep reinforcement learning leveraging symbolic planning. In *Proceedings of the AAAI Conference on Artificial Intelligence* (Vol. 33, pp. 2970–2977).
78. Vaezipoor, P., Li, A. C., Icarte, R. A. T., & McIlraith, S. A. (2021). LTL2action: Generalizing LTL instructions for multi-task RL. In *International Conference on Machine Learning* (pp. 10497–10508).
79. Yang, F., Lyu, D., Liu, B., & Gustafson, S. (2018). PEORL: Integrating symbolic planning and hierarchical reinforcement learning for robust decision-making. In *Proceedings of the 27th International Joint Conference on Artificial Intelligence* (pp. 4860–4866). AAAI Press.
80. Van der Pol, E., Worrall, D., van Hoof, H., Oliehoek, F., & Welling, M. (2020). MDP homomorphic networks: Group symmetries in reinforcement learning. *Advances in Neural Information Processing Systems, 33*, 4199–4210.
81. Höpner, N., Tiddi, I., & van Hoof, H. (2022). Leveraging class abstraction for commonsense reinforcement learning via residual policy gradient methods. In *International Joint Conference on Artificial Intelligence, IJCAI 2022* (pp. 3050–3056).
82. Gao, J., Chen, S., Li, X., & Zhang, J. (2022). Transient voltage control based on physics-informed reinforcement learning. *IEEE Journal of Radio Frequency Identification, 6*, 905–910.
83. Christiano, P. F., Leike, J., Brown, T., Martic, M., Legg, S., & Amodei, D. (2017). Deep reinforcement learning from human preferences. In *Advances in neural information processing systems* (p. 30).
84. Kaplan, R., Sauer, C., & Sosa, A. (2017). Beating atari with natural language guided reinforcement learning. arXiv preprint arXiv:1704.05539.
85. Knox, W. B., & Stone, P. (2009). Interactively shaping agents via human reinforcement: The tamer framework. In *Proceedings of the International Conference on Knowledge Capture* (pp. 9–16).
86. Charpentier, B., Senanayake, R., Kochenderfer, M., & Günnemann, S. (2022). Disentangling epistemic and aleatoric uncertainty in reinforcement learning. arXiv preprint arXiv:2206.01558.

87. Hüllermeier, E., & Waegeman, W. (2021). Aleatoric and epistemic uncertainty in machine learning: An introduction to concepts and methods. *Machine Learning, 110*(3), 457–506. https://doi.org/10.1007/s10994-021-05946-3

88. Palminteri, S., & Lebreton, M. (2021). Context-dependent outcome encoding in human reinforcement learning. *Current Opinion in Behavioral Sciences, 41*, 144–151. https://doi.org/10.1016/j.cobeha.2021.06.012

89. Nylin, M., Westberg, J. J., & Lundberg, J. (2022). Reduced autonomy workspace (raw)—An interaction design approach for human automation cooperation. *Cognition, Technology & Work, 24*(2), 261–273. https://doi.org/10.1007/s10111-021-00674-1

90. Bondi, E., Koster, R., Sheahan, H., Chadwick, M., Bachrach, Y., Cemgil, A. T., et al. (2022). Role of human-AI interaction in selective prediction. *Proceedings of the AAAI Conference on Artificial Intelligence, 36*, 5286–5294.

91. van den Bosch, K., Schoonderwoerd, T., Blankendaal, R., & Neerincx, M. (2019). Six challenges for human-AI co-learning. In *Adaptive Instructional Systems: First International Conference, AIS 2019, Held as Part of the 21st HCI International Conference, HCII 2019, Orlando, FL, USA, July 26–31, 2019, Proceedings* (pp. 572–589).

92. Hayes, C. F., Rădulescu, R., & Bargiacchi, E. (2022). A practical guide to multi-objective reinforcement learning and planning. *Autonomous Agents and Multi-Agent Systems, 36*(26). https://doi.org/10.1007/s10458-022-09564-8

93. Parasuraman, R., Sheridan, T. B., & Wickens, C. D. (2000). A model for types and levels of human interaction with automation. *IEEE Transactions on Systems, Man, and Cybernetics Part A, Systems and Humans, 30*(3), 286–297. Retrieved from http://www.ncbi.nlm.nih.gov/pubmed/11760769

94. Vicente, K. J., Christoffersen, K., & Pereklita, A. (1995). Supporting operator problem solving through ecological interface design. *IEEE Transactions on Systems, Man, and Cybernetics, 25*, 529–545.

95. Borst, C., Flach, J. M., & Ellerbroek, J. (2015). Beyond ecological interface design: Lessons from concerns and misconceptions. *IEEE Transactions on Human-Machine Systems, 45*(2), 164–175. https://doi.org/10.1109/THMS.2014.2364984

96. Nachreiner, F., Nickel, P., & Meyer, I. (2006). Human factors in process control systems: The design of human–machine interfaces. *Safety Science, 44*(1), 5–26. https://doi.org/10.1016/j.ssci.2005.10.019

97. Marot, A., Kelly, A., Naglic, M., Barbesant, V., Cremer, J., Stefanov, A., & Viebahn, J. (2022). Perspectives on future power system control centers for energy transition. *Journal of Modern Power Systems and Clean Energy, 10*(2), 328–344. https://doi.org/10.35833/MPCE.2021.000687

98. Amokrane-Ferka, K., Marot, A., Meddeb, M., Dussartre, M., Crochepierre, L., Rozier, A., Renoir, N., Gosselin, S., Girod, H., Khouadjia, M., et al. (2024). Framework for human and AI assistant bidirectional interaction applied to industrial system operations.

99. Metzinger, T. (2019). Ethics washing made in Europe. *Der Tagesspiegel*. https://www.tagesspiegel.de/politik/ethics-washing-made-in-europe-5937028.html

100. Ryan, M. (2020). In AI we trust: Ethics, artificial intelligence, and reliability. *Science and Engineering Ethics, 26*, 2749.

101. Zanotti, G., Petrolo, M., Chiffi, D., & Schiaffonati, V. (2024). Keep trusting! A plea for the notion of trustworthy AI. *AI & Society, 39*(6), 2691–2702.

102. Nickel, P. J., & Vaesen, K. (2012). Risk and trust. In S. Roeser, R. Hillerbrand, M. Peterson, & P. Sandin (Eds.), *Handbook of risk theory* (pp. 857–870). Springer. https://doi.org/10.1007/978-94-007-1433-5_34

103. Royal Society (Great Britain). (1983). *Risk assessment: Report of a Royal Society study group*. Royal Society.

104. Hansson, S. O. (2009). From the casino to the jungle: Dealing with uncertainty in technological risk management. *Synthese, 168*, 423–432. https://doi.org/10.1007/s11229008-9447-0
105. Nordström, M. (2022). Ai under great uncertainty: Implications and decision strategies for public policy. *AI & Society, 37*, 1703–1714. https://doi.org/10.1007/s00146022-01393-1
106. van de Poel, I. (2016). An ethical framework for evaluating experimental technology. *Science and Engineering Ethics, 22*, 667–686.
107. Prunkl, C. (2022). Human autonomy in the age of artificial intelligence. *Nature Machine Intelligence, 4*, 99.

CyclOps: Leveraging Semantic Technologies for AI and Data Life Cycle Management and Governance

Monica Caballero, Anna Queralt, Diego Calvanese, Alexandros Oikonomidis,
Chukwuemeka Muonagor, Alberto Abella, Piero Campalani,
Benjamin Cogrel, Gabriele Tassi, Julia Palma, Alex Barceló,
Saeed Talebzadeh, Ross Little, Peio Oiz, Florencia Pérez,
Alexandros Kalafatelis, Maria Font, Sergi Nadal, Alexandros Nizamis,
Admela Jukan, Simone Martin Marotta, Ricardo Simón-Carbajo,
Javier Conejero, and Dolores Ordóñez

Abstract Artificial intelligence (AI) and data-driven applications are transforming industry and research across domains. Despite research advances, the management of the AI and data life cycle in organizations remains largely manual and ad hoc. At the same time, data spaces initiatives are emerging, providing new data-sharing opportunities but posing new challenges for organizations seeking to leverage all

M. Caballero (✉) · M. Font
NTT DATA, Barcelona, Spain
e-mail: monica.caballero.galeote@nttdata.com; maria.fontsanchez@nttdata.com

A. Queralt · S. Nadal
Universitat Politècnica de Catalunya, BarcelonaTech, Barcelona, Spain
e-mail: anna.queralt@upc.edu; sergi.nadal@upc.edu

D. Calvanese
Free University of Bozen-Bolzano, Ontopic s.r.l., Bolzano, Italy
e-mail: diego.calvanese@unibz.it

A. Oikonomidis · A. Nizamis
Centre for Research and Technology, Hellas—Information Technologies Institute (CERTH/ITI), Thessaloniki, Greece
e-mail: aleoikon@iti.gr; alnizami@iti.gr

C. Muonagor · A. Jukan
Technische Universität Braunschweig, Braunschweig, Germany
e-mail: c.muonagor@tu-braunschweig.de; a.jukan@tu-braunschweig.de

A. Abella
FIWARE Foundation, Berlin, Germany
e-mail: alberto.abella@fiware.org

P. Campalani
Eurac Research—Center for Climate Change and Transformation, Bolzano, Italy
e-mail: piero.campalani@eurac.edu

© The Author(s) 2026
E. Curry et al. (eds.), *Artificial Intelligence, Data and Robotics*,
https://doi.org/10.1007/978-3-032-10561-5_14

B. Cogrel
Ontopic s.r.l., Bolzano, Italy
e-mail: benjamin.cogrel@ontopic.ai

G. Tassi · S. M. Marotta
Expert.AI, Naples, Italy
e-mail: gtassi@expert.ai; smarotta@expert.ai

J. Palma · R. Simón-Carbajo
CeADAR—Ireland's Centre for Artificial Intelligence, University College Dublin,
Dublin, Ireland
e-mail: julia.palma@ucd.ie; ricardo.simoncarbajo@ucd.ie

A. Barceló · J. Conejero
Barcelona Supercomputing Center, Barcelona, Spain
e-mail: alex.barcelo@bsc.es; javier.conejero@bsc.es

S. Talebzadeh
DataCalculus, Tallinn, Estonia
e-mail: saeed@datacalculus.com

R. Little
ATOS, Madrid, Spain
e-mail: ross.little@eviden.com

P. Oiz · D. Ordóñez
AnySolution SL, Palma de Mallorca, Spain
e-mail: poiz@anysolution.eu; dom@anysolution.eu

F. Pérez
Avoris Corporación Empresarial, Palma de Mallorca, Spain
e-mail: mf.perez@avoristravel.com

A. Kalafatelis
Four Dot Infinity, Athens, Greece
e-mail: alkalafatelis@fourdotinfinity.com

their potential. Tackling these challenges, this chapter describes the new framework proposed by the CyclOps project, which aims to enable interoperable and trustworthy automatic management, governance, and maintenance of the entire data life cycle for large-scale volumes of data generated in heterogeneous distributed sources. This supports the development and deployment of AI-based applications across both business and research contexts. CyclOps operationalizes the end-to-end data life cycle, placing at its core knowledge graphs, an established semantic formalism to represent data and metadata adhering to the FAIR Principles, while capturing relevant information to improve reproducibility, traceability, and explainability of the AI results. This core layer is complemented with tools for the automation of data management tasks; distributed data processing; AI tools, algorithms, and models; data space interoperability; and a human-centric interface. The chapter presents the key innovative propositions of CyclOps and its underlying technologies and illustrates them through selected use cases that highlight the general applicability of the approach.

Keywords Semantic technologies · Data interoperability · Data life cycle · AI life cycle · Knowledge graphs · Automation

1 Introduction

Advanced data-driven applications that use artificial intelligence (AI) techniques to extract value from data have shown the power to reshape industries at all levels [1], steering innovation, increasing the efficiency, and unlocking new business opportunities. Modern data-intensive ecosystems are shifting from modelling well-defined domains with a few structured data sources, toward the scale of hundreds and thousands of evolving sources, each providing abundance of data, in a variety of forms. Simultaneously, data spaces initiatives are emerging, as a new framework to support data sharing and exchange across data ecosystems in such a new data-based economy [2]. The European Strategy for Data has defined the ultimate objective of creating a single market for data, ensuring Europe's global competitiveness and data sovereignty.[1]

In this context, the ability to crossmatch and cross-reference heterogeneous data sources could be a major competitive advantage for organizations. However, there are several aspects that still ultimately limits the ability to leverage data as a strategic asset and drive business value for all kinds of players in the industry.

On the one hand, well-known challenges related to data integration to enable later analysis, such as ensuring data quality, consistency, and interoperability, as well as the technical expertise and skills required to solve these problems, not only remain but are even more difficult to address. On the other hand, data spaces add additional requirements on aspects such as access control, usage policies, usage of open standards for interoperability, provenance, and traceability, etc., on top of the already complex traditional problems. While initiatives such as the Data Spaces Support Centre (DSSC) focus on defining the necessary building blocks for interoperability and governance within these data spaces, a complementary challenge remains for individual organizations: managing the complex data and AI life cycle internally to process data that has been consumed from a data space or to prepare data for sharing.

Consequently, the need for systematic governance of the end-to-end data and AI life cycle has recently become more evident [3]. For the sake of clarity, by systematic governance, in this chapter, we mean the holistic framework of policies, processes, and technologies used to manage and control the entire data and AI life cycle, from origin to deployment. This includes ensuring the traceability and lineage of all assets, enforcing rules for data quality and access control, and automating the complex workflows that connect data sources to AI-based services. From now on, we will use governance to refer to this comprehensive systematic approach.

Despite the progress made on defining foundations for data systematic governance, the adoption of its principles is scarce in practice; most organizations today do not carry out the operationalization of data governance (i.e., the definition of the data processes and flows that implement its principles) in a systematic way and define their own data architecture and associated data flows that best suit their

---

[1] https://digital-strategy.ec.europa.eu/en/policies/strategy-data

requirements [4]. This is known to be an arduous and complex task requiring a high degree of specialization and many resources. There are solutions available from big vendors [5] (e.g., Collibra, Alation, IBM, or Informatica) that offer full suites including tools for governance, metadata management, data catalogue, etc., as well as data science and AI platforms in the market [6] (e.g., Dataiku or Databricks), and even a wider variety of products focused and specialized on specific aspects of the data and AI life cycle (data discovery, pipeline automation, etc.). Still, most of these solutions require a significant effort in cost and resources for its deployment, as they are often designed for large enterprises with mature data infrastructure and are too complex or expensive for many small or medium-sized organizations. In addition, a significant part of the process needs to be developed in an ad hoc fashion, or they do not offer a full coverage of both the data and AI life cycle, from governance to the design and execution of data flows and the deployment of the AI model.

Therefore, the current governance of the data life cycle in organizations, from the data sources into valuable insights, remains largely manual in many cases, which is not only impractical but also results in the creation of data and information silos, as the structure, meaning, and policies related to the data sources remain hidden or implicit.

Tackling these challenges, this chapter describes the new framework proposed by the CyclOps project,[2] a 3-year project funded by the European Commission, having the automation of the end-to-end data and AI life cycle as its main goal. The solution described relates also to the response to the need for solutions for the integration of data life cycle, architectures, and standards for complex data cycles and/ or human factors as part of the implementation of the AI, Data, and Robotics Partnership [7].

The CyclOps framework manages the life cycle of data and AI assets within an organization, from the consumption of data from multiple sources to its meaningful integration and processing to the serving of data, models, and services that are then ready for consumption by or publication to data spaces. CyclOps addresses the abovementioned limitations by providing a modular, open, and lightweight framework that integrates data and AI life cycle management using semantic technologies. It focuses on reducing entry barriers for organizations of different sizes by automating repetitive and complex tasks, reusing shared data and models through interoperable standards to reduce integration costs.

The proposed approach has at its core knowledge graphs [8], an established semantic formalism to represent data and metadata adhering to the FAIR principles, while capturing relevant information to improve reproducibility, traceability, and explainability of the AI results. Additionally, CyclOps incorporates tools for the automation of data management tasks; distributed data processing; AI tools, algorithms, and models; data space interoperability; and a human-centric interface.

The CyclOps framework is mainly addressed to industrial and public services' developers and data scientists: users that are responsible for implementing and

---

[2] https://www.cyclopsproject.eu/

managing the data pipelines that take data from diverse sources and process it to generate specialized AI models or other insights, which will be the basis for new services and products to be offered in a variety of sectors. These users traditionally deal with the technical aspects of pipeline orchestration and data integration, tasks that often require significant time and expertise. CyclOps alleviates this burden by operationalizing and automating large portions of the workflow, allowing developers to focus on more critical, higher-level tasks. They can use the framework to define pipelines and execute the models, ultimately delivering actionable insights with less manual effort. Another important user group includes knowledge scientists and domain experts, who typically manage the creation and handling of semantic assets, such as ontologies, vocabularies, and data models, which are crucial for data integration and interoperability. CyclOps minimizes the need for their manual input to handle complex integration processes by automating much of the metadata generation, data connection, and governance-related processes. As a result, the framework makes it easier for organizations with limited technical resources and/or extensive domain expertise to obtain valuable insights from data.

The CyclOps framework consists of four interconnected layers, as depicted in Fig. 1, the Knowledge Layer, the User Intent Layer, the Runtime Layer, and the Interoperability Layer. Each layer serves a specific purpose, facilitating a seamless workflow that considers the user-in-the-loop, from data exploration and pipeline generation to governance.

At the core of CyclOps lies the Knowledge Layer, aimed at operationalizing and automating data governance throughout the data and AI life cycle. This layer relies on an ontology-based approach supported by semantic technologies to provide a unified view of data from diverse sources, including structured, semi-structured, and unstructured data in an Integrated Knowledge Base (IKB). The User Intent Layer is the entry point for users, where their needs and objectives are captured and translated into actionable steps within CyclOps. It processes the intents behind the user input and, supported by the Knowledge Layer, transforms them into pipelines for execution in the subsequent layers. The Runtime Layer handles the execution of the pipelines received from the User Intent Layer. It manages data ingestion and preprocessing tasks, as well as AI algorithms and models, and transparently drives the execution of the pipelines in distributed environments. Finally, the Interoperability Layer is responsible for defining and managing the conditions for data interoperability. It addresses the legal, organizational, semantic, and technical requirements necessary for enabling data exchange with data spaces.

The following sections of this chapter present the different layers of CyclOps solution, the key innovative propositions, and its underlying technologies. Later, this chapter introduces how the layers interact and how each component could be utilized in a generic usage workflow and present four use cases in four different domains selected to showcase the generalizability of the framework.

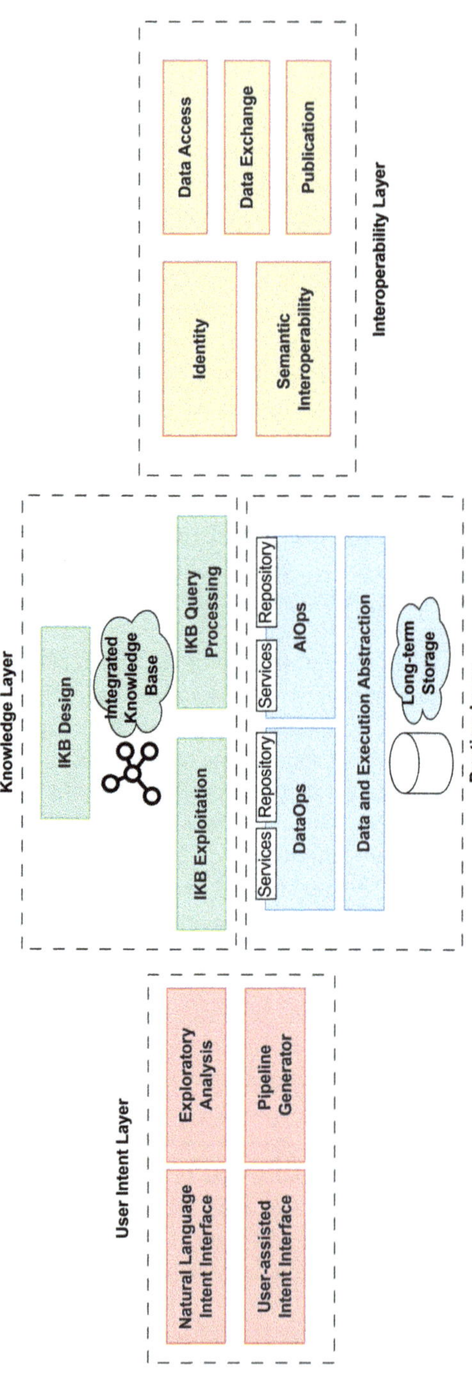

Fig. 1 CyclOps framework

2 Knowledge Layer

The Knowledge Layer in CyclOps serves as the core of the framework for AI and data life cycle management and governance, to which all other components of the CyclOps framework relate. It provides a unified view of data from diverse sources by means of a knowledge graph, seamlessly integrating structured, semi-structured, and unstructured data. Additionally, it enables components within CyclOps to write and consume annotations about the processes being executed, which ensure their interoperability and traceability, and enable automation of several tasks. As such, this layer is a backbone for data operations and decision-making. It implements an Integrated Knowledge Base (IKB) based on the Virtual Knowledge Graph (VKG) approach, which relies on an ontology mapped to the data sources and provides query processing mechanisms over the data sources that exploit the domain knowledge encoded in the ontology.

The core functionalities within the Knowledge Layer include the design and mapping of ontologies, efficient query processing over VKGs, and the extraction and categorization of concepts from text for enabling advanced queries. Additionally, it supports explainability for AI models, thus promoting transparency and trust in AI-driven processes.

2.1 IKB Design Tools

The IKB Design Tools allow the user to build in a semi-automatic way an ontology and the mapping to the data sources connected to the CyclOps framework. While for the design of the ontology we can rely on well-assessed methodologies [9–11], this is not the case for the mapping component, where effective design methodologies and tools support are still missing. The IKB Design Tool mapping component semi-automatically bootstraps a candidate ontology of the IKB and generates mappings between the data sources and such ontology. It makes use of novel techniques based on mapping patterns [12], which rely on different types of information. On the one hand, so-called schema-driven patterns exploit schema-level information, notably structural constraints (such as keys and foreign keys) and linguistic information related to schema element names. On the other hand, data-driven patterns are based on an analysis of the content of data sources and of specific elements therein, by relying on machine learning approaches, so as to extract from the data mapping-relevant information. The adopted techniques are similar in spirit to those used for schema matching in databases [13], but they are complicated by the fact that mappings in the VKG setting not only make use of possibly complex queries over the data sources but also use templates to construct the object identifiers (IRIs) for the ontology objects that virtually populate the knowledge graph. The methods adopted in the IKB Design Tools are semi-automatic, since the bootstrapped candidate mappings constitute for the designer useful information and hints for possible matches, to be then interactively refined and improved. Indeed, once the mapping and the ontology have been bootstrapped, the component offers a user interface where to refine the mapping and the ontology. It also enables visualizing the resulting structure of the produced knowledge graph.

2.2 IKB Query Processing

The IKB Query Processing component allows the efficient processing of queries over the IKB. Such queries are typically issued by the other components in the CyclOps framework that need to access the information stored in the IKB, and they are expressed in SPARQL or SQL. They are either directly answered or just translated into SQL queries ready to be executed over the data sources. The latter case aims at improving performance by minimizing the number of data conversions. The implementation relies on the open-source VKG engine *Ontop* [14, 15] and on its commercial extension *Ontopic Server*. The latter complements *Ontop* with a SQL connector implementing the wire protocol of PostgreSQL, enabling querying of the IKB not only via SPARQL but also via the more popular SQL language, thus facilitating the interoperation with the other CyclOps components.

2.3 IKB Concept Extraction and Categorization Engine

This component bridges the gap between unstructured textual data and structured knowledge representation, enabling more precise, context-aware, and complex querying functionalities. Moreover, this component lays the grounds for the explainability of the query results, every time textual data is involved in the generation of an answer. The IKB Concept Extraction and Categorization Engine plays a role in facilitating data interoperability and traceability by extracting concepts from text in alignment with the ontologies defined in the IKB. This engine allows for the precise identification of where specific results appear in the text, ensuring a structured and transparent process. Furthermore, it categorizes documents according to a customizable taxonomy, making it adaptable to different domains and use cases. This functionality enhances the research potential by providing organized and traceable outputs. The tool relies on transforming ontological concepts into structured data representations and using them to direct a large language model (LLM) into extracting structured information from unstructured textual data. The results of this process enable automated reasoning, structured information retrieval, advanced querying capabilities, and traceability and explainability of query results.

To ensure consistency in data extraction and representation, the ontology represented in the IKB is parsed to dynamically generate structured schemas. These schemas act as a bridge between conceptual knowledge and real-world unstructured data sources, facilitating the alignment of extracted information with predefined entity structures. The process involves:

1. Ontology interpretation: identifying key concepts (classes) and their properties (relations, attributes) to ensure structured knowledge extraction
2. Schema generation: translating each concept into a structured data model while maintaining relationships and constraints across different sources
3. Type inference: mapping datatype properties (such as text, numerical values, and dates) to structured fields in the data model to ensure format alignment

This process lays the foundation for accurate entity recognition and interoperability between datasets. The structured extraction pipeline works in synergy with the ontology, ensuring that extracted information adheres to predefined schemas. By leveraging ontological definitions, the system guarantees consistency in entity representation, making it possible to interconnect extracted information with the IKB.

Once text is processed for potential entities, a validation mechanism ensures that only relevant and accurate extractions are retained. This involves lexical matching, which checks if extracted entities explicitly appear in the text; fuzzy matching, which identifies variations of known entities that may be expressed differently; and linguistic normalization, which handles morphological variations such as plural and singular forms to improve recognition. These validation techniques ensure that extracted data is reliable and contextually relevant to the ontology.

To enable advanced queries, extracted entities must be mapped to structured templates that conform to the ontology's schema. Each identified entity is categorized under a specific conceptual framework, where entities are assigned to their corresponding classes within the ontology and extracted attributes are associated with entity properties, ensuring semantic alignment. The structured output is then integrated into the broader knowledge base, allowing for cross-referencing and enrichment of existing data. The explainability is provided by accompanying information regarding the position of the extracted entities and related properties in the text. This structured representation facilitates interoperability between different datasets and enhances data usability.

2.4 Explainability of AI Models

CyclOps incorporates a Multimodal Explainable AI component (MXAI) to provide its users means to understand how decisions are made by the models implemented within the framework. Through the MXAI, CyclOps helps identify different available tools for explainability of multimodal AI models in a structured way, according to the user's intent, the data, and algorithms used. The MXAI component enables transparency in AI decision-making by integrating models, datasets, and metadata to generate interpretable insights.

MXAI provides both model-specific and model-agnostics methods, offering diverse methods to generate clear, interpretable insights into model behavior [16].

The process begins with a user query, which is processed by the User Intent Layer to extract intent and determine the required explanation. MXAI serves as the core of the system, retrieving the appropriate AI model for which explainability is required from the AI Models Repository, and the corresponding dataset used for training the model from DataOps (see next section). Additionally, it gathers metadata from the IKB, including the model architecture, features details, and relevant dataset information. By integrating models, datasets, and metadata, MXAI provides a structured, data-driven explanation framework for AI systems. Innovation in this component relies in providing an environment for the implementation of general safeguards (explanation, contestation, human-in-the-loop) that allows the implementation of ad hoc safeguards for each application.

3   Runtime Layer

The Runtime Layer of CyclOps provides users with a complete execution environment to manage the entire life cycle of data and AI workflows, ensuring integration between data ingestion, processing, AI model deployment, and distributed execution. It consists of three core components, DataOps, AIOps, and Data and Execution Abstraction (DEA), each contributing to the automated and scalable processing of heterogeneous data sources.

The subsequent sections provide a detailed description of DataOps, AIOps, and DEA, specifying their individual functionalities and interdependencies within the CyclOps framework.

3.1   DataOps

The DataOps component enables users to ingest and prepare data from various sources, including external data spaces and internal repositories, making it accessible to AIOps for its processing. To this aim, it provides a set of repositories and services to support data management and preprocessing tasks within pipelines. The repositories provide functions to be used as operators in data pipelines (e.g., data preparation, cleaning, or integration), while the services are tools that consider the user-in-the-loop and allow CyclOps's users to get added-value services based on DataOps functionalities (e.g., discover relatedness measures or quality constraints).

The DataOps Repositories provide functions that enable the ingestion of data from various sources through different modalities, such as single/batch file(s) upload, upload through APIs, or ingestion of streaming data through Kafka, streamlining the overall data collection process. Additional functions are provided to perform some essential tasks upon ingestion of the data, including cleaning, removing outliers, and transformation/normalization, to ensure consistency and quality in the input data. By standardizing the data, managing missing values, and applying necessary transformations, it is ensured that the data is appropriately preprocessed for subsequent analysis and ML model training.

The DataOps Services include services for data discovery, data quality rules discovery, data curation, and data augmentation. These services are either data based, which directly interact with samples of raw data, or semantic based, which leverage the domain knowledge in the IKB.

The Data Discovery service finds similar or related datasets that can be integrated and propagated to analysis tasks. As one of the most intensively discussed data management challenges for big data, dataset selection or data discovery tries to find a subset of relevant datasets from the ones available in a data lake. Existing approaches search for datasets with similar attribute names or overlapping instance values. The challenges include finding joinable tables [17, 18] or semantically related tables [19]. At the data level, CyclOps adopts the novel approach of using compact representations of datasets and their attributes in the form of profiles, which allow to scale up the discovery process [20]. In this way, this service profiles the data to find datasets that can be effectively crossed (e.g., via union or join operations) to enrich the input dataset, finding columns whose values are not only from the same domain but also have a similar distribution and granularity. At the semantic level, data discovery consists in finding similarities among semantic datasets, using ontological elements to automatically identify mappings with the highest score.

The Data Quality Rules Discovery is a data-based service that automatically extracts the integrity constraints that hold in a raw dataset, preventing the user from manually defining rules to ensure the consistency of the data. Examples of constraints can be identifiers, functional dependencies, order constraints, or general constraints in the domain that guarantee that data is consistent before analysis.

DataOps Services also provide curation and augmentation methods for dataset enrichment. The Semantic-based Data Curation service aims to ensure the quality of the data, by using SHACL shapes, evaluating the semantic resources with FAIR principles. It discovers implicit constraints to detect wrong elements or duplicities, and considers FAIR principles to curate the data, guaranteeing that it is unique, normalized, and not duplicated. Finally, the Data Augmentation service enhances AI model robustness by generating additional synthetic data based on the original training dataset. This process expands the dataset by applying transformations to create new variations of the data. By increasing data diversity, the service helps the model generalize better, improving its performance and resilience to unseen data during inference.

As a whole, the DataOps component can be seen as a set of functionalities that perform the necessary steps to transform the raw data sources into data assets. By leveraging semantic mappings and metadata integration through the IKB, DataOps enhances data interoperability, preparing it for advanced AI-driven workflows.

3.2 AIOps

AIOps automates and optimizes AI-driven workflows, ensuring model selection, training, deployment, and monitoring [21]. It integrates multiple repositories and services to manage the full AI life cycle, enhancing scalability, efficiency, and adaptability. While several off-the-shelf solutions exist to handle part of AIOps (e.g., MLFlow [22], Kubeflow,[3] or AWS SageMaker[4]), CyclOps implements a novel approach to AIOps that tightly interconnects with the IKB leveraging on the results of DataOps and its VKG.

Following the same approach as DataOps, the AIOps component offers a set of repositories and services. The repositories include the AI Models Repository, storing trained models with versioning for traceability and reproducibility [23], allowing retrieval for inference and further fine-tuning. It also incorporates the Optimization Functions Repository contains automated hyperparameter tuning and model compression, improving efficiency in training and deployment phases. As for services, the Feature Engineering component offers tools for analyzing and transforming dataset variables prior to model training, such as correlation analysis or feature importance estimation, annotating its outputs in the IKB to ensure interpretability and traceability. The AI Marketplace provides a unified interface for exploring algorithms, models, and optimization functions across CyclOps, enabling filtering of algorithms and including a recommendation system that suggests relevant algorithms based on the user intent. The AI Model Protection and Monitoring services continuously assess data drift and model performance degradation, enabling proactive intervention through anomaly detection. The Model Deployment service integrates with the DEA component for a scalable and distributed execution, ensuring smooth real-time and batch inference workflows. These capabilities align with automated end-to-end CI/CD AI life cycle pipelines [24, 25], which are crucial for deploying AI models in production while maintaining explainability, accountability, and responsibility [26].

All these AIOps functionalities are exposed to users through the CyclOps Lab, which acts as the central development and interaction environment. Although not a repository or service itself, the Lab integrates and orchestrates all AIOps components, enabling users to build, execute, and monitor AI workflows through a unified interface.

CyclOps Lab combines interactive notebooks, visual pipeline composition tools, decentralized collaboration features, and semantic integration mechanisms, all designed to support robust and transparent AI workflows. Users can define workflows combining data ingestion, preprocessing, algorithm execution, and postprocessing using a drag-and-drop interface. These workflows are stored as versioned, reusable assets and are automatically annotated with task type, I/O schema, and component lineage. All outputs are aligned to the IKB, ensuring that AI results are

---

[3] Kubeflow: Machine Learning Toolkit for Kubernetes. https://www.kubeflow.org/

[4] Amazon SageMaker: Developer Guide. https://docs.aws.amazon.com/sagemaker/latest/dg/

not only reproducible but also semantically interoperable. This environment provides the backbone for building complex pipelines that remain explainable and traceable.

In addition to development, CyclOps Lab also supports model deployment, enabling users to seamlessly transition from experimentation to production. Trained models can be serialized in formats such as ONNX,[5] Pickle, or TensorFlow SavedModel and wrapped into standardized containers for deployment.

Workflows developed in CyclOps Lab can be executed locally or distributed across computing environments by means of the DEA component. Results, whether intermediate or final, can be inspected, compared, and exported. The Lab also provides visual tools for monitoring pipeline status, tracking resource usage, and diagnosing failures or anomalies. AIOps interacts closely with other CyclOps components: DataOps supplies preprocessed data for model training, while IKB enhances explainability and interpretability through metadata. DEA ensures parallel execution and computational scalability, while the User Intent Layer triggers model selection and training based on user-defined workflows.

By automating AI life cycle management, AIOps ensures that AI-driven processes within CyclOps remain robust, effective, and seamlessly integrated across the entire system.

Together, AIOps components establish a complete, modular framework for AI development, deployment, and monitoring within CyclOps. Through their integration in the CyclOps Lab, and alignment with the IKB, they enable semantically transparent workflows that are fully traceable, explainable, and reusable. The system's modularity ensures that individual components—whether repositories or services—can evolve independently while remaining interoperable. With AIOps in place, the execution of these AI workflows now depends on robust, scalable, and infrastructure-agnostic mechanisms—requirements that are addressed by the Data and Execution Abstraction module.

3.3 *Data and Execution Abstraction (DEA)*

The Data and Execution Abstraction (DEA) module ensures good performance and scalability without sacrificing programmability nor introducing any steep adoption costs. The integration of DEA within CyclOps represents a paradigm shift in how AI-driven data-intensive applications interact with distributed data environments. It allows for the efficient execution of complex workflows, ensuring that both business and research communities can leverage large-scale, distributed data in a seamless and intelligent manner.

This module is designed to seamlessly bridge the gap between DataOps and AIOps by providing a unified computing environment. It facilitates efficient and

---

[5] https://onnx.ai

scalable data management and processing pipelines while abstracting the complexities of distributed infrastructures. That is, the DEA simplifies the complexity of the underlying infrastructure, making the execution of pipelines over a distributed environment transparent to CyclOps users.

In order to achieve these goals, the DEA module employs a dual-layered approach to data and execution management. On the one hand, a Large-Scale Data Management offers a unified interface to the datasets and distributed data sources. This enables a uniform treatment of data regardless of its physical location within the infrastructure. On the other hand, a Distributed Execution Engine facilitates parallel execution of data transformations, AI model training, and data analysis tasks across distributed environments. This engine abstracts the underlying infrastructure, decoupling the application logic from the execution logistics. By integrating task-based execution paradigms, the module optimizes resource allocation and improves performance efficiency.

To achieve these capabilities, DEA integrates state-of-the-art technologies that streamline distributed data access and distributed computation. The two software pillars for this endeavor are dataClay [27] for the Data Management aspect and COMPSs [28] for the Distributed Execution. These technologies enable in situ data processing, avoiding unnecessary data movement, which is particularly relevant in high-performance data analytics and AI workflows. The abstraction layer provided by DEA empowers developers to define scalable and efficient data pipelines without deep knowledge of the underlying distributed computing mechanisms.

A key innovation within the DEA is the in-depth integration of data management and distributed execution. These building blocks are often used for large-scale data processing, but a profound interaction between them can collectively enhance data handling and computational efficiency at a scale.

4 User Intent Layer

The User Intent Layer receives and interprets user intents, either expressed in natural language or in a structured format. This layer translates and processes user requirements to configure and generate data processing pipelines, which automate the movement and transformation of data between source and target systems. Additionally, it facilitates the exploration of output data, enabling a streamlined analysis and insight generation process.

The User Intent Layer is divided into four components: The Natural Language Intent Interface, the User-assisted Intent Interface, the Pipeline Generator, and the Exploratory Analysis modules. The Natural Language Intent Interface parses users' high-level desires expressed in natural language and extracts the actual intention and valuable information in the request from factors like sentence structure, context, etc., in a structured format, so that the Pipeline Generator or the Exploratory Analysis Engine are capable to understand. Alternatively, the User-assisted Intent Interface allows advanced users—who can provide precise inputs and instructions based on their expertise—to select processes or input keywords to guide the generation of specific pipelines by the Pipeline Generator. The Exploratory Analysis provides users with a platform through which they can interact with their desired data, retrieved by interacting with the IKB.

4.1 Intent Interface and Translator

The Intent Interface and Translator enables the input of intents in high-level (natural language) or in low-level (selection of pipelines) formats. It includes two intent interfaces: the User-assisted and the Natural Language Intent Interfaces.

The Natural Language Intent Interface is an interface through which the user keys in their intents in a high-level (natural language) format. To enter the intents in high level, the user is directed to a Natural Language Processing (NLP) Chat. The NLP Chat provides a chat interface that is able to gather as much information as possible from the user, based on knowledge of predefined required fields. These required fields contain the core parameters that are expected to be captured in the user intents. The NLP Chat communicates with the user while ensuring that the important information captured in the required fields, which also relate to the user intents, are extracted from the user. It then passes the gathered information to an Info Retrieval component. Info Retrieval retrieves the important parameters from the conversation received from the NLP Chat based on the same defined required fields. The important parameters are also put in a structured format by Info Retrieval.

Intents can also be entered in low-level format through the User-assisted Intent Interface where the user can select processes and generate their desired data or execution pipeline with the help of the Pipeline Generator, without having to pass through the NLP Chat or the Info Retrieval component. This functionality is meant for CyclOps users who have knowledge of what they want and how it could be achieved. This component also acts as the interface through which the user explores and interacts with the data retrieved through the Exploratory Analysis Engine.

The job of this collection of components is to gather as much information from the user as possible and extract the important fields from this conversation, thereby converting the intent into a structured intent.

In intent-based systems (IBS), LLMs have been applied for intent translation [29], which is one of the core functionalities of IBS. This technology is also adopted in the intent interface component. In detail, LLMs are applied for interaction with users to capture user intents in natural language and further translate these intents to a structured format. This process involves the extraction of the needed information from the natural language conversation between the user and the NLP Chat.

4.2   Exploratory Analysis Engine

The Exploratory Analysis Engine allows users to interact with the data available in the CyclOps framework leveraging the IKB. This is achieved through a sub-component within the User-assisted Intent Interface through which the user explores or interacts with the retrieved data. This component also allows for the policies compiled from the user natural language inputs by the Intent Compiler to be converted to queries for retrieving the data from the IKB. That is, the Exploratory Analysis Engine converts the policies in JSON to SPARQL or SQL queries through which data querying is fulfilled.

This component directly interacts with the Knowledge Layer. The user inputs in natural language are translated into SPARQL or SQL queries with the aid of existing knowledge provided by the Knowledge and/or Runtime Layer(s). A response is given to the user based on the available data. This response can also be visualized by the user.

There are few technologies that support the automation of the conversion of JSON data to SPARQL queries, which is one major function of this component. LLMs have also been applied for automatic SPARQL query generation from texts [30]. Nguyen et al. [31] proposed a translator to convert REST/JSON request messages to SPARQL commands based on an ontology of their data. In CyclOps, this module implements a JSON-to-SPARQL translation engine to transform JSON into SPARQL queries for the IKB.

4.3   Pipeline Generator

The Pipeline Generator compiles the user intents received from the User-assisted Intent Interface or the Natural Language Intent Interface into pipeline orchestration commands and delivers them to the DEA component in the Runtime Layer for its execution.

It contains two major sub-components: the Intent Compiler and the Pipeline Orchestrator. The Intent Compiler parses the outputs of the intent interface and verifies and generates policies from the intents. The Orchestrator manages and dictates the actions to be performed in setting up data processing execution pipelines encompassing functions and algorithms in AIOps and DataOps.

For the pipeline generation, this component retrieves the pipeline components metadata from the IKB and implements a categorization framework domiciled in the Knowledge Layer for assigning algorithms to policies. This component adopts an intent-2-algorithm categorization framework that provides a means to categorize policies retrieved by the Pipeline Orchestrator according to the algorithms required to execute the ML tasks contained in those policies.

5   Interoperability Layer

The interoperability layer acts as a bridge between CyclOps and external ecosystems, enabling secure and structured data flows to and from data spaces and service marketplaces. The interoperability layer in CyclOps facilitates interaction with external data sources such as data spaces and service marketplaces. It ensures smooth data retrieval and publication while maintaining security and compliance with regulatory frameworks and legal provisions from the sources.

This layer supports identity management through Self-Sovereign Identity (SSI) mechanisms, extensively described in the next section, enabling authentication, and access control for external resources. Additionally, it integrates semantic interoperability tools to standardize data formats and align them with common ontologies, ensuring efficient data exchange across platforms. This layer introduces several innovations, including the use of W3C verifiable credentials[6] for accessing external data spaces, leveraging Keycloak[7] and other components for authentication, and implementing pysmartdatamodels[8] for semantic data transformation.

5.1   Identity Management

In CyclOps, the identity management component follows a Self-Sovereign Identity approach as envisaged by the Data Spaces Business Alliance (DSBA) Technical Convergence guidelines [32], aiming to align with the converging approaches from EBSI/ESSIF, IDSA, eIDAS, and GAIA-X.

CyclOps delivers a state-of the-art identity solution for data space as envisaged by the DSBA Technical Convergence built on a Trust Framework supported by W3C Verifiable Credentials and Decentralized identifiers following the EBSI Conformance Guidelines.[9]

The identity component is made up of several services that are deployed in the different participant organization domains depending upon their role in a data space. It is therefore important to note the primary participants as described by GAIA-X,[10] are Consumers, Providers, and Operators, where the latter role is responsible for the Trust Anchor Framework for the Data Space, which governs the trusted participants and Issuers for the ecosystem.

The identity management component is made up of a set of component services:

- SSI Wallet: A mobile app for storing, managing, and presenting Verifiable Credentials (VCs). It supports OID4VCI for Receiving credentials from trusted organizations (aligned with EBSI standards). It furthermore supports OID4VP for presenting credentials to data space providers and marketplaces.
- SSI Agent (Issuer): Issues VCs to members of consumer and provider organizations, with capabilities such as verifying presentations for provider organizations and issuing VCs to employees for role-based authorization in a data space.
- VC Verifier: Verifies VCs for authentication and authorization. Key features include providing endpoints for SIOP-2/OIDC4VP-compliant authentication flows; and exchanging VCs for JWT tokens to enable downstream authorization and authentication within Data Provider DSC components.
- Credential Configuration Service: Used to configure the credentials required to access them with associated trusted issuer and partner lists.
- Trusted Issuer List (TIL): Manages trusted issuers for a data space by implementing an EBSI Trusted Issuers Registry.
- Trusted Participant List: Part of the Verifiable Data Registry (Dataspace Registry). It registers consumer and provider organizations as participants in a data space.

5.2 Data Interoperability with Data Spaces

The interoperability layer in CyclOps ensures seamless data exchange with external data spaces by implementing structured mechanisms for data retrieval, publication, and compliance.

When retrieving data, CyclOps users provide credentials by using SSI Wallet, as explained in the previous section, which interacts with the data space to generate the authorization to operate in the data space. With this authorization, the data exchange module will be allowed to fetch the requested data, store it in a context broker, and return it to the user in a standardized format.

For data publication, the process involves VCs, enriching metadata via semantic interoperability tools and registering datasets in the data space's catalogue. The data exchange agent plays a crucial role in transforming data into NGSI-LD format (a specific variety of JSON format), ensuring compatibility with external platforms. By integrating automated semantic mapping and standardized data models, this component enhances the Findability, Accessibility, Interoperability, and Reusability (FAIR) principles, promoting more efficient and transparent data sharing across diverse digital ecosystems.

The interoperability layer includes several semantic facilities that provides:

1. Access to more than 160,000 attributes definitions compiled from more than 1000 data models, as well as the access to the data types of these attributes or its semantic model when available, the units of an attribute when available, and the list of all attributes for a specific data model.

2. The type of NGSI attribute (Property, Relationship, or GeoProperty). The last two attributes are especially relevant because the Relationship one allows the exploration of related entities and the GeoProperty one defines geographical resources, either a point, multipoint, line, multiline, polygon, or multi-polygon.
3. The generation of random payloads compliant with any of the data models in different formats.
4. The ingestion or update of data compliant with a data model in a context broker (e.g., coming from a sensor).
5. The generation of a data model compatible with SQL databases.
6. Facilities to find a data model when there is limited information about the name.
7. Access to all the metadata about the data model (e.g., the source of the data model, if it is derived from an existing standard or ontology, the general description, its title, etc.).

The data interoperability layer also integrates a data exchange broker, to enable the data sharing by the users without compromising the security of their systems. The broker is compliant with the NGSI-LD standard defined at ETSI. This standard extends previous NGSI versions by incorporating Linked Data (LD) principles, allowing for a more structured and interoperable representation of information. An NGSI-LD broker is designed to manage and facilitate access to context information in distributed environments, particularly as the temporal storage when accessing and retrieving information from data spaces but also for other environments like smart cities, Industry 4.0, and IoT applications. It is capable of the storage, retrieval, and exchange of dynamic, semantically enriched data using the NGSI-LD standard. The broker supports CRUD (Create, Read, Update, Delete) operations on context entities, which represent real-world objects or logical business objects, and allows subscription mechanisms where clients can receive notifications about changes in specific entities according to their defined conditions. This is particularly important in case of interaction with data spaces because it could eventually simplify and reduce the number of requests to the data space. One of its key features is context expansion, where relationships between entities can be traversed and resolved using linked data references by using a specific type of attribute named relationships. It also provides temporal querying, which enables tracking the evolution of context data over time, crucial for historical analysis and predictive analytics. Through JSON-LD encoding, it ensures compatibility with semantic Web technologies, fostering seamless data exchange across domains. By supporting federated architectures, it allows multiple brokers to collaborate, ensuring horizontal scalability and cross-domain interoperability.

The broker is a software component that can be found in several open-source implementations in the FIWARE catalogue, like Orion-LD,[11] Stellio,[12] and Scorpio.[13]

---

[11] https://github.com/FIWARE/context.Orion-LD/tree/develop/doc/manuals-ld

[12] https://stellio.rtfd.io/

[13] https://scorpio.rtfd.io/

6   Foreseen Usage of CyclOps and Use Cases

6.1   A Generic Usage Workflow

In a generic case, data scientists will use CyclOps for running advanced analytical tasks that span data management and analysis operations to be consumed and/or published in data spaces. The resultant workflow for this general usage is depicted in Fig. 2. This workflow assumes that CyclOps has already been configured to access the data sources, either data spaces or private data stores, and that the subset of data required for the analysis has previously been ingested in the Long-term Storage.

The workflow starts with the user expressing the analytical intent as natural language or via a user-assisted user interface (step ①). In both cases, from user intents, the Pipeline Generator component automatically defines the pipeline that satisfies such intent (step ②). The input data of the pipeline is that stored in the Long-term Storage, obtained provided by the IKB through its Virtual Knowledge Graph (step ③). Once data are fetched, the pipeline (which is a combination of DataOps and AIOps functions, steps ④ and ⑤) is executed on a distributed infrastructure leveraging the Data and Execution Abstraction (DEA) component. In the process of executing it, the pipeline itself will generate final or intermediate results to be stored back to the Long-term Storage component and annotations to the IKB about its execution (steps ⑥). Then, once a pipeline has been executed, its output can also be passed to the Interoperability Layer, which will publish the generated data or models to the target data space (step ⑦). Additionally, we consider that Knowledge Scientists can scrutinize the content of the IKB concerning the execution of pipelines, i.e. via the IKB Exploitation services, they can obtain assessment of explainability aspects (step ⑧), which are ultimately translated to queries over the IKB (step ⑨).

6.2   Use Cases in Different Domains

To fully demonstrate the potential of the CyclOps framework, four use cases have been selected from very diverse areas of industry and research. Indeed, the variety of the selected use cases is what best expresses the wide applicability of the framework. The first use case joins the European Data Space for Mobility, Transport, and Tourism (EONA-X) and the European Data Space for Tourism (DEPLOYTOUR), currently in deployment phase. The second use case revolves around the Green Deal Dataspace (GDDS) and its exploitation for storing information tailored to Disaster Risk Reduction and Climate Change Adaptation. The third use case is focused on the Public Procurement Data Space (PPDS), the first data space under development with direct involvement of the European Commission. Finally, the fourth use case is applied on a Manufacturing Data Space, an embryonic data space for offset

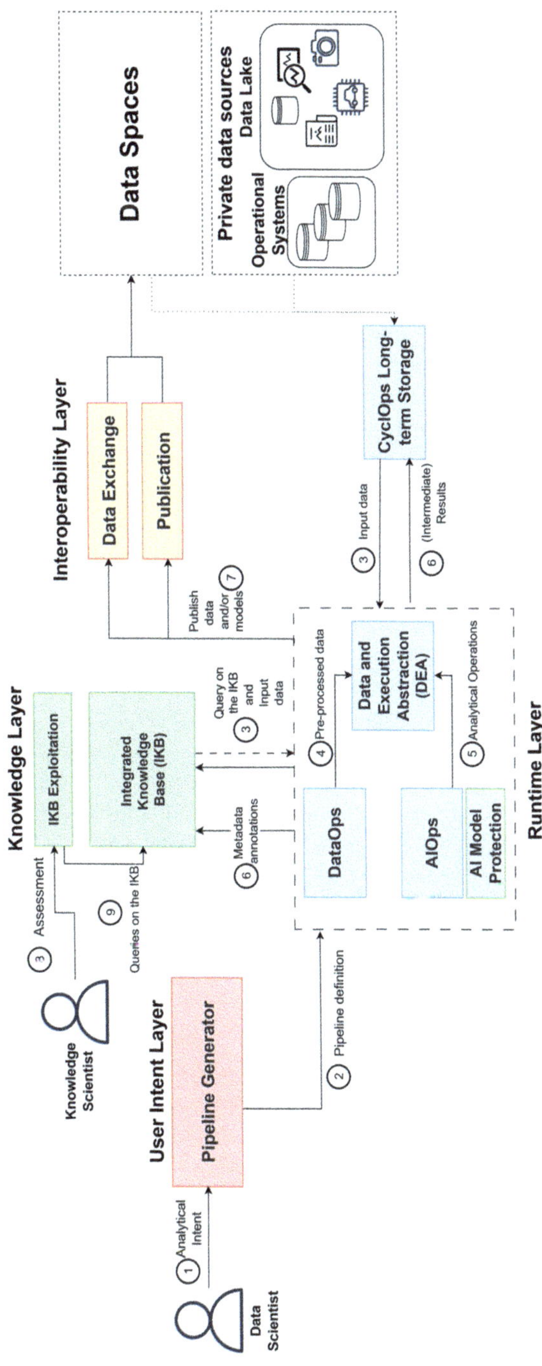

Fig. 2 CyclOps generic usage workflow

printing. Next, we provide a detailed description of each use case and how CyclOps can be applied for solving challenges nowadays.

6.2.1 Use Case on Tourism

Tourism is a complex sector composed by a high number of entities (public and private) that must collaborate to ensure the sustainability, competitiveness, and resilience of the sector [33, 34]. Data has become a key issue to ensure a personalized tourism experience while supporting decision-making [35].

This use case's objective is to create added-value services that provide a seamless and integrated experience to tourists, based on the information on the way they move and consume services during their journey. By leveraging advanced data processing and AI-driven insights, it aims to enhance decision-making, improve service quality, and create a seamless experience for both tourists and industry stakeholders.

Focusing on the hospitality and leisure sector, as well as mobility actors, this use case serves a wide range of users. Destination Management Organizations (DMOs) can use the insights to analyze and optimize tourist flows, ensuring better resource allocation and improved visitor experiences. Tourism professionals, including hotels, tour operators, and service providers, can refine their offerings based on real-time data. Research entities and academic institutions can access structured information to study tourism trends, sustainability, and visitor behavior [36]. Finally, tourists themselves benefit from personalized recommendations, enhanced services, and a more efficient travel experience.

This use case requires the integration of heterogeneous data from various sources to construct a comprehensive view of the tourist journey. On the one hand, it relies on common European data spaces, which provide structured datasets on tourism and mobility, the Common European Tourism Data Space (ETDS)[14] and the European Mobility Data Space (EMDS).[15] On the other hand, private data sources might contribute with valuable insights from hotel reservations, transportation records, social media interactions, and contextual factors like events and/or weather conditions. This poses several challenges concerning semantic interoperability, data fragmentation, and required distributed computing, among others.

In this use case, the expected workflow and pipelines to be implemented as are follows. The different intents begin with the collection of data from multiple sources, including data spaces, mobility networks, accommodation providers, digital platforms, etc. Once gathered, by using CyclOps, the data undergoes cleaning and normalization, and an associated unified knowledge base is created. This ensures consistency, reliability, and interoperability across different datasets. Relationships between the data points can then be established, enabling a holistic understanding of tourist behavior, preferences, and movement patterns. CyclOps will allow

---

[14] https://deploytour.eu/

[15] https://deployemds.eu/

real-time capabilities (update and analysis for dynamic insights), scale to accommodate future data sources and increasing volumes, tourist data collection, and sharing throughout. CyclOps will also allow to optimize the global process of offering and providing leisure and tourism services for destinations.

The CyclOps framework addresses the functional, operational, and technical necessities of this use case. Advanced processing techniques can be applied to extract meaningful insights. Natural language processing (NLP) models can be used to analyze tourist feedback from reviews, surveys, and social media interactions, identifying sentiment trends and areas for improvement. AI-driven pattern recognition techniques can be developed to detect common behaviors, helping tourism professionals anticipate the demand and adjust their services accordingly.

On top of CyclOps, this use case can build spatial analysis tools for mapping mobility routes, allowing for the optimization of transportation networks and the reduction of congestion in key areas, as well as interactive dashboards and reports to present the insights obtained to use case-end users. These dashboards visualize key metrics such as mobility trends, visitor satisfaction levels, and personalized recommendations, enabling stakeholders to make data-driven decisions. Furthermore, to ensure accessibility and interoperability, this use case plans to make insights also available through APIs, allowing tourism organizations and businesses to integrate them into their own systems, tourism apps, or third-party systems.

6.2.2   Use Case on Climate Change

Climate change is pushing national and local authorities to employ efficient and pragmatic actions to both reduce the greenhouse gas emissions and mitigate the risks related to the changes in climate in the short and medium term [37]. A correct assessment of climate risk is essential for understanding its impacts, and thus, it must be underpinned by empirical evidence and data available to drive dynamically the adaptation measures based on the combination of different so-called adaptation pathways.

In this context—also referred to as Climate Change Adaptation (CCA)—several challenges and sources of uncertainties currently hinder an effective and impactful understanding of the key risks and their complex relationships, both from an environmental and societal perspective [38]. The knowledge from the qualitative and quantitative sources of data and information is sparse, fragmented, and often hard to integrate; uncertainties propagate and must be accounted for, and the pace at which the increasingly available Earth Observation (EO) data coming from satellites does not yet translate to equivalent available information [39].

Within this use case, the objective is to use the CyclOps framework as a key bridging technology for CCA. Revolving around the so-called Impact Chains [40] (a conceptual framework based on cause-effect chains for climate risk assessment), the use case will demonstrate how a well-structured knowledge could turn heterogeneous sources into an actionable structure with a "hot and dry" risk scenario that will focus on three different (and correlated) hazards—droughts, heat waves, and wildfires—in the inter-regional Adige River Basin in the north of Italy.

By properly encoding and linking spatially explicit quantitative datasets with formal vocabularies and ontologies in the CyclOps-integrated knowledge base, the world of data-intensive processes and algorithms can get closer to the world of policymakers and climate risk practitioners in general. With the intermediation of CyclOps, the factors and processes leading to specific climate risks that are represented in the impact chains can turn into interactive triggers of information to be visualized and used to support decisions. Together with CyclOps' language models capabilities and its registry of AI algorithms, complex models can be trained and executed on the recommended catalogues of data, on demand, and then stored in the Green Deal Data Space, to make it findable and foster its re-use.

6.2.3 Use Case on Public Procurement

The proposed use case in this domain is the definition and implementation of a forecasting and scenario assessment tool to evaluate and estimate the participation and success of small and medium enterprises (SMEs) in public procurement procedures. The Directive 2014/24/EU aims to enhance SME participation in public procurement. SME participation in public procurement is still very limited compared to their role in national economies, where they represent 99% of all businesses [41]. Several barriers make it difficult for SMEs to participate in and win tenders [42].

This use case aims to identify the limiting factors faced by SMEs when accessing public procurement and provide AI-based tools to assess the impact of modifying them. These tools would support both the EU and national policymakers on the promotion of SME in public procurement, as well as contracting authorities willing to enhance the participation of SME in their tendering procedures, and SME themselves in the identification of those procedures where they are most likely to succeed.

Currently, the implementation of the use case requires different profiles, from the public procurement expert to knowledge engineers for the management of the linked data life cycle and data scientists for the AI models development, among others. These users face different challenges concerning the usage of the public procurement data and the ability to analyze it and automatize the data life cycle process, where CyclOps can be of great support.

In detail, the main challenges and how CyclOps can be leveraged to overcome them are the following.

The public procurement data sources (Public Procurement Data Space[16] and MiTender[17]) have different formats although containing similar information. The Knowledge Layer and DataOps modules in CyclOps will help the knowledge engi-

---

[16] https://single-market-economy.ec.europa.eu/single-market/public-procurement/digital-procurement/public-procurement-data-space-ppds_en

[17] https://mitender.es

neers to transform the data from different data sources into a format that allows to merge and compare data, in a faster way. They will help as well review the data to ensure its quality and reliability, considering different aspects such as completeness, validity, consistency, and accuracy; with this, the number of human errors will be significantly decreased, providing a better outcome of the process.

The AI models chosen for this specific use case depend on the most relevant data within the datasets. This is a hard task since the success of the use case and the relevance of the results will depend on that. CyclOps automatizes this decision-making process and facilitates the implementation and execution of the model in a distributed manner. Thus, the data scientist can also focus on enhancing the performance of the model through the proper validation of it.

The management of a large volume of data is also a challenge for the developers as it requires a balance of well-defined architecture pointing toward an efficiency-wise implementation. CyclOps will provide such a powerful architecture that will maximize the results with the minimal use of resources.

6.2.4   Use Case on Manufacturing

This use case aims to apply CyclOps capabilities to the manufacturing domain of Offset Printing (OP). While the European printing industry holds a significant role in EU economy (i.e., with an annual turnover of 88 billion Euro), it currently faces significant challenges attributed to the rise of global competition, costs, and environmental regulations [43].

In detail, companies within the sector generate vast amounts of structured and unstructured data daily, which remain underutilized during their entire life cycle, affecting real-time decision-making and raw-material consumption. Furthermore, this lack of digitalization also leads to product deficiencies, detected at the final quality control, hindering timely corrective actions, while machine downtime disruptions further reduce operational efficiency [44]. In addition, the sectors defective production has significant economic and environmental costs, with an OP press emitting 7100 metric tons of CO_2 and wasting 230 metric tons of paper annually, highlighting the need for more proactive and efficient processes. Therefore, leveraging the generated data presents an opportunity for the OP companies to optimize their production processes and support EU's data economy vision [45].

CyclOps can help address the aforementioned challenges, transforming the daily generated data into actionable insights, enabling companies to meet the complex market demands and regulatory requirements. In detail, leveraging the DataOps component, factory users can ensure the integration and curation of high-quality data generated from various factory sources, thus maintaining data integrity throughout operations. By integrating their data within CyclOps, companies can utilize the AIOps component to train new or existing ML models available on the CyclOps marketplace (e.g., factory optimization, predictive maintenance, etc.). Toward this end, CyclOps's DEA can manage the large-scale data management as well as the distributed deployment for ML models. With the interoperability layer, users can securely share data, making it easier to collaborate with partners, suppliers, and research organizations, creating a manufacturing data space environment, as well as leverage data from existing data spaces. Additionally, the User Intent Layer empowers factory operators to explore real-time production data. By transforming complex data into actionable insights, this interface helps identify inefficiencies and guides adjustments to machine settings and workflows, enabling companies to reduce product defects and material waste, thereby enhancing both operational efficiency and environmental sustainability [46].

6.3 Going Beyond the Use Cases: Implications and Challenges

The previous use case descriptions illustrate the diverse applicability and potential benefits of the CyclOps framework across various domains. Here, we aim to provide an overall discussion of challenges that might arise when generalizing such approach and deploying it in other use cases.

High-quality ontologies and semantic resources. At the heart of CyclOps lies the IKB, whose accuracy and utility heavily depend on the availability of high-quality, domain-relevant ontologies and reliable metadata. However, crafting or adapting ontologies requires domain expertise, and available ontologies often lack completeness, formal rigor, or contextual alignment with operational data. CyclOps facilitates semi-automated ontology and mapping generation (e.g., using mapping patterns and bootstrapping techniques); however, the quality of these assets remains a foundational bottleneck.

Semantic interoperability across ecosystems. The increasing participation in federated data environments and data spaces (e.g., those promoted by DSSC and GAIA-X) requires robust semantic interoperability. CyclOps addresses this through its Interoperability Layer and use of standards such as NGSI-LD and pysmartdatamodels. Still, aligning data from heterogeneous sources—each potentially using different data models, vocabularies, and policies—remains a significant technical and organizational challenge. Achieving semantic interoperability not only requires technical infrastructure but also collective agreement on shared conceptualizations, which may evolve over time and vary across domains.

Tailoring AIOps to domain-specific needs. CyclOps provides an extensive AIOps module with capabilities for feature engineering, model selection, optimization, and deployment. However, real-world AI implementations often require domain-specific models, tailored performance metrics, and contextual constraints (e.g., fairness in public procurement or geospatial robustness in climate applications). CyclOps addresses this through its AI Marketplace, yet such environment must be continuously updated and maintained to incorporate the specific needs of new use cases.

Automating processes over heterogeneous and custom data formats. Despite support for varied data ingestion mechanisms (e.g., API uploads, streaming via Kafka), many organizations still rely on legacy systems. The diversity of formats hinders automation, especially when structural metadata is missing or unreliable. CyclOps attempts to bridge this gap through schema inference and the use of SHACL for data curation, but the integration of custom data formats into automated workflows often requires manual intervention or the development of specific adapters.

7 Conclusions

The CyclOps framework is presented in this chapter as a new solution for facilitating the development of AI-based data-driven applications for all players, business, and research alike, addressing a critical gap in the management of the end-to-end data and AI life cycle. It does so by providing automated solutions for integrating and analyzing both machine- and human-generated data from heterogeneous data sources and enabling the sharing and exchange of data, models, and services from and for data spaces.

The use of well-defined and semantically rich metadata in the form of a KG in the IKB, at the core of the framework, enhances data interoperability and allows for automation. The IKB approach also incorporates innovative components providing traceability and explainability capabilities to the system. All this information is leveraged by the Runtime layer components, allowing the management and distributed execution of pipelines and providing and ensuring integration between data ingestion, processing, AI model deployment, and distributed execution.

CyclOps also embraces the human-in-the-loop concept, providing a human-centric entry point to all features through the User Intent layer, which also includes capabilities that allow the user to express intents in a flexible manner. This makes the framework user-friendly and accessible to non-experts, ensuring that even small businesses with limited technical resources can benefit from CyclOps. Finally, the Interoperability layer provides all functionalities required for the interaction with external data sources such as data spaces and service marketplaces, implementing structured mechanisms for data retrieval, publication, compliance, and identity management.

CyclOps builds on existing and novel standards, models, and architectures and complements or expands them as necessary to foster interoperability, data sharing, and exchange. It does so with a set of interconnected tools offering a variety of functionalities to adapt to a great diversity of use cases, which allows further versatile scientific research, without restricting them to specific vertical domains. In this sense, to serve as an inspiration for the reader, four use cases in different domains (tourism, climate change, public procurement, and manufacturing) are introduced, which present different challenges that could be addressed by the described CyclOps solution.

CyclOps is being developed in the framework of a research and innovation project. At the time of writing this chapter, CyclOps is reaching the finalization of its first version, which will be validated within the use cases presented. More innovations and features will be further developed in the next months, dealing with aspects such as privacy, regulatory compliance, and AI model robustness, among others.

Acknowledgments This chapter describes work undertaken in the context of the CyclOps project, funded by the European Union under GA No. 101135513. Views and opinions expressed are however those of the author(s) only and do not necessarily reflect those of the European Union or the European Commission as the granting authority. Neither the European Union nor the granting authority can be held responsible for them.

References

1. Rashid, A. B., & Kausik, M. A. K. (2024). AI revolutionizing industries worldwide: A comprehensive overview of its diverse applications. *Hybrid Advances, 7*, 100277. https://doi.org/10.1016/j.hybadv.2024.100277
2. Curry, E., Scerri, S., & Tuikka, T. (2022). *Data spaces: Design, deployment and future directions*. Springer Nature.
3. Nadal, S., Jovanovic, P., Bilalli, B., et al. (2022). Operationalizing and automating data governance. *Journal of Big Data, 9*, 117. https://doi.org/10.1186/s40537-022-00673-5
4. Nadal, S., et al. (2022). Operationalizing and automating data governance. *Journal of Big Data, 9*(1), 117.
5. Jaffri, A., Popa, A., Krensky, P., Hare, J., Bhati, R., Hassanlou, M., & Zhang, T. (2024, June 17). 2024 Gartner Magic Quadrant for data science and machine learning platforms.
6. De Simoni, G., Raj, A., Chien, M., & Kennedy, S. (2025, January 7). Gartner, Magic Quadrant for data and analytics governance platforms. ID: G00807073.
7. Curry, E., Heintz, F., Irgens, M., Smeulders, A. W., & Stramigioli, S. (2022). Partnership on AI, data, and robotics. *Communications of the ACM, 65*(4), 54–55.
8. Hogan, A., Blomqvist, E., Cochez, M., D'amato, C., De Melo, G., Gutierrez, C., Kirrane, S., Gayo, J. E. L., Navigli, R., Neumaier, S., Ngomo, A.-C. N., Polleres, A., Rashid, S. M., Rula, A., Schmelzeisen, L., Sequeda, J., Staab, S., & Zimmermann, A. (2022). Knowledge graphs. *ACM Computing Surveys, 54*(4), 71, 37 p. https://doi.org/10.1145/3447772
9. Guarino, N., & Welty, C. A. (2009). An overview of OntoClean. In *Handbook on ontologies* (pp. 201–220). Springer. https://doi.org/10.1007/978-3-540-92673-3_9
10. Davies, K., Keet, C. M., & Lawrynowicz, A. (2019). More effective ontology authoring with test-driven development and the TDDonto2 tool. *International Journal on Artificial Intelligence Tools, 28*(7), 1950023:11–1950023:25. https://doi.org/10.1142/S0218213019500234
11. Flores, J., Rabbani, K., Nadal, S., Gómez, C., Romero, O., Jamin, E., & Dasiopoulou, S. (2024). Incremental schema integration for data wrangling via knowledge graphs. *Semantic Web, 15*(3), 793–830.
12. Calvanese, D., Gal, A., Lanti, D., Montali, M., Mosca, A., & Shraga, R. (2023). Conceptually-grounded mapping patterns for virtual knowledge graphs. *Data & Knowledge Engineering, 145*, 102157. https://doi.org/10.1016/J.DATAK.2023.102157
13. Shvaiko, P., & Euzenat, J. (2005). A survey of schema-based matching approaches. *Journal on Data Semantics*, 146–171. https://doi.org/10.1007/11603412_5
14. Calvanese, D., Cogrel, B., Komla-Ebri, S., Kontchakov, R., Lanti, D., Rezk, M., Rodriguez-Muro, M., & Xiao, G. (2017). Ontop: Answering SPARQL queries over relational databases. *Semantic Web, 8*(3), 471–487. https://doi.org/10.3233/SW-160217

15. Xiao, G., Lanti, D., Kontchakov, R., Komla-Ebri, S., Kalayci, E. G., Ding, L., Corman, J., Cogrel, B., Calvanese, D., & Botoeva, E. (2020). The virtual knowledge graph system Ontop. In *ISWC 2020* (pp. 259–277). https://doi.org/10.1007/978-3-030-62466-8_17
16. Sun, S., An, W., Tian, F., Nan, F., Liu, Q., Liu, J., Shah, N., & Chen, P. (2024). A review of multimodal explainable artificial intelligence: Past, present and future. https://doi.org/10.48550/arXiv.2412.14056.
17. Fernandez, R. C., et al. (2018). Aurum: A data discovery system. In *ICDE 2018*.
18. Bogatu, A., et al. (2020). Dataset discovery in data lakes. In *ICDE 2020*.
19. Dong, Y., et al. (2021). Efficient joinable table discovery in data lakes: A high-dimensional similarity based approach. In *ICDE 2021*.
20. Flores, J., et al. (2021). Towards scalable data discovery. In *EDBT 2021*.
21. Dang, Y., et al. (2020). AIOps: Real-world challenges and research innovations. In *IEEE/ACM 42nd International Conference on Software Engineering: Companion Proceedings (ICSE-Companion)*.
22. Zaharia, M., et al. (2018). Accelerating the machine learning lifecycle with MLflow. *IEEE Data Engineering Bulletin, 41*(4), 39–45.
23. Schelter, S., et al. (2018). Automating large-scale machine learning model management. In *VLDB* (Vol. 11(12), pp. 1543–1557).
24. Sculley, D., et al. (2015). Hidden technical debt in machine learning systems. In *NeurIPS*.
25. Amershi, S., et al. (2019). Software engineering for machine learning: A case study. In *ICSE-SEIP*.
26. Garg, S., Pundir, P., Rathee, G., Gupta, P. K., Garg, S., & Ahlawat, S. (2022). On continuous integration/continuous delivery for automated deployment of machine learning models using MLOps. arXiv:2202.03541. https://doi.org/10.48550/arXiv.2202.03541.
27. Martí, J., Queralt, A., Gasull, D., Barceló, A., Costa, J. J., & Cortes, T. (2017). Dataclay: A distributed data store for effective inter-player data sharing. *Journal of Systems and Software, 131*, 129–145.
28. Sala, B., Maria, R., Conejero, J., Diaz, C., Ejarque, J., Lezzi, D., Gomis, F.-J. L., Cortés, C. R., & Pardell, R. S. (2015). Comp superscalar, an interoperable programming framework. *SoftwareX, 3*, 32–36.
29. Mekrache, A., & Ksentini, A. (2024). LLM-enabled intent-driven service configuration for next generation networks. In *2024 IEEE 10th International Conference on Network Softwarization (NetSoft), Saint Louis, MO, USA* (pp. 253–257). https://doi.org/10.1109/NetSoft60951.2024.10588881
30. Arazzi, M., Ligari, D., Nicolazzo, S., & Nocera, A. (2025). Augmented knowledge graph querying leveraging LLMs.
31. Nguyen, K. M., Nguyen, T. H., & Huynh, X. H. (2016). Automated translation between RESTful/JSON and SPARQL messages for accessing semantic data. In *2016 International Conference on Electronics, Information, and Communications (ICEIC), Danang, Vietnam* (pp. 1–4). https://doi.org/10.1109/ELINFOCOM.2016.7562981
32. Data Spaces Business Alliance. (2022, September). Technical convergence discussion document, v1.0.1. Retrieved from https://data-spaces-business-alliance.eu/dsba-releases-technical-convergence-discussion-document/
33. Gretzel, U., Sigala, M., Xiang, Z., & Koo, C. (2015). Smart tourism: Foundations and developments. *Electronic Markets, 25*(3), 179–188. https://doi.org/10.1007/s12525-015-0196-8
34. European Commission. (2022). Transition pathway for tourism. Retrieved from https://single-market-economy.ec.europa.eu/sectors/tourism/transition-pathway-tourism_en
35. Werthner, H., Koo, C., Gretzel, U., & Lamsfus, C. (2015). ICTs in tourism: Challenges and new solutions. *Information Technology & Tourism, 15*(1), 1–4. https://doi.org/10.1007/s40558-015-0022-3
36. European Travel Commission (ETC). (2021). Exploring consumer travel preferences after COVID-19. Retrieved from https://etc-corporate.org/reports/
37. Calvin, K., Dasgupta, D., Krinner, G., Mukherji, A., Thorne, P. W., Trisos, C., et al. (2023). In Core Writing Team, H. Lee, & J. Romero (Eds.), *IPCC, 2023: Climate change 2023:*

Synthesis report, summary for policymakers. Contribution of Working Groups I, II and III to the sixth assessment report of the Intergovernmental Panel on Climate Change (pp. 1–34). IPCC. Retrieved from https://www.ipcc.ch/report/ar6/syr/

38. Arribas, A., Fairgrieve, R., Dhu, T., et al. (2022). Climate risk assessment needs urgent improvement. *Nature Communications, 13,* 4326. https://doi.org/10.1038/s41467-022-31979-w
39. Hain, L. I., Kölbel, J., & Leippold, M. (2021, April 19). Let's get physical: Comparing metrics of physical climate risk. https://doi.org/10.2139/ssrn.3829831
40. Zebisch, M., Terzi, S., Pittore, M., Renner, K., & Schneiderbauer, S. (2022). Climate impact chains—A conceptual modelling approach for climate risk assessment in the context of adaptation planning. In C. Kondrup et al. (Eds.), *Climate adaptation modelling* (Springer climate). Springer. https://doi.org/10.1007/978-3-030-86211-4_25
41. Ploeger, A., & Prutsch, M. J. Fact sheets on the European Union—2025. Small and medium-sized enterprises. Retrieved from https://www.europarl.europa.eu/factsheets/en/sheet/63/small-and-medium-sized-enterprises
42. Morais, C., & Santos, A. B. (2024). Lessons for a fairer, more transparent, and more competitive public procurement in the EU. *Journal of Innovation Management, 12*(1), 1.
43. Kalafatelis, A. S., Trochoutsos, C., Giannopoulos, A. E., Angelopoulos, A., & Trakadas, P. (2023, January). A stacking ensemble learning model for waste prediction in offset printing. In *Proceedings of the 2023 10th International Conference on Industrial Engineering and Applications* (pp. 267–272).
44. Spantideas, S. T., Giannopoulos, A. E., Kapsalis, N. C., Angelopoulos, A., Voliotis, S., & Trakadas, P. (2022). Towards zero-defect manufacturing: Machine selection through unsupervised learning in the printing industry. In *CEUR Workshop Proceedings*. ISSN: 1613-0073. Retrieved from http://ceur-ws.org
45. Kalafatelis, A. S., Nomikos, N., Angelopoulos, A., Trochoutsos, C., & Trakadas, P. (2023, January). An effective methodology for imbalanced data handling in predictive maintenance for offset printing. In *International Conference on Mechatronics and Control Engineering* (pp. 89–98). Springer.
46. Trochoutsos, C., Kalafatelis, A. S., & Trakadas, P. Transforming offset printing with digital twins and AI: Insights from research initiatives. In *Proceedings of the 12th International Symposium GRID 2024*. Department of Graphic Engineering and Design, Faculty of Technical Sciences, University of Novi Sad.

Intelligent Underwater Perception: Current Trends and Future Directions

Nicoletta Risi, Gabriel Guimarães Carvalho, Olivier Vasseur, Phil Reiter, and Tom De Schepper

Abstract Underwater systems have witnessed an increasing demand for enhanced sensing technologies in a wide range of applications, ranging from pipes inspection and ground exploration to hull inspections and rescue operations. Meanwhile, recent technological advances in automation, robotics, and AI have set the stage for a new generation of autonomous underwater systems with increased levels of autonomy. However, compared to research on, for instance, terrestrial or aerial sensing and autonomous systems, remote underwater environments present additional unique challenges due to factors such as low-lighting conditions, constrained power resources, and limited connectivity. This chapter on intelligent underwater perception discusses an area with growing interest from both academia and industry. Building upon previous surveys, we consider state of the art from an academic and industry perspective, with the goal of bringing the views of AI, data, and sensing technologies into one overview. We find that the latest technological advancements in hardware substrates, alongside emergent AI pipelines and cross-fertilization from the neighboring automotive sector, are pointing toward a disruptive change in the underwater ecosystem and have the potential to deliver unprecedented levels of intelligence for next-generation underwater perception systems.

Keywords Underwater intelligent perception · Sensing technologies · Edge AI

This chapter relates to the Sensing and Perception enablers for the European AI, Data and Robotics Partnership [1].

N. Risi (✉) · G. G. Carvalho · P. Reiter · T. De Schepper
imec, Kapeldreef, Leuven, Belgium
e-mail: Nicoletta.Risi@imec.be; Gabriel.GuimaraesCarvalho@imec.be; Phil.Reiter@imec.be; Tom.DeSchepper@imec.be

O. Vasseur
Faculty of Applied Engineering, IDLab—imec—University of Antwerp, Antwerp, Belgium
e-mail: Olivier.Vasseur@imec.be

E. Curry et al. (eds.), *Artificial Intelligence, Data and Robotics*,
https://doi.org/10.1007/978-3-032-10561-5_15

1   Introduction

The underwater domain is one of the real frontiers that await real human exploration. The ocean covers approximately 71% of the surface of the earth and is the key enabler of life on earth. Yet, the ocean depths are still, for a large majority, unexplored, and it is said we know more about the surface of the Moon than about the oceans. For instance, in 2024, only about 26% of the global seafloor has been mapped in detail (using multibeam sonar systems from the surface) [2, 3], despite the interest in oceanography for many centuries [4].

However, this scenario is about to change. Today, the extremely complex underwater sensing ecosystem is at a pivotal point that will soon deliver groundbreaking technological advances across various oceanographic applications. Enhanced *underwater intelligence* is already at the core of next-generation underwater systems and is witnessing an increasing interest from academic research, as well as industry key players [5].

The market demand for enhanced underwater intelligence systems is rising in several domains. Key areas are underwater sensing and robotics. It is anticipated that advances in underwater sensing will lead to a novel generation of truly intelligent vehicles with unprecedented levels of autonomy in underwater environments [6]. Moreover, according to the latest reports, the global underwater robotics market size reached USD 4.7 billion in 2024 and is projected to exhibit a growth rate (CAGR) of 10.26% by 2033 [7]. This steep increase stems from growing research and industrial activities in several applications [8], with the largest market segment being underwater commercial exploration.

For instance, the increasing shortage of fossil fuels on land necessitates harvesting resources on the ocean floor [8]. With the harsh and risky environmental conditions of deep waters hindering human intervention, improving the level of autonomy of underwater vehicles naturally becomes an interesting and viable solution to tackle those missions. Other driving applications are in the defense, security, and safety space where there are growing concerns on underwater threats and sabotage [9]. In this context, the latest advancements in underwater sensing, as well as computing technology, are enhancing security and level of autonomy in, among others, search-and-rescue and inspection operations (e.g., for ship-hull, mines, and cable inspection) [10]. Finally, an increasing demand for underwater intelligence stems from the rising interest in marine and ocean research, where complex and time-consuming experiments call for advanced sensing technologies to enhance the quality of data captured and reduce costs.

Such missions, involving not only shallow but also deep-sea environments, typically require underwater systems endowed with the ability to run complex tasks, such as localization, mapping, detection and tracking, navigation, and often communication, with the on-board sensors suite and limited compute power. Thus, within this so-called "see, think, act" cycle of robot operations [8], underwater intelligence stems from the successful implementation of a variety of *tasks*, embedded on static, as well as moving, *platforms*, endowed with different *sensing technologies* applied to several *domains* (Table 1).

Within this highly complex ecosystem, enhanced perception is at the foundation of the robot's ability to make sense of the surrounding environment and, as such, becomes a vital asset for truly intelligent underwater systems. Thus, this chapter presents an overview of current state of the art and solutions toward intelligent underwater systems, with a focus on *underwater perception*. As the "intelligent" nature of underwater perception entails an active component, this work elaborates on sensing technologies for active perception tasks. Specifically, the main focus is on *acoustic* and *optical* sensors, typically mounted on *unmanned underwater vehicles* (*UUVs*) to scan the environment, identify landmarks, and understand the surroundings (e.g., for *search-and-rescue* operations or underwater *inspection*). While these sensors can also be used for so-called geophysical navigation, they are also substantially different from those involved in the realm of communication (e.g., radio communication), resource exploration, and monitoring of essential ocean variables (e.g., conductivity, temperature, and depth) [2]. Building up on previous surveys, we consider state-of-the-art work from academic and industry perspectives, with the goal of bringing the views of AI, data, and sensing technologies into one overview. For a deeper dive into specific aspects of intelligent underwater systems, the authors refer the reader to Table 2, with a collection of review papers focusing on a subset of specific building blocks in the field.

The structure of the chapter is as follows: First, we discuss the unique challenges that make underwater perception so complex (Sect. 2) and discuss some of the current sensing technologies and industry key players (Sect. 3). Next, we present an overview of various simulation environments (Sect. 4). These simulators are key to compensate for the cumbersome acquisition of real-world data in harsh underwater environments. We finalize this chapter by discussing the frontier of underwater intelligent perception, with an overview of future research challenges and interesting links to the automotive sector (Sect. 5).

Table 1 The underwater intelligence ecosystem

Task	Platforms	Sensing technology	Domains
Perception[a]	Unmanned underwater vehicles[a]	Acoustic[a]	Search and rescue[a]
Navigation	Critical undersea infrastructures	Optical[a]	Inspection[a]
Communication		Magnetic/electric	Defense[a]
Localization		Quantum/ gravitational	Resource exploration
Joint operations		Bionic sensing	Seafloor mapping
			Marine research

[a] Out of the many building blocks of underwater intelligence, these will be the main focus of this chapter

Table 2 Reviews on intelligent underwater systems

Reference	Domains	Platforms	Sensing technology	Processing methods	Simulators
Whitt et al. (2020) [11]	✓	✓		✓	
Cong et al. (2021) [12]			✓		
Sun et al. (2021) [2]	✓		✓		
Kahraman-Bacher (2021) [13]			✓	✓	
Forti et al. (2022) [6]	✓		✓		
Wang et al. (2022) [14]				✓	
Er et al. (2023) [15]				✓	
Huy et al. (2023) [8]	✓	✓	✓	✓	
Wang et al. (2023) [16]				✓	
Ciuccoli et al. (2024) [17]					✓
Ioannou et al. (2024) [10]	✓	✓			
Chen et al. (2024) [18]				✓	

2 The Unique Challenges of Intelligent Underwater Perception

Compared to research on terrestrial or aerial sensing and autonomous systems, enhancing the quality of perception in remote underwater systems poses remarkable technical challenges [19].

First, environmental factors such as biofouling (mainly in shallow water), corrosion, current speed, and extreme pressure and temperature conditions (mainly in deep water) demand unique system-level requirements for the target sensors suite [20].

Moreover, the environmental conditions faced by underwater systems inherently lower the sensing quality. For instance, the interactions between photons and water molecules result in light attenuation, reflection, refraction, and diffusion effects, with red light having stronger attenuation than blue and green spectra [2]. Indeed, while red and violet light disappear within 10 m ranges, blue and green wavelengths can travel up to around 30 m ranges [8]. Such attenuation effects are detrimental for optical sensors and are worsened under turbid-water conditions, due to the light interactions with particles suspended in water. Additionally, the signal quality is also usually affected by water turbulence, which contributes to signal attenuation and distortion [21]. Sound waves, in turn, are affected by factors such as temperature and pressure, as well as environmental noise in seawater (e.g., boats, waves, biological noise, and, for on-shore applications, harbor infrastructures) [2].

Finally, from a compute-platform standpoint, the underwater environment demands strict low-power, limited-bandwidth and limited-memory requirements. Indeed, remotely operated as well as fully autonomous, tetherless, underwater vehicles are typically deployed in hardly accessible areas, with limited power sources and limited room for human intervention.

Thus, enhancing intelligent perception capabilities in such unique and critical scenarios requires a joint effort from industry players and academic institutions to deliver novel sensing technologies, as well as digital frameworks to support software simulations of the underwater environment. An overview of the current state-of-the-art technology in this field is presented in the following sections, with a focus on acoustic and optical sensors mounted on UUVs.

3 Intelligent Underwater Perception

3.1 *Unmanned Underwater Vehicles*

Unmanned underwater vehicles (UUVs) are submersible vehicles capable of operating without human occupants, also referred to as underwater drones. UUVs are typically categorized in terms of their levels of autonomy (LoA). In the lowest level of autonomy, the UUV behavior is fully controlled via external sources by a trained operator. These remotely operated vehicles (ROVs) have an umbilical cable, or tether, that physically links the ROV to a control station and its power supply, normally at surface level, and through which data is transmitted. The presence of a tether limits the maneuverability and the mission range, being the major disadvantage of such a UUV. With a much larger LoA, there are autonomous underwater vehicles (AUVs), which are self-propelled and untethered, addressing the drawbacks presented by ROVs. AUVs can operate without real-time control by human operators for extended periods of time and greater depths. Table 3 summarizes the key aspects of ROVs and AUVs.

Beyond ROVs and AUVs, the future of underwater exploration lies in groundbreaking technologies that enhance autonomy, adaptability, and efficiency. The next generation of UUVs, designed for longer and more complex missions with greater precision, includes:

- Hybrid ROV (HROV) [10]: Hybrid ROVs, or HROVs, offer real-time control when needed and autonomous functions for distant explorations, behaving as an AUV or a ROV depending on the application. They can operate both tethered/untethered. A well-known representative of an HROV is the Nereus [22] vehicle, lost in 2014 at a depth of 9.9 km at the Kermadec Trench.

Table 3 UUV categories comparison

Feature	ROV	AUV
Control	Human	Autonomous
Tether	Present	Absent
Power	Remotely supplied	Onboard
Operation	Real-time needs	Programmable missions
Application	Maintenance, repair, inspection	Mapping, surveying

- Intervention AUV (I-AUV) [10]: Intervention AUVs are AUVs with increased autonomy for intervention tasks. They can interact with the environment, manipulating objects and performing maintenance. For examples of different I-AUV concepts, we refer to [23].
- Bio-inspired/biomimetic robots: Operating autonomously or remotely, they mimic marine life for efficient mobility and energy efficiency. Examples of fish-inspired robots, such as MIT's RoboTuna and SoFi, can be found in [24], and the concept of snake-inspired robots can be found in [25].

3.2 *Underwater Sensors*

Collecting environmental information on UUVs involves a variety of sensing techniques, often used jointly to leverage complementary features and limitations. The core technologies behind underwater sensing can be clustered in four main classes: acoustic, optical, electromagnetic, quantum/gravitational and bionic sensing (Table 1). Autonomous underwater vehicles typically solve perception tasks, such as object detection, with a combination of onboard optical and acoustic sensors [2, 8, 10, 12], which are often well represented in current underwater software simulators.

Underwater acoustic sensing methods comprise a variety of sonar-based sensors. Forward-looking sonar (FLS), for instance, consists of a single sonar beam, often directed forward, to detect objects in a localized illuminated area [10]. Multibeam sonar, instead, is typically used to generate a map of the seabed by simultaneously transmitting a fan of beams and by extracting depth and directional information from the returning soundwaves [2, 10, 12]. Better suited for search and rescue operations or object search is instead the side-scan sonar (SSS) [2, 10, 12]. These sensors are typically mounted on the side of a vehicle and emit directional pulse acoustic signals with a wide vertical beam angle and a short horizontal beam angle. Finally, the sub-bottom profiler (SBP) is an acoustic sensor used to analyze the seabed structure by sending pulses to the seabed and measuring the soundwave reflection and refraction that occur at the interface of different media [2].

Compared to acoustic sensing, optical sensors, such as passive optical cameras, pose additional environmental challenges arising from the rapid underwater light attenuation, especially in turbid water. Although the use of active artificial light is often considered as a viable strategy to compensate for the low-lighting conditions, this often comes at the cost of additional backscattering effects due to the interaction between the light source and water particles [8]. Alternatively, laser or LiDAR technology is typically used to compensate for the shortcomings of passive optical sensors [26]. More recently, hyperspectral imaging (HSI) and polarized cameras are also being explored as promising candidates to mitigate the seawater attenuation, by capturing a wider spectral range. A beautiful demonstration of such principles has already been found in nature, in the eyes of the mantis shrimp, which can capture visible UV and polarized light by means of clusters of photoreceptors cells called ommatidia. Interestingly, inspired by this remarkable feature, a team of researchers at the North Caroline State University, has developed a new type of optical, mantis shrimp-inspired multispectral and polarization-sensitive (SIMPOL) sensor [27], which can capture simultaneously hyperspectral and polarized light. Although the development of such technology is still in an early stage, a successful implementation of these principles could unleash a novel generation of optical sensors with enhanced underwater sensitivity. Indeed, the potential of using polarized light in underwater environments has already been shown, for instance, in the context of underwater navigation [28].

Given their complementary features, optical and acoustic sensors are often combined with sensor fusion techniques to maximize the quality of perception [8]. Acoustic sensors offer higher accuracy and longer detection ranges, but higher costs and lower resolution, which makes them good candidates for object detection at long distances, especially in harsh water conditions since acoustic signals are less attenuated in water. Optical sensors, by contrast, offer higher resolution but shorter detection ranges, due to underwater light attenuation, which makes them better candidates for small object detection and robot grasping and maneuvers. Active optical sensors, such as laser or LiDAR, are often used also for marine monitoring, bathymetry, and seafloor mapping.

An overview of some commercial underwater acoustic and optical sensing devices is shown in Fig. 1, sorted with respect to largest depth rating (i.e., deployment depth) and depth range.[1] The corresponding raw data is reported in Table 4. Our industry review confirms that acoustic sensors exhibit the longest detection

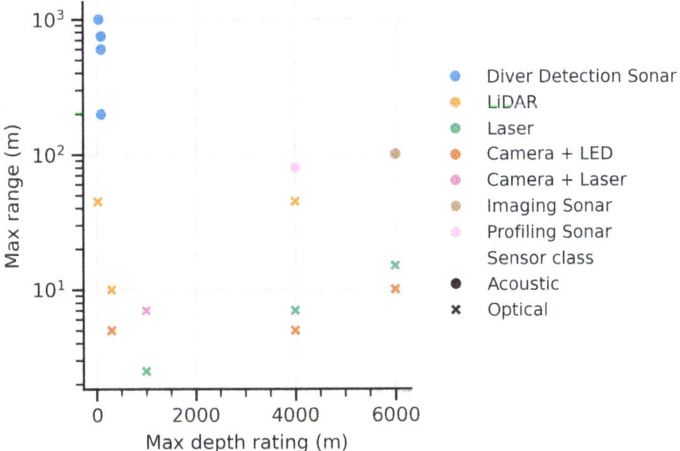

Fig. 1 Depth range vs max depth rating of some commercial, active underwater sensors

---

[1] Since the depth range of passive sensors, such as optical cameras or hydrophones, depends mostly on the water attenuation of acoustic and electromagnetic waves, only active sensors are considered here.

Table 4 Companies and active underwater sensing technologies

Company	Technology	Device	Max depth rating (m)	Max range (m)
NORBIT	Diver detection sonar	GuardPoint70	55	1000
NORBIT	Diver detection sonar	GuardPoint100	100	750
NORBIT	Diver detection sonar	GuardPoint200	100	600
NORBIT	Diver detection sonar	GuardPoint400	100	200
3DAtdepth	LiDAR	SL4 LiDAR	4000	45
3DAtdepth	LiDAR	SL6 LiDAR	4000	45
3DAtdepth	LiDAR	SL4n LiDAR	30	45
Fraunhofer	LiDAR	ULi	300	10
Voyis	Laser	Insight Nano	1000	2.5
Voyis	Laser	Insight Micro	4000	7
Voyis	Laser	Insight Pro	6000	15
Voyis	Camera + LED	Observer & Nova Micro-5MP	1000	7
Voyis	Camera + LED	Observer & Nova Micro-4k	1000	7
Voyis	Camera + LED	Observer & Nova Pro	6000	10
Voyis	Camera + LED	Recon—Small AUVs	300	5
Voyis	Camera + LED	Recon—Medium AUVs	1000	7
Voyis	Camera + laser	Recon LS	1000	7
Voyis	Camera + LED	Discovery Camera	300	5
Voyis	Camera + LED	Discovery Stereo	4000	5
Impact Subsea	Imaging sonar	ISS360HD	6000	100
Impact Subsea	Profiling sonar	ISP360	4000	80

range. Out of them, the diver detection sonar is confined within small depth rating ranges due to its specific application in rather shallower water. Moreover, while the sensitivity of passive optical sensors is higher at small depth ranges, due to underwater light attenuation, combining optical sensors with active illumination allows to increase both depth range and rating. See, for instance, the Observer & Nova Pro from the Canadian company Voyis, which leverages an actively cooled imaging sensor paired with up to 700,000-lumen lighting to deliver images at up to 10 m range and 6000 m depth ratings.

In addition to gathering information from the outside world, underwater vehicles typically rely on proprioceptive sensors to track their current state, e.g., Inertial Measurement Units (IMUs), Doppler Velocity Log (DVL), and positioning systems. Given their key role in navigation, path planning, and simultaneous localization and mapping (SLAM), these sensors are also typically supported in simulation frameworks (see Table 5).

Table 5 Comparison of underwater robotics simulators

	UWSim [30]	UUV [32]	DAVE [34]	MARUS [35]	HoloOcean [38]	UNav-Sim [43]	Stonefish [45]
Year	2012	2016	2022	2022	2022	2023	2025
Sim engine (R: rendering, P: physics)	OpenScene graph (R), bullet physics (P)	Gazebo (R,P)	Gazebo (R,P)	Unity 3D (R,P)	UE4 (R), PhysX (P)	UE5 (R), AirSim (P)	OpenGL (R), bullet physics (P)
ROS	ROS 1	ROS 1	ROS 1	ROS 1,2	ROS 2	ROS 1,2	ROS 2
Maintained (git commit in 2024)	No	No	No	Yes	Yes	Yes	Yes
Requirements	GPU/graphics card	CPU	CPU/GPU	GPU	GPU	GPU	GPU
Robotics domain	Marine	Underwater	Underwater	Marine	Underwater	Underwater	Underwater
Dataset annotation				✓		✓	✓
Optical and acoustic sensors							
Camera	✓	✓	✓	✓	✓	✓	✓
Depth camera							✓
Event-based camera							✓
Thermal camera							✓
LiDAR			✓	✓			✓
Multibeam sonar	✓	✓	✓		✓		✓
Forward-looking sonar			✓		✓		✓
Side-scan sonar				✓	✓		✓
Doppler velocity log	✓	✓	✓	✓	✓		✓
Rendering quality	Low	Low	Low	Moderate	High	High	High
Physics quality	Low	Low	Moderate	Moderate	Low	Moderate	High

4 Underwater Simulators

Real-world underwater experiments are extremely costly due to the expensive equipment and the harsh, risky environmental conditions. Thus, in addition to enhancing the sensing quality of underwater technology, a significant effort of both industry and researcher is devoted to the development of more accurate and realistic software simulations, or digital twins (DTs), of the underwater environment [17]. Building a digital counterpart of the physical world has two major applications.

First, it provides a testbed to validate new sensors and actuator models under a variety of controlled environmental conditions, prior to deployment to the real world.

Moreover, it serves as a powerful strategy for data generation to mitigate one of the biggest roadblocks in the field, i.e., the scarcity of large datasets to train emergent AI models. Indeed, today, the latest advancements in edge-computing sensing and processing technologies are delivering promising solutions to tackle the low-power, low-bandwidth constraints of underwater systems by computing data directly on the edge. In addition to enhancing the system energy efficiency, these approaches mitigate the need to communicate large amounts of data to a central server for downstream processing above water, which would otherwise be extremely power hungry and technically challenging. However, deploying truly robust AI-powered systems compatible with emergent energy-efficient architectures for underwater perception needs large-scale and annotated data from underwater scenarios. Thus, leveraging underwater simulation frameworks for data generation proves a promising and viable solution to address the scarcity of data.

As the digital counterpart of a physical system, such simulators must fulfill a set of critical requirements [17].

- First, they should support simulation of UUVs and their interaction with the external environment. This entails a graphic engine for rendering underwater agents and the surrounding environment, as well as a physics engine to simulate robot dynamics and interactions with the environment.
- To enable reliable digital twinning, simulators should prioritize a high level of physical fidelity while preserving a reasonable level of graphical rendering. For instance, they should provide modeling physical interactions between rigid bodies, as well as hydrodynamics and hydrostatic effects, ideally under both clean and turbid water conditions.
- Finally, to mitigate the lack of real-world data, they should support the simulation of a wide range of underwater sensors, as well as a robust framework for collecting and annotating large datasets, and potentially the possibility to interact with AI simulation engines, such as OpenAI's gym [29].

Below, we present an overview of the most used, current open-source underwater simulators. For a more detailed analysis on the current ecosystem of digital twins for the underwater environment, the authors refer the reader to [17]. A direct comparison with respect to some key software features is provided in Table 4.

4.1 UUV and UWSim

The UnderWater Simulator (UWSim) [30], developed on the OpenSceneGraph [31] and Bullet graphics and physics engines, has been extensively used in a variety of applications. Despite its versatility, it functions primarily as a visualization tool due to its inability to accurately simulate the dynamics and hydrodynamics of vehicles or their manipulator dynamics.

The unmanned underwater vehicle (UUV) simulator [32], built on the Gazebo [33] platform, prioritizes high physical fidelity over advanced rendering quality. It includes a package of Gazebo plugins with integrated simple hydrodynamics and Robot Operating System (ROS) nodes to control UUVs. However, it lacks support for simulating manipulator hydrodynamics and buoyancy, and its development is limited to ROS Melodic.

Although UUV and UWSim were once the most widely used underwater simulators, both are now discontinued.

4.2 DAVE

Building on the UUV simulator, the DAVE Aquatic Virtual Environment (DAVE) simulator [34] expands capabilities by supporting a wider range of underwater sensors, ROVs, tools, and scenarios. It integrates real-world, high-resolution bathymetry, ocean current data, and models for vehicles and objects, along with their physical degradation. The simulator supports various sensors, including the DVL, an underwater 3D pulse LiDAR, modeled after the SL3 Subsea LiDAR Laser, an Ultra Short Baseline (USBL) system for navigation and positioning, and a multibeam FLS. However, DAVE currently lacks support for underwater cameras, and the rendering quality provided by the Gazebo simulator remains unrealistic.

4.3 MARUS

The Marine Robotics Unity Simulator (MARUS) simulator [35] is built on the Unity3D physics engine [36], which has been enhanced to incorporate water body physics, including buoyancy and currents, as well as simulations of various sensors, visualization tools, dataset generation, and automatic annotation. It supports a wide range of sensors [37], such as a 9-DOF IMU, GNSS, Pose, Camera, Depth, Ranging, 360° LiDAR, Imaging Sonar, Acoustic Modem, RF devices, DVL, and Automatic Identification System (AIS). Additionally, by means of a gRPC interface, the simulator is compatible with various backends, including ROS 1 and 2, MATLAB, and LabView. Moreover, it includes tools for dataset generation and automatic annotation.

4.4 HoloOcean

HoloOcean [38] builds on the open-source reinforcement learning and robotics simulator Holodeck [39]. As such, it leverages Unreal Engine 4 (UE4) [40] and PhysX [41] as rendering and physics engine, respectively, with a Python API that facilitates easy installation and contribution. It includes a few example environments and agents and supports multi-agent missions, making it versatile for a range of applications. One of its notable features is the implementation of an efficient imaging sonar. The simulator supports a variety of sensors [42], including a DVL, IMU, depth sensor, camera, GPS, range finder, and pose sensor.

4.5 UNav-Sim

UNav-Sim [43] is the first underwater simulator built on Unreal Engine 5 (UE5), offering high-quality rendering capabilities. It leverages open-source AirSim [44] extensions and supports integration with ROS, enabling the use of an autonomous vision-based navigation stack. Currently, the simulator provides support for GPS, IMU, distance sensors, and cameras. In addition, it comes with an OpenAI gym environment that allows for benchmarking learning-based navigation algorithms. Similar to HoloOcean, based on Unreal Engine, UNav-Sim prioritizes high-quality rendering over physical realism.

4.6 Stonefish

Stonefish [45] is a lightweight simulator that uses a custom rendering engine implemented directly in OpenGL, without relying on any external graphics engine. The simulator is composed of two main components: a C++ library that provides the core simulation framework and a ROS package that offers a sensor message interface, simulator nodes, and launch files.

Key features of Stonefish include the computation of buoyancy and hydrodynamics based on the actual geometry of objects, as well as realistic rendering that accounts for controllable light sources, absorption, scattering, and air-light effects. The simulator supports a diverse set of sensors [46]: rotary encoder, torque, force-torque, link sensors, accelerometer, gyroscope, IMU, odometry, GPS, compass, pressure, DVL, Inertial Navigation System (INS), profiler, multibeam sonar, color camera, depth camera, FLS, mechanical scanning imaging sonar, and side-scan sonar.

Interestingly, a new version of Stonefish has just been released [47], featuring enhancements in a few key aspects. Pertaining to the sensor suite, notable additions include the integration of new GPU-accelerated sensors and improved sonar simulation, the implementation of visual light communication, and the introduction, for the first time, of event-based cameras [48]. Designed to respond asynchronously to local changes in the light intensity, these sensors exhibit low-latency response, high dynamic range, and low power consumption. Moreover, as the event-based paradigm of these sensors extends also to asynchronous neuromorphic processing systems [49, 50], event-based cameras are often interfaced with dedicated, energy-efficient architectures. Overall, these features make them highly promising to address some of the key shortcomings of underwater sensing, i.e., low lighting conditions and limited power supply. Moreover, the recent rising interest in event cameras in the automotive sector [51] has boosted the development of novel event-based algorithms for perception tasks, such as detection, tracking, and optic flow. Thus, coupling simulated event cameras from Stonefish with emergent event-based models has the potential to unleash a new generation of low-latency and more energy-efficient edge-computing systems in the underwater domain [52, 53].

Exceptional advancements in Stonefish are not only limited to the sensor suite. They also include AI-compatible tools. For instance, as introduced by the MARUS and UNav-Sim simulators, the latest release includes an automatic annotation tool for segmentation and object detection, which are crucial to leverage software simulators for dataset generation. In addition, to ease the integration with reinforcement learning tools for the development of learning-based methods, the authors included Python bindings to expose Stonefish data to OpenAI Gym, thereby removing the need for a higher-latency ROS interface.

Building upon its already impressive first software release, which stands out for high physics and rendering quality, the recent advancements in the Stonefish simulator position it as the leading open-source tool for underwater simulation.

5 The Frontier of Intelligent Underwater Perception

The latest trend in enhancing intelligent underwater perception is paving the way for next-generation autonomous underwater agents, which will soon deliver more modular, easier to operate, and, most importantly, adaptive and energy-efficient systems. As covered in the prior sections, new sensing technologies are currently being proposed (e.g., hyperspectral). In parallel, as the digital counterpart of underwater sensing is expected to reflect the complexity of the real-world environment, the current growth in levels of autonomy of underwater vehicles is driving a significant research effort in the development of more realistic and reliable underwater simulators. Yet, compared to the more mature terrestrial, aerial and space robotics applications, the realm of intelligent underwater sensing is just about to thrive and is still far from reaching its full potential. This scenario unfolds exciting frontiers in underwater technologies, as well as the potential for cross-fertilization from research in neighboring fields, such as the automotive sector.

From a technology standpoint, increasing the computing capabilities as close as possible to the sensor, also known as "near/in-sensor computing" [54], is the real frontier of truly intelligent edge systems. Its deployment for underwater sensing has the potential to introduce a disruptive increase in levels of autonomy for three main reasons. First, moving computation closer to the sensory terminal allows to compress input signals and to extract higher-level features in situ, thereby reducing the movement of large amounts of data across long distances. Such a bandwidth optimization becomes even more crucial in underwater environments, due to limited communication and where cybersecurity is becoming a critical issue [6]. Moreover, especially if combined with emergent neuromorphic sampling and computing schemes, near/in-sensor computing can lead to huge power savings, as already demonstrated in other domains. For instance, the neuromorphic, event-driven, level-crossing analog-to-digital converter (LCADC) is shown to lead to more than one order of magnitude less power consumption for data transmission and storage with biomedical signals, such as electroencephalograms [55]. Thus, the recent introduction of event-based sensors in the ecosystem of underwater simulators is expected to drive a novel research path toward edge-computing devices for underwater sensing [47, 52, 53]. Finally, enhancing in situ computation unlocks the potential to endow underwater systems with more adaptive sensing, or "self-calibrating" capabilities, which is extremely crucial given the harsh and difficult-to-access conditions where UUVs typically operate. Not only does this entail enhancing the system ability to cope with environmental variability, as well as sensors drift, but it also mitigates the complexity of sensor fusion techniques, which rely on accurate sensors calibration.

Aside from the innovation in edge-computing hardware, there is a necessary effort to provide non-compute-intensive perception software. In fact, the low-power requirement of underwater detection and the need for real-time execution in some applications make it a key feature for perception software to be lightweight. The advent of deep learning (DL) in object detection has brought accuracy improvements compared to traditional machine learning (ML) techniques, especially for environments characterized by complex visual features [13, 14]. On the downside, such methods often come at a high computational cost, making research into lightweight models an important and promising field.

A promising research direction to enhance sensing efficiency and quality of underwater intelligent systems is to drive inspiration from biology. In addition to the abovementioned mantis-inspired HSI, remarkable examples are the artificial lateral line and whiskers sensors [12]. The former ones are inspired by the biological lateral line, the sensing organ of the fish responsible for perceiving changes in the water flow. When navigating in dark environments, these sensors are a powerful resource to make sense of the surroundings. Similarly, inspired by their biological counterpart, artificial whisker sensors can be used for accurate perception of the surrounding flow.

While the underwater intelligence domain is still heavily underexplored, it is interesting to notice that underwater and autonomous driving systems share multiple underlying technical challenges. Thus, leveraging insights from the automotive sector promises to be a powerful direction worth investigating to tackle some of the outstanding challenges in the realm of underwater intelligence.

For instance, one current critical milestone in the field is the effective integration of multiple sensory modalities. In this regard, recent advancements in DL techniques for autonomous vehicles have encouraged innovative ways to ease the alignment of multiple sensor modalities without the need for large, synchronized datasets across all sensors. The use of modality translation, for example, has been explored to predict the observation of a specific sensor based on the output of other sensors. This is especially useful in conditions where the sensor of interest is unavailable or unable to provide accurate data. For example, the sensing capabilities of LiDAR sensors are degraded in environments containing a high concentration of suspended particles (e.g., smoky air for automotive or turbid water for underwater), where, instead, radar or sonar would perform better. In such degraded optical conditions, translating radar measurements into LiDAR-like measurements could allow to generate data that contains the information of the radar-perceived environment while still featuring a high point cloud density like a LiDAR could provide. A DL framework of this type for autonomous vehicles was proposed in [56].

As concerns underwater software simulations, one of the major roadblocks is the complexity of underwater environments, which requires adequate physics modeling to understand the dynamics of interactions between underwater agents and the surroundings. Future directions in this regard include supporting simulations of different water conditions (e.g., water turbidity, currently supported in Stonefish only), weather conditions, as well as novel sensing technologies. While some emergent devices are already being introduced (e.g., event cameras in Stonefish), supporting novel sensing technologies often entails introducing additional complexity in physics modeling and is therefore a challenging task in the domain of digital twins. Simulating HSI, for instance, and generating realistic sensor data requires addressing two roadblocks: (1) supporting multiple spectral bands outside of the visible range with the underlying rendering engine and (2) extending the physics engine to capture the interactions between the underwater environment and such electromagnetic waves. This is a well-known problem in the neighboring automotive ecosystem, where current digital twins are not yet equipped to support the next generation of sensors. For instance, the latest version of the open-source simulator CARLA [57] does not support hyperspectral sensors. Yet, the use of HSI is currently gaining attention for enhancing autonomous driving systems [58]. In this domain, the complexity of simulating emerging sensing technologies, combined with the scarcity of data, is typically addressed with physics-based sensor modeling and by leveraging the power of AI [59]. For instance, the integration of modality translation techniques (e.g., for RGB-to-thermal [60] or sonar and radar to LiDAR [56]) with digital twins of autonomous driving systems promises to be a powerful tool to augment the simulated scenarios of interest with next-generation sensors [59]. If extended to the underwater domain, the use of AI models to mitigate the shortcomings of current physics and rendering engines can pave the way for AI-enriched pipelines for underwater intelligent sensing.

In essence, while still lagging behind the more mature terrestrial, aerial, and space robotics applications, the underwater intelligence domain has effectively reached a pivotal point, and its biggest outstanding quests are being unfolded. In this realm, the latest technological advancements in hardware substrates, alongside emergent AI pipelines and cross-fertilization from the neighboring automotive sector, are pointing toward a disruptive change in the underwater ecosystem and have the potential to deliver unprecedented levels of intelligence for next-generation underwater perception systems.

Acknowledgments This research received funding from the Flemish Government under the "Onderzoeksprogramma Artificiële Intelligentie (AI) Vlaanderen" program (Flanders AI Research).

References

1. Curry, E., Heintz, F., Irgens, M., et al. (2022). Partnership on AI, data, and robotics. *Communications of the ACM, 65*, 54–55. https://doi.org/10.1145/3513000
2. Sun, K., Cui, W., & Chen, C. (2021). Review of underwater sensing technologies and applications. *Sensors, 21*, 7849. https://doi.org/10.3390/s21237849
3. How much of the ocean has been explored? Ocean Exploration Facts: NOAA Office of Ocean Exploration and Research. Retrieved February 24, 2025, from https://oceanexplorer.noaa.gov/facts/explored.html
4. Sonnewald, M., Lguensat, R., Jones, D. C., et al. (2021). Bridging observations, theory and numerical simulation of the ocean using machine learning. *Environmental Research Letters, 16*, 073008. https://doi.org/10.1088/1748-9326/ac0eb0
5. Zhao, Q., Peng, S., Wang, J., et al. (2024). Applications of deep learning in physical oceanography: A comprehensive review. *Frontiers in Marine Science, 11*. https://doi.org/10.3389/fmars.2024.1396322
6. Forti, N., d'Afflisio, E., Braca, P., et al. (2022). Next-Gen intelligent situational awareness systems for maritime surveillance and autonomous navigation [point of view]. *Proceedings of the IEEE, 110*, 1532–1537. https://doi.org/10.1109/JPROC.2022.3194445
7. Research and Markets. Underwater robotics market by type, application, and region 2025-2033. Retrieved February 24, 2025, from https://www.researchandmarkets.com/reports/5947058/underwater-robotics-market-type-application
8. Huy, D. Q., Sadjoli, N., Azam, A. B., et al. (2023). Object perception in underwater environments: A survey on sensors and sensing methodologies. *Ocean Engineering, 267*, 113202. https://doi.org/10.1016/j.oceaneng.2022.113202
9. Loik, R. (2024). Undersea hybrid threats in strategic competition: The emerging domain of NATO–EU defense cooperation. *Journal on Baltic Security, 10*, 1–25. https://doi.org/10.57767/jobs_2024_008
10. Ioannou, G., Forti, N., Millefiori, L. M., et al. (2024). Underwater inspection and monitoring: Technologies for autonomous operations. *IEEE Aerospace and Electronic Systems Magazine, 39*, 4–16. https://doi.org/10.1109/MAES.2024.3366144
11. Whitt, C., Pearlman, J., Polagye, B., et al. (2020). Future vision for autonomous ocean observations. *Frontiers in Marine Science, 7*, 697. https://doi.org/10.3389/fmars.2020.00697
12. Cong, Y., Gu, C., Zhang, T., & Gao, Y. (2021). Underwater robot sensing technology: A survey. *Fundamental Research, 1*, 337–345. https://doi.org/10.1016/j.fmre.2021.03.002
13. Kahraman, S., & Bacher, R. (2021). A comprehensive review of hyperspectral data fusion with LiDAR and SAR data. *Annual Reviews in Control, 51*, 236–253. https://doi.org/10.1016/j.arcontrol.2021.03.003

14. Wang, N., Wang, Y., & Er, M. J. (2022). Review on deep learning techniques for marine object recognition: Architectures and algorithms. *Control Engineering Practice, 118*, 104458. https://doi.org/10.1016/j.conengprac.2020.104458

15. Er, M. J., Chen, J., Zhang, Y., & Gao, W. (2023). Research challenges, recent advances, and popular datasets in deep learning-based underwater marine object detection: A review. *Sensors, 23*, 1990. https://doi.org/10.3390/s23041990

16. Wang, N., Chen, T., Liu, S., et al. (2023). Deep learning-based visual detection of marine organisms: A survey. *Neurocomputing, 532*, 1–32. https://doi.org/10.1016/j.neucom.2023.02.018

17. Ciuccoli, N., Screpanti, L., & Scaradozzi, D. (2024). Underwater simulators analysis for digital twinning. *IEEE Access, 12*, 34306–34324. https://doi.org/10.1109/ACCESS.2024.3370443

18. Chen, L., Huang, Y., Dong, J., et al. (2024). Underwater object detection in the era of artificial intelligence: Current, challenge, and future.

19. Pydyn, A., Popek, M., Janowski, Ł., et al. (2024). Between water and land: Connecting and comparing underwater, terrestrial and airborne remote-sensing techniques. *Journal of Archaeological Science: Reports, 53*, 104386. https://doi.org/10.1016/j.jasrep.2024.104386

20. Skålvik, A. M., Saetre, C., Frøysa, K.-E., et al. (2023). Challenges, limitations, and measurement strategies to ensure data quality in deep-sea sensors. *Frontiers in Marine Science, 10*, 1152236. https://doi.org/10.3389/fmars.2023.1152236

21. Baykal, Y., Ata, Y., & Gökçe, M. C. (2022). Underwater turbulence, its effects on optical wireless communication and imaging: A review. *Optics & Laser Technology, 156*, 108624. https://doi.org/10.1016/j.optlastec.2022.108624

22. Bowen, A., Yoerger, D., Taylor, C., et al. (2009). The Nereus hybrid underwater robotic vehicle. *Underwater Technology, 28*, 79–89. https://doi.org/10.3723/ut.28.079

23. Ridao, P., Carreras, M., Ribas, D., et al. (2014). Intervention AUVs: The next challenge. *IFAC Proceedings Volumes, 47*, 12146–12159. https://doi.org/10.3182/20140824-6-ZA-1003.02819

24. Li, J., Li, W., Liu, Q., et al. (2024). Current status and technical challenges in the development of biomimetic robotic fish-type submersible. *Ocean-Land-Atmosphere Research, 3*, 0036. https://doi.org/10.34133/olar.0036

25. Pettersen, K. Y. (2017). Snake robots. *Annual Reviews in Control, 44*, 19–44. https://doi.org/10.1016/j.arcontrol.2017.09.006

26. Zhou, G., Li, C., Zhang, D., et al. (2021). Overview of underwater transmission characteristics of oceanic LiDAR. *IEEE Journal of Selected Topics in Applied Earth Observations and Remote Sensing, 14*, 8144–8159. https://doi.org/10.1109/JSTARS.2021.3100395

27. Altaqui, A., Sen, P., Schrickx, H., et al. (2021). Mantis shrimp–inspired organic photodetector for simultaneous hyperspectral and polarimetric imaging. *Science Advances, 7*, eabe3196. https://doi.org/10.1126/sciadv.abe3196

28. Cheng, H., Chen, Q., Zeng, X., et al. (2023). The polarized light field enables underwater unmanned vehicle bionic autonomous navigation and automatic control. *Journal of Marine Science and Engineering, 11*, 1603. https://doi.org/10.3390/jmse11081603

29. Brockman, G., Cheung, V., Pettersson, L., et al. (2016). OpenAI Gym.

30. Dhurandher, S. K., Misra, S., Obaidat, M. S., & Khairwal, S. (2008). UWSim: A simulator for underwater sensor networks. *Simulation, 84*, 327–338. https://doi.org/10.1177/0037549708096606

31. OpenSceneGraph. OpenSceneGraph. Retrieved February 27, 2025, from https://openscenegraph.github.io/openscenegraph.io/openscenegraph.io/

32. Manhães, M. M. M., Scherer, S. A., Voss, M., et al. (2016). UUV simulator: A Gazebo-based package for underwater intervention and multi-robot simulation. In *OCEANS 2016 MTS/IEEE Monterey* (pp. 1–8).

33. Koenig, N., & Howard, A. (2004). Design and use paradigms for Gazebo, an open-source multi-robot simulator. In *2004 IEEE/RSJ International Conference on Intelligent Robots and Systems (IROS) (IEEE Cat. No.04CH37566)* (Vol. 3, pp. 2149–2154).

34. Zhang, M. M., Choi, W.-S., Herman, J., et al. (2022). Dave aquatic virtual environment: Toward a general underwater robotics simulator. In *2022 IEEE/OES Autonomous Underwater Vehicles Symposium (AUV)* (pp. 1–8). IEEE.
35. Lončar, I., Obradović, J., Kraševac, N., et al. (2022). MARUS—A marine robotics simulator. In *OCEANS 2022, Hampton Roads* (pp. 1–7).
36. Juliani, A., Berges, V.-P., Teng E., et al. (2020). Unity: A general platform for intelligent agents.
37. marus-core/Scripts/Sensors at dev, MARUSimulator/marus-core. Retrieved March 5, 2025, from https://github.com/MARUSimulator/marus-core/tree/dev/Scripts/Sensors
38. Potokar, E., Ashford, S., Kaess, M., & Mangelson, J. G. (2022). HoloOcean: An underwater robotics simulator. In *2022 International Conference on Robotics and Automation (ICRA)* (pp. 3040–3046). https://doi.org/10.1109/ICRA46639.2022.9812353
39. Greaves, J., Robinson, M., Walton, N., et al. (2018). Holodeck: A high fidelity simulator.
40. Unreal Engine. The most powerful real-time 3D creation tool. Retrieved February 26, 2025, from https://www.unrealengine.com/en-US/home
41. NVIDIAGameWorks/PhysX (2025).
42. Sensors—HoloOcean 1.0.0 documentation. Retrieved March 5, 2025, from https://byu-holoocean.github.io/holoocean-docs/UE5.3_Prerelease/holoocean/sensors.html
43. Amer, A., Álvarez-Tuñón, O., Uğurlu, H. İ., et al. (2023). UNav-Sim: A visually realistic underwater robotics simulator and synthetic data-generation framework. In *2023 21st International Conference on Advanced Robotics (ICAR)* (pp. 570–576). IEEE.
44. Shah, S., Dey, D., Lovett, C., & Kapoor, A. (2018). AirSim: High-fidelity visual and physical simulation for autonomous vehicles. In M. Hutter & R. Siegwart (Eds.), *Field and service robotics* (pp. 621–635). Springer International Publishing.
45. Cieślak, P. (2019). Stonefish: An advanced open-source simulation tool designed for marine robotics, with a ROS interface. In *OCEANS 2019—Marseille* (pp. 1–6).
46. Sensors—Stonefish 1.4.0 documentation. Retrieved March 5, 2025, from https://stonefish.readthedocs.io/en/latest/sensors.html
47. Grimaldi, M., Cieslak, P., Ochoa, E., et al. (2025). Stonefish: Supporting machine learning research in marine robotics.
48. Gallego, G., Delbrück, T., Orchard, G., et al. (2022). Event-based vision: A survey. *IEEE Transactions on Pattern Analysis and Machine Intelligence, 44*, 154–180. https://doi.org/10.1109/TPAMI.2020.3008413
49. Roy, K., Jaiswal, A., & Panda, P. (2019). Towards spike-based machine intelligence with neuromorphic computing. *Nature, 575*, 607–617. https://doi.org/10.1038/s41586-019-1677-2
50. Liu, S.-C., Delbruck, T., Indiveri, G., et al. (2014). *Event-based neuromorphic systems*. John Wiley & Sons.
51. Shariff, W., Dilmaghani, M. S., Kielty, P., et al. (2024). Event cameras in automotive sensing: A review. *IEEE Access, 12*, 51275–51306. https://doi.org/10.1109/ACCESS.2024.3386032
52. Dadson, N. K. N., & Barbalata, C. Marine event vision: Harnessing event cameras for robust object detection in marine scenarios.
53. AliAkbarpour, H., Moori, A., Khorramdel, J., et al. (2024). Emerging trends and applications of neuromorphic dynamic vision sensors: A survey. *IEEE Sensors Reviews, 1*, 14–63. https://doi.org/10.1109/SR.2024.3513952
54. Zhou, F., & Chai, Y. (2020). Near-sensor and in-sensor computing. *Nature Electronics, 3*, 664–671. https://doi.org/10.1038/s41928-020-00501-9
55. Safa, A., Van Assche, J., Alea, M. D., et al. (2022). Neuromorphic near-sensor computing: From event-based sensing to edge learning. *IEEE Micro, 42*, 88–95. https://doi.org/10.1109/MM.2022.3195634
56. Balemans, N., Anwar, A., Steckel, J., & Mercelis, S. (2024). LiDAR-BIND: Multi-modal sensor fusion through shared latent embeddings. *IEEE Robotics and Automation Letters, 9*, 9159–9166. https://doi.org/10.1109/LRA.2024.3457384
57. Dosovitskiy, A., Ros, G., Codevilla, F., et al. (2017). CARLA: An open urban driving simulator. In *Proceedings of the 1st Annual Conference on Robot Learning* (pp. 1–16). PMLR.

58. Shah, I. A., Li, J., Glavin, M., et al. (2024). Hyperspectral imaging-based perception in autonomous driving scenarios: Benchmarking baseline semantic segmentation models.
59. imec. SENSAI: Digital twin for next-gen automotive sensors. Retrieved February 28, 2025, from https://www.imec-int.com/en/expertise/artificial-intelligence/sensai-digital-twin-next-gen-automotive-sensors
60. Wang, K., Ravaglia, L., Longo, R., et al. (2024). Increasing the diversity in RGB-to-thermal image translation for automotive applications. In *2024 IEEE sensors* (pp. 1–4).

Brain-to-Speech: Prosody Feature Engineering and Transformer-Based Reconstruction

Mohammed Salah Al-Radhi, Géza Németh, Andon Tchechmedjiev, and Binbin Xu

Abstract This chapter presents a novel approach to brain-to-speech (BTS) synthesis from intracranial electroencephalography (iEEG) data, emphasizing prosody-aware feature engineering and advanced transformer-based models for high-fidelity speech reconstruction. Driven by the increasing interest in decoding speech directly from brain activity, this work integrates neuroscience, artificial intelligence, and signal processing to generate accurate and natural speech. We introduce a novel pipeline for extracting key prosodic features directly from complex brain iEEG signals, including intonation, pitch, and rhythm. To effectively utilize these crucial features for natural-sounding speech, we employ advanced deep learning models. Furthermore, this chapter introduces a novel transformer encoder architecture specifically designed for brain-to-speech tasks. Unlike conventional models, our architecture integrates the extracted prosodic features to significantly enhance speech reconstruction, resulting in generated speech with improved intelligibility and expressiveness. A detailed evaluation demonstrates superior performance over established baseline methods, such as traditional Griffin-Lim and CNN-based reconstruction, across both quantitative and perceptual metrics. By demonstrating these advancements in feature extraction and transformer-based learning, this chapter contributes to the growing field of AI-driven neuroprosthetics, paving the way for assistive technologies that restore communication for individuals with speech impairments. Finally, we discuss promising future research directions, including the integration of diffusion models and real-time inference systems.

Keywords Brain AI · Neural signals · iEEG · Prosody feature · Transformer model

---

M. S. Al-Radhi (✉) · G. Németh
Department of Telecommunications and Artificial Intelligence, Budapest University of Technology and Economics, Budapest, Hungary
e-mail: malradhi@tmit.bme.hu; nemeth@tmit.bme.hu

A. Tchechmedjiev · B. Xu
EuroMov Digital Health in Motion, University of Montpellier, IMT Mines Alès, Montpellier, France
e-mail: andon.tchechmedjiev@mines-ales.fr; binbin.xu@mines-ales.fr

E. Curry et al. (eds.), *Artificial Intelligence, Data and Robotics*,
https://doi.org/10.1007/978-3-032-10561-5_16

453

1   Introduction

Speech is a fundamental means of human communication, and its loss due to neurological disorders can severely impact quality of life. Brain-to-speech (BTS) technology, which aims to synthesize speech directly from neural activity, has emerged as a promising solution for individuals with speech impairments [1–4]. Among various neural recording techniques [5], intracranial electroencephalography (iEEG) provides high spatial and temporal resolution, making it an ideal candidate for reconstructing speech from brain signals [6]. However, despite recent advances in deep learning and neural decoding [7–10], achieving natural and intelligible speech synthesis remains a significant challenge.

A major limitation in current BTS research is the inadequate modeling of prosody [11, 12], which encompasses rhythm, stress, and intonation (key elements that contribute to speech naturalness and intelligibility). Traditional speech synthesis techniques primarily focus on spectral envelope reconstruction [13], often neglecting these essential prosodic attributes, resulting in robotic and monotonous speech. Furthermore, deep learning models struggle with the inherent variability in neural signals [14, 15], limiting their ability to generalize across subjects and speech conditions [16, 17].

Another critical challenge is the accurate reconstruction of phase information, which is essential for high-fidelity speech synthesis. Traditional vocoding techniques, such as the Griffin-Lim algorithm, introduce phase inconsistencies, leading to perceptual distortions [18, 19]. Although neural vocoders have demonstrated promising improvements, their adaptation to iEEG-based speech synthesis remains underexplored.

To address these challenges, this chapter introduces a novel BTS framework that integrates prosody-aware feature extraction with a transformer-based speech reconstruction model. The proposed approach employs a wavelet-based feature extraction pipeline to capture both spectral and prosodic characteristics of iEEG signals, enhancing their representation for speech decoding. In addition, a specialized transformer encoder architecture is designed to control the extracted prosodic features, significantly improving speech synthesis quality by generating more natural and meaningful speech. To further enhance phase reconstruction, an iterative harmonic phase correction mechanism is introduced, ensuring greater harmonic consistency and minimizing perceptual distortions. The effectiveness of this framework is evaluated against state-of-the-art iEEG-to-speech synthesis models, demonstrating superior performance in terms of intelligibility and naturalness.

Beyond the immediate contributions to BTS synthesis, this work provides insights into the broader applications of brain-computer interface (BCI) research, particularly in the development of neural speech prostheses. By advancing prosody-aware feature engineering and transformer-based reconstruction methods, this chapter contributes to the growing field of AI-driven neuroprosthetics, paving the way for future innovations in assistive communication technologies. Moreover, potential directions for future research include the integration of diffusion-based generative models and real-time inference strategies, which could further enhance the feasibility and practicality of BTS systems. Additionally, this chapter aligns with the objectives of the AI, Data, and Robotics Partnership [20], contributing to advancements in AI-driven speech neuroprostheses and data-driven methodologies for brain-computer interfaces.

The remainder of this chapter is organized as follows: Sect. 2 reviews related work in brain-to-speech synthesis, focusing on feature extraction, deep learning models, and vocoding techniques. Section 3 presents the proposed methodology, detailing the prosody feature engineering pipeline and the transformer-based reconstruction model. Section 4 describes the experimental setup, including datasets, evaluation metrics, and baseline comparisons. Section 5 discusses results and provides insights into model performance. Finally, Sect. 6 outlines future research directions and concludes the chapter.

2 Related Work

The field of brain-to-speech (BTS) synthesis has evolved significantly over the past decade, supported by advances in neural signal processing, deep learning, and speech synthesis technologies. Early studies primarily focused on linear mappings between neural activity and acoustic features such as formants or spectral envelopes [21], which provided foundational insights but lacked the capacity to model the complex temporal and spectral structures inherent in natural speech. With the rise of deep learning, more sophisticated models have been developed to decode speech directly from neural signals, improving intelligibility and expressiveness.

2.1 Existing Deep Learning Models for Brain-to-Speech

Several deep learning architectures have been explored to map neural recordings to speech features. Convolutional neural networks (CNNs) [22] have been widely used due to their ability to capture local spatial and spectral structures in iEEG data. CNNs have shown improved accuracy in reconstructing speech by focusing on features like high-gamma power and band-specific energy distributions [23].

Recurrent neural networks (RNNs), particularly long short-term memory (LSTM) networks [24], have also been employed to model temporal dependencies in neural signals, which are essential for prosody and rhythm. However, RNNs often struggle with long-range dependencies and generalization across speakers and sessions. Sequence-to-sequence (Seq2Seq) models [25], enhanced with attention mechanisms, have demonstrated more robust performance by learning both local and global temporal relationships. Despite their benefits, many Seq2Seq approaches still lack explicit prosody modeling, limiting the expressiveness of the synthesized speech.

2.2 Feature Extraction Techniques in iEEG-to-Speech Studies

Accurate representation of neural features is critical in BTS systems. Prior studies have commonly used empirical features such as high-gamma activity, band-limited power, and phase-amplitude coupling (PAC) to capture speech-relevant neural dynamics [23]. High-gamma activity (70–170 Hz) has been associated with articulatory processing, while slower oscillations in the theta and beta ranges have been linked to prosodic elements and motor planning [9].

To improve temporal resolution and multi-scale analysis, wavelet-based methods have been introduced. Discrete Wavelet Transform (DWT), for example, allows decomposition of iEEG signals into time-frequency representations that can capture both transient articulatory and sustained prosodic components [26]. Furthermore, cross-frequency coupling (CFC) metrics, particularly PAC, have been leveraged to understand hierarchical interactions between frequency bands in speech production [27].

2.3 Prior Work on Prosody Representation in BTS Systems

Prosody, which includes intonation, stress, and rhythm, is crucial for producing natural-sounding speech, yet it has historically been underrepresented in BTS literature. Many earlier models focused solely on spectral envelope reconstruction, often yielding robotic outputs lacking emotional tone or natural flow.

Recent research has begun to address this by incorporating prosodic cues directly into neural decoding pipelines. Some studies have attempted to predict fundamental frequency (F0) from neural recordings using supervised learning techniques [28], while others have explored additional prosodic features such as energy, duration, shimmer (amplitude perturbation), and phase variability [29]. However, consistent integration of these features into BTS models remains limited, and more comprehensive approaches to prosody-aware decoding are still needed.

2.4 Neural Vocoding Methods Applied in BTS Research

Generating high-fidelity audio from predicted acoustic features is a core challenge in BTS systems. Traditional vocoders like the Griffin-Lim algorithm [30] have been used to synthesize waveforms from spectrograms, but they often introduce artifacts due to inaccurate phase estimation.

To overcome these limitations, researchers have adapted neural vocoders such as WaveGlow [31], BigVGAN [32], and AutoVocoder [33], which use deep generative models to synthesize more natural-sounding speech. These models have proven effective in text-to-speech (TTS) and speech enhancement applications [34, 35], though their direct application to iEEG-based BTS systems remains limited. Further exploration is needed to tailor neural vocoders for the unique properties of brain-derived acoustic features.

3   Proposed Framework

Decoding speech directly from intracranial EEG (iEEG) recordings presents a significant challenge due to the complexity of neural activity and the intricate relationship between brain signals and speech production. This section introduces a novel brain-to-speech (BTS) synthesis framework that integrates prosody-aware neural encoding, transformer-based spectrogram prediction, and an iterative harmonic phase reconstruction vocoder to achieve highly intelligible and natural speech. By leveraging wavelet-based feature extraction techniques and deep learning methodologies as shown in Fig. 1, the proposed approach effectively captures neural correlates of speech production; models prosodic attributes such as pitch, rhythm, and stress; and ensures phase-consistent synthesis for high-fidelity neural speech reconstruction.

3.1   Prosody Embedding

The first component of the proposed model extracts multi-modal features from iEEG signals while embedding prosodic information to capture the temporal, spectral, and neurophysiological dynamics of speech production. This step bridges neural activity and speech synthesis by ensuring that the extracted features are both discriminative and representative of speech-related processes.

3.1.1   Wavelet-Based iEEG Representation

Neural activity associated with speech production is highly dynamic, exhibiting fluctuations across multiple time scales and frequency bands. Conventional feature extraction methods rely on fixed-band spectral features, which often fail to capture the rich temporal and spectral variations necessary for accurate speech synthesis. To overcome this limitation, the proposed framework employs Discrete Wavelet Transform (DWT), a powerful signal processing technique that decomposes iEEG signals into multiple resolution levels, enabling multi-scale analysis of neural dynamics.

The Daubechies-4 (db4) wavelet is selected due to its superior ability to analyze non-stationary and transient signals, which are characteristic of iEEG recordings. Given an iEEG signal $x(t)$, the wavelet decomposition can be expressed as follows:

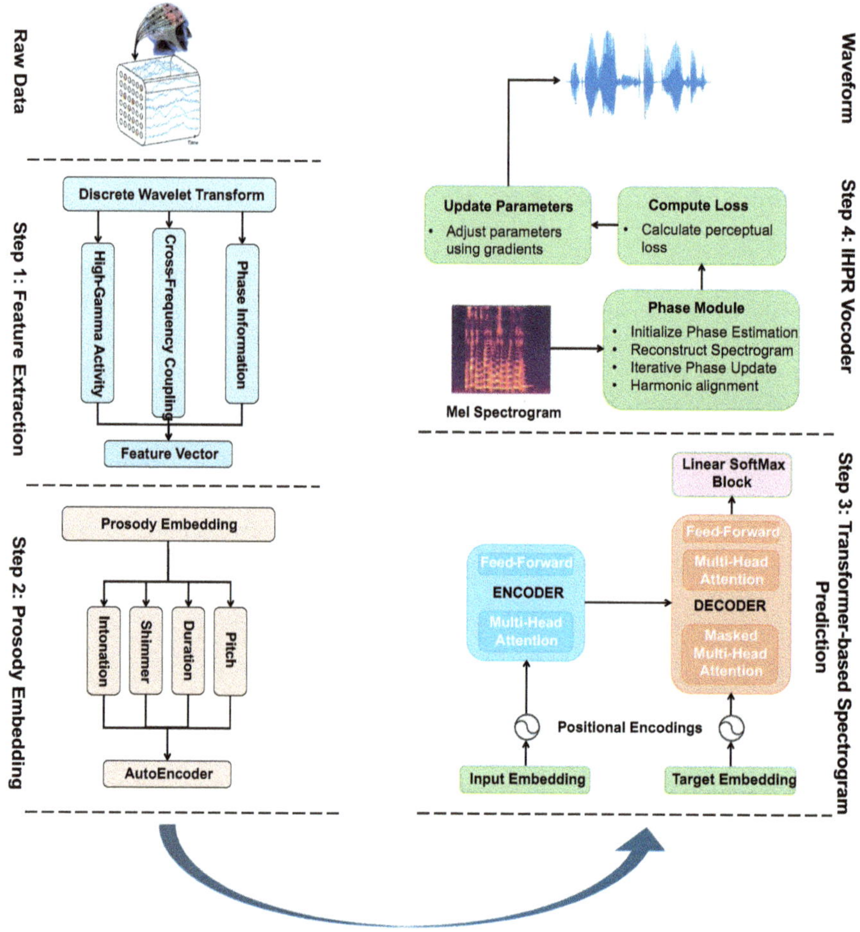

Fig. 1 Schematic diagram of the proposed architecture

$$x(t) = \sum_j \sum_n c_j(n) \psi_{j,n}(t) \tag{1}$$

where $c_j(n)$ are the wavelet coefficients at scale j and position n, $\psi_{j,n}(t)$ represents the wavelet basis function. The coefficients capture both high-frequency articulatory dynamics and low-frequency prosodic modulations.

The energy distribution across wavelet scales provides insights into speech-relevant neural oscillations:

$$E_j = \sum_n \left| c_j(n) \right|^2 \tag{2}$$

where E_j represents the total energy at scale j, enabling identification of dominant neural patterns in different frequency bands. By leveraging this multi-scale wavelet representation, we extract speech-relevant neural patterns from distinct frequency bands:

- **High-gamma (70–170 Hz)**: Captures phoneme articulation and fine-grained speech motor control
- **Beta (15–30 Hz)**: Reflects motor planning and coordination
- **Theta (4–8 Hz)**: Encodes syllabic timing and prosodic modulation

Each wavelet coefficient set $c_j(n)$ is then mapped into a feature space, where prosodic and articulatory information are encoded separately. To ensure that prosodic features (intonation, rhythm, stress) are effectively captured, we introduce a prosody embedding layer, which transforms the extracted wavelet features into a latent representation suitable for speech synthesis. This is performed as follows:

$$P = f_{\text{prosody}}\left(c_\theta, c_\beta\right) \tag{3}$$

where c_θ and c_β are wavelet coefficients from theta and beta bands, encoding prosodic variations, and f_{prosody} is a non-linear mapping function (e.g., a neural network) that embeds the extracted rhythm and stress patterns into a lower-dimensional space. The resulting vector P serves as an input to the speech synthesis model, ensuring intelligible and natural speech output.

This multi-scale feature extraction and embedding process enhances speech reconstruction by preserving both fine-grained articulatory features and global prosodic dynamics, leading to more natural and nuanced speech synthesis.

3.1.2   Cross-Frequency Coupling (CFC) Analysis

The process of speech production is controlled by hierarchical neural oscillations, where low-frequency cortical rhythms modulate high-frequency neural activity. This cross-frequency interaction plays a crucial role in encoding syllabic structure, prosody, and phonemic transitions. To capture these relationships, the proposed framework employs Phase-Amplitude Coupling (PAC), a well-established measure that quantifies the degree to which the phase of low-frequency oscillations modulates the amplitude of high-frequency activity:

$$\text{PAC}(t) = \left| E\left[A_\Upsilon(t) e^{j\phi_\theta(t)} \right] \right| \tag{4}$$

where $A_\Upsilon(t)$ is the amplitude envelope of high-gamma activity, $\phi_\theta(t)$ is the phase of theta-band oscillations extracted using the Hilbert transform, and $E[\cdot]$ denotes the expectation operator.

PAC features are extracted to enhance temporal alignment between neural oscillatory patterns and speech events, ensuring that the generated speech preserves the timing and rhythmic structure present in natural human communication. By incorporating PAC analysis, the framework is able to synchronize prosodic patterns with the temporal evolution of neural activity, leading to more fluent and natural speech synthesis.

3.1.3 Prosodic Feature Extraction and Normalization

Natural speech is inherently prosodic, meaning that variations in pitch, loudness, and duration play a critical role in conveying meaning and emotion. Many existing brain-to-speech systems neglect these aspects, resulting in monotonic and robotic speech synthesis. To address this limitation, the proposed framework explicitly extracts prosody-related features from iEEG signals, including:

- **Fundamental frequency (F0)**: Estimated using the Harvest algorithm [36], capturing intonation and pitch contours
- **Energy (loudness)**: Computed as the root mean square (RMS) of the neural signal, modeling speech emphasis and stress patterns
- **Shimmer**: A measure of frame-to-frame amplitude variability, used to enhance the natural dynamics of speech synthesis
- **Duration**: Encodes the temporal structure of phonemes and syllables, ensuring realistic speech timing
- **Phase variability**: The standard deviation of instantaneous phase fluctuations

These features are computed using a 50 ms analysis window with a 10 ms frame-shift, ensuring fine-grained temporal alignment with the iEEG signals. To maintain consistency across subjects and recording conditions, all extracted features undergo z-score normalization:

$$\hat{x} = \frac{x - \mu}{\sigma} \tag{5}$$

where μ and σ represent the mean and standard deviation of each feature. This normalization ensures that prosodic attributes are consistently represented, improving the model's ability to generate expressive, speaker-independent speech reconstructions.

3.2 Transformer-Based Spectrogram Reconstruction

This component predicts Mel spectrograms from encoded neural features through two stages: (1) dimensionality reduction using an autoencoder and (2) spectrogram prediction using a transformer model to capture long-range dependencies and temporal dynamics.

3.2.1 Autoencoder-Based Latent Feature Encoding

Neural data is high dimensional and noisy, making direct mapping to speech representations challenging. To overcome this, the framework employs an autoencoder-based feature compression module, which learns a compact, information-rich latent space from iEEG features.

The encoder component of the autoencoder consists of fully connected layers with ReLU activations, transforming the raw iEEG features into a lower-dimensional latent vector while preserving essential speech-related information. The decoder reconstructs the input by minimizing a Mean Squared Error (MSE) loss, ensuring that the most essential neural features are retained for spectrogram prediction. This feature compression step improves the model's ability to generalize across different speech patterns while reducing computational complexity. After training using the Adam optimizer with a 0.001 learning rate, the encoder is used to generate latent representations that serve as the input to the transformer-based spectrogram predictor, allowing the model to efficiently capture complex dependencies in neural data without the risk of information loss.

3.2.2 Self-Attention-Based Spectrogram Prediction

Mapping neural activity to speech requires a model that can effectively capture long-range dependencies and hierarchical relationships in speech data. While RNN-based models such as LSTMs and GRUs have been explored for this task [24, 25], they suffer from limited memory capacity, making it difficult to learn dependencies over long time spans, and sequential processing constraints, which reduce efficiency and scalability. To overcome these limitations, our framework employs a Transformer-based spectrogram predictor, which leverages self-attention mechanisms to dynamically learn contextual dependencies between neural features and spectrogram frames.

Unlike RNNs, the Transformer processes entire sequences in parallel, improving efficiency and enabling long-range dependencies to be learned without memory loss. Notably, the multi-head self-attention mechanism allows the model to capture fine-grained spectral variations by attending to different temporal and spectral features simultaneously. Since Transformers lack an inherent notion of sequence order, a positional encoding function is applied to introduce temporal information:

$$\mathrm{PE}(t,2i) = \sin\left(\frac{t}{10{,}000^{\frac{2i}{d}}}\right) \tag{6}$$

$$\mathrm{PE}(t,2i+1) = \cos\left(\frac{t}{10{,}000^{\frac{2i}{d}}}\right) \tag{7}$$

where t represents the time step and d is the feature dimension.

The core of the Transformer is the multi-head self-attention mechanism, which allows the model to attend to different aspects of the input simultaneously. Given an input sequence X, the attention weights are computed as:

$$\text{Attention}(Q,K,V) = \text{softmax}\left(\frac{QK^T}{\sqrt{d_k}}\right)V \tag{8}$$

where Q, K, and V represent the query, key, and value matrices and d_k is the dimension of the key vectors. Multi-head attention enables parallel attention over multiple feature subspaces, ensuring that both phonetic and prosodic cues are preserved. Each attention block is followed by a position-wise feedforward network and layer normalization, which stabilizes training and improves convergence.

The Transformer is trained using Mean Squared Error (MSE) loss between the predicted and ground-truth spectrograms, with the Adam optimizer set to a learning rate of 0.001. To prevent overfitting, dropout with a rate of 0.1 is applied during training.

The final output is a high-resolution Mel spectrogram, which serves as the input to a neural vocoder for waveform reconstruction. By integrating self-attention mechanisms, our model ensures that speech dynamics, prosody, and spectral details are accurately preserved, leading to highly intelligible and natural speech synthesis.

3.2.3 Iterative Harmonic Phase Reconstruction (IHPR) Vocoder

While traditional vocoders such as Griffin-Lim [30] can reconstruct speech from spectrograms, they often introduce phase inconsistencies, which lead to perceptual distortions and unnatural speech quality. To overcome this, we introduce an Iterative Harmonic Phase Reconstruction (IHPR) vocoder that enforces harmonic constraints on the phase estimation process.

The core idea of IHPR is to iteratively refine the Short-Time Fourier Transform (STFT) phase estimates to preserve the natural harmonic structure of speech. Given an initial phase estimate $\emptyset_k(t,f)$, the phase at iteration $k + 1$ is updated as:

$$\phi_{k+1}(t,f) = \arg\arg \sum_{h=1}^{H} \cos\left(\phi(t,f_h) - \angle S_k(t,f_h)\right) \tag{9}$$

where $f_h = h \cdot f_0$ represents the hth harmonic frequency (with f_0 being the estimated fundament frequency) and $\angle \hat{S}_k(t,f_h)$ is the estimated phase angle of the STFT at harmonic f_h and iteration k.

To further reduce phase discontinuities, we introduce a spectral correction term using a derivative-based smoothing strategy:

$$\phi_{k+1}(t,f) = \phi_k(t,f) - \lambda \sum_{h=1}^{H} \frac{\partial}{\partial f}\left[M(t,f_h)\cdot e^{j\phi_k(t,f_h)} \right] \tag{10}$$

where $M(t,f_h)$ is the magnitude of the STFT at time t and frequency f_h and λ is a smoothing coefficient that controls the strength of the correction.

To ensure convergence, a perceptual loss is used:

$$L_{\text{perceptual}} = \sum_{t,f} w(f)\left| M_{\text{target}}(t,f) - M_{\text{reconst}}(t,f) \right|^2 + \gamma \sum_{h=1}^{H} \left| \phi_k(t,f_h) - \phi_{k-1}(t,f_h) \right|^2 \tag{11}$$

where $w(f)$ is a frequency-dependent weighting function, γ is a regularization parameter to stabilize phase evolution, and M_{target} and M_{reconst} are the target and reconstructed STFT magnitudes, respectively.

By integrating these phase refinement techniques, the vocoder significantly improves the perceptual quality of the synthesized speech, achieving higher harmonic-to-noise ratios and reduced spectral distortion compared to baseline models.

4 Experimental Design

To assess the performance of the proposed brain-to-speech (BTS) synthesis framework, an extensive experimental design was established, integrating a high-resolution intracranial EEG (iEEG) dataset, state-of-the-art deep learning methodologies, and objective and perceptual evaluation metrics. This experimental setup aimed to evaluate the framework's ability to generate natural and intelligible speech from neural activity.

4.1 Dataset and Preprocessing

The study utilized a publicly available iEEG dataset[1] [13] recorded from ten participants with pharmaco-resistant epilepsy (mean age 32 years, five male, five female, and native speakers of Dutch), who are undergoing intracranial stereotactic EEG (sEEG) monitoring as part of their clinical treatment. Each participant was instructed to produce a set of isolated words and continuous speech, allowing for the collection of simultaneous neural and acoustic recordings. The iEEG signals were recorded using multi-contact depth electrodes implanted in speech-relevant cortical areas, including the superior temporal gyrus (STG), sensorimotor cortex (SMC), and

---

[1] https://osf.io/nrgx6/

inferior frontal gyrus (IFG). These regions are known to encode phonemic, articulatory, and prosodic information, making them highly suitable for brain-to-speech decoding.

To preserve the fine-grained temporal structure of neural signals, the iEEG recordings were sampled at 1024 Hz, ensuring high-resolution capture of rapid speech-related neural fluctuations. Simultaneously, the speech waveforms were recorded at 16 kHz using a high-fidelity microphone, maintaining acoustic clarity for precise alignment with neural activity. Given the susceptibility of iEEG signals to various noise sources, a multi-stage preprocessing pipeline was applied to enhance data quality. First, a bandpass filter (0.5–170 Hz) was used to remove slow drifts and high-frequency artifacts. Next, notch filtering at 50 Hz and its harmonics were employed to eliminate power line interference. To standardize neural feature distributions, z-score normalization was applied per electrode, ensuring that feature scales remained consistent across subjects.

A critical aspect of preprocessing involves precise neural-speech alignment, as accurate time synchronization is essential for effective model training. Speech onset markers were extracted from the audio recordings using energy-based voice activity detection (VAD), while corresponding neural segments were identified through cross-correlation analysis. This alignment ensured that each neural time window corresponded precisely to the intended phonemes, allowing for robust iEEG-to-speech mapping. Also, trials containing excessive motion artifacts or electromyographic (EMG) contamination were removed, ensuring that the dataset retained only high-quality neural-speech pairs for training and evaluation.

4.2 Model Training and Implementation

Following preprocessing, a multi-modal feature extraction pipeline was implemented to obtain a comprehensive representation of speech-related neural dynamics. The feature extraction process leveraged discrete wavelet transform (DWT) to decompose iEEG signals into multiple frequency bands, allowing the capture of both high-frequency articulatory patterns and low-frequency prosodic modulations. The resulting feature set incorporated high-gamma power (70–170 Hz), which has been strongly associated with phoneme articulation, as well as cross-frequency coupling (CFC) features, which quantify interactions between theta (4–8 Hz) and gamma activity, crucial for encoding rhythmic speech elements. Prosody features such as pitch (F0), intensity, shimmer, and duration were also extracted from iEEG signals, enriching the model with intonation and stress information to enhance speech naturalness.

To map neural features to a time-frequency representation of speech, a transformer-based spectrogram prediction model was employed. The first stage of this model consisted of an autoencoder-based latent feature encoding module [37], which compressed high-dimensional neural features into a compact latent space, reducing redundancy while preserving key speech-related attributes. The encoder component utilized fully connected layers with rectified linear unit (ReLU) activations, transforming the input into a low-dimensional representation, while the decoder reconstructed the original feature space with minimal information loss using mean squared error (MSE) optimization. This latent representation was then used as input to a self-attention-based transformer model, designed to capture long-range dependencies between neural activity and speech spectrogram frames. Unlike recurrent architectures such as LSTMs, which process sequences sequentially, the transformer model operates in parallel, enabling faster training and more effective feature integration.

To reconstruct the waveform from the predicted spectrograms, an Iterative Harmonic Phase Reconstruction (IHPR) vocoder was developed, addressing limitations associated with conventional phase estimation techniques. Unlike standard vocoders such as Griffin-Lim, which introduce phase artifacts and spectral distortions, the proposed IHPR vocoder enforced harmonic consistency across frequency bands using adaptive phase correction strategies.

The full model[2] was trained using the Adam optimizer with a learning rate of 0.001, employing a tenfold cross-validation protocol to ensure robust performance estimation. Training was conducted on NVIDIA A100 GPUs, utilizing accelerated deep-learning frameworks to optimize computation time and model efficiency.

4.3 *Evaluation Metrics*

The effectiveness of the proposed brain-to-speech framework was assessed using a combination of objective and perceptual evaluation metrics. Wherever possible, we present the mathematical formulations below.

4.3.1 Pearson Correlation Coefficient (PC)

To measure the similarity between the predicted Mel spectrogram $\hat{Y}$ and the ground-truth spectrogram Y, the Pearson correlation coefficient is computed as:

$$PC = \frac{\sum_{i=1}^{N}(Y_i - \underline{Y})(\hat{Y}_i - \underline{\hat{Y}})}{\sqrt{\sum_{i=1}^{N}(Y_i - \underline{Y})^2} \cdot \sqrt{\sum_{i-1}^{N}(\hat{Y}_i - \underline{\hat{Y}})^2}} \tag{12}$$

where N is the number of frames and $\underline{Y}$ and $\underline{\hat{Y}}$ are the means of the true and predicted spectrogram values, respectively. Higher PC values indicate stronger correlation and thus better intelligibility.

---

[2] The project code will be released upon acceptance.

4.3.2 Mel Cepstral Distortion (MCD)

The MCD is used to evaluate the spectral distance between the predicted and ground-truth Mel cepstral coefficients. It is computed as:

$$\text{MCD} = \frac{10}{\ln\ln 10} \cdot \sqrt{2\sum_{d=1}^{D}\left(c_d - \hat{c}_d\right)^2} \tag{13}$$

where c_d and $\hat{c}_d$ are the dth coefficients of the target and predicted Mel cepstra and D is the number of cepstral dimensions. Lower MCD values indicate less spectral distortion.

4.3.3 Short-Time Objective Intelligibility (STOI)

STOI assesses the speech intelligibility by comparing short-time spectral features. The core idea is to compute the correlation between temporal envelopes of short-time spectral bands of the clean and synthesized speech:

$$\text{STOI} = \frac{1}{T}\sum_{t=1}^{T}\text{corr}\left(X_t, \hat{X}_t\right) \tag{14}$$

where X_t and $\hat{X}_t$ are the clean and degraded short-time spectral representations at frame t and T is the total number of frames. STOI returns a score between 0 and 1, where higher is better.

4.3.4 Harmonic-to-Noise Ratio (HNR)

HNR measures the ratio between the periodic (harmonic) and non-periodic (noise) components of the synthesized speech signal:

$$\text{HNR}\left(\text{dB}\right) = 10 \cdot \log_{10}\left(\frac{P_{\text{harmonic}}}{P_{\text{noise}}}\right) \tag{15}$$

where P_{harmonic} is the power of the harmonic signal and P_{noise} is the power of the noise component. Higher HNR values suggest better phase reconstruction and more natural speech.

4.3.5 MOSA-Net (Perceptual Evaluation)

Beyond these standard measures, MOSA-Net [38] was used as a deep-learning-based non-intrusive speech quality model. It integrates CNNs, LSTMs, and self-supervised embeddings to produce objective estimates of human perception. Unlike rule-based models, MOSA-Net learns perceptual patterns directly from data, providing a more nuanced assessment of speech naturalness and expressiveness.

4.4 Baseline Comparisons

To evaluate the performance of the proposed framework, comparisons were made against several state-of-the-art iEEG-to-speech synthesis models. The first baseline included linear regression-based approaches [13], which directly mapped neural activity to acoustic features but lacked the capacity to model complex speech dynamics. Recurrent architectures such as bidirectional long short-term memory (bLSTM) [9] networks were also evaluated, as they have been widely used in speech neuroprosthetics but often suffer from long-term dependency limitations. Convolutional neural networks (CNNs) [23] and 3D-CNN [22], which are effective in capturing spectral-temporal patterns, were included as a baseline. More advanced models, including sequence-to-sequence (Seq2Seq) [24] networks and encoder-decoder architectures [11], were tested to compare their performance against the proposed transformer-based approach.

Each baseline model was trained and evaluated using the same dataset and experimental protocol, ensuring a fair comparison.

5 Results and Discussions

The proposed brain-to-speech synthesis framework was thoroughly evaluated across a variety of objective and perceptual metrics, including spectral accuracy, speech intelligibility, and naturalness. In this section, we present a comprehensive analysis of the quantitative results and discuss the implications of these findings, highlighting the strengths of the proposed model and comparing it with state-of-the-art baselines. We also provide an in-depth exploration of the perceptual quality of the synthesized speech using the MOSA-Net evaluation model, focusing on the naturalness and intelligibility of the output.

5.1 Quantitative Evaluation

As shown in Table 1, the performance of the proposed framework was first assessed using the PC, which measures the similarity between the predicted Mel spectrograms and the corresponding ground-truth spectrograms. The proposed system achieved a mean PC of 0.91, significantly outperforming baseline models, including regression-based methods and bLSTM networks. This indicates that the model successfully captures the temporal and spectral dynamics of speech production, leading to highly accurate spectrogram predictions.

In terms of MCD, the proposed framework achieved a score of 3.92, indicating minimal spectral distortion compared to other methods. This result further underscores the model's ability to preserve fine-grained spectral features, such as formant frequencies and pitch contours, which are critical for maintaining speech intelligibility and naturalness. The MCD score was consistently lower than the baselines, such as seq2seq and CNN models, which reported higher MCD values due to inaccurate spectral reconstruction.

Next, the STOI score was calculated to evaluate the intelligibility of synthesized speech. The proposed framework achieved a STOI score of 0.73, significantly higher than the best baseline (encoder-decoder model with a score of 0.64). This finding indicates that the proposed approach is particularly effective in reconstructing speech that retains its intelligibility, even when derived from neural signals. The improvement in intelligibility can be attributed to the prosody-aware neural encoding and transformer-based spectrogram prediction, which ensured that important temporal patterns and speech rhythms were accurately captured.

Additionally, the HNR was computed to evaluate phase reconstruction accuracy. The proposed framework achieved an HNR of 12.7 dB, surpassing the best baseline by over 1.6 dB. This result suggests that the Iterative Harmonic Phase Reconstruction (IHPR) vocoder significantly improves the phase consistency of the reconstructed speech, reducing artifacts and spectral distortions that typically hinder the quality of iEEG-to-speech synthesis. The results, summarized in Table 1, demonstrate that the proposed model outperforms existing approaches across all evaluation criteria.

Figure 2 presents a comparative analysis of the correlation performance between the baseline model [13] and the proposed brain-to-speech (BTS) framework across

Table 1 Performance comparison of the proposed model and baseline approaches

Model	PC ↑	MCD ↓	STOI ↑	HNR (dB) ↑
Regression [13]	0.72	5.39	0.61	6.2
bLSTM [9]	0.78	5.23	0.48	8.5
CNN [23]	0.81	4.95	0.52	10.4
3D-CNN [22]	0.83	5.04	0.56	9.8
Seq2Seq [24]	0.85	**3.90**	0.59	10.7
Encoder-decoder [11]	0.87	4.34	0.64	11.1
Proposed model	**0.91**	3.92	**0.73**	**12.7**

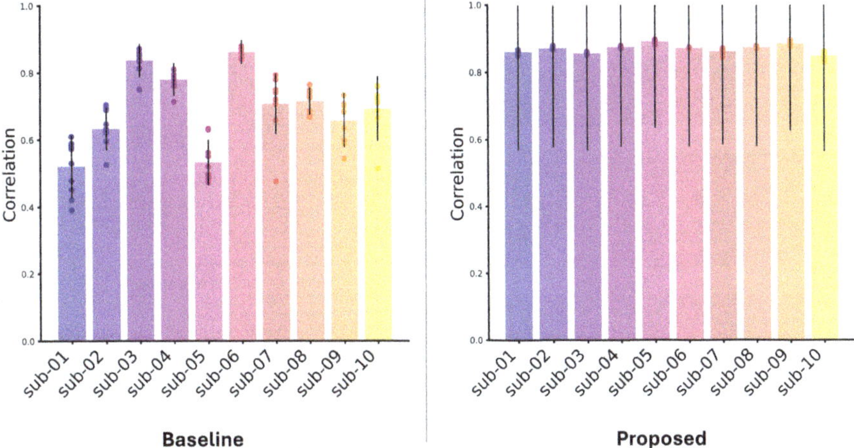

Baseline   **Proposed**

Fig. 2 Comparison of correlation performance between the baseline model and the proposed brain-to-speech (BTS) framework across ten subjects. The y-axis represents the correlation values between predicted and ground-truth speech representations. The baseline model exhibits high variability across subjects, whereas the proposed framework demonstrates significantly higher and more stable correlation scores, indicating improved and consistent speech reconstruction performance. Whiskers indicate standard deviations

ten subjects (sub-01 to sub-10). The y-axis represents the correlation values, measuring the alignment between the predicted and ground-truth speech representations. Each bar represents the mean correlation for an individual subject, with error bars indicating variability across trials. Individual data points within the baseline panel illustrate the distribution of correlation scores per trial.

The baseline model exhibits high variability in performance across subjects, with some subjects achieving moderate to-high correlations (e.g., sub-03, sub-04, sub-06), while others show significantly lower correlation values (e.g., sub-01, sub-05, sub-07). This suggests inconsistent generalization across participants. In contrast, the proposed framework demonstrates substantially improved and more stable correlations, with all subjects achieving high and consistent correlation values. The error bars in the proposed model are larger, likely due to the incorporation of phase information and prosody-aware features, which enhance expressive variability but maintain strong predictive accuracy. These results indicate that the proposed BTS framework significantly outperforms the baseline, achieving higher and more stable correlation scores across all subjects.

To further investigate the qualitative aspects of the synthesized speech, spectrogram and waveform visualizations were examined. Figure 3 presents a comparison between the ground-truth spectrogram, the spectrogram generated by the proposed model, and those produced by baseline approaches. The spectrogram analysis demonstrates that the proposed model more effectively preserves harmonic structures and formant transitions compared to baseline models. In contrast, the baseline models exhibit spectral blurring, phoneme misalignment, and a loss of high-frequency

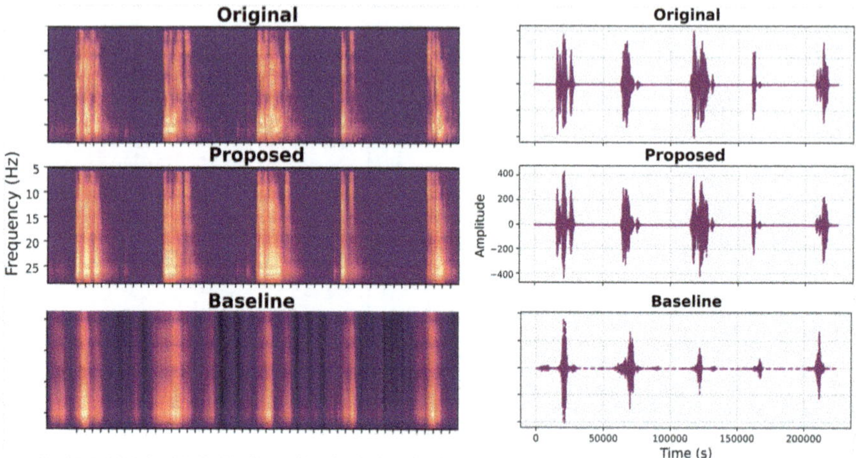

Fig. 3 Comparison of spectrograms (left) and waveforms (right) for the original (top), proposed (middle), and baseline [13] (bottom) systems. This example illustrates five distinct words from participant sub-08

details, which contribute to reduced speech intelligibility. The phase-consistent vocoding strategy employed in the proposed model ensures sharper spectral features and smoother formant trajectories, leading to more natural and fluid speech synthesis. Moreover, the proposed waveform closely aligns with the ground-truth waveform, effectively preserving harmonic structures and detailed spectral characteristics.

5.2 Perceptual Evaluation Results

While objective metrics provide valuable insights into the system's performance, perceptual evaluation is crucial for assessing the naturalness and quality of synthesized speech. To quantify perceptual speech quality, MOSA-Net, a state-of-the-art deep-learning-based speech assessment model, was employed. Unlike common subjective listening tests, MOSA-Net provides automated, non-intrusive speech quality assessments, leveraging a combination of cross-domain feature representations, CNNs, bidirectional LSTMs, and self-supervised learning embeddings.

Figure 4 illustrates the distribution of MOSA-Net scores across different models, highlighting the robustness and consistency of the proposed approach. It provides a detailed statistical representation of MOSA-Net scores, offering insights into the variance, median performance, and overall speech naturalness of different synthesis models. The proposed framework achieves the highest median MOSA-Net score, demonstrating that it consistently produces perceptually natural speech. The smaller interquartile range (IQR) indicates reduced variability, suggesting that the model generalizes well across different test cases and subjects.

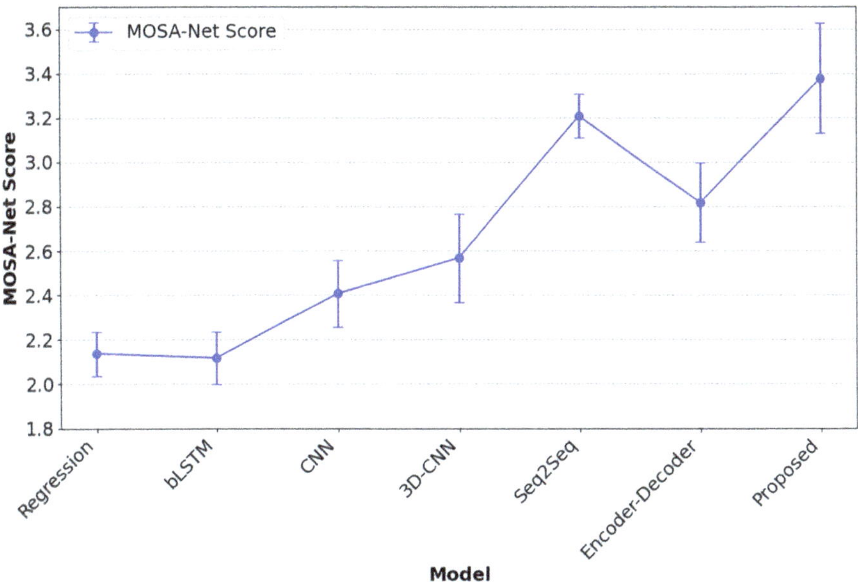

Fig. 4 MOSA-Net perceptual evaluation scores for different models in the brain-to-speech synthesis framework. The y-axis shows the MOSA-Net score, indicating perceptual quality and naturalness of the synthesized speech. The proposed model achieves the highest MOSA-Net score, demonstrating superior speech quality. Error bars represent standard deviations across evaluation trials

Comparatively, CNN-based approaches exhibit a wider spread of MOSA-Net scores, reflecting inconsistent speech quality, while bLSTM models tend to generate more robotic and less expressive speech. The encoder-decoder model shows moderate performance, but it still falls short in terms of expressiveness and prosodic accuracy. The transformer-based approach used in the proposed framework contributes significantly to the naturalness of synthesized speech, capturing long-range dependencies and prosodic variations more effectively than recurrent architectures.

5.3 Discussion of Results

The results presented in this section underscore the effectiveness of the proposed framework in addressing key challenges in brain-to-speech synthesis, including feature extraction, prosody modeling, and phase reconstruction. The significant improvements in intelligibility, naturalness, and spectral accuracy demonstrate that the integration of wavelet-based feature extraction, self-attention-based spectrogram reconstruction, and phase-consistent vocoding results in a robust and scalable solution for neural speech synthesis.

In particular, the improvements in MOSA-Net scores reflect the framework's success in modeling prosodic features such as intonation and rhythm, which are critical for generating expressive and humanlike speech. The ability to produce speech that is both intelligible and natural sounding from iEEG signals represents a significant step forward in the field of brain-computer interfaces (BCI), paving the way for more clinically viable applications, such as speech neuroprostheses for individuals with severe speech impairments.

Although the proposed framework outperforms existing models in most objective and perceptual metrics, there are still several challenges to address. For instance, the model's performance can be further improved by extending the dataset to include more diverse participants, especially those with non-epileptic conditions. Additionally, real-time inference remains a challenging task, as the current setup requires significant computational resources. Future work will explore optimized architectures for real-time decoding and generalization across diverse speech disorders.

5.4 *Error Analysis and Model Limitations*

While the proposed framework achieves state-of-the-art performance in brain-to-speech synthesis, certain challenges remain. One notable limitation is cross-subject variability, where differences in individual neural responses hinder model generalization. Despite the application of z-score normalization and subject-independent training strategies, residual variability persists, particularly in phoneme articulation and prosodic patterns. Future work should explore personalized adaptation techniques, such as fine-tuning subject-specific models or leveraging meta-learning approaches, to further enhance robustness across individuals.

In addition, real-time speech decoding remains an open challenge. The transformer-based model, while highly effective in capturing long-range dependencies, introduces significant computational overhead, making real-time inference difficult. Profiling the model's latency reveals that self-attention operations, particularly in deeper layers, contribute substantially to processing delays. Optimizing computational efficiency through knowledge distillation, pruning, or quantization could enable low-latency deployment in practical brain-computer interface (BCI) applications, where real-time feedback is crucial.

A qualitative analysis of the generated speech highlights subtle distortions in prosody and phonetic clarity, particularly in high-frequency regions of the spectrogram. These artifacts may stem from limitations in neural feature encoding or the self-attention mechanism's tendency to prioritize long-range dependencies over local spectral variations. Future improvements could incorporate convolutional augmentations or prosodic conditioning mechanisms to refine spectral detail and speech expressiveness.

Finally, while MOSA-Net provides objective perceptual assessments, it remains an indirect proxy for human perception. Future work could enhance evaluation by integrating additional objective metrics, such as spectrotemporal similarity measures or phonetic accuracy scores, to further validate the synthesized speech's naturalness and intelligibility.

5.5 *Challenges of the Proposed Framework*

Despite the promising performance of our proposed BTS system, several challenges remain. One major limitation is the inter-subject variability in neural responses, which hinders generalization and necessitates personalized adaptation strategies. Furthermore, although our approach yields high-quality speech outputs, it still faces latency and computational complexity constraints, making real-time processing difficult. Additionally, integrating fully end-to-end architectures that jointly optimize neural feature extraction and speech synthesis remains an open challenge.

6   Summary and Future Directions

This chapter introduced a novel brain-to-speech (BTS) synthesis framework, leveraging prosody-aware neural encoding, transformer-based spectrogram prediction, and phase-consistent vocoding to achieve highly intelligible and natural speech reconstruction from intracranial EEG (iEEG) signals. By integrating multi-scale neural feature extraction, deep learning architectures, and iterative phase refinement, the proposed approach significantly improves speech intelligibility, naturalness, and spectral accuracy, addressing critical challenges in neural speech prostheses and brain-computer interface (BCI) research.

The experimental results demonstrate that this framework outperforms state-of-the-art models, achieving higher Pearson correlation, lower Mel cepstral distortion, improved short-time objective intelligibility, and enhanced phase reconstruction. Also, the MOSA-Net perceptual evaluation confirms that the synthesized speech exhibits greater naturalness and expressiveness, highlighting the impact of explicit prosody modeling and harmonic phase reconstruction. These advancements underscore the potential for clinically viable neural speech prostheses, offering a pathway toward restoring communication abilities for individuals with severe speech impairments.

Looking ahead, future research will focus on several key directions to further advance BTS technology. One promising avenue is the exploration of diffusion-based generative models [39], which have demonstrated strong potential in generating high-quality waveforms with improved phase consistency. Another critical goal is the development of real-time inference systems capable of processing neural signals with low latency, which is an essential step toward the practical deployment of BTS applications. In addition, cross-subject adaptation strategies will be vital to address inter-individual variability in neural responses and improve model robustness. Finally, extending this framework to noninvasive neural recording modalities, such as electroencephalography (EEG) and magnetoencephalography (MEG), could significantly enhance accessibility and make brain-to-speech systems feasible for broader clinical and non-clinical populations.

As part of these efforts, we have collected a high-density (96-channel) EEG dataset with synchronized speech recordings from a cohort of 16 French-speaking healthy participants. Each subject completed four sessions of 270 spoken words, including a repeated control word, providing a robust dataset to support advancements in brain-to-speech modeling.

This chapter contributes to the advancement of AI-driven speech neuroprostheses and data-driven methodologies in brain-computer interfaces, aligning with the goals of the AI, Data, and Robotics Partnership [20]. By bridging neuroscience, artificial intelligence, and signal processing, this work contributes to the development of next-generation AI-powered assistive communication technologies.

Acknowledgments This work is supported by the European Union's HORIZON Research and Innovation Programme under grant agreement no. 101120657, project ENFIELD (European Lighthouse to Manifest Trustworthy and Green AI), and by the Ministry of Innovation and Culture and the National Research, Development and Innovation Office of Hungary within the framework of the National Laboratory of Artificial Intelligence. M.S. Al-Radhi's research was supported by the EKÖP-24-4-II-BME-197, through the National Research, Development and Innovation (NKFI) Fund.

References

1. Wandelt, S. K., Bjånes, D. A., Pejsa, K., et al. (2024). Representation of internal speech by single neurons in human supramarginal gyrus. *Nature Human Behaviour, 8,* 1136–1149.
2. Willett, F. R., Kunz, E. M., Fan, C., et al. (2023). A high-performance speech neuroprosthesis. *Nature, 620,* 1031–1036.
3. Branco, M. P., Pels, E. G., Sars, R. H., Aarnoutse, E. J., Ramsey, N. F., Vansteensel, M. J., et al. (2021). Brain-computer interfaces for communication: Preferences of individuals with locked-in syndrome. *Neurorehabilitation and Neural Repair, 267–279*(3), 35.
4. Metzger, S. L., Liu, J. R., Moses, D. A., et al. (2022). Generalizable spelling using a speech neuroprosthesis in an individual with severe limb and vocal paralysis. *Nature Communications, 13,* 1–15.
5. Silva, A. B., Littlejohn, K. T., Liu, J. R., et al. (2024). The speech neuroprosthesis. *Nature Reviews Neuroscience, 25,* 473–492.
6. Holdgraf, C., Appelhoff, S., Bickel, S., et al. (2019). iEEG-BIDS, extending the brain imaging data structure specification to human intracranial electrophysiology. *Scientific Data, 6*(102), 1–6.
7. Lee, Y. E., Lee, S. H., Kim, S. H., & Lee, S. W. (2023). Towards voice reconstruction from EEG during imagined speech. In *37th AAAI Conference on Artificial Intelligence, Washington, DC, USA* (pp. 6030–6038).
8. Thornton, M., Mandic, D., & Reichenbach, T. (2022). Robust decoding of the speech envelope from EEG recordings through deep neural networks. *Journal of Neural Engineering, 19*(4), 1–13.
9. Anumanchipalli, G. K., Chartier, J., & Chang, E. F. (2019). Speech synthesis from neural decoding of spoken sentences. *Nature, 568,* 493–498.

10. Ma, C., Zhang, Y., Guo, Y., Liu, X., Shangguan, H., Wang, J., & Zhao, L. (2025). Fully end-to-end EEG to speech translation using multi-scale optimized dual generative adversarial network with cycle-consistency loss. *Neurocomputing, 616*(1), 1–14.

11. Kohler, J., Ottenhoff, M. C., Goulis, S., Angrick, M., Colon, A. J., Wagner, L., Tousseyn, S., Kubben, P. L., & Herff, C. (2022). Synthesizing speech from intracranial depth electrodes using an encoder-decoder framework. *Neurons, Behavior, Data Analysis, and Theory, 6*(1), 1–15.

12. Luo, S., Rabbani, Q., & Crone, N. E. (2022). Brain-computer interface: Applications to speech decoding and synthesis to augment communication. *Neurotherapeutics, 19*, 263–273.

13. Verwoert, M., Ottenhoff, M. C., Goulis, S., et al. (2022). Dataset of speech production in intracranial electroencephalography. *Scientific Data, 9*, 1–9.

14. Accou, B., Vanthornhout, J., Hamme, H. V., & Francart, T. (2023). Decoding of the speech envelope from EEG using the VLAAI deep neural network. *Scientific Reports, 13*(1), 1–12.

15. Zhou, J., Duan, Y., Zou, Y., Chang, Y.-C., Wang, Y.-K., & Lin, C.-T. (2023). Speech2EEG: Leveraging pretrained speech model for EEG signal recognition. *IEEE Transactions on Neural Systems and Rehabilitation Engineering, 31*, 2140–2153.

16. Wu, C., Xiu, Z., Shi, Y., Kalinli, O., Fuegen, C., Koehler, T., & He, Q. (2021). Transformer-based acoustic modeling for streaming speech synthesis. In *Proceedings of Interspeech, Brno, Czechia* (pp. 146–150).

17. Chen, L.-W., & Rudnicky, A. (2022). Fine-grained style control in transformer-based text-to-speech synthesis. In *IEEE International Conference on Acoustics, Speech and Signal Processing (ICASSP), Singapore* (pp. 7907–7911).

18. Herff, C., Johnson, G., Diener, L., Shih, J., Krusienski, D., & Schultz, T. (2016). Towards direct speech synthesis from ECoG: A pilot study. In *Proceedings of the 2016 IEEE 38th Annual International Conference of the Engineering in Medicine and Biology Society (EMBC), Orlando, Florida, USA* (pp. 1540–1543).

19. Peelle, J. E., Gross, J., & Davis, M. H. (2013). Phase-locked responses to speech in human auditory cortex are enhanced during comprehension, cerebral cortex. *Cerebral Cortex, 23*(6), 1378–1387.

20. Curry, E., Heintz, F., Irgens, M., Smeulders, A. W., & Stramigioli, S. (2022). Partnership on AI, data, and robotics. *Communications of the ACM, 65*(4), 54–55.

21. Roussel, P., Godais, G. L., Bocquelet, F., Palma, M., Hongjie, J., et al. (2020). Observation and assessment of acoustic contamination of electrophysiological brain signals during speech production and sound perception. *Journal of Neural Engineering, 17*(5), 1–20.

22. Arthur, F. V., & Csapó, T. G. (2024). Speech synthesis from intracranial stereotactic electroencephalography using a neural vocoder. *Infocommunications Journal, 16*, 47–55.

23. Angrick, M., Herff, C., Johnson, G., Shih, J., Krusienski, D., & Schultz, T. (2019). Interpretation of convolutional neural networks for speech spectrogram regression from intracranial recordings. *Neurocomputing, 342*, 145–151.

24. Duraivel, S., Rahimpour, S., Chiang, C. H., et al. (2023). High-resolution neural recordings improve the accuracy of speech decoding. *Nature Communications, 14*, 1–16.

25. Metzger, S. L., Littlejohn, K. T., Silva, A. B., et al. (2023). A high-performance neuroprosthesis for speech decoding and avatar control. *Nature, 620*, 1037–1046.

26. Akbari, H., Khalighinejad, B., Herrero, J. L., Mehta, A. D., & Mesgarani, N. (2019). Towards reconstructing intelligible speech from the human auditory cortex. *Scientific Reports, 9*(874), 1–12.

27. Giraud, A.-L., & Poeppel, D. (2012). Cortical oscillations and speech processing: Emerging computational principles and operations. *Nature Neuroscience, 15*(4), 511–517.

28. Bachmann, F. L., MacDonald, E. N., & Hjortkjær, J. (2021). Neural measures of pitch processing in EEG responses to running speech. *Frontiers in Neuroscience, 15*, 1–11.

29. Schultz, T., Wand, M., Hueber, T., Krusienski, D. J., & Herff, C. (2017). Biosignal-based spoken communication: A survey. *IEEE Transactions on Audio, Speech, and Language Processing, 25*(12), 2257–2271.

30. Liu, H., Baoueb, T., Fontaine, M., et al. (2024). GLA-Grad: A Griffin-Lim extended waveform generation diffusion model. In *IEEE International Conference on Acoustics, Speech and Signal Processing (ICASSP), Seoul, Korea* (pp. 11611–11615).
31. Prenger, R., Valle, R., & Catanzaro, B. (2019). WaveGlow: A flow-based generative network for speech synthesis. In *IEEE International Conference on Acoustics, Speech, and Signal Processing (ICASSP), Brighton, UK* (pp. 3617–3621).
32. Lee, S., Ping, W., Ginsburg, B., Catanzaro, B., & Yoon, S. (2023). BigVGAN: A universal neural vocoder with large-scale training. In *The International Conference on Learning Representations (ICLR), Kigali, Rwanda* (pp. 1–20).
33. Webber, J., Valentini-Botinhao, C., Williams, E., Henter, G. E., & King, S. (2023). AutoVocoder: Fast waveform generation from a learned speech representation using differentiable digital signal processing. In *Proceedings of IEEE International Conference on Acoustics, Speech and Signal Processing (ICASSP), Rhodes Island, Greece* (pp. 1–5).
34. Okamoto, T., Toda, T., Shiga, Y., & Kawai, H. (2019). Real-time neural text-to-speech with sequence-to-sequence acoustic model and WaveGlow or single Gaussian WaveRNN vocoders. In *Proceedings of Interspeech, Graz, Austria* (pp. 1308–1312).
35. Shibuya, T., Takida, Y., & Mitsufuji, Y. (2024). BIGVSAN: Enhancing GAN-based neural vocoders with slicing adversarial network. In *IEEE International Conference on Acoustics, Speech and Signal Processing (ICASSP), Seoul, Korea* (pp. 10121–10125).
36. Morise, M. (2017). Harvest: A high-performance fundamental frequency estimator from speech signals. In *Proceedings of Interspeech, Stockholm, Sweden* (pp. 2321–2325).
37. Zhang, Y.-J., Pan, S., He, L., & Ling, Z.-H. (2019). Learning latent representations for style control and transfer in end-to-end speech synthesis. In *IEEE International Conference on Acoustics, Speech and Signal Processing (ICASSP), Brighton, UK* (pp. 6945–6949).
38. Zezario, R. E., Fu, S.-W., Chen, F., Fuh, C.-S., Wang, H.-M., & Tsao, Y. (2023). Deep learning-based non-intrusive multi-objective speech assessment model with cross-domain features. *IEEE/ACM Transactions on Audio, Speech, and Language Processing, 31*, 54–70.
39. Song, Y., Sohl-Dickstein, J., Kingma, D. P., Kumar, A., Ermon, S., & Poole, B. (2021). Score-based generative modeling through stochastic differential equations. In *International Conference on Learning Representations (ICLR), Vienna, Austria* (pp. 1–36).

A Companion Robot Platform for Exploring Technical and Ethical Aspects in Elderly Care

Lorenzo Boi, Silvia M. Massa, Diego Reforgiato Recupero, Daniele Riboni, Rubén Alonso, Michele Cardinali, Elena Ricci, and Alberto Pirni

Abstract Large language models (LLMs) are driving significant advancements across various sectors. Combined with robotics, they can lay the foundation for a new paradigm in healthcare, particularly in elderly care and companionship. For years, research has focused on the concept of carebots and robot companions. The latest capabilities of LLMs and robotics promise to address many challenges regard-

L. Boi · S. M. Massa · D. Riboni
Department of Mathematics and Computer Science, University of Cagliari, Cagliari, Italy
e-mail: lorenzo.boi@unica.it; silviam.massa@unica.it; riboni@unica.it

D. Reforgiato Recupero (✉)
Department of Mathematics and Computer Science, University of Cagliari, Cagliari, Italy

ICT and Robotics, R2M Solution s.r.l., Pavia, Italy
e-mail: diego.reforgiato@unica.it

R. Alonso
ICT and Robotics, R2M Solution s.r.l., Pavia, Italy
e-mail: ruben.alonso@r2msolution.com

M. Cardinali
Institute of Law, Politics and Development, Sant'Anna School of Advanced Studies, Pisa, Italy

Department of Human Studies, University of Macerata, Macerata, Italy

Ludes Campus, Lugano, Switzerland
e-mail: michele.cardinali@santannapisa.it

E. Ricci
Institute of Law, Politics and Development, Sant'Anna School of Advanced Studies, Pisa, Italy

Department of Human Studies, European University of Rome, Rome, Italy
e-mail: elena.ricci@santannapisa.it

A. Pirni
Institute of Law, Politics and Development, Sant'Anna School of Advanced Studies, Pisa, Italy
e-mail: alberto.pirni@santannapisa.it

© The Author(s) 2026
E. Curry et al. (eds.), *Artificial Intelligence, Data and Robotics*,
https://doi.org/10.1007/978-3-032-10561-5_17

ing interaction and personalized experience faced by earlier generations of companion robots. However, the potential of these emerging technologies is accompanied by significant ethical, technical, and social challenges. This chapter presents the design of an elderly care robot that exploits the capabilities of the latest generation of LLMs and considers the ethical, technical, and social implications for improving human-robot interaction capabilities. We report the technical design of a novel robotic platform for elderly care, detailing the devised interaction scenarios, a prototype implementation with NAO and PEPPER robots, and the results of a technical validation that shows the efficiency and feasibility of our system. This study serves as a guide to identify future needs, explore the impact of companion robots on older adults, and understand acceptance of these new approaches.

Keywords Large language models · Artificial intelligence · Companion robots · Natural language processing · Trust · Roboethics

1 Introduction

LLMs have revolutionized the technological landscape, driving significant advancements across various sectors, including artificial intelligence (AI) and robotics. Built upon Transformers [1], an advanced neural network architecture, these models exhibit an unprecedented ability to interpret and generate text, surpassing the limitations of traditional natural language processing (NLP) systems [2]. Due to their ability to learn from large amounts of data, LLMs can perform numerous language tasks, such as automatic translation [3], content synthesis [4], answering questions [5], and code generation [6]. However, their impact is not limited to linguistic processing. Their integration with robotic systems has opened up new opportunities for developing autonomous and social robots, significantly enhancing the interaction between humans and machines [7].

The synergy between LLMs and robotics is transforming the relationship between technology and humans, promoting more personalized, accessible, and efficient healthcare [8]. Robots equipped with advanced language capabilities can engage with patients naturally, gathering symptom information, answering questions, and supporting healthcare professionals during diagnosis and treatment [9]. These advancements enhance medical resource utilization and offer a more personalized and reassuring experience [10].

In the past, many conversational agents in healthcare employed finite-state or frame-based dialogue management strategies, limiting the ability to have natural, adaptive conversations [11]. These limitations hindered customization and the ability to respond to the individual needs of users, particularly older adults. The advent of LLMs offers a promising avenue to overcome these weaknesses. However, the use of LLMs in companion robots for older adults is not without challenges. Irfan et al. [12] highlight several problems encountered when integrating LLMs into conversational robots. These include frequent interruptions in conversations, slow or repetitive responses, inconsistent interactions, language barriers, hallucinations, and out-of-date information. These problems can cause frustration, confusion, and worry in the elderly, potentially impeding the effectiveness of robotic companions.

By leveraging the advanced features of the latest versions of LLMs and robotics, a variety of innovative use cases have emerged across multiple domains. Emerging key applications include:

- Interaction and preliminary triage: Robots can collect details about patients' symptoms, record essential information, and respond promptly to common questions. This support lightens the burden on healthcare staff, improving efficiency during the initial stages of diagnosis [13].
- Elderly assistance: Humanoid robots equipped with LLMs monitor physical health, remind patients to take medications, provide emotional support, and send alerts in an emergency. These functionalities promise to improve the quality of life for the elderly, ensuring greater safety and autonomy [14].
- Rehabilitation: Specialized robots guide patients during rehabilitation exercises, providing personalized feedback to improve treatment effectiveness and engagement [15].
- Psychological support and autism therapy: Robots equipped with advanced language capabilities can provide assistance to individuals with psychological or cognitive needs by engaging in seemingly empathetic conversations. In particular, they can play a significant role in helping children with autism develop social skills [16].
- Feedback collection and monitoring: Robots equipped with LLMs can interview patients, collecting detailed data on their physical and mental state to optimize future treatments [17].

Companion robots and carebots represent a rapidly growing sector, designed especially to support and provide companionship to older adults [18].

The former provides emotional support, social interaction, and companionship; the latter assists with physical care tasks, medical monitoring, and well-being management. With their advanced language comprehension, these types of robots respond to users' emotional and daily needs, reducing feelings of isolation and promoting greater autonomy. Features such as empathetic conversations, personalized reminders, and daily health monitoring make them essential tools for improving quality of life in home settings. However, despite the progress, using LLMs in the healthcare sector raises important safety and ethical concerns. The sensitive data processed and the vulnerability of LLMs to manipulations or errors could result in undesirable behavior or misleading information. Therefore, it is essential to develop rigorous verification protocols and alignment strategies to ensure the safe and ethical use of these technologies [19, 20].

A major challenge in using companion robots for elderly care is ensuring their successful adoption. Indeed, older adults are often unfamiliar with new technologies, and their seamless interaction with robots may be disrupted by skepticism or cognitive and physical barriers. While encouraging trust and emotional bonds with robots is crucial for engagement, these connections carry significant risks. Indeed, over time, robotic companions may replace (rather than complement) human interaction, increasing social isolation. Additionally, excessive reliance on robots for emotional support and daily tasks may determine dependency, limit autonomy, and reduce opportunities for engagement with family, caregivers, and friends. In order to mitigate these risks, designs must look for an appropriate balance between technological assistance and human-centered care. The most appropriate designs are those that promote healthy technology habits, foster connections with real people, and support group and social activities.

The Italian project TRI-TECH ("TRust in Technology: How to Assess and Improve RoboT-User Interaction in Elderly Care Integrating EtHical, Technical and Social Variables") aims to address these challenges by adopting a multidisciplinary approach. In particular, the project aims to explore the impact of companion robots on the lives of elderly individuals, with a particular focus on assessing the ethical implications through various indicators. Various tailored functionalities are implemented, which aim at fully understanding the participants' requests and demonstrating empathy in every stage of the human-robot interaction. The goal is to break down traditional human-robot barriers, creating a more natural and human-like interaction that fosters smooth and reassuring communication, ultimately improving the overall user experience.

This chapter aligns with the core objectives of the AI, Data, and Robotics Partnership [21] by introducing a novel robotic platform designed to enable user-friendly interaction between robots and elderly individuals, leveraging LLMs and carefully designed interaction flows to promote human-centric and accessible healthcare solutions. Starting from a critical review of current robotic platforms and related open challenges, both on the technical and ethical side, we designed a system architecture that leverages innovative AI tools to stimulate the interaction between the older adult and the robot. In order to evaluate the usability and ethical considerations in the usage of robotic systems by elderly people, we designed different human-robot interaction scenarios, which were implemented using the PEPPER and NAO humanoid robots. We also conducted a technical validation of the system, which showed the efficiency and feasibility of our proposed solution. We also conducted a technical validation of the system, which showed the efficiency and feasibility of our proposed solution.

The main contributions of this work are the following:

1. We critically review and discuss the state of the art in human-robot interaction, considering practical, technical, and ethical challenges.
2. We introduce the architecture of a novel robotic platform specifically designed to explore the technical and ethical implications of human-robot interaction in elderly care.
3. We explain how we designed and implemented the interaction flow between the robot and the senior in different use case scenarios.
4. We present two prototype implementations using NAO and PEPPER robots, together with the results of a technical validation of our system.

The rest of this chapter is organized as follows: Sect. 2 reviews the related work about human-robot interaction. Section 3 highlights the key challenges related to the definition of a companion robot for personal assistance addressed to elderly people. Section 4 presents our system's architecture and implementation. Section 5 focuses on practical interaction and its implications. Section 6 discusses ethical issues and their impact. Future research directions are outlined in Sect. 7, where opportunities for further exploration are emphasized. Finally, Sect. 8 concludes the chapter and summarizes the main implications.

2　Related Works

In recent years, robotic systems have significantly advanced in their intelligence and ability to engage in natural interactions with humans [22]. The desired result would be a humanoid robot that communicates seamlessly, recognizes objects, understands actions, and adapts its responses to the user's needs. These capabilities have been significantly enhanced by integrating ontologies with NLP techniques [23–25]. For instance, the NAO robot was able to discuss various topics and execute commands through the work of different authors [26–28]. Furthermore, Fukuda's framework allowed robots to identify objects through verbal interaction, fostering a communication interface that is more intuitive and that resembles human interaction [29]. The idea that the interaction should be similar to human-to-human interaction is one of the primary goals of multimodal interaction [30]. In fact, the field of multimodal interaction is grounded in the understanding that human communication is inherently multimodal. Principles of multimodality not only facilitate fluidity and engagement but also help reduce errors in interaction, adapt to potential user limitations, and allow users to choose their preferred mode of interaction at any given moment. Another relevant work in the field of human-robot interaction was carried out by Markievicz, which used supervised machine learning to interpret natural language commands, allowing actions to be classified and voice instructions to be converted into executable tasks [31].

AI and robotics are revolutionizing their field and making their way into healthcare and elderly companionship. Research projects like Triage-Bot [32] illustrate how automation and AI can optimize healthcare delivery. Developed by Sharma et al., Triage-Bot leverages AI-driven tools like facial recognition, automatic speech recognition, and photoplethysmography to collect and process patient data. Integrated with Electronic Medical Records, it automatically classifies patients according to the Canadian Triage and Acuity Scale (CTAS), enhancing efficiency and access to care.

Moreover, in the field of social robotics, several projects have aimed to develop robotic companions capable of long-term interaction and emotional engagement with humans. The FLASH robot,[1] designed within the European LIREC project,[2] focused on integrating emotional expression, behavioral consistency, and modular design to enhance human-robot interaction [33]. The MARIO project[3] focused on active and healthy aging, designing a robot aimed at exploring its use as companionship and to reduce loneliness and social isolation in dementia patients [34]. These works demonstrated the potential for robotic companions to establish meaningful and enduring relationships with users by combining advanced hardware with affective control systems. Arunachalam et al., in 2024, presented a work that proposed an AI-driven human-robot interaction system using recurrent neural network (RNN) models to provide adaptive assistance and intelligent companionship for the elderly [32]. The significance of this work lies in enhancing the accuracy and responsiveness of robotic caregivers in geriatric settings by leveraging high-quality and diversified datasets. This technology aims to manage medication, monitor health, and improve the autonomy and quality of life of the elderly by integrating IoT and AI.

3 Technical and Design Challenges

Designing a companion robot for personal assistance for elderly people presents a range of technical and design challenges. In addition to meeting complex functional requirements, the robot must ensure a positive, empathetic, and personalized user experience. An impactful aspect, yet challenging, is the integration with an LLM, since it enhances NLP capabilities. However, the success of the robot also depends on its ability to consider and enhance user perception, which influences key factors such as acceptance, trust, and engagement. In the following, we illustrate how the robot integrates advanced technologies, such as LLMs, to provide real-time, empathetic, and personalized interactions. We consider challenges like optimizing computational efficiency and ensuring scalability through modular design. Finally, we explore strategies to enable multimodal communication for natural interactions and foster trust and acceptance through empathy and user feedback and adaptability to evolving needs.

In designing a carebot or a companion robot to assist elderly people, it is necessary to ensure that it provides practical support but also emotional engagement and adaptability. The key factors that should be considered during the development include:

- Execution of various functions: the robot must handle a wide range of tasks, from daily practical assistance (e.g., medication reminders) to entertainment features, such as simple games (e.g., riddles or "guess the word"), storytelling, and the possibility to play songs or videos. Moreover, the robot must include func-

---

[1] https://spectrum.ieee.org/strange-polish-robot-

[2] https://cordis.europa.eu/project/id/215554/it

[3] http://www.mario-project.eu/portal/

tionalities that encourage the elderly person to engage in beneficial activities, such as stretching routines or mindfulness exercises. Integrating an LLM allows for contextual and personalized understanding and response to the user's needs, balancing complexity and ease of use.

- Speed of execution and fluidity of interaction: a quick, natural, and efficient interaction is crucial for conveying empathy and responsiveness, key aspects for building trust and improving the user experience. The LLM represents the core of the robot's linguistic intelligence, but its processing requires significant computational resources. Optimizing the model's inference pipeline is essential to reduce latency and response times while ensuring smooth and scalable performance. The user perception is closely tied to these aspects; indeed, a system that responds quickly and accurately is perceived as more reliable, enhancing the user's sense of security and satisfaction.

- Scalability and integration of new features: to keep the robot resilient over time, a modular design is needed that allows for the addition of new functionalities and system customization for specific contexts. The ability of the robot to evolve according to the needs of the elderly person reinforces the sense of usefulness and engagement, fostering a long-term emotional bond and improving the user's perception.

- Multimodality of the interaction and adaptation to context and user's sensations: the robot must interpret the context and respond appropriately to the user's emotions. Multimodality enables a natural interaction while allowing the robot to supplement the information received through one mode with data from others (e.g., better understanding voice commands through gestural information) or to better understand the context of the interaction. The integration of the LLM with sensors (e.g., cameras or microphones) enables the combination of linguistic and physical cues for a more complete understanding. This adaptation not only improves the quality of interactions but also reinforces the perception of the robot as an empathetic and attentive companion, capable of responding to the elderly person's emotional needs.

- Demonstration of empathy: the robot's ability to demonstrate empathy is central to building an emotional bond with the user. Thanks to the LLM, the robot can generate responses that simulate an empathetic attitude. However, this function must be carefully designed to avoid biases or inappropriate responses, ensuring that the tone remains respectful and consistent with the user's expectations. Perceived empathy is what transforms the robot from a mere technological assistant into a life companion. Also, in relation to multimodality, multimodal outputs can be provided, offering more information through gestures, voice, or visual effects that allow the robot to seem more natural and facilitate communication.

- User feedback collection: collecting feedback is crucial for improving the robot's functionalities and personalizing its responses. However, the chosen method must respect the elderly person's cognitive and physical capabilities, avoiding intrusive interactions. Well-designed feedback mechanisms also enhance the perception of the robot as a system attentive to the user's needs, increasing trust and acceptance. Additionally, the robot provides empathetic feedback based on

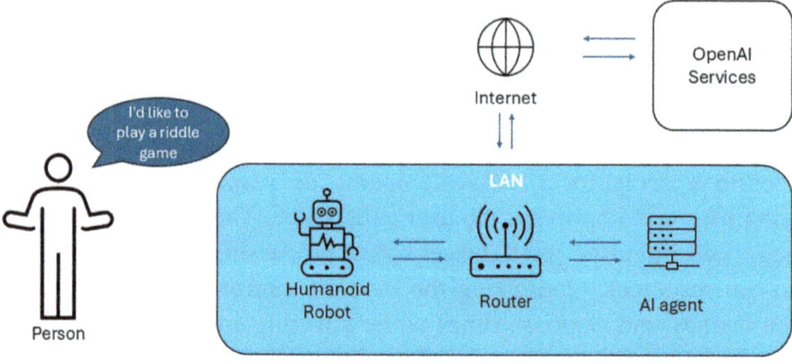

Fig. 1 System architecture

context, using personalized movements that combine body language, eye color, and voice tone. This approach makes interactions even more natural, human, and responsive, strengthening the emotional bond between the robot and the elderly person.

4 System Architecture

The architecture of our proposed human-robot interaction system is based on the employment of LLMs and programmable robots. This configuration allows robots to understand and respond to complex commands, both physical and cognitive, in a natural and intuitive manner. The architecture is illustrated in Fig. 1.

In this section, we will describe its main components and their functioning.

4.1 Main Components

The system architecture is composed of four fundamental elements:

- Humanoid robots: For this study, we selected the NAO[4] and PEPPER[5] humanoid robots. Both robots are equipped with a variety of sensors, including gyroscopes, microphones, cameras, and tactile sensors, enabling multimodality and an advanced perception of their surroundings. Additionally, both robots are fully programmable and capable of performing complex movements, making them ideal for the intended purpose. The main difference between the two lies in their physical and functional configuration. PEPPER lacks legs but moves using

---

[4] https://corporate-internal-prod.aldebaran.com/en/nao

[5] https://corporate-internal-prod.aldebaran.com/en/pepper

wheels and features an integrated touchscreen tablet. In contrast, NAO is equipped with legs and walks but does not include a tablet.

- Language model: We conducted preliminary evaluations to determine the most suitable LLM for our system's specific requirements. In our informal benchmarking, we considered several state-of-the-art models, including GPT-4 (OpenAI), Gemini (Google), LLaMA (Meta), and Mistral. Our primary selection criteria included support for the Italian language, fluency in generating natural and contextually appropriate responses, fast response times, and ease of integration via APIs. Based on these factors, GPT-4 demonstrated the most reliable performance, particularly in handling Italian inputs and delivering coherent, user-friendly replies. Hence, we adopt OpenAI's GPT-4 language model [35] to process user requests and generate appropriate responses or actions. GPT-4 can understand natural language and respond to general questions or interpret specific commands, adapting to different linguistic variants and contexts. Importantly, the system's architecture is model-agnostic by design: the underlying LLM can be replaced with minimal effort by any other model that supports OpenAI-compatible APIs, such as Gemini via a wrapper, or even local models via Ollama. For well-known alternatives that do not natively support this interface, only minor code adjustments are required, typically limited to a few lines to adapt the API endpoints and response parsing logic.
- AI agent: The AI agent is designed to leverage the advanced capabilities of an LLM model, ensuring efficient and seamless management of interactions between the user and the robot. The architecture consists of the following main modules:

  1. Functionality request recognition module: It analyzes user requests to identify and activate the most suitable functionalities available in the system.
  2. Robot gesture selection module: It processes contextual data to select and execute the most relevant movements according to the specific situation.
  3. User data tracking module: It manages user-related information, including preferences, interaction history, and details helpful in personalizing the experience.
  4. Log tracking module: It records and monitors all system activities, including user-robot interactions, providing valuable data for debugging, optimization, and performance analysis.

4.2 Detailed Architecture and Data Flow

In this section, we provide a detailed description of the data flow within the system architecture:

1. Reception of voice input: The interaction begins with a voice instruction provided by the user to the humanoid robot (e.g., "I would like to play a guessing

game"). The audio signal is captured through microphones integrated into the robot (NAO or Pepper), as depicted in Fig. 2.

2. Audio transmission and textual transcription: After receiving the audio signal, the robot transmits it over the local network (LAN) for transcription. The specific transmission flow depends on the robot model used:

   (a) NAO robot: The robot sends the audio signal to the AI agent server, which forwards it to the Speech-to-Text (STT) transcription service utilizing OpenAI's Whisper model, as represented in Fig. 3.
   (b) Pepper robot: The Pepper robot sends the audio signal directly to the STT service, receives the transcribed text, and subsequently forward it to the AI agent, as illustrated in Fig. 4.

3. Intent classification and response generation: Upon receiving the textual transcription of the voice input, the AI agent employs a LLM, such as OpenAI's GPT-4, to perform a detailed semantic classification of the user's intent. The LLM analyzes the received text by comparing it with a predefined set of available functionalities (e.g., play games, listen to music, perform exercises, answer general questions) and selects the functionality that best matches the user's request as illustrated in Fig. 5.

The response generation can follow two distinct modalities:

1. Static response (predefined): The AI agent selects a pre-existing, hardcoded response stored in the robot. The most appropriate response is selected based on the recognized intent, as depicted in Fig. 6.
2. Dynamic response: When a personalized and original response is required, the AI agent utilizes LLM, such as OpenAI's GPT, to generate content specifically tailored to the interaction context, as illustrated in Fig. 7.
3. Transmission of response to the robot and user interaction: The generated response, whether static or dynamic, is transmitted back to the humanoid robot, which converts it into vocal output using an integrated text-to-speech (TTS) system. Verbal communication is accompanied by appropriate body movements from the robot to enhance interaction and natural communication with the user.

Fig. 2 Reception of voice input

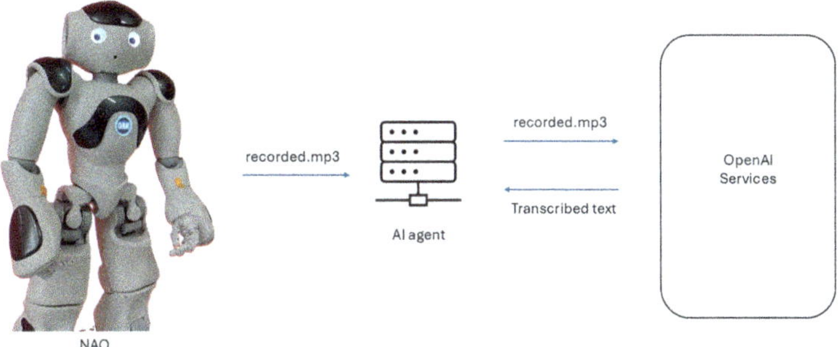

Fig. 3 Transcription of vocal input on NAO

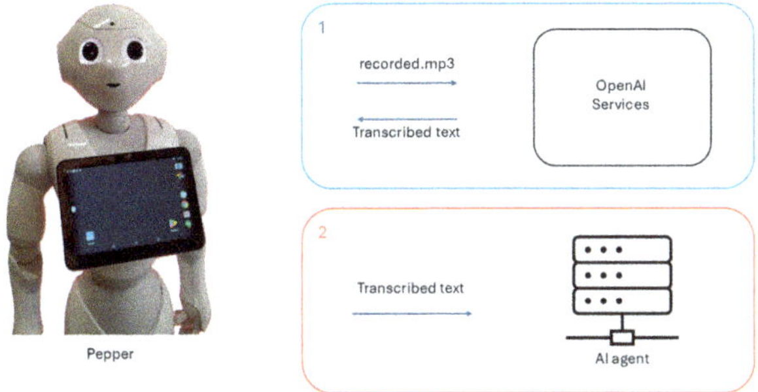

Fig. 4 Transmission of the transcribed text to the AI agent

Fig. 5 Intent classification

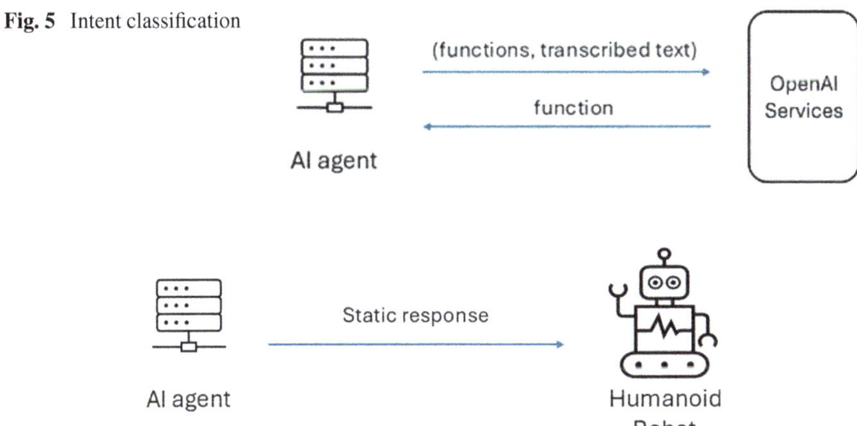

Fig. 6 Sending a static response to the robot

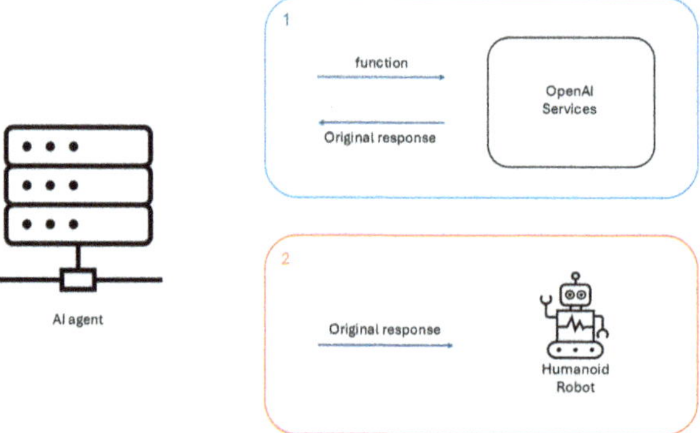

Fig. 7 Generating and sending a static response to the robot

4.3   Integration and Communication

The communication between the various components of the system occurs through a local area network (LAN), where the robots interact with a central unit responsible for managing the core functions and integrating with GPT-4 and the AI agent. The exchange of requests and responses is implemented using the REST architecture over the HTTP protocol, ensuring efficient data transmission. This approach promotes smooth and natural robot-user interaction while significantly reducing perceived conversation latency.

4.4   Technical Evaluation

We tested our robotic infrastructure by measuring the response times of its various modules to evaluate the system's efficiency and responsiveness. The most computationally expensive modules of the system are the following:

- Speech-to-text, for transcribing the words spoken by the elderly
- Text generation, used in cases where it is necessary to produce a response to the elderly
- Input classification, that is, the analysis of the speech-to-text output to understand the elderly's intentions

To evaluate the system's performance, the response times of each of these components were measured during real-world interaction with the robot. The collected data includes repeated measurements over 20 sessions of interaction. Table 1 reports the average and standard deviation of the recorded times for each module. Results show that the execution times are feasible for our application scenarios. The most expensive task is text generation, which may take considerable time for particular

Table 1 Response times of system components

	μ (s)	σ (s)
Input classification	1.43	1.10
Speech to text	1.47	0.47
Text generation	4.58	1.50

tasks, especially when the LLM is asked to invent a story based on a theme chosen by the elderly.

Additionally, we conducted preliminary tests involving colleagues and individuals external to the project. This approach allowed us to gather unbiased feedback, free from implementation-related influence, which helped identify potential issues and improve the overall interaction with the robot.

4.5   Advantages of the Proposed Architecture

The proposed architecture offers several advantages:

- Precise natural language processing and classification: Integration with an LLM enables the system to extract relevant information from user inputs typically expressed in non-technical language, ensuring effective system operation.
- Scalability and adaptability: Due to its modular design, the system allows for easy integration of new features and modifications to the AI agent's parameters, facilitating future adjustments to the interaction and enabling continuous system evolution.
- Multimodality: Given the capabilities of robots and the flexibility of LLMs, the architecture supports the customization of different forms of user communication. Multimodality also streamlines the robot's ability to respond in a more versatile and engaging way.
- Relationality: The architecture is designed to foster meaningful human-robot relationships, prioritizing the development of connections that address emotional and social needs of elderly persons. By promoting interactions that prevent the isolation often experienced in caregiving contexts, robots would enhance psychological and relational well-being. This focus on relational engagement ensures that the technology transcends its role as a tool, becoming a trusted companion that integrates seamlessly into the social fabric, ultimately improving the quality of life for individuals in need of care.
- Stability of ethical framework: The proposed architecture consolidates a robust ethical framework, with trust as its cornerstone value. Designed with context-related usage in mind, it ensures that every deployment of the system is preceded by careful consideration of the situational needs and potential impacts. By embedding trust as a central tenet, it supports ethical decision-making while fostering reliable and meaningful interactions, establishing a standard for responsible and inclusive innovation. This framework embodies a twofold role. On the

one hand, it can be considered as a foundational-orientative guide for the work of robotic engineers, shaping the programming process to align with ethical principles. On the other, it offers the opportunity for stabilizing a common ground of inter-disciplinary cooperation and fostering trans-disciplinary achievements, from both a technological and social point of view.

5 User Interaction

The system operates through four main phases to ensure smooth and personalized interaction between the user and robots. Each phase is accompanied by complete tracking of conversations through the Log Tracking Module, which records and monitors every exchange of information between the robot and the user. The four main phases that ensure the system's proper functioning are discussed below.

5.1 User Basic Information

The first phase involves the robot acquiring personalized information through a series of questions posed to the user. These questions primarily concern the user's name, the name the user wishes to give to the robot, and their preference regarding the formality of the conversation (more or less formal). The collected parameters are stored in the User Data Tracking Module, allowing the system to adapt to the user's needs. Based on these preferences, the robot will begin addressing the user by name and use appropriate language, creating a more natural and engaging experience. Figure 8 illustrates the flowchart of the User Basic Information Configuration process, highlighting the sequential steps from posing questions to the user to adapting the robot's behavior based on the collected preferences.

5.2 Actual Interaction

Once the information are set, the system begins actively interacting with the user, inviting them to perform various activities. The Functionality Request Recognition Module analyzes the user's requests and activates the most suitable features. Specifically, the robot can:

- Play songs or videos (only in the case of PEPPER)
- Propose simple games, such as riddles or "guess the word," allowing the user to test the robot
- Tell themed stories
- Guide the user through basic physical exercise routines or mindfulness sessions

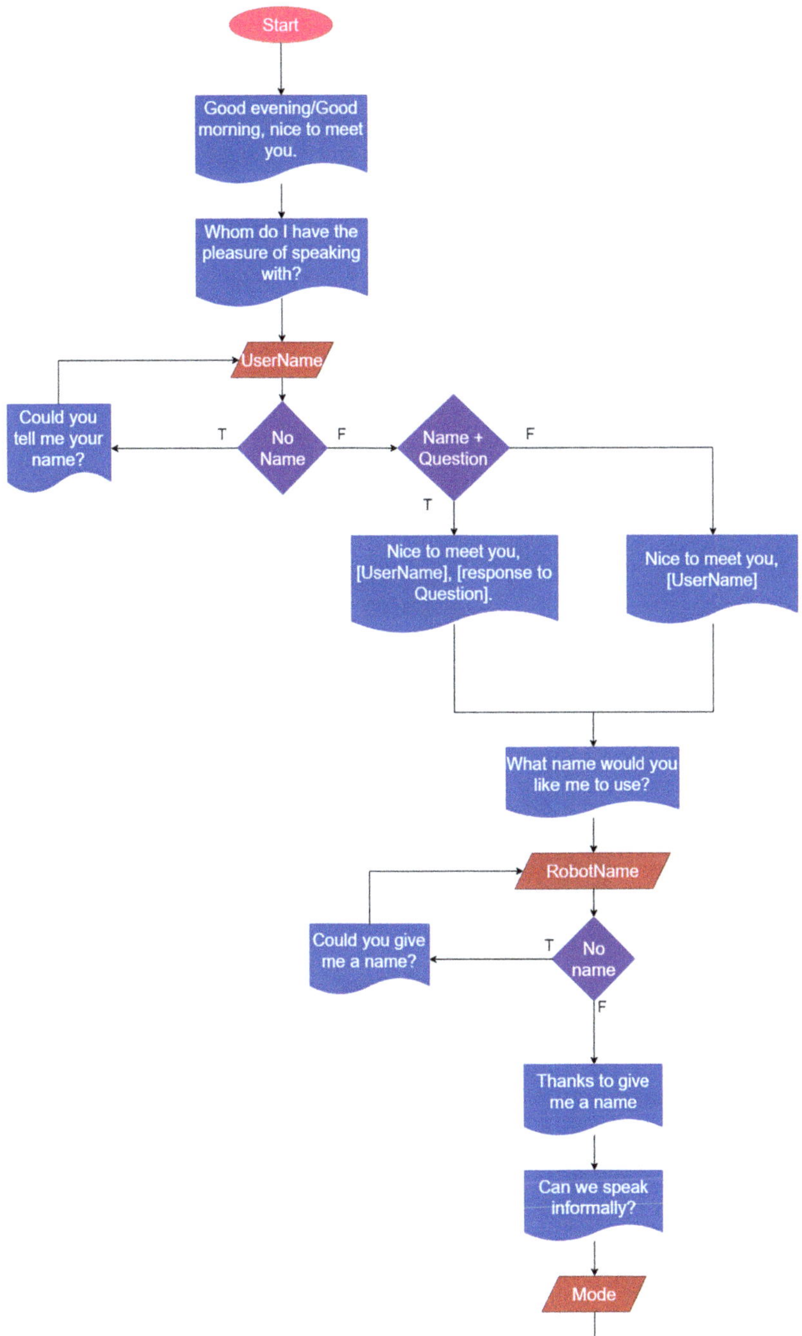

Fig. 8 User basic information configuration flowchart

- Respond to general questions or comments from the user (e.g., "Who was Napoleon?" or "Tell me a joke")

During each activity, the robot will ask the user whether they enjoyed the experience, fostering engagement and empathy. In Fig. 9, the feedback request can be observed when the user asks the robot to tell a story. After the story is told, the question "Did you enjoy the story?" is asked.

Furthermore, the robot will react emotionally and gesturally based on the context: for instance, if the user expresses sadness, the robot will respond with appropriate gestures, thanks to the Robot Gesture Classification Module.

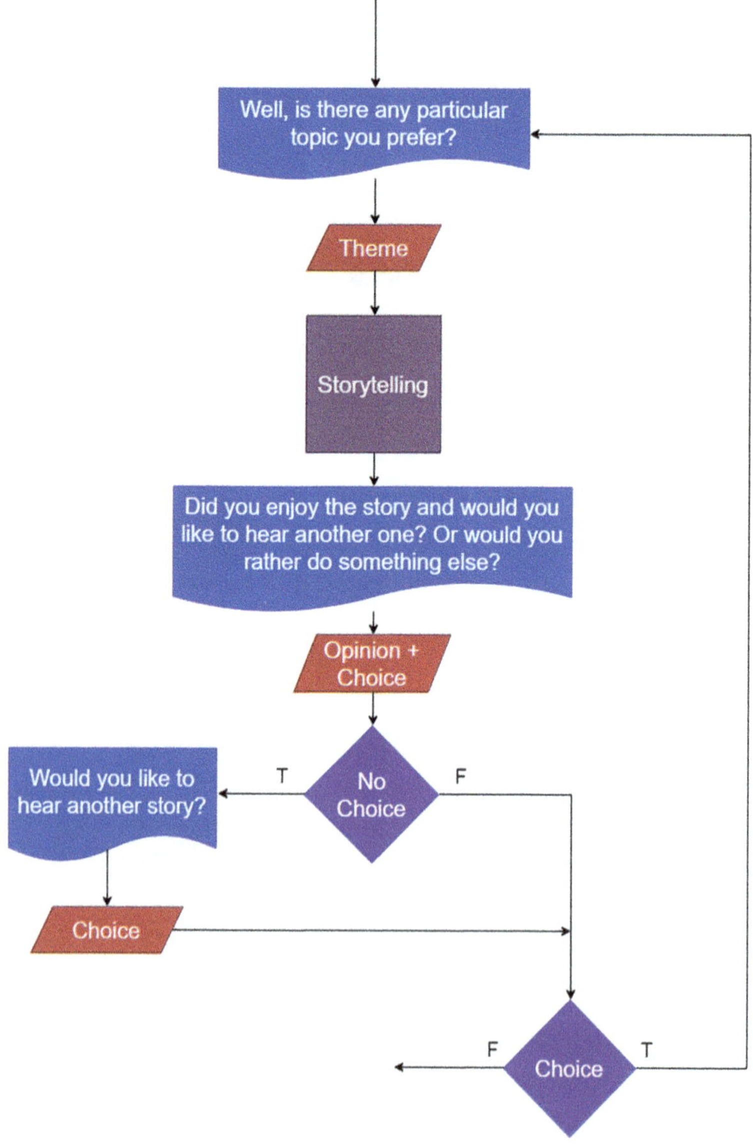

Fig. 9 Storytelling flowchart

5.3 Event Reminders

The system is designed to track important events and remind the user about them, encouraging the elderly user to perform activities necessary for their well-being. For example, periodically, the robot will remind the user to drink water or do some physical exercise, ensuring that the user follows daily routines and does not forget essential tasks.

5.4 Continued Interaction

After the initial interaction phase, the robot will continue encouraging the user to engage in activities, maintaining active involvement. If the user shows no interest or provides no feedback to continue, the robot will wait for new requests. The user can stop using the robot anytime, interrupting the interaction. The system is designed to be flexible and respect the user's wishes, ensuring a personalized and non-intrusive experience. This modular and interactive approach allows the system to dynamically respond to the user's needs, enhancing the overall interaction experience and fostering continuous, respectful, and engaging communication.

5.5 System at Work

The TRI-TECH project aims to evaluate and enhance the interaction between elderly individuals and companion robots, focusing on key aspects such as trust, acceptance, and perceived utility of these technologies. The proposed architecture will be employed in an experimental setting, involving 1-h usage sessions with groups of elderly participants recruited from care facilities and healthcare environments. Recruited seniors will interact with two robots, shown in Fig. 10: NAO and PEPPER. A video showcasing the Pepper robot interacting with a user, narrating a story, and utilizing the mentioned architecture is available at https://youtu.be/kaR-vO7YFEGs. A video showcasing NAO proposing a set of physical exercises in available at https://youtube.com/shorts/rcd99D44Wpo.

Participants will follow a structured process: initially, they will complete a preliminary questionnaire to assess their expectations and predispositions. Subsequently, after their experience with the robot, they will fill out a second questionnaire to provide detailed observations and feedback on their perceptions and satisfaction levels. During the sessions, qualitative and quantitative data on the interaction between participants and the robot will be collected. These data, derived from textual input provided by users, will include patterns of verbal communication, emotional expressions, and the level of engagement evident in their responses. The methodological approach also involves simulating real-life scenarios, such as

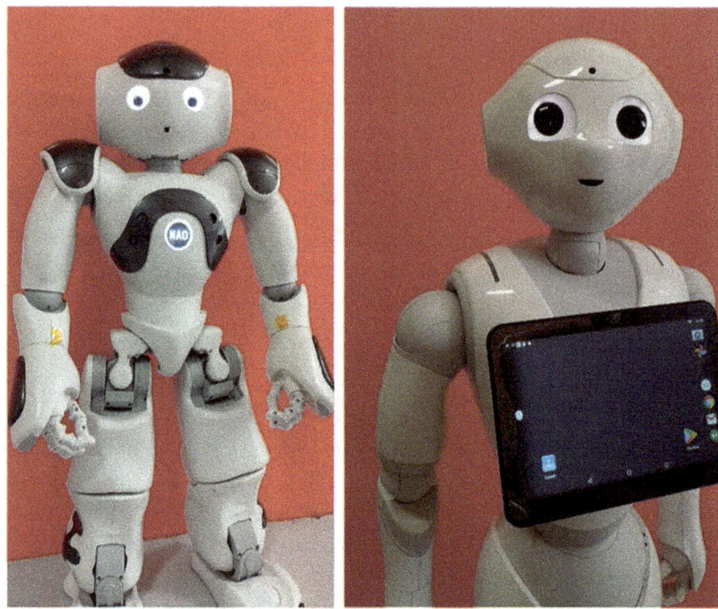

Fig. 10 The NAO and PEPPER robots used in our prototypes

companionship and health monitoring, to ensure that the analyzed interactions reflect realistic and meaningful situations for the target demographic. This element ensures that the findings can be applied to practical and relevant contexts in the daily lives of elderly individuals. The integration of ethical, technical, and social variables is essential for a comprehensive analysis of the impact of robots on the quality of life of the elderly. The collected data will be used to investigate critical aspects such as trust in the robot, the degree of acceptance, and the effectiveness of interaction modalities in promoting user well-being and comfort. The results of this study will guide the optimization of the design, functionality, and communication methods of companion robots, fostering their large-scale adoption. Ultimately, the project aims to improve elderly care by providing reliable, useful technologies that seamlessly integrate into their daily lives, thereby contributing to significant advancements in care and social inclusion.

6　Ethical Issues

In the sections above, we examined the technical aspects involved in the project; we now turn our attention to the ethical implications and concerns. The rapid advancement of digital technologies has resulted in the increasing utilization of intelligent systems and autonomous machines across diverse domains, encompassing healthcare, industry, education, and transportation. Within these contexts, trust has been identified in the scientific literature as a crucial factor [36] significantly influencing the acceptance, implementation, and efficacy of robots and computerized systems in general [37].

6.1   Mapping Trust in Human-Robot Interaction

Generally speaking, trust constitutes a cornerstone of interpersonal relationships, encompassing both human-to-human and human-to-robot interactions, particularly when these interactions involve elements of reciprocity (where applicable) or shared experiences. The presence of trust exerts a significant influence on the quality of relationships and reveals a complex dynamic: as trust increases, relationships tend to become more robust, authentic, and resilient to dialectical tensions in a generative manner, facilitating the resolution of conflicts without descending into exclusionary or adversarial dynamics. Conversely, in the absence of trust, relationships often assume a defensive profile characterized by caution or suspicion, frequently necessitating precautionary or contractual measures to mitigate perceived risks. These considerations assume even greater significance in interactions with artificial intelligence systems or robotic entities [38].

Fostering trust can become particularly critical when it comes to robots because, unlike human beings, robots lack the innate social and emotional cues that naturally elicit trust. Humans are predisposed to trust other human beings due to shared experiences, emotional understanding, and the capacity for reciprocal empathy [39]. These qualities are deeply rooted in our evolutionary and cultural frameworks and make human-to-human trust more intuitive and accessible.

In contrast, robots often lack these relatable attributes. They do not share our biological makeup, cultural experiences, or emotional depth, which makes it harder for us to perceive them as trustworthy agents. It is no coincidence that robots are often designed to resemble humans in appearance and behavior. It has been proven that anthropomorphism plays a crucial role in fostering trust and facilitating interaction [40, 41].

6.2   Preliminary Ethical Framing of Emotional States

Research in human-robot interaction has shown that the more a robot appears capable of understanding and replicating human emotional states, the more likely users are to perceive it as reliable and approachable [42]. This suggests that we tend to feel more comfortable with and trust things that resemble us as human beings.

Moreover, robots are perceived as operating due to complex, often opaque, systems of rules and algorithms that are not immediately understandable to most users. This "black box" nature of AI systems can further erode trust, as users may struggle to predict or fully comprehend a machine's actions and decisions [43–45]. It should also be noted that the perceived trustworthiness of these systems is often contingent on the reliability of their information, performance, and capabilities as demonstrated in practice. Furthermore, from the user's perspective, a robotic entity will be considered "worthy of trust" only to the extent that it is not perceived as threatening, obstructive, or deceptive [46, 47]. Instead, it must be seen as a mechanical entity capable of promoting advantageous and productive outcomes while remaining sustainable from the perspective of human relationships.

6.3 Considering Intercultural Ethical Complexity

Finally, an additional layer of complexity arises from the cultural context. For example, the elderly population in Europe has historically exhibited a cautious attitude toward artifacts such as robots and artificial intelligence systems, contrasting with more accepting attitudes observed in Asian or Eastern populations [48]. Historical, political, socio-cultural, symbolic, anthropological, mythical, and religious factors shape the perception of robots, thereby influencing the degree of trust accorded to them [49]. For instance, some individuals may perceive a robot as a mere service tool, enabling the delegation of burdensome tasks, thereby inspiring a level of trust comparable to that traditionally reserved for a tool like a hammer.

As previously discussed, numerous factors have the potential to undermine human-robot trust, which becomes particularly significant in the context of companion robots. Examples include elderly care, assistance for individuals with disabilities, and the promotion of general well-being. Indeed, some scholars emphasize that the relationship between trust and use can be captured by the phrase "no trust, no use," underscoring its fundamental importance [50]. This presents a significant challenge: if care robots are designed to combat issues such as isolation and loneliness but fail to inspire trust, they risk becoming underutilized despite their potential benefits.

For this reason, the TRI-TECH project places a strong emphasis on investigating this construct, which, in the context of personalized companion robots (PCRs), can be conceptualized as encompassing more than merely an emotion, a perception of reliability, or technical competence in performing specific tasks. Rather, these dimensions are interrelated and necessitate a holistic, integral approach that transcends a narrow focus on the immediate goals the system achieves. Given these complexities that characterize human-robot interactions, strategies to enhance trust in these systems are crucial and can take various forms. These include investigating how humans perceive robots or identifying feasible modes of interaction, designing robots guided by the principles of "Ethics by Design," and ensuring that robots embody the very properties that constitute trust.

6.4 Enabling Multifaceted Trust

The TRI-TECH project integrates several properties identified as critical to the value of trust while addressing these dimensions. Specifically, by refining communicative exchanges, it pays attention to aspects such as proxemics, gestures consistent with semantic content, and ensuring that robots perform "checks" to verify the accuracy of their actions. We expect these characteristics will enhance the trust that elderly users may place in these systems. More concretely, we focus on designing actions embedded with specific sub-values that constitute trust and programming them into robots.

The selected values included autonomy and vulnerability—concepts previously explored in existing literature [51–53]—as well as relationality and meaningfulness, which are recognized by the Ethics of Care as indispensable components of human and caregiving contexts. Integrating these values into robot behavior aims to create interactions that inspire trust and strengthen user-robot relationships.

In summary, fostering trust in intelligent systems, particularly in sensitive and vulnerable environments, requires a multidimensional and culturally aware approach. Trust is not merely an auxiliary attribute but a fundamental element that influences the efficacy and sustainability of human-robot interactions. By incorporating design principles that align with human values and needs, we can establish a foundation for robotic systems that are not only technologically reliable but also socially and ethically integrated into the fabric of human relationships. This integration is essential to ensure their widespread acceptance and effective utilization, ultimately maximizing their potential to enhance users' well-being.

7 Implementation Barriers and Future Research Directions

The integration of robotics and LLMs as a tool for healthcare presents several challenges that need to be addressed, along with opportunities for further improvement. At the time of writing, LLM models often present problems such as generating false, obsolete, general, or unreliable information in response to questions. Currently, while the interaction flow is predefined by the systems designers, the dialogue of the robot is generated by an LLM based on a set of defined prompts. We have carefully crafted the prompts in order to minimize unexpected outputs or hallucinations from the LLM. Despite our efforts, occasional unexpected or hallucinated outputs remain a known limitation of current LLMs. In order to mitigate this issue, in future work, we plan to investigate the use of Retrieval Augmented Generation methods to drive the LLM outputs by means of curated external knowledge [54]. This approach would ensure more accurate, relevant, and contextually appropriate responses. During the experiments, a variety of interaction information is collected, enabling the identification of weaknesses in the robot's current functionality and the areas that require improvement. By analyzing user interaction patterns, it will be possible to refine responses, incorporate the most required functions, and optimize overall performance. Advanced machine-learning techniques will allow the robot to adapt and evolve based on real-world usage.

A further technical issue in the dialogue interaction with the robot regards the variability in elderly speech along with privacy and data security concerns. Indeed, in order to optimize speech-to-text, it is useful to exploit powerful AI tools that are commonly executed on cloud infrastructures due to their computational complexity. Of course, relying on external infrastructures for speech-to-text transcription, as done in our current implementation, exposes the sensitive communications of the elderly to unintended access by third parties. In future work, we will investigate the use of compact yet effective AI models for executing speech-to-text transcription on a dedicated secure server inside our system infrastructure. However, in scenarios where local computational resources or a private cloud infrastructure are available, the proposed system can be adapted seamlessly to operate within these environments, thereby ensuring greater control over data privacy and security.

Future work also includes seamless integration with various medical devices, such as smartwatches, blood pressure monitors, and smart scales.

These devices can provide valuable health information, enabling the robot to assess the health status of elderly users more accurately. By incorporating such devices, the robot can offer personalized health advice, fostering a more profound sense of trust and strengthening the relationship between the user and the companion robot. This integration is a significant step toward creating a more holistic and supportive caregiving solution.

From an ethical standpoint, two aspects warrant particular attention as key future research directions: the emotional and relational needs of elderly users and the potential risks associated with the use of companion robots. With regard to the first issue, it should be noted that the use of companion robots necessitates a critical evaluation of the modes of reception or expectations that arise from users, as well as the ways in which the presence of robots reconfigures daily practices or the meanings they assume within the experiences of the elderly population. On one hand, the presence of assistive and companion robots, as demonstrated in the literature, can reduce levels of loneliness, stress, and depression, offering users communicative and interactive exchanges that mitigate feelings of isolation. Conversely, there is a tangible risk of delegating the entirety of relational tasks to robots or assuming that they can replicate the complexity of human relationships. Human connections are inherently multifaceted, comprising components of value, semantics, perception, and shared worldviews, as well as profound emotional resonance. To address these challenges, several measures warrant consideration. For example, companion robots should be designed with the ability to recognize and adapt to the emotional and relational needs of elderly users, as well as their individual preferences. Specifically, advancements in technology should enable companion robots to be engineered with the capacity to identify and respond to the nuanced emotional and relational requirements of each user. This could encompass mechanisms for detecting subtle emotional cues or anticipating specific needs, thereby ensuring that their responses are both appropriate and aligned with their preferences. Moreover, when a robot is unable to meet the subject's needs, it should be equipped to effectively communicate this limitation to the relevant individuals, such as family members or caregivers, thereby facilitating the efficient transfer of responsibility for further support. A second consideration pertains to the design and deployment of companion robots and related artificial intelligence technologies, ensuring that these systems are firmly grounded in ethical principles that prioritize transparency and fairness. For instance, it would be crucial to identify the most effective strategies for informing elderly users about potential risks while critically assessing the contexts in which such information is both necessary and genuinely beneficial for the individual. These risks may include misinterpretations or unintended deception that could arise during interactions with companion robots. Proactively addressing these challenges is essential to managing expectations realistically and preventing ethical oversights, thereby ensuring the preservation of dignity and autonomy. Moreover, companion robots should not function in isolation but should be integrated into a comprehensive network of care that encompasses human caregivers, family members, and social connections. Implementing this integration necessitates meticulous understanding and planning to ensure that robots are seamlessly incorporated into the caregiving ecosystem. Through this approach, robots can complement human relationships rather than supplant them, fostering collaboration rather than engendering dependence.

Finally, there is a need for longitudinal studies to comprehensively evaluate older adults' perceptions and experiences with the companion robot.

These studies will help assess the emotional and functional impact of the robot over long periods. By addressing these challenges and undertaking these initiatives, it would be possible to develop a truly reliable, adaptable, and user-centered robotic companion that caters to the unique needs of older adults.

8   Conclusions

Integrating LLMs with robotics represents a considerable opportunity to enhance healthcare and elderly care, providing personalized, empathetic, and efficient assistance. Several initiatives have shown that the development of companion robots can improve the quality of life of older people by addressing their emotional, physical, and social needs. The TRI-TECH project aims to assess and address the ethical issues related to human-robot interaction by creating a companion robot capable of performing various tasks, from daily reminders to advanced multimodal interactions and adapting to users' preferences, creating meaningful, humanlike connections. The proposed system architecture, which combines the advanced NAO and PEPPER humanoid robots with the capabilities of LLM GPT-4, enables continuous communication and personalized interaction and lays the foundation for future advances in robotic assistance.

Key challenges, such as latency optimization, user trust enhancement, and ethical and safe use assurance, emphasize the importance of continuous research and development. Ethical considerations are crucial in this context, as the interaction between robots and users involves sensitive data and complex emotional dynamics. Future research should prioritize this aspect along with efforts to perfect technical capabilities, ensuring that robots not only assist but also respect and support their users.

The continued evolution of companion robots will require interdisciplinary collaboration across technical, medical, and social domains. By leveraging advanced technologies such as *RAG* and integrating with wearable medical devices, the proposed system will be able to provide increasingly reliable and contextually relevant care. Long-term studies will be crucial to assess user acceptance and identify areas for improvement, contributing to the development of truly adaptive and user-centered solutions.

In conclusion, the combination of robotics and LLM paves the way for a new era of intelligent care. If carefully designed and ethically deployed, companion robots can redefine elderly care, fostering independence, improving well-being, and creating a more inclusive and humane healthcare system. By addressing the challenges and following future research directions, we can realize the vision of robots as reliable partners in improving the lives of the elderly and transforming the wider healthcare landscape.

Acknowledgments This work represents a preliminary outcome of the TRI-TECH project ("Trust in Technology: How to Assess and Improve RoboT-User Interaction in Elderly Care Integrating Ethical, Technical and Social Variables," coord. by A. Pirni), funded under the National Recovery and Resilience Plan (PNRR).

The project, identified by CUP B83C22004800006, is supported within Mission 4 "Education and Research," Component 2 "From Research to Business," Investment 1.3, financed by the European Union, NextGenerationEU. Funding details are specified in the official directives: D.D. 343/2024 (pdf), D.D. 365/2024 (rectification of Article 3.3, Table 1, Annex D, pdf), and D.D. 468/2024 (pdf), which extended the submission deadline to April 10, 2024, at 12:00.

The authors collaborated closely in discussing, planning, writing, and revising all parts of the chapter. Specifically, Lorenzo Boi, Silvia M. Massa, Diego Reforgiato Recupero, Daniele Riboni, and Rubén Alonso contributed to the writing of Sects. 1, 2, 3, 4.1, 4.2, 4.3, 5, and 7. Michele Cardinali, Alberto Pirni, and Elena Ricci focused on Sects. 4.3, 6, and 7.

References

1. Vaswani, A., Shazeer, N., Parmar, N., Uszkoreit, J., Jones, L., Gomez, A. N., Kaiser, L., & Polosukhin, I. (2017). Attention is all you need. In I. Guyon et al. (Eds.), *Advances in neural information processing systems* (Vol. 30). Curran Associates, Inc..
2. Obinwanne, T., & Brandtner, P. (2024). Enhancing sentiment analysis with gpt-a comparison of large language models and traditional machine learning techniques. In A. K. Nagar et al. (Eds.), *Intelligent sustainable systems* (pp. 187–197). Springer.
3. Zhang, B., Haddow, B., & Birch, A. (2023). Prompting large language model for machine translation: A case study. In *Proceedings of the 40th International Conference on Machine Learning, ICML'23*. Retrieved from JMLR.org

4. Zhang, T., Ladhak, F., Durmus, E., Liang, P., McKeown, K., & Hashimoto, T. B. (2024). Benchmarking large language models for news summarization. *Transactions of the Association for Computational Linguistics, 12*(1), 39–57.
5. Kamalloo, E., et al. (2023). Evaluating open-domain question answering in the era of large language models. In A. Rogers et al. (Eds.), *Proceedings of the 61st Annual Meeting of the Association for Computational Linguistics, Toronto, Canada, July 2023* (Long Papers) (Vol. 1, pp. 5591–5606). Association for Computational Linguistics.
6. Wang, J., & Chen, Y. (2023). A review on code generation with LLMS: Application and evaluation. In *2023 IEEE International Conference on Medical Artificial Intelligence (MedAI)* (pp. 284–289).
7. Kim, C. Y., Lee, C. P., & Mutlu, B. (2024). Understanding large-language model (LLM)-powered human-robot interaction. In *2024 19th ACM/IEEE International Conference on Human-Robot Interaction (HRI)* (pp. 371–380).
8. Kim, K., Windle, J., Christian, M., Windle, T., Ryherd, E., Huang, P.-C., Robinson, A., & Chapman, R. (2024). Framework for integrating large language models with a robotic health attendant for adaptive task execution in patient care. *Applied Sciences, 14*(21), 9922.
9. Wang, D., & Zhang, S. (2024). Large language models in medical and healthcare fields: Applications, advances, and challenges. *Artificial Intelligence Review, 57*, 299.
10. Browne, R., et al. (2024). Reflective dialogues with a humanoid robot integrated with an LLM and a curated NLU system for positive behavioral change in older adults. *Electronics, 13*(22), 4364.
11. Laranjo, L., et al. (2018). Conversational agents in healthcare: A systematic review. *Journal of the American Medical Informatics Association, 25*(9), 1248–1258.
12. Irfan, B., et al. (2025). Between reality and delusion: Challenges of applying large language models to companion robots for open-domain dialogues with older adults. *Autonomous Robots, 49*(1), 9.
13. Frosolini, A., Catarzi, L., Benedetti, S., Latini, L., Chisci, G., Franz, L., Gennaro, P., & Gabriele, G. (2024). The role of large language models (LLMS) in providing triage for maxillofacial trauma cases: A preliminary study. *Diagnostics, 14*(8), 839.
14. Cavallaro, A., Perillo, F., Romano, M., Sebillo, M., & Vitiello, G. (2024). Social robot in service of the cognitive therapy of elderly people: Exploring robot acceptance in a real-world scenario. *Image and Vision Computing, 147*, 105072.
15. Albanese, G. A., et al. (2024). Robotic systems for upper limb rehabilitation in multiple sclerosis: A SWOT analysis and the synergies with virtual and augmented environments. *Frontiers in Robotics and AI, 11*, 1335147.
16. Peca, A. (2016). Robot enhanced therapy for children with autism disorders: Measuring ethical acceptability. *IEEE Technology and Society Magazine, 35*(6), 54–66.
17. Han, S., Min, X., Lao, J., & Liang, Z. (2023). Collecting patient feedback as a means of monitoring patient experience and hospital service quality—Learning from a government-led initiative. *Patient Preference and Adherence, 17*(2), 385–400.
18. Weiss, A., & Hannibal, G. (2018). What makes people accept or reject companion robots? A research agenda. In *Proceedings of the 11th PErvasive Technologies Related to Assistive Environments Conference, PETRA'18* (pp. 397–404). Association for Computing Machinery.
19. Jain, N., Schwarzschild, A., Wen, Y., Somepalli, G., Kirchenbauer, J., Chiang, P. Y., Goldblum, M., Saha, A., Geiping, J., & Goldstein, T. (2023). Baseline defenses for adversarial attacks against aligned language models.
20. Zou, A., Wang, Z., Carlini, N., Nasr, M., Zico Kolter, J., & Fredrikson, M. (2023). Universal and transferable adversarial attacks on aligned language models.
21. Curry, E., Heintz, F., Irgens, M., Smeulders, A. W., & Stramigioli, S. (2022). Partnership on AI, data, and robotics. *Communications of the ACM, 65*(4), 54–55.
22. Umbrico, A., Orlandini, A., & Cesta, A. (2020). An ontology for human-robot collaboration. *Procedia CIRP, 93*, 1097–1102. In *53rd CIRP Conference on Manufacturing Systems 2020*.

23. Maynard, D., Li, Y., & Peters, W. (2008). NLP techniques for term extraction and ontology population. In *Ontology learning and population: Bridging the gap between text and knowledge* (Vol. 167(6), pp. 107–127).

24. Recupero, D. R., & Boi, L. (2024). Towards seamless human robot dialogue through a robot action ontology. In B. Sartini et al. (Eds.), *Joint Proceedings of the ESWC 2024 Workshops and Tutorials Co-located with 21st European Semantic Web Conference (ESWC 2024), Hersonissos, Greece, May 26–27, 2024* (CEUR Workshop Proceedings) (Vol. 3749). Retrieved from CEUR-WS.org

25. Recupero, D. R., & Spiga, F. (2020). Knowledge acquisition from parsing natural language expressions for humanoid robot action commands. *Information Processing & Management, 57*(6), 102094.

26. Alonso, R., Bonini, A., Recupero, D. R., & Spano, L. D. (2022). Exploiting virtual reality and the robot operating system to remote-control a humanoid robot. *Multimedia Tools and Applications, 81*(11), 15565–15592.

27. Alonso, R., et al. (2020). A flexible and scalable social robot architecture employing voice assistant technologies. In B. N. De Carolis et al. (Eds.), *Proceedings of the Workshop on Adapted intEraction with SociAl Robots, cAESAR 2020, Cagliari, Italy, March 17, 2020* (CEUR Workshop Proceedings) (Vol. 2724, pp. 36–40). Retrieved from CEUR-WS.org

28. Kobayashi, S., et al. (2011). Intelligent humanoid robot with Japanese Wikipedia ontology and robot action ontology. In *Proceedings of the 6th International Conference on Human-Robot Interaction, HRI'11* (pp. 417–424). Association for Computing Machinery.

29. Fukuda, H., Mori, S., Kobayashi, Y., Kuno, Y., & Kachi, D. (2013). Object recognition for service robots through verbal interaction based on ontology. In G. Bebis et al. (Eds.), *Advances in visual computing* (pp. 395–406). Springer.

30. Oviatt, S. (1996). Multimodal interfaces for dynamic interactive maps. In *Proceedings of the SIGCHI Conference on Human Factors in Computing Systems* (pp. 95–102).

31. Markievicz, I., Kapociute-Dzikiene, J., Tamosiunaite, M., & Vitkute-Adzgauskiene, D. (2015). Action classification in action ontology building using robot-specific texts. *Information Technology and Control, 44*, 6. https://doi.org/10.5755/j01.itc.44.2.7322

32. Sharma, D., Rashno, E., Zulkernine, F., El Khodary, E., Beninger, M., Almeida, R., Tao, J., Alaca, F., & Elgazzar, K. (2024). Triage-bot: An assistive triage framework. In *2024 IEEE International Conference on Digital Health (ICDH)* (pp. 138–140).

33. Kedzierski, J., Kaczmarek, P., Dziergwa, M., & Tchon, K. (2015). Design for a robotic companion. *International Journal of Humanoid Robotics, 12*(1), 1550007.

34. D'Onofrio, G., Sancarlo, D., Raciti, M., Burke, M., Teare, A., Kovacic, T., Cortis, K., Murphy, K., Barrett, E., Whelan, S., et al. (2019). Mario project: Validation and evidence of service robots for older people with dementia. *Journal of Alzheimer's Disease, 68*(4), 1587–1601.

35. OpenAI, Achiam, J., Adler, S., Agarwal, S., et al. (2024). GPT-4 Technical Report

36. Dario, P., Ciuti, G., Pirni, A., Capasso, M., & Bisconti, P. (2022). Social robots between trust and deception: The impact on institutions and practices. *Frontiers in Artificial Intelligence and Applications, 366*, 677–682.

37. Sanders, T., Kaplan, A., Koch, R., Schwartz, M., & Hancock, P. A. (2019). The relationship between trust and use choice in human-robot interaction. *Human Factors, 61*(4), 614–626.

38. Fabris, A. (2020). Can we trust machines? The role of trust in technological environments. In A. Fabris (Ed.), *Trust* (pp. 123–135). Springer International Publishing.

39. Decety, J. (2014). *The neuroevolution of empathy and caring for others: Why it matters for morality* (pp. 127–151). Springer International Publishing.

40. Hancock, P. A., Billings, D. R., Schaefer, K. E., Chen, J. Y. C., de Visser, E. J., & Parasuraman, R. (2011). A meta-analysis of factors affecting trust in human-robot interaction. *Human Factors, 53*(5), 517–527.

41. Hancock, P. A., Kessler, T. T., Kaplan, A. D., Brill, J. C., & Szalma, J. L. (2021). Evolving trust in robots: Specification through sequential and comparative meta-analyses. *Human Factors, 63*(7), 1196–1229.

42. Kolomaznik, M., Petrik, V., Slama, M., & Jurik, V. (2024). The role of socio-emotional attributes in enhancing human-AI collaboration. *Frontiers in Psychology, 15*, 1369957.
43. Belisle-Pipon, J.-C., Monteferrante, E., Roy, M.-C., & Couture, V. (2023). Artificial intelligence ethics has a black box problem. *AI and Society, 38*(4), 1507–1522.
44. Hassija, V., Chamola, V., Mahapatra, A., Singal, A., Goel, D., Huang, K., Scardapane, S., Spinelli, I., Mahmud, M., & Hussain, A. (2024). Interpreting black-box models: A review on explainable artificial intelligence. *Cognitive Computation, 16*(1), 45–74.
45. von Eschenbach, W. J. (2021). Transparency and the black box problem: Why we do not trust AI. *Philosophy & Technology, 34*(4), 1607–1622.
46. Akalin, N., Kristoffersson, A., & Loutfi, A. (2022). Do you feel safe with your robot? Factors influencing perceived safety in human-robot interaction based on subjective and objective measures. *International Journal of Human-Computer Studies, 158*, 102744.
47. Hancock, P. A., Billings, D. R., & Schaefer, K. E. (2011). Can you trust your robot? *Ergonomics in Design, 19*(3), 24–29.
48. Brohl, C., Nelles, J., Brandl, C., Mertens, A., & Nitsch, V. (2019). Human-robot collaboration acceptance model: Development and comparison for Germany, Japan, China and the USA. *International Journal of Social Robotics, 11*(5), 709–726.
49. Haring, K. S., Mougenot, C., Ono, F., & Watanabe, K. (2014). Cultural differences in perception and attitude towards robots. *International Journal of Affective Engineering, 13*(3), 149–157.
50. Schaefer, K. E., Chen, J. Y. C., Szalma, J. L., & Hancock, P. A. (2016). A meta-analysis of factors influencing the development of trust in automation: Implications for understanding autonomy in future systems. *Human Factors, 58*(3), 377–400.
51. Pirni, A., Balistreri, M., Capasso, M., Umbrello, S., & Merenda, F. (2021). Robot care ethics between autonomy and vulnerability: Coupling principles and practices in autonomous systems for care. *Frontiers in Robotics and AI, 8*, 654298.
52. Pirni, A., & Carnevale, A. (2013). The challenge of regulating emerging technologies: A philosophical framework. In *Law and technology: The challenge of regulating technological development* (Vol. 1, pp. 59–75). Pisa University Press.
53. Pirni, A., et al. (2017). Sostenibilità etica dei personal care robot. Linee per un inquadramento preliminare. In A. Pirni (Ed.), *Il post-umano realizzato. Orizzonti di possibilita e sfide per il nostro tempo* (Nuova Corrente) (Vol. LIX, pp. 133–151). gennaio-giugno.
54. Gao, Y., Xiong, Y., Gao, X., Jia, K., Pan, J., Bi, Y., Dai, Y., Sun, J., Wang, M., & Wang, H. (2024). Retrieval augmented generation for large language models: A survey.

Human-Centric and Intelligent Robotics in Healthcare: Integrating IoT, AI, and Beyond

A'adel Sayyahi , Seyed Enayatallah Alavi , and Morteza Jaderyan

Abstract Since the integration of robotics into the field of healthcare, considerable potential has arisen to transform the delivery of medical care, making services more precise, efficient, and accessible. Robotics has begun making the surgical, rehabilitation, and patient care realms its own. With the arrival of IoT and deep learning, the potential of these hybrid systems is accentuated with seamless connectivity, real-time data analysis, and intelligent decision-making. While IoT allows robotic devices to interact with other healthcare systems, deep learning enables robots to process massive datasets and learn from experience to better their performance. These two fields together are setting the stage for smarter, adaptive, and highly efficient healthcare solutions.

As the demand for healthcare increases, the solutions through robotics using IoT and deep learning algorithms solve problems ranging from the lack of healthcare resources for minimally invasive procedures to the management of chronic diseases. For example, an IoT-enabled robotic system could monitor patient conditions in real time while adapting robot functions based on predictive analytics, whereas deep learning can identify patterns from more intricate datasets that lead to swifter and more accurate diagnoses. By bringing together these recently developed technologies, healthcare robotics is in for a dramatic restructuring of the generally accepted practice of medicine, whereby new solutions will emerge to enhance patient outcomes and streamline clinical workflows.

Keywords Robotics in healthcare · Internet of Things (IoT) · Deep learning · Surgical assistance · Rehabilitation robotics · Patient monitoring · Real-time analytics · Chronic disease management · Data security

A. Sayyahi · S. E. Alavi (✉) · M. Jaderyan
Department of Computer Engineering, Faculty of Engineering, Shahid Chamran University of Ahvaz, Ahvaz, Iran
e-mail: A-Sayyahi@stu.scu.ac.ir; Se.Alavi@scu.ac.ir; m.jaderyan@scu.ac.ir

E. Curry et al. (eds.), *Artificial Intelligence, Data and Robotics*,
https://doi.org/10.1007/978-3-032-10561-5_18

505

1 Introduction

The chapter relates to the technological integration objectives of the AI, Data, and Robotics Partnership [1], particularly in advancing smart healthcare systems through robotics, IoT, and deep learning. The entry of robotics in the realm of healthcare marks a paradigm shift, having helped overcome ever-persistent limitations to service deliveries. Robot technology has begun to impact surgical intervention, patient care, and administrative operation, radically changing the clinician-patient-side interaction and the performance of very delicate medical procedures. This technological change arose from the increasing need for accuracy, operational efficiency, and patient outcomes in aging populations and growing numbers of chronic illnesses.

Healthcare robots are developed to support a wide array of functions, including surgical assistance, rehabilitation, and medication management. Robotic surgical systems, notably the da Vinci Surgical System, have transformed minimally invasive procedures by offering superior precision, dexterity, and visualization. These systems facilitate complex surgeries with reduced patient trauma, shorter recovery times, and fewer postoperative complications [2]. This impact is especially evident in fields such as urology, gynecology, and general surgery, where surgical accuracy is critical. The key components and capabilities of the da Vinci system, including the surgeon's console and dexterous instruments, are illustrated in Fig. 1.

Making a surgical intervention is just one aspect of the possibility. Robo-systems have more relevance in rehabilitation, providing modes of adaptive and person-specific treatment. Rehabilitation robots permit highly individualized treatment promising to parallel individual recovery trajectories and, hence, therapeutic outcome improvement while simultaneously reducing physical and mental demands on human therapists. In addition to that, robots managing medications at home guarantee adherence to prescriptions as well as enable auto-dispensing with the spotting of

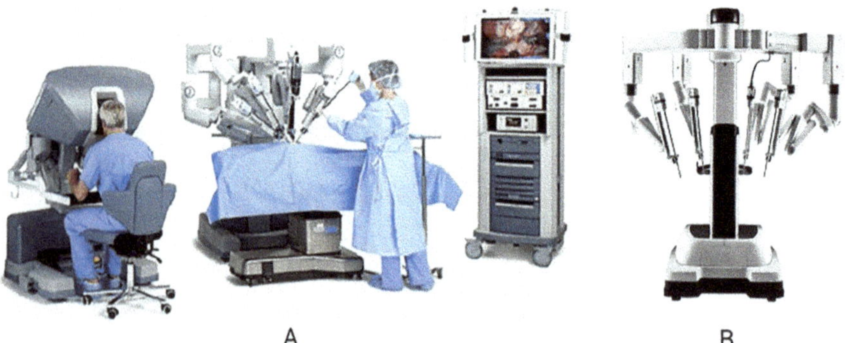

Fig. 1 The da Vinci surgical system: components and capabilities in minimally invasive surgery. (**a**) The system includes a master console and a teleoperated patient-side robot. (**b**) Dexterous instruments enable fine manipulation inside the body. (*Source*: This image is taken from an open-access journal licensed under a Creative Commons Attribution 4.0 International; Gumbs et al. [3])

Fig. 2 Key applications of robotics and AI in enhancing patient-centered healthcare. (*Source*: AI generated chart created with Napkin Inc., Napkin AI (Beta Version), 2024)

irregularities to lessen the workload of health workers [4]. Healthcare robotics receive tremendous impetus with the integration of IoT and deep learning. With the IoT, systems enable real-time acquisition of data and inter-device communications, hence supporting individualized care solutions.

In a complementary manner, deep learning models work on a large scale on medical data that enables the recognition of patterns and predictive diagnosis, hence enhancing clinical decision-making and individualized treatment planning. Recent advances range from diagnostic imaging to robot-assisted biopsy and intelligent prosthetics, thus impacting clinical effectiveness and moving toward patient-centered care. These developments are shown in Fig. 2.

Complementary powers are harnessed through IoT and deep learning. While IoT maintains real-time connections among devices, deep learning uses smart analytics to command robotic actions [5, 6]. In the rehabilitation scenario, IoT sensors capture highly detailed data regarding movement and send it to deep learning models for dynamic alteration of therapies [7]. In surgical scenarios, IoT enables system integration, while deep learning enables anatomical recognition and risk prediction tasks [8]. This makes remote robotic surgeries possible, where IoT maintains communication fidelity and deep learning maintains surgical precision [9]. The transformative impact on surgical robotics is presented in Fig. 3.

The synergy of IoT and deep learning is key to realizing a data-driven, efficient, and patient-centric future healthcare ecosystem. These technologies will automate mundane tasks and supplement decision support services with gross improvement in the quality of delivery of care [10].

1.1 Background: Rise of Robotics in Healthcare

In the past few years, robotization in medical treatment has significantly impacted the practice of healing through the improvement of precision, efficiency, and patient outcomes. Robotic systems have been working alongside human beings in such

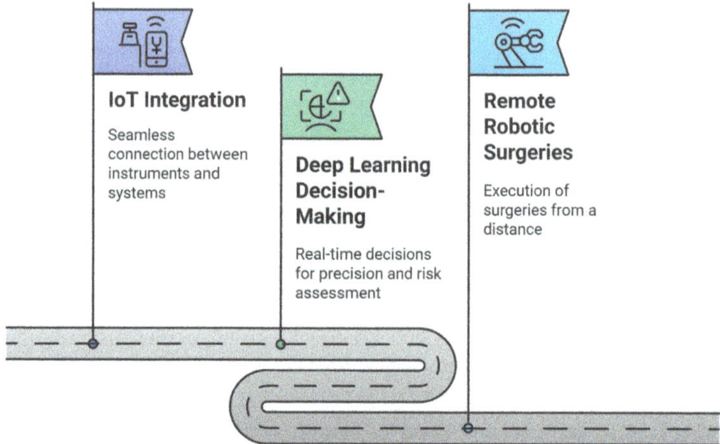

Fig. 3 Integration of IoT, deep learning, and robotics for advanced remote surgical systems. (*Source*: AI generated chart created with Napkin Inc., Napkin AI (Beta Version), 2024)

domains as surgery, rehabilitation, telemedicine, and other types of patient assistance to put an end to long-standing challenges while adapting to evolving clinical needs.

The da Vinci Surgical System and CMR Surgical's Versius are two examples of robotic platforms that allow careful manipulation during minimally invasive procedures, reducing patient trauma and recovery time, with applications in urology, gynecology, and general surgery, among others, thereby improving the accuracy of surgical operations and increasing the safety of patients in general. There has also been an improvement in rehabilitation robotics. Robotic exoskeletons and assistive robots would provide individualized therapy that adapts to the particular needs of a patient so as to recover their mobility and independence. For instance, the ARI robot, developed by the National Robotarium, assists patients in performing rehabilitation exercises, alleviating the scarcity of physiotherapists and improving therapy outcomes.

Internet of Things (IoT) and artificial intelligence (AI) integration have further empowered healthcare robotics. Such IoT-enabled systems allow them to capture real-time data and communicate, while AI algorithms offer intelligent decision-making and predictive analytics. The resulting adaptability and responsiveness of robotic systems find utility in various healthcare environments. Socially assistive robots (SARs) have been involved in rendering emotional support, patient monitoring, and assistance with daily activities, especially for the elderly and persons with chronic conditions. These robots thus contribute to greater patient engagement as well as compliance with treatment plans [11]. Implementation costs of these technologies are rather high, requiring special training on their operation; ethical concerns have also been posed with respect to how patients interact with robotic systems. Future work and research answer all these concerns, thus leaving robotics as a key instrument in the development of patient-centered healthcare.

1.2   Current State of Healthcare Robotics

Surgical robotics represents one of the most transformative applications of robotics in medicine. These systems have revolutionized traditional surgical procedures, offering solutions to mitigate risks such as excessive blood loss, infections, and postoperative complications [12]. By enabling minimally invasive techniques, surgical robots, such as the da Vinci Surgical System, provide surgeons with enhanced precision, dexterity, and control. These features allow for the execution of complex procedures that were once deemed too risky or technically challenging.

The benefits of robotic-assisted surgeries extend beyond the operating room. Studies have demonstrated that patients undergoing robot-assisted procedures often experience shorter hospital stays, reduced recovery times, and lower overall complications compared to traditional surgical methods. These advantages significantly enhance patient outcomes and satisfaction [13]. Furthermore, robotic systems facilitate surgeon training through simulation modules, enabling the acquisition of advanced skills in a controlled environment. This capability underscores the broader educational impact of robotics on the surgical field. Surgical robotics represents one of the most transformative applications of robotics in medicine. These systems have revolutionized traditional surgical procedures, offering solutions to mitigate risks such as excessive blood loss, infections, and postoperative complications [12]. The benefits extend to shorter hospital stays, reduced recovery times, and lower overall complications for patients [13].

1.3   Integration of IoT and Deep Learning in Robotics

The convergence of the Internet of Things (IoT), robotics, and deep learning is a key step in the digital transformation of healthcare systems. This integration produces intelligent autonomous platforms to improve clinical outcomes, streamline healthcare workflows, and reduce cognitive load from medical professionals. Through this synergy of technologies, healthcare may bring in new standards for precision, adaptability, and personalized care delivery.

1.3.1   Smart Ecosystems Enabled by IoT and Robotics

Being formed within interconnected networks of devices that exchange and analyze data in real-time, IoT-enabled robotic systems are embedded with sensors and communication modules that allow them to interact with other digital health infrastructures, such as wearable biosensors, imaging systems, and hospital information networks. These robots, for instance, can gather physiological parameters such as heart rate, oxygen saturation, and blood pressure from the patients in an ICU and send the acquired data to a centralized Patient Data Management System (PDMS)

[14]; then the PDMS processes the data and generates alerts against critical deviations to translate into a timely clinical intervention. Conventional monitoring in an ICU is "human-in-the-loop," wherein medical personnel interpret patient data collected by bedside monitors and central monitors before forwarding it to the PDMS for analysis. With IoT, this process gets streamlined to automate data acquisition and enhance safety through decision-making at an algorithmic real-time level.

1.3.2   Enhancing Robotic Intelligence Through Deep Learning

Deep learning algorithms can further enhance medical robot functionality by enabling the robots to analyze and extract clinically significant insights from large and complex datasets. They can use some very deep neural networks to pick up subtle deviations that would have eluded rule-based systems for recognition purposes. Hence, robots endowed with deep learning techniques may be able to read diagnostic images, such as X-rays, MRIs, or ultrasounds, with an accuracy nearly on par, and sometimes surpassing, with human radiologists [13]. This in turn builds confidence in diagnosis and speeds up clinical decision-making, thereby enhancing patient care and improving efficient utilization of resources.

1.4   Chapter Structure and Research Approach

The chapter begins with an introduction comprising three sections and carries on with a multilayered approach. Section 2 addresses the technical, ethical, legal, and economic challenges that hinder a broader adoption of robotics in healthcare. Section 3 explores current developments and solutions, including the application of deep learning to create adaptive robotic systems and the utilization of IoT scenarios for real-time data transmission and informed decision-making. Further examples are given in terms of real-world applications, case studies, and models for human-robot collaboration to marry theory with practice. Section 4 considers future research directions, with emphasis on emerging trends, infrastructure expansion, and policy considerations. Section 5 ends the chapter by summarizing and synthesizing key findings. The methodology involves performing a critical review of relevant and recent literature while providing specific selected case studies and conceptual modeling, ensuring both theoretical and practical relevance. Figure 4 visually summarizes the structure of the chapter in a concise manner.

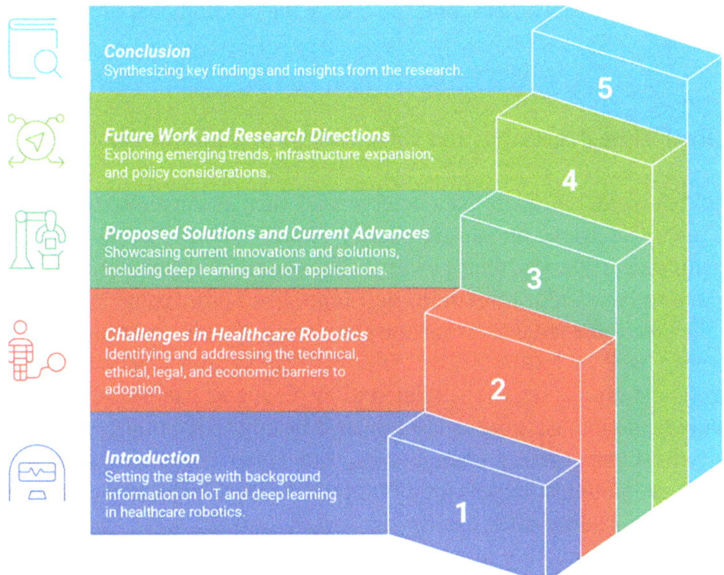

Fig. 4 Chapter structure and research approach. (*Source*: AI generated chart created with Napkin Inc., Napkin AI (Beta Version), 2024)

2   Challenges in Healthcare Robotics

The infusion of robotics into healthcare represents a historic advancement that improves patient care, streamlines operations, and delivers better outcomes through IoT, AI, and deep learning. Galactic magnates stand uncertain in the face of barriers littering the technical, privacy, ethical, legal, social, and economic fields. This session then covers these barriers, emphasizing the need for interdisciplinary cooperation, common frameworks, and novel innovations that could allow robotics to fully evolve into a patient-centric platform. Issues are considered from the perspectives of technical obstacles, data privacy and security, ethical and legal issues, and cost-related challenges: current research and real-world examples back up this discussion.

2.1   Technical Barriers

The effective deployment of robotics in healthcare is conditioned on overcoming several major technical hurdles that affect the performance, reliability, and integration of the system. Key barriers are the shortcomings of real-time processing, edge computing, and system interoperability, whose seamless operation is indispensable in the macroscopic dynamic scenario that a clinical environment represents. Hardware design, data processing, and standardized protocols must move forward to circumvent unmovably these issues to enable the development of robust and scalable systems.

2.1.1 Real-Time Processing and Edge Computing Limitations

In the field of medicine, apart from emergencies such as imprecise cardiac arrest or trauma, timely decisions remain imperative on the other end of the series for adequate medical care. Despite their power, traditional cloud systems encounter latency periods due to bandwidth limitations and centralized processing. Edge computing tries to alleviate these issues through the processing of data closer to its data source: on the robot itself or locally via an edge gateway system. This means latency is minimized, and critical decisions related to medical care can be made in real time. For example, nursing-assistive robots powered by edge computing can immediately process patient data, detect emergencies, and either trigger an alert or initiate CPR. Edge computing's implementation in these robots reduces the time taken for emergency response by 40%, reshaping critical care delivery. The intraoperative systems on edge computing also permit surgeons to engage in real-time feedback and accordingly modify their actions on an ongoing basis to maximize precision, thereby mitigating surgical risks [15]. Figure 5 shows the contribution of edge computing in real-time decision-making for healthcare. The diagram displays IoT-enabled devices (wearables, surgical robots) transmitting data to local edge servers as opposed to farther-away clouds. For example, a nursing robot processes ECG signals locally to detect cardiac anomalies and instantly alerts clinicians, skipping

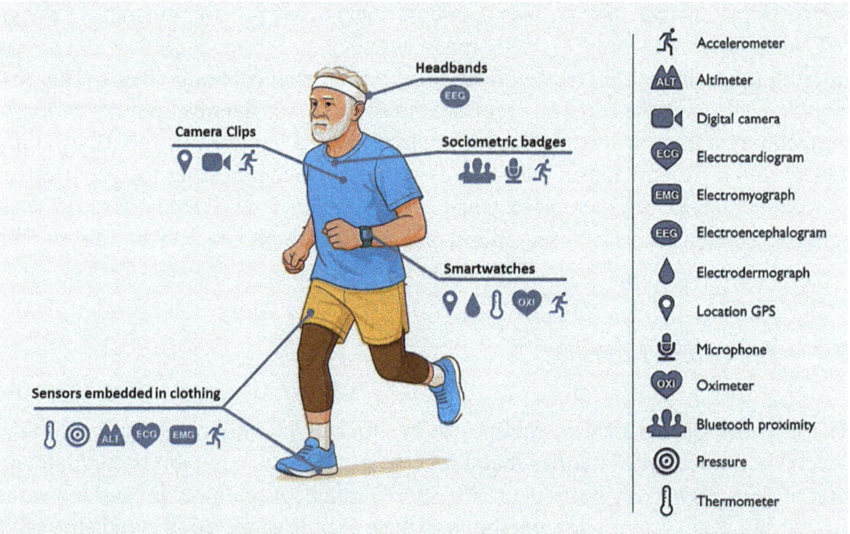

Fig. 5 Examples of wearables and the associated medical capabilities. (*Source*: AI generated illustration of an elderly athletic male element created with OpenAI, ChatGPT-4.0, 2025)

sudden delays entailed by cloud processing. This visual aligns with the examples described above, highlighting how edge systems enable rapid responses in emergencies and intraoperative scenarios.

2.1.2 System Interoperability

The convergence of robotics, IoT, and deep learning in healthcare necessitates sturdy interoperability existing side by side for the hardware, communication protocols, and data analytics systems for smooth workflow integrated into the clinical environment. Interoperability allows robotic systems to interface with existing healthcare infrastructure such as electronic health records (EHR), therefore enabling data exchange in real time and facilitating personalized care [16]. This section places before you sensor integration, data transmission protocols, deep learning analytics, and real time, the four pertinent faces of interoperability, including their roles and challenges within integrated healthcare systems:

1. **Hardware Design and Sensor Integration (IoT Sensors)**

   - IoT-enabled robots use sensors like the **AD8232 (ECG), MAX30100** (heart rate and oxygen), and **MLX90614** (non-contact temperature) to monitor patient vitals continuously and noninvasively. These sensors enhance patient comfort and hygiene, with collected data sent to remote servers for analysis.

2. **Data Transmission and Communication Protocols**

   - IoT sensor data is sent to central servers via protocols like **MQTT** or **HTTP**, often using low-power **NodeMCU** modules with **Wi-Fi**. Standardized protocols ensure seamless integration with existing healthcare systems.

3. **Deep Learning Models and Data Analysis**

   - Deep learning models analyze IoT sensor data (e.g., ECG, heart rate, temperature) to detect conditions like arrhythmias or fever. By combining inputs from multiple sensors, they provide comprehensive health assessments.

4. **Real-Time Reporting and Alerts**

   - The system generates real-time vital sign reports, flags abnormalities, and alerts healthcare providers. It may also suggest treatment adjustments based on data analysis.

By combining IoT sensors with deep learning, this framework facilitates real-time health monitoring, accurate analysis, and timely response, ultimately leading to improved patient outcomes. Figure 6 provides a detailed illustration of the proposed system and its components.

The interoperable framework allows for personalized healthcare, where data exchange between systems is essential so that interventions can be tailored to the needs of individual patients. For example, a robotic system can start adapting

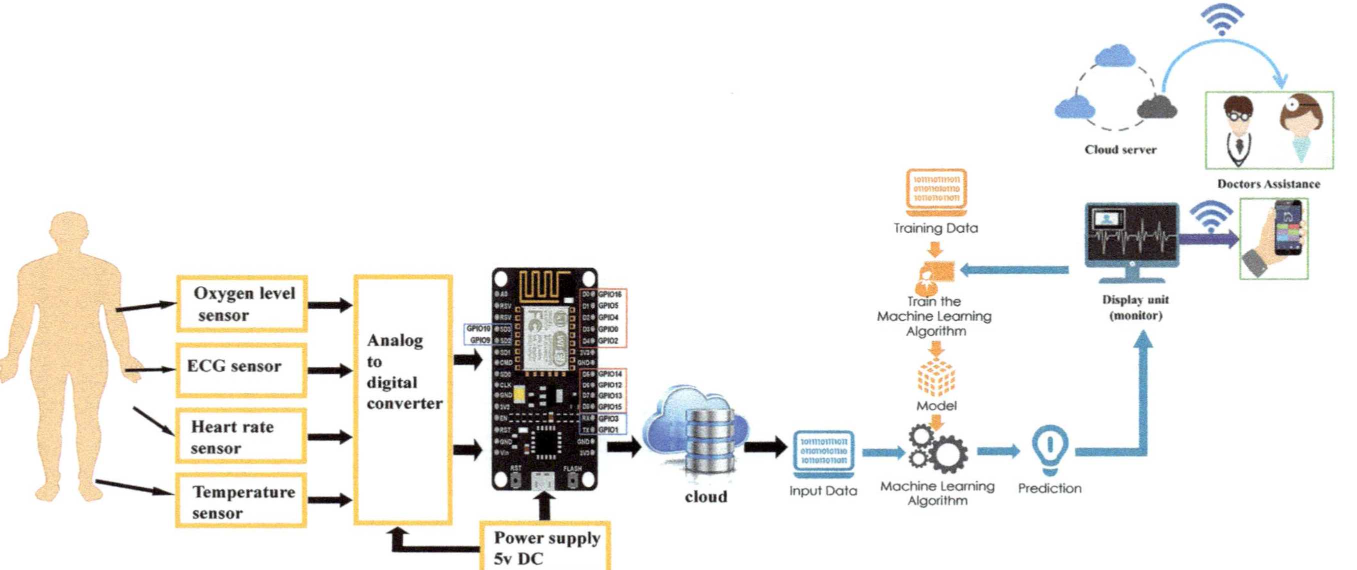

Fig. 6 Remote health monitoring and detection using IoT sensors and deep learning algorithms. This framework utilizes IoT sensors and deep learning algorithms to analyze patient data, enabling remote health monitoring and early detection of health issues for improved healthcare outcomes. (*Source*: Islam, M. R., Kabir, M. M., Mridha, M. F., Alfarhood, S., Safran, M., & Che, D. (2023). Deep Learning-Based IoT System for Remote Monitoring and Early Detection of Health Issues in Real-Time. *Sensors*, 23(11), 5204. https://doi.org/10.3390/s23115204)

monitoring parameters according to a patient's medical history to maximize treatment efficacy [16]. However, interoperability still requires overcoming certain challenges, like fragmented standards, fragmentation of systems, and limited training of clinicians. Training programs with an interdisciplinary approach covering robotics, IoT, and data analytics must be developed to arm healthcare professionals with working knowledge of these technologies [17]. With these barriers solved, interoperable robotics may be put to effective use in improving patient outcomes and satisfaction with operational efficiency.

2.2 Data Privacy and Security Issues

IoT-enabled robotics in healthcare have raised major issues in the field of data privacy and security. Considering the sensitive nature of medical data and the interconnectedness of IoT systems, these aspects give birth to higher threats of cyberspace and unauthorized access. Problems of this domain require dealing with countermeasures placed legally and from a technical standpoint greatest of standards allowing protection of patient information and wearing on the retainment of trust by the public sector in healthcare robotics.

2.2.1   IoT Vulnerabilities

IoT-enabled robotics in healthcare are vulnerable to cyber threats due to the sensitive nature of patient data and the critical functions they perform. Encryption, authentication, and intrusion detection systems are essential to protect data integrity and prevent unauthorized access. Regular security audits and updates are critical to mitigate vulncrabilities, while standardized frameworks ensure interoperability and security across diverse platforms [15]. Research emphasizes the importance of robust security measures, such as encryption protocols and access controls, to safeguard patient information [18]. Table 1 summarizes key risks, including unauthorized access, cyberattacks, and inconsistent security practices.

Table 1 Security concerns and solutions in IoT-enabled robotics

Aspect	Description	Implications	Potential solutions
Data privacy	IoT devices handle sensitive patient data	Breaches compromise confidentiality	Encryption protocols, access controls
Cyber threats	Vulnerability to hacking and malware	Potential system manipulation	Multilayered security, audits
Interoperability	Lack of standardized security protocols	Creates security gaps	Standardized frameworks
Accountability	Unclear liability in security breaches	Legal and ethical dilemmas	Clear accountability guidelines

2.2.2   Medical Data Protection Regulations

The use of AI and IoT in healthcare raises complex ethical and legal questions, particularly regarding accountability and decision-making. As AI systems gain autonomy, determining liability for errors or adverse outcomes becomes challenging. Clear guidelines and accountability frameworks are essential to balance improved patient care with trust in these technologies [19]. Table 2 highlights the need for responsibility frameworks, explainable AI, and unbiased training data to ensure fairness and trust in AI-driven healthcare systems.

2.3   Ethical, Legal, and Social Concerns

The adoption of robotics in healthcare has deep ethical, legal, and societal considerations, the tackling of which is crucial for establishing trust with the patient and acceptance by society at large. These concern the ethical design in the human-robot interaction and how responsibility and liability are assigned in any given robotic intervention issue at the core of aligning these technologies with both patient needs and regulatory directives.

2.3.1   Human-Robot Interaction Ethics

The current rehabilitation robotics system faces operational constraints because existing laws prevent its broad implementation which Fig. 7 shows. The first hurdle is that the high cost of the devices acts as the primary deterrent, with an exoskeleton such as the Lokomat Pro priced from $150,000 to $250,000, thereby placing the solution beyond the reach of many healthcare institutions, especially those in low-resource settings [20]. Most organizations lack staff members who possess the necessary training for robotic therapy operations. A 2022 survey found that 60% of rehabilitation centers reported an inability to train staff in robotic therapy protocols [21]. The ReStore Exo-Suit interfaces with EHR systems because interoperability challenges, which include system connection difficulties, create significant barriers to its operation [22].

Table 2 Security concerns and solutions in IoT-enabled robotics

Aspect	Description	Implications	Potential solutions
Autonomy	AI makes decisions without human oversight	Difficult to assign responsibility for errors	Establish clear AI accountability guidelines
Transparency	Opaque AI algorithms	Reduced trust, legal challenges	Implement explainable AI (XAI) frameworks
Bias	AI may reflect biases in training data	Ethical concerns, inequitable patient care	Use diverse and representative datasets

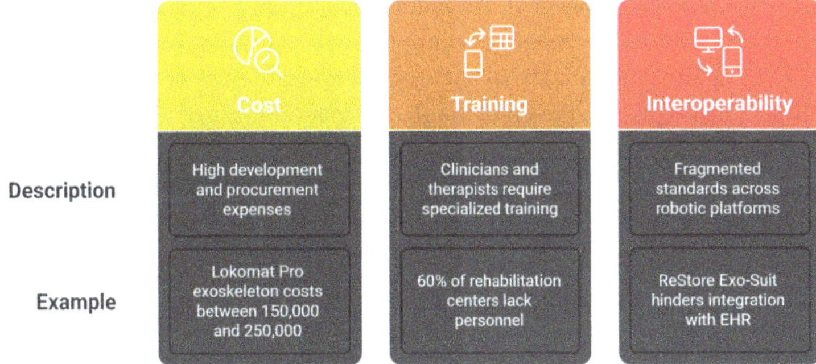

Fig. 7 Barriers to rehabilitation robotics adoption, illustrating cost, training, and interoperability challenges with examples

Innovations are attempting to cope with these challenges. Open-source and 3D-printed exoskeletons are cutting production costs by up to 60%, thus making these machines more affordable [21]. Meanwhile, the AMIKA system uses machine learning to adapt therapy protocols based on the actual performance of the intervention, thereby enhancing the ability to predict the most effective interventions by 30% for diseases such as Parkinson's [23]. Such technologies bring a bright light of hope to badly underserved populations, in particular, pediatric and geriatric patients who have scant options for traditional modeling of therapy.

Eventually, the fusion of healthcare with robotics must be achieved via three strategic anchors: ethical codesign to ensure development centered on patients, cost reduction for greater access, and establishing standards for interoperability with existing healthcare infrastructures [23, 24].

2.4 Cost, Maintenance, and Adoption Barriers

Despite the possibility of robotics and IoT bringing changes in health, economic and operational issues stand in their way. High costs, maintenance requirements, and retraining of the workforce prevent adoption, especially in poor regions. These factors need to be addressed by employing innovative financial models, cost and benefit assessments, and complete training programs.

2.4.1 Financial Sustainability

The heavy initial investment in advanced robotics and IoT systems, along with the subsequent upkeep costs, can limit access to medium and small healthcare facilities. Front-end development comes at the highest level of design and deployment costs, which is practically prohibitive for implementation at smaller hospitals. Customization to suit higher needs escalates costs, and this is detrimental to scalability. Maintenance involves frequent updates and repairs, thus becoming a cost overburden, whereas frequent technological upgrades call for heavy investments all over again. And yet, sharp and comprehensive cost-benefit analyses reveal the true long-term worth achieved through improved outcomes and efficiencies that these have shown to bear. To enable more access, different funding models can be considered, such as public-private partnerships, lease models, and subsidization. Modular design can cut down on customization costs, whereas AI-supported predictive maintenance could reduce maintenance costs, and a thorough cost-benefit analysis could support the decision regarding an upgrade [25].

2.4.2 Workforce Training Needs

The rapid advancement in technology is challenging healthcare providers in their ability to stay trained in the IoT-enabled robotic and AI systems that require specialized skills. Promoting a culture of lifelong learning is of paramount importance to adjust to such technologies and to offer the best care [26]. Along those lines, the targeted training programs presented in Table 3 can help tackle the skill gaps, resistance to change, and lack of resources that hinder providers from fully exploiting these innovations.

Collaboration among the stakeholders to solve problems of cost, security, and ethics, and training is required when implementing robotics, IoT, and deep learning technologies into healthcare. Standardized frameworks and security protocols, and an emphasis placed on training and education to instill confidence among patients and caregivers, could ensure this technology is said to be safe, secure, and trustworthy.

Table 3 Security concerns and solutions in IoT-enabled robotics

Aspect	Description	Implications	Potential solutions
Skill gaps	Healthcare professionals lack advanced skills	Reduced effectiveness and misuse of technology	Continuous education and training programs
Resistance to change	Reluctance to adopt new technologies	Slower adoption and underutilization	Promote a culture of lifelong learning and hands-on training
Resource constraints	Limited time and resources for training	Difficulty in keeping up with advancements	Online training modules and dedicated learning time

3   Proposed Solutions and Current Advances

The earlier session addressed the multifaceted challenges of integrating robotics, IoT, and deep-learning technologies in healthcare. These challenges include matters of real-time processing and edge computing constraints, issues of interoperability of systems, vulnerabilities within the IoT infrastructure, questions regarding medical data safeguarding, ethics of interaction with humans by robots, and finally matters of funding and workforce training. Building on this foundation, the current session delves into preeminent solutions and advancements trying to resolve these barriers to augment the efficacy, access, and ethical application of these technologies. The session then looks into the integration of IoT and deep learning into robotics, practically demonstrated by case studies, human-robot cooperation models, and the evolution of emotionally intelligent and socially assistive robots, each carrying a part of the transformation of healthcare.

3.1   *Integration of IoT and Deep Learning in Robotics*

The merging of IoT and deep learning to operate with robotic systems holds great promise in rising beyond the currently existing technological pitfalls. In this section, two crucial sub-areas are explored: smart data transmission and adaptive learning and decision-making to power the efficiency and adaptability of robotic applications in healthcare.

3.1.1   Smart Data Transmission

The IoT infrastructure lays the foundation for the smooth extraction and transmission of sensor-generated data, equipping robotic systems to gain a perception of and a lucid understanding of their surroundings with supreme precision. Integrated sensors, such as IMUs, cameras, and environmental sensing, generate an endless stream of data that IoT networks gather and process in real time. As Mouradian et al. commented, IoT's ability to service large, distributed sensor networks allows robots to simultaneously track numerous complex variables, like user movements, environmental impediments, and physiological indicators [27]. Figure 8 presents an example of such a system via remote patient monitoring, where biosensors and wearable devices enabled by IoT collect crucial health-related signals, such as heart rate, movement patterns, and medication adherence. The data is then transmitted using microcontrollers and gateways to cloud platforms for real-time analysis. This type of instant processing becomes a quintessential element in applications that require immediate response, like robotic exoskeletons preventing a gait pattern impairment or in assistive robots that have to adjust their navigation through an unstructured environment.

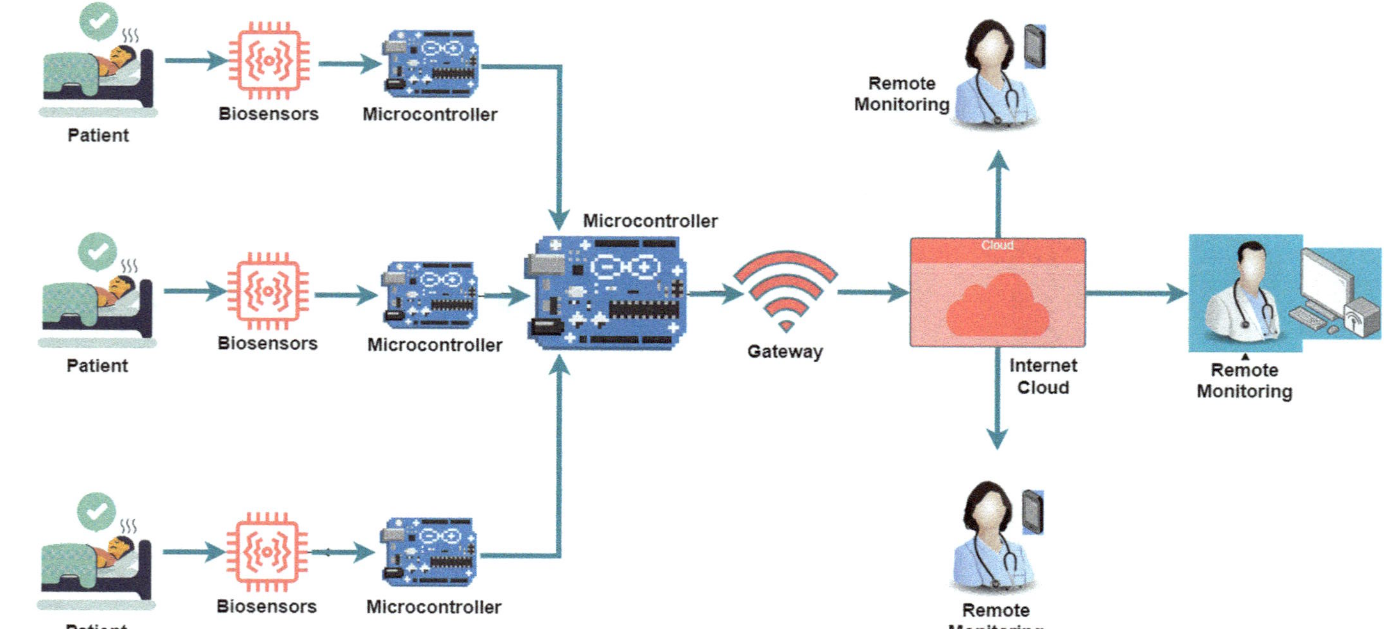

Fig. 8 Remote monitoring via IoT healthcare systems. The framework depicts biosensors and wearable devices collecting patient data, transmitted through microcontrollers and gateways to cloud platforms. This enables real-time health monitoring and robotic interventions, particularly vital in remote or underserved regions. (*Source*: Uddin, R., & Koo, I. (2024). Real-Time Remote Patient Monitoring: A Review of Biosensors Integrated with Multi-Hop IoT Systems via Cloud Connectivity. *Applied Sciences*, 14(5), 1876. https://doi.org/10.3390/app14051876)

3.1.2 Adaptive Learning and Decision-Making

The integration of IoT with robotic systems fosters dynamic behavioral adaptation through the application of machine learning and predictive analytics. In detail, by exploiting cloud-based IoT platforms, robots benefit from huge computational resources for analyzing a large amount of historical and real-time data in order to make better decisions. Kamilaris and Botteghi highlight that due to their IoT capabilities, robots can enhance their working skills through user interaction, thereby refining the tasks they perform, such as object manipulation or personalized therapeutic programs [15]. As an example, in the case of rehabilitation, the robot could adjust the resistance level in real time during the exercise, depending on muscle fatigue from IoT-connected electromyography (EMG) sensors, such that its capacity remains aligned with the needs of the user and is continuously effective in a dynamic environment. Figure 9 demonstrates the adaptation framework of IoT-enabled robotics, with highlights of how a closed-loop system works, integrating sensing, analysis, and actuation. Visual data related to patient movement during rehabilitation are time-stamped, observed by the camera, and processed through the IMAGE-Net image processing network IoT system to infer user intent. For example, it may be presumed that the VIDEO-Net detects that the patient intends that robot resistance ought to be diminished due to fatigue; thereupon, actuators, such as motors of a robotic exoskeleton, are enabled to make fine adjustments by diminishing resistance or modifying limb trajectories accordingly. In the rehabilitation environment, Video-Net detects the fine movement patterns of fatigue and immediately actuates adjustments to benefit the therapeutic efficacy. This interconnection shows how IoT helps build context-aware adaptability, which aligns with the conclusions drawn in [15] by Kamilaris and Botteghi concerning the IoT-based evolution of robotics and allows robots to dynamically anticipate and adapt to user needs.

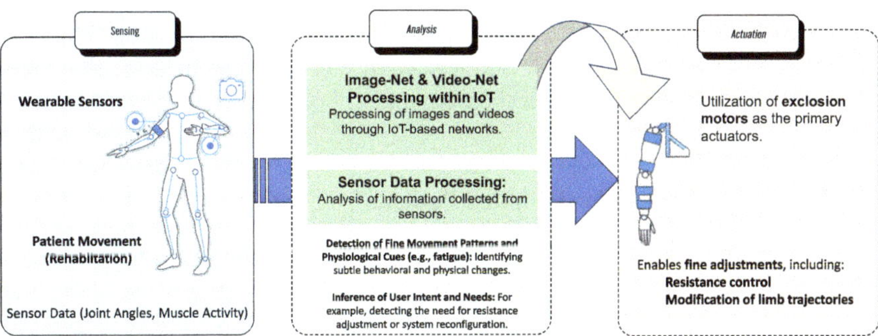

Fig. 9 IoT-enabled adaptive robotics framework. The diagram outlines the closed-loop process: real-time visual data acquisition (camera view), intention inference via VIDEO-Net, and responsive actuation, enabling context-aware robotic adaptability. (*Source*: AI generated illustration of a human element equipped with wearable sensors, created with Google, Gemini (Banana model), 2025. The overall image and flowchart were designed by the author)

3.2 Practical Applications and Case Studies

This section evaluates the practical implementation of robotics, IoT, and deep learning through specific case studies, demonstrating their impact across diverse healthcare domains.

3.2.1 Surgical Robotics

The introduction of robotic surgical systems under the da Vinci Surgical System and Medtronic Hugo brands has profoundly affected minimally invasive surgery MDT, enabling operations with less than a millimeter precision. They utilize technologies that go beyond the limits of human dexterity to significantly improve the outcomes in critical surgeries, such as prostatectomies and hysterectomies. Their successful application is dependent upon the design of three principles: articulation of instruments, imaging technologies, and tactile feedback. The systems incorporate articulating instruments fitted with wristed tools that provide seven degrees of freedom, thus enabling surgeons to imitate natural hand movements within constrained anatomical spaces. Because they require navigating delicate tissues in nerve-sparing prostatectomies, these capabilities are crucial. Moreover, the advanced imaging system uses 3D HD cameras to deliver magnified real-time sight to the surgeon on the surgical site, thereby enhancing anatomical clarity and supporting precise dissection. In gynecological surgeries, for example, this advanced imaging lets surgeons view and preserve critical structures like the uterine arteries.

The advancement of haptics presents these robots with added functionality in force sensing to reproduce tactile sensations. Through haptics, surgeons feel tissue resistance during dissection and suturing, using touch as well as sight. Such feedback is more necessary for tasks requiring accurate motor control, such as colorectal anastomosis, where too much force will cause complications to arise during recovery. These technologies articulated together with imaging and haptic feedback have been reported through clinical studies to reduce intraoperative errors by 70% [15]. Table 4 compares the best robotic surgical platforms regarding their specific capabilities and applications in a clinical environment.

3.2.2 Assistive Devices for Disabled Patients

Assistive systems for disabled patients represent an extraordinary amalgamation of robotics, technology, and design centered on user needs. Some physical, sensory, or cognitive needs of disabled persons must be of intervention to allow physical impediments that hinder independence. This section thus reviews the use of robotics for assisting patient-specific needs and considers innovations in assistive technologies, especially toward their acceptance within healthcare ecosystems.

Table 4 Comparative analysis of robotic surgical systems

Feature	da Vinci Surgical System	Medtronic Hugo	Verb surgical
Articulation range	360° instrument rotation	270° rotation	340° rotation
Imaging resolution	3D HD (1080p)	4K ultra HD	3D 4K with AI overlay
Haptic feedback	Partial (force sensing)	Not available	Advanced haptic sensors
Specialized applications	Urology, gynecology	General surgery	Oncology, cardiology
IoT integration	Limited	Real-time data sharing	AI-driven predictive analytics

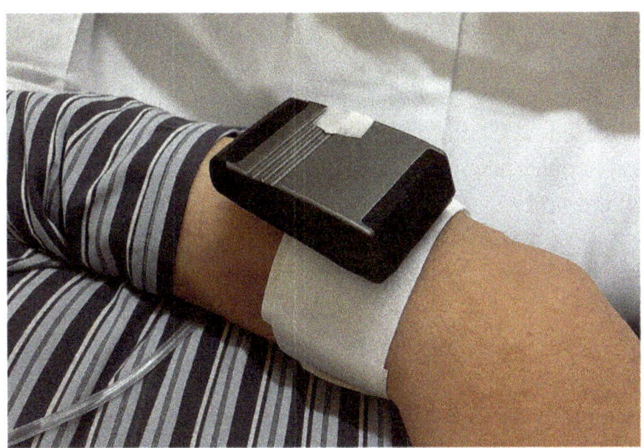

Fig. 10 Portable pager from the Signolux system. The Signolux system's portable pager, utilized to notify patients of critical pump events following left ventricular assist device (LVAD) implantation. (*Source*: Spiliopoulos, S., Hergesell, V., & Dapunt, O. (2019). Left ventricular assist device therapy in a patient with hearing and speech disabilities. *The Journal of Thoracic and Cardiovascular Surgery*, *157*(1), e1–e2. https://doi.org/10.1016/j.jtcvs.2018.05.048)

Addressing the Needs of Disabled Patients Through Robotics

Robotic systems came forth to equip gaps in access and care for disabled persons. These systems employ advanced automated mechanisms, sensor technologies, and artificial intelligence to give solutions tailored to an individual's daily needs and rehabilitation processes. Another instance of this is the use of the Signolux system in the post-surgical care of a 74-year-old male with hearing and speech disabilities. After the patient underwent implantation of an LVAD, he used a portable Figure 10 to get real-time notifications regarding pump events, thus ensuring that he responded to critical alarms promptly [28]. This case demonstrates that assistive robotics can partly or wholly overcome sensory limitations to facilitate patient safety in high-stakes clinical scenarios.

IoT-Driven Robotics for Rehabilitation

We once considered assistive robotics as a new way of diagnosing problems among people with disabilities; IoT is now enabling continuous communication between devices, caregivers, and healthcare providers. Robots enabled with multimodal sensors such as motion sensors and biometric sensors check on patients' activities and health parameters in real time for adaptive interventions. Within rehab settings, for example, joystick-based robotic exoskeleton applications can switch support levels dynamically in line with the patient's progress for enhanced recovery trajectories. Kamilaris and Botteghi emphasize that the IoT aids collaborations between assistive devices such as smart wheelchairs and environmental sensors to create responsive care ecosystems [15]. Such an interconnected approach finds more usage in rehabilitation, where precision and adaptability weigh heavier on recovery.

When dealing with classic robots, ethics mean to safeguard user autonomy. Excessive reliance upon robotic assistance may, in fact, lead to decreased agency on the patient's part. Hence, Gherardini et al. recommended co-design frameworks that involve disabled people during the development phases, guaranteeing that the solutions speak to the dignity and preferences of the user [24]. This approach is applied in the designing of some wearable assistive devices, in which ergonomic and aesthetic customization both help prevent stigmatization and enhance the functionality itself.

3.2.3 Innovations in Assistive Technologies and Their Integration

The modern system is more assistive than a horizontal appliance through integration with broader healthcare systems via advanced interoperability protocols, collaborative ecosystems, and stakeholder-driven paradigms. In this section, three core innovations are further developed, the Web of Things (WoT), stakeholder collaboration models, and semantic technologies, with an emphasis on their practical applications.

Web of Things (WoT): Advancing Interoperability

The Web of Things (WoT) serves as a set of extension capabilities that build upon those of the Internet of Things (IoT), enabling application-layer interoperability among robotic components, services, and users. WoT addresses some inherent limitations of IoT and enables the easy integration of Web technologies into robotic systems. By bringing in standard Web protocols, WoT enables safe and efficient discovery, orchestration, and use of services and data, thus putting robotic systems to greater functional use.

Whereas the Internet of Things (IoT) remains an infrastructure upon which the Web of Things (WoT) extends activities, WoT aims for application-layer interoperability between robotic entities, services, and users. WoT alleviates some of the inherent limitations of IoT and allows some of the Web technologies to be conveniently interfaced with robotic systems. By providing standard Web-based protocols, WoT enables robotic systems to have greater functional use through safe and efficient discovery, orchestration, and use of services and data.

The early instances of Web-enabled robots, such as MERCURY (1994) [29], TELE-GARDEN (1995) [30], and XAVIER (2002) [31], explored the possibilities of remote teleoperation and basic Web interaction. Whereas these pioneer systems showed what could be achieved in Web-connected robotics, their functionalities were confined to predefined tasks with minimalistic interfaces. Modern WoT approaches, however, enhanced greatly by RESTful APIs and semantic technologies, bring forth advanced interaction paradigms, including robot-to-robot (R2R), robot-to-human (R2H), and machine-to-machine (M2M) communications, vastly broadening the domain of Web-enabled robotics. The overall benefits of WoT toward robotics are summarized in Table 5.

Such advances are seen to implement major improvements in assistive technologies: robotic systems respond in real time and are adaptive to the users' needs. In the adoptive environment of WoT, robotic systems gain better interoperability, scalability, and adaptability, thereby fostering applications for innovative and effective assistive solutions. One must also think about ethical considerations when developing assistive systems so that these technologies do not destroy but rather enhance the autonomy of disabled persons. Research on the co-design of ADs also stresses the need to engage the user in the design phase to achieve a solution that respects his/her dignity and preferences while remaining practical [24]. That approach and attitude must facilitate the users in trusting and accepting assistive technologies.

Remote Monitoring and Telepresence

The use of a robotic telepresence facility can be a lifesaver in emergency and critical care contexts. For instance, telepresence robots can be fitted with high-definition cameras and audio systems so that specialists can perform remote diagnosis and

Table 5 Enhanced perspectives of key stakeholders on assistive technology

Benefit	Description	Example
Standardized data formats	Exports sensor data in widely used formats for compatibility across platforms	Robots sharing data in JSON or XML for easy integration
Interoperable APIs	Exposes robotic functions via APIs accessible with common programming languages	Managing robot motion through a REST API
Enhanced semantic understanding	Uses semantic Web tech to interpret and reason about data for better decision-making	Categorizing "obstacles" or "paths" for navigation
Web-based robotic mashups	Combines robotic and Web services to access contextual Web knowledge for smarter behavior	Adjusting cleaning operations based on weather APIs
Cloud integration	Connects robots to cloud systems for real-time data processing and storage	Streaming video to the cloud for AI object recognition

Table 6 Comparison of traditional telemedicine and robotic telepresence

Feature	Traditional telemedicine	Robotic telepresence
Real-time interaction	Limited to video calls	Enhanced with mobility and presence
Accessibility	Requires patient-side equipment	Robot can independently access patient area
Specialist collaboration	Video or text-based communication	Real-time, robot-mediated collaboration
Emergency care application	Delays due to static setup	On-the-spot diagnosis and intervention
Patient engagement	Passive interaction	Interactive with physical presence

consultation, enabling interventions to prevent life from being endangered [32]. This feature is an exquisite technology for increasing access to specialized care and effectively cutting the distance between patients and healthcare providers in far-away regions. Table 6 elucidates the comparison of key features and benefits of robotic telepresence with traditional telemedicine to stress the added value that robotics brings to healthcare delivery.

Also, to maximize clinical efficiency, robotic telemedicine provides a bridge for enhanced communication among teams of healthcare providers as well as for coordination. For instance, a case telepresence robot acts as a medium of communication for multidisciplinary case discussions wherein specialists working remotely collaborate quite readily. Such an integration into telemedicine allows for more cohesive patient care, subsequently reducing the chance of miscommunication and timely diagnosis and treatment of the patient with a whole lot of accuracy.

3.3 Human-Robot Collaboration Models

This section explores frameworks that enhance collaboration between humans and robots, optimizing clinical workflows and surgical assistance.

3.3.1 Enhancing Clinical Workflow

With the increasing acceptance of robotics, the applications vary widely, from drug delivery and sterilization to operative surgical assistance. In a smart-hospital environment, these robots must work together for optimum efficiency. Using IoT-driven management systems, these robots coordinate using advanced algorithms for resource allocation: swarm intelligence, reinforcement learning, etc. These algorithms identify priorities, allocate resources, and ensure smooth cooperation across different robotic systems. For example, in a situation where there is insufficient staff, the drug-delivery robots autonomously reroute themselves to the critical care units so that medications are delivered on time to patients. The sterilization robots can adapt their cleaning schedules based on real-time information from IoT sensors so that the adverse MIC risks are minimized, and surgical robots work with diagnostic devices and IoT systems to coordinate their movements and actions in the operative room with more precision and efficiency.

3.3.2 Enhanced Human-Robot Collaboration

The Internet of Things facilitates the smooth integration between human intent and robotic execution via an intuitive control interface. Wearable IoT such as intelligent gloves or eye-tracking systems make it possible for users to give their commands with ease. For example, as shown in Fig. 11a, b, the discussed IoT-based assistive system includes two subsystems: (i) one is a stationary assistive system for bedridden patients and (ii) a mobile assistive system for wheelchair users. Both subsystems use eye-tracking technology to permit those with extreme motor impairments to control home appliances, communicate with caregivers, or even drive a wheelchair autonomously. The stationary system incorporates IoT-enabled sensors to recognize gaze patterns to dim the lights or sound alarms; meanwhile, the autonomous wheelchair is being steered via the mobile system using an eye-tracking controller for directional commands on a real-time basis. The IoT assures minimal latency communication between the user's eye movements, robotic actuators, and external devices, allowing for fluid interchange between the subsystems with respect to the patient's contextual scene, e.g., switching between bed and wheelchair environments. A two-way interaction between alternative living environments shows why IoT forms an essential cog in developing themselves as responsive user-centric robotic solutions that dynamically unfold within such diverse living environments [8].

3.3.3 Collaborative Surgical Assistance

With the optimization of joint surgical assistance in surgical conditions, optimization of cooperation between manual surgeons and robotic systems is achieved. Henceforth, research efforts should go into building intuitive interfaces that reduce the human-robot interaction cognitive burden. One-nineteenth of the emphasis should be placed on the whole communication mechanism; it should follow a standard framework of RESTful API or, preferably, an IoT-based architecture that ensures efficient low-latency data dataflow between robotic systems and their human operators. At the same time, robotic systems must be able to dynamically adapt to the healthcare professionals' and patients' needs and provide adaptability according to surgical context and patient condition. AFP advances allow for highly natural and efficient command inputs, thus enabling surgeons to interact with robotic systems through voice- or text-based instructions. Similarly, HCI research will provide a basis for the development of interfaces that improve situational awareness and maintain precise control during surgical procedures. These outcomes

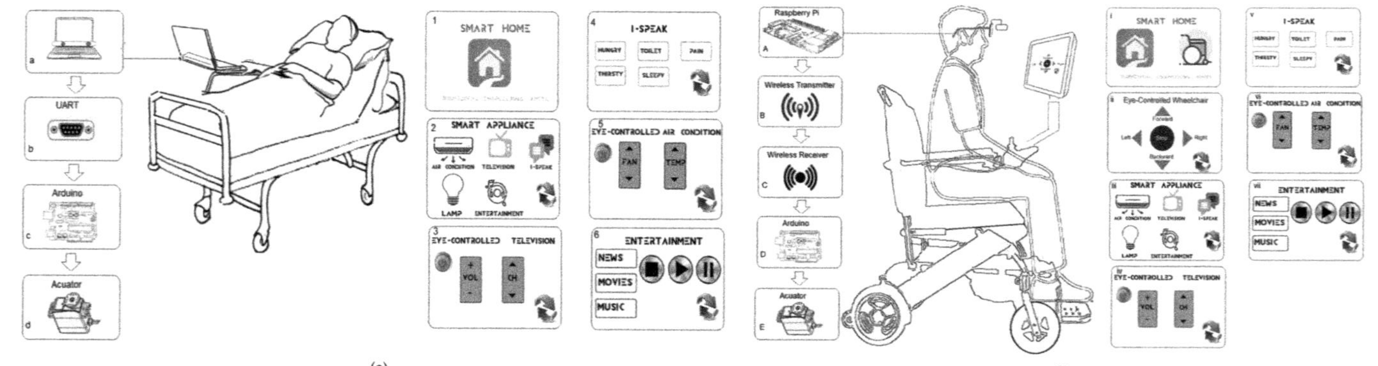

Fig. 11 Smart home system based on eye-tracking technology. The smart home assistive system integrates two subsystems: (**a**) a **stationary system** using IoT and eye-tracking for bedridden patients to control appliances and communicate and (**b**) a **mobile system** enabling wheelchair users to navigate and interact with devices via eye-tracking and IoT. Unified by low-latency cloud connectivity, these systems ensure seamless transitions between environments, delivering adaptive, user-centric autonomy and connectivity. (*Source*: Paing, M. P., Juhong, A., & Pintavirooj, C. (2022). Design and Development of an Assistive System Based on Eye Tracking. *Electronics*, 11(4), 535. https://doi.org/10.3390/electronics11040535)

are thus necessary for creating a complete human-robot partnership, thus recommending improved surgical outcomes and patient safety.

3.4 Emotionally Intelligent and Socially Assistive Robots

This section addresses the psychological benefits of robotics, focusing on mental health support and social interaction for isolated patients.

3.4.1 Mental Health Support

The realm of socially assistive robots (SARs) is one in which they tend to the emotional and psychological needs of the patients, complementing the physical healthcare application of the robots. Being companion robots, they provide opportunities for engaging in interactions, which are conducive to better mental health among the elderly. Studies suggest that SARs have been instrumental in alleviating symptoms of depression and anxiety by imparting a feeling of purpose and social engagement, thereby counteracting feelings of isolation [33]. This method ensures that both aspects of health, mental and physical, are cared for in healthcare.

From here, the latest advances in artificial intelligence have given birth to adaptive, highly interactive SARs. Machine learning techniques are involved so that these robots can consider prior interactions and adjust their answers and actions accordingly with respect to the needs of a given individual, hence retaining a great therapeutic effect for longer periods. This adaptability has greatly increased their ability to provide emotional support consistently, thereby making them truly valuable tools in today's mental healthcare.

3.4.2 Social Interaction for Isolated Patients

Socially assistive robots (SARs) play an essential role in rehabilitation to aid people with physical disabilities or those with mobility impairments, especially elderly people recovering from strokes or falls. The robot NAO, the most studied in human-robot interaction, fosters therapy adherence by providing companionship and context-tailored support during rehabilitation exercises [34]. In cardiac rehabilitation, Lara et al. [35] detail patient-robot interfaces that combine physiological monitoring and social interaction. Sensors like the Zephyr HxM ensure reliability of data such as training heart rate and recovery heart rate, while the social interaction encourages patient use of the system through real-time feedback. Figure 12 shows a detailed architecture of such patient-robot interfaces that elegantly integrate sensor technology and interactive components into a complete and effective rehabilitation solution [34].

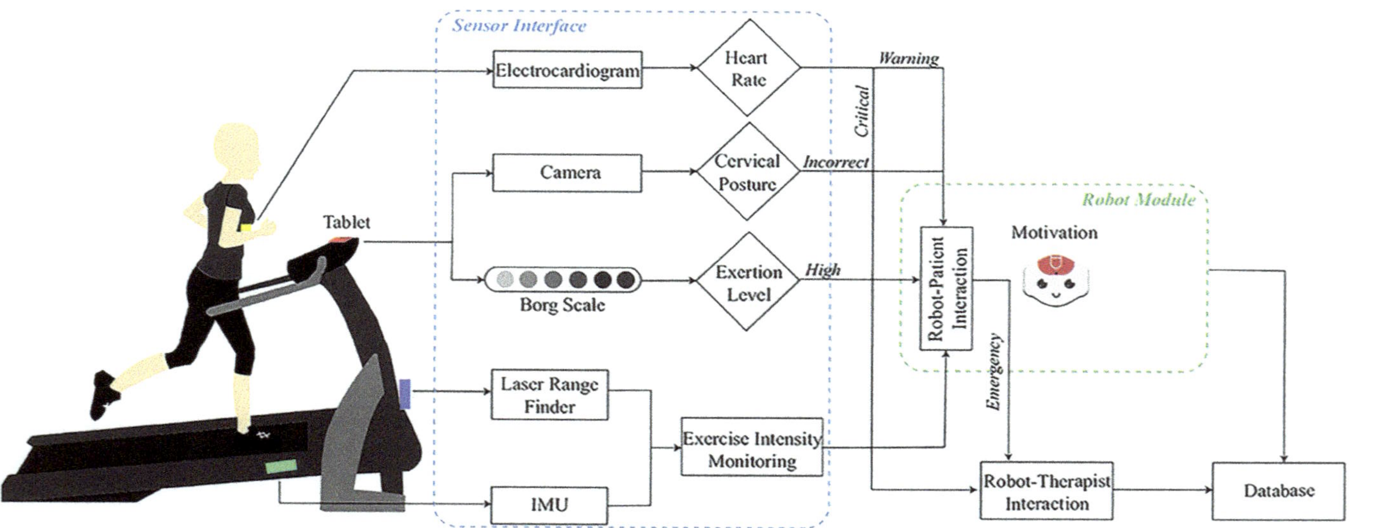

Fig. 12 A diagram of the patient-robot interface used in the cardiac rehabilitation program. As shown in figure, a set of sensors were integrated to measure each parameter. (*Source*: Céspedes, N., Irfan, B., Senft, E., Cifuentes, C. A., Gutierrez, L. F., Rincon-Roncancio, M., et al. (2021). A socially assistive robot for long-term cardiac rehabilitation in the real world. *Frontiers in Neurorobotics*, 15, 633248. https://doi.org/10.3389/fnbot.2021.633248)

On another occasion, SARs take care of emotional and psychological needs for older adults, either those living by themselves or in care homes, thereby greatly reducing social isolation and loneliness [33]. Such robots enhance living and well-being outright by fostering meaningful human interactions. In a holistic approach to medicine, their adaptability and customizable features make them invaluable.

4 Future Work and Research Directions

This section outlines emerging avenues in healthcare robotics, IoT-enabled care environments, and the associated policy and ethical challenges. It points out trends like customized robotic systems for patients, AI-driven diagnostic tools, the growth of IoT monitoring and electronic health record integration in smart hospitals, and the necessity for consistent standards and data-sharing systems.

4.1 Emerging Trends in Healthcare Robotics

The integration of robotics, IoT, and artificial intelligence (AI) in healthcare is rapidly creating a new area of innovation in medical practices and patient care. Innovations are revitalizing existing systems while creating new applications to address some of the most acute problems in the delivery of healthcare. Here, we look at some of the most promising trends and applications in healthcare robotics: neuromorphic computing, digital twins, federated learning, AR, VR, and ethical and regulatory frameworks. These technologies provide unique opportunities to enhance healthcare outcomes, make work processes easy, and make sure robotics are responsibly applied in clinical settings.

Neuromorphic Computing
Neuromorphic computing replicates human brain neurons to achieve its goal which enables advancements in AI and robotics. The system allows real-time handling of intricate data streams which supports decision-making processes in fast-changing settings such as operating rooms. The use of neuromorphic architectures in surgical robots improves their ability to respond to situations while maintaining high operational control during autonomous procedures. The ability of neuromorphic computing to improve healthcare mobile robots demonstrates its value for tasks needing quick calculations and large data handling and high adaptability [36].

Digital Twins
Digital twins are virtual models of real objects. The healthcare sector is embracing digital twins for simulations relating to patient conditions and treatment outcomes. By modeling various scenarios, the effects of different interventions can be predicted by healthcare providers for more personalized and better care. Digital twin systems working alongside IoT devices may monitor and collect data continuously,

granting much-needed insight into evolving patient health and the efficacy of treatment procedures. The use of digital twins in the clinical setup will weigh heavily on decisions and could potentially influence clinical outcomes, especially for complicated cases such as chronic disease management and surgical planning [37].

Federated Learning

This approach to machine learning aims at putting all institutions together to co-train AI models without disclosing patient-health-related data. Disregarding critical athlete data privacy considerations, the technology has its limitations because the AI algorithms could have been trained better, and maybe more generalizable. Using data from different sources has been regarded as the federated learning approach to make AI systems in healthcare better, including diagnostic tools and predictive analytics. Ahmad et al. stated that federated learning holds promise for making AI applications in healthcare more accurate and effective while maintaining respect for data protection laws [38].

Augmented Reality (AR) and Virtual Reality (VR)

In conjunction with AR and VR, robotics and IoT are set to take training and education in healthcare to a whole new level. With immersive technologies recreating surgical procedures and patient interactions, they've taken medical education quite a step forward. Phupattanasilp and Tong provide evidence supporting the use of AR in the betterment of healthcare providers' training, emphasizing its potential in skill acquisition and retention [39]. AR and VR technologies will have a significant role in healthcare education to prepare the workforce of the future as they become more widespread.

Ethical and Regulatory Frameworks

Until the moment when AI and robotics are found to be inseparable from clinical practice, there will be a demand for land development around issues of ethics related to accountability, transparency, and patient autonomy. The creation of ethical frameworks and regulatory guidelines will drastically modulate the further edification of healthcare robotics. Among themselves, stakeholders that would include healthcare providers, technology developers, and regulatory bodies will have to cooperate to set guidelines that responsibly encourage innovation, to the detriment of patient rights.

4.1.1 Personalized Robotic Systems

The future of robotics in healthcare is poised for potentially transformational advancements with the help of emerging technologies and applications. As the integration of robotics, IoT, and artificial intelligence (AI) advances, opportunities are opening up to improve patient care, create efficiency in operations, and tackle persistent challenges in healthcare delivery. This section discusses new trends and applications in healthcare robotics and future directions and potential areas of research that will drive the development of the next generation of medical technologies.

4.1.2 AI-Driven Diagnostic Tools

CNNs are important in medical image processing, mimicking the human visual system for extracting hierarchical features from complex images like MRI, CT, and dermatoscopic scan images. These models surpass skin cancer diagnostic accuracies from radiologists by noticing variations in subtle patterns such as irregular lesion borders, thereby improving precision [40]. Other state-of-the-art models used in such tasks include ResNet, U-Net, and Transformer-based models, with U-Net designed in 2015 by Ronneberger et al. achieving significant success in biomedical image segmentation, including tumor detection in brain MRI scans. The U-shape design of U-Net involves an encoder-decoder with skip connections between the four downsampling and upsampling stages that help in restoring high-resolution information lost during pooling and achieved an accuracy of above 95% on the BraTS dataset due to effective feature fusion [41]. Figure 13 illustrates this architecture, highlighting its robust performance in multiscale segmentation tasks.

Recurrent neural networks (RNNs) and their architectures, especially LSTM and GRU, are excellent for sequentializing medical data, such as ECG, EEG, and usual trends of vital signs [41]. Being able to learn long-term temporal dependencies, LSTMs forecast ICU patient deterioration with an accuracy of 89%, taking real-time blood pressure and oxygen saturation data to examine delayed effects, such as the delayed effect of oxygen desaturation and the development of septic shock. Appendix 1 shows the architectures compared with their features, applications, and limitations in medical robotics and data managements.

4.2 Expanding IoT Infrastructure in Smart Hospitals

The Internet of Things (IoT) is transforming healthcare delivery inside the walls of a hospital by interconnecting medical devices, wearables, and data systems. This expansion has enabled the realization of real-time monitoring and data-centric decision-making. The main focus of innovation remains on pervasive patient monitoring and ensuring seamless data integration into electronic records.

4.2.1 Real-Time Health Monitoring Systems

The future smart hospitals will have massively distributed IoT sensor networks for, among other things, continuous patient and environmental monitoring. Wearable devices such as smartwatches and patches, together with ambient sensors, can assess vital signs, mobility, and environmental variables like room temperature and

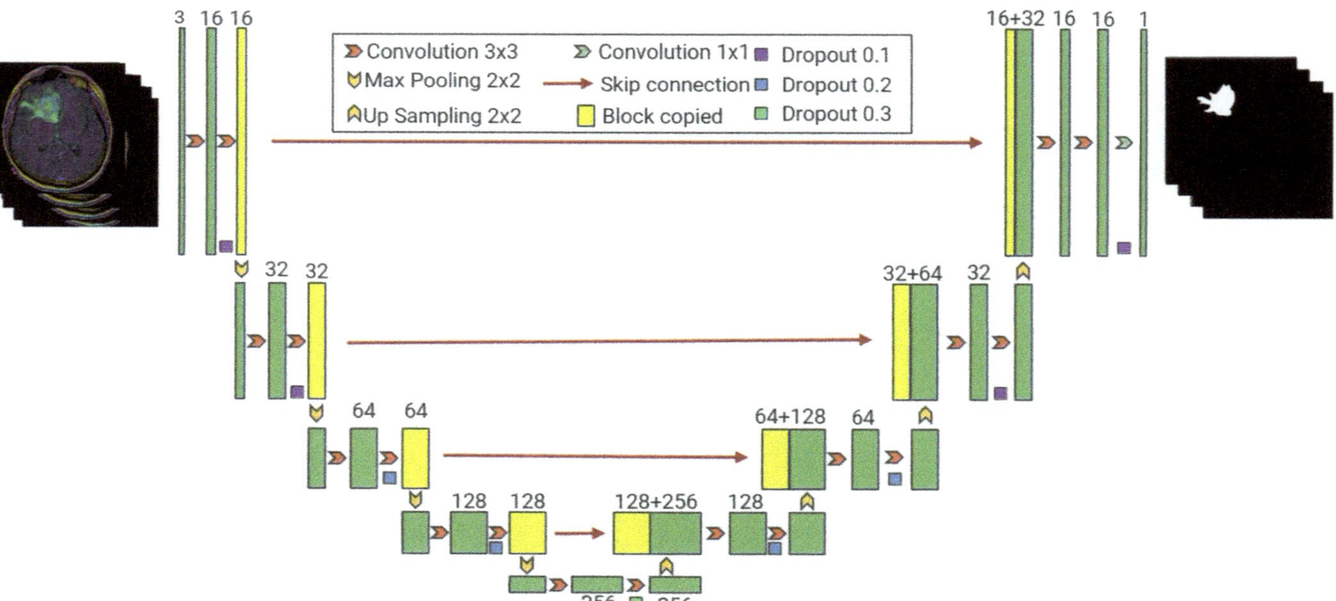

Fig. 13 U-Net architecture for brain tumor segmentation. This figure illustrates the U-Net architecture's encoder-decoder design with skip connections. The encoder (contraction path, left side) progressively downsamples input images to extract hierarchical features, while the decoder (expansion path, right side) upsamples these features to reconstruct segmentation masks. Skip connections (horizontal arrows) fuse encoder outputs with decoder inputs at equivalent resolution levels, preserving spatial details lost during pooling. This integration allows the model to localize lesions (e.g., tumors) with pixel-level precision, even in low-data scenarios. The architecture's efficiency and accuracy have made it a benchmark for medical image segmentation tasks. (*Source*: Sun, H., Yang, S., Chen, L. et al. Brain tumor image segmentation based on improved FPN. *BMC Med Imaging* 23, 172 (2023). https://doi.org/10.1186/s12880-023-01131-1)

air quality on a continuous basis. When swift clinical deterioration occurs, such as with sepsis or respiratory failure, rapid intervention improves patient outcomes and saves lives. Current studies emphasize the importance of the IoT in healthcare, denoting that they offer secured real-time remote monitoring facilities, which improve the quality of life [42]. The wireless wearable sensors keep transmitting ECG, blood pressure, and oxygen saturation data to monitoring stations non-stop while raising alarms to the staff when abnormalities are observed, while IoT-based monitors in ICUs track multiple patients at once in tandem with AI-based early warning scoring. In environments with less urgent care, IoT robots and ambient sensors identify patient locations and detect falls, thus enabling early interventions.

Now, this theory shines in telehealth as well, where home-monitoring systems send data to hospital networks to curb readmission rates. The envisaged continuous monitoring network seeks to render health monitoring accessible from home, thus liberating the hospital. Realization of this will call for an IoT-based infrastructure with ultra-reliable wireless networks and 5G/6G connectivity, plus processor-intensive analytics. Future developments should target cheap, disposable, or implantable sensors, better batteries, and wearables integrated with AI for predictive alerts, triggering a next generation in responsive and personalized hospital care.

4.2.2 Integration with Electronic Health Records (EHRs)

Integration of IoT-generated data into hospital information systems and electronic health records to present a complete view of the patient to clinicians is the challenge. Presently, the industry's ongoing attempts focus on interlinking IoT devices and continuous monitoring data with EHR platforms through standard interfaces like HL7/FHIR and through a middleware solution. Essentially, this would keep a record of every single sensor reading or even alerts of the patient ever recorded into that patient's own EHR; in other words, longitudinal records of vital signs and activities become available for analytics and decision support, e.g., trends in daily weight or blood sugar measured by home devices for chronic disease management.

Data interoperability is a major obstacle cited by stakeholders considering the use of big data from wearable technology, whereas poor interoperability has been cited as hindering the ability to integrate with the various health data [43]. Most EHR systems use proprietary formats and do not have standard methods to import IoT streams, which calls for adopting common standards such as IEEE 11073 and HL7/FHIR in conjunction with cloud-based architectures for data exchange. Besides being secure and guaranteeing privacy, IoT devices gather sensitive information that must comply with health privacy regulations when combined with EHRs. Studies have revealed fragmented systems and inconsistent standards are key to EHR interoperability issues, a root already troubling IoT [44]. In the upcoming years, however, the focus shall be cloud-native platforms and distributed ledger technologies such as blockchain for securely linking IoT data with the EHR, providing secured networks, and a well-standardized way of integration for actionable insight to be deployed by both clinicians and AI tools.

4.3 Policy, Regulation, and Ethical Frameworks

The rapid development of healthcare robotics and IoT distill many legal and ethical complex issues into a single topic. Strict laws and standards should be established to ensure safety, data governance, and international data flows. Once looked at carefully, two main aspects are highlighted: technical/ethical standardization and policies for cross-border data sharing.

4.3.1 Standardization

While healthcare robotics is evolving, some important questions about what must be researched further to fully realize the promises of these technologies will be posed. Viewing the subject for a moment from improving human-robot collaboration and scaling to accessibility and more ethical standpoints, the new generation of research will have a great deal to contribute in determining the very nature of how health will be delivered. To depict key future areas of research graphically, Fig. 14 presents the key themes that must be addressed, including enhanced human-robot collaboration, available and scalable robotic systems, interoperability standards,

Fig. 14 Future research directions for integrating robotics, AI, and IoT into healthcare: from collaboration and scalability to ethical frameworks and immersive training technologies. (*Source*: AI generated chart created with Napkin Inc., Napkin AI (Beta Version), 2024)

predictive analytics driven by AI, explainable and ethical AI, and utilizing immersive technologies for training healthcare workers. This visual encapsulation offers a structured overview of the interlinked areas where research and innovation can significantly impact the effectiveness, fairness, and integration of robotics and AI into modern healthcare systems.

4.3.2 Cross-Border Data-Sharing Policies

Healthcare data globalization presents a considerable legal challenge, especially when data are shared across borders for multi-site research, or in telemedicine, with privacy regulations that differ by region, such as the GDPR in Europe, HIPAA in the United States, or some framework in Asia. Stakeholders stress that the incompatibility of privacy laws in differing jurisdictions occurs at the forefront of hindering international collaboration, highlighting the lack of appropriate legal protections and the incompatibilies of regulations for transferring data [45]. Remaining as arguably the most important issue is the lack of uniform data protection practices, calling for collaborative efforts to try and address these issues.

Future policymaking will be oriented toward setting up agreements and frameworks that will pave the way for lawful data exchange, including harmonizing standards via mutual adequacy decisions or worldwide agreements and implementing technical safeguards such as encryption and anonymization to meet strict regulatory requirements. Proposals include bolstering existing legislation and promoting the creation of transparent inter-jurisdictional contracts, such as the European Union–United States Data Privacy Framework, to harmonize divergent legal systems. Evolving legal frameworks would be imperative since healthcare robotics frequently requires patient data derived from various sources, such as wearables and imaging, which may cross national borders. Ethical points raised in the review weigh in favor of an international balancing act, through bodies like the WHO and OECD and via bilateral agreements, to mitigate privacy concerns against innovation [46]. Future work will investigate federated learning architectures with data staying in situ while models are exchanged and consent management based on blockchain for cross-border data exchange security, thus creating a trusted and interoperable data ecosystem for global healthcare robotics.

5 Conclusion

The combination of robotics, the Internet of Things (IoT), and deep learning is revolutionary in healthcare, as it provides data-theoretic, exact, and patient-centered solutions. The integration is opening new doors in breaking down clinical challenges, advancing treatments, and promoting the accessibility and equity of care. In this conclusion, we review the major insights examined in the chapter and offer recommendations toward practical implementation, impact, and expected research interest.

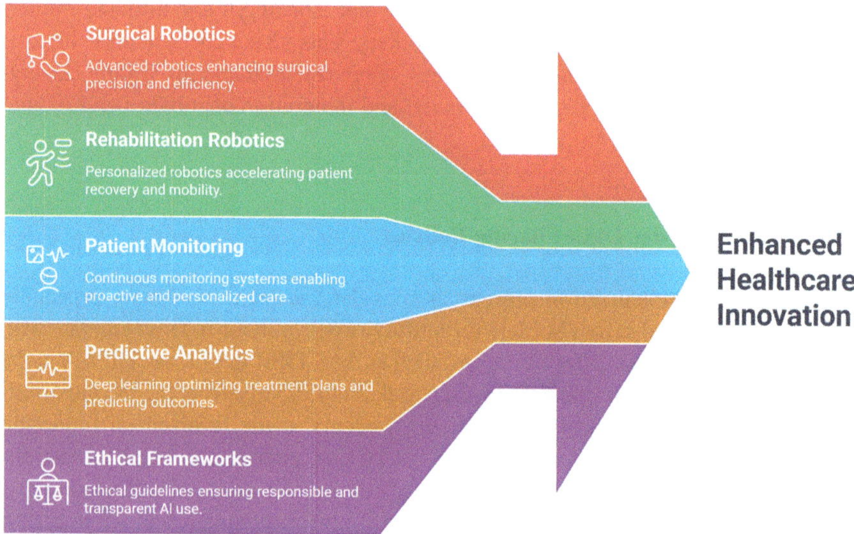

Fig. 15 Illustration of key innovations in surgical robotics, rehabilitation, patient monitoring, predictive analytics, and ethical frameworks enhancing healthcare. (*Source*: AI generated chart created with Napkin Inc., Napkin AI (Beta Version), 2025)

5.1   Summary of Key Findings

The convergence of robotics, IoT, and deep learning has ushered in a new era of healthcare innovation, offering unprecedented opportunities to enhance patient outcomes, streamline clinical workflows, and address long-standing challenges in healthcare delivery. In Fig. 15, we summarize the key areas where these technologies have demonstrated transformative potential.

5.2   Final Thoughts on Implementation and Impact

The integration of robotics, IoT, and deep learning into healthcare does present challenges. But they are nothing compared to the potential that offers solutions to some of the oldest medical problems:

1. **Resource Limitations**: With its possibilities of automation and workflow efficiency, robotics and IoT can reduce the load upon health systems so that more resources can be geared toward patient care. This becomes more essential for resource-constrained settings since such is the lack of access to newer medical technologies [47].

2. **Chronic Disease Management**: IoT-enabled monitoring systems and AI-based predictive analytics can radically change the way chronic diseases are managed. By providing a continuous assessment of patients' health, the interventions will be very prompt, allowing treatment plans to be adjusted to their needs for better outcomes and reduced health costs [15, 38].

3. **Minimally Invasive Procedures**: The precision and dexterity of robot-assisted surgical operations have made minimally invasive procedures much more accessible and effective. These measures collectively help shorten patients' recovery time, thereby reducing potential complications and providing a higher quality of care [36, 37].

4. **Equity and Accessibility**: Among the most notable societal effects of these technologies is their potential to amplify access to care for underserved populations. Telehealth and AI-enabled chatbots can, therefore, breach the walls constituted by the patient-provider distance, enabling a remote or resource-deficient individual to procure timely and appropriate care [48].

5. **Future Research Directions**: However, advancing healthcare robotics still requires extensive research. Future research should concentrate on facilitating better human-robot interaction, scalability, and accessibility and creating generic standards for interoperability. Likewise, the incorporation of immersive technologies, like AR and VR, into healthcare education will be very important for preparing the workforce of tomorrow [39].

The integration of robotics, IoT, and deep learning into healthcare is metamorphosing medical services into intelligent, adaptive, smart, and efficient solutions for modern-day health challenges. These technologies have exhibited the potential to act from surgical state robots, patient monitoring, and rehabilitations to certainty for better outcomes, operational efficiency, and resource limitations. To harness this potential requires a situation addressed against cost issues, concerns for data security, overcoming interoperability problems, and grappling with ethical issues.

Taking into account the interventions and lessons learned in real applications and identifying the future directions of research, the healthcare industry can tap the full potential of these transformative technologies. The multi-party collaboration between healthcare providers, technology developers, and policymakers will strongly drive innovation address barriers to accessibility and keep patient rights protected. Looking into the future, a robotics/IoT/deep-learning integration would continue to be a key element of the evolving landscape of a more efficient, more equitable, and patient-focused healthcare system. Appendix 2 provides a glimpse of the transformative possibilities of IoT and deep learning applied to healthcare robotics while capturing the comparative benefits, key challenges, and future directions for each application.

Acknowledgments   This work is supported in part by Shahid Chamran University of Ahvaz under grant number **SCU.EC1403.450.**

Appendix 1: Comparative Analysis of Deep Learning Algorithms for Medical Data Management and Intelligent Decision-Making

Algorithm	Key features	Medical applications	Advantages	Limitations
CNN	– Hierarchical feature extraction from images – Ideal for structured data (e.g., MRI)	– Skin cancer detection – Tumor segmentation	– High accuracy in image processing – Outperforms traditional methods	– Requires large labeled datasets– Sensitive to image noise
RNN/ LSTM/GRU	– Sequential data processing – Long-term memory retention	– ICU patient deterioration prediction – EEG/ECG signal analysis	– Models temporal dependencies – Handles continuous data streams	– High computational complexity – Struggles with very long-term dependencies
U-Net	– Encoder-decoder structure with skip connections – Pixel-level segmentation	– Brain tumor segmentation in MRI – Skin lesion identification	– High accuracy with limited data – Recovers spatial details	– Hyperparameter tuning required – Limited 3D image processing
Transformer	– Attention mechanisms – Parallel data processing	– Medical text report analysis – Disease progression prediction	– Captures non-local relationships – Highly scalable	– Computationally intensive – Low interpretability

Appendix 2: Transformative Potential of IoT and Deep Learning in Healthcare Robotics

Application	Key benefits	Challenges	Future directions
Surgical robotics	Enhanced precision, reduced operation times, and minimally invasive procedures	High costs and technical complexity	Develop cost-effective solutions and improve interoperability
Rehabilitation robotics	Faster recovery times and improved patient mobility	High costs and need for specialized training	Focus on modular designs and low-cost systems for broader accessibility
Patient monitoring	Continuous tracking of vital signs and early detection of health anomalies	Data security and privacy concerns	Implement robust encryption and federated learning for secure data sharing

Application	Key benefits	Challenges	Future directions
Predictive analytics	Personalized treatment plans and optimized patient outcomes	Ethical concerns related to AI decision-making	Develop ethical AI frameworks and explainable AI (XAI) methods
Ethical frameworks	Ensures responsible use of AI and robotics in healthcare	Balancing innovation with ethical considerations	Collaborate with regulatory bodies to establish guidelines for ethical AI use

References

1. Curry, E., Heintz, F., Irgens, M., Smeulders, A. W., & Stramigioli, S. (2022). Partnership on AI, data, and robotics. *Communications of the ACM, 65*(4), 54–55.
2. Okamura, A. M., Matarić, M. J., & Christensen, H. I. (2010). Medical and health-care robotics. *IEEE Robotics & Automation Magazine, 17*(3), 26–37.
3. Gumbs, A. A., De Simone, B., & Chouillard, E. (2020). Searching for a better definition of robotic surgery: Is it really different from laparoscopy? *Mini-invasive Surgery, 4,* 90.
4. Turjamaa, R., Vaismoradi, M., & Kangasniemi, M. (2023). Older home care clients' experiences of digitalisation: A qualitative study of experiences of the use of robot for medicines management. *Scandinavian Journal of Caring Sciences, 37*(2), 561–570.
5. Naeem, S., Ali, A., Memon, K., Bavluwala, M., Shinde, U. P., & Patil, A. V. (2023). A review of flexible high-performance supercapacitors for the Internet of Things (IoT) and artificial intelligence (ai) applications. *Energy and Thermofluids Engineering, 3,* 1–9.
6. Siripurapu, S., Darimireddy, N. K., Chehri, A., Sridhar, B., & Paramkusam, A. V. (2023). Technological advancements and elucidation gadgets for healthcare applications: An exhaustive methodological review-part-II (robotics, drones, 3D-printing, Internet of Things, virtual/ augmented and mixed reality). *Electronics, 12*(3), 548.
7. Long, V. N., & Hoang, N. A. (2017, June). Development of IoT based lower limb exoskeleton in rehabilitation. In *2017 14th International Conference on Ubiquitous Robots and Ambient Intelligence (URAI)* (pp. 824–826). IEEE.
8. Qian, C., & Ren, H. (2025). Deep reinforcement learning in surgical robotics: Enhancing the automation level. In *Handbook of robotic surgery* (pp. 89–102). Academic Press.
9. Minopoulos, G. M., Memos, V. A., Stergiou, K. D., Stergiou, C. L., & Psannis, K. E. (2023). A medical image visualization technique assisted with AI-based haptic feedback for robotic surgery and healthcare. *Applied Sciences, 13*(6), 3592.
10. Gyrard, A., Tabeau, K., Fiorini, L., Kung, A., Senges, E., De Mul, M., et al. (2023). Knowledge engineering framework for IoT robotics applied to smart healthcare and emotional well-being. *International Journal of Social Robotics, 15,* 1–28.
11. Aymerich-Franch, L., & Ferrer, I. (2023). Socially assistive robots' deployment in healthcare settings: A global perspective. *International Journal of Humanoid Robotics, 20*(1), 2350002.
12. Giri, S., & Sarkar, D. K. (2012). Current status of robotic surgery. *Indian Journal of Surgery, 74,* 242–247.
13. Jain, S., & Gautam, G. (2015). Robotics in urologic oncology. *Journal of Minimal Access Surgery, 11*(1), 40–44.
14. Kosa, G., Morozov, O., Lehmann, A., Pargger, H., Marsch, S., & Hunziker, P. (2023). Robots and intelligent medical devices in the intensive care unit: Vision, state of the art, and economic analysis. *IEEE Transactions on Medical Robotics and Bionics, 5*(1), 2–17.
15. Kamilaris, A., & Botteghi, N. (2020). The penetration of Internet of Things in robotics: Towards a web of robotic things. *Journal of Ambient Intelligence and Smart Environments, 12*(6), 491–512.

16. Dahl, T. S., & Kamel Boulos, M. N. (2013). Robots in health and social care: A complementary technology to home care and telehealthcare? *Robotics, 3*(1), 1–21.

17. Łukasik, S., Tobis, S., Kropińska, S., & Suwalska, A. (2020). Role of assistive robots in the care of older people: Survey study among medical and nursing students. *Journal of Medical Internet Research, 22*(8), e18003.

18. Cakmak, M., & Ozerhan, Y. (2022). The effects of Internet of Things on the transportation cost management: A study of logistics company. *Journal of Business Economics and Finance, 11*(3), 130–142.

19. BiBi, Z., Wang, G., Xu, L. D., Thompson, M., Mir, R., Nyikos, J., et al. (2017). IoT-based system for communication and coordination of football robot team. *Internet Research, 27*(2), 162–181.

20. Pradhan, B., Bharti, D., Chakravarty, S., Ray, S. S., Voinova, V. V., Bonartsev, A. P., & Pal, K. (2021). Internet of Things and robotics in transforming current-day healthcare services. *Journal of Healthcare Engineering, 2021*(1), 9999504.

21. Open Source Robotics Foundation. (2023). Affordable exoskeletons for global health. Retrieved from https://www.openrobotics.org/foundation

22. Hassan, A. A., Tutuncu, K., Abdullahi, H. O., & Ali, A. F. (2023). IoT-based smart health monitoring system: Investigating the role of temperature, blood pressure and sleep data in chronic disease management. *Instrumentation Mesure Métrologie, 22*(6), 231–240.

23. Sivaparthipan, C. B., Muthu, B. A., Manogaran, G., Maram, B., Sundarasekar, R., Krishnamoorthy, S., et al. (2020). Innovative and efficient method of robotics for helping the Parkinson's disease patient using IoT in big data analytics. *Transactions on Emerging Telecommunications Technologies, 31*(12), e3838.

24. Gherardini, F., Petruccioli, A., Dalpadulo, E., Bettelli, V., Mascia, M. T., & Leali, F. (2020). A methodological approach for the design of inclusive assistive devices by integrating co-design and additive manufacturing technologies. In *Intelligent Human Systems Integration 2020: Proceedings of the 3rd International Conference on Intelligent Human Systems Integration (IHSI 2020): Integrating People and Intelligent Systems, February 19–21, 2020, Modena, Italy* (pp. 816–822). Springer International Publishing.

25. Nambiappan, H. R., Arboleda, S. A., Lundberg, C. L., Kyrarini, M., Makedon, F., & Gans, N. (2022). Mina: A robotic assistant for hospital fetching tasks. *Technologies, 10*(2), 41.

26. Zakhama, A., Charrabi, L., & Jelassi, K. (2019). Intelligent selective compliance articulated robot arm robot with object recognition in a multi-agent manufacturing system. *International Journal of Advanced Robotic Systems, 16*(2), 1729881419841145.

27. Mouradian, C., Jahromi, N. T., & Glitho, R. H. (2018). NFV and SDN-based distributed IoT gateway for large-scale disaster management. *IEEE Internet of Things Journal, 5*(5), 4119–4131.

28. Spiliopoulos, S., Hergesell, V., & Dapunt, O. (2019). Left ventricular assist device therapy in a patient with hearing and speech disabilities. *The Journal of Thoracic and Cardiovascular Surgery, 157*(1), e1–e2.

29. Goldberg, K., Mascha, M., Gentner, S., Rothenberg, N., Sutter, C., & Wiegley, J. (1995, May). Desktop teleoperation via the world wide web. In *Proceedings of 1995 IEEE International Conference on Robotics and Automation* (Vol. 1, pp. 654–659). IEEE.

30. Goldberg, K., Santarromana, J., Bekey, G., Gentner, S., Morris, R., Wiegley, J., & Berger, E. (1995). The telegarden. In *Proceedings of the ACM SIGGRAPH* (pp. 135–1140).

31. Simmons, R., Fernandez, J., Goodwin, R., Koenig, S., & O'Sullivan, J. (1998). Xavier: An autonomous mobile robot on the web.

32. Lee, H., Kim, J., Kim, S., Kong, H. J., & Ryu, H. (2019). Investigating the need for point-of-care robots to support teleconsultation. *Telemedicine and e-Health, 25*(12), 1165–1173.

33. Pino, M., Boulay, M., Jouen, F., & Rigaud, A. S. (2015). "Are we ready for robots that care for us?" Attitudes and opinions of older adults toward socially assistive robots. *Frontiers in Aging Neuroscience, 7*, 141.

34. Céspedes, N., Irfan, B., Senft, E., Cifuentes, C. A., Gutierrez, L. F., Rincon-Roncancio, M., et al. (2021). A socially assistive robot for long-term cardiac rehabilitation in the real world. *Frontiers in Neurorobotics, 15*, 633248.

35. Lara, J. S., Casas, J., Aguirre, A., Munera, M., Rincon-Roncancio, M., Irfan, B., et al. (2017, July). Human-robot sensor interface for cardiac rehabilitation. In *2017 International Conference on Rehabilitation Robotics (ICORR)* (pp. 1013–1018). IEEE.

36. Luo, H., & Yin, Q. (2021). Distributed optimization for mobile robots under mobile edge computing environment. *Complexity, 2021*(1), 8342610.

37. Zhang, J., Luo, H., & Xu, J. (2022). Towards fully BIM-enabled building automation and robotics: A perspective of lifecycle information flow. *Computers in Industry, 135*, 103570.

38. Ahmad, A., Fahmideh, M., Altamimi, A. B., Katib, I., Albeshri, A., Alreshidi, A., et al. (2021). Software engineering for IoT-driven data analytics applications. *IEEE Access, 9*, 48197–48217.

39. Phupattanasilp, P., & Tong, S. R. (2019). Augmented reality in the integrative Internet of Things (AR-IoT): Application for precision farming. *Sustainability, 11*(9), 2658.

40. Badr, N. G., & Dankar, M. (2022). Assistive healthcare robotics–challenges in nursing service innovation: Critical review. *ITM Web of Conferences, 41*, 02002.

41. Cheikh, I., Aouami, R., Sabir, E., Sadik, M., & Roy, S. (2022). Multi-layered energy efficiency in LoRa-WAN networks: A tutorial. *IEEE Access, 10*, 9198–9231.

42. Abdulmalek, S., Nasir, A., Jabbar, W. A., Almuhaya, M. A., Bairagi, A. K., Khan, M. A. M., & Kee, S. H. (2022). IoT-based healthcare-monitoring system towards improving quality of life: A review. *Healthcare, 10*(10), 1993.

43. Canali, S., Schiaffonati, V., & Aliverti, A. (2022). Challenges and recommendations for wearable devices in digital health: Data quality, interoperability, health equity, fairness. *PLOS Digital Health, 1*(10), e0000104.

44. Carlos Ferreira, J., Elvas, L. B., Correia, R., & Mascarenhas, M. (2024). Enhancing EHR interoperability and security through distributed ledger technology: A review. *Healthcare, 12*(19), 1967.

45. Chan, H. Y., Toh, H. J., & Lysaght, T. (2024). Cross-jurisdictional data transfer in health research: Stakeholder perceptions on the role of law. *Asian Bioethics Review, 16*(4), 663–682.

46. Elendu, C., Amaechi, D. C., Elendu, T. C., Jingwa, K. A., Okoye, O. K., Okah, M. J., et al. (2023). Ethical implications of AI and robotics in healthcare: A review. *Medicine, 102*(50), e36671.

47. Valner, R., Masnavi, H., Rybalskii, I., Põlluäär, R., Kõiv, E., Aabloo, A., et al. (2022). Scalable and heterogenous mobile robot fleet-based task automation in crowded hospital environments—A field test. *Frontiers in Robotics and AI, 9*, 922835.

48. Seng, K. P., Ang, L. M., & Ngharamike, E. (2022). Artificial intelligence Internet of Things: A new paradigm of distributed sensor networks. *International Journal of Distributed Sensor Networks, 18*(3), 15501477211062835.

Future Directions

Roadmaps and Agendas for Research and Innovation in Artificial Intelligence, Data, and Robotics

Paulo Figueiras, Giorgos Ioannou, Charalambos Lambri, Sangheeta Reji,
Elena Mossali, Martina Imarisio Neviani, Sotiris Koussouris,
Nefeli Bountouni, Marina Da Bormida in Cugurra, Robert Hellbach,
Dimitris Bibikas, Philip O'Brien, Carlos Agostinho,
and Ricardo Jardim-Gonçalves

P. Figueiras · C. Agostinho (✉) · R. Jardim-Gonçalves
UNINOVA—Centre of Technology and Systems (CTS), FCT Campus, Caparica, Portugal
e-mail: paf@uninova.pt; ca@uninova.pt; rg@uninova.pt

G. Ioannou · C. Lambri
Department of Computer Science, University of Cyprus, Nicosia, Cyprus
e-mail: ioannou.george@ucy.ac.cy; lambri.charalambos@ucy.ac.cy

S. Reji
Digital Public Services, Fraunhofer Institute for Open Communication Systems,
Berlin, Germany
e-mail: sangeetha.reji@fokus.fraunhofer.de

E. Mossali · M. I. Neviani
Consorzio Intellimech, Bergamo, Italy
e-mail: elena.mossali@intellimech.it; martina.imarisio@intellimech.it

S. Koussouris · N. Bountouni
Suite5 Data Intelligence Solutions Ltd., Limassol, Cyprus
e-mail: sotiris@suite5.eu; nefeli@suite5.eu

M. D. B. in Cugurra
S&D Consulting Europe, Milan, Italy
e-mail: marina.cugurra@sdconsulting-eu.com

R. Hellbach
BIBA-Bremer Institut für Produktion und Logistik GmbH at the University of Bremen,
Bremen, Germany
e-mail: hel@biba.uni-bremen.de

D. Bibikas
Zenith Gas & Light SA, Thessaloniki, Greece
e-mail: d.bimpikas@zenith.gr

P. O'Brien
Walton Institute, South East Technological University, Waterford, Ireland
e-mail: Philip.OBrien@waltoninstitute.ie

© The Author(s) 2026

E. Curry et al. (eds.), *Artificial Intelligence, Data and Robotics*,
https://doi.org/10.1007/978-3-032-10561-5_19

Abstract Artificial intelligence (AI), data, and robotics are transformative technologies that drive innovation across industries and address critical societal challenges. This chapter outlines priority areas, including data processing for AI, synthetic data generation, autonomous systems, human-centric robotics, and explainable, science-guided, and adaptive AI, just to name a few. It emphasizes the need for robust frameworks to ensure ethical AI development, addressing concerns around bias, fairness, and accountability. The roadmaps and agendas will highlight cross-disciplinary collaboration, leveraging synergies among AI, big data analytics, and robotics. This chapter advocates foundational research in adaptive, explainable, science-guided AI while emphasizing applied research for real-world deployment. It provides a strategic vision to align public and private sector investments, strengthen Europe's global competitiveness, and promote trust and inclusivity in AI, data, and robotics technologies.

Keywords Roadmaps · Research and innovation agendas · Artificial intelligence · Data · Robotics

1 Introduction

Artificial intelligence (AI), data, and robotics are intertwined technological domains [1] that have been driving innovation and industrial transformation for at least two decades and will continue to do so. Their convergence is accelerating, reshaping sectors such as healthcare, manufacturing, and smart cities. Recent advancements, including generative AI, synthetic data generation, edge computing, and human-centric robotics, demonstrate the rapid evolution of these fields. Their growing influence demands coordinated research efforts, ensuring responsible deployment and alignment with societal and economic priorities.

This chapter aims to outline research and innovation pathways and prospects for the coming years, pointing out some of the research priorities and define potential research and innovation agendas and roadmaps in these domains, addressing several aspects such as research priorities, ethical and societal considerations, and future collaboration enterprises and investment strategies, and highlighting the transformative role of AI, data, and robotics in the near future. It will provide a strategic vision to align public and private sector investments, strengthen Europe's global competitiveness, and promote trust and inclusivity in AI, data, and robotics technologies.

1.1 The Role of AI, Data, and Robotics in Innovation

AI, data, and robotics are three technological domains that have emerged as truly transformative technologies and have and still are reshaping several sectors of our society. And although these fields are independent, they are also deeply connected, each playing a crucial role in enabling smarter, more efficient, and autonomous systems, transforming industries, driving economic growth, and addressing societal challenges.

AI encompasses a wide range of technologies and methodologies that enable machines to perform tasks that typically require human intelligence, including learning, reasoning, problem-solving, perception, and decision-making. Recent advances in AI-related fields, such as machine learning (ML), deep learning (DL), neural networks, and, more recently generative AI, have dramatically improved AI's capabilities and innovation potential across almost all industry and society domains. Examples can be found in several sectors, such as the healthcare and life sciences sector, particularly in the case of AI-powered diagnostics [2], in the finance and banking sector, as in the case of AI-driven fraud detection [3], or in the manufacturing and industry, namely, related to predictive maintenance, quality control [4], etc.

Robotics integrates AI, data, and automation to create intelligent machines capable of performing physical tasks with precision, adaptability, and autonomy or of supporting humans in physically demanding tasks and contributing to optimized ergonomics. Recent advancements in robotic perception, collaborative robotics, human-centric robotics, and AI-based control systems have significantly expanded the role of robotics in both industrial and service-oriented domains. Key innovation areas include industrial automation and Smart Factories, such as, for instance, human-centric and collaborative robotics [5], autonomous vehicles and drones [6], or medical and assistive robotics [7], just to name a few examples.

Data is the glue between AI and robotics. More than that, data represents the lifeblood of AI and robotics, serving as the foundation for training models and for cognitive robotics while enabling real-time decision making in both domains. The advent of true intelligent systems is only possible due to the rapid growth of big data and cloud and edge computing. Beyond these fields, recent years have seen great advances in terms of how to process, manage, and capitalize on different types of data:

1. Structured data, namely, automated data engineering with AI, graph databases and knowledge graphs, and federated learning.
2. Unstructured data, as in the case of foundation models for text and other multi-modal data, self-supervised learning, and other generative AI–based management solutions.
3. Real-time streaming data, especially focusing on edge AI and federated learning for real-time processing [8], event-driven AI, and digital twins.
4. Synthetic data, for instance, generative adversarial networks (GANs), differentially private synthetic data, and AI-augmented labelling and data generation [9], just to point out some examples.

As AI, data, and robotics continue to evolve, they present unprecedented opportunities for scientific discovery, economic growth, and societal progress. Further, the synergy between these fields will drive next-generation intelligent systems that can process vast amounts of information, learn from their environments, and autonomously adapt to new challenges. However, responsible development is critical to ensuring these technologies align with human values, ethics, and regulatory frameworks [10].

1.2 The Need for Agendas and Roadmaps

Despite their potential advantages, AI, data, and robotics present several technical, ethical, and regulatory challenges. AI models require vast amounts of high-quality data, which is often not easy to obtain, leading to data scarcity and biases, as well as high maintenance, processing, and storage costs. Additionally, many AI models lack transparency due to their black-box nature, making their results difficult to interpret and use in systematic decision-making processes. These challenges also amplify biases and lead to discriminatory outcomes in predictions and classifications. Thus, future developments of AI should adhere to strong legal, societal, and ethical standards to ensure responsible use.

Regarding robotics, it is crucial to ensure that robots make safe, ethical, and efficient decisions in dynamic environments and can work seamlessly with humans in shared spaces. Energy efficiency is another key challenge for future robotics systems, as developing power-efficient robots is essential for extending their operational capabilities. Moreover, the integration of AI with robotics will require global standards to govern the deployment of intelligent robotics in critical applications. On the side of data, besides the data quality and bias challenges already mentioned, data security and privacy remain huge challenges. Ensuring compliance with regulations such as General Data Privacy Regulation (GDPR), establishing data governance and creating policies for data collection, sharing, and access across organizations and borders will be critical.

Europe has lagged behind leading innovative countries, particularly in relation to generative AI and cloud-based data infrastructure [11]. The rapid pace of AI, data, and robotics innovation presents both opportunities and challenges. Without a coordinated roadmap, advancements may occur in a fragmented manner, leading to issues such as duplication of efforts, missed opportunities for synergies, and unbalanced funding distributions, due to disproportionate attention or interest in some areas, leaving critical but less-publicized domains underfunded. This is partly due challenges in European research and innovation landscape, including bureaucratic hurdles, unbalanced funding, and the tension between ethical approaches to technological development, data, and robotics innovation [12]. One good example of such a coordinated roadmapping agenda is the Artificial Intelligence, Data and Robotics Association's Strategic Research, Innovation and Deployment Agenda (ADRA SRIDA) [10]. ADRA [13], a pan-European network fostering collaboration between academia, industry, and policymakers, seeks to build on the foundations of Europe's ambition to lead globally in AI, data, and robotics, by boosting the revenue potential of business models while also delivering broader benefits to society.

As AI, data, and robotics continue to evolve, their transformative potential can only be fully realized through a strategic and coordinated approach. Research and innovation roadmaps are essential to guide advancements, prioritize key areas, address ethical concerns, and ensure long-term societal benefits. These roadmaps outline key research priorities, technological milestones, and policy actions needed to achieve medium- to long-term progress, aligning research and innovation efforts across sectors. They also provide a strategic framework for challenge prioritization that collaboratively addresses ethical, regulatory, and technical barriers by fostering interdisciplinary partnerships between research institutions, industries, and governmental bodies.

1.3 Approach for Roadmap and Agenda Development

Developing comprehensive research and innovation roadmaps requires a structured and multi-stakeholder approach that aligns with technological trends, societal needs, policy regulations, and global economic objectives while ensuring ethical considerations and sustainability. This section outlines a systematic approach to guide roadmap development, covering its core principles, phases, and key stakeholder engagement strategies. The starting point for the approach that serves as a basis for this chapter is the AI-DAPT project's Research Agenda. AI-DAPT [14] is an EU-funded innovative research project focused on addressing critical challenges in AI deployment, particularly related to data utilization, model reliability, and adaptability. It aims at reinstating the pure data-related work in its rightful place in AI and at reinforcing the generalizability, reliability, trustworthiness, and fairness of AI solutions. The AI-DAPT Research Agenda is closely related to the ADRA missions, goals, and so-called Big Tickets (BTs) in AI, data, and robotics, described in the ADRA SRIDA [10].

Starting from the AI-DAPT Research Agenda, the approach addresses the identification of key research and innovation priorities and stakeholders, establishes the technological milestones for these priorities through a set of agendas and roadmaps for each key topic (Sect. 2), delves through future ethical and societal challenges and how they can be mitigated (Sect. 3), and defines several pathways concerning business, policy and regulatory frameworks, and investment and funding strategies to guide the future innovations in the fields of AI, data, and robotics (Sect. 4), all along mapping the intrinsic relations to the ADRA SRIDA's specific challenges and Big Tickets (BTs).

The selection of research priorities, challenges, and future pathways is driven by the need to address both foundational challenges and transformative opportunities in the symbiotic relationship between AI, data, and robotics, directly related to the AI, Data, and Robotics Partnership [15] and specifically to the ADRA SRIDA. As these technologies evolve, they must be guided by structured research agendas that push innovation boundaries while ensuring ethical alignment, practical applicability, and societal benefit.

Data is a critical foundation for AI systems, yet real-world data is often incomplete, biased, or difficult to access—limiting AI's ability to generalize effectively.

This makes research into data processing and synthetic data essential, not only for improving AI robustness but also for developing fairer, privacy-conscious systems. Advancing autonomy with a human-centric focus is another priority. As AI powered robotics and autonomous systems take on complex roles in sectors like manufacturing, healthcare, and public services, they must be designed to work safely and intuitively alongside people, fostering trust and adaptability. Likewise, explainable, science-guided, and adaptive AI is increasingly vital in high-stakes areas such as medicine, law, and policymaking, where transparency and accountability are essential.

With technology shifting toward decentralized and real-time use cases, edge AI and IoT have become pivotal. By moving computation closer to the data source—whether in smart cities, autonomous vehicles, or industrial systems—AI can deliver faster, more efficient decisions with lower energy use and less dependence on the cloud. AI-driven decision support systems are also becoming central in finance, healthcare, and security, offering data-driven insights that enhance, rather than replace, human judgment. These systems must be context-aware, explainable, and resilient. Beyond the technical dimension, these research priorities are deeply linked with ethical and societal challenges. This chapter underscores the importance of developing AI and robotics systems that are fair, trustworthy, and aligned with European values while also emphasizing the need for supportive policy frameworks, smart investments, and workforce development to ensure widespread, inclusive benefits.

2 Key Research Priorities, Agendas, and Roadmaps

2.1 Data Processing and Synthetic Data Creation

The demand for high-quality data is creating bottlenecks in acquisition, pre-processing, and augmentation workflows. In many situations, synthetic data, which are artificially generated mimicking real-world patterns, has emerged to solve this issue. In this section, the challenges, emerging solutions, and the transformative potential of synthetic data will be analyzed. This challenge aligns closely with ADRA SRIDA's BT1 [10], which calls for innovations in data handling and preprocessing methods to deliver high-performance AI in real-world contexts.

Accurate, up-to-date, and well-labelled data is the foundation and fuel to building an AI/ML system. Frequently, in large AI projects, datasets are reused or recycled, which poses a challenge, while ad hoc processing brings challenges in data acquisition [16]. Data marketplaces often lack transparency in pricing, format standardization, and metadata—forcing engineers through inefficient procurement processes involving multi-vendor negotiations and laborious cleaning. Along with the inefficiency of this process, data acquisition systems should be constantly evolving and adapting to infrastructure changes.

Nowadays, there are several challenges in data acquisition, resulting in data scarcity. In many situations, data collection faces logistical, ethical, and financial barriers. For example, medical imaging datasets are constrained by privacy laws (e.g., GDPR). Traditional anonymization techniques like k-anonymity often degrade data utility, creating a bottleneck between compliance and model performance. These privacy-driven data scarcity challenges align with BT6 [10], which promotes the adaption and standardization of Privacy-Enhancing Technologies (PETs) and a compliance-by-design approach. Moreover, they align with BT3 [10], which promotes secure multi-party computation and federated learning platforms within sensitive and regulated data environments, such as the European Health Data Space to facilitate data sharing while preserving privacy and security.

During the pre-processing phase, automated pipelines address data quality challenges through ML solutions. Techniques like federated learning enable collaborative model training across decentralized datasets without raw data sharing, preserving privacy while improving generalization. For example, IBM's LAB (Large-scale Alignment for chatBots) framework uses synthetic data to train large language models (LLMs), reducing reliance on human annotations [17].

Although data augmentation techniques are used to tackle data scarcity and improve model performance, it does not come without challenges. In many cases, data augmentation may amplify biases. Therefore, transformations applied to datasets should strike a balance between enriching the data and maintaining its authenticity. Another challenge is computational efficiency, as augmenting large datasets requires additional processing power and time, which slows down training or would require more powerful hardware. Finally, most developments in augmentation techniques are in image domain datasets, with techniques for tabular, textual, and time series being limited [18]. Advancing bias-aware, frugal augmentation methods and embedding fairness checks into augmentation pipelines aligns with BT1's [10] call for unbiased and resource-efficient ADR techniques and to BT6's [10] requirement for integrated fairness check and correction tooling in all data workflows.

Synthetic data generation uses AI to create privacy-compliant, scalable datasets. In healthcare, synthetic patient records enable rare disease research without compromising confidentiality. For instance, Google worked with healthcare providers to create synthetic datasets that simulate patient medical records, allowing AI systems to be trained without violating patient privacy [19]. Despite its promise, synthetic data introduces novel challenges. In some cases, there are statistical mismatches between synthetic and real data that degrade model performance. Moreover, synthetic data generators trained on biased sources can systematize discrimination [20]. As synthetic data becomes AI's primary training fuel by 2030 [21], its responsible deployment will determine whether AI evolves as an equitable tool or an amplifier of existing disparities.

2.2 Autonomous Systems and Human-Centric Robotics

Robots are already widely utilized across various sectors in Europe, and the integration of AI could expand the scope of automation even further. This expansion could overcome traditional challenges, enhance productivity, and bolster Europe's competitiveness in both its internal market and on a global scale [22]. Advances in AI and machine learning have enhanced autonomous systems' ability to make context-sensitive decisions in real time and adapt via feedback across all operation functions. This is possible thanks to more capable, reliable, and flexible hardware but also to advanced software embedding new foundation models and technologies [23]. Therefore, the systems changed from rule-based robotics, based on individual programming, "if…then" instructions, and a low flexibility, to training-based robotics, which can learn skills via reinforcement learning (RL) in a trial-and-error approach, up to context-based, autonomous robotics, built on robotics foundation models (RFMs) and with a reduced effort for the programming [24]. This transition is in line with one of ADRA's key priorities, which highlights the need for groundbreaking technological foundations in autonomy, high performance, and predictability to ensure robust and scalable autonomous systems (BT1 ADRA SRIDA [10]). These advancements let robots navigate dynamic environments, optimize tasks, and adjust to unpredictable situations with increased efficiency and safety.

Another critical area of research is human-robot interaction (HRI), which relies on intuitive communication, transparency, and adaptability. The human-centric approach, in fact, is one of the three pillars of the Industry 5.0 at European level and refers to all technologies, including robotics and AI [25]. Today, many models of collaborative robots are available on the market. However, the performance achievable with collaborative robots is still limited. Advancements in natural language processing (NLP), computer vision, multimodal interaction, and explainable AI allow robots to better understand and respond to humans and to be more transparent, predictable, and user-friendly. These advancements support the development of effective and trustworthy general-purpose AI, as outlined in ADRA's agenda, which includes generative AI, continual learning, and explainability as key enablers for human-centric robotics (BT2 ADRA SRIDA [10]). Additionally, safety mechanisms are fundamental to fostering trust and acceptance in collaborative environments.

This is just the beginning: the field of autonomous systems and human-centric robotics is evolving rapidly, and greater advancements are expected for a broader range of applications. Some of the main key research priorities in this field include:

1. Rethinking current processes involving robotics, since Europe leads in robot production but must boost deployment to stay competitive and enhance efficiency, security, and sovereignty.
2. Leveraging on robotics potentialities to solve the problems of modern society, such as the demographic shift. In this case, robots can be used to maximize productivity despite the workforce shrinking and ageing that will be faced in the next decades.

3. Supporting the green transition to reduce the environmental impact and promote the adoption of circular economy.
4. Enhancing adaptability in dynamic environments through advanced algorithms that utilize sensor data and AI to predict operators' behavior and to reconfigure its tasks to adapt to changes in production to minimize downtime.
5. Promoting the adoption of methodologies to allow effective HRI and its analysis. Intuitive communication, through multi-modal systems and natural language algorithms, will simplify the operators' work.
6. Equipping the robot with technologies to handle complex tasks and the use of sensors, appropriate control techniques, and also cameras and AI can make the robot more efficient and effective.
7. Simplifying robot programming, introducing supporting software for the assisted programming, lowering the needed competences and costs, and promoting an easy robot reconfiguration.
8. Integrating robotics in the factory production line and systems, because a safe and efficient system presupposes integrated collaborative automation, providing an aggregate and holistic answer to support industrial decisional processes.
9. Aligning with the new regulatory framework, emphasizing the use of ethical AI in robotics and favoring a human-centered design. Such innovations reflect ADRA's strategic focus on next-generation smart embodied robotic systems, particularly in areas like soft robotics, configurability, and lifelong learning to enhance adaptability and efficiency in dynamic environments. In this regard, ADRA also emphasizes the importance of research and tools that go beyond regulatory compliance to ensure safety, trust, and privacy in autonomous systems (BT4 and BT6 ADRA SRIDA [10]).

These key research priorities are moving from theory to practice. While robots have been used in manufacturing for over 60 years, collaborative robots have only gained traction in the last decade, becoming more affordable and more accessible to small and medium-sized enterprises (SMEs) through simplified programming. Furthermore, AI integration has expanded their scope of use. For example, BMW is testing humanoid robots in assembly processes, reusing existing workstations originally designed for human operators [26]. Meanwhile, in the healthcare sector, robotic-assisted surgery has evolved significantly: today's systems are more sophisticated, enabling multiple reconfigurable robots to collaborate seamlessly with surgical teams [7].

These priorities echo the vision presented in the ADRA SRIDA [10], which calls for an integrated and forward-looking approach to building robust, trustworthy, and human-centric autonomous systems across Europe. According to this, it appears evident that the main objective in autonomous systems and robotics for Europe is to maintain the leadership, avoiding that they will be developed, commercialized, and sourced outside Europe.

2.3 *Explainable, Science-Guided, and Adaptive AI*

Given AI's influence in sectors such as healthcare and critical infrastructure, a paradigm shift toward models prioritizing transparency, scientific rigor, and adaptability has become essential. Moving beyond performance metrics alone, future AI research must focus on developing systems that can explain their reasoning, incorporate scientific domain knowledge, and adapt to changing environments while maintaining human oversight. This research direction is particularly crucial for Europe's goal of maintaining global research impact in AI while achieving strategic autonomy in trustworthy ADR technologies [10]. These advances address key challenges from ADRA SRIDA's Big Tickets, particularly BT2 (effective and trustworthy general-purpose ADR), BT5 (ADR technology for the sciences), and BT6 (research, innovation, and tools for compliance).

The immediate priority is establishing standardized frameworks for evaluating and implementing explainability across various AI applications. Current post hoc explanation techniques like Local Interpretable Model-agnostic Explanations (LIME) [27] and SHapley Additive exPlanations (SHAP) [28] provide valuable insights but face limitations with complex deep learning architectures. Research should focus on developing domain-specific XAI evaluation metrics that align with regulatory requirements and stakeholder needs, addressing BT2's challenge of creating explainable and robust AI while supporting the EU AI Act's transparency requirements. Healthcare applications demand different explainability standards than automated financial systems, supporting both BT2's evaluation and auditing requirements and BT6's compliance-by-design ADR systems that can handle evolving legal requirements [10]. The integration of semantic knowledge graphs with AI models shows promise in enhancing reasoning capabilities and providing natural language explanations that stakeholders can understand, contributing to BT2's trusted reasoning and statistical inference capabilities challenge.

In parallel, science-guided AI approaches that incorporate domain knowledge and physical constraints into model architectures represent a critical research direction [29]. Unlike purely data-driven approaches, science-guided models embed established scientific principles within their structure, improving generalizability and interpretability. Physics-informed neural networks that integrate differential equations into their loss functions exemplify this approach, demonstrating superior performance in modelling complex physical systems while requiring less training data, addressing BT5's challenge of building AI capabilities for scientific research through advances in simulation, emulation, and methods to embed domain knowledge. These approaches have shown promise in climate modelling, materials science, and biomedical applications where first principles guide model development, supporting BT5's goal of enabling causality and exploration of cause-effect relationships in scientific processes.

As AI becomes more explainable and science-guided, future research will focus on integrating these capabilities with adaptive learning frameworks. Modular AI architectures will separate reasoning from adaptation, ensuring transparency while enabling continuous learning, directly addressing BT2's continual/incremental learning challenge [10]. A key challenge will be handling concept drift, where data relationships shift over time, requiring uncertainty quantification and robust validation methods to trigger human oversight when necessary [30]. This connects to BT6's life-cycle management requirements for adaptive AI systems, particularly the challenge of incremental and evolutionary qualification where there is no strict separation between design and operational phases due to AI's dynamic nature [10]. Institutions will need governance structures for monitoring and auditing these evolving systems.

Science-guided AI will increasingly integrate multi-scale and multi-physics models, bridging microscopic and macroscopic phenomena. In fields like drug discovery and materials science, AI can connect molecular interactions to system-level effects, accelerating innovation while maintaining scientific validity. This supports ADRA's vision of AI-enabled research assistants capable of synthesizing literature and data while addressing BT5's goal of creating next-generation analytical tools to simulate complex systems [10]. These advances address BT5's cross-disciplinary collaboration challenge by developing AI methods capable of translating findings across domains and supporting collaboration of expert teams. Over the next decade, AI systems will evolve toward dynamic explainability, tailoring insights based on user expertise. These models will simulate counterfactual scenarios, enabling causal reasoning that goes beyond correlations to uncover true cause-and-effect relationships [31].

A major goal is to develop neuro-symbolic AI, blending neural networks' pattern recognition with symbolic reasoning's logical transparency [32]. This advancement supports BT2's vision of effective and trustworthy general-purpose ADR by combining symbolic and neural representations in ADRA's principled approach to combining reason and learning [10]. Future AI could even propose scientific theories and design experiments, working alongside human researchers in what BT5 envisions as AI-enabled laboratory assistants. However, progress depends on balancing model complexity with explainability, establishing standardized evaluation metrics, and fostering interdisciplinary collaboration between AI experts and domain scientists.

In healthcare, AI will create adaptive diagnostic systems that integrate medical knowledge and evolving disease data, offering clinically relevant explanations [2]. In climate science, adaptive AI will improve extreme weather predictions while clarifying uncertainty and causal factors. In manufacturing and energy, predictive maintenance systems will combine physics-based modelling with real-time learning, reducing failures and optimizing efficiency while incorporating energy-efficient designs that address sustainability concerns. The financial sector will benefit from AI that adapts to market shifts while maintaining explainable risk assessments. These applications demonstrate the convergence of challenges across BT2, BT5, and BT6, where systems must be simultaneously trustworthy, scientifically rigorous, and compliant with regulatory frameworks [10].

Future research in AI should focus on quantum-enhanced modelling, collaborative and brain-inspired computing, and privacy-preserving approaches like federated learning. Addressing data governance, sustainability, and energy efficiency is essential, aligning with BT6's priorities. Progress will depend on continued investment in foundational research, real-world application development, and strong governance to ensure transparency and trust. This integrated effort can position Europe as a leader in advancing AI for science and society.

2.4 AI for Edge Computing and IoT

Edge AI enables near real-time data processing at the network edge, reducing latency, bandwidth usage, and dependence on centralized cloud computing. This shift is particularly impactful in applications where low-latency decision-making is essential, such as autonomous systems, industrial automation, and real-time monitoring.

Across various sectors, IoT devices generate vast amounts of data from sensors, cameras, smart devices, and connected infrastructure. By leveraging AI-powered edge computing, organizations can enable predictive analytics, anomaly detection, and autonomous control directly at the data source. In healthcare, this means real-time patient monitoring and diagnostics [2]; in smart manufacturing, it enables predictive maintenance and process optimization [4]; in transportation, it supports traffic flow optimization and vehicle autonomy [6]; while in the energy sector, edge AI can assist in the decentralization of energy management, optimizing power consumption and integrating renewable energy sources while ensuring grid stability [33]. By reducing reliance on centralized processing, edge AI enhances efficiency, security, and responsiveness, driving innovation and operational improvements across several industries.

The challenges and opportunities of edge AI for IoT are closely aligned with several ADRA SRIDA Big Tickets [10]:

1. Advancing foundational technologies to enable efficient, autonomous, and real-time processing in resource-constrained environments (BT1)
2. Addressing interoperability, data governance, and secure model integration across diverse edge and IoT infrastructures (BT3)
3. Ensuring privacy, security, and regulatory compliance in decentralized AI deployments through robust tools and methods (BT6)

On a closer view, the deployment of AI at the edge is enhancing efficiency, resilience, and adaptability across various sectors, enabling real-time decision-making and predictive intelligence where it is most needed. Traditional centralized systems often struggle with dynamic, decentralized environments, such as smart grids [33], industrial automation [25], and autonomous transportation [6]. AI models deployed at the edge can process real-time data streams from IoT devices, sensors, and connected infrastructure, enabling predictive maintenance, fault detection, and adaptive resource allocation. For instance, AI-enhanced industrial monitoring systems can predict equipment failures and optimize workflows, reducing downtime and improving operational efficiency. In smart buildings, edge AI can autonomously regulate heating, ventilation, and air conditioning (HVAC) systems, adjusting to occupancy patterns, environmental conditions, and energy demand to enhance comfort and sustainability [33].

The use of federated learning techniques further enhances edge AI by enabling decentralized model training without exposing sensitive user data, addressing privacy, security, and data governance concerns across industries [34]. A practical implementation of AI-driven IoT in smart infrastructure is seen in automated transportation systems, where edge AI algorithms analyze real-time traffic data, weather conditions, and historical mobility patterns to optimize traffic signals, reduce congestion, and enhance road safety [35]. Additionally, in the Smart Energy domain, AI-based predictive analytics help forecast the day-ahead power market price, communicating price signals to decentralized smart heating devices [14].

Despite the significant advantages of AI at the edge, several challenges must be addressed to fully unlock its potential across industries. One of the primary concerns is computational constraints, as many edge devices have limited processing power, memory, and battery life. To overcome this, researchers are developing efficient AI model compression techniques and lightweight neural networks that reduce computational overhead while maintaining high performance. Another critical issue is data privacy and security. Since edge AI processes data locally, rather than relying on centralized cloud computing, it is crucial to implement robust encryption techniques, secure access controls, and decentralized authentication methods to protect sensitive consumer and industrial data.

Interoperability is another major challenge, as the lack of standardized communication protocols and AI frameworks makes it difficult to integrate AI-driven edge computing across diverse IoT devices and platforms. Standardization efforts are necessary to ensure seamless compatibility between different manufacturers and technologies, enabling scalable and flexible deployments [36]. Additionally, energy efficiency remains a key consideration for AI inference at the edge. Running AI models on battery-powered or resource-constrained devices requires optimized algorithms and hardware accelerators that minimize power consumption while maintaining real-time processing capabilities. This is particularly important for applications in remote or off-grid environments, where energy availability is limited.

Looking ahead, advancements in federated learning and reinforcement learning algorithms are expected to further enhance AI-driven edge computing for IoT applications. These innovations will enable more intelligent, autonomous, and resilient systems capable of dynamically adapting to changing conditions in real time. By addressing these challenges, AI at the edge can become a foundational technology for smarter, more secure, and more sustainable digital ecosystems across multiple sectors.

2.5 *AI-Driven Decision Support Systems*

AI-driven Decision Support Systems (DSS) leverage big data, ML, explainable AI (XAI), generative AI, and human-AI collaboration to provide accurate, timely, and interpretable recommendations, enhancing decision-making processes across healthcare, finance, Industry 4.0, and other societal and industrial areas. The future of AI-driven DSS lies in building trustworthy, adaptive, and scalable AI solutions that empower decision-makers across industries and by combining future advances in several technological domains in an interdisciplinary research endeavor, which brings together AI, cognitive science, and decision theory, on one side, and, on the other, ethical AI frameworks that ensure responsible transparent and trustworthy DSS deployments.

These advances can be divided into three major research and innovation pillars. First, human-centered AI models and algorithms for decision support will entail the development of fair, unbiased, interpretable white-box AI models that are explainable and trustworthy [37]. This will lead to decision-making processes that may truly depend on AI-driven recommendations and transparent explanations about them. Although some current large language models present reasoning capabilities, the evolution of neuro-symbolic AI to enhance reasoning and inference [32] will not only enable better and more autonomous recommendations but also further means of explainability and transparency for future models used in DSS. Further, future research on adaptive AI, such as traditional reinforcement learning, evolutionary ML and DL [38], transfer learning and meta-learning models [39], and AI-powered optimization techniques for multi-criteria decision-making [40] will allow future DSS to adapt and make increasingly adequate recommendations toward bigger and more broad contexts. Finally, the design of human-in-the-loop systems for shared decision-making, considering cognitive load [41], and trust calibration [42] in AI-assisted decisions will provide the adequate amount and quality of information within each recommendation to effectively support decision-making processes. The pillar's specific challenges are closely related to several ADRA SRIDA's BTs [10]:

1. Reflecting deep foundational research aiming to improve reasoning, learning, and decision-making capabilities in AI systems (BT1)
2. Emphasizing the goal of turning complex data into actionable, trustworthy recommendations (BT2)
3. Supporting the development of systems that foster user confidence and ethical use (BT5)

Second, novel data and knowledge management processes, such as multi-source, real-time data integration and fusion techniques or the development of semantic knowledge graphs that effectively integrate and contribute to AI model reasoning and inference [43], will work as the information backbone of AI-driven DSS, since information is key to, on one hand, provide increasingly accurate recommendations and, on the other, achieve true transparency and explainability in AI-driven systems. Synthetic data will play a fundamental role in future AI-driven DSS to overcome scarcity of data while preserving data privacy, while the integration of novel federated learning techniques will allow for privacy-preserving decision support across organizations and entities [44]. This pillar is closely related to the specific objectives of ADRA SRIDA's BT2, by supporting the development of generalizable, reasoning-capable, and transparent AI systems using semantic knowledge graphs, synthetic data, and federated learning, and BT3, by addressing real-time data integration, multi-source fusion, and privacy-preserving methods like federated learning.

Finally, real-world deployments and scalability improvements of AI-driven DSS are essential for future research and innovation undertakings. These real-world scalable deployments should be based on both cloud-native and edge AI architectures [8] and robust AI validation and benchmarking methodologies for domain-specific DSS. Furthermore, these deployments should also account for the integration of AI-driven DSS with IoT infrastructures and digital twins for real-time monitoring and decision-making support [45]. These real-world deployments will lead, in a medium to long term, to the development of fully autonomous, self-learning DSS [46], with seamless integration with quantum-based AI for increasingly complex decision modelling [47], supporting AI-driven, multi-agent decision ecosystems that autonomously optimize their own policies and strategies. These systems will unlock unprecedented levels of efficiency, precision, and adaptability, reshaping future industrial and societal decision-making processes. This pillar is closely connected to BT1, since its vision of self-learning, scalable, and performance-optimized decision-support systems falls directly under the scope of foundational technological breakthroughs; BT2, as it aims at general-purpose, complex, and trustworthy AI systems that can operate adaptively across diverse domains; and toward BT3's focus on operationalization, governance, and technical integration, which is echoed in this need for real-world, scalable deployment and coordinated decision ecosystems.

3 Ethical and Societal Considerations

3.1 Bias, Fairness, and Accountability

While AI offers transformative potential, its capacity requires rigorous frameworks for bias mitigation, fairness assurance, and accountability enforcement. Bias in AI systems originates from three primary sources: data bias, algorithmic bias, and human decision bias [48]. The challenge of bias in these three forms are closely related to several ADRA SRIDA's BTs [10]:

1. Developing models that minimize social biases by design (BT1)
2. Avoiding bias toward the underprivileged that may be less represented in datasets (BT2)
3. Ensuring transparency of training data for both legal reasons as well as trustworthiness (BT6).

Data bias occurs when training datasets are not representative of the population, leading to skewed outcomes. For instance, facial recognition systems have been shown to misidentify individuals from minority groups due to underrepresentation in training data. Such biases are intensified by selection bias (e.g., excluding rural populations in medical datasets) and measurement bias (e.g., using zip codes as proxies for socioeconomic status). Model architectures and optimization objectives may inadvertently prioritize majority groups, thus producing algorithmic bias. For instance, when data imbalanced exists in the nature of a problem, some algorithms often optimize for majority-class accuracy at the expense of minority groups. Human judgments embedded in labelling processes or model design choices introduce subjectivity and human decision bias. The Correctional Offender Management Profiling for Alternative Sanctions recidivism algorithm controversy exemplifies this: human-generated risk assessments encoded systemic biases against black defendants, which were then codified into the algorithmic system [48].

The challenge in ensuring fairness in decision-making comes from the fact that there is no universal definition of fairness. Specifically, BT2 [10] mentions that contradictory definitions of biases exist, making the relevant metrics used depending on the goal. To tackle biases, literature mainly splits fairness in the following three groups:

1. Individual fairness: Give similar predictions to similar individuals [49]. For example, loan applicants with identical credit histories should receive equivalent approval probabilities.
2. Group fairness: Treat different demographic groups equally. For this, metrics like demographic parity (equal approval rates) or equalized odds (matching false positive/negative rates) can be used.
3. Procedural fairness: Focuses on equitable decision-making processes rather than outcomes. This includes transparency in model logic and contestability mechanisms allowing users to challenge decisions.

Several mitigation strategies have been proposed to optimize fairness in AI. Diverse data collection can help ensure that training datasets are diverse and representative. When possible, collecting more diverse data can mitigate any biases from the trained models. Increasing participation from marginalized communities during data collection can help mitigate existing biases [48]. When diverse data collection is not possible, creating synthetic data using GANs can create balanced datasets. Moreover, techniques during model training have been proposed. For example, adversarial debiasing trains models to simultaneously optimize accuracy and minimize bias detection [50]. Moreover, fairness metrics can be directly integrated into loss functions [51].

To obtain quantitative measures of fairness, several tools exist to guide researchers in constructing unbiased models. For example, the open-source AI Fairness 360 by IBM provides fairness metrics, as well as algorithms, to mitigate bias [52]. Furthermore, for enabling explainability and therefore understanding "black box" models, SHAP values can be utilized for decision rationales, though limitations persist in explaining DL systems. Utilizing such tools is aligned with BT6, emphasizing on developing tools for compliance and auditability. Moreover, the output of these tools scientifically support the explainability and interpretability of AI systems, supporting BT5 [10]. Fairness is closely linked to the AI accountability, which demands to put in place mechanisms to ensure responsibility for AI systems and their outcomes, both before and after their development, deployment, and use. Accountability, which is acknowledged as one of the seven principles the European Union seeks to be ascribed to ethically founded AI, goes hand in hand with risk management, especially regarding risk to safety and ethical principles, including fundamental rights. Accountability rotates around auditability, which refers to the enablement of the internal or external assessment of algorithms, data, and design processes. Reports or records on actions or decisions contributing to a certain system outcome, together with the capability to respond to the consequences of such an outcome, are essential to guaranteeing accountable and trustworthy AI systems. Algorithmic audits and Impact Assessment, involving the scrutinizing of the inputs, processes, and outputs of AI systems to identify and mitigate biases, are tools to detect discrimination and other biases in AI systems. Thereby, accountability pertains to the idea that AI systems should be developed, deployed, and used such that responsibility for bad or harmful outcomes can be assigned to liable parties. Therefore, an underlying building block of accountability is liability. In case something goes wrong with an AI system, and it produces harm or damage, there should be someone responsible.

The AI Act refers to the "provider accountability," meaning that the individuals or organizations developing, deploying, or operating AI systems are held responsible for their actions. In line with its risk-based approach, the AI Act sets forth the conformity assessment procedure for the high-risk AI systems, meant as the process of verifying and/or demonstrating that a high-risk AI system complies with the requirements provided under Title III, Chapter 2 of the Act, which refer to risk management system, data governance, technical documentation, record-keeping, transparency and provision of information, human oversight, accuracy, robustness, and cybersecurity [53].

3.2   Robustness in AI, Data, and Robotics

Robustness, when referring to AI or robotics systems, or to their convergence, lies on ensuring that systems operate as expected, maintaining their stability and performance, under a wide range of real-world conditions and disturbances, thus minimizing the risk of failure and improving their overall reliability and safety. Although robustness is one of the key trust dimensions for AI and robotics, there is still lack in relevant research, while insufficient measures toward ensuring robustness at production may lead to potentially harmful situations and hurt the trust of humans toward AI. The way toward creating a robust environment of trust is through the implementation of a multifaceted strategy that has as key elements:

1. Explainability and transparency
2. Engagement and collaboration through open-source initiatives
3. Use of synthetic data for privacy and bias mitigation
4. The integration of robust validation methodologies and tools for reliability

Trust goes hand in hand with understanding. Explainability and interpretability are seen as facilitators of trust and robustness, as on the one hand they enable users to understand the reasoning of decisions while allowing them to validate the results and identify pitfalls. Common approaches like LIME and SHAP have gone from desk to practice, with their widespread use to unravel the mechanisms behind the decisions of complex models, and the roadmaps for future research XAI directions have already been identified [54]. Future areas for XAI include the advancement of current XAI methods for application on deep learning models; the contextualization of explainability to specific contexts, circumstances, or even user preferences; and the integration of human feedback, for example, through interactive XAI interfaces. Some additional research directions include bias-free and privacy-preserving XAI, the development of domain-specific XAI-evaluation metrics, the combination of explanatory approaches, and more. These directions are closely knit to ADRA SRIDA's [10] BT1 that involves the improvement of XAI interfaces for trustworthiness by design and fostering human-AI collaboration and BT2 by aiming to address the explainability and privacy trade-off through privacy-preserving XAI techniques.

Code openness and sharing are vital in ensuring security and robustness through collaboration. Numbers illustrate the current big trend toward democratizing AI through open-source initiatives: Hugging Face hosting more than 50,000 free access AI models, over a million downloads of the open-source ML Library TensorFlow, over 100,000 GitHub repositories related to open-source AI, and over 100 models are spawning daily. Future roadmaps for open-source AI development include the integration of distributed learning and decentralized deployment for enhanced privacy-preservation and security and democratization of access to AI even for resource-constrained organizations like SMEs through the development of less resource-demanding models. This profound movement, however, needs to address various challenges to ensure high-quality, smooth collaboration and ethical use, which will most likely be addressed through the emergence of stronger security protocols and regulations and the establishment of unified standards. AI Model Marketplaces should also adapt and evolve in the upcoming period [55] to meet security, model evolution lineage tracking, and other challenges. The discussed directions toward code openness and decentralization are aligned with various aspects of ADRA SRIDA's [10] BTs, including BT1, toward the development of trustworthy and generalizable AI systems; BT2, fostering interoperability and integration through decentralization; and BT3, ensuring adherence to ethical and legal aspects with the appropriate approaches that promote responsible AI and privacy preservation.

Synthetic data in AI and robotics systems are seen as the answer to the problems of data scarcity and overcoming privacy concerns, especially for data-hungry model training operations. One of the emerging trends in synthetic data generation for AI is real-time generation with the purpose of feeding continuous learning systems that adapt dynamically to changes. With regard to the quality of generated data per se, and with the use of generative AI for synthetic data generation tasks becoming more and more popular, the improvement of the fidelity and controllability of generative systems and the development of standardized evaluation and contamination protocols and tools are within identified research axes [56]. Advancements in the quality, fidelity, and contamination analysis of synthetic data generation systems act as direct or indirect enablers of various ADRA SRIDA's [10] BTs. Specifically, they improve reliability and quality of ADR systems utilizing synthetic data (BT1), ensure fitness of synthetic datasets for interoperability and reuse (BT2), and support the validation of the absence of bias, privacy (BT3) and security (BT4) issues in the generated data.

Finally, the automation of AI and robotics system validation and monitoring will ensure robustness and further strengthen trust. From model validation during and after training, up to model and system monitoring during production, issues like accuracy, fairness, data drift, model decay, and compliance are in the microscope and are identified before it is too late. Although existing model validation frameworks and tools already showcase continuous monitoring functionalities, a future advancement in the area will include turning observations into actions, through enhanced integration with automated retraining systems for model updates based on observed performance. Another emerging trend in AI validation and automation is XAI monitoring; an approach that goes beyond traditional model performance monitoring and focuses also on the interpretability and explainability aspects of the systems in scope, thus ensuring that the production systems meet high explainability standards. Streamlined and automated AI/XAI and robotics system validation aligns with ADRA SRIDA's [10] BT1 (ensuring explainability, performance, fairness, and trust), BT2 (enabling the delivery of reusable and robust components), BT3 (automating regulatory and compliance checks), and BT4 (detecting vulnerabilities and validating explanation security).

3.3 Ensuring Societal Impact, Inclusivity, Safety, and Liability

As AI, data, and robotics become more pervasive and ubiquitous, shaping critical aspects of society and industry, ranging from healthcare to finance and manufacturing, it is increasingly essential to ensure that future innovations in these domains, their individual applications, or potential synergies between them and other fields reach and serve all individuals equitably, enabling safe operation at all times, and provide clear liability frameworks. Although all these aspects are related to already-discussed problematics, such as bias, fairness, accountability, or robustness, it is important to emphasize the need to address inclusivity, safety, and legal responsibility, both today and in the future, to foster public trust, ethical deployment, and long-term sustainability of future innovations in these domains. In fact, addressing these three pillars—inclusivity, safety, and liability—under the umbrella of AI, data, and robotics, will bring both opportunities and challenges.

Regarding inclusivity, these domains may increase the risks of widening economic inequalities and reinforcing systemic biases. Workforce displacement may be one of the biggest challenges of the next decades due to application of AI-, data- and robotics-related innovations [57], while it can also create new opportunities in terms of new AI-driven jobs and reskilling and upskilling programs and investments [57] to ensure equitable workforce transitions. Bias in AI-driven decision-making is already a big challenge nowadays but can be another increasingly problematic challenge in the future, as AI systems could perpetuate biases in hiring or law enforcement applications. The need for diverse, unbiased datasets that are subject to regular audits is essential to prevent discrimination [58]. Further, accessibility to these life-changing technologies must be equal and universal, especially for disabled and marginalized communities. In fact, these technologies can provide greater direct societal impacts to people that need greater social support if they are designed with universal accessibility in mind, supporting multiple languages, cultural contexts, and different disabilities [59].

On the other hand, and due to the potential future pervasiveness of autonomous AI- and robotics-based systems, particularly in critical infrastructure, public services, and everyday lives, ensuring safety is a top societal challenge. For instance, autonomous systems in high-risk environments, such as in AI-driven healthcare, autonomous vehicles, and industrial robots, will require rigorous safety validation and fail-safe mechanisms to prevent accidents [60]. In cases in which AI and robotics play greater roles in education healthcare and elderly care, for instance, it is crucial to ensure that these systems have greater levels of emotional intelligence, ethical behaviors, and user safety toward a more ethical human-AI-robotics interaction. Moreover, cybersecurity-related threats in AI and robotics technologies and the rise of autonomous weapons and predictive policing raise future ethical and safety concerns about misuse and accountability. Future safety frameworks are key to regulate security and military applications and to enforce adequate encryption, adversarial defenses, and threat detection processes.

Lastly, as AI and robotics take on more autonomous functions, legal and ethical dilemmas concerning accountability and liability become more pressing. Future liability concerns will shape how society trusts, adopts, and regulates AI-driven decisions. One pressing matter is the view of AI as a legal entity [61], and future debates on whether AI can be held accountable for decisions, especially in cases of errors or harm. The perspective of AI as a legal entity will have further implications, such as the introduction of global AI laws and regulatory bodies or specialized insurance models for AI risks and liabilities. Nevertheless, the focus should be in proactive measures toward the transparency and explainability of the decisions taken by AI and automated systems, ensuring algorithmic accountability and that such decisions will be auditable, interpretable, and challengeable.

3.4 AI and Sustainability

AI is increasingly recognized as a key enabler of sustainability, driving efficiency, resilience, and responsible resource management across multiple industries. By leveraging AI-driven predictive analytics, automation, and optimization, organizations can minimize their environmental impact while improving operational efficiency and long-term sustainability. AI's role in sustainability is twofold: it supports the transition to more resource-efficient systems and enhances responsible consumption through intelligent, data-driven decision-making [62].

Across various sectors, AI-powered sustainability solutions are transforming operations by optimizing resource use, reducing waste, and enabling circular economy practices. In manufacturing, AI enhances process efficiency, predictive maintenance, and material recycling, reducing industrial waste [63]. In transportation, AI-driven traffic management and route optimization help lower fuel consumption and emissions, while autonomous and electric vehicle technologies contribute to a greener mobility future [6]. Finally, in the energy sector, AI can facilitate smart grid management, demand-response optimization, and renewable energy forecasting, among others, ensuring the efficient integration of distributed energy resources and AI-driven sustainability solutions are gradually expanding to industrial operations, where machine learning algorithms optimize energy use, minimize waste, and enhance circular economy practices [64].

AI and sustainability goals align closely with several ADRA SRIDA Big Tickets:

1. BT1, by emphasizing the need for energy-efficient AI technologies and optimized algorithms to reduce the environmental footprint of large-scale AI systems
2. BT3, through the call for interoperability, standardization, and integration of AI into existing sustainability frameworks across sectors
3. BT6, by addressing regulatory, ethical, and compliance challenges related to sustainable AI deployment

Together, these tickets support the development of responsible, efficient, and scalable AI systems that drive the transition toward greener, more resilient industries and societies. As AI continues to evolve, its applications in sustainability will play a crucial role in building more resilient, resource-efficient, and environmentally conscious industries and societies. The transition toward more sustainable and efficient AI-driven and autonomous systems requires intelligent control mechanisms to manage fluctuations in resource availability and demand. AI-driven solutions can optimize supply and demand balancing by forecasting consumption patterns, resource availability, and operational needs, and ML algorithms can analyze historical and real-time data, ensuring optimal efficiency and minimal reliance on unsustainable practices [65].

Despite its transformative potential, AI-driven sustainability solutions present several challenges that must be addressed to ensure their effectiveness and long-term viability. One key concern is the energy consumption of AI models, particularly the training of large-scale machine learning algorithms, which requires significant computational power and contributes to a growing carbon footprint. To mitigate this, advancements in energy-efficient AI algorithms, carbon-aware computing, and hardware optimization are essential to balance AI's sustainability benefits with its own resource demands [66]. Another challenge is interoperability and standardization. The integration of AI into existing sustainability frameworks requires standardized protocols that allow seamless communication between different systems, industries, and stakeholders. Without a unified approach, AI-driven sustainability efforts risk fragmentation and inefficiency, limiting their ability to scale across sectors such as transportation, urban infrastructure, and resource management [67]. Additionally, regulatory and ethical compliance remains a critical factor. AI solutions must align with sustainability regulations, ethical guidelines, and transparency requirements, ensuring equitable access to AI-driven environmental solutions.

AI is expected to play an increasingly pivotal role in climate change mitigation, renewable energy expansion, and sustainable resource management. Emerging AI techniques, such as reinforcement learning and federated learning, offer enhanced efficiency in managing distributed networks and optimizing large-scale sustainability initiatives. However, widespread adoption, ethical governance, and accessibility must remain priorities. Collaboration between policymakers, researchers, and industrial stakeholders will be essential to ensuring AI's potential is fully realized in a way that supports economic resilience, environmental responsibility, and long-term sustainability goals [62].

3.5 *AI Literacy and Public Perception*

While AI, data, and robotics become increasingly integrated into daily life, the need for AI literacy grows. AI literacy refers to the public's ability to understand, critically evaluate, and interact with AI-driven technologies in an informed and responsible way. Without widespread AI education, public mistrust, misinformation, and ethical concerns could hinder innovation and adoption. Furthermore, reskilling and upskilling processes are necessary to empower future workers in the use of AI, robotics, and autonomous systems [57].

The European AI Act, entered into force on August 1, 2024, introduces a regulatory requirement for AI literacy, set out in Article 4 of the AI Act. This requirement for AI literacy is among the first provisions of the AI Act, which came into effect as of February 2, 2024 [53]. Article 4 states that providers and deployers of AI systems shall take measures to ensure, to their best extent, a sufficient level of AI literacy of their staff and other persons dealing with the operation and use of AI systems on their behalf, considering their technical knowledge, experience, education, and training and the context the AI systems are to be used in, and considering the persons or groups of persons on whom the AI systems are to be used.

Recital 20 clarifies that it is paramount that the AI literacy should equip providers, deployers, and affected people with the necessary notions to make informed decisions regarding AI systems, to grab the greatest advantages from AI systems while safeguarding fundamental rights. Furthermore, it provides insight regarding the need to consider the relevant context, which might have an impact on such necessary notions and might entails the understanding of the correct application of technical elements during the AI system's development phase, as well as the measures to be applied during its use and the knowledge necessary to understand how AI-driven decisions might have an impact on the affected people. Pursuant to the AI Act, AI literacy should provide all relevant AI value-chain actors with the insights required to ensure the appropriate compliance and its correct enforcement.

Since the wide implementation of AI literacy measures can pave the way to sustain the consolidation and innovation path of trustworthy AI in the Union, the European AI Board is expected to promote AI literacy tools, public awareness, and understanding of the benefits, risks, safeguards, rights, and obligations in relation to the use of AI systems. Voluntary codes of conduct to advance AI literacy among people dealing with the development, operation, and use of AI should be developed. In the same direction, EU AI Office, considering that several organizations have anticipated and prepared themselves to the entry into application of the AI Act and with the aim of encouraging learning and exchange among providers and deployers of AI systems, gathered some current practices among the pledgers of the AI Pact with the purpose of creating a living repository to provide examples on ongoing AI literacy practices [68]. Such a list of practices is non-exhaustive and will be updated with further practices on a regular basis.

These challenges have deep implications for future public perception on these technologies. For instance, public perception of AI is shaped by media narratives, real-world experiences, and trust in institutions, while for robotics, cinema productions have been prolific in showing the potential hazards brought by future robots and autonomous systems. Negative portrayals—such as fears of job displacement, loss of human control, bias—can lead to public resistance and stricter regulations, slowing innovation. Conversely, overestimating technologies' capabilities could create unrealistic expectations, leading to misuse and disappointment.

4 Collaboration and Strategic Investments

4.1 Synergies and Public-Private Partnerships Toward AI, Big Data, and Robotics Competitiveness

AI, big data, and robotics are deeply interconnected, working together to drive technological advancements across industries. AI relies on vast amounts of data for pattern recognition and decision-making, while robotics applies AI in real-world scenarios, enhancing automation and adaptability. As AI models become more sophisticated and computing power expands, the seamless integration of these technologies continues to unlock new opportunities for efficiency, safety, and innovation. Despite their potential, many industries still rely on conventional methods, and developing intelligent, reliable robots for unpredictable environments remains a challenge. In Europe, AI-powered robotics could revolutionize critical fields, particularly where human intervention is limited. However, achieving this requires a regulatory framework that not only promotes innovation but also accelerates adoption through targeted financial and legislative support. The 2021 revision of the Coordinated Plan on AI emphasizes the importance of collaboration among the European Commission, member states, and private stakeholders to strengthen Europe's leadership in trustworthy AI. Similarly, the EU AI Act introduces regulations that ensure AI and robotics development align with ethical standards, fostering responsible deployment.

While regulation is necessary, it must be balanced with investment to avoid stifling innovation. Europe's strong focus on ethical AI sets it apart as a leader in responsible technology, but rigid frameworks may drive companies to regions with more flexible policies. Additionally, Europe faces significant competition from the United States and China, where private investments far exceed public funding in the EU. The long-term nature of robotics development, often requiring 15–20 years from research to market deployment, further complicates the landscape, making sustained investment and strategic planning essential for maintaining competitiveness. To address these challenges, the EU should prioritize private investment, fund key innovations, and streamline regulations to reduce barriers in AI and robotics development. A holistic approach is needed—one that includes long-term research funding, workforce training, clear policies, and strategies for technology adoption [69]. Encouraging interdisciplinary collaboration will also be crucial in maximizing the potential of these technologies and fostering innovation at regional, national, and European levels.

At the regional level, initiatives that connect businesses, research institutions, and technology developers can help integrate AI-driven robotics into industries, raising awareness of its potential and accelerating adoption. Public-private partnerships between universities and companies are particularly valuable, as they enable the translation of research into practical applications, strengthening regional industries and contributing to Europe's broader technological landscape [70]. On a national scale, interdisciplinary cooperation is essential for scaling up technological advancements and aligning them with strategic industrial objectives. Robotics and AI can enhance efficiency, improve production processes, and ensure that European manufacturers remain globally competitive. Well-designed national policies will be key to achieving these goals [71].

At the European level, collaboration across borders is crucial for advancing AI and robotics. Initiatives like EUROBIN [72], funded by the EU, demonstrate the power of cooperative research and development, bringing together experts from various fields. The European Robotics Forum (ERF) [73] serves as another important platform, allowing researchers, entrepreneurs, and policymakers to exchange ideas and explore opportunities for technological advancement. A vital force in Europe's robotics strategy is euRobotics [69], which has played a key role in shaping AI and robotics development. During the Horizon 2020 program, it partnered in the SPARC Public-Private Partnership, the world's largest civilian robotics research initiative. More recently, euRobotics has helped establish ADRA, which is instrumental in driving AI and robotics innovation, ensuring that Europe remains globally competitive while maintaining a strong commitment to responsible technology development. In this context, ADRA SRIDA's [10] BTs for 2025–2027 provide a clear roadmap for advancing Europe's leadership in AI, data, and robotics through foundational technologies, trustworthy systems, interoperable ecosystems, next-generation robotics, scientific applications, and compliance-focused innovation. Through these initiatives, Europe is actively strengthening its robotics ecosystem, ensuring that innovation continues to thrive. By fostering collaboration, balancing regulation with investment, and supporting interdisciplinary research, Europe can maintain its leadership in AI-powered robotics while addressing both industrial and societal challenges in a strategic and responsible manner.

4.2 Policy, Regulation, and Funding Strategies

Policy regulation and funding strategies are pivotal in steering the ethical development and deployment of AI, data, and robotics. A comprehensive approach involves establishing robust policy frameworks, implementing effective funding mechanisms, and addressing critical regulatory considerations to align technological advancements with societal values and ethical standards.

Globally, nations are formulating AI strategies that emphasize ethical considerations. For instance, China's ethical guidelines for AI stress maintaining human oversight and accountability in AI applications. Similarly, the UK's National AI Strategy underscores the significance of aligning AI development with public trust and ethical norms [74]. These frameworks aim to balance innovation with ethical responsibility, ensuring that AI technologies serve the public good without infringing on individual rights. Strategic funding is essential to promote research and development in ethical AI. National and international funding bodies, such as the European Union's Horizon Europe program, allocate substantial resources to AI research, emphasizing projects that incorporate ethical considerations and societal impact assessments. Additionally, public-private partnerships emerge as effective models to fund initiatives that prioritize ethical AI development, fostering collaboration between governments, academia, and industry.

Effective regulation is crucial to mitigate risks associated with AI and robotics. The European Union's AI Act proposes a risk-based approach, categorizing AI applications based on their potential harm and imposing stricter requirements on high-risk systems [75]. This includes mandates for transparency, accountability, and human oversight. Moreover, the Act introduces conformity assessments to ensure compliance with ethical standards before AI systems enter the market [76]. In the United States, agencies like the National Institute of Standards and Technology (NIST) are developing frameworks to manage AI risks, promoting standards that uphold safety and trustworthiness in AI applications [77].

In conclusion, the convergence of well-defined policy frameworks, targeted funding strategies, and comprehensive regulatory measures is essential to foster the ethical development and deployment of AI, data, and robotics. By prioritizing ethical considerations alongside technological innovation, societies can harness the benefits of these technologies while safeguarding fundamental values and rights.

4.3 Skills and Literacy Development and Workforce Upskilling

The integration of AI in education is essential to equip the workforce with the skills needed for the rapidly evolving digital economy. AI is increasingly shaping industries, automating routine tasks, and redefining job roles. However, a significant skills gap exists, with only 15% of workers receiving AI-related training, while 42% recognize the need to upskill [78]. Due to the given potential, it is essential to distinguish a twofold perspective on a rapidly evolving AI landscape and a need for targeting both (1) future workforce (e.g., students, vocational learners, even pupils) and (2) current workforce (e.g., workers, technicians, operators). This approach combines early AI education and lifelong learning opportunities by upskilling and reskilling, ensuring AI literacy across different career stages [79].

In the case of (1), introducing AI concepts at an early educational stage is pivotal in preparing future generations for an AI-infused world. Initiatives like Estonia's national program to teach AI skills to high school students, in collaboration with tech companies, aim to cultivate critical thinking and AI awareness among youth [80]. Additionally, research indicates that AI literacy is essential for children's development, enabling them to interact effectively with AI technologies and understand their implications [81]. Furthermore, studies have shown that early exposure to AI education can enhance students' understanding of complex concepts, thereby reducing future skills gaps. These efforts highlight the necessity of embedding AI education within early learning curricula to equip students with the skills required for future job markets. Further, AI education must be embedded across all levels and pathways of learning, including vocational training [82]. Vocational learners need practical AI applications in fields such as manufacturing, logistics, and healthcare, integrating AI-powered simulations, robotics, and predictive maintenance tools into their training. Dual training systems should incorporate AI-related apprenticeships, where students apply AI tools in real-world job settings. Additionally, partnerships between educational institutions and industries must be strengthened, offering AI-driven internships and on-the-job training. Finally, AI literacy must also include awareness of ethical AI use, data security, and bias prevention, ensuring responsible adoption of AI technologies in professional environments.

On the other hand, for (2), AI upskilling is critical to ensure job security and career advancement as industries transition to AI-driven processes. Operators, workers, and technicians must be trained in AI-enhanced tools, such as predictive maintenance, process automation, and real-time data analytics. AI literacy programs should be tailored to industry-specific needs, equipping employees with practical knowledge of AI-integrated machinery, quality control algorithms, and robotics-assisted production. Modular, flexible AI training, including online courses, workplace learning, and hands-on workshops, will help workers transition into AI-augmented roles, e.g., by Google [83]. Since new skills are emerging slower in the current workforce, incentives should be established to meet the demand in future AI technologies. Companies should establish AI mentorship programs, ensuring workers receive continuous support while adapting to AI applications. Additionally, government and industry-funded reskilling initiatives should focus on workers at risk of job displacement, offering alternative AI-driven career paths. Promoting lifelong AI education through corporate learning platforms will ensure that workers stay competitive and contribute effectively to an AI-driven economy.

AI literacy and workforce upskilling are crucial for adapting to AI-driven job transformations. First, AI literacy enables workers to understand automation, data analytics, and machine learning, fostering collaboration with AI rather than displacement. Second, upskilling programs provide practical AI skills, problem-solving abilities, and adaptability, ensuring job market resilience. Continuous education further enhances critical thinking, ethical AI awareness, and human-AI interaction, driving productivity, innovation, and sustainable employment in the digital economy.

5 Conclusion and Future Directions

This chapter has outlined a structured research and innovation roadmap for AI, data, and robotics, emphasizing the interconnected nature of these technologies and their potential to drive transformative change across industries and society while mapping specific challenges and objectives to the ADRA SRIDA [10]. The synergy between AI-driven analytics, data infrastructures, and robotic automation is creating new opportunities for autonomy, efficiency, and decision-making in complex environments. By addressing key research priorities—ranging from data processing, autonomous systems, and explainable AI to edge computing and AI-driven decision support—this chapter expects to lay the groundwork for more adaptive, efficient, and trustworthy intelligent systems. At the same time, ethical and societal considerations such as bias mitigation, transparency, inclusivity, and sustainability must remain at the forefront to ensure that a holistic approach that advances research, innovation, and responsible deployment across all three domains is achieved.

Looking ahead, research and innovation must focus on several crucial directions. First, advancing AI's ability to generalize and adapt across different environments will be key. Future systems need to become more resilient, self-learning, and context aware, allowing them to operate effectively in dynamic real-world conditions. Alongside this, hybrid AI architectures that integrate neurosymbolic AI, causal reasoning, and quantum computing will push the boundaries of decision-making capabilities while maintaining interpretability and trust. Second, high-quality, interoperable, and secure data infrastructures are crucial for developing reliable AI models, predictive analytics, and autonomous decision-making systems. Future research must focus on improving data governance, privacy-preserving techniques, synthetic data generation, and real-time data processing to ensure that AI and robotics operate with trustworthy and actionable insights. Finally, the next generation of robotic systems must be more adaptable, autonomous, and capable of interacting seamlessly with human operators and dynamic environments. Research priorities should include advancements in cognitive robotics and human-robot collaboration, allowing robots to function in unstructured and unpredictable conditions, and integrating real-time AI processing, sensor fusion, and edge computing will enable robots to perceive, reason, and act with greater precision and autonomy.

While the technological possibilities are immense, a strong regulatory and innovation framework is essential to accelerate deployment. The European Union has taken significant steps in shaping trustworthy AI, data sovereignty, and robotics innovation, but further regulatory clarity, targeted funding, and interdisciplinary cooperation are needed to maintain global competitiveness. The challenge lies in striking a balance between stringent ethical AI principles and fostering an innovative, friendly environment, ensuring that companies and research institutions can thrive without excessive bureaucratic hurdles. Bridging the gap between research, policy, and industry adoption is crucial for ensuring that AI, data, and robotics technologies are effectively integrated into real-world applications. Strengthening public-private partnerships will play a pivotal role, enabling collaboration between governments, academia, and industry leaders to accelerate technological progress and maintain global competitiveness. Moreover, long-term investments in education, skills development, and interdisciplinary cooperation will ensure that Europe has the necessary talent pipeline to support future AI and robotics innovation.

By following this strategic roadmap, AI, data, and robotics can transition from being disruptive technologies to becoming integral enablers of economic resilience, societal well-being, and industrial competitiveness. The challenge ahead is not just about advancing these technologies but about ensuring they are trusted, inclusive, and aligned with Europe's long-term innovation vision.

Acknowledgments The authors acknowledge the contribution of the European Commission-funded Horizon Europe research project AI-DAPT (Grant agreement ID: 101135826) for the development of the research and innovation agendas and roadmaps proposed in this chapter.

References

1. Soori, M., Arezoo, B., & Dastres, R. (2023). Artificial intelligence, machine learning and deep learning in advanced robotics, a review. *Cognitive Robotics, 3*, 54–70.
2. Aamir, A., Iqbal, A., Jawed, F., Ashfaque, F., Hafsa, H., Anas, Z., Oduoye, M. O., Basit, A., Ahmed, S., Rauf, S. A., Khan, M., & Mansoor, T. (2024). Exploring the current and prospective role of artificial intelligence in disease diagnosis. *Annals of Medicine and Surgery, 86*(2), 943–949.
3. Luo, B., Zhang, Z., Wang, Q., Ke, A., Lu, S., & He, B. (2024). AI-powered fraud detection in decentralized finance: A project life cycle perspective. *ACM Computing Surveys, 57*(4), 1–38.

4. Johanesa, T. V. A., Equeter, L., & Mahmoudi, S. A. (2024). Survey on AI applications for product quality control and predictive maintenance in Industry 4.0. *Electronics, 13*(5), 976.

5. Keshvarparast, A., Battini, D., Battaia, O., & Pirayesh, A. (2024). Collaborative robots in manufacturing and assembly systems: Literature review and future research agenda. *Journal of Intelligent Manufacturing, 35*, 2065–2118.

6. Jahani, H., Khosravi, Y., Kargar, B., Ong, K.-L., & Arisian, S. (2024). Exploring the role of drones and UAVs in logistics and supply chain management: A novel text-based literature review. *International Journal of Production Research, 63*, 1–25.

7. Barua, R. (2024). Innovations in minimally invasive surgery: The rise of smart flexible surgical robots. In *Emerging technologies for health literacy and medical practice* (pp. 110–131). IGI Global.

8. Gill, S. S., Golec, M., Hu, J., Xu, M., Du, J., Wu, H., Walia, G. K., Murugesan, S. S., Ali, B., Kumar, M., Ye, K., Verma, P., Kumar, S., Cuadrado, F., & Uhlig, S. (2025). Edge AI: A taxonomy, systematic review and future directions. *Cluster Computing, 28*, 18.

9. Akkem, Y., Biswas, S. K., & Varanasi, A. (2024). A comprehensive review of synthetic data generation in smart farming by using variational autoencoder and generative adversarial network. *Engineering Applications of Artificial Intelligence, 131*, 107881.

10. Heintz, F., Belbachir, N., & Curry, E. (2024). *Strategic research, innovation, and deployment agenda 2025–2027*. ADRA.

11. van Ufford, H. Q. (2025). *Regulate or stagnate: Why the EU must lead on AI*. European Council of Foreign Relations.

12. European Court of Auditors. (2024). *EU artificial intelligence ambition—Stronger governance and increased, more focused investment essential going forward*. European Union.

13. ADRA Association. (2024). *ADRA—The AI Data Robotics Association*. European Union. Retrieved from https://adr-association.eu/

14. AI-DAPT Consortium. (2024). AI-DAPT European project. Retrieved from https://www.ai-dapt.eu/

15. Curry, E., Heintz, F., Irgens, M., Smeulders, A. W. M., & Stramigioli, S. (2022). Partnership on AI, data, and robotics. *Communications of the ACM, 65*(4), 54–55.

16. Whang, S. E., Roh, Y., Song, H., & Lee, J.-G. (2023). Data collection and quality challenges in deep learning: A data-centric AI perspective. *The VLDB Journal, 32*, 791–813.

17. Brodsky, S. (2024, August 20). *Examining synthetic data: The promise, risks and realities*. IBM. Retrieved from https://www.ibm.com/think/insights/ai-syntheticdata

18. Maharana, K., Mondal, S., & Nemade, B. (2022). A review: Data pre-processing and data augmentation techniques. *Global Transitions Proceedings, 3*(1), 91–99.

19. Kosinski, M. (2025, January 14). Why synthetic data is the future of AI: Exploring the shift from real-world datasets. Retrieved from https://www.1950.ai/post/whysynthetic-data-is-the-future-of-ai-exploring-the-shift-from-real-world-datasets

20. Binici, K., Pham, N. T., Mitra, T., & Leman, K. (2022). Preventing catastrophic forgetting and distribution mismatch in knowledge distillation via synthetic data. In *IEEE/CVF Winter Conference on Applications of Computer Vision (WACV 2022), Waikoloa, HI, USA*.

21. Brasseur, A. (2024, August 28). *The synthetic data revolution: How does it fuel AI?* AXA Venture Partners. Retrieved from https://www.axavp.com/the-synthetic-datarevolution-how-does-it-fuel-ai/

22. World Economic Forum. (2025). *Frontier technologies in industrial operations: The rise of artificial intelligence agents*. World Economic Forum.

23. Tang, C., Abbatematteo, B., Hu, J., Chandra, R., Martín-Martín, R., & Stone, P. (2025). Deep reinforcement learning for robotics: A survey of real-world successes. *Annual Review of Control, Robotics, and Autonomous Systems, 8*, 153.

24. Hu, Y., Xie, Q., Jain, V., Francis, J., Patrikar, J., Keetha, N., Kim, S., Xie, Y., Zhang, T., Fang, H.-S., Zhao, S., Omidshafiei, S., Kim, D.-K., Agha-Mohammadi, A.-A., Sycara, K., Johnson-Roberson, M., Batra, D., Wang, X., Scherer, S., Wang, C., Kira, Z., Xia, F., & Bisk, Y. (2023). Toward general-purpose robots via foundation models: A survey and metaanalysis. arXiv:2312.08782.

25. Panagou, S., Neumann, W. P., & Fruggiero, F. (2024). A scoping review of human robot inter-action research towards Industry 5.0 human-centric workplaces. *International Journal of Production Research, 64*(3), 974–990.
26. BMW Group. (2024, August 6). Successful test of humanoid robots at BMW Group Plant Spartanburg. Retrieved from https://www.bmwgroupwerke.com/spartanburg/en/news/2024/successful-test-of-humanoid-robots.html
27. Ribeiro, M. T., Singh, S., & Guestrin, C. (2016). Why should I trust you? Explaining the predictions of any classifier. In *22nd ACM SIGKDD International Conference on Knowledge Discovery and Data Mining (KDD 2016), San Francisco, CA, USA.*
28. Lundberg, S. M., & Lee, S.-I. (2017). A unified approach to interpreting model predictions. In *31st International Conference on Neural Information Processing Systems (NIPS 2017), Long Beach, CA, USA.*
29. Karpatne, A., Atluri, G., Faghmous, J. H., Steinbach, M., Banerjee, A., & Ganguly, A. (2017). Theory-guided data science: A new paradigm for scientific discovery from data. *IEEE Transactions on Knowledge and Data Engineering, 29*(10), 2318–2331.
30. Gama, J., Žliobaitė, I., Bifet, A., Pechenizkiy, M., & Bouchachia, A. (2014). A survey on con-cept drift adaptation. *ACM Computing Surveys, 46*(4), 1–37.
31. Pearl, J. (2019). The seven tools of causal inference, with reflections on machine learning. *Communications of the ACM, 62*(3), 54–60.
32. Bhuyan, B. P., Ramdane-Cherif, A., Tomar, R., & Singh, T. P. (2024). Neuro-symbolic artifi-cial intelligence: A survey. *Neural Computing and Applications, 36*, 12809–12844.
33. Feng, C., Wang, Y., Chen, Q., Ding, Y., Strbac, G., & Kang, C. (2021). Smart grid encounters edge computing: Opportunities and applications. *Advances in Applied Energy, 1*, 100006.
34. Li, P., Zhang, H., Wu, Y., Qian, L., Yu, R., & Niyato, D. (2024). Filling the missing: Exploring generative AI for enhanced federated learning over heterogeneous mobile edge devices. *IEEE Transactions on Mobile Computing, 23*(10), 10001–10015.
35. Din, I. U., Almogren, A., & Rodrigues, J. J. P. C. (2024). AIoT integration in autonomous vehi-cles: Enhancing road cooperation and traffic management. *IEEE Internet of Things Journal, 11*(22), 35942–35949.
36. Lee, E., Seo, Y.-D., Oh, S.-R., & Kim, Y.-G. (2021). A survey on standards for interoperabil-ity and security in the internet of things. *IEEE Communications Surveys & Tutorials, 23*(2), 1020–1047.
37. Ali, S., Abuhmed, T., El-Sappagh, S., Muhammad, K., Alonso-Moral, J. M., Confalonieri, R., Guidotti, R., Del Ser, J., Díaz-Rodríguez, N., & Herrera, F. (2023). Explainable artificial intelligence (XAI): What we know and what is left to attain trustworthy artificial intelligence. *Information Fusion, 99*, 101805.
38. Zhan, Z.-H., Li, J.-Y., & Zhang, J. (2022). Evolutionary deep learning: A survey. *Neurocomputing, 483*, 42–58.
39. Hospedales, T., Antoniou, A., Micaelli, P., & Storkey, A. (2022). Meta-learning in neural net-works: A survey. *IEEE Transactions on Pattern Analysis and Machine Intelligence, 44*(9), 5149–5169.
40. Černevičienė, J., & Kabašinskas, A. (2022). Review of multi-criteria decision-making meth-ods in finance using explainable artificial intelligence. *Frontiers in Artificial Intelligence, 5*, 827584.
41. Wang, X., Zhang, Z., & Jiang, Q. (2024). The effectiveness of human vs. AI voice-over in short video advertisements: A cognitive load theory perspective. *Journal of Retailing and Consumer Services, 81*, 104005.
42. Naiseh, M., Simkute, A., Zieni, B., Jiang, N., & Ali, R. (2024). C-XAI: A conceptual frame-work for designing XAI tools that support trust calibration. *Journal of Responsible Technology, 17*, 100076.
43. DeLong, L. N., Mir, R. F., & Fleuriot, J. D. (2024). Neurosymbolic AI for reasoning over knowledge graphs: A survey. *IEEE Transactions on Neural Networks and Learning Systems, 36*, 1–21.

44. Wei, H., Wang, D., Ouyang, X., Wan, J., Liu, J., & Li, T. (2024). Multimodal federated learning: Concept, methods, applications and future directions. *Information Fusion, 112*, 102576.
45. Rakshit, P., Saha, N., Nandi, S., & Gupta, P. (2024). Artificial intelligence in digital twins for sustainable future. In *Transforming industry using digital twin technology* (pp. 19–44). Springer.
46. Gorobchenko, O., Holub, H., & Zaika, D. (2024). Theoretical basics of the self-learning system of intelligent locomotive decision support systems. *Archives of Transport Quarterly, 71*(3), 169–186.
47. How, M.-L., & Cheah, S.-M. (2024). Forging the future: Strategic approaches to quantum AI integration for industry transformation. *AI, 5*(1), 290–323.
48. Akter, S., McCarthy, G., Sajib, S., Michael, K., Dwivedi, Y. K., D'Ambra, J., & Shen, K. (2021). Algorithmic bias in data-driven innovation in the age of AI. *International Journal of Information Management, 60*, 102387.
49. Kusner, M., Loftus, J., Russell, C., & Silva, R. (2017). Counterfactual fairness. In *31st Conference on Neural Information Processing Systems (NIPS 2017), Long Beach, CA, USA.*
50. Zhang, B. H., Lemoine, B., & Mitchell, M. (2018). Mitigating unwanted biases with adversarial learning. In *AAAI/ACM Conference on AI, Ethics, and Society (AIES 2018), New Orleans, LA, USA.*
51. Serna, I., Morales, A., Fierrez, J., & Obradovich, N. (2022). Sensitive loss: Improving accuracy and fairness of face representations with discrimination-aware deep learning. *Artificial Intelligence, 305*, 103682.
52. Bellamy, R. K. E., Dey, K., Hind, M., Hoffman, S. C., Houde, S., Kannan, K., Lohia, P., Martino, J., Mehta, S., Mojsilovic, A., Nagar, S., Ramamurthy, K. N., Richards, J., Saha, D., Sattigeri, P., Singh, M., Varshney, K. R., & Zhang, Y. (2019). AI Fairness 360: An extensible toolkit for detecting and mitigating algorithmic bias. *IBM Journal of Research and Development, 63*(4/5), 4:1–4:15.
53. Demetzou, K., & Rovilos, V. (2023). *Conformity assessments under the proposed EU AI act: A step-by-step guide.* Future of Privacy Forum and OneTrust LLC.
54. Mathew, D. E., Ebem, D. U., Ikegwu, A. C., Ukeoma, P. E., & Dibiaezue, N. F. (2025). Recent emerging techniques in explainable artificial intelligence to enhance the interpretable and understanding of AI models for human. *Neural Processing Letters, 57*, 16.
55. Qian, M., Musa, A. A., Biswas, M., Guo, Y., Liao, W., & Yu, W. (2025). Survey of artificial intelligence model marketplace. *Future Internet, 17*(1), 35.
56. Liu, R., Wey, J., Liu, F., Si, C., Zhang, Y., Rao, J., Zheng, S., Peng, D., Yang, D., Zhou, D., & Dai, A. M. (2024). Best practices and lessons learned on synthetic data. arXiv:2404.07503.
57. Wang, K.-H., & Lu, W.-C. (2025). AI-induced job impact: Complementary or substitution? Empirical insights and sustainable technology considerations. *Sustainable Technology and Entrepreneurship, 4*(1), 100085.
58. Jibril, M., & Florentina, T. A. (2024). Governing AI in hiring: An effort to eliminate biased decision. In *New frontiers in artificial intelligence, JSAI-isAI* (pp. 49–63).
59. Kubullek, A.-K., van Ledden, S., & Dogangün, A. (2024). A collection of standards-based recommendations for sustainable, social, accessible robots and Systems in Public Spaces—A systematic review and derivation of unified equality requirement descriptions. In *12th International Conference on Human-Agent Interaction (HAI 2024), Swansea, UK.*
60. Yampolskiy, R. V. (2025) On monitorability of AI. *AI and Ethics, 5*, 698–707.
61. Kostenko, O. M., Bieliakov, K. I., Tykhomyrov, O. O., & Aristova, I. V. (2024). "Legal personality" of artificial intelligence: Methodological problems of scientific reasoning by Ukrainian and EU experts. *AI and Society, 39*, 1683–1693.
62. Rolnick, D., Donti, P. L., Kaack, L. H., Kochanski, K., Lacoste, A., Sankaran, K., Ross, A. S., Milojevic-Dupont, N., Jaques, N., Waldman-Brown, A., Luccioni, A. S., Maharaj, T., Sherwin, E. D., Mukkavilli, S. K., Kording, K. P., Gomes, C. P., Ng, A. Y., Hassabis, D., Platt, J. C., Creutzig, F., Chayes, J., & Bengio, Y. (2022). Tackling climate change with machine learning. *ACM Computing Surveys, 55*(2), 1–96.

63. Cheah, C. G., Chia, W. Y., Lai, S. F., Chew, K. W., Chia, S. R., & Show, P. L. (2022). Innovation designs of Industry 4.0 based solid waste management: Machinery and digital circular economy. *Environmental Research, 213*, 113619.
64. European Commission. (2024). *AI and generative AI: Transforming Europe's electricity grid for a sustainable future*. Directorate-General for Communications Networks, Content and Technology—European Commission.
65. Ahmad, T., Madonski, R., Zhang, D., Huang, C., & Mujeeb, A. (2022). Data-driven probabilistic machine learning in sustainable smart energy/smart energy systems: Key developments, challenges, and future research opportunities in the context of smart grid paradigm. *Renewable and Sustainable Energy Reviews, 160*, 112128.
66. Strubell, E., Ganesh, A., & McCallum, A. (2020). Energy and policy considerations for modern deep learning research. In *AAAI Conference on Artificial Intelligence (AAAI 2020), New York, NY, USA*.
67. Sarkadi, S., Tettamanzi, A. G. B., & Gandon, F. (2022). Interoperable AI: Evolutionary race toward sustainable knowledge sharing. *IEEE Internet Computing, 26*(6), 25–32.
68. EU AI Office—European Union. (2025). *Living repository to foster learning and exchange on AI literacy*. Directorate-General for Communications Networks, Content and Technology—European Commission.
69. euRobotics. (2024). *A unified vision for European robotics. A strategy for innovation, growth and societal impact*. euRobotics aisbl.
70. Ingenarius, Lda. (2025). *CRIARTE—Construção com Robótica Inteligente e Arquitetura Revolucionária de Tecnologias Emergente*. COMPETE 2030. Retrieved from https://www. p2030criarte.com/
71. SIRI—Associazione Italiana Robotica e Automazione. (1975). SIRI—Associazione Italiana Robotica e Automazione. Retrieved from https://www.robosiri.it/
72. euROBIN Consortium. (2022). *euROBIN*. European Commission. Retrieved from https:// www.eurobin-project.eu
73. European Robotics Forum. (2025). *European Robotics Forum*. AIM Group International. Retrieved from https://eu-robotics.net/european-robotics-forum/
74. Segate, R. V., & Daly, A. (2024). Encoding the enforcement of safety standards into smart robots to harness their computing sophistication and collaborative potential: A legal risk assessment for European Union policymakers. *European Journal of Risk Regulation, 15*(3), 665–704.
75. European Parliament. (2023, June 8). EU AI act: First regulation on artificial intelligence. Retrieved from https://www.europarl.europa.eu/topics/en/article/20230601STO93804/ eu-ai-act-firstregulation-on-artificial-intelligence
76. Novelli, C., Hacker, P., Morley, J., Trondal, J., & Floridi, L. (2024). A robust governance for the AI act: AI office, AI board, scientific panel, and National Authorities. *European Journal of Risk Regulation, 16*, 1–25.
77. National Institute of Standards and Technology. (2023). *Artificial intelligence risk management framework*. NIST.
78. Cedefop—European Union. (2025). *Skills empower workers in the AI revolution*. Publication Office of the European Union.
79. O'Brien, K., & Downie, A. (2024, October 15). *Upskilling and reskilling for talent transformation in the era of AI*. IBM.
80. Thornhill, J., & Milne, R. (2025, February 26). Estonia launches AI in high schools with US tech groups. *Financial Times*.
81. Anderson, J. (2024, October 2). *The impact of AI on children's development*. Harvard Graduate School of Education.
82. Rott, K. J., Lao, L., Petridou, E., & Schmidt-Hertha, B. (2022). Needs and requirements for an additional AI qualification during dual vocational training: Results from studies of apprentices and teachers. *Computers and Education: Artificial Intelligence, 3*, 100102.
83. Google. (2025). *Grow with Google—AI training overview*. Google. Retrieved from https:// grow.google/ai/

The Future of Digital Twins in Europe

Emerging Trends, Strategic Foresight, and Innovation Pathways

Victor Alonso Ramos, Gema Antequera García, Iria Galiñanes Romero, Celia Minguet Requeni, Sergio Gusmeroli, Iddo Bante, Jose Ramón Sierra, Nikolaos Nikolakis, Mila Koeva, Katalin Kovacs, Gorka Aguirre, and Ibon Ocaña

Abstract This chapter explores the future of digital twin (DT) technologies in Europe, identifying emerging trends, strategic challenges, and innovation pathways toward 2030 and beyond. Building on the foundational work of the AI, Data and Robotics Association (ADRA) Digital Twin Topic Group, it provides a forward-looking analysis of how DTs can transform European industry and society. The

V. A. Ramos (✉) · G. A. García · I. G. Romero · C. M. Requeni
CTAG Automotive Technology Center of Galicia, O Porriño, Spain
e-mail: victor.alonso@ctag.com; gema.antequera@ctag.com; iria.galinanes@ctag.com; celia.minguet3892@ctag.com

S. Gusmeroli
Politecnico di Milano, Milan, Italy
e-mail: sergio.gusmeroli@polimi.it

I. Bante · M. Koeva
Univerity of Twente, Enschede, The Netherlands
e-mail: i.bante@utwente.nl; m.n.koeva@utwente.nl

J. R. Sierra
LiveM, Zaragoza, Spain
e-mail: jrsierra@livem.eu

N. Nikolakis
Laboratory for Manufacturing Systems, University of Patras, Patras, Greece
e-mail: nikolakis@lms.mech.upatras.gr

K. Kovacs
Innomine, Budapest, Hungary
e-mail: katalin.kovacs@innomine.com

G. Aguirre
IDEKO, Elgoibar, Spain
e-mail: gaguirre@ideko.es

I. Ocaña
Ceit-IK4, San Sebastian, Spain
e-mail: iocana@ceit.es

581

E. Curry et al. (eds.), *Artificial Intelligence, Data and Robotics*,
https://doi.org/10.1007/978-3-032-10561-5_20

chapter highlights Europe's unique opportunity to lead in ethical, sustainable, and value-driven DT adoption, distinct from the investment-heavy approaches of global competitors. Key enablers such as AI integration, data governance, frugal computing, talent development, and regulatory alignment are discussed in depth. A multiphase technology roadmap is proposed, outlining short-, mid-, and long-term priorities, including DTs' convergence with mechatronics, AR/VR, and predictive intelligence. Finally, the chapter calls for coordinated European action to foster open innovation ecosystems and establish Europe as a global reference in trusted, human-centric digital twin technologies.

Keywords Artificial intelligence · Manufacturing · Advanced robotics · Internet of Things · Data analytics

1   Introduction

Digital twins (DTs) are rapidly evolving beyond their origins as simulation and monitoring tools to become foundational elements in the transformation of Europe's digital, industrial, and societal systems. These virtual counterparts of physical entities are increasingly viewed as engines of systemic innovation, capable of integrating real-time data, artificial intelligence (AI), and advanced analytics to support more adaptive, sustainable, and resilient infrastructures. As such, digital twins are poised to play a decisive role in advancing the EU's twin transitions, digital and green, while also contributing to strategic priorities such as technological sovereignty and industrial competitiveness.

This second chapter adopts a forward-looking perspective to explore how Europe can shape and lead the future of digital twin development. Building on current achievements and strategic initiatives, such as those promoted by ADRA, the European Data Strategy, and Horizon Europe, it provides a prospective roadmap for consolidating Europe's position as a trusted, value-driven leader in DT innovation. It does so with full awareness of global dynamics and the urgent need to differentiate Europe's approach from that of other major players such as the United States and China, who prioritize massive investment over ethical and sustainable governance.

The *Draghi Report* on EU competitiveness has underscored the importance of investing in key enabling technologies, including digital twins, to address the widening productivity gap and secure Europe's economic future. Furthermore, the Strategic Research, Innovation, and Deployment Agenda (SRIDA) developed under ADRA emphasizes the critical role of DTs in fostering cross-sector digital transformation aligned with European values, namely, trustworthiness, human centricity, and sustainability.

In contrast to the previous chapter, which focused on the current state of the art and use cases, this chapter seeks to answer a different question: How can Europe anticipate, shape, and lead the next generation of digital twins over the coming decade? It addresses this question through the lens of strategic foresight, identifying both technological and structural priorities across the short, medium, and long term.

In particular:

- **Section 2** analyses the core challenges and strategic capabilities required to position Europe as a leader in digital twin innovation.
- **Section 3** sets out actionable recommendations for advancing secure, interoperable, and inclusive DT ecosystems.
- **Section 4** proposes a technology roadmap with short-, mid-, and long-term goals to guide Europe's efforts toward trusted, value-based digital transformation.

2   DT: A Breakthrough Europe Cannot Miss

After extensive technological analysis and European Commission research and innovation efforts, the ADRA Digital Twin Topic Group has identified key challenges and competencies needed to position Europe at the forefront of digital twin (DT) adoption and innovation.

2.1   Redefining Competitiveness in the EU

The convergence of DTs with emerging technologies such as AI, IoT, and machine learning is redefining the boundaries of industrial and societal innovation. DTs enable real-time monitoring, simulation, and optimization, offering solutions to critical European challenges including sustainability, energy management, and digital transformation.

The inherent flexibility and adaptability of DTs open opportunities for more responsive and efficient systems. This enables industries to optimize operations, anticipate failures, and minimize waste, accelerating innovation and driving improvements in:

- Improving performance and reducing resource consumption across manufacturing, energy, and transportation
- Designing systems capable of responding dynamically to changing conditions
- Enabling real-time, automated decisions for processes such as predictive maintenance, supply chain management, and autonomous operations

2.2 Addressing Critical Challenges in a European Way

2.2.1 Ethics and European Values

DT development in Europe must address challenges arising from global data sources and AI models. Much training data originates outside Europe, potentially introducing biases inconsistent with European values. This is particularly significant for urban planning, healthcare, and public services. In particular:

- **Data diversity**: Use of European-specific data sources to ensure DTs reflect privacy, data sovereignty, and ethical standards
- **European-centric models**: Development of models and frameworks that prioritize transparency, fairness, and inclusivity

2.2.2 Frugality and Energy Efficiency

Scaling DTs presents significant energy consumption challenges. Simulations and data processing can be resource intensive, so Europe must prioritize:

- **Frugal AI and optimized data usage**: Prioritizing solutions that require less data without sacrificing effectiveness
- **Energy-efficient systems**: Emphasizing low-energy, high-performance models to avoid unsustainable energy use
- **Resource optimization**: Leveraging underutilized computing resources, including IoT and edge devices

2.2.3 Physical Aspect and Real-World Integration

For DTs to be effective, they must integrate seamlessly with physical systems, requiring detailed knowledge of physical dynamics, sensor networks, and communication protocols.

- **System optimization**: Accurate representation of real-world dynamics to enhance reliability and performance
- **Human-system interaction**: User-centric DTs that are accessible and responsive across different national and cultural contexts
- **Multi-system coordination**: Managing real-time monitoring and predictive maintenance across complex, interconnected systems (e.g., smart cities, industry 4.0)
- **Interoperability**: Establishing communication standards to enable effective data exchange between devices, robots, and sensors

2.3 *Urgency for Europe to Move Forward*

A clear mission for Europe is to create a robust DT ecosystem that integrates AI, data science, and robotics, engaging stakeholders across sectors. This collaborative, multidimensional approach can directly support up to 10 of the 17 Sustainable Development Goals (SDGs) outlined in the EU's 2030 Agenda [1].

3 Our Strategic Recommendations

Inside the ADRA DT Topic Group, we opened a discussion on strategic recommendations to improve EU competitiveness. The future of DTs is uncertain. We do not know exactly how it will affect our lifestyles, but Europe should manage to be in a situation of being able to take advantage of the benefits that this technology will offer in the future. If Europe wants to get to a privileged position in this technology, we should avoid using the same strategies of China and the USA of massive investment but clearly define our own objectives and the paths and actions to achieve them, always considering Europe's strengths and weaknesses.

Our strategic recommendations for pursuing this objective are listed hereafter:

1. **Adopt Blue Ocean Strategy**: Rather than replicating investment-driven approaches seen in China or the USA, Europe should focus on value innovation by identifying new application domains for DTs, such as cultural heritage, sustainable agriculture, and public sector transformation [2]. This "Blue Ocean Strategy" encourages Europe to lead in market spaces that remain underexplored globally, leveraging its diversity in research, regulation, and societal priorities to develop differentiated, locally relevant DT solutions.

2. **Develop, attract, and retain talent**: To ensure a robust pipeline of expertise, Europe must invest in comprehensive education and training on DTs, AI, and robotics across all levels, from secondary schools to advanced university programs. Integration of DT concepts in school curricula will foster early familiarity and career interest, while advanced courses will create high-skill specialists. To counteract brain-drain, the EU should develop incentives for talent retention, including attractive research opportunities, career pathways, and competitive working conditions.

3. **Ensure security**: As DTs are increasingly deployed in critical infrastructure, cybersecurity must be a priority. Blockchain technology offers robust, decentralized solutions for data integrity and traceability in DT ecosystems [3]. EU-wide initiatives, such as the Cybersecurity Act (Regulation 2019/881) and the NIS2 Directive, provide frameworks for standardizing certification, risk management, and incident response, coordinated by ENISA [4].

4. **Fully integrate hardware and software**: Achieving seamless integration between digital and physical systems is vital for DT effectiveness. This requires unified system architectures, interoperable communication protocols, advanced

sensors, and robust cybersecurity practices. Standardization and open-source solutions will play a critical role in fostering cross-platform compatibility.

5. **Advance mechatronics and real-time control**: Ongoing progress in mechatronics, the fusion of mechanical, electrical, and software engineering, is essential to realize DTs that can autonomously interact with and adapt to physical environments in real time. Establishing shared standards for data exchange and control will unlock greater interoperability and scalability.

6. **Explore novel AI and data approaches**: Europe should invest in explainable, energy-efficient AI models specifically designed for DTs, to ensure both transparency and sustainability. Prioritizing research in novel AI techniques can drive system performance while maintaining alignment with European values of trustworthiness and privacy.

7. **Build a collaborative community and digital commons**: To drive sustained innovation, Europe must foster collaborative platforms and digital commons for shared data, models, and tools. Initiatives such as challenge competitions, open datasets, and libraries of virtual assets will support knowledge transfer and accelerate adoption across sectors.

4 Technology Roadmap for Europe

The ADRA DT Topic Group has developed a Technology Roadmap for Digital Twins (DTs) in Europe, structured around short-, mid-, and long-term objectives. The roadmap aligns with the Strategic Research, Innovation, and Deployment Agenda (SRIDA) [5] and addresses both technical and structural priorities for the EU. In particular, key strategic priorities include:

- Advancing software platforms and professional training to accelerate DT adoption
- Developing standards for interoperability, security, and data governance
- Fostering multi-sector collaboration and robust digital infrastructure
- Ensuring ethical, sustainable, and responsible use of DTs

4.1 Short-Term Goals (2–3 Years)

Short Term Objective: Accelerating the Implementation of DT Systems and Opening Up to More Complex Applications, Up to 2–3 Years
To accelerate the development of DT technology, it is needed to capitalize on and expand the current state of the art by enabling integrators to leverage these advancements through a structured offering of software toolchains. These toolchains should utilize comprehensive modelling frameworks and advanced analytics to address

challenges that currently limit the broad application of DTs, such as real-time data integration, scalability, and system interoperability.

Key issues that need to be addressed in the initial phase include simplifying the DT development process, reducing the time and economy costs associated with deploying these systems, and enhancing the interoperability of DTs between all the involved technologies. One of the principal characteristics of DT systems is their inherent adaptability, which is essential for customizing applications to meet specific industry needs.

The expected outcome is the broader adoption by industries of advanced DT technology, encompassing enhanced functionalities, greater adaptability of systems, and accelerated deployment across sectors including SMEs. This will facilitate more dynamic and efficient operations, enabling companies to respond swiftly to changing market demands and operational challenges. Besides, 24% of aerospace and defense organizations are prioritizing DTs to optimize full product life cycle operations, with an additional 50% planning to do so within the next 1–2 years.

4.1.1 Technology

Technology readiness levels of the below detailed are expected to increase from the current range of 3–4 to 7–8 within the next 2–3 years.

- Data acquisition and integration: Seamless integration of heterogeneous data sources, including IoT devices, legacy systems, and external data streams, is essential. The diversity in data types and formats poses a significant challenge for real-time data integration and synchronization. This may be mitigated by developing standardized protocols and interfaces that facilitate robust data integration. Finally, it is required to have a good IoT infrastructure to support the development of a DT, so the need to upgrade machines and factories or digitize existing machines so that they are all IoT compatible is also very important. On the other hand, we must be able to check if the acquired data is correct. The data is the basis of the DT, so we must be sure about its robustness.
- Scalability and infrastructure: In order to accommodate the expansive data and computational needs of DTs, establishing scalable infrastructure is an aspect to consider. This involves developing and implementing cloud and edge computing solutions that can dynamically scale to meet the fluctuating demands of DT applications.
- Cybersecurity and data privacy: Ensuring the cybersecurity and privacy of DTs is paramount, particularly in the context of increasing data breaches and the new stringent compliance requirements (like NIST, ISO/IEC 27001 y IEC 62443). Implementing advanced cybersecurity measures such as state-of-the-art encryption, secure access controls, and conducting regular security audits is crucial. In this context, studying technologies like Blockchain makes a lot of sense.
- Interoperability between systems: Enhancing interoperability among various industrial systems and DTs is critical for achieving a seamless operational

ecosystem. This requires the *development* and adoption of industry-wide standards for DT interoperability.

- Digital twin spaces: To enhance interoperability between systems, it would be especially valuable to create common frameworks and shared spaces where standardized digital twins can interact. This would enable seamless communication among digital twins from different companies, optimizing production and the entire supply chain—from raw materials to the end user.

4.1.2 Structural

Communication, Education, and Training Develop and implement clear communication strategies that articulate the benefits, capabilities, and operational requirements of DT technology to all stakeholders. This includes creating informational materials, case studies, and regular updates that keep all parties informed about the latest developments, best practices, and success stories. Effective communication will ensure that the value proposition of DTs is well understood and embraced across organizational levels. According to the scientific literature, the potential for predictive maintenance and operational optimization are consistently postulated as major benefits of DT technology, so it may be pertinent to firstly highlight these benefits to the stakeholders.

Establish targeted educational programs focused on DT technologies. These programs should provide theoretical knowledge as well as practical skills in designing, implementing, and managing DTs. Incorporating interdisciplinary approaches that cover IT, engineering, data analytics, and industry-specific applications will prepare a skilled workforce capable of advancing DT technology. It would be useful collaborating with local clusters, projects incubators, or other similar organizations to develop these trainings and achieve a good engagement.

In a similar way, to organize regular engagement sessions such as webinars, seminars, and conferences where industry leaders, technology experts, and academic researchers can share insights, discuss challenges, and explore new opportunities in the field of DTs. These sessions not only enhance learning and knowledge sharing but also foster collaborations that can lead to innovative uses and improvements in DT technology. It would be useful to promote and collaborate with their own DT technology providers and their partners at webinars and conferences. For instance, Siemens and its partner, CT Advantek, is continually offering webinars where their own Siemens experts explain and show their products and services.

Ultimately, to develop certification and accreditation programs that validate expertise in DT technology. These programs will standardize the qualifications for professionals working with DTs, ensuring high standards of practice and instilling trust among enterprises looking to adopt this technology.

4.2 Mid-Term Goals (3–5 Years)

Over the next 3–5 years, DT technology is poised to undergo transformative advancements. As an integral component of Industry 4.0, DTs are expected to evolve beyond their current deployments in manufacturing, aerospace, and automotive sectors, making significant inroads into areas like healthcare, urban planning, energy, and agriculture. This expansion is driven by the technology's ability to integrate seamlessly with the Internet of Things (IoT) and leverage vast amounts of big data to create dynamic, real-time simulations of physical systems and assets.

As DTs become more sophisticated, incorporating advanced machine learning algorithms and AI, they will shift from merely descriptive tools to predictive and prescriptive systems. This evolution will enable organizations to not only monitor and replicate the performance of physical assets but also anticipate future challenges and optimize processes proactively. The increased accuracy and predictive capabilities of DTs will enhance decision-making the most.

Furthermore, in conjunction with advances in user interface technologies, such as augmented and virtual reality, DTs will become more accessible and useful to a wider range of professionals, including those without deep technical expertise. This inclusivity will further drive the adoption and utility of DTs, solidifying their role in digital transformation strategies in numerous fields.

4.2.1 Technology

Technology readiness levels of the below detailed are expected to increase from the current range of 2–3 to 5–6 in 3–5 years' time.

- **Advanced AI and machine learning integration**: The integration of advanced AI algorithms and machine learning techniques will transform DTs into intelligent models capable of predictive analytics, anomaly detection, and autonomous decision-making (according to some papers, if the DT does not act into the reality, it would be considered a Shadow Twin more than a DT). Techniques such as reinforcement learning and real-time data processing will enable DTs to anticipate and mitigate potential issues before they arise.
- **Augmented reality (AR) and virtual reality (VR) integration**: The incorporation of AR and VR technologies will significantly enhance the visualization and interaction capabilities of DTs. These technologies will provide immersive experiences, allowing stakeholders to interact with DTs in a three-dimensional space. This will make complex data and simulations more accessible and understandable.
- **Security enhancements**: Constantly developing robust cybersecurity frameworks tailored for DTs will be essential. New cyberattacks are continually renewing and creating, so it is needed that cybersecurity follows this pace, by addressing potential vulnerabilities and building trust in the technology's

reliability and security. These enhancements will ensure wider adoption and utilization of DTs.

4.2.2 Structural

- **Industry-wide standardization**: Establishing and promoting industry-wide standards for DT development, deployment, and operation will ensure consistency and interoperability across different sectors. As the DT becomes established in one sector, such as automotive, it will become easier to implement in the other sectors.
- **Pilot programs and demonstration projects**: At this mid-term time, the adoption of the DT would be enough to have various successful projects that serve as an example of what a DT can do. **Launching pilot programs and demonstration projects in various industries will help showcase the benefits and ROI of DTs**, encouraging wider adoption. These projects should focus on practical applications, highlighting how DTs can improve efficiency, reduce costs, and enhance decision-making.
- **Focus on user experience**: At this moment, the technological fundamentals of a DT would be more developed, so the **efforts can be made in the direction of the user experience**. Enhancing the user experience by developing intuitive and accessible interfaces for DT technology will be essential. This involves leveraging AR/VR technologies and user-centric design principles to make DTs more user-friendly (with no need of technical operators) and effective for a wide range of stakeholders.

4.3 *Long-Term Goals (5–7 Years)*

Over the long term, spanning the next 5–6 years, DT technology is poised to undergo profound transformations where DTs are not just a complement to physical system but integral components that drive innovation and efficiency and have a total focus on sustainability.

Firstly, we can expect DTs to become deeply embedded in the operational fabric of organizations, going beyond traditional uses in manufacturing and automotive industries to become ubiquitous in sectors like healthcare, urban planning, and even agriculture. At this moment, the integration with underlying technologies such as AI, machine learning, and IoT will be maximized, and the DT must be a total predictive tool.

Moreover, as these technologies mature, DTs will increasingly interact with each other within and across industries, creating a mesh of interconnected systems that share data and insights. This interconnectivity will enhance the collective intelligence of DTs, leading to more comprehensive solutions to global challenges such as climate change, resource scarcity, and urban congestion.

In the longer term, the focus will likely shift from individual DT deployments to holistic ecosystems where DTs operate as dynamic and integral parts of a connected world.

4.3.1 Technology

Technology readiness levels of the below detailed are expected to increase from the current range of 1–2 to 4–5 in 5–6 years' time.

- **Toward interconnected systems**: As DT technologies mature, they are poised to transform into a network of interconnected systems that communicate and collaborate across various industries. This mesh of interconnected DTs will enable seamless data sharing and insights exchange, amplifying the collective intelligence of the systems involved. By integrating data from diverse sources such as traffic patterns, weather systems in smart cities, and energy consumption metrics in smart grids, DTs can learn from a broader dataset, enhancing their accuracy and the sophistication of their analyses. This not only improves the functionality of individual systems but also fosters a unified, intelligent ecosystem capable of more nuanced insights and actions.

  This enhanced interconnectivity will lead to robust solutions for global challenges such as climate change, resource scarcity, and urban congestion. For example, DTs across energy, agriculture, and urban planning sectors can synergize to optimize resource use and reduce environmental impacts. An energy grid's DT could dynamically source more power from renewables in response to predicted demand spikes, aided by weather condition data from environmental DTs. Similarly, interconnected DTs in urban settings can manage utilities, public transport, and emergency services in a cohesive manner, ensuring optimal city living conditions by adjusting operations in real time based on shared data and predictive analytics. For example, in the future, a DT of a factory can plan its production based on the information acquired by the DT of the logistics of the supplier, and the supplier could plan its logistics of the day and optimum routes based on the information given by a DT of the city's traffic, etc. The network could be infinite.

  To enhance these interconnections and ensure full integration, a clear and specific standardization should be proposed within the staff. As suggested in the short term, a dedicated space for digital threads—similar to how data spaces are being created—should be developed across Europe and for all industries. This would help connect the entire value chain and enable its simulation and optimization, reducing costs and the carbon footprint.
- **Sustainability focus**: Over the long term, the sustainability focus of DT technology is going to be on the top of the priorities, particularly as global pressures related to environmental sustainability intensify. By enabling precise monitoring and control of resources, DTs can drive significant reductions in waste and

optimize the use of energy and raw materials across industries [6]. With Destination Earth (DestinE), the EU established the goal to generate a highly accurate, complete DT of the Earth by 2030 [7]. Furthermore, DTs will play a pivotal role in enhancing the resilience of urban and natural systems against the impacts of climate change. On a larger scale, DTs of entire ecosystems can facilitate more informed conservation and restoration decisions, helping preserve biodiversity and ensure the sustainability of natural resources (European DT Ocean as example). By integrating environmental data into the operational algorithms of DTs, businesses and governments can not only achieve better economic outcomes but also contribute to broader environmental goals, such as those outlined in the Paris Agreement and the Sustainable Development Goals. Besides, 57% of organizations identified sustainability as a key motivator behind their investments in DTs. Organizations have achieved an average improvement of 16% in their sustainability metrics thanks to DTs.

4.3.2 Structural

In the long term, the purposes of DTs are to facilitate the implementation of the EU's Green Deal—in line with internationally binding climate and environmental targets.

Moreover, 75% of European businesses are expected to be fully digitized by 2030, according to the Europe's Digital Decade's digital targets 2030.

In line with the interconnectivity of sectors mentioned above, among the EU's most ambitious objectives is also the creation of a DT of the Ocean (DTO), which exemplifies the cross-sectorality of this technology.

Ultimately, what is clear is that DT will be widely adopted by many different sectors and companies; thus, according to Accenture's Technology Trends 2022 report, the global DT market was valued at $3.21 billion in 2020 and is expected to grow to $184.5 billion by 2030. The projected growth of the global DT market is significant, increasing from €16.55 billion in 2024 to an estimated €242.11 billion by 2032. This represents a compound annual growth rate (CAGR) of 39.8% throughout the forecast period. DT Market Size, Share and Growth Forecast, Report, Juanuary 2026.

The worldwide market for DTs is set to increase from about €16.42 billion in 2024 to €240.11 billion by 2032, with an annual growth rate of 39.8%. Manufacturing is expected to be the fastest-growing sector in this market.

5 Conclusions

Digital twin (DT) technologies are rapidly emerging as a foundational layer of Europe's future digital infrastructure. Far beyond their original applications in engineering or manufacturing, DTs are now positioned to act as intelligent, adaptive systems capable of transforming how we plan, operate, and optimize complex socio-technical environments. This chapter has provided a forward-looking road-map to guide Europe in harnessing the transformative potential of DTs in alignment with its broader strategic objectives.

In the coming years, DTs are expected to evolve into fully integrated, autonomous, and interconnected platforms, moving from passive digital replicas to active agents of systemic change. They will be central to achieving strategic goals such as climate neutrality, industrial competitiveness, resilience in public services, and citizen-centric digital governance. Their growing role in the EU's digital and green transitions calls for coordinated action at multiple levels: technological, structural, and strategic.

From a technological standpoint, the priorities include advancing real-time data integration, ensuring interoperability across platforms and sectors, incorporating explainable and frugal AI, and embedding robust cybersecurity frameworks. The ability of DTs to simulate, predict, and respond in real time to dynamic conditions will be essential to tackle critical challenges in areas such as mobility, energy, logistics, and health.

Structurally, Europe must address key enablers such as education, training, and workforce development. Building a pipeline of skilled professionals is crucial to avoid brain-drain and maintain innovation capacity. Likewise, fostering inclusive ecosystems through open platforms, shared resources, and cross-sector dialogue will be essential to ensure equitable access and broad adoption of DT technologies across regions and industries.

Strategically, the chapter emphasizes the need for Europe to chart its own course. Rather than replicating foreign models based on scale and capital intensity, the EU should leverage its comparative advantages: regulatory strength, societal trust, and a strong public research base. The proposed "Blue Ocean" approach advocates for innovation in underexplored but high-impact areas such as sustainable agriculture, cultural heritage, and public services, where Europe can lead by example and create new market space globally.

The proposed roadmap, structured into short-, medium-, and long-term priorities, offers a concrete plan for consolidating Europe's leadership in digital twin development. It highlights the importance of aligning research and industrial strategies with ethical and environmental imperatives while also responding to global competition and evolving societal needs.

Ultimately, the development and deployment of digital twins offer a unique opportunity to reimagine how Europe designs its infrastructures, governs its resources, and empowers its citizens. If developed responsibly and inclusively, DTs can become a key interface between data, people, and decision-making, bridging physical and digital realms in ways that enhance sustainability, transparency, and social value.

At this pivotal moment, Europe must act decisively. With the right investments, policy frameworks, and collective vision, DTs can help ensure that the next phase of digital transformation is not only innovative but also equitable, resilient, and aligned with European values.

Acknowledgments The content of this chapter originates from the last years of workshops, discussions, and webinars inside the ADRA's Digital Twin Topic Group, where experts, projects, and industrial pilots have actively participated.

References

1. https://commission.europa.eu/strategy-and-policy/sustainable-development-goals/eu-and-united-nations-common-goals-sustainable-future_en
2. Kim, W. C., & Mauborgne, R. A. (2014). *Blue Ocean strategy, expanded edition: How to create uncontested market space and make the competition irrelevant.* Harvard Business Review Press.
3. Mendling, J., Weber, I., Aalst, W. V. D., Brocke, J. V., Cabanillas, C., Daniel, F., Debois, S., Ciccio, C. D., Dumas, M., Dustdar, S., & Gal, A. (2018). Blockchains for business process management-challenges and opportunities. *ACM Transactions on Management Information Systems (TMIS), 9*(1), 1–16.
4. https://www.enisa.europa.eu/topics/product-security-and-certification/cybersecurity-certification-framework
5. https://adr-association.eu/srida
6. Digitalising the energy system—EU action plan, COM (2022) 552 final. Retrieved from https://eur-lex.europa.eu/legal-content/EN/TXT/?uri=CELEX%3A52022DC0552&qid=1666369684560
7. Garske, B., Holz, W., & Ekardt, F. (2024). Digital twins in sustainable transition: Exploring the role of EU data governance. *Frontiers in Research Metrics and Analytics: Research Policy and Strategic Management, 9.* https://doi.org/10.3389/frma.2024.1303024

Advancing Industrial Collaboration: The Next Generation of Human-Robot Interaction

Anibal Reñones, Mireya de Diego, Francisco Melendez, Zoi Arkouli, Nikos Dimitropoulos, Christos Gkrizis, Sotiris Makris, Fernando Castaño, Wael M. Mohammed, Rodolfo E. Haber, and Leire Bastida

Abstract This chapter examines the transformative impact of artificial intelligence (AI), data, and robotics on human-centric manufacturing, in alignment with the European Union's Horizon Europe objectives of digital competitiveness and sustainability. It presents the contributions of three EU-funded projects (ARISE, FORTIS, and JARVIS) focused on advancing industrial human-robot interaction (HRI) through technological innovation and cross-sector collaboration. ARISE introduces an open-source middleware integrating FIWARE and ROS2, designed to enhance industrial interoperability, scalability, and AI-driven decision-making. The project validates its solutions through Testing and Experimentation Facilities (TEFs) and an FSTP program supporting SMEs in deploying HRI applications. FORTIS emphasizes adaptive robotic systems that enhance multimodal human-robot inter-

A. Reñones (✉) · M. de Diego
Industry 4.0 Research Area, Fundacion CARTIF, Boecillo, Spain
e-mail: aniren@cartif.es; mirdie@cartif.es

F. Melendez
FIWARE Foundation, Berlin, Germany
e-mail: francisco.melendez@fiware.org

Z. Arkouli · N. Dimitropoulos · C. Gkrizis · S. Makris
Laboratory for Manufacturing Systems and Automation, University of Patras, Patras, Greece
e-mail: arkouli@lms.mech.upatras.gr; dimitropoulos@lms.mech.upatras.gr;
gkrizis@lms.mech.upatras.gr; makris@lms.mech.upatras.gr

F. Castaño · R. E. Haber
Centre for Automation and Robotics (CAR), Spanish National Research Council (CSIC),
Arganda del Rey, Spain
e-mail: fernando.castano@car.upm-csic.es; rodolfo.haber@car.upm-csic.es

W. M. Mohammed
FAST-Lab, Faculty of Engineering and Natural Sciences, Tampere University,
Tampere, Finland
e-mail: wael.mohammed@tuni.fi

L. Bastida
eServices, TECNALIA, Basque Research and Technology Alliance (BRTA), Derio, Spain
e-mail: leire.bastida@tecnalia.com

E. Curry et al. (eds.), *Artificial Intelligence, Data and Robotics*,
https://doi.org/10.1007/978-3-032-10561-5_21

595

action, leveraging AI-driven cognition, digital twins, and predictive maintenance to optimize industrial workflows. JARVIS advances intuitive human-robot interaction by implementing cognitive mechatronics, AI-enhanced multi-modal interfaces for human-centric social collaboration, and advanced digital twins. The JARVIS toolset will be demonstrated in four large-scale industrial pilots across the energy, automotive, and aeronautics sectors, addressing diverse challenges in HRI for complex tasks. Together, these projects extend the frontiers of HRI, fostering safer, more efficient, and adaptable robotics solutions. By validating AI, data, and robotics integration at scale, they contribute to the evolution of Industry 5.0, ensuring that robotics systems complement human labor while promoting trust, safety, and sustainability in industrial environments.

Keywords Human-robot interaction · Artificial intelligence · Smart manufacturing · Industrial automation · Middleware for robotics · Circular economy in robotics · Testing and experimentation facilities

1 Introduction

The growing demand for flexible and reconfigurable production has stimulated the development of collaborative robots with varying payloads (3–170 kg), mobility, and integrated perception and safety systems enabling human robot coexistence. Researchers have demonstrated the benefit for human-robot collaboration in several applications, e.g., in the automotive and white goods sector [1]. However, the industry still seems reluctant in adopting hybrid production solutions. AI-based solutions and particularly perception systems will enhance the benefit of such solutions, making them gradually more appealing for industrial players [2].

Despite the significant progress in automation and perception technologies, humans remain an indispensable production resource [3]. This in turn underscores operator acceptance as a determining factor for successful deployment [4] of any automation solution. Collaborative AI and virtual reality (VR) simulations can support the optimization of workplace layouts and improving human-robot collaboration by design [5], while multi-user immersive VR frameworks can support the implementation of participatory task design, enhancing ergonomics and reducing operator discomfort or feelings of intimidation and distrust [6]. Cooperating robots have been tested to support manufacturing applications [7], with the benefit of human-robot cooperation being particularly visible in flexible manufacturing [8], especially in assembly tasks [9]. At the same time, different technologies such as wearables and the use of artificial intelligence for human-aware robot motion have been proposed [10].

Building on these technological advancements, three collaborative research projects with a common aim have been launched. As part of the strategic agenda set by the AI, Data, and Robotics Partnership [11], ARISE, JARVIS, and FORTIS seek to explore innovative approaches to human-robot collaboration. These three research projects have been funded under the same Horizon Europe call "Industrial leadership in AI, Data and Robotics—advanced human robot interaction,"[1] tasking them with the mandate of *demonstrating the added value of integrating AI, data, and robotics technologies through large-scale validation scenarios reaching critical mass and mobilizing the user industry.* Following is presented a short overview of the projects that tackle different aspects of collaborative robotics, from middleware development to adaptive robotic control and immersive task design, highlighting their unique contributions to advancing human-robot interaction (HRI) [12] in industrial environments. Accelerating industrial transformation while ensuring that AI and robotics complement manual labor rather than replace it is the key challenge addressed in these projects [13].

FORTIS addresses key challenges in human-robot interaction (HRI) by developing adaptive, human-centric robotic systems. These systems are designed to enhance collaboration, safety, and productivity in physically demanding sectors such as construction, infrastructure services, and manufacturing. One significant challenge is ensuring effective communication between humans and robots. FORTIS tackles this by implementing multimodal interfaces that integrate voice, gesture, and visual cues, allowing robots to interpret and respond to human behaviors dynamically. This approach fosters trust and seamless collaboration. Another challenge is adapting to diverse and changing environments. FORTIS employs digital twin technology [14] to create virtual replicas of physical systems, enabling real-time monitoring and adaptation without compromising human privacy. To validate these solutions, FORTIS conducts pilot programs across various sectors, demonstrating the practical benefits of its HRI technologies in real-world scenarios.

JARVIS advances human-robot collaboration by addressing key industrial adoption barriers such as low collaborative performance, programming complexity, and technology acceptance. Its framework is built on four pillars: user-centric social interaction design, cognitive mechatronics, AI-enhanced interfaces for intuitive communication and control, and trustworthy AI. The JARVIS toolset focuses on dynamic robot adaptation, situational awareness, and ergonomic safety, ensuring seamless interaction by considering human intentions, task constraints, and dynamic robot surroundings.

ARISE contributes to industrial HRI by developing an open framework that accelerates human-robot collaboration through open-source technologies and human-centric design methodologies. It addresses operational performance and cost-effective HRI implementation across industries while ensuring solutions integrate ethical AI principles, intuitive interaction interfaces, and adaptive learning mechanisms. ARISE empowers innovators and solution providers by offering methodological tools that facilitate the development of scalable, ethical, and efficient robotic solutions.

---

[1] https://cordis.europa.eu/programme/id/HORIZON_HORIZON-CL4-2023-DIGITAL-EMERGING-01-02/en

The following sections provide a detailed overview of the key technological advances driving collaborative robotics, followed by an in-depth look at the specific contributions of the ARISE, FORTIS, and JARVIS projects. We then explore how these projects are fostering innovation and boosting business transformation across industry, before concluding with a discussion on the lessons learned and future directions for human-robot interaction in industrial environments.

2 Key Technological Advances in Collaborative Robotics

2.1 *Architectural Foundations for Collaborative Robotics Systems*

Collaborative robotics and human-robot interaction (HRI) are at the forefront of modern industrial automation, enabling seamless cooperation between humans and robots in shared workspaces [15]. The software architecture of collaborative robotics systems is the system's backbone and plays a crucial role in ensuring flexibility, scalability, and interoperability across industrial environments. These systems require a robust and modular infrastructure that enables real-time data exchange, adaptive control, and seamless communication between robots, humans, and industrial applications. A well-defined architecture must support distributed computing, low-latency decision-making, and standardized interfaces to accommodate diverse hardware and software components. By adopting standardized and modular architectural frameworks, industries can accelerate the deployment of robotic solutions while ensuring long-term sustainability and ease of maintenance.

ARISE follows an open and modular approach to architecture, leveraging the FIWARE ecosystem and ROS2 framework to enable seamless interoperability between robotics platforms, industrial applications, and AI-based control systems. FIWARE provides standardized open-source components for context-aware data management, facilitating efficient communication between robots, sensors, and enterprise systems. Meanwhile, ROS2 ensures real-time robotic control, adaptability, and scalability, allowing for robust and flexible system integration. This combination enables ARISE to create highly responsive and intelligent robotic environments that can be deployed across different industrial sectors.

JARVIS adopts a modular and scalable architecture that combines AI-driven decision-making, cognitive mechatronics, and multimodal human-robot interaction (through gestures, voice, physical contact, etc.) to enable context-aware and human-centric collaboration in industrial settings. Built on ROS2 middleware, the architecture facilitates interoperability across heterogeneous hardware and software platforms. The JARVIS architecture is designed to orchestrate complex processes by processing different data pipelines, deep learning models, and reinforcement learning agents for perception, reasoning, and successful task execution. The JARVIS toolset is designed for reusability, enabled by Docker-based microservices deployment, which ensures portability across edge, cloud, and on-premise infrastructures. The framework dynamically adapts robot behavior based on operator state, task context, and environment feedback. By fusing perception, cognition, and communication layers, JARVIS enables robots to interpret human intent, autonomously adapt task execution, and maintain fluent collaboration in hybrid teams. Security, privacy, and trustworthy AI principles are embedded within the architecture, ensuring compliance with industrial safety standards and regulatory frameworks.

Similarly, FORTIS's architecture for collaborative robotics systems is designed to ensure safe, efficient, and human-centric human-robot interactions (HRI) across industries such as construction, manufacturing, and infrastructure services. The system prioritizes three core needs: human-centered, technical, and business requirements. From a human perspective, FORTIS emphasizes user experience, safety, and trust by integrating multimodal communication (voice, gestures, and haptics), real-time cognitive and physical state monitoring, and adaptive interfaces that allow seamless interaction between workers and robots. Technically, the architecture is modular, scalable, and interoperable, leveraging AI, IoT, and digital twins to facilitate real-time decision-making, environmental awareness, and predictive maintenance. The ROS-based middleware enables communication across robots, humans, and other industrial systems while supporting multi-robot collaboration, sensor integration, and dynamic task allocation in complex environments. Business-wise, FORTIS aligns with operational efficiency and cost-effectiveness by integrating smoothly with enterprise systems such as ERP and MES, ensuring compliance with international safety standards and fostering the gradual adoption of robotics within existing workflows. The hybrid cloud-edge deployment ensures robust operation even in connectivity-constrained environments, making FORTIS an intelligent, adaptable, and industry-ready framework for collaborative robotics that enhances productivity while maintaining worker trust and safety.

Despite their different focus areas, ARISE, JARVIS, and FORTIS converge on common architectural needs such as modularity, real-time responsiveness, interoperability, and human-centric design, which are essential for scalable and trustworthy deployment of collaborative robotics across industrial sectors.

2.2 Collaborative Robotics and Human-Robot Interaction

Besides software architecture, technology enablers such as collaborative robots and interfaces are crucial to achieving natural HRI. Recent advancements in HRI have focused on multimodal interaction, incorporating gesture, voice, and touch control to enhance communication and efficiency. These developments aim to create intuitive and adaptive robotic systems that can respond dynamically to human behavior, fostering a more natural and productive working relationship.

Safety is a critical component of collaborative robotics, ensuring that humans and robots can work together without risk. Cutting-edge technologies such as real-time motion prediction [16], force-limiting actuators, and advanced sensor systems have significantly improved safety standards in close-proximity collaboration. These measures not only protect operators but also facilitate smoother interaction, reducing the need for physical barriers and increasing the overall efficiency of robotic deployments in industrial collaborative environments.

The widespread adoption of collaborative robots also depends on trust and acceptance among workers. Ensuring transparency in robotic decision-making, providing user-friendly interfaces, and incorporating feedback from operators are essential strategies to build confidence in robotic systems. Social and ergonomic considerations are equally important in designing robots that enhance worker well-being rather than disrupt established workflows.

ARISE contributes to this field by developing AI-driven multimodal interface systems that facilitate seamless and natural human-robot interaction, enhancing responsiveness and adaptability. Additionally, ARISE utilizes mixed-reality interfaces to enable more intuitive and immersive human-robot interactions, improving operator situational awareness and control in industrial environments. These approaches contribute to safer, more efficient, and human-centered robotic solutions.

JARVIS places human experience, safety, and interaction intuitiveness at the core of its toolset design. It addresses major adoption barriers such as the complexity of robot programming, the lack of fluidity in human-robot collaboration, and the mental workload imposed on operators by introducing multimodal interfaces including speech, gesture, XR-based teleoperation, and physical contact. JARVIS minimizes the need for explicit commands and fosters natural interactions by enhancing the robot's ability to understand human intent, monitor task progress, and dynamically adapt its behavior, thanks to machine learning-based perception systems, digital twins, and intelligent mechatronics. JARVIS embeds Social Sciences and Humanities (SSH) methodologies to assess and improve technology acceptance, inclusivity, and trust, ensuring that robotic systems are not only technically capable but also aligned with human values and industrial ergonomics.

FORTIS integrates multimodal communication, including gesture recognition, voice commands, and haptic feedback, to create intuitive and natural interactions between humans and robots. The system leverages AI-driven perception tools such as action recognition, anomaly detection, and 3D modelling to enhance real-time understanding of human intent and environmental context, ensuring seamless collaboration. Safety in close-proximity human-robot collaboration is a core priority, addressed through adaptive control systems, real-time monitoring of human physiological and behavioral states, and compliance with international safety standards. The use of wearable and non-wearable sensors allows continuous tracking of human movements, alerting robots to potential hazards and dynamically adjusting robot behavior to prevent accidents. The FORTIS architecture further supports trust building by enabling digital twins for simulation and training, allowing workers to familiarize themselves with robotic systems before full deployment. Additionally, the system incorporates resource allocation optimization and AI-driven decision support to balance automation with human oversight, ensuring that workers remain actively engaged rather than displaced.

Collectively, ARISE, JARVIS, and FORTIS demonstrate how the integration of multimodal interaction, adaptive intelligence, and human-centered design can advance trustworthy and efficient human-robot interaction (HRI) across diverse industrial environments.

2.3 Support for SMEs in Manufacturing

Small and medium-sized enterprises (SMEs) play a critical role in the manufacturing ecosystem but often face significant barriers in adopting advanced robotics and AI solutions. Challenges such as high initial investment costs, lack of technical expertise, and difficulties in integrating new technologies into existing workflows hinder widespread adoption. To address these challenges, research initiatives have been developing tailored strategies that facilitate SME engagement, lower entry barriers, and ensure the practical applicability of automation solutions. Additionally, research projects often include funding mechanisms to support third parties, providing SMEs with access to innovation funding through effective and simple procedures. This enables SMEs to focus on technology development rather than administrative complexity. Consequently, the projects emphasize lower TRL technology development, allowing SMEs to focus on the application of such technologies.

The main case studies for SMEs will be obtained from the open calls to integrate their technologies within the ARISE, JARVIS, and FORTIS ecosystem, fostering innovation through financial and technical support. Specifically, the ARISE project focuses on enabling SMEs to integrate robotics and AI solutions seamlessly by leveraging an open-source middleware approach. By utilizing a modular and scalable architecture built on FIWARE and ROS2, ARISE ensures that SMEs can access interoperable and adaptable robotic solutions without high upfront costs. Additionally, ARISE's Testing and Experimentation Facilities (TEFs) provide SMEs with a controlled environment to test and validate their solutions before full deployment, reducing risk and optimizing performance. The project also offers technical guidance, training, and collaborative opportunities to enhance SME participation in Industry 5.0 transformation.

JARVIS provides SMEs with funding, mentorship, and access to its reusable toolset to comprehensively support the wider adoption of human-robot collaboration through structured open calls. JARVIS offers two types of open calls: external pilots, where consortia can bring their own solutions and use cases aligned with the project's objectives, and co-development pilots, where new solutions are developed for the JARVIS pilots, which include the manufacturing of aircraft components, assembly of hybrid car batteries, nuclear station decommissioning, and offshore energy infrastructure inspection and maintenance. The case studies from the open calls fall within two main domains: agile manufacturing and intelligent inspection and maintenance. Depending on the type of open call, SMEs are granted access to human-robot interaction interfaces, planners, and perception systems along with expert guidance to integrate these tools into their use cases and enable the implementation of social human-robot interaction.

The FORTIS project engages SMEs through an open call, encouraging collaboration between businesses, universities, and research organizations to develop and integrate innovative solutions into the FORTIS ecosystem. The project offers financial support, mentoring, and technical guidance to help SMEs overcome adoption barriers. Furthermore, FORTIS provides clear eligibility criteria, structured funding mechanisms, and milestone-based evaluations to maximize impact. By fostering an ecosystem that promotes knowledge sharing, interoperability, and human-centric AI development, the initiative accelerates the deployment of trustworthy and efficient robotics solutions. A mentorship and evaluation process with predefined sprints ensure continuous support, enabling SMEs to refine and validate their solutions with expert guidance.

Through strategic funding mechanisms, modular technology development, and structured open call processes, the ARISE, FORTIS, and JARVIS projects collectively aim to accelerate the adoption of AI-driven robotics in SMEs. By addressing financial, technical, and procedural barriers, these projects ensure that SMEs can seamlessly integrate advanced automation solutions, contributing to a more innovative and competitive manufacturing ecosystem.

2.4 Interoperability and Standardization

Interoperability and standardization are fundamental to ensuring that robotic systems can seamlessly communicate and integrate across diverse industrial environments. To address this aspect, the European Digital Industrial Platform for Robotics (EDIPR)[2] brings an excellent foundational framework and is shared as a common baseline framework across FORTIS, JARVIS and ARISE projects.

To make progress in interoperability aspects across architectures and components, the architectural elements of the EDIPR framework are particularly relevant to the aforementioned initiatives. By building on these elements, the HRI innovation landscape can benefit from a better alignment and harmonization across middleware tools, communication protocols, and data models. These three elements, when conveniently implemented, combined, and architected together, are considered by these initiatives the essential building blocks to support the cross-platform compatibility, scalability, and replicability of advanced solutions for robotics and AI-driven automation. The final aim is to reduce development complexity and promote cost-effective implementations and deployments, increasing their potential for achieving broader adoption.

However, interoperability and standardization are intrinsically linked, especially in the fields of robotics, sensing, and automation, where a high degree of

---

[2] https://robmosys.eu/wiki/

fragmentation in protocols, data models, and architectures creates significant barriers to seamless integration [17]. To address this challenge, it is essential to take a step-by-step approach, developing incremental, interoperable solutions that progressively bridge these gaps by showcasing successful approaches and producing reusable patterns and reference models, paving the way for broader standardization.

The initiatives behind this chapter are also aligned in selecting objectives that contribute to interoperability and standardization. These projects, as innovation actions with broader goals and objectives, are unlikely to delve enough into problems and produce curated contributions to formal standardization at levels such as ISO. However, they seek to promote and facilitate the alignment of component implementation with these standards, fostering convergence. Thus, the joint focus of these projects is more on following the path built by de facto standards that already have some level of acceptance and analyzing their applicability in the most relevant use cases for HRI solutions. For this purpose, the previous contributions from other projects within the ADRA ecosystem, such as DIH2[3] or SHOP4CF,[4] as well as from standardization communities, European standardization support initiatives like HSBooster.eu, and the entire innovation ecosystem around leading technologies for robotics middleware, such as OPC-UA and ROS2, have been decisive.

Building on the relevant use cases determined by the state of the art in HRI and the aforementioned contributions from the related innovation ecosystems, three main contexts of the robotics-based solutions are analyzed:

- First, the embedded context, which deals with the components that are either embedded or tightly coupled to the control plane of the robotic platform.
- Second, the operational context, which deals with the interactions between robotics platforms and external devices or systems that are meant to support humans in physical operations.
- Last but not least, the analytical context, which focuses on enabling real-time integration with the operational context and seamless data interoperability across IT systems to produce powerful data visualizations, valuable data-driven insights, and accelerate decision-making.

In the embedded context, ROS4HRI[5] (Robot Operating System for Human-Robot Interaction) extends the capabilities of ROS2 by introducing standardized interfaces and data structures specifically designed for human-robot collaboration [18]. It defines a set of ROS2 topics that handle multimodal interaction data, including speech recognition, facial expression detection, and hand gesture tracking. Additionally, it represents key human-related concepts such as body joints (e.g., shoulder, elbow, wrist), gaze direction, and proximity to the robot. This structured

---

[3] https://www.dih-squared.eu/

[4] https://shop4cf.eu/

[5] https://ros4hri.github.io/ros4hri-tutorials/

approach enables robots to process and react to human actions in real time [19], improving adaptability and responsiveness in manufacturing settings.

In the operational context, where robots should be seen as any other asset or machine responsible for implementing a task, the role of standardized APIs and shared data models is crucial to ensure their seamless integration across distributed components and third-party systems (e.g., ERP, MES, SCADA). FIWARE Smart Data Models [20] and the FIWARE NGSI-LD protocol [21] provide a structured framework for harmonizing information from multiple sources, including human status, robot behaviors, and environmental conditions, enabling true interoperability between different toolkits and robotic systems.

The final aim is to enhance their integration in the analytical context that covers interactions with industrial IoT platforms, AI-driven decision-making systems, and cloud-based analytics. This enhanced interoperability allows more complex system architectures in the operational context to employ structured human, robot, and human-robot interaction models.

2.5 *AI and Data-Driven Robotics*

Artificial Intelligence (AI) and data-driven robotics are transforming industrial automation with the promise of enabling robots to operate with greater autonomy, adaptability, and efficiency [22]. AI will enhance robotic autonomy and behavior adaptation through real-time decision-making and learning mechanisms, allowing robots to adjust to dynamic environments. In human-robot interaction, AI-driven models will improve communication, perception, and responsiveness, fostering seamless collaboration. Generative AI [23] is expected to contribute by creating synthetic scenarios for training, workflow optimization, and predictive maintenance. Additionally, AI will play a crucial role in translating human interactions into machine-executable commands, ensuring intuitive and effective control of robotic systems in industrial settings.

Machine learning, generative AI, and reinforcement learning are leveraged in manufacturing within the FORTIS project framework to enhance human-robot collaboration, optimize resource allocation, and improve decision-making. Machine learning techniques enable robots to learn from human behaviors, environmental conditions, and task demands, leading to adaptive automation that enhances efficiency and safety [24]. Generative AI is utilized for synthesizing human-robot interaction scenarios, predicting potential failures, and optimizing workflows by simulating various operational conditions [25]. This allows for improved planning and reduced downtime in manufacturing environments. Reinforcement learning further refines robotic behavior by allowing systems to learn through trial and error, adjusting their actions dynamically based on real-world feedback.

JARVIS advances beyond traditional frameworks by integrating advanced AI algorithms into real-world industrial pilots to overcome constraints of conventional robotics systems such as limited robot intelligence, which in turn restricts robot adaptability, context awareness, human intention interpretation, and the overall system's performance. JARVIS employs Deep Learning and Reinforcement Learning as core enablers of real-time reasoning, decision-making autonomy, and robot behavior adaptation. The project takes advantage of multimodal sensor data processing to allow robots to predict human actions, dynamically adapt to their environment, optimize task execution based on continuous learning cycles, and actively refine robot behavior, ensuring a seamless, efficient, and ergonomic collaboration with human operators. JARVIS couples its AI-driven technologies with Social Sciences and Humanities (SSH) methodologies to facilitate the development of inclusive, trust-driven, and adaptive interactions while accounting for industrial robot ethics [26]. Furthermore, JARVIS extends human-robot intellectual interaction by integrating generative AI to generate adaptive workflows, enhance robot communication strategies, and personalize interaction models based on operator preferences and workload conditions [27].

FORTIS project integrates advanced artificial intelligence (AI), adaptive learning mechanisms, and real-time human-robot interaction (HRI) to optimize autonomy and efficiency in dynamic environments such as construction, manufacturing, and infrastructure services. The use of AI-driven cognition and decision-making allows robots to process contextual data, understand human behavior, and make informed decisions without direct supervision. Key components include the AI-based Human Cognition Toolkit, which models human behavior and cognitive states, and the Multi-Robot Dynamic Reconfigurability Toolkit, which enables robots to adapt to real-time environmental changes. Additionally, FORTIS employs digital twin technology for predictive modelling and optimization, ensuring that robots can learn from past interactions to improve future performance. These developments contribute to more autonomous robotic systems capable of seamlessly integrating into complex workflows while prioritizing safety, adaptability, and human collaboration [28].

2.6 *Sustainability and Circular Economy in Robotics*

Sustainability and circular economy principles are becoming increasingly relevant in industrial robotics, aiming to reduce waste, optimize resource efficiency, and extend the life cycle of materials and components. By integrating AI-driven planning, predictive maintenance, and energy-efficient task execution, robotic systems can contribute to more sustainable manufacturing and industrial processes. Circular economy models emphasize the importance of reusing, remanufacturing, and recycling materials, reducing the environmental impact of production and decommissioning activities.

ARISE addresses sustainability and circular economy efforts by defining key challenges within its Financial Support to Third Parties (FSTP) program, a funding initiative that enables SMEs, research institutions, and startups to develop and test innovative robotics solutions in real-world industrial environments. These challenges focus on enhancing robotic solutions for disassembly, remanufacturing, and resource-efficient automation. ARISE promotes the use of open-source middleware to facilitate adaptive and modular robotics that can be easily integrated into sustainable industrial workflows. One of the primary circular economy challenges addressed by ARISE is the so-called Collaborative robotics for dismantling and assembly of high-value products [29]. This challenge focuses on the development of robotic systems capable of efficiently disassembling complex products for reuse and remanufacturing, reducing waste and enhancing sustainability in manufacturing.

JARVIS addresses sustainability and circular economy objectives through real-world use cases that demonstrate how HRI can contribute to resource efficiency, reduced environmental impact, and safer working conditions. Among its four pilot domains, the nuclear decommissioning most clearly exemplifies how robotic systems guided by human operators can contribute to sustainable dismantling processes. In this high-risk environment, JARVIS enables the teleoperated disassembly of large and potentially contaminated components, reducing operator exposure while facilitating the controlled separation and recovery of valuable materials. The project's open call mechanism reinforces this direction through a dedicated challenge on robotic remanufacturing and disassembly, inviting external solutions that advance HRI in support of circular manufacturing practices. Furthermore, the JARVIS toolsets including AI-driven planning, digital twins, intuitive robot programming, and adaptive perception are designed to be reusable and easily extendable to disassembly and remanufacturing scenarios. These technologies facilitate energy-efficient task execution, anomaly detection, and workload balancing, ensuring minimal waste and maximum resource efficiency. By embedding these capabilities within a modular HRI framework, JARVIS advances both the technological and operational enablers of circular economy models in high-impact industrial contexts.

FORTIS project leverages robotics and AI to enhance efficiency, reduce waste, and optimize resources across construction, manufacturing, and infrastructure. By integrating robotic automation, real-time monitoring, and AI-driven decision-making, FORTIS minimizes material waste and energy consumption while improving precision in industrial operations. AI-powered quality control and predictive maintenance further contribute to reducing defects, scrap, and rework costs, leading to more sustainable manufacturing practices. A key aspect of FORTIS is its focus on robotic disassembly and remanufacturing to support circular economy models. The project employs AI-driven human-robot collaboration to optimize material reuse, reduce waste, and enhance resource efficiency. Digital twins and AI toolkits enable real-time monitoring, ensuring sustainable operations and adaptive material handling. These innovations are particularly impactful in industries like construction and railways, where automated material reconfiguration and sustainable logistics play a crucial role. FORTIS also demonstrates these advancements through real-world case studies in the construction, manufacturing, and rail industries. For instance, Garcia Garcia, a construction company, integrates robotic systems for material transportation and sandwich panel assembly, improving workplace safety and production efficiency. In the railway sector, Vias applies robotics for the maintenance of railway fastenings, enhancing worker safety and operational reliability. In manufacturing, Arçelik (Beko) explores robotics for assembly line automation, focusing on energy-efficient processes and minimizing environmental impact.

3 Specific Contributions of ARISE, JARVIS, and FORTIS Projects

3.1 ARISE Project

The ARISE project (Agile, human-centric, and Real-tIme enabled open SourcE technologies advancing industrial HRI in Europe[6,7]) introduces an All-in-One Middleware designed to bridge the gap between operational technologies (OT) and information technologies (IT) within industrial Human-Robot Interaction (HRI) applications. This middleware integrates ROS2 and FIWARE, ensuring seamless communication between high-performance, real-time OT networks and IT-driven analytics systems. By extending NGSI-LD brokers with native DDS (Data Distribution Service) support, ARISE enables efficient data exchange across industrial automation platforms [30].

The middleware developed in ARISE addresses key industrial barriers by harmonizing diverse communication protocols and architectures. The aim of the middleware is to ensure (Fig. 1):

- Real-time data handling from OT devices within broader IT infrastructures
- Bidirectional communication, allowing insights from analytics and digital twins to influence OT processes dynamically
- Open standards compliance, enhancing scalability and modularity across industrial settings

ARISE has developed a dedicated version of Orion-LD, a data brokering component that natively supports ROS2 and FIWARE (NGSI-LD) interfaces. This integration allows for flexible interaction between OT and IT domains, providing seamless access to data streams from robotic applications, sensors, and industrial equipment [31].

The integration of ROS2 and FIWARE is key to achieving interoperability in industrial settings:

- FIWARE: Provides an open-source framework of components supporting context-aware data management, facilitating seamless integration with industrial applications, real-time analytics, and IIoT-based operations.

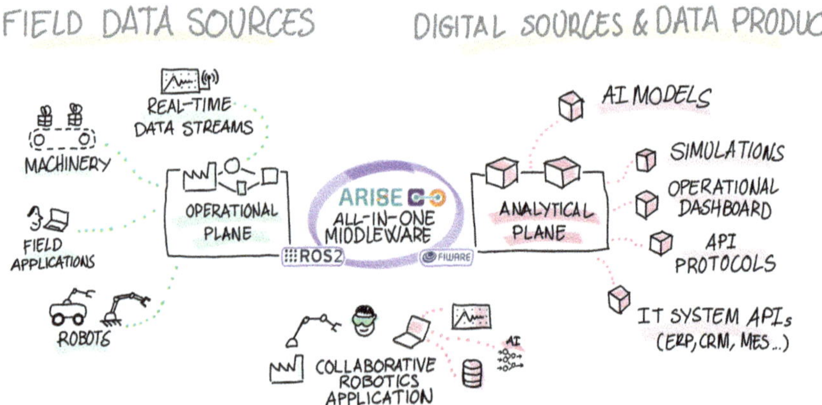

Fig. 1 Conceptual scheme of ARISE middleware connecting the operational plane with the analytical plane in industrial collaborative robotic applications

- ROS2: Serves as the operational middleware for robotic systems, ensuring deterministic behavior, safety compliance, and real-time performance.
- Binding approach: The middleware acts as a translator between DDS topics and NGSI-LD entities, allowing bidirectional data flow between robotic applications and industrial analytics platforms.

To validate and refine the middleware, ARISE has defined eight industrial challenges, which are defined as key industrial and technological barriers that must be addressed to advance human-robot interaction (HRI) in manufacturing and other relevant sectors. These challenges serve as focal points for research, development, and validation efforts, ensuring that the solutions proposed align with industry needs, sustainability goals, and interoperability requirements. Last but not least, the challenges are also seen as inspirational scenarios that the applicants of the ARISE funding program (see Sect. 4) must adhere to to ensure alignment with the project's overarching goals, technological focus, and industrial relevance. These are the ARISE eight challenges:

1. Dismantling and assembly of high-value products [29]: The goal is to automate repetitive or hazardous tasks, enhance efficiency, reduce risks associated with hazardous materials, and ensure worker safety, improving overall productivity. Integration of mixed reality (XR) provides enhanced situational awareness and robust voice command recognition for robot coordination to guide actions in real time. The focus is seamless human-robot collaboration, reducing manual handling and increasing operational efficiency.

2. Complex product picking in industrial warehouses [32]: This challenge addresses the need for efficient and accurate picking operations, reducing manual labor and improving the handling of delicate products; includes real-time 3D object localization to identify various products and voice-based user interaction to

facilitate human-robot collaboration and ensure effective communication between them.

3. Flexible collaborative robots [33]: The goal is a reprogrammable robotics workstation that adapts to variable production requirements, ensuring plug-and-play modularity for hardware and software components. The robot recognizes objects, picks up, and places them in trays based on pre-programmed sequences, ensuring safety, adaptability, and efficient collaboration to optimize production processes.

4. Smart programming [33]: The challenge aims to provide intuitive programming interfaces and AI algorithms to ensure flexibility, allowing operators to adapt the workstation's behavior without requiring extensive coding skills. This approach enhances adaptability to different products while preserving operator's autonomy in agile manufacturing environments, enabling operators to assemble kits by selecting products.

5. Enhanced robot functionality through multimodal HRI interactions [34]: The challenge aims to revolutionize human-robots interactions by leveraging advanced multimodal communications methods, including verbal cues, gestures, and visual feedback, particularly in assisted living contexts. The robot assesses the user's needs, guides them through exercises using multimodal communication, and adapts the session based on the user's progress and feedback.

6. Fetch and carry tasks in healthcare environments [35]: The challenge focuses on developing robotic systems capable of autonomously performing fetch and carry tasks within healthcare settings. The main goal is to alleviate the workload of medical staff, allowing them to concentrate on direct patient care, developing robust interaction flows, building trust among end-users, and creating a safe social navigation system.

7. HRI for improving the efficiency of workers in high precision flexible tasks [36]: This challenge focuses on deploying to enhance the efficiency of workers engaged in high-precision flexible tasks within manufacturing environments. By leveraging robotics, the aim is to streamline the workspace setup process, reducing overall setup time and improving operational efficiency. This process involves real-time 3D object detection to identify tools and components, robotic arm trajectory planning to execute precise movements, and also autonomous mobile robot navigation for seamless transportation.

8. HRI for improving ergonomics in high precision tasks [37]: This challenge focuses on the design of a setup considering the human physical properties in order to respect operator's ergonomics and well-being. Operators can verbally command the cobot to make further adjustments for comfort, including speech command detection for understanding operator instructions and ergonomics-aware positioning systems to ensure optimal comfort and reduce strain during taks. To ensure the challenges address critical gaps in robotics, AI, and industrial automation, they are tested and refined through real-world applications in ARISE, so-called Testing and Experimentation Facilities (TEFs). These are facilities with state-of-the-art robotics and automation technologies that provide a controlled environment for validating innovative human-robot interaction (HRI) solutions, AI-driven automation, and industrial interoperability. The TEFs

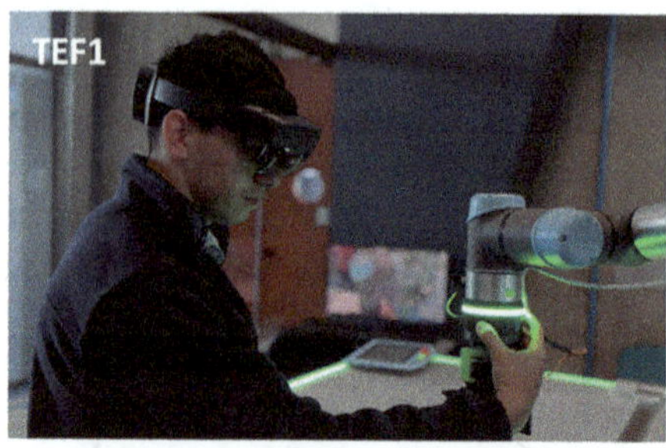

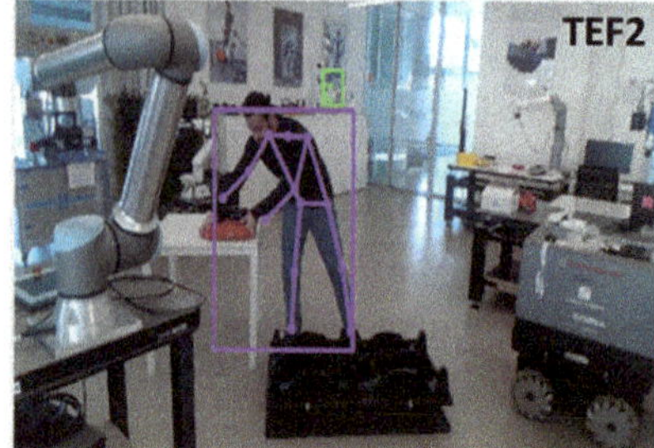

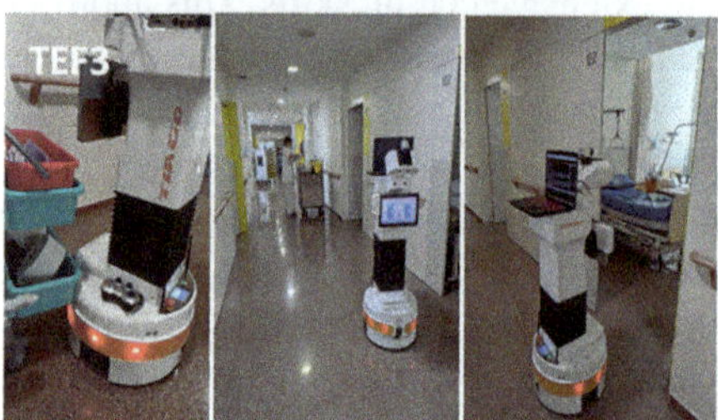

Fig. 2 Testing and experimental facilities of ARISE project

enable rigorous testing of robotic applications under close-to-real-world conditions (Fig. 2):

- TEF1: Experimental robotic cell for complex picking and dismantling applications
- TEF2: Re-programmable cobots for flexible manufacturing
- TEF3: Revolutionizing healthcare with human-centered robotics
- TEF4: Leading the way toward human-centric zero-defect manufacturing

3.2 *FORTIS Project*

The FORTIS project[8,9] offers a solution featuring a modular and scalable system architecture that integrates various toolkits and components. The Robot Gateway and Robotic Intelligibility Toolkit facilitates seamless robotic system integration, while the Human-Centered Data Collection Toolkit captures human activity and behavior for adaptive responses. The Multi-Robot Dynamic Reconfigurability Toolkit ensures coordination and adaptability in shared spaces, and the Digital Twin enables real-time simulation and optimization of human-robot interactions, while the Resource Allocation and Optimization Toolkit optimizes task distribution among human workers and robots. Additionally, the AI-Based Human Cognition Toolkit enhances robots' contextual understanding and decision-making capabilities.

---

[8] https://cordis.europa.eu/project/id/101135707

[9] https://fortis-project.eu/

FORTIS deploys a hybrid cloud-edge model ensuring robust real-time processing even in connectivity-limited environments. Middleware based on the Robot Operating System (ROS) facilitates real-time communication among system components. The interaction framework employs multimodal adaptive communication, incorporating voice, gestures, and text to enable seamless human-robot interaction.

The three pilots in the FORTIS project—construction, manufacturing, and infrastructure services—serve as practical testbeds for implementing its human-centric robotic solutions. In construction, AI-powered robots assist workers with heavy lifting and precision assembly, reducing workplace injuries and enhancing productivity. The manufacturing sector benefits from intelligent automation, where robots collaborate with human operators to optimize production lines and minimize errors. In infrastructure services, autonomous robotic systems are introduced for maintenance tasks, ensuring efficiency and safety in high-risk environments such as railways and large-scale public infrastructure projects (Fig. 3).

Applying the FORTIS solution across these pilots highlights its adaptability and transformative potential in the industry. By integrating AI, data analytics, and robotics into real-world scenarios, the project demonstrates how human-robot collaboration can enhance efficiency while maintaining a human-centric approach. The scalability of FORTIS allows for future expansion into additional sectors, such as logistics and healthcare, where similar human-robot interactions can drive operational improvements and safety enhancements. This underscores FORTIS's commitment to advancing the next generation of AI-driven industrial transformation.

A fundamental principle of FORTIS is the human-centered design approach, incorporating Design Thinking to align user needs with technological advancements. These principles ensure that AI and robotics systems are designed with workers in mind, promoting a smooth and safe transition to human-robot collaboration.

HRI is a multidisciplinary field requiring insights from social sciences and humanities to address ethical, psychological, and societal factors. FORTIS promotes a shift from traditional automation to collaborative frameworks where robots function as cooperative and predictive entities, responding in real time to human needs. A thorough understanding of how humans perceive and interact with robots informs the system design, ensuring usability and acceptance.

3.3   *JARVIS Project*

The JARVIS project[10,11] is designed to address the critical challenges hindering the industrial adoption of human-robot interaction (HRI), which persist despite research advancement. Specifically, JARVIS consortium aims to address:

---

[10] https://cordis.europa.eu/project/id/101135708

[11] https://www.jarvis-project.eu/

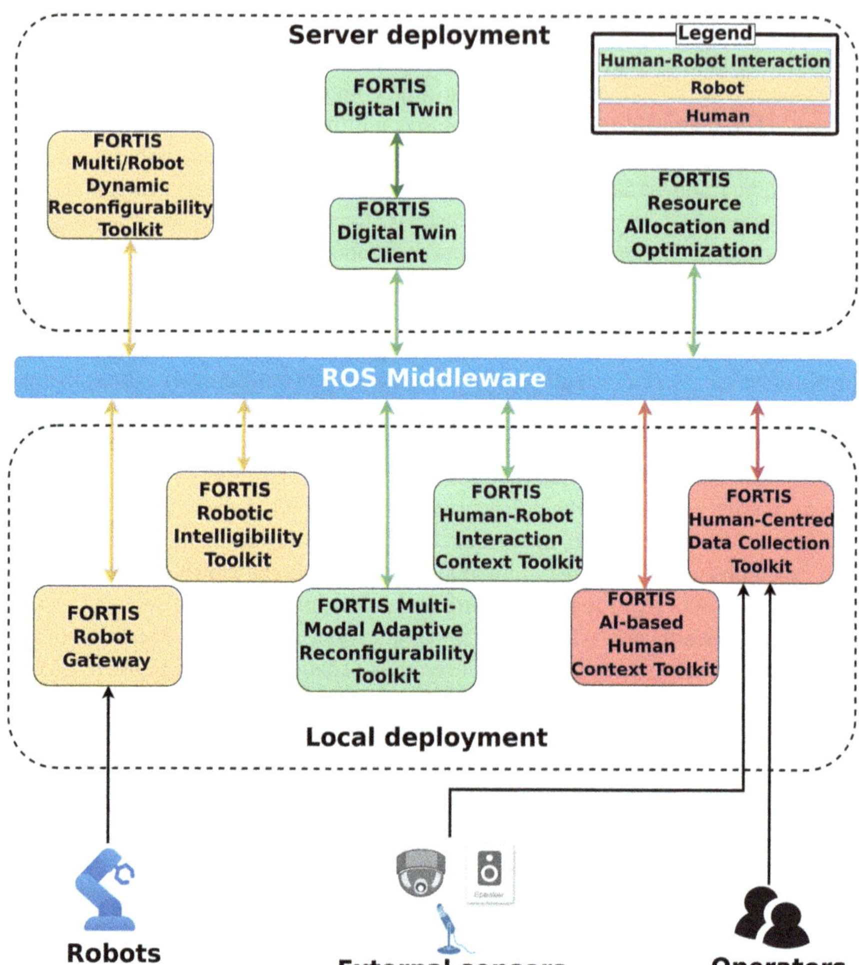

Fig. 3 FORTIS architecture, communication middleware, and toolkits

1. The limited cognition and intelligence of current robotic systems, which can coexist with humans but lack the ability to predict human actions and proactively adapt
2. The low performance of collaborative operations, where outdated safety measures and rigid action sequencing restrict robot speed and efficiency
3. The lack of collaboration fluidity, as operators are forced to adapt to the constraints of robotic systems rather than the other way around
4. The complexity of robot programming, which requires specialized expertise, preventing flexible, intuitive programming approaches that could benefit from operators' tacit knowledge

These limitations can significantly impact the systems performance in complex industrial environments where human behavior, task variability, and workspace dynamics cannot be easily pre-programmed or predicted. Traditional systems often rely on fixed interaction models and rule-based logic, which fail to accommodate the variability and contextual richness of human-robot collaboration. Moreover, the frequent absence of integrated perception and reasoning mechanisms restricts the robot's ability to interpret intent, adapt to deviations, or adjust in real time, ultimately hindering trust, usability, and performance.

In this context, JARVIS is built on four technological pillars to overcome these limitations. The first is user-centric social interaction design, ensuring intuitive, human-friendly collaboration by reducing operator strain, cognitive load, and interaction friction. The second is cognitive mechatronics, which embeds adaptive intelligence into robotic systems to enhance real-time perception, learning, and human-intention prediction. The third pillar focuses on AI-enhanced interfaces for seamless communication, control, and programming, facilitating natural and multimodal human-robot interaction. Finally, JARVIS integrates security, privacy, and trustworthy AI principles to ensure compliance with regulatory frameworks and promote reliable and responsible automation.

The JARVIS framework, as illustrated in the figure, leverages ROS2 middleware for scalability, modularity, and interoperability across diverse industrial environments [38]. The system features AI-driven work plan optimization, dynamically adapting task execution based on calculated key performance indicators to ensure process efficiency. A centralized AI engine orchestrates task execution, decision-making, and coordination while relying on structured data management through MongoDB. Multimodal human-robot interaction is facilitated by immersive XR interfaces, enabling real-time teleoperation, process monitoring, and adaptive control. Additionally, perception-driven autonomy is achieved through AI-powered computer vision and sensor fusion, allowing robots to autonomously navigate, recognize objects, and adapt to changing environments. JARVIS is designed as a containerized solution using Docker-based microservices, ensuring cross-platform deployment across edge, cloud, and on-premise infrastructures (Fig. 4).

The system's capabilities will be demonstrated in four industrial pilot applications, each highlighting JARVIS's adaptability to complex and unstructured environments. In aeronautics assembly, AI-driven robots will support the precise manufacturing of composite aircraft components, reducing rework and optimizing material usage. The automotive battery assembly pilot will focus on improving the safe and efficient manipulation of hybrid vehicle battery packs, minimizing manual intervention in hazardous environments. In nuclear decommissioning, JARVIS will enable teleoperated robots to dismantle and remove large plant components, significantly reducing operator exposure to radiation while ensuring precise and controlled disassembly. Lastly, in offshore energy maintenance, XR-based teleoperation will allow remote intervention in offshore facilities, ensuring that robots can inspect and maintain critical infrastructure with minimal human risk (Fig. 5).

JARVIS promotes user-centric social human-robot collaboration, which aims to augment human capabilities in complex tasks. With the integration of AI [39], cognitive mechatronics, and immersive human-robot interfaces, JARVIS aims to transform industrial automation by making robots more adaptable, intuitive, and easy to

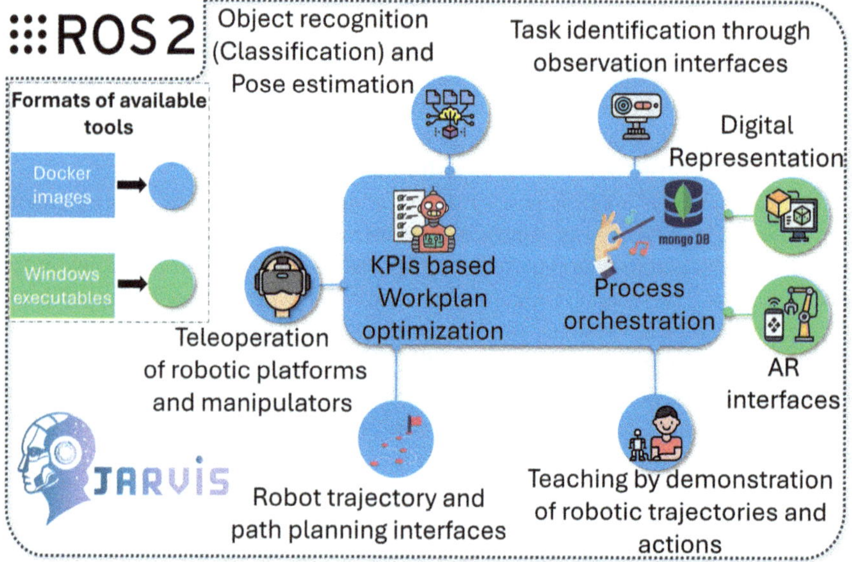

Fig. 4 JARVIS architecture, communication middleware, and tools

Fig. 5 JARVIS vision for (**a**) aircraft seat production, (**b**) car battery assembly, (**c**) nuclear decommissioning, and (**d**) offshore inspection and maintenance

integrate into human workflows. With its four key focus areas and real-world testing across different industries, JARVIS sets a new benchmark for reliable, scalable, and efficient human-robot collaboration in demanding work environments.

4   Boosting Business and Innovation for the Industry

Small and medium-sized enterprises (SMEs) play a crucial role in technological advancement but often face significant challenges in accessing funding opportunities. Traditional funding mechanisms can be complex and time-consuming, misaligned with the rapid and iterative nature of SME-driven innovation. To bridge this gap, Financial Support to Third Parties (FSTP) calls serve as a streamlined instrument to provide accessible and efficient funding, allowing SMEs to focus on developing and validating cutting-edge solutions without administrative burdens. These calls facilitate quicker technology development cycles, ensuring that innovations can transition smoothly from research to market applications.

Beyond its immediate research objectives, ARISE project is committed to fostering a long-term community that extends beyond the project's duration. Through its open calls, ARISE supports SMEs, research institutions, and developers in testing and validating cutting-edge robotic solutions in real-world industrial environments. By providing access to Testing and Experimentation Facilities (TEFs), ARISE ensures that innovations can be rigorously evaluated under practical conditions, improving their readiness for market adoption. Additionally, ARISE is building an ecosystem that promotes continuous collaboration between industry players, academia, and technology providers. The project actively engages in knowledge-sharing activities, networking events, and joint initiatives with other European robotics and AI projects. By establishing an open-source repository of middleware solutions, standardized interfaces, and best practices, ARISE aims to create a foundation that future innovators can build upon, ensuring the sustainability and growth of the robotics industry well beyond the project's timeline.

FORTIS project aims to drive business innovation and industrial progress through its two Open Calls, focusing on human-robot interaction (HRI) solutions. The initiative is designed to attract and support SMEs, startups, and research institutions working on innovative technologies that can enhance collaboration between humans and robots in various industrial and non-industrial settings. By integrating AI-based cognition, multimodal communication, and dynamic robot adaptation, FORTIS seeks to develop more effective and user-friendly robotic systems that improve safety, productivity, and efficiency in workplaces. The program provides funding, mentorship, and structured evaluation processes to ensure high-quality technological advancements that align with market needs.

To boost business and market deployment, FORTIS open calls provide financial and strategic support for the development and integration of robotic solutions within the FORTIS ecosystem. By funding projects that align with specific toolkits such as AI-based cognition, robotic intelligibility, and environment analysis, FORTIS ensures that selected innovations meet industry requirements and are ready for market adoption.

FORTIS also prioritizes strong collaboration between industry and research institutions to create a sustainable innovation ecosystem. The structured selection process, including external evaluation and expert mentoring, guarantees that only high-potential projects receive funding. Additionally, the project emphasizes human-centric, robot-centric, and trust-based interactions to enhance market trust in robotic solutions. By ensuring technological robustness and seamless integration into existing industrial environments, FORTIS accelerates commercialization, making robotics more accessible and beneficial for businesses, ultimately leading to increased market competitiveness and industry transformation.

The JARVIS project actively promotes business innovation through targeted open calls designed to bring advanced human-robot interaction (HRI) technologies closer to industrial adoption. Its Financial Support to Third Parties (FSTP) mechanism supports two distinct open call tracks: external pilots, in which consortia propose solutions and use cases aligned with JARVIS objectives, and co-development pilots, where new HRI technologies are co-created for the project's industrial pilots. These open calls provide equity-free funding, mentorship, and access to the JARVIS toolset including reusable AI-enhanced planning modules, cognitive mechatronics, multimodal interaction interfaces, and digital twins. Through these instruments, JARVIS facilitates the development, validation, and deployment of scalable HRI solutions tailored to real-world industrial needs. In particular, the project incentivizes innovations that address the barriers to collaborative robotics adoption in complex and safety-critical environments, with a strong focus on usability, trust, and open source and interoperability. This approach ensures that the outcomes of the open calls are not only technically robust but also aligned with long-term industry demands and societal expectations.

Collectively, the open call mechanisms of ARISE, FORTIS, and JARVIS demonstrate a shared commitment to fostering an innovation-friendly environment for SMEs, startups, and research actors. By lowering entry barriers, accelerating development cycles, and offering structured pathways from experimentation to market deployment, these initiatives contribute to a more dynamic, inclusive, and competitive European robotics ecosystem.

5 Conclusions

The three projects share a common vision of advancing AI-driven robotics within the framework of Industry 5.0.

Their collective efforts complement the comprehensive ecosystem that has been established by the European Commission to support AI innovation, facilitate industrial adoption, and strengthen cross-sector collaboration. Specifically, Horizon Europe, ADRA, Networks of Excellence Centers, and Testing and Experimentation Facilities (TEFs) provide critical infrastructure for developing, validating, and deploying AI-driven solutions across diverse industries while promoting the Industry 5.0 principles, emphasizing human-centric AI, sustainable manufacturing, and trustworthy automation.

Focusing on HRI, the integration of AI, data, and robotics in Industry 5.0 has demonstrated significant advancements by fostering safer, more efficient, and adaptable automation solutions. ARISE, JARVIS, and FORTIS showcase further progress in HRI, thanks to key innovations in middleware interoperability, AI-driven perception, cognitive mechatronics, and multimodal interfaces, and facilitate the validation of such technologies across diverse industrial applications.

The projects collectively emphasize the importance of standardization, real-world validation, and SME empowerment to drive industrial adoption. ARISE has developed an All-in-One Middleware leveraging ROS2 and FIWARE to enhance data exchange, interoperability, and scalability, while its Testing and Experimentation Facilities (TEFs) and Financial Support to Third Parties (FSTP) program accelerate technology transfer. JARVIS has developed a robust set of tools to support multimodal human-robot interaction, aiming at promoting the adoption of user-centric context-aware HRI. The developed JARVIS toolset will be validated in four large-scale industrial pilots from the automotive, aeronautics, and energy sectors. At the same time, JARVIS, to further support wider industrial adoption of human-robot interaction, provides FSTP to SMEs along with mentoring and access to the JARVIS toolset. These tools are to be integrated in cooperation with third-party beneficiaries and applied in real-world use cases defined by the external pilots themselves. FORTIS demonstrated the value of modular robotics for enhancing safety and efficiency. It emphasized adaptability and human-centric design as critical success factors. These insights provide a foundation for developing comprehensive industrial policies.

Despite these advancements, challenges remain in interoperability, workforce acceptance, regulatory clarity, and cost barriers, particularly for SMEs. To foster large-scale adoption, standardized frameworks, AI safety regulations, and funding mechanisms must support seamless AI-robotics integration. Future research should focus on cross-sector expansion, enhanced human-machine collaboration, and the alignment of European initiatives to ensure sustainable and inclusive robotics innovation.

In summary, the collaborative synergies among ARISE, JARVIS, and FORTIS have underscored the critical importance of establishing standardized middleware frameworks, specifically through ROS-based systems combined with FIWARE technologies. This standardization has significantly improved interoperability, allowing seamless real-time data exchange between diverse operational technologies (OT) and IT platforms, thereby reducing complexity and cost barriers that typically hinder the adoption of ADR solutions. By validating these solutions through extensive pilot applications and structured open calls aimed particularly at SMEs, the projects demonstrated the practical viability of integrating sophisticated robotic precision with human adaptability. Furthermore, this integration fosters trust, promotes worker safety and well-being, and supports sustainability and circular economy practices through optimized resource usage and efficient robotic operations.

Ultimately, the collaborative experiences from ARISE, JARVIS, and FORTIS underline the need for continued cross-sector engagement, inclusive technological innovation, and strategic support mechanisms.

Acknowledgments This research was partly funded by the ARISE, JARVIS, and FORTIS projects, which have received funding from the European Union's Horizon Europe Research and Innovation Programme under grant agreement nos. 101135784, 101135708, and 101135707, respectively.

References

1. Makris, S., Michalos, G., Dimitropoulos, N., Krueger, J., & Haninger, K. (2024). Seamless human–robot collaboration in industrial applications. In *CIRP novel topics in production engineering* (Vol. 1, pp. 39–73). Springer.
2. Chryssolouris, G., Alexopoulos, K., & Arkouli, Z. (2023). *A perspective on artificial intelligence in manufacturing* (Vol. 436, pp. 1–135). Springer.
3. Arkouli, Z., Michalos, G., Kokotinis, G., & Makris, S. (2024). Worker-centered evaluation and redesign of manufacturing tasks for ergonomics improvement using axiomatic design principles. *CIRP Journal of Manufacturing Science and Technology, 55*, 188–209.
4. Zhang, C., Wang, Z., Zhou, G., Chang, F., Ma, D., Jing, Y., Cheng, W., Ding, K., & Zhao, D. (2023). Towards new-generation human-centric smart manufacturing in Industry 5.0: A systematic review. *Advanced Engineering Informatics, 57*. https://doi.org/10.1016/j.aei.2023.102121
5. Arkouli, Z., Tompoulidis, I., Kontos, M., Michalos, G., & Makris, S. (2024). Collaborative human-centered design of manufacturing tasks: A multi-user immersive VR experience. *Procedia CIRP, 128*, 597–602.
6. Arkouli, Z., Tompoulidis, I., Dimitropoulos, N., Michalos, G., & Makris, S. (2024). Collaborative AI & immersive VR simulation for workplace layout optimization in manufacturing applications. *Procedia CIRP, 130*, 336–341.
7. Matheson, E., Minto, R., Zampieri, E. G., Faccio, M., & Rosati, G. (2019). Human–robot collaboration in manufacturing applications: A review. *Robotics, 8*(4), 100.
8. Makris, S. (2021). *Cooperating robots for flexible manufacturing* (pp. 2293–2300). Springer.
9. Krüger, J., Lien, T. K., & Verl, A. (2009). Cooperation of human and machines in assembly lines. *CIRP Annals, 58*(2), 628–646.
10. Dimitropoulos, N., Togias, T., Zacharaki, N., Michalos, G., & Makris, S. (2021). Seamless human–robot collaborative assembly using artificial intelligence and wearable devices. *Applied Sciences, 11*(12), 5699.
11. Curry, E., Heintz, F., Irgens, M., Smeulders, A. W. M., & Stramigioli, S. (2022). Partnership on AI, data, and robotics. *Communications of the ACM, 65*(4), 54–55. https://doi.org/10.1145/3513000
12. Wani, S., & Nemade, G. (2023). The synergy between artificial intelligence and robotics. *International Journal of Scientific Engineering and Research, 11*(10), 9–12.
13. Schmidpeter, B., & Winter-Ebmer, R. (2021). Automation, unemployment, and the role of labor market training. *European Economic Review, 137*, 103808. https://doi.org/10.1016/j.euroecorev.2021.103808. ISSN: 0014-2921.
14. Marasigan, J., Albert, J., & Wong, Y. (2024). Adaptive robotics: Integrating robotic simulation, AI, image analysis, and cloud-based digital twin simulation for dynamic task completion. In *International Conference on Human-Computer Interaction* (pp. 262–271). Springer.
15. Michalos, G., Karagiannis, P., Dimitropoulos, N., Andronas, D., & Makris, S. (2022). Human robot collaboration in industrial environments. In *The 21st century industrial robot: When tools become collaborators* (pp. 17–39). Springer.

16. Katsampiris-Salgado, K., Dimitropoulos, N., Gkrizis, C., Michalos, G., & Makris, S. (2024). Advancing human-robot collaboration: Predicting operator trajectories through AI and infrared imaging. *Journal of Manufacturing Systems, 74*, 980–994.
17. Schlechtriem, M. (2022). Interoperability: A key challenge for the robotics industry. HowToRobot. Retrieved from https://howtorobot.com/expert-insight/interoperability
18. GitHub. (2024). ROS4HRI framework. Retrieved from https://github.com/ros4hri
19. Séverin, L., et al. (2020). ROS for human-robot interaction. https://doi.org/10.48550/arXiv.2012.13944
20. GitHub. (2024). FIWARE smart data models. Retrieved from https://github.com/smart-data-models/
21. ETSI. (2024). Context Information Management (CIM); NGSI-LD API, RGS/CIM-009v181.
22. Kunze, L. et al. (2018). Artificial intelligence for long-term robot autonomy: A survey. arXiv [Cs.RO]. Retrieved from http://arxiv.org/abs/1807.05196
23. Thaker, R. (2024). Generative AI and robotics: From large language models to intelligent human-robot interaction and task planning. *International Journal of Innovative Research in Engineering and Multidisciplinary Physical Sciences, 12.* https://doi.org/10.5281/zenodo.14001479
24. Mohsen, S., Arezoo, B., & Dastres, R. (2023). Artificial intelligence, machine learning and deep learning in advanced robotics, a review. *Cognitive Robotics, 3*, 54–70.
25. Aristeidou, C., Dimitropolus, N., & Michalos, G. (2024). Generative AI and neural networks towards advanced robot cognition. *CIRP Annals, 73*(1), 21–24.
26. Fletcher, S. R., & Webb, P. (2017). Industrial robot ethics: The challenges of closer human collaboration in future manufacturing systems. In *A World with Robots: International Conference on Robot Ethics: ICRE 2015* (pp. 159–169). Springer.
27. Dimitropoulos, N., Papalexis, P., Michalos, G., & Makris, S. (2023). Advancing human-robot interaction using AI—A large language model (LLM) approach. In *European Symposium on Artificial Intelligence in Manufacturing* (pp. 116–125). Springer.
28. Morandini, S., et al. (2024). Human factors and emerging needs in aerospace manufacturing planning and scheduling. *Cognition, Technology & Work, 27*, 1–19.
29. Youtube. (2024). TEF1 Challenge 1 (CARTIF) experimental robotic cell for complex picking and dismantling applications. Retrieved from https://youtu.be/pVYA1EUMzQ0?si=fsfo7De-wBFFCdzT
30. ARISE Consortium. (2024). D2.1 NGSI-LD Context Broker with native DDS support.
31. ARISE Consortium. (2024). ARISE 1st OC—Technical guidelines. Retrieved from https://s3.amazonaws.com/fundingbox-sites/gear%2F1729262917453-ARISE+1st+OC+-+Technical+Guidelines.pdf
32. Youtube. (2024). TEF1 Challenge 2 (CARTIF) experimental robotic cell for complex picking and dismantling applications. Retrieved from https://youtu.be/ECLC7G9yq5k
33. Youtube. (2024). TEF2 Challenges 3/4 (INTELLIMECH) re-programmable co-bots for flexible manufacturing. Retrieved from https://youtu.be/OtXSwHZ7k9Q
34. Youtube. (2024). TEF3 Challenge 5 (PAL Robotics) enhancing HRI in healthcare via multimodal interaction. Retrieved from https://youtu.be/IoRsV484LuA
35. Youtube. (2024). TEF3 Challenge 6 (PAL Robotics) enhancing HRI in healthcare via multimodal interaction. Retrieved from https://youtu.be/-87R32HHaGI
36. Youtube. (2024). TEF4 Challenge 7 (Politecnico di Milano) human centric zero-defect manufacturing. Retrieved from https://youtu.be/R1yeb38ggEI
37. Youtube. (2024). TEF4 Challenge 8 (Politecnico di Milano) human centric zero-defect manufacturing. Retrieved from https://youtu.be/eWuR9BwIS54
38. Ginting, M. F., et al. (2021). CHORD: Distributed data-sharing via hybrid ROS 1 and 2 for multi-robot exploration of large-scale complex environments. *IEEE Robotics and Automation Letters, 6*(3), 5064–5071.
39. Makris, S., Alexopoulos, K., Michalos, G., Arkouli, Z., Papacharalampopoulos, A., Stavropoulos, P., et al. (2023). Artificial intelligence in manufacturing white paper prepared by the artificial intelligence in manufacturing network-AIM-NET.

Human-AI Interaction and Visualization Perspectives on ADR

Kostiantyn Kucher, Magnus Bång, and Jonas Lundberg

Abstract Recent advances in artificial intelligence (AI), data, and robotics (ADR) have pushed the boundaries of the benchmark performance of the respective methods and have already started to change the landscape in various application domains. Some of these domains are mission critical, with control activity that must match running processes. For those domains, a number of questions and challenges related to safety, robustness, and trustworthiness of AI and ADR methods and models still remain open, especially in the scenarios involving human operators. In this chapter, we provide an overview of human-centered perspectives on ADR with an emphasis on human-AI interaction, interactive visualization, and visual analytics. We explain the relationship of these fields to the related disciplines and fields, including human factors and human-computer interaction. We introduce the readers to basic concepts from these fields and discuss how the prior work fits with ADR principles, focusing on examples in visualization for explainable AI, cognitive systems engineering for joint human-AI control, and evaluation approaches for human-AI decision support systems. We argue that the techniques and frameworks proposed in these human-centered fields can and should be integrated with ADR methods.

Keywords Human-AI interaction · Information visualization · Visual analytics · Vis4ML · Joint human-AI control

K. Kucher (✉) · J. Lundberg
Department of Science and Technology, Linköping University, Norrköping, Sweden
e-mail: kostiantyn.kucher@liu.se; jonas.lundberg@liu.se

M. Bång
Department of Computer and Information Science, Linköping University, Linköping, Sweden
e-mail: magnus.bang@liu.se

E. Curry et al. (eds.), *Artificial Intelligence, Data and Robotics*,
https://doi.org/10.1007/978-3-032-10561-5_22

1 Introduction

The remarkable progress achieved by researchers and practitioners in artificial intelligence (AI) and robotics over the past decade has benefitted from access to larger datasets, better hardware and infrastructure, and more ingenious algorithms. New workflows and paradigms have emerged with an emphasis on making use of existing models that are often very powerful but also very costly to train and highly complex with respect to their structure and behavior. In order to facilitate further progress in AI, data, and robotics (ADR)—and to enable successful applications of the respective advances across various domains and scenarios, including mission-critical ones—a number of human-centered challenges must be addressed. While it may seem at first glance that focusing on humans is not so relevant for the disciplines of AI and robotics, counter-arguments are quite straightforward. Stakeholders at various stages of the underlying theoretical research, implementation, deployment, and regulation for AI and robotics are humans: researchers, developers, decision-makers, technical support staff, operators, end users, general public, authorities, etc. (as illustrated in a simplified way in Fig. 1). These stakeholders typically require a certain level of understanding and insight (ranging from high-level familiarity to in-depth expertise) in order to approve, use, and trust the corresponding ADR tools or robots. Regulations existing in specific fields and disciplines under specific jurisdictions can impose explicit requirements for such ADR systems. For instance, the recent EU AI Act lists a number of requirements for human oversight of high-risk AI systems in Article 14,[1] including the requirements for supporting monitoring, interpretation, and intervention in the operations of such systems [1].

Thus, it becomes clear that the concerns related to various forms of interaction and teaming across humans and ADR systems are highly relevant for most (if not all) stages of ADR design, development, and deployment. Such concerns have been studied across several disciplines and fields, from applied psychology and human factors to autonomous systems and data science. More specifically, researchers and developers may benefit from the techniques that provide overview and support interactive exploration of the data and models in detail, especially for machine learning (ML) scenarios, including deep learning (DL) and large language models (LLMs); system operators and decision-makers require carefully designed user interfaces that facilitate rather than hinder their activities; and further stakeholders as well as the general public may benefit from high-level summaries that convey the main points about the respective ADR systems without excessive detail and clutter. To explore the possibilities for addressing these challenges, we focus on the concepts and methods from human-computer interaction, information visualization, and visual analytics relevant to ADR.

This chapter relates to several concerns among cross-sectorial AI, data, and robotics technology enablers of the AI, Data, and Robotics Partnership [2],

---

[1] https://artificialintelligenceact.eu/article/14/; last accessed on June 1, 2025.

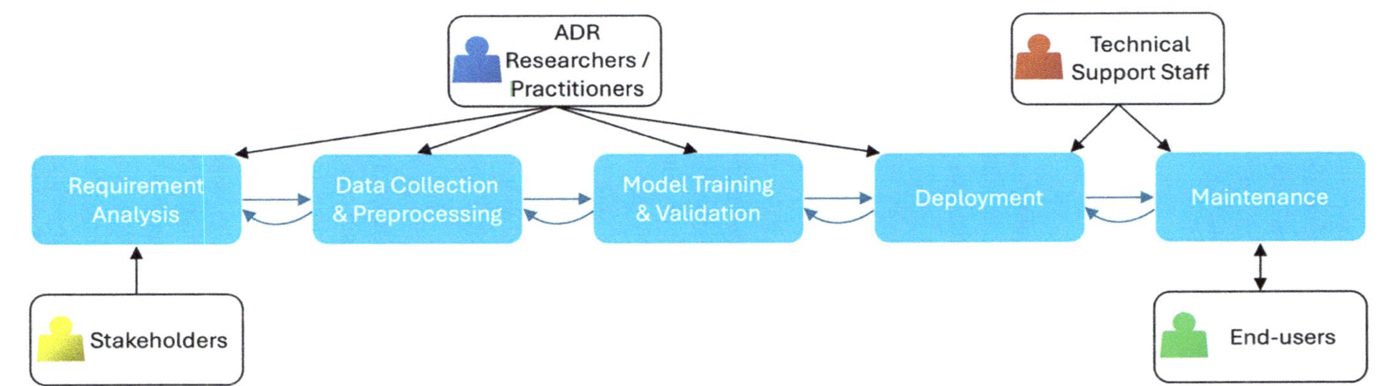

Fig. 1 A simplified model of ADR solution construction, deployment, and maintenance, including several key actors. While the focus of ADR research and applications may lie mainly on technical aspects, the involvement of humans in various roles alongside the pipeline (especially considering end-users) necessitates the involvement of human-centered methods and perspectives

including action and interaction, but also systems, methodologies, hardware, and tools. The rest of this chapter is organized as follows: in the next section, we start by introducing the basic concepts from the fields of human factors and human-computer interaction and then information visualization and visual analytics. Afterward, we discuss three topics that illustrate the potential of human-computer interaction and visualization applications for and interactions with ADR. Finally, we outline several potential directions for future work and conclude this chapter.

2 Background

In this section, we introduce the fields of human-computer interaction and information visualization (as well as several directly related fields) that is necessary for a further discussion of human-centered techniques and findings relevant to ADR in the next section.

2.1 Human Factors and Human-Computer Interaction

The discipline of *human factors* (or *ergonomics*) focuses on interactions within systems comprising humans and their environments, typically in the context of work (professional duties) performed by humans [3], but also services received by humans [4]. The environments include physical or virtual artefacts, organizational structures related to activities, and social structures related to other people. The focus of human factors research is thus to jointly improve the performance of the complete system, typically by fitting the environment to the human. Two major system outcomes can be achieved in this way through systems design: performance (productivity, quality, effectiveness, efficiency, etc.) and well-being (health and safety, satisfaction, learning, etc.) [4].

While the discipline of human factors has its roots in earlier research from the nineteenth century, it is generally considered that the contemporary discipline emerged in the late 1940s [3]. While the main concerns of human factors were originally related to *physical* ergonomics, with time, focus was also put on *cognitive* and *organizational* ergonomics. Cognitive aspects (especially from not only the reactive, summative perspective but also more proactive and potentially prescriptive one) have also been the subject of the field of cognitive systems engineering that is defined by Militello et al. as "an approach to the design of technology, training, and processes intended to manage cognitive complexity in sociotechnical systems" [5]. Relationship between these and further disciplines and fields is complicated by their interdisciplinary nature and existence of a rich lexicon of related terms and topics [6]. While the traditional human factors research and applications focusing on physical ergonomics had their roots in medical and biomechanical fields, the focus on cognitive aspects required methods from applied psychology, systems engineering, and design.

Eventually, the discipline (or field) of *human-computer interaction* emerged on the intersection of human factors and computer science. Ebert et al. describe the goals and scope of human-computer interaction as "to improve the interaction between users and computing devices in the sense that this interaction should become more user-friendly and better adapted to the needs and capabilities of the users and the capabilities of the device" [7]. Dix mentions a number of disciplines and fields as overlapping with and helping define human computer interaction, but specifically, computer science, psychology, and ergonomics, afterward joined by social sciences due to the interest in sociotechnical design challenges [8]. One of the important topics in human-computer interaction is the design and evaluation of user interfaces. The methods from cognitive systems engineering can be applicable and helpful in this regard when considering support for *decision-making* in complex sociotechnical systems, in particular, the *ecological interface design* [9, 10]. This methodology relies on *work domain analysis* of the respective domain problem with a distinction between lower-level physical information and higher-level, more abstract functional information, which is relevant to the notion of *abstraction hierarchy*. Our previous work elaborated on the applications of this methodology for the air traffic control domain. Our initial interactive interface contribution for supporting "what-if" and "what-else" probes [11] was followed up by a study applying work domain analysis. The latter included design and evaluation of two ecological user interfaces for dealing with conflicts in air traffic control while aiming for safety, performance, and efficiency as functional purposes [12].

The development and adoption of technology has been naturally affecting human factors and human-computer interaction, initially with respect to physical but eventually also cognitive and organizational ergonomics. Bainbridge discussed a number of "ironies of automation"—challenges arising, among others, from the assumptions that replacing human operators with fully automated systems would only lead to benefits in the short and long term [13]. The issues that could ensue, however, include the loss of vigilance while monitoring a long-running process; losing the skills over a longer period of time; and reducing (or even eliminating) the opportunities to train new staff members. With recent advances in artificial intelligence, new ironies have emerged [14]. In addition to these concerns, the issue of *trust* in automation should also be considered [15], as both over- and under-trust can lead to suboptimal and even catastrophic results in mission-critical scenarios. While the notion of trust has been debated across numerous studies and fields, the definition by Lee and See has the benefit of being operationalizable for both research and application purposes: according to them, trust is "the attitude that an agent will help achieve an individual's goals in a situation characterized by uncertainty and vulnerability" [15]. Further concerns relevant to human factors in human-automation collaboration scenarios include *cognitive load* [16] and *situation awareness* [17, 18]. The latter can be described as a dynamic process that comprises the ongoing and continuously updated comprehension of the situation at large (*frames*), more specific details affecting the situation (*implications*), elements of the environment (*objects*), and the *event horizon* of actual and potential developments [18].

In order to evaluate/validate particular designs and techniques, human factors and specifically human-computer interaction researchers and practitioners rely on a multitude of experimental methods that can be traced back to behavioral and social sciences [19]. Such experimental methods can be characterized by the dimensions of precision, generalizability, and realism: for instance, a laboratory experiment can provide a high degree of precision, but it will lead to a trade-off with realism and potentially generalizability. Both quantitative and qualitative methods have their use in human-computer interaction [20], while one of the most common target measures is *usability*. Usability can be defined and evaluated alongside three aspects [21]:

Effectiveness (Are the users able to complete their tasks? How many errors they make?)
Efficiency (How much time does it take for the users to complete their tasks?)
Satisfaction (What is the users' opinion/attitude toward the tool?) [21]

While effectiveness and efficiency are often measured as part of task-based user studies (typically laboratory or perhaps field experiments), satisfaction is usually gauged by interviews and questionnaires with Likert-scale items, with several prominent examples being the 10-item System Usability Scale [22] or the 2-item UMUX-LITE [23]. The knowledge about such methodologies may be beneficial when designing ADR experiments involving human participants or judges; and they are also often relevant to evaluation of visualization techniques and tools, as discussed next.

2.2 *Information Visualization and Visual Analytics*

The field of information visualization (InfoVis) has emerged in 1990s on the intersection of research and applications in scientific visualization, computer graphics, human-computer interaction (especially graphical user interfaces), cognitive science, data science, knowledge discovery in databases, cartography, graphic design, color science, and further fields. The historical roots of InfoVis can be traced back for several centuries, including examples of thematic and topological maps, mathematical diagrams and statistical charts, as well as infographics handcrafted for printed media with illustratory purposes. However, the identity and special focus of the InfoVis field are strongly associated with a general workflow for visually representing and interacting with abstract data (typically multivariate/multidimensional and often nonspatial in contrast to scientific visualization). The commonly accepted definition of the scope of InfoVis is "The use of computer-supported, interactive, visual representations of abstract data to amplify cognition" [24]. As Stuart Card put it, "The purpose of information visualization is to amplify cognitive performance, not just to create interesting pictures. Information visualizations should do for the mind what automobiles do for the feet" [25].

The overall design of InfoVis techniques and tools typically follows the InfoVis Reference Model [24] or similar frameworks: the tool allows the user to load their data, and then the data is preprocessed and mapped to certain *visual representations* and rendered in one or multiple *views*. The user is usually provided with a number of *interaction* options to investigate the data using the corresponding views, adjust them, or even change the representation. The options and concerns associated with each of the individual steps of the design process according to such a framework, their interactions/relationships, series of such interactions, and the overall context of use constitute the space of research and application challenges for the InfoVis field. For instance, the question of choosing the optimal (or at least more suitable than baseline alternatives) visual representation for a particular data type and scale with further constraints (e.g., the conventions existing for graphical representations and notations in a specific discipline/field such as biology) is a rather typical research problem in InfoVis, addressed through a combination of theoretical and empirical methods.

The complete process of designing, implementing, and validating InfoVis techniques and tools should be carried out in a human-centered way while being based on the existing theoretical models and empirical evidence accumulated in this field. One of the methodologies commonly adopted in InfoVis is the nested blocks model [26] that defines several nested steps for the iterative design and validation process, from domain problem characterization to abstraction to design of individual visual representations and interactions. Another prominent example is the data-users-tasks design triangle model [27] that highlights the respective concerns and their interactions. Finally, Sedlmair et al. describe the design study methodology for InfoVis suitable for applied research and collaborations across and beyond academia [28].

Continuing this discussion in a top-down fashion, the next crucial step of the design process is to define the intended *user tasks* that should drive the resulting design, as high-level tasks of exploratory or confirmatory data analyses would typically require a different design than storytelling/narrative visualization. Considering low-level tasks, the classical InfoVis workflow describing a mix of user tasks and corresponding interactions is known as the Visual Information Seeking Mantra: "Overview first, zoom and filter, then details-on-demand" [29]. Besides these four tasks, Shneiderman's taxonomy also includes further tasks such as "relate," "history," and "extract." Another prominent taxonomy focuses on ten low-level analysis tasks [30]: "retrieve value," "filter," "compute derived value," "find extremum," "sort," "determine range," "characterize distribution," "find anomalies," "cluster," and "correlate."

Analysis and design of individual visual representations can be approached from the perspective of combining basic *marks* (graphical elements such as points or lines) with further visual *channels* (graphical properties such as position, color, shape, curvature, orientation/tilt, size, texture, or motion) [31, 32], typically for a given spatial substrate (such as a 2D Cartesian coordinate system). For example, switching line segments into individual dots would turn a line plot into a scatterplot, while varying the area of dots based on the underlying data would turn a scatterplot into a bubble plot. It should be mentioned that InfoVis techniques rely "by default" on 2D rather than 3D representations when the latter is not justified [32], especially for standard 2D computer monitor screens. The motivation for this includes distortion, occlusion, and navigation issues as well as lack of performance benefits for many data types and tasks, based on prior studies. When properly motivated and used in right contexts (including immersive environments), 3D may have its benefits, of course.

Since the purpose of InfoVis is not only to provide static visual representations of the data (such as diagrams and plots created using spreadsheet or scientific computation software, for instance) but generally also to facilitate further engagement with such representations, the next step of the design process is to consider *interactions* that should be supported for the user to facilitate their tasks. High-level interaction strategies in InfoVis [33] include support for *navigation* through the data space, *direct manipulation* (allowing the user select and adjust data items via their visual representations, rather than requiring database queries or scripting/coding), and *human/agent interaction*, which has become an active area of research and applications over the past years due to the development of conversational agents and natural language interfaces [34, 35]. Considering more low-level interaction techniques, the taxonomy by Shneiderman described above [29] is complemented by the interaction taxonomy by Yi et al.: "select," "explore," "reconfigure," "encode," "abstract/elaborate," "filter," and "connect" [36]. As an example of "abstract/elaborate," *geometric zooming* allows the user to simply adjust the scaling when exploring representations such as scatter plots, while *semantic zooming* may change the visual representation based on the level of detail, e.g., display additional details and labels when zooming in (this technique is thus also related to the "encode" category). Further, multiple *focus + context* techniques that apply distortions (such as the fisheye lens) provide a more salient representation of the selected data subset ("focus") while still supporting the overview of further data ("context") in the same view [37].

While the discussion above was mainly centered on users and tasks, the data-related concerns also play a crucial role in the choice of appropriate visual representations and interactions: "The expressiveness principle dictates that the visual encoding should express all of, and only, the information in the dataset attributes" [32]. In other words, the visual representation should not "invent" additional dimensions or relations (such as typical 3D bar charts with the depth attribute not based on the underlying data) or mislead the users by hinting at such non-existent relations and patterns due to poor visual encoding choices. At the same time, while it is possible to directly map one, two, or even three data attributes/dimensions to the respective 2D/3D spatial substrate, the choice of visual representation for more than three attributes at a time is generally not a trivial task—and in the general case, certain trade-offs must be made that compromise the "… express all of, and only, the information" part of the expressiveness principle. Suggestions for the choice of visual representations of low-dimensional data as well as some "special" data types (such as geospatial data) are provided by Shneiderman as well as Heer et al. [29, 38]. Further, in-depth discussions and online survey browsers for hundreds of existing techniques and tools have also been contributed over the years in the InfoVis field.

For instance, our work on text visualization [39] describes the design space of respective techniques and tools, as depicted in Fig. 2. The respective online survey, the TextVis Browser,[2] not only provides access and links to the respective publications but also hosts a collection of links to similar resources focusing on the topics such as visualization of trees, multivariate graphs/networks, etc. Another prominent collection to mention is the Visualization Resources Web site[3] [40].

The strategies for representation of high-dimensional data [41] can be mapped back to the elements of the InfoVis Reference Model [24], for instance, *dimensionality reduction* [42] could be applied early on to lay out the source data entries as points in a 2D or 3D space, with the assumption that the entries similar in the original high-dimensional data space will be positioned close to each other in the resulting projection space. This strategy could be generally applied for any high-dimensional dataset—however, the quality of the resulting projection as well as the ability of users to actually interpret the resulting point cloud layouts depends on multiple factors (algorithm/model properties, applicability for specific data and tasks, etc.). Alternative strategies include the use of *multiple coordinated views* [43] that usually rely on *brushing* + *linking* interactions: as the user directly selects items in one view such as a scatterplot, e.g., using a rectangular or lasso selection (the "brushing" step), the other views are dynamically updated to highlight, filter, change pan or zoom level, etc. accordingly (the "linking" step). For example, this strategy could be applied alongside a *scatterplot matrix* as yet another general approach for any high-dimensional dataset. This is a powerful strategy; however, it requires more space to juxtapose several views and potentially imposes the *change blindness* risk for a larger number of views and more complex and larger datasets. Finally, the visual metaphor itself could be designed in a nontrivial way that facilitates one or several tasks with such high-dimensional data, including iconic *glyphs* [44] or representations of a single data entry as a polyline in *parallel coordinates* [45]; the rather popular *radar chart* representation (also known as a *Kiviat diagram, star chart, spider chart*, etc.) is a variation of the parallel coordinate plot with a radial axis layout, for instance.

Last but not least, the stage of *evaluation/validation* of InfoVis approaches is crucial for ensuring that novel ideas or combined applications of existing techniques reach the intended goals. InfoVis relies on the body of knowledge of evaluation in human-computer interaction described above, ranging from formal methods to usability studies to expert reviews [46]. Isenberg et al. provide a systematic review of the evaluation methods used in InfoVis studies [47], while the works by Lam et al. and Elmquist and Yi provide guidelines for designing and conducting evaluations according to typical scenarios and patterns [48, 49].

While this description provides an overview of the field of InfoVis, the reader may have noticed that the focus of the discussion so far has been on the careful mapping from user requirements and data characteristics to the design and evaluation of

---

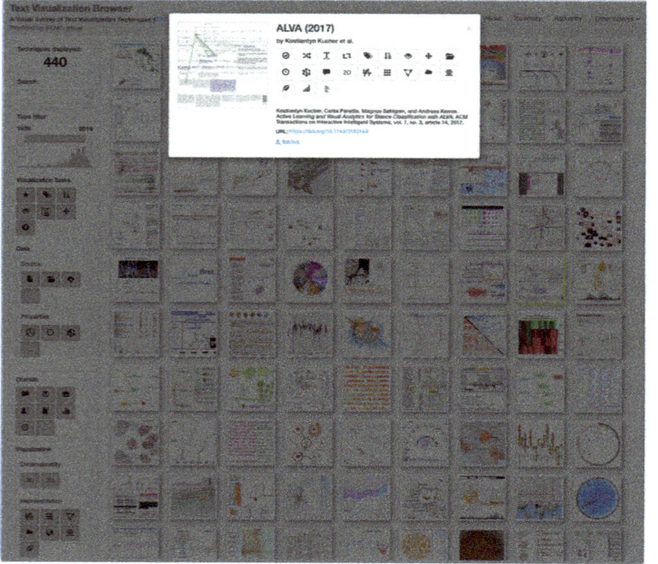

Fig. 2 TextVis Browser, a manually curated online visual survey of text visualization techniques. Left: on clicking one of the thumbnail images in the grid, the user is presented with a modal dialog with details about the respective visualization technique, including bibliographic details as well as categorization in the respective design space. Right: the overview of the complete categorized dataset included in the survey browser, providing a glimpse at the patterns existing within and across the respective design space dimensions/categories

interactive visualization techniques and tools with only cursory mentions of *computational* methods. Indeed, while the traditional InfoVis approaches relied on a rather direct mapping process from data (traditionally, from databases or static files) to visual representations, eventually the challenges related to processing larger, more complex, and heterogeneous datasets for broader and more demanding tasks gave way to the concept of *visual analytics*. Keim et al. provide the following definition for this field: "Visual analytics combines automated analysis techniques with interactive visualizations for an effective understanding, reasoning and decision making on the basis of very large and complex data sets" [50]. While the boundary between visual analytic and InfoVis fields and methods is often fuzzy, the following Visual Analytics Mantra by Keim et al. illustrates the respective focus: "Analyze First, Show the Important, Zoom, Filter and Analyze Further, Details on Demand" [50]. The need to accommodate analysts in the world with more complex and less predictable data and models was discussed by Amar and Stasko around the same time as a follow-up to their earlier work [30, 51], with the focus on supporting high-level analytic activities such as "complex decision-making, especially under uncertainty," "learning a domain," "identifying the nature of trends," and "predicting the future." The later Knowledge Generation Model by Sacha et al. describes the relationships across elements and activities that take place within the software (data, computational model, interactive visualization affordances) and beyond (the human analyst's steps for exploring the data, formulating and verifying hypotheses, and eventually generating new knowledge about the domain problem) [52]. Visual analytics has become an important field over the past two decades, and its role especially in the context of ADR-related topics and concerns cannot be overstated. The recent discussion of grand challenges in visual analytic applications, for instance, included closer intertwining of visual analytics and AI, providing guidance (which requires mixed-initiative approaches relying on AI), and explainability in complex visual analytic systems [53], which opens up opportunities for mutual benefits for research and applications involving visual analytics and ADR.

Turning to industrial and societal uptake of InfoVis, novel tool prototypes as well as commercial products have been developed within or close to the visualization research community [54], such as Tableau, complemented by a variety of *business intelligence and analytics* tools that integrate some level of support for visual representation and interaction techniques with data preprocessing and computational models [55], such as Microsoft Power BI or IBM Cognos. Design, implementation, and evaluation of such tools aimed at professionals with a varying degree of expertise in visualization and computational disciplines pose a set of additional challenges. The existing work on this genre of *visualization dashboards* includes lessons learned [56], design spaces [57], and actionable design patterns [58] from visualization experts.

3 Human-Centered Perspectives on ADR

In this section, we discuss three topics based on the existing literature as well as our own work that highlight different constellations of human-computer interaction, visualization, and ADR concerns.

3.1 Human-Computer Interaction and Visualization for Explainable AI

One of the directions of research and applications in human-computer interaction, InfoVis, and visual analytics that is highly relevant to ADR is support for various stages of AI (especially ML/DL) model training and deployment pipeline. Some of the prominent examples of the work in visual analytics focusing on ML (rather than later DL) approaches include interactive support for construction, debugging, and analysis of application of decision trees [59], support vector machine classifiers for text retrieval [60], inter-active ensemble learning for video analytics [61], or visual interactive labeling [62].

Our own prior work on supporting the active learning-based multi-label data annotation and classifier training processes for research on sentiment and stance in text data led to designing and applying a similar approach with our collaborators, researchers in linguistics and computational linguistics [63]. ALVA is a visual analytic tool that provides user interfaces for annotation, visual analyses of the annotated data and annotation process itself, as well as monitoring of the classification model performance over the active learning training rounds. The Web-based tool supports several user roles with permissions to access and make annotations, control the active learning process, or access the visual analytic interfaces. The focus of the latter is on analyzing the progress of the annotation process as well as identifying patterns in the data annotated so far, as displayed in Fig. 3. By using a novel visual representation titled *CatCombos* ("category combinations"), our collaborators were able to identify interesting cases and patterns of interaction between several semantic categories of stance in the annotated text data, such as expressions of uncertainty, hypotheticals, and prediction. Similar approaches are sought after by researchers and practitioners in computational linguistics, as understanding and supporting the data annotation/labeling process contributes in a major way to the success of resulting models. Considerations for interdisciplinary collaborations in visual text analytics [64] can thus be extended for other areas of ADR in the future.

The rapid development of AI approaches across academia and industry over the past decade has also affected the fields of human-computer interaction and visualization, with new challenges and research directions emerging to support the research and applications of such approaches. The rising interest for *interpretable* (with the model internals understandable to humans) and *explainable* (with the model behavior and individual decisions being described and motivated for humans)

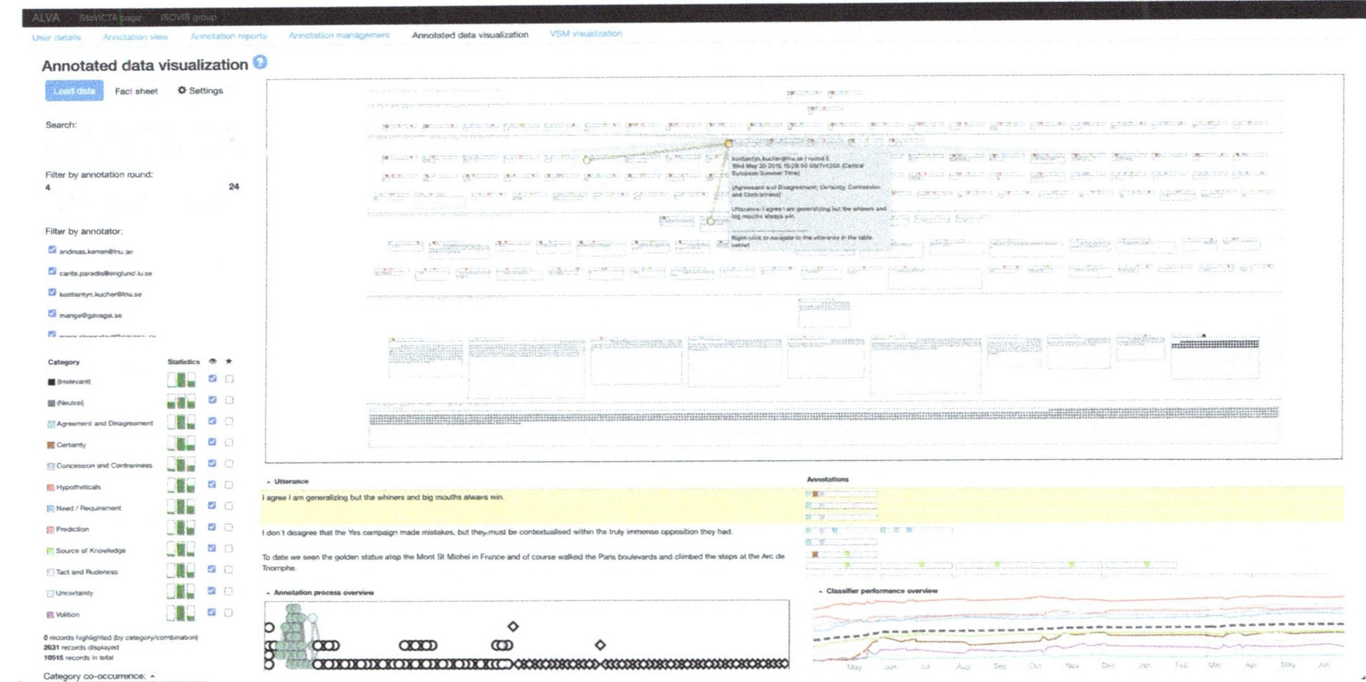

Fig. 3 The user interface of *ALVA* used for the visual analysis of annotated data, annotation process, and active learning classifier training for multi-label text classification

AI [65] has led to the initiatives on human-computer interaction research agenda for explainable, accountable, and intelligent systems [66], as well as the human-centered artificial intelligence agenda [67]. Amershi et al. proposed a set of 18 generally applicable guidelines for human-AI interaction applicable not only for research prototypes but also industrial products, divided into several main categories according to the training and use stages/situations: initial, ongoing, AI error/failure, and long-term use [68]. Liao et al. contributed an explainable AI question bank collected via discussions with industry design practitioners on the users' needs for explanations and explainability [69]. These guidelines and resources are relevant for ongoing and future work in ADR, as they elaborate on the constraints and preferences for AI models and tools that are used and/or overseen by humans.

The work within the InfoVis and especially visual analytic fields relevant to interpretable and explainable AI has resulted in a number of interactive visual approaches to support human-centered machine learning [70, 71], explainable deep learning [72], and trust in AI [73]. Several prominent examples to mention here include the *What-If Tool* for interactive model probing and counterfactual reasoning [74], *AttentionViz* for a global view on Transformer attention patterns [75], and the *explAIner* framework for generating visual explanations and suggestions for model steering [76].

Our own work in this area has resulted in design spaces and interactive survey browsers focusing on the topics of enhancing the trust in ML models with the use of visualization[4] [77, 78] and the use of data embedding approaches in visual analytics[5] [79]. These resources (see Fig. 4) position the existing work and allow researchers and practitioners alike to find the existing visual analytic techniques supporting tasks such as ML model construction, debugging, quality/bias control, interactive exploration and comparison of the embedding space, as well as representation and explanation of model decisions. Furthermore, we contributed visual analytic techniques and tools such as *StackGenVis* for training stacking ML classifiers [80] and *EEVO* for text similarity analyses using ensembles of embeddings [81].

There is potential for further advances in the area by combining human-computer interaction, visual analytics, and ADR involving the challenges of interpretability, explainability, and trustworthiness of AI—not only within academia but also with further societal and industrial collaborations. This is highlighted by our recent conceptual framework on this topic [82] as well as a study on explainable AI interfaces for operators in process industries [83].

---

[4] https://trustmlvis.lnu.se; last accessed on June 1, 2025.

[5] https://va-embeddings-browser.ivis.itn.liu.se/; last accessed on June 1, 2025.

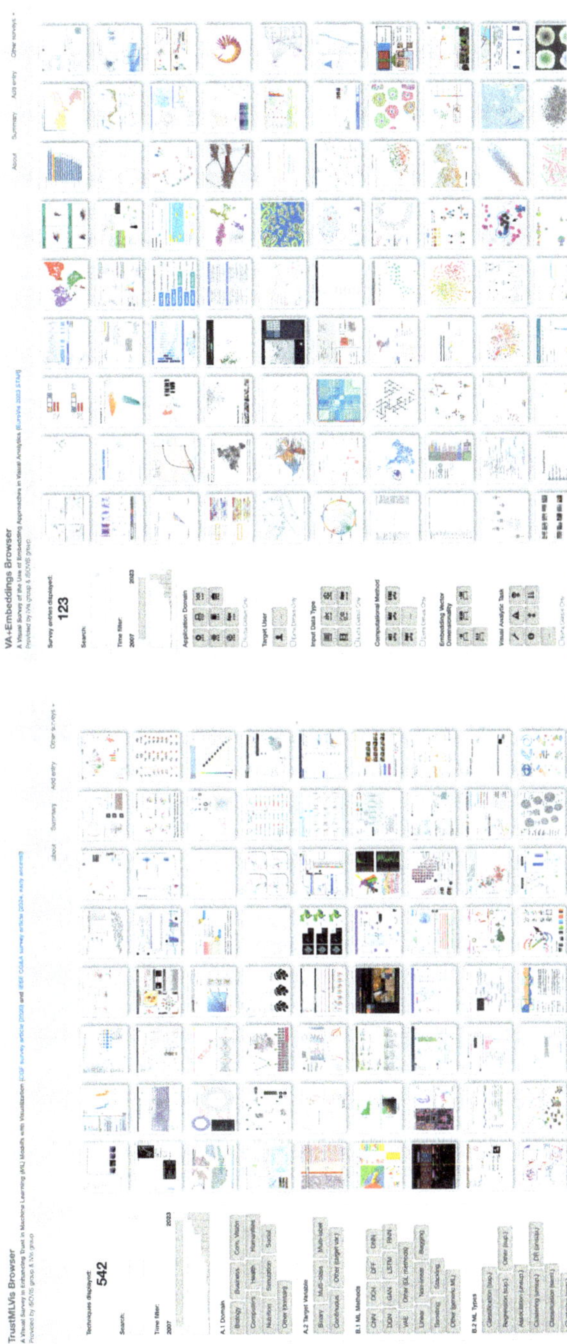

Fig. 4 Further online survey browsers relevant to ADR topics. Left: TrustMLVis Browser, currently hosting a dataset with 542 categorized visualization techniques aimed at enhancing trust in Machine Learning (ML) models. Right: VA + Embeddings Browser, currently hosting a dataset with 123 categorized visualization techniques relevant to the use of data embedding methods at various stages of the visual analytic pipeline

3.2   Joint Human-AI Control

Some of the "ironies of automation" described above [13] remained open over the course of decades, resulting in the *automation conundrum* associated with the change of *level of automation* (LOA) from completely manual human control toward full automation [17]: "as more autonomy is added to a system, and its reliability and robustness increase, the lower the situation awareness of human operators and the less likely that they will be able to take over manual control when needed." When considering AI instead of traditional automation techniques, new ironies emerge: according to Endsley, for instance, "The more intelligent and adaptive the AI, the less able people are to understand the system" and "The more capable the AI, the poorer people's self-adaptive behaviours for compensating for shortcomings" [14]. Requirements for explainability, human interaction and oversight, training and skill retention, and joint testing of human-AI systems are mentioned among alleviations for these challenges [14].

Relying on the notions of abstraction hierarchy [10], situation awareness [18], and LOA [17], our own work proposed the *Joint Control Framework* (*JCF*) [84] that describes human-machine (including automation and AI) interaction scenarios with a focus on *cognitive joints*, moments in time and space where human and machine work together. JCF relies on three key components:

1. Process mapping that results in definition and categorization of relevant subjects, objects, and effects for the scenario at hand.
2. The framework of *Levels of Autonomy in Cognitive Control* (*LACC*) that defines six levels (1–2: *how?*, 3–4: *what?*, 5–6: *why?*) of autonomy and interaction in relation to situational awareness concepts such as frames and the event horizon.
3. JCF *Score*, a notation for temporal description resembling the musical notation used for sheet music. This score uses six horizontal lines to represent six LACC levels. Separate groups of lines can be used for separate subjects or objects, similarly to separate staffs used for different instruments in sheet music.

While the framework and score can be applied on their own, the ongoing work on a prototype of *JCF Editor* depicted in Fig. 5 facilitates transcription and analyses of complex use case scenarios recorded with the existing human-interaction tools and users. In this example, future drone traffic is analyzed from the perspective of an operator who needs to inspect planned traffic before approving it [85]. Looking at Fig. 5, we immediately can see that the blue decision joints of the human subject are much higher up than their green and orange action and decision joints. This means that the subject must make inferences in abstraction, i.e., consequences of implemented plans on higher goals—and then back again from those higher-level goals to what adjustments on lower levels that must be made to fulfil the goals.

The relevance of this line of work to ADR lies in the ability to formalize and analyze complex scenarios including multiple subjects and objects over time, such as one or several operators and one or several automation/AI agents. The methodology can be applied for evaluation of existing techniques and planning of future

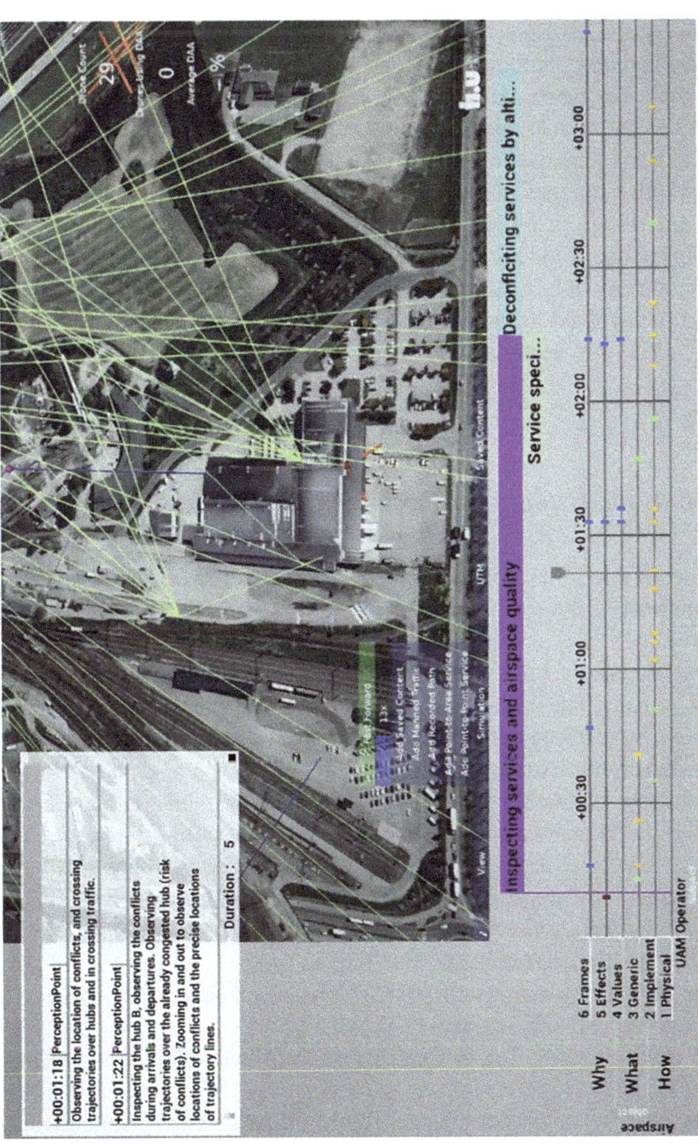

Fig. 5 The user interface of JCF Editor used for annotating and analyzing the data from a study on unmanned aerial vehicle traffic coordination [85]. Blue markers represent decision joints between human operators and the traffic process, green markers represent actions, and orange markers represent perception joints. The markers represent what the joint is about, e.g., is it about a plan (L3) or a goal (L5), and is this goal or plan explicitly represented so that it can be perceived and/or manipulated? Level differences between perception-decision-action mean that the subject does the interpretation themselves, e.g., cognitive or computational load

studies. A relevant example here involves the work on human-in-the-loop AI for future, advanced air traffic management systems combining manned and unmanned vehicles and services [86]. To accommodate the long-term planning and design for such eventual future systems (in the air traffic control domain and beyond), Nylin et al. described the concept of Reduced Autonomy Workspace that combines the notions of LOA and LACC to describe the extent of divided and collaborative work between subjects, including automation/AI agents [87]. The term "reduced autonomy" refers to the cases when the automation/AI agent needs to hand over some degree of control to the human operator (e.g., due to uncertainty or accountability requirements), and this handover should be adapted to accommodate the operator (e.g., checking whether the operator is highly occupied at the moment). Nylin et al. also introduced the design of glyphs for similar real-time scenarios in future systems that visually encodes time remaining to make the necessary decision [88]. The development of actual models, agents, and autonomous systems within the scope of ADR that would be involved in such traffic management or unmanned traffic scenarios provides clear opportunities for future contributions and collaborations across fields.

3.3 *Evaluation for Human-AI Decision Support Systems*

Our final topic is relevant to the context of the ongoing *AI4REALNET* project[6] that aims to achieve human-in-the-loop decision-making for co-learning between AI and humans. The project focuses on application scenarios with critical network infrastructure systems (electricity, railway, and air traffic management) that are traditionally operated by humans. The project is tasked with not only purely technical (such as novel reinforcement learning methods for complex networks) but also sociotechnical goals to enhance transparency, safety, explainability, and acceptance of the developed AI solutions by human stakeholders (in particular, industrial operators). More specifically, the human-centered evaluations planned within the project are intended to assess the social-technical decision quality; AI acceptability, trust, and trustworthiness; human user experience; AI and human learning curves; task allocation balance; and long-term consequences of AI assistants.

As discussed above, the concerns related to human-centered AI and ML [67, 68, 71] are relevant for such scenarios in ADR. More specifically, Hoffman et al. describe the 7-item Explanation Goodness Checklist (to be used by other researchers/experts), the corresponding 7-item Explanation Satisfaction Scale with Likert items (to be used by study participants), and the 8-item XAI Trust Scale with Likert items (also to be used by study participants) [89]. Similar questionnaires are also used with respect to further (X)AI properties in the published work, and they are certainly useful; however, the work by Buçinca et al. indicated that proxy tasks and

---

[6]https://ai4realnet.eu/; last accessed on June 1, 2025.

subjective measurements may be misleading in evaluating explainable systems; thus, experimental design for such tasks can require careful planning [90]. The work by Nauta et al. defined 12 properties for explanations such as *correctness, completeness*, or *compactness* and provided first steps toward systematic evaluation of these properties by combining computational methods (e.g., randomly perturbing the internals of the model and checking whether the generated explanation would change) as well as user studies [91], which may also be relevant for this project.

It is also interesting to note within the context of ADR, especially robotics, that some of the prior work focusing on human-robot interaction discusses measurements and evaluation methods applicable for human-AI interaction involving software AI agents. Nikolaidis et al. evaluate human-robot *mutual adaption*, including direct measurements of human participants' self-confidence in their ability to complete the task as well as their assessment of trustworthiness of the robot (and several further measurements in separate experiments) [92]. Hoffman discusses both subjective (based on self-reported questionnaire results) and objective metrics of *fluency* in human-robot interaction; the latter include the percentage of concurrent activity, the human's idle time, the robot's functional delay, and the robot's idle time [93]. These measures are interesting to consider for other human-AI interaction scenarios; however, for various tasks involving monitoring of ongoing activities, measures such as idle time may not be relevant. Perception of robots and other AI agents by humans is also the subject of the Artificial-Social-Agent Questionnaire [94]. The resulting short version of the questionnaire comprises 24 Likert-scale items that range from the appearance of the agent to its usability, social presence, and perceived user-agent alliance. Finally, Hauptman et al. discuss an evaluation of perceptions of *adaptive* autonomous agents for human-AI teaming [95]. In their study, AI teammates could be either considered manually set to a particular LOA or considered able to adapt their LOA dynamically. The results of the factorial survey indicate that predictability and degree of team experience can enable higher levels of autonomy in such scenarios. These studies as well as further work related to sociotechnical evaluation of human-AI systems will be relevant to the evaluations within the scope of the *AI4REALNET* project, and we hope that the outcomes of the project (including experimental protocols) will nurture future work in ADR as well as interdisciplinary collaborations.

4 Conclusions and Opportunities

In this chapter, we have outlined the basic concepts as well as several scenarios and prior results in human-AI interaction and visualization that are relevant to ADR. We have described methods and techniques that originate from these fields and would be directly relevant and helpful for ADR researchers and practitioners. *We conclude that combining these would have broad impact into several areas of concern such as developing user interfaces and visual analytic solutions to facilitate ADR development, debugging, and monitoring and analyzing the existing and potential scenarios of multi-agent interactions.* We especially see promise in addressing the involvement of human stakeholders by approaching the evaluation of ADR solutions as sociotechnical problems, with human-centered concerns being as important as model performance.

Besides these direct applications of human-centered methods, we would also like to mention the following challenges and directions where collaborations between researchers and practitioners in ADR and human-centered fields have the potential to make impact:

4.1 Reaching Out to Audiences in Critical Application Fields

As discussed in this chapter, operators in highly specialized fields such as air traffic control or process industries are trained to follow protocols that emphasize safety and predictability. Introducing black-box ADR methods directly at the level of full automation for such users and fields may cause short-term and especially long-term issues including loss of situational awareness, under−/over-trust, and deskilling. The study with decision-makers from the industry by Bedué and Fritzsche shows that AI adoption intention among such respondents is affected by several dimensions of trust (including *access to knowledge, transparency, explainability,* and *reliability*) as well as perceived benefits and risks [96]. Peres et al. highlight the requirements for keeping human in the loop using human-machine technology, achieving data availability and quality, and ensuring interpretability and trust for AI as major challenges for industrial AI research and adoption, especially considering higher LOA [97]. Making use of human-centered methods to introduce ADR techniques and models may thus be the best way to find common ground with such stakeholders and users.

4.2 Reaching Out to Wider Audiences

Interestingly, not only audiences in highly specialized domains but also the general public can benefit from collaborations across human-centered methods and ADR, considering the ubiquity and impact of ADR technologies on the society and everyday life. Sinderman et al. describe the Attitude Towards Artificial Intelligence Scale based on five Likert-scale items with questions such as "I fear AI" and "AI will benefit humankind" [98]. Human-centered methods can thus be helpful for introducing and educating the general public on various aspects of ADR, which would allow them to form informed opinions.

4.3 Supporting Data-Centric and Human-Centric AI Paradigms

One of the recent developments directly relevant to ADR is the emergence of the *data-centric* AI paradigm that shifts the focus from algorithm/model performance toward the role of data in the AI life cycle; as Gröger put it, "There is no AI without data" [99]. Sambasivan et al. discuss the risks rising from cascades of negative, downstream effects from data issues and highlight the role of data as a first-class citizen in AI [100]. Jarrahi et al. describe the principles for data-centric AI, which explicitly mention "Human-Centeredness of 'Data Work,'" "AI as a Sociotechnical System," and "Continuous and Substantive Interactions Between AI and Domain Experts," which align with our recommendations for deeper collaboration between researchers in human-centered fields and ADR [101].

4.4 Preparing Appropriate Responses to ADR Cybersecurity and Privacy Challenges

Peres et al. highlight the role of cybersecurity and privacy for industrial AI [97]; however, we would argue that ADR applications in further critical application fields must make use of all feasible strategies to ensure the highest degree of cybersecurity. West and Aydin discuss how the process of aligning AI models with user preferences and values can be manipulated by malicious actors [102]. Machine learning models used for dimensionality reduction to represent data can also be attacked and manipulated in some scenarios, as discussed in our recent work [103]. There is thus room for extensive research and innovation on the intersection of ADR, human-centered methods, and cybersecurity as well as privacy.

4.5 Contributing Novel ADR-Driven Interfaces and Interactions

Finally, besides applying human-centered methods for ADR, ADR methods could also be highly beneficial for human-computer interaction and visualization research and applications. The existing work on automation in visualization [104], guidance [105], adaptive visualization [106], and further opportunities of integrating ML and AI in visualization [107–109] provides us with clear evidence for the interest existing for such collaborations among our fields.

Acknowledgments The work on this chapter was carried out within the scope of the AI4REALNET project. AI4REALNET has received funding from European Union's Horizon Europe Research and Innovation Programme under Grant Agreement No. 101119527 and from the

Swiss State Secretariat for Education, Research and Innovation (SERI). Views and opinions expressed are however those of the authors only and do not necessarily reflect those of the European Union and SERI. Neither the European Union nor the granting authority can be held responsible for them.

References

1. Enqvist, L. (2023). 'Human oversight' in the EU artificial intelligence act: What, when and by whom? *Law, Innovation and Technology, 15*(2), 508–535.
2. Curry, E., Heintz, F., Irgens, M., Smeulders, A. W., & Stramigioli, S. (2022). Partnership on AI, data, and robotics. *Communications of the ACM, 65*(4), 54–55.
3. Karwowski, W. (2005). Ergonomics and human factors: The paradigms for science, engineering, design, technology and management of human-compatible systems. *Ergonomics, 48*(5), 436–463.
4. Dul, J., Bruder, R., Buckle, P., Carayon, P., Falzon, P., Marras, W. S., Wilson, J. R., & Van der Doelen, B. (2012). A strategy for human factors/ergonomics: Developing the discipline and profession. *Ergonomics, 55*(4), 377–395.
5. Militello, L. G., Dominguez, C. O., Lintern, G., & Klein, G. (2010). The role of cognitive systems engineering in the systems engineering design process. *Systems Engineering, 13*(3), 261–273.
6. Hoffman, R. R., Feltovich, P. J., Ford, K. M., Woods, D. D., Klein, G., & Feltovich, A. (2002). A rose by any other name… would probably be given an acronym. *IEEE Intelligent Systems, 17*(4), 72–80.
7. Ebert, A., Gershon, N. D., & van der Veer, G. C. (2012). Human-computer interaction: Introduction and overview. *KI-Künstliche Intelligenz, 26*, 121–126.
8. Dix, A. (2017). Human–computer interaction, foundations and new paradigms. *Journal of Visual Languages and Computing, 42*, 122–134.
9. Bennett, K. B., & Flach, J. (2019). Ecological interface design: Thirty-plus years of refinement, progress, and potential. *Human Factors, 61*(4), 513–525.
10. Vicente, K. J. (2002). Ecological interface design: Progress and challenges. *Human Factors, 44*(1), 62–78.
11. Zohrevandi, E., Westin, C. A., Lundberg, J., & Ynnerman, A. (2022). Design and evaluation study of visual analytics decision support tools in air traffic control. *Computer Graphics Forum, 41*(1), 230–242.
12. Zohrevandi, E., Westin, C. A., Vrotsou, K., & Lundberg, J. (2022). Exploring effects of ecological visual analytics interfaces on experts' and novices' decision-making processes: A case study in air traffic control. *Computer Graphics Forum, 41*(3), 453–464.
13. Bainbridge, L. (1982). Ironies of automation. *IFAC Proceedings Volumes, 15*(6), 129–135.
14. Endsley, M. R. (2023). Ironies of artificial intelligence. *Ergonomics, 66*(11), 1656–1668.
15. Lee, J. D., & See, K. A. (2004). Trust in automation: Designing for appropriate reliance. *Human Factors, 46*(1), 50–80.
16. Kosch, T., Karolus, J., Zagermann, J., Reiterer, H., Schmidt, A., & Woźniak, P. W. (2023). A survey on measuring cognitive workload in human-computer interaction. *ACM Computing Surveys, 55*(13s), 283.
17. Endsley, M. R. (2017). From here to autonomy: Lessons learned from human–automation research. *Human Factors, 59*(1), 5–27.
18. Lundberg, J. (2015). Situation awareness systems, states and processes: A holistic framework. *Theoretical Issues in Ergonomics Science, 16*(5), 447–473.
19. McGrath, J. E. (1995). Methodology matters: Doing research in the behavioral and social sciences. In *Readings in human–computer interaction* (pp. 152–169). Morgan Kaufmann.

20. Purchase, H. C. (2012). *Experimental human-computer interaction: A practical guide with visual examples*. Cambridge University Press.
21. Frøkjær, E., Hertzum, M., & Hornbæk, K. (2000). Measuring usability: Are effectiveness, efficiency, and satisfaction really correlated? In *Proceedings of the CHI Conference on Human Factors in Computing Systems* (pp. 345–352). ACM.
22. Brooke, J. (2013). SUS: A retrospective. *Journal of Usability Studies, 8*(2), 29.
23. Lewis, J. R., Utesch, B. S., & Maher, D. E. (2013). UMUX-LITE: When there's no time for the SUS. In *Proceedings of the CHI Conference on Human Factors in Computing Systems* (pp. 2099–2102). ACM.
24. Card, S. K., Mackinlay, J., & Shneiderman, B. (Eds.). (1999). *Readings in information visualization: Using vision to think*. Morgan Kaufmann.
25. Card, S. (2007). Information visualization. In *The human-computer interaction handbook: Fundamentals, evolving technologies and emerging applications* (2nd ed., pp. 509–543). CRC Press.
26. Munzner, T. (2009). A nested model for visualization design and validation. *IEEE Transactions on Visualization and Computer Graphics, 15*(6), 921–928.
27. Miksch, S., & Aigner, W. (2014). A matter of time: Applying a data–users–tasks design triangle to visual analytics of time-oriented data. *Computers & Graphics, 38*, 286–290.
28. Sedlmair, M., Meyer, M., & Munzner, T. (2012). Design study methodology: Reflections from the trenches and the stacks. *IEEE Transactions on Visualization and Computer Graphics, 18*(12), 2431–2440.
29. Shneiderman, B. (1996). The eyes have it: A task by data type taxonomy for information visualizations. In *Proceedings of the IEEE Symposium on Visual Languages* (pp. 336–343). IEEE.
30. Amar, R., Eagan, J., & Stasko, J. (2005). Low-level components of analytic activity in information visualization. In *Proceedings of the IEEE Symposium on Information Visualization* (pp. 111–117). IEEE.
31. Görg, C., Pohl, M., Qeli, E., & Xu, K. (2007). Visual representations. In *Human-Centered Visualization Environments: GI-Dagstuhl Research Seminar, Dagstuhl Castle, Germany, March 5–8, 2006, Revised Lectures* (pp. 163–230). Springer.
32. Munzner, T. (2014). *Visualization analysis and design*. CRC Press.
33. Fikkert, W., D'Ambros, M., Bierz, T., & Jankun-Kelly, T. J. (2007). Interacting with visualizations. In *Human-Centered Visualization Environments: GI-Dagstuhl Research Seminar, Dagstuhl Castle, Germany, March 5–8, 2006, Revised Lectures* (pp. 77–162). Springer.
34. Hoque, E., & Islam, M. S. (2025). Natural language generation for visualizations: State of the art, challenges and future directions. *Computer Graphics Forum, 44*(1), e15266.
35. Voigt, H., Alaçam, Ö., Meuschke, M., Lawonn, K., & Zarrieß, S. (2022). The why and the how: A survey on natural language interaction in visualization. In *Proceedings of the Conference of the North American Chapter of the Association for Computational Linguistics: Human Language Technologies* (pp. 348–374). ACL.
36. Yi, J. S., ah Kang, Y., Stasko, J., & Jacko, J. A. (2007). Toward a deeper understanding of the role of interaction in information visualization. *IEEE Transactions on Visualization and Computer Graphics, 13*(6), 1224–1231.
37. Kosara, R., Hauser, H., & Gresh, D. L. (2003). An interaction view on information visualization. In *EuroGraphics'03—State of the art reports*. Eurographics.
38. Heer, J., Bostock, M., & Ogievetsky, V. (2010). A tour through the visualization zoo. *Communications of the ACM, 53*(6), 59–67.
39. Kucher, K., & Kerren, A. (2015). Text visualization techniques: Taxonomy, visual survey, and community insights. In *Proceedings of the IEEE Pacific Visualization Symposium* (pp. 117–121). IEEE.
40. Liu, X., Alharbi, M. S., Chen, J., Diehl, A., Rees, D., Firat, E. E., Wang, Q., & Laramee, R. S. (2023). Visualization resources: A survey. *Information Visualization, 22*(1), 3–30.

41. Liu, S., Maljovec, D., Wang, B., Bremer, P. T., & Pascucci, V. (2016). Visualizing high-dimensional data: Advances in the past decade. *IEEE Transactions on Visualization and Computer Graphics, 23*(3), 1249–1268.
42. Espadoto, M., Martins, R. M., Kerren, A., Hirata, N. S., & Telea, A. C. (2019). Toward a quantitative survey of dimension reduction techniques. *IEEE Transactions on Visualization and Computer Graphics, 27*(3), 2153–2173.
43. Roberts, J. C. (2007). State of the art: Coordinated & multiple views in exploratory visualization. In *Proceedings of the International Conference on Coordinated and Multiple Views in Exploratory Visualization* (pp. 61–71). IEEE.
44. Borgo, R., Kehrer, J., Chung, D. H., Maguire, E., Laramee, R. S., Hauser, H., Ward, M., & Chen, M. (2013). Glyph-based visualization: Foundations, design guidelines, techniques and applications. In *EuroGraphics'13—State of the art reports* (pp. 39–63). Eurographics.
45. Inselberg, A. (1985). The plane with parallel coordinates. *The Visual Computer, 1*, 69–91.
46. Carpendale, S. (2008). Evaluating information visualizations. In *Information visualization: Human-centered issues and perspectives* (pp. 19–45). Springer.
47. Isenberg, T., Isenberg, P., Chen, J., Sedlmair, M., & Möller, T. (2013). A systematic review on the practice of evaluating visualization. *IEEE Transactions on Visualization and Computer Graphics, 19*(12), 2818–2827.
48. Elmqvist, N., & Yi, J. S. (2015). Patterns for visualization evaluation. *Information Visualization, 14*(3), 250–269.
49. Lam, H., Bertini, E., Isenberg, P., Plaisant, C., & Carpendale, S. (2011). Empirical studies in information visualization: Seven scenarios. *IEEE Transactions on Visualization and Computer Graphics, 18*(9), 1520–1536.
50. Keim, D., Andrienko, G., Fekete, J. D., Görg, C., Kohlhammer, J., & Melançon, G. (2008). Visual analytics: Definition, process, and challenges. In *Information visualization: Human-centered issues and perspectives* (pp. 154–175). Springer.
51. Amar, R. A., & Stasko, J. T. (2005). Knowledge precepts for design and evaluation of information visualizations. *IEEE Transactions on Visualization and Computer Graphics, 11*(4), 432–442.
52. Sacha, D., Stoffel, A., Stoffel, F., Kwon, B. C., Ellis, G., & Keim, D. A. (2014). Knowledge generation model for visual analytics. *IEEE Transactions on Visualization and Computer Graphics, 20*(12), 1604–1613.
53. Wu, A., Deng, D., Chen, M., Liu, S., Keim, D., Maciejewski, R., Miksch, S., Strobelt, H., Viégas, F., & Wattenberg, M. (2023). Grand challenges in visual analytics applications. *IEEE Computer Graphics and Applications, 43*(5), 83–90.
54. Satyanarayan, A., Lee, B., Ren, D., Heer, J., Stasko, J., Thompson, J., Brehmer, M., & Liu, Z. (2019). Critical reflections on visualization authoring systems. *IEEE Transactions on Visualization and Computer Graphics, 26*(1), 461–471.
55. Behrisch, M., Streeb, D., Stoffel, F., Seebacher, D., Matejek, B., Weber, S. H., Mittelstädt, S., Pfister, H., & Keim, D. (2018). Commercial visual analytics systems—Advances in the big data analytics field. *IEEE Transactions on Visualization and Computer Graphics, 25*(10), 3011–3031.
56. Froese, M. E., & Tory, M. (2016). Lessons learned from designing visualization dashboards. *IEEE Computer Graphics and Applications, 36*(2), 83–89.
57. Sarikaya, A., Correll, M., Bartram, L., Tory, M., & Fisher, D. (2018). What do we talk about when we talk about dashboards? *IEEE Transactions on Visualization and Computer Graphics, 25*(1), 682–692.
58. Bach, B., Freeman, E., Abdul-Rahman, A., Turkay, C., Khan, S., Fan, Y., & Chen, M. (2022). Dashboard design patterns. *IEEE Transactions on Visualization and Computer Graphics, 29*(1), 342–352.
59. van den Elzen, S., & van Wijk, J. J. (2011). BaobabView: Interactive construction and analysis of decision trees. In *Proceedings of the IEEE Conference on Visual Analytics Science and Technology* (pp. 151–160). IEEE.

60. Heimerl, F., Koch, S., Bosch, H., & Ertl, T. (2012). Visual classifier training for text document retrieval. *IEEE Transactions on Visualization and Computer Graphics, 18*(12), 2839–2848.

61. Höferlin, B., Netzel, R., Höferlin, M., Weiskopf, D., & Heidemann, G. (2012). Interactive learning of ad-hoc classifiers for video visual analytics. In *Proceedings of the IEEE Conference on Visual Analytics Science and Technology* (pp. 23–32). IEEE.

62. Bernard, J., Zeppelzauer, M., Sedlmair, M., & Aigner, W. (2018). VIAL: A unified process for visual interactive labeling. *The Visual Computer, 34*, 1189–1207.

63. Kucher, K., Paradis, C., Sahlgren, M., & Kerren, A. (2017). Active learning and visual analytics for stance classification with ALVA. *ACM Transactions on Interactive Intelligent Systems, 7*(3), 14.

64. Kucher, K., Sultanum, N., Daza, A., Simaki, V., Skeppstedt, M., Plank, B., Fekete, J. D., & Mahyar, N. (2022). An interdisciplinary perspective on evaluation and experimental design for visual text analytics: Position paper. In *Proceedings of the IEEE Workshop on Evaluation and Beyond—Methodological Approaches for Visualization* (pp. 28–37). IEEE.

65. Gilpin, L. H., Bau, D., Yuan, B. Z., Bajwa, A., Specter, M., & Kagal, L. (2018). Explaining explanations: An overview of interpretability of machine learning. In *Proceedings of the IEEE International Conference on Data Science and Advanced Analytics* (pp. 80–89). IEEE.

66. Abdul, A., Vermeulen, J., Wang, D., Lim, B. Y., & Kankanhalli, M. (2018). Trends and trajectories for explainable, accountable and intelligible systems: An HCI research agenda. In *Proceedings of the CHI Conference on Human Factors in Computing Systems*. ACM.

67. Shneiderman, B. (2020). Human-centered artificial intelligence: Reliable, safe & trustworthy. *International Journal of Human-Computer Interaction, 36*(6), 495–504.

68. Amershi, S., Weld, D., Vorvoreanu, M., Fourney, A., Nushi, B., Collisson, P., Suh, J., Iqbal, S., Bennett, P. N., Inkpen, K., & Teevan, J. (2019). Guidelines for human-AI interaction. In *Proceedings of the CHI Conference on Human Factors in Computing Systems*. ACM.

69. Liao, Q. V., Gruen, D., & Miller, S. (2020). Questioning the AI: Informing design practices for explainable AI user experiences. In *Proceedings of the CHI Conference on Human Factors in Computing Systems*. ACM.

70. Andrienko, N., Andrienko, G., Adilova, L., & Wrobel, S. (2022). Visual analytics for human-centered machine learning. *IEEE Computer Graphics and Applications, 42*(1), 123–133.

71. Sperrle, F., El-Assady, M., Guo, G., Borgo, R., Chau, D. H., Endert, A., & Keim, D. (2021). A survey of human-centered evaluations in human-centered machine learning. *Computer Graphics Forum, 40*(3), 543–568.

72. La Rosa, B., Blasilli, G., Bourqui, R., Auber, D., Santucci, G., Capobianco, R., Bertini, E., Giot, R., & Angelini, M. (2023). State of the art of visual analytics for explainable deep learning. *Computer Graphics Forum, 42*(1), 319–355.

73. Beauxis-Aussalet, E., Behrisch, M., Borgo, R., Chau, D. H., Collins, C., Ebert, D., El-Assady, M., Endert, A., Keim, D. A., Kohlhammer, J., & Oelke, D. (2021). The role of interactive visualization in fostering trust in AI. *IEEE Computer Graphics and Applications, 41*(6), 7–12.

74. Wexler, J., Pushkarna, M., Bolukbasi, T., Wattenberg, M., Viégas, F., & Wilson, J. (2020). The What-If Tool: Interactive probing of machine learning models. *IEEE Transactions on Visualization and Computer Graphics, 26*(1), 56–65.

75. Yeh, C., Chen, Y., Wu, A., Chen, C., Viégas, F., & Wattenberg, M. (2024). AttentionViz: A global view of transformer attention. *IEEE Transactions on Visualization and Computer Graphics, 30*(1), 262–272.

76. Spinner, T., Schlegel, U., Schäfer, H., & El-Assady, M. (2020). explAIner: A visual analytics framework for interactive and explainable machine learning. *IEEE Transactions on Visualization and Computer Graphics, 26*(1), 1064–1074.

77. Chatzimparmpas, A., Kucher, K., & Kerren, A. (2024). Visualization for trust in machine learning revisited: The state of the field in 2023. *IEEE Computer Graphics and Applications, 44*(3), 99–113.

78. Chatzimparmpas, A., Martins, R. M., Jusufi, I., Kucher, K., Rossi, F., & Kerren, A. (2020). The state of the art in enhancing trust in machine learning models with the use of visualizations. *Computer Graphics Forum, 39*(3), 713–756.

79. Huang, Z., Witschard, D., Kucher, K., & Kerren, A. (2023). VA + Embeddings STAR: A state-of-the-art report on the use of embeddings in visual analytics. *Computer Graphics Forum, 42*(3), 539–571.

80. Chatzimparmpas, A., Martins, R. M., Kucher, K., & Kerren, A. (2020). StackGenVis: Alignment of data, algorithms, and models for stacking ensemble learning using performance metrics. *IEEE Transactions on Visualization and Computer Graphics, 27*(2), 1547–1557.

81. Witschard, D., Jusufi, I., Martins, R. M., Kucher, K., & Kerren, A. (2022). Interactive optimization of embedding-based text similarity calculations. *Information Visualization, 21*(4), 335–353.

82. Kucher, K., Zohrevandi, E., & Westin, C. A. (2025). Towards visual analytics for explainable AI in industrial applications. *Analytics, 4*(1), 7.

83. Zhang, Y., Methnani, L., Brorsson, E., Zohrevandi, E., Darnell, A., & Kucher, K. (2025). Designing explainable and counterfactual-based AI interfaces for operators in process industries. In *Proceedings of the International Joint Conference on Computer Vision, Imaging and Computer Graphics Theory and Applications* (GRAPP, HUCAPP and IVAPP) (Vol. 1, pp. 831–841). SciTePress.

84. Lundberg, J., & Johansson, B. J. (2021). A framework for describing interaction between human operators and autonomous, automated, and manual control systems. *Cognition, Technology & Work, 23*(3), 381–401.

85. Lundberg, J., Arvola, M., & Palmerius, K. L. (2021). Human autonomy in future drone traffic: Joint human–AI control in temporal cognitive work. *Frontiers in Artificial Intelligence, 4*, article 704082.

86. Lundberg, J., Bång, M., Johansson, J., Cheaitou, A., Josefsson, B., & Tahboub, Z. (2019). Human-in-the-loop AI: Requirements on future (unified) air traffic management systems. In *Proceedings of the IEEE/AIAA Digital Avionics Systems Conference*. IEEE.

87. Nylin, M., Johansson Westberg, J., & Lundberg, J. (2022). Reduced autonomy workspace (RAW)—An interaction design approach for human-automation cooperation. *Cognition, Technology & Work, 24*(2), 261–273.

88. Nylin, M., Lundberg, J., Bång, M., & Kucher, K. (2024). Glyph design for communication initiation in real-time human-automation collaboration. *Visual Informatics, 8*(4), 23–35.

89. Hoffman, R. R., Mueller, S. T., Klein, G., & Litman, J. (2023). Measures for explainable AI: Explanation goodness, user satisfaction, mental models, curiosity, trust, and human-AI performance. *Frontiers in Computer Science, 5*, 1096257.

90. Buçinca, Z., Lin, P., Gajos, K. Z., & Glassman, E. L. (2020). Proxy tasks and subjective measures can be misleading in evaluating explainable AI systems. In *Proceedings of the ACM International Conference on Intelligent User Interfaces* (pp. 454–464). ACM.

91. Nauta, M., Trienes, J., Pathak, S., Nguyen, E., Peters, M., Schmitt, Y., Schlötterer, J., Van Keulen, M., & Seifert, C. (2023). From anecdotal evidence to quantitative evaluation methods: A systematic review on evaluating explainable AI. *ACM Computing Surveys, 55*(13s), 295.

92. Nikolaidis, S., Hsu, D., & Srinivasa, S. (2017). Human-robot mutual adaptation in collaborative tasks: Models and experiments. *International Journal of Robotics Research, 36*(5–7), 618–634.

93. Hoffman, G. (2019). Evaluating fluency in human–robot collaboration. *IEEE Transactions on Human-Machine Systems, 49*(3), 209–218.

94. Fitrianie, S., Bruijnes, M., Li, F., Abdulrahman, A., & Brinkman, W. P. (2022). The artificial-social-agent questionnaire: Establishing the long and short questionnaire versions. In *Proceedings of the ACM International Conference on Intelligent Virtual Agents*. ACM.

95. Hauptman, A. I., Schelble, B. G., McNeese, N. J., & Madathil, K. C. (2023). Adapt and overcome: Perceptions of adaptive autonomous agents for human-AI teaming. *Computers in Human Behavior, 138*, 107451.

96. Bedué, P., & Fritzsche, A. (2022). Can we trust AI? An empirical investigation of trust requirements and guide to successful AI adoption. *Journal of Enterprise Information Management, 35*(2), 530–549.

97. Peres, R. S., Jia, X., Lee, J., Sun, K., Colombo, A. W., & Barata, J. (2020). Industrial artificial intelligence in Industry 4.0—Systematic review, challenges and outlook. *IEEE Access, 8*, 220121–220139.
98. Sindermann, C., Sha, P., Zhou, M., Wernicke, J., Schmitt, H. S., Li, M., Sariyska, R., Stavrou, M., Becker, B., & Montag, C. (2021). Assessing the attitude towards artificial intelligence: Introduction of a short measure in German, Chinese, and English language. *KI-Künstliche Intelligenz, 35*(1), 109–118.
99. Gröger, C. (2021). There is no AI without data. *Communications of the ACM, 64*(11), 98–108.
100. Sambasivan, N., Kapania, S., Highfill, H., Akrong, D., Paritosh, P., & Aroyo, L.M. (2021). "Everyone wants to do the model work, not the data work": Data cascades in high-stakes AI. In *Proceedings of the CHI Conference on Human Factors in Computing Systems*. ACM.
101. Jarrahi, M. H., Memariani, A., & Guha, S. (2023). The principles of data-centric AI. *Communications of the ACM, 66*(8), 84–92.
102. West, R., & Aydin, R. (2025). The AI alignment paradox. *Communications of the ACM, 68*(3), 24–26.
103. Fujiwara, T., Kucher, K., Wang, J., Martins, R. M., Kerren, A., & Ynnerman, A. (2025). Adversarial attacks on machine learning-aided visualizations. *Journal of Visualization, 28*(1), 133–151.
104. Domova, V., & Vrotsou, K. (2023). A model for types and levels of automation in visual analytics: A survey, a taxonomy, and examples. *IEEE Transactions on Visualization and Computer Graphics, 29*(8), 3550–3568.
105. Collins, C., Andrienko, N., Schreck, T., Yang, J., Choo, J., Engelke, U., Jena, A., & Dwyer, T. (2018). Guidance in the human–machine analytics process. *Visual Informatics, 2*(3), 166–180.
106. Yanez, F., Conati, C., Ottley, A., & Nobre, C. (2025). The state of the art in user-adaptive visualizations. *Computer Graphics Forum, 44*(1), e15271.
107. Endert, A., Ribarsky, W., Turkay, C., Wong, B. W., Nabney, I., Blanco, I. D., & Rossi, F. (2017). The state of the art in integrating machine learning into visual analytics. *Computer Graphics Forum, 36*(8), 458–486.
108. Wang, Q., Chen, Z., Wang, Y., & Qu, H. (2022). A survey on ML4VIS: Applying machine learning advances to data visualization. *IEEE Transactions on Visualization and Computer Graphics, 28*(12), 5134–5153.
109. Wu, A., Wang, Y., Shu, X., Moritz, D., Cui, W., Zhang, H., Zhang, D., & Qu, H. (2022). AI4VIS: Survey on artificial intelligence approaches for data visualization. *IEEE Transactions on Visualization and Computer Graphics, 28*(12), 5049–5070.

GPSR Compliance
The European Union's (EU) General Product Safety Regulation (GPSR) is a set
of rules that requires consumer products to be safe and our obligations to
ensure this.

If you have any concerns about our products, you can contact us on

ProductSafety@springernature.com

In case Publisher is established outside the EU, the EU authorized
representative is:

Springer Nature Customer Service Center GmbH
Europaplatz 3
69115 Heidelberg, Germany

www.ingramcontent.com/pod-product-compliance
Ingram Content Group UK Ltd.
Pitfield, Milton Keynes, MK11 3LW, UK
UKHW021023080726
473054UK00004B/260